Foodservice Organizations

A Managerial and Systems Approach

SIXTH EDITION

Mary B. Gregoire
Director, Food and Nutrition Services
Rush University Medical Center
Professor, Clinical Nutrition
and Health Systems Management
Rush University

Marian C. Spears (deceased)
Professor Emeritus
Kansas State University

Upper Saddle River, New Jersey 07458

Library of Congress Cataloging-in-Publication Data
Mary B. Gregoire
Foodservice organizations : a managerial and systems approach / Mary B. Gregoire, Marian C. Spears.—6th ed.
p. cm.
Includes bibliographical references and index.
ISBN 0-13-193632-8
1. Food service management. I. Spears, Marian C. II. Title.
TX911.3.M27S69 2006
647.95068--dc22

2006010691

Editor-in-Chief: Vernon R. Anthony
Senior Editor: Eileen McClay
Assistant Editor: Maria Rego
Director of Manufacturing and Production: Bruce Johnson
Managing Editor: Mary Carnis
Production Liaison: Janice Stangel
Design Coordinator: Miguel Ortiz
Manufacturing Manager: Ilene Sanford
Manufacturing Buyer: Cathleen Petersen
Interior Design & Formatting: Pine Tree Composition, Inc.
Production Editor: Jessica Balch (Pine Tree Composition, Inc.)
Cover Designer: Joseph DePinho, DePinho Design
Cover Image: Steven Hunt, Getty—The Image Bank
Executive Marketing Manager: Ryan DeGrote
Senior Marketing Coordinator: Elizabeth Farrell
Marketing Assistant: Les Roberts
Printer/Binder: Hamilton Printing Company
Cover Printer: Phoenix Color

Pearson Education, Ltd.
Pearson Education Australia PTY. Limited
Pearson Education Singapore Pte. Ltd.
Pearson Education North Asia Ltd.
Pearson Education Canada, Ltd.
Pearson Educación de Mexico, S.A. de C.V.
Pearson Education–Japan
Pearson Education Malaysia, Pte. Ltd.

10 9 8 7 6 5 4
ISBN 0-13-193632-8

Contents

Part 2 Transformation: Functional Subsystems

Chapter 4 Food Product Flow and Kitchen Design 85

Chapter 5 Procurement 119

Chapter 6 Food Production 201

Chapter 7 Distribution and Service 281

Chapter 8 Safety, Sanitation, and Maintenance 305

Part 3 Transformation: Management Functions and Linking Processes

Preface

The sixth edition of *Foodservice Organizations: A Managerial and Systems Approach* is dedicated to the life of and foodservice management legacy left by Dr. Marian Spears. Dr. Spears passed away in May 2004, shortly after the publication of the fifth edition. She was a life-long advocate of the importance of foodservice management as a component of the dietetics profession. Her career included years of foodservice administration experience, first in hospitals and then in nursing homes, prior to completing her Ph.D. and shifting to a career in dietetics and foodservice management education. She served as department chair at Kansas State University prior to her retirement. In 1989, the American Dietetic Association recognized her with the Marjorie Hulsizer Copher Award, the highest honor given to one of its members. I am honored to have had the chance to work with Dr. Spears for nearly 25 years. She was the inspiration and mentor every professional dreams of having in his or her career; I was fortunate to have that dream be a reality.

The foodservice systems model, originally developed by Dr. Allene Vaden, has provided the framework for this text since the first edition was published in 1985. The model has withstood the test of time and remains an innovative conceptualization for describing a foodservice operation. Organizing the text around this model provides a unique design for this textbook compared with other foodservice management texts. The material in each chapter provides detailed information on how managers can efficiently and effectively transform the human, material, facility, and operational inputs of the system into outputs of meals, customer satisfaction, employee satisfaction, and financial accountability.

Foodservice Organizations provides a blending of theory and practice. The text is guided by a belief that effective foodservice managers must have an understanding of the empirical base that can be used to better manage their operation. Each chapter attempts to provide a blending of quoted research and the practical application of that research.

The foodservice and hospitality industries continue to grow. Each year new job opportunities become available for graduates. Students entering the field come from programs focusing on dietetics, foodservice management, and hospitality management. The basic principles for effectively managing a foodservice operation remain the same for all, and, thus, this text can meet the needs of students in a variety of programs. It was written primarily for junior- and senior-level students and also as a resource for graduate students and instructors. The text was designed as one which could be used for multiple courses, thus reducing the financial burden on students who purchase new textbooks each semester.

Every effort was made to keep the text short by providing quick reviews of information and discussions of the applications of this information. Extensive reference lists and websites at the end of each chapter provide sources of additional information that can be used by students and instructors to expand discussion of topics introduced in the text.

Organization of Foodservice Organizations

The foodservice systems model serves as the conceptual framework for the book. Part 1 focuses on describing the Foodservice Systems Model. Concepts of the model are explained in depth. In Part 2, the Functional Subsystems (procurement; production; distribution and service; and safety, sanitation, and maintenance) of the transformation process are discussed. Part 3 focuses on the Management Functions and Linking Processes of the transformation process. Information on management, leadership, communication, and decision making is included as well as discussions on human resource management, financial management, and marketing. The last section, Part 4, focuses on Outputs of the System and includes methods for evaluating the effectiveness of the system outputs.

Each chapter contains margin notes with definitions of key terms. A foodservice management–related professional organization is showcased in each chapter to create awareness among students of the many professional opportunties that exit. A glossary of approximately 500 key terms is included at the end of the text. Each chapter contains an extensive bibliography and list of websites that can provide additional information about the chapter material. Each chapter also includes a summary of key points, study questions, and ideas for class projects.

New to This Edition

Feedback from those who have used this text was very helpful in its revision. Several new topics and features have been added, including

- An introduction to strategic thinking and strategic planning has been added to Chapter 1.
- Updated information on process improvement and measurement of quality has been added to Chapter 2.
- Additional information on foodservice layout and design information has been added to Chapter 4.
- Updated information on Hazard Analysis Critical Control Points (HACCP) and food safety based on *Food Code 2005* has been added to Chapter 8.
- Current theories in management and leadership have been added to Chapters 9 and 10.
- Information on intra- and cross-cultural communication has been added to Chapter 11.
- A professional organization profile is included at the end of each chapter.

For Instructors

Several support materials have been developed to accompany this text. An *Instructor's Manual,* which includes answers to the chapter study questions and exam questions is available from Prentice Hall. PowerPoint slides can be downloaded from the book's website at www.prenhall.com/spears. The text *Exploring Quantity Food Production and Service Through Problems,* by Lieux and Luoto (2000), provides excellent problem-based learning exercises designed to accompany the subject matter presented in this text. The hospital foodservice case study *Inlet Isles,* by Allen-Chabot, Curtis, and Blake (2001), provides an excellent case with problem sets that could be used to supplement several of the chapters in this text. Also included is a TestGen, which allows instructors to customize, save, and generate classroom tests.

For Students

The companion website, www.prenhall.com/spears, contains support materials you might find helpful as you work with the material in this text. The website includes PowerPoint slides that accompany each chapter and can serve as an outline of the key points in each chapter. Also included on the website are practice questions with information about why a particular answer may be correct or incorrect.

Acknowledgments

Work on a text such as this requires support from many individuals, and I would like to recognize their work and support. This edition of the text would not have been completed were it not for the extreme understanding, support, and encouragement provided by my husband, Wayne; my daughter, Theresa; and my son, Jonathan. In our more than 30 years of marriage, Wayne has been the "spouse extraordinare," allowing me to work late and on weekends on projects such as this text while he manages to keep things running smoothly at home while working full time himself. The words *thank you* hardly convey my appreciation and gratitude to him. Theresa was my girl Friday for the textbook and instructor's guide materials. Recently graduated from college, she was able to provide a "college student" set of eyes to improve the presentation of material. She checked website accuracy and references and helped develop study questions and answers. Every mom should be so lucky to have such a great assistant.

I also want to thank Diane Schweitzer, Foodservice Director for the School City of Hammond, a program recognized as a District of Excellence by the School Nutrition Association. Diane provided valuable information on workplace ergonomics and kitchen design for inclusion in this edition.

I appreciate the suggestions made by the reviewers, including Elizabeth Caffman, University of Central Arkansas; Virginia Cantrell, San Francisco State University; and Kevin Sauer, Kansas State University. I tried to incorporate all of their suggestions as I made revisions in the text.

Finally, I want to thank the staff at Prentice Hall for their help in guiding this publication to its latest revision. Thanks also to those who served as copy and production editors, for their review and preparation of the text for publication.

Mary B. Gregoire

About the Authors

Marian C. Spears, Ph.D., R.D., was Professor Emeritus and former chair of the Department of Hotel, Restaurant, Institution Management and Dietetics at Kansas State University. She held bachelor's and master's degrees from Case Western Reserve University, and a Ph.D. from the University of Missouri–Columbia. She had nearly 20 years of professional practice before entering academia, including positions as manager of a commercial cafeteria, chief dietitian of a children's home, and chief dietitian of a private hospital, all in Cleveland, Ohio, and Associate Director of Dietetics in St. Louis, Missouri. Her academic experiences included a faculty position at the University of Arkansas, Fayetteville, and serving as director of the Food Systems Management Coordinated Program in Dietetics at the University of Missouri–Columbia. During their years of residence in Arkansas, she and her husband maintained an extensive consulting practice in the design and operation of hospital foodservice facilities. Dr. Spears authored and coauthored numerous publications. In 1989, she received the Marjorie Hulsizer Copher Award, the highest honor conferred on a member by the American Dietetic Association. She passed away in May 2004.

Mary B. Gregoire, Ph.D., R.D., F.A.D.A., C.H.E. is Director of Food and Nutrition Services at Rush University Medical Center in Chicago and Professor in the Departments of Clinical Nutrition and Health Systems Management at Rush University. She has more than 25 years of experience as an administrator in both education and foodservice operations. Her career includes positions as professor and chair of apparel, educational studies, and hospitality management at Iowa State University, associate foodservice director and internship director at Rush University Medical Center, associate director of research at the National Food Service Management Institute, graduate program director at Kansas State University, and foodservice director at Jasper County Hospital. She has been an active researcher in the area of foodservice and hospitality management and has published numerous articles related to various aspects of foodservice management. Dr. Gregoire has her bachelor's and master's degrees from North Dakota State University and her Ph.D. from Kansas State University. She holds distinction as a charter fellow of the American Dietetic Association and is a Certified Hospitality Educator.

CHAPTER 1

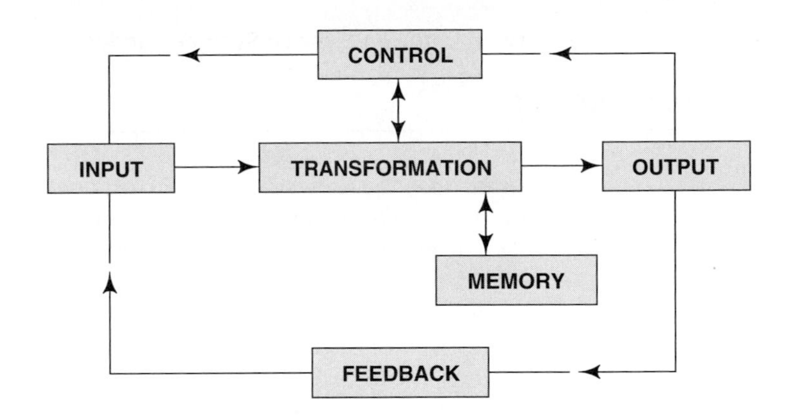

Systems Approach to a Foodservice Organization

Enduring Understanding

- Foodservice operations are open systems that transform inputs into outputs.
- Every decision made will impact a foodservice operation in many ways.
- The same or similar output can be achieved using different inputs.
- Factors in the environment impact the foodservice system in profound ways.
- Strategic management is critical to the success of a foodservice operation.
- The foodservice industry is diverse.

Learning Objectives

After reading and studying this chapter, you should be able to:

1. Define systems terms such as interdependency, dynamic equilibrium, and equifinality.
2. Analyze foodservice operations using the foodservice systems model.
3. Identify inputs and outputs of the foodservice systems model.
4. Discuss the transformation process.
5. Analyze ways in which factors in the environment impact the foodservice system.
6. Discuss steps in the strategic management process.
7. Describe foodservice operations in the foodservice industry.

Welcome to the exciting world of foodservice management. To guide your exploration of this field, the foodservice systems model approach to the management of foodservice organizations will be used. Viewing a foodservice operation as a system provides a way to understand the interrelatedness of work that goes on in that foodservice operation and will help you be a more effective manager. In this chapter, we discuss systems theory concepts and introduce you to the foodservice systems model. Each of the following chapters in this text then expands on various components of this foodservice systems model. You also will learn about the importance of strategic management for a foodservice operation and the many opportunities that exist for managers in the foodservice industry.

THE SYSTEMS CONCEPT

Systems may be viewed as closed or open, based on the amount of interaction with their environment. Foodservice operations are viewed as open systems.

The application of systems concepts has been used to facilitate problem solving and decision making for managers. The systems approach focuses on the totality of the organization rather than its processes or parts. It considers the impact of both the internal and external environment on the organization.

Model: Conceptual simplification of a real situation in which extraneous information is excluded and analysis is simplified.

Several **models** of foodservice systems have been published in trade and professional literature. These models enabled managers, suppliers, and others to evaluate current practices and the impact of proposed changes on the foodservice operation (David, 1972; Freshwater, 1969; Gue, 1969; Konnersman, 1969; Livingston, 1968; Martin, 1999).

Before 1960, analytical fact-finding approaches were used to examine organizations. The systems era started in the 1960s and began a focus on synthesis, the act of combining separate parts into a conceptual whole. Effective managers must be capable of coordinating complex organizations by focusing on interactions and interrelationships of components and subsystems of the organization to ensure all are working together to achieve the organization's goals.

System: Collection of interrelated parts or subsystems unified by design to obtain one or more objectives.

A **system** is defined as a collection of interrelated parts or subsystems unified by design to obtain one or more objectives. Luchsinger and Dock (1976) listed fundamental implications of the term system.

- A system is designed to accomplish an objective.
- Subsystems of a system have an established arrangement.
- Interrelationships exist among the elements.
- Flow of resources through a system is more important than basic elements.
- Organization objectives are more important than those of the subsystems.

The systems approach to management is simply keeping the organization's objectives in mind throughout the performance of all activities. It requires a communication network and coordination among all parts of the organization. Decisions and actions by the manager in one area of the operation will affect others.

THE ORGANIZATION AS A SYSTEM

Input: Any human, physical, or operational resource required to accomplish objectives of the system.

Transformation: Action or activity to change inputs into outputs.

Output: Result of transforming input into achievement of a system's goal.

The basic systems model of an organization is shown in Figure 1.1. The major parts of a system include: input, transformation, and output, as shown in the model.

The **input** of a system may be defined as any human, physical, or operational resource required to accomplish objectives of the system. **Transformation** involves any action or activity used in changing input into output, such as activities involved in production of food. The **output** is the result from transforming the input, and it represents achievement of the system's goal. For example, the primary

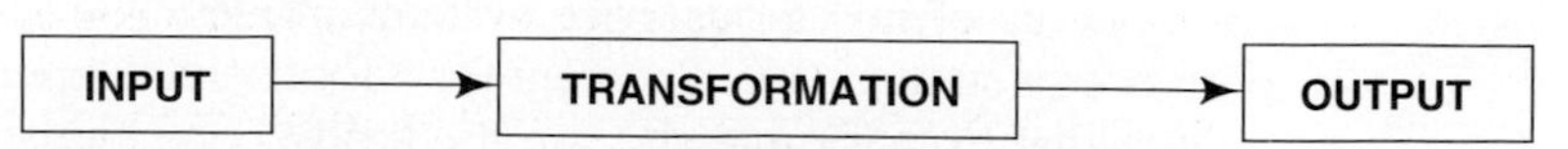

Figure 1.1. Basic systems model of an organization.

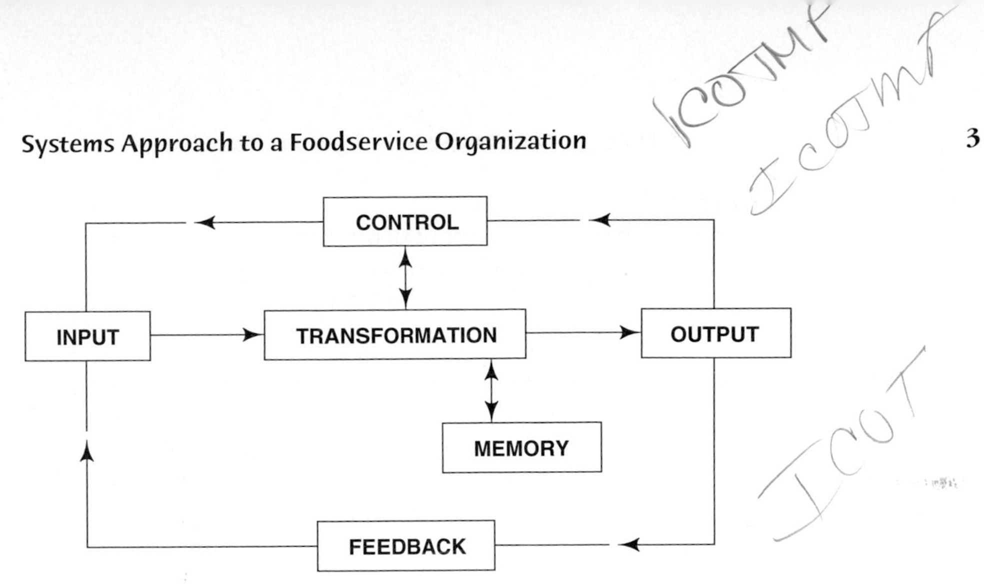

Figure 1.2. Expanded systems model of an organization.

output in a foodservice system is the production of the desired quantity and quality of food to meet customers' needs.

The expanded systems model of an organization includes four additional parts: control, memory, environmental factors, and feedback (Figure 1.2). Internal and external control provides guidance for the system. Internal control consists of plans including the goals and objectives of the organization, standards, and policies and procedures. External control consists of local, state, and federal regulations and contracts with outside companies.

The **control** element performs three functions in a system. It ensures that resources are used effectively and efficiently in accomplishing organizational objectives; it ensures that the organization is functioning within legal and regulatory constraints; and it provides standards to be used in evaluation of operations.

Memory: All stored information that provides historical records of a system's operations.

Memory includes all stored information and provides historical records of the system's operations. Analysis of past records can assist the manager in making plans and avoiding repetition of past mistakes. Rapid advances in computer technology are revolutionizing the memory capability of all types of systems. Rather than rely on filing cabinets for storage of information, managers increasingly are relying on computers for rapid access to records.

Environmental factors: Things outside the system that can impact the operation of the system.

Environmental factors are things which occur outside of the foodservice system yet impact some component of the system. Environmental factors may include technological innovation, globalization, competition, changing demographics, and political changes.

Feedback: Processes by which a system continually receives information from its internal and external environment.

Feedback includes those processes by which a system continually receives information from its internal and external environment. If used, feedback assists the system in adjusting to needed changes. For instance, feedback from customers' comments could be valuable information to the manager. Organizations without effective feedback mechanisms become relatively closed systems and may go out of business.

CHARACTERISTICS OF OPEN SYSTEMS

Open systems: Organizations that are in continual interaction with the environment.

An **open system** has a number of unique characteristics:

- interdependency of parts, leading to integration and synergy
- dynamic equilibrium

- equifinality
- permeable boundaries
- interface of systems and subsystems
- hierarchy of the system

Interdependency: Each part of the system affects performance of other parts of the system.

Interdependency is the reciprocal relationship of the parts of a system; each part mutually affects the performance of the others. This characteristic emphasizes the importance of viewing the organization as a whole rather than the parts in isolation. For example, in a foodservice system a decision to purchase a new piece of automated equipment may affect the menu, type of food purchased, and employee schedules.

Interaction among units of an organization is implied by interdependency. Units do not operate in a vacuum but continually relate with others. For example, for the organization to function as an effective system, the purchasing department must interact with the production unit and advertising with the sales department. The result of effective interaction is **integration,** in which the parts of the system share objectives of the entire organization. Integration leads to **synergy,** meaning that the units or parts of an organization working together may have greater impact than each of them operating separately.

Integration: Parts are blended together into a unified whole.

Synergy: Working together can create greater outcomes than working individually.

Dynamic equilibrium: Continuous response and adaptation of a system to its internal and external environment.

Dynamic equilibrium, or steady state, is the continuous response and adaptation of a system to its internal and external environment, which includes all the conditions, circumstances, and influences affecting the system. To remain viable, an organization must be responsive to social, political, and economic pressures. A foodservice director continually must evaluate cost and availability of food, labor, and supplies and advances in new technology. Change then is required to adapt to these new conditions and maintain viability. Feedback is important in maintaining dynamic equilibrium.

Equifinality: Same or similar output can be achieved by using different inputs or by varying the transformation process.

The term **equifinality** is applied to the organization as a system. It means that a same or similar output could be achieved by using different inputs or by varying the transformation processes. In other words, various alternatives may be used to attain similar results. In a foodservice organization, a decision to change from conventional to convenience foods will affect inputs and the transformation processes; however, a similar output, meals for a given clientele, will be achieved from these different inputs and processes.

Permeability of boundaries is the characteristic of an open system that allows the system to be penetrated or affected by the changing external environment. **Boundaries** define the limits of a system, and permeability allows the system to interact with the environment. For example, a hospital constantly interrelates with the community, other healthcare institutions, and government agencies, all of which are part of the external environment.

Boundaries: Limits of a system that set the domain of organizational activity.

The concept of boundaries among levels of a system, between subsystems, or between systems is a rather nebulous one. The walls between subsystems cannot be rigid, however, if an organization is to be effective. For example, activities of the food production and service units provide boundaries for each subsystem. Despite separate realms of activity, the boundaries between the two subsystems need to be highly permeable because the goal of the foodservice system is to satisfy the customer. The production and service subsystems, therefore, must be interdependent to meet this goal.

Interface: Area where two systems or subsystems come in contact with each other.

The area of interdependency between two subsystems or two systems is often referred to as the **interface.** The example just cited illustrates the interface between the production and service subsystems of a foodservice organization. The overall organizational system has many interfaces with other systems such as suppliers, government agencies, community organizations, and unions.

A point of friction often occurs when two moving parts come together. Similarly, the interface between two subsystems within an organization is likely to be characterized by tension. Whyte (1948), in a landmark study of the restaurant industry, and later Slater (1989) identified the area between the front and back of the house as a point of maximum tension between waitstaff and cooks. In hospitals, a classic example of interface is patient tray service. Dietetic and nursing services are in direct contact with each other, and conflict frequently occurs. An age-old argument in many hospitals has been, "Who is responsible for clearing the patient's bedside table at mealtime?" These interface areas often require special attention by managers.

An organization can be described as having three different levels. The model shown in Figure 1.3 was adapted from that proposed initially by Parsons (1960) and modified by Petit (1967) and Kast and Rosenzweig (1985) as a way of conceptualizing the organization as a system.

The internal level, also called the technical core or production level, is where goods and services of the organization are produced. The organizational level, in which coordination and services for the technical operations are provided, is responsible for relating the technical and policy-making levels. Policy making, the third level, is primarily responsible for interaction with the environment and long-range planning. An example is the corporate headquarters for a chain of restaurants. Although all three levels have permeable boundaries and environmental interaction, the degree of permeability increases from the technical to the policy-making level.

Hierarchy: Characteristic of a system that is composed of subsystems of a lower order and a suprasystem of a higher order.

Another characteristic of a system is **hierarchy.** A system is composed of subsystems of lower order; the system is also part of a larger suprasystem. In fact, the ultimate system is the universe. For purposes of analysis, however, the largest unit with which one works generally is defined as the system, and the units become subsystems. A subsystem, a complete system in itself but not independent, is an interdependent part of the whole system.

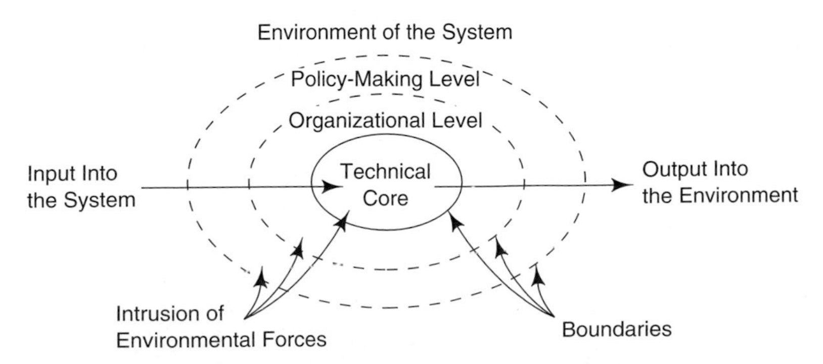

Figure 1.3. Levels of the organization.
Source: Adapted from "A Behavioral Theory of Management" by T. A. Petit, 1967, *Academy of Management Journal, 10*(4), p. 346. Reprinted by permission.

For example, a hospital may be viewed as a system; dietetic services, nursing, radiology, and other departments are considered subsystems. By the same token, a college or university may be viewed as a system, and academic units and student services as two of the subsystems. Often in colleges and universities the foodservice department is a component of the student services division. One could, however, analyze foodservice departments in a hospital and college in more detail and thus view them as systems; the units within the foodservice would then become the subsystems.

A FOODSERVICE SYSTEMS MODEL

A foodservice systems model (Figure 1.4) was developed to illustrate applications of systems theory to a foodservice organization. An examination of the model reveals that it is based on the basic systems model of an organization (Figure 1.1), which includes input, transformation, and output. The additional components of

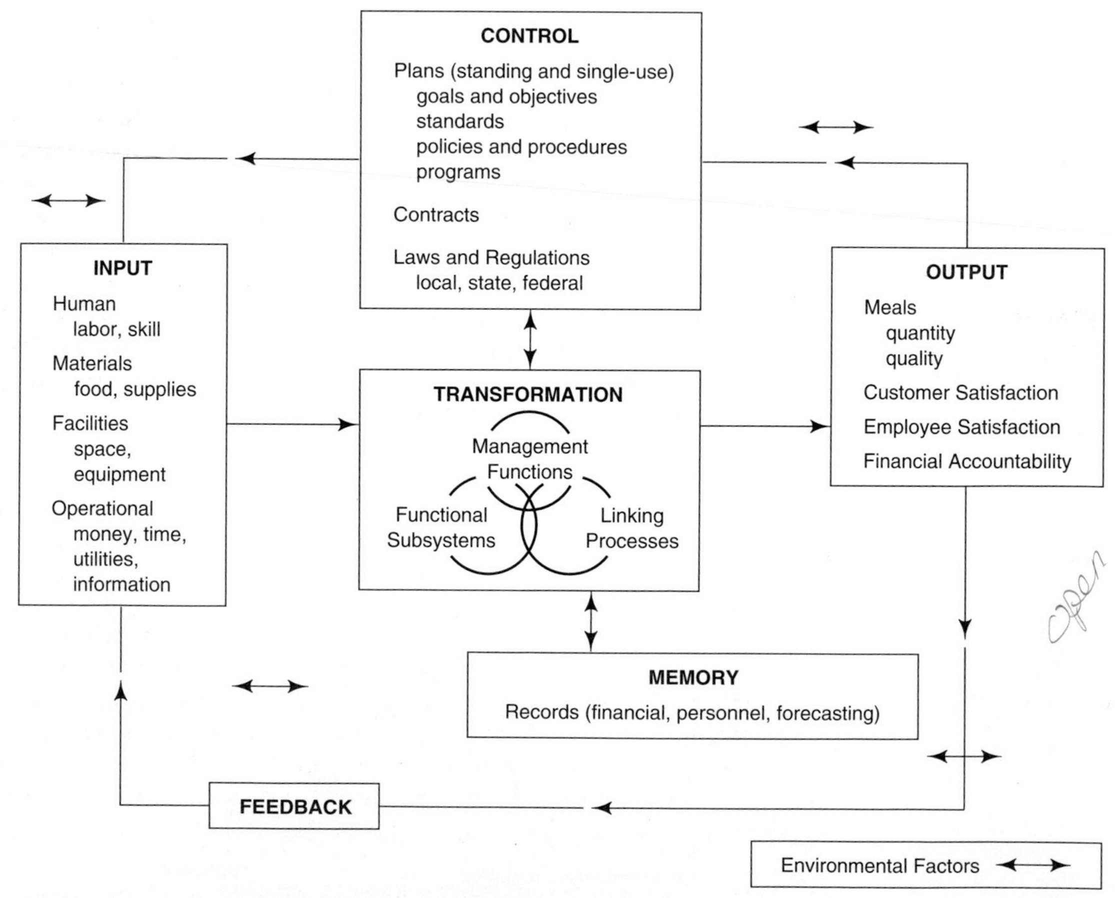

Figure 1.4. A foodservice systems model.
Source: Adapted from *A Model for Evaluating the Foodservice System,* by A. G. Vaden, 1980. Manhattan, KS: Kansas State University. © 1980. Used by permission.

control, memory, environmental factors, and feedback, which are from the expanded systems model of an organization (Figure 1.2), are integral parts of the foodservice systems model.

Arrows in the model represent the flow of materials, energy, and information throughout the foodservice system. Gaps in the arrows from output to input on the periphery of the model represent the permeability of the boundaries of the foodservice system and reflect the environmental interaction inherent in the effectiveness of the system. The bidirectional arrows represent environmental interactions, both internal and external to the system.

The **inputs** of the foodservice system are the human and physical resources that are transformed to produce the output. Traditionally, these resources have been referred to as men, materials, money, and minutes. This traditional definition has been expanded by defining the following four types of **resources:**

- **Human:** labor and skills
- **Materials:** food and supplies
- **Facilities:** space and equipment
- **Operational:** money, time, utilities, and information

Input requirements are dependent upon and specified by the objectives and plans of the organization. For example, the decision to open a full-service restaurant serving fine cuisine rather than a limited-menu operation with carryout service would have a major impact on type and skill of staff, food and supplies for production of menu items, capital investment, and type of foodservice facility and layout.

In the foodservice systems model (Figure 1.4), **transformation** includes the functional subsystems of the foodservice operation, managerial functions, and linking processes. These are all interdependent parts of transformation that function in a synergistic way to produce the output of the system. Each will be discussed in further detail in future chapters of this text.

Subsystem: Complete system within itself that is part of a larger system.

The **functional subsystems** of a foodservice system (Figure 1.5) are classified according to their purpose and may include procurement, production, distribution and service, and sanitation and maintenance. Depending on the type of foodservice system, the subsystems within the system may vary.

The type of system determines the characteristics and activities of the subsystems. In the example given previously, the full-service restaurant serving fine cuisine would have a more sophisticated and elaborate production unit than that of the limited-menu restaurant. Distribution and service in hospitals and many schools represent a very complex and difficult subsystem to control; the appropriate food at the correct temperature and quality must be delivered to patients and students in many locations. The distribution and service subsystem is the most important difference between restaurants and onsite foodservice operations. Food contractors providing meals for several airlines face the complexities of different menus and schedules, varying numbers of passengers, and problems such as delayed and canceled flights. Designing subsystems to meet the unique characteristics of these various foodservice organizations requires a systems approach in which the overall objectives of the organization are considered along with interrelationships among parts of the system.

Management functions, an integral component of the transformation element, are performed by managers to coordinate the subsystems in accomplishing the system's objectives. In this text, management functions include planning, organizing,

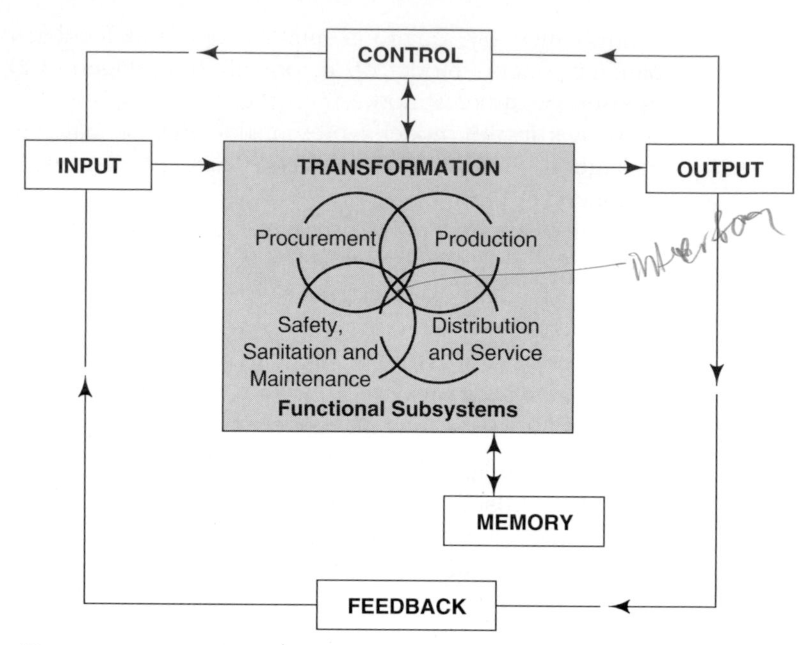

Figure 1.5. Functional subsystems of a foodservice system.

staffing, directing, and controlling. These functions are used to manage the operation, including human resources, finances, and marketing.

The **linking processes** of decision making, communication, and balance are needed to coordinate the characteristics of the system in the transformation from inputs to outputs. **Decision making** is defined as the selection by management of a course of action from a variety of alternatives. **Communication,** which is the vehicle for transmitting decisions and other information, includes oral and written forms. **Balance** refers to management's ability to maintain organizational stability under shifting technological, economic, political, and social conditions.

The **outputs** are the goods and services that result from transforming the inputs of the system; they express how objectives are achieved. The primary output in the foodservice system is meals in proper quantity and quality. In addition, customer and employee satisfaction and financial accountability are desired outcomes.

Traditionally, textbooks in this field have stated that the objective of a foodservice is to produce the highest possible quality food at the lowest possible cost. In this text, however, the objective of production of food is to satisfy the expectations, desires, and needs of customers, clients, or patients. A customer at an office snack bar, for example, might be content with a grilled cheese sandwich and tomato soup; that evening at an upscale restaurant, however, the customer will have quite different expectations of the cuisine.

Customer satisfaction is closely related to the types and quality of food and services provided and to customer expectations. For example, a college student, pleased with pizza on the luncheon menu of a college residence hall, would be unhappy if that same item were served at a special function of a social fraternity at a country club, even though in both instances the product may be of high quality. The student's expectations in these two situations are quite different.

Employee satisfaction is another important output of the foodservice system. Management should be concerned about the satisfaction of their employees. Managers also should be concerned about assisting employees in achieving and coordinating personal and organizational objectives. The effectiveness of any system, in large measure, is related to the quality of work done by the people staffing the organization.

Financial accountability is an output applicable to either a for-profit or not-for-profit foodservice organization. A foodservice manager must control costs in relation to revenues regardless of the type of operation. In the profit-making organization, a specific profit objective generally is defined as a percent of income. In a not-for-profit organization, the financial objective may be to generate a certain percentage of revenues in excess of expenses to provide funds for renovations, replacement costs, or expansion of operations.

Control encompasses the goals and objectives, standards, policies and procedures, and programs of the foodservice organization. The menu is considered the most important internal control of a foodservice system. The menu controls food and labor costs, type of equipment needed, customer and employee satisfaction, and profit. All plans, however, are internal controls of the system and may be either standing or single use. Standing plans are those used repeatedly over a period of time and updated or reviewed periodically for changes. Single-use plans are those designed to be used only one time for a specific purpose or function.

A cycle menu is an example of a standing plan. For example, a hospital may have a 2-week menu cycle that is repeated throughout a 3-month seasonal period. Many restaurants use the same menu every day; some might add a daily special. Various types of organizational policies also are examples of standing plans. The menu for a special catered function, however, is an example of a single-use plan. A particular single-use plan may provide the basis for a subsequent event of a similar type, but is not intended to be used in its exact form on a second occasion.

Contracts and various local, state, and federal laws and regulations are other components of control. Contracts are either internal or external controls. Internal controls may be for security, pest control, and laundry services; legal requirements are externally imposed controls on the foodservice system. The foodservice manager must fulfill various contractual and legal obligations to avoid litigation. For example, in constructing a foodservice facility, local, state, and federal building and fire codes must be followed in both design and construction. New federal regulations were imposed with the passage of the Americans with Disabilities Act, which requires that public accommodations and private businesses serving the public must remove barriers that interfere with access to the facilities and services provided (National Restaurant Association, 1992). Controls, then, are standards for evaluating the system, and they provide the basis for the managerial process of controlling.

Memory stores and updates information for use in the foodservice system. Inventory, financial, forecasting, and personnel records and copies of menus are among the records that management should maintain. Review of past records provides information to management for analyzing trends and making adjustments in the system.

Environmental factors include things such as technological innovation, globalization, competition, changing demographics, government regulations, and other factors that are external to the foodservice operation. They involve how a foodservice relates to and interacts with customers, employees, government officials, vendors, crop growers, food distributors, truckers, health inspectors, and thousands of other influences affecting its operation. These environmental factors often require that organizations be flexible, willing to change, quality conscious, and customer focused if they are to be successful. **Environmental scanning** is the term used to describe the search for and acquisition of information about events and trends external to the organization.

Environmental scanning: Search for and acquisition of information about events and trends external to the organization.

Feedback provides information essential to the continuing effectiveness of the system and for evaluation and control. As stated earlier, a system continually receives information from its internal and external environment that, if used, assists the system in adapting to changing conditions. Effective use of feedback is critical to maintaining viability of the system. A few examples of feedback that a foodservice manager must evaluate and use on a regular basis are comments from customers, plate waste, patronage, profit or loss, and employee performance and morale.

STRATEGIC MANAGEMENT

A systems approach to managing a foodservice operation involves creative and intuitive **strategic thinking.** Such thinking synthesizes information from the internal and external environments to create an integrated perspective for guiding the organization into the future (Mintzberg, 1994). According to Harrison and Enz (2005), strategic thinking has the following characteristics:

- Intent focused—has vision for where the organization is/should be going
- Comprehensive—views organization as part of larger system
- Opportunistic—takes advantage of unanticipated opportunities
- Long-term oriented—goes beyond here and now and looks into the future
- Builds on past and present—learns from past; recognizes constraints of present
- Hypothesis driven—evaluates creative ideas in a sequential process

Strategic thinking is a major component of a process termed **strategic management.** According to Coulter (2005), managing strategically means developing and implementing strategies that assist an organization in maintaining a **competitive advantage** that sets it apart from others in the industry. A competitive advantage might mean doing something that other companies do not do (i.e., hospitals that provide room service), doing things better than others (i.e., Taco Bell's minimization of kitchen space through use of central commissaries to reduce store overhead), or doing something others cannot do (i.e., Kentucky Fried Chicken's recipe and equipment for cooking chicken).

Competitive advantage: Characteristic(s) of a company that distinguish it from others.

Steps in the Strategic Management Process

Although authors differ on the actual number of steps that occur in the strategic management process, all agree that the process involves analysis of the company and its environment, creation and implementation of strategies to move a company toward its goals, and evaluation of progress. The process is circular in nature as information from the evaluation is fed back into the analysis step and the steps repeated. Figure 1.6 shows the steps in the strategic management process. Although

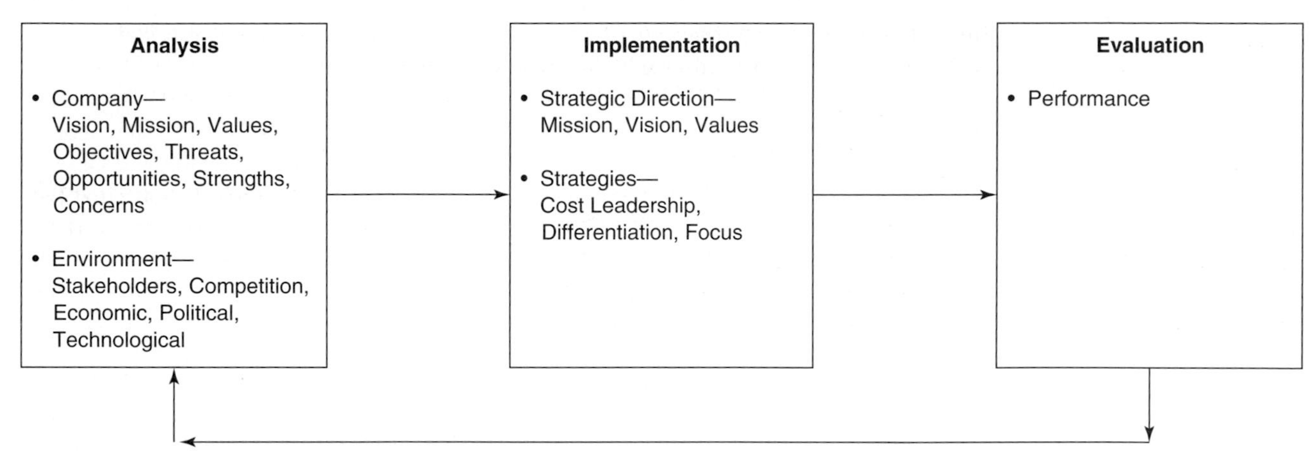

Figure 1.6. Strategic management process.

Stakeholder: Individuals or groups who are significantly affected by or can significantly influence a company's decisions.

Competitor: Another organization selling a similar product/service to the same market segment.

Vision: Statement of where a company wants to be in the future.

Mission statement: Describes what a company does; differentiates it from others.

once thought to be a process only used in for-profit organizations, strategic management has become a critical component for all foodservice organizations regardless of profit status.

Analysis

Analysis includes a review, and revision if needed, of a company's vision, mission, values, and objectives. The analysis should determine a company's external threats and opportunities and identify strengths and concerns. A careful analysis is needed of the company's **stakeholders** and **competitors** and the economic, political, and technology environment. As shown in Figure 1.7, a company's **vision** statement expresses in broad terms where the company wants to be in the future. It builds on the company's values and identifies its purpose. The **mission statement** provides more focus and describes what the company does. Harrison and Enz (2005) indicate that a mission statement directs decision making and resource allocation,

Vision

The Office of Food and Nutrition Services, Fairfax County Public Schools, will assist in developing excellence in each and every student. In an atmosphere that values diversity and human resources, we will be the best Food and Nutrition Services program in the nation by providing students food and nutrition knowledge, skills, and values they will need for a healthy body in an ever-changing global society.

Mission

We will provide nutritious meals that reflect current research and meet the recommended dietary allowances for school-age children. We will provide nutrition education in the classroom with the school cafeteria serving as the nutrition laboratory. We will serve as a nutrition education resource to the community.

Figure 1.7. Vision and mission statements for Fairfax county public schools food and nutrition services.
Source: http://www.fcps.edu/fs/food

inspires higher levels of performance and pride, communicates organization purpose and values, and enhances organizational reputation.

Implementation

Strategies: Decisions and actions to assist a company meet its objectives.

The second step in the strategic management process, implementation, involves the determination of strategic direction for the company and the creation and implementation of **strategies** to help a company gain competitive advantage. The most widely quoted author on strategy development, Michael Porter (1985, 1990), proposed three bases for strategies:

- **Cost leadership**—being the lowest cost provider of a product/service for a broad target market
- **Differentiation**—providing product/service that is unique, that customers value, and that customers are willing to pay a higher price for
- **Focus**—using a cost leadership or differentiation strategy to target a specific, limited-sized market niche

Once an organization's strategies are determined, implementation can begin. According to David (2005), managers may need to do several things to implement strategies effectively: change organizational structures, link performance evaluation and pay to strategies, create an organizational climate supportive of change, and adapt or modify production processes.

Evaluation

Evaluation, the final step in the strategic management process, involves assessing whether changes have occurred in the organization's internal or external strategic position and determining whether the organization is progressing satisfactorily toward achieving its objectives. An effective strategy evaluation program should be economical; provide meaningful, accurate information in a timely manner; and identify factors that have led to an organization's current position. Results of this evaluation will help managers decide if the organization is moving toward its goals as planned or if changes are needed to help ensure success.

THE FOODSERVICE INDUSTRY

The foodservice industry is exciting because it is in a constant state of change. More and more women are entering the workforce and diversity in the workplace is becoming more prevalent. Customers are being heard and are given more choices in the market. Employees are becoming more empowered. The importance of food safety is being recognized. This list is just a beginning. The industry offers many challenging and rewarding career opportunities. Foodservice operations commonly are categorized as either **commercial** or **onsite foodservices.** The commercial segment includes foodservices in which selling food for profit is the primary activity of the business; the onsite (sometimes referred to as noncommercial or institutional) segment provides foodservice as a secondary activity for the business in which the foodservice is located. These segments are not clear-cut, and each segment has a few characteristics of the other segments. Business in some of these foodservice operations is dependent upon the economy. Current information about the foodservice industry is available at the National Restaurant Association Web site: www.restaurant.org.

Commercial Segment

Commercial foodservice: Foodservice operations in which sale of food is the primary activity and a profit is desired.

The **commercial foodservice** segment includes a broad range of restaurants (from limited-service to fine dining), lodging, food and beverage, recreation and sports, and convenience stores. The number of commercial foodservice operations and total sales in these operations continues to increase.

Limited-Service, Limited-Menu

Limited-service, limited-menu restaurants (sometimes referred to as quick-service or fast-food) were designed to provide a limited number of food items to a customer in a relatively short period of time. Often the customer orders food at a counter and pays for it before eating. These restaurants are targeting working professionals and parents who want to have a meal served quickly at a low price. Many restaurants have created a new concept referred to as "fast/casual," "adult fast food," or "quality quick service." These companies combine the speed and convenience of fast food with food quality and exciting decor at a price between the two.

Other restaurants that offer an upscale menu and environment combine them with some quick-service techniques.

Full-Service Restaurants

Full-service restaurants provide waited table service for customers. Guests are greeted and seated by a host/hostess and orders taken and delivered by waitstaff. Payment occurs after the meal is completed. A tip is typically given for the service provided by the waitstaff member. The style and ambiance of full-service restaurants varies greatly from casual to fine dining.

Casual Dining Restaurants. Casual dining restaurants are designed to attract middle-income individuals who enjoy dining out but do not want the formal atmosphere and high price found in a fine dining restaurant (Chon & Sparrows, 2000). The atmosphere is casual, the mood relaxed, and the price midrange at these restaurants. Some may have themes, such as Olive Garden or Red Lobster. Others offer more varied menus such as Applebee's, Bennigan's, and T.G.I. Friday's. Entrée items in these restaurants typically are less than $12.

Fine Dining Restaurants. Fine dining restaurants, often referred to as "white tablecloth" restaurants, are characterized by a high level of attentive table service, expensive-looking furnishings and décor, and fine cuisine (Chon & Sparrows, 2000). Staff members in these restaurants work to create a memorable dining experience that communicates elegance and attention to every need of the guest. Prices paid for a meal in a fine dining restaurant often can exceed $100 per person. Chez Panisse in Berkeley, Charlie Trotter's in Chicago, and Le Bernardin in New York City are examples of well-known fine dining restaurants.

Hotel and Motel Restaurants

A hotel's food and beverage department is an exception if the profit exceeds 20% while the rooms division has profit margins of 75 to 80% (Payne, 1998). Operating a full-service restaurant is expensive, and guests for some reason prefer not to eat in a traditional hotel restaurant (Andorka, 1998a,b). Food courts in hotels have become a popular way to meet the needs of the customer while keeping the investment at low levels. "People don't usually want to eat in a traditional hotel restaurant" said Shuster, foodservice director for a large chain of hotels. She did an analysis of her company's full-service hotels and found that the operators were having real problems keeping their restaurants profitable. She found that 40 to 50% of their investments were on food costs and 45 to 50% on labor. Making a profit was impossible. In 1995, a program was launched for adding food courts in place of restaurants to 18 hotels around the country. Food costs were slashed to an average of 32% and labor costs to an average of 28% of their investments. This allowed hotels to provide foodservice options that better meet guests' needs without a large investment of capital (Andorka, 1998a,b).

Food and beverage operations in hotels usually have longer hours of service than do independent restaurants. Hotels serve three meals a day, seven days per week. Operating a medium-priced, a fine dining, and perhaps a theme restaurant inside one full-service hotel has been common in major U.S. cities. These operations require multiple kitchens with duplicate staffing, expanded managerial controls, and higher costs for distribution of food and supplies and maintenance of the foodservice operation. Hotel dining operations are labor intensive but room service, usually advertised as quick-service, is the most labor-intensive service in the entire hotel. Labor is a hotel's greatest cost, representing as much as 50% of all expenses and being responsible for 25% of all revenue. Most food and beverage revenues are based on the number of rooms sold for a night. The American Hotel and Lodging Association is the professional organization for these managers. Information about the organization and the industry can be found at www.ahma.com.

The bed-and-breakfast (B&B) segment of the hospitality industry, long popular in European private homes and country inns, has expanded in the United States. An increasing number of B&Bs are being opened as primary businesses; a major change is the addition of restaurants offering service at periods other than breakfast and to customers other than overnight guests (Lanier & Berman, 1993). More information about B&Bs can be found at the Professional Association of Innkeepers International's Web site: www.paii.org.

Country Club Restaurants

Country clubs usually have the challenge of running foodservice outlets from snack bars to fine dining restaurants (Apfel, 1998b). In addition to having an 18-hole golf course, tennis courts, and pool, customers will take advantage of the foodservice facilities that may include dining at the 19th hole grill, a wonderful Sunday buffet brunch displayed on two 30-foot tables in the dining room, or a snack and cold beverage by the pool. Individualized customer service occurs regularly in this segment. Foodservice staff often prepare a member's favorite recipe even if it is not on the menu or bring members their favorite cocktails before they ask for them.

Country clubs typically feature four dining concepts: an informal grill used mostly for breakfast and lunch before or after a game of golf or tennis; an upscale dining room; a banquet facility for corporate meetings, weddings, and dances; and a snack bar. Country clubs may be classified as not-for-profit, but their financial objectives likely will be to generate sufficient funds for repair, maintenance, and upgrading facilities. Members pay dues, making the cost of menu items lower than for-profit operations. Club members have top priority for use of banquet facilities, but nonmembers may be encouraged to use them because they subsidize the facility.

Country club members are really a hybrid of customer and owner. The market is small, between 200 and 1,000 customers per club, much smaller than for-profit restaurants. Outstanding customer service is required because no one wants to upset customers/owners.

The professional association for club managers is the Club Managers Association of America (CMAA). More information can be found at their Web site: www.cmaa.org.

Airport Restaurants

Airlines are decreasing onboard foodservice by serving snacks such as pretzels and peanuts and cans of cold beverages and cups of hot coffee. In the 1970s, airport restaurants were popular and many were considered as fine dining by local residents. As Americans became casual, so did airport restaurants (Crecca, 1997). In the 1980s, airports had bars and a couple of food items. In the 1990s, airports added popular brands of food for sale to airline customers. Hamburgers or hot dogs on buns, pizza, Mexican tortillas, frozen yogurt, and other quick foods were found in most airline terminals. In the late 1990s, airports began contracting with local restaurants to operate properties in their facilities. The Old Philadelphia Tavern in the Philadelphia airport and The Chicago Tap Room and Berghoff Café in Chicago's O'Hare airport are two examples.

Each airport operates differently (Crecca, 1997). Operators either bid on their own or enter into a franchising arrangement with a major concession operator. Restaurants rely on the airlines to attract people to their concourse. Most airport space is priced high like shopping mall space. Limited space and unusual space configurations require special equipment and higher costs. Kitchen and dining space must be downsized and menus are limited. Most airports require foodservice providers to cover all day, from early morning to late at night, and offer a take-out option primarily for airline crews. Often airline passengers have limited time to order and eat. Such speed requires top-notch employees. Kitchen staff should be cross-trained in various preparation tasks, and servers must be trained for speed. Restaurant operators need to know flight schedules and departure times. Each airport operates differently and has different needs and requirements. Increased airport security following the 9/11 terrorist attacks in the United States has changed how many airports and airlines operate their foodservice venues. Because of the long lines that often exist for passengers to process through the security checkpoints at airports, many passengers arrive earlier. When going through security takes less time than anticipated, these passengers then have more time in the gate area so many airports have increased the number of foodservice venues available to passengers inside the secured areas of the airport.

Cruise Ship Dining

Today, a cruise is considered one of the best vacations a person can take. There is no such thing as a typical cruise passenger (Cruise With Us, 1998). All kinds of people take cruises, of all ages, from all walks of life—singles, couples, and families. Cruise ships have a reputation for service of excellent food. No limit is set on what you choose or how much you eat. The cost of the food is included in the price of the cruise. For those who want a low-calorie meal or a special diet, they are available. Most ships have an early and late seating in their dining room and offer late evening buffets. Some ships have theme restaurants, such as Italian, Chinese, Japanese, or Southwestern, in addition to the main dining area. Formal attire for dinner is a cruise tradition. Most cruise ships also offer more casual options for meals such as coffee shops and pool-side cafes.

Zoos

Foodservice is becoming a profit center in several zoos across the country, and operators are upgrading the food eaten by visitors (Walkup, 1998). As zoo tax dollars become scarcer, zoo managers are looking at ways to increase revenue. Zoo directors are researching how foodservice can be upgraded to boost revenues. Brookfield Zoo in a western suburb of Chicago self-operates the Ituri Café in its Habitat Africa exhibit. In addition, three more traditional year-round restaurants and several seasonal food outlets are in the zoo.

Many zoos use contract foodservice companies to provide the food options in their facilities. Service Systems Associates and K-M Concessions are contract companies that specialize in food concessions and retail shopping for zoos and aquariums. Their clients include the Denver, Los Angeles, Detroit, and Houston zoos. In addition to themed concessions such as the Hungry Elephant Food Court and the Safari Grille, the group also provides catering services for a variety of fund-raising and exhibit-opening types of events.

Museums

Fine dining foodservice operations are becoming more prevalent in large museums. Although snack food options have been common in many museums, the addition of a seated serve restaurant is a relatively new trend. The restaurant at the Philadelphia Museum of Art, for example, hosts wine-tasting events and serves a champagne brunch on weekends. The Art Institute of Chicago employs an executive chef for its Garden Restaurant and hosts jazz performances there during the summer months. The St. Louis Art Museum worked with Chef Wolfgang Puck to open Puck's and Puck's Café in its facility. The Metropolitan Museum of Art in New York City offers a variety of dining options, from self-service dining in its Cafeteria to European-style waiter service in its Petrie Café.

Sports Events

Americans are spending more of their leisure time close to home visiting theme parks, sports events, and national parks. More and more foodservices are operated by contract recreation companies. Orlando-based Disney World and Universal Studios, however, are examples of theme parks that operate their own foodservices.

The Georgia Dome stadium has an in-house catering company, MGR Food Services (Georgia Dome, 1998). It operates 15 concession stands, Sideline Sports Grill (a sports bar), the InZone Restaurant (a fine dining operation), and Miller's Lodge, a microbrew pub. In the stands, traditional fare, such as hot dogs, popcorn, and beer, is sold. Three food courts have many well-known Atlanta restaurants with familiar foods. All foodservices are open to the public 1 1/2 hours before every Falcons game, and customers are advised to come early for "seatgating" (coolers, food, or beverages may not be brought into the Georgia Dome).

A number of contract companies specialize in servicing sports events. Contractors have launched furious bidding wars for lucrative ballpark concessions. Hot dogs, popcorn, peanuts in the shell, ice cream, soda, and beer have long been the items sold most often at ballparks. Today, however, fans can leave work and go straight to the ballpark to eat dinner that could include deli sandwiches, pizza, fajitas, frozen yogurt, and cheesecake.

Convenience Stores

According to the National Association of Convenience Stores (NACS), a **convenience store** is a retail business with primary emphasis placed on providing the public a convenient location to purchase quickly from a wide array of consumable products (predominantly food) and gasoline. Convenience store sales were $394.7 billion in 2004. Foodservice is the second largest instore product sold and generates more than 30% of gross profit dollars for convenience stores (NACS online, 2005). Based on research, six convenience store formats were identified as trends in the industry (What Is a Convenience Store?, 2005).

- **Kiosk.** Less than 800 square feet intended to provide additional revenue beyond gasoline sales. Sells only tobacco, beverages, snacks, and confectioneries, no groceries. Parking only at the gas pumps. Typical customers are transients and locals stopping in to buy gasoline.
- **Mini.** Usually 800 to 1,200 square feet in size, with emphasis on gasoline sales. Grocery selection is usually sparse and the only food is prepared sandwiches. Parking is often only at the pumps. Usually open from 18 to 24 hours and customers usually only buy gas.
- **Limited selection.** Range from 1,500 to 2,200 square feet and are becoming more numerous. Gasoline and store sales are profitable. They have a broader product mix and grocery offering; a simple foodservice, for example, may offer hot dogs, nachos, popcorn, and other snacks and fast food. Gasoline buyers are normally the biggest customer base. Striped parking and extended hours are common.
- **Traditional.** About 2,400 to 2,500 square feet, offering a product mix that includes dairy, bakery, snack foods, beverages, tobacco, grocery, health and beauty aids, confectionery, as well as gasoline sales. Other possible items are prepared foods to go, fresh or frozen meats, various products, and limited produce items. Usually have 6 to 12 striped parking spaces and are open 24 hours a day. These are normally owned by convenience store chains, but oil companies also are building them.
- **Expanded.** The number of stores that have 2,800 to 3,600 square feet is growing fast. Stores have more shelving for grocery products and more room for fast-food operations and seating. A greater percentage are using the space

to take advantage of the high profit margins in fast foods. These stores usually carry regular convenience store items. Parking is important, with most having about 10 to 20 spaces. Hours are extended.

- **Hyper.** These are very large stores, 4,000 to 5,000 square feet, and usually offer many products and services arranged in departments. For example, such stores may offer a bakery, a sit-down restaurant area, and a pharmacy. Many sell gas. The number of employees per shift can be large, particularly if a small restaurant is added. The number of parking spaces is substantial and hours are extended. Some stores are mini-truck stops, which affects product mix and customer base.

Onsite Segment

Onsite foodservice: Foodservice operations in which sale of food is secondary to the goal of the organization; typically not-for-profit.

The **onsite foodservice** segment includes hospitals, schools, college, and universities, correctional facilities, and military operations. These foodservice operations provide meals primarily for those directly involved with the facility such as patients, students, prisoners, and employees. Some visitors to these facilities also may be served.

Hospitals

Hospital foodservice operations provide food for both in-patients and out-patients and their family and friends. Employee meal service and retail sales are becoming increasingly important. **Managed care,** providing care under a fixed budget, has put pressure on hospitals to control costs (Managed Care Terms, 1998). Many hospital foodservice operators are remodeling space so it is a viable destination, adding brand names to menus, and providing street-side entrances to lure outside customers (Best Change Practices, 1998). Patient census counts are declining and foodservice managers are streamlining menus, staffing only one or two shifts, and relying on more convenience foods. According to Silverman et al. (2000), hospital foodservice directors are being expected to increase revenues. As a result, food kiosks, retail bakeries, and coffee carts are becoming more commonplace in hospitals. Hospital foodservice is now expected to be a revenue producer rather than a cost center and employee benefit.

Several professional organizations provide support for hospital foodservice managers. They include the American Dietetic Association, American Society for Healthcare Service Administrators, Dietary Managers Association, and Healthcare Foodservice Management. Information on these organizations can be found at www.eatright.org, www.ashfsa.org, www.dmaonline.org, and www.hfm.org.

Schools

The National School Lunch Program (NSLP) is a federally assisted meal program operating in nearly 100,000 public and nonprofit private schools and residential child care institutions (National School Lunch Program, 2005). It was the first federally funded meal program for children. The NSLP provides nutritionally balanced, low-cost, or free lunches to more than 26 million children each school day. The fiscal year 2005 appropriation for the NSLP was more than $7 billion. In 1946, as President Harry Truman signed the National School Lunch Act establishing the meal program, he said, "Nothing is more important in our national life than the welfare of our children and proper nourishment comes first in attaining this welfare."

Several additional federal programs have been added since that time to provide additional nutritional support for feeding children. The School Breakfast Program (SBP) began as a pilot project in 1966 and was made permanent in 1975. The SBP provides nutritionally balanced, low-cost or free breakfasts to children; more than 8 million in 2004. The Special Milk Program, which provides milk free or at a reduced cost to children, also was started in 1966. The Summer Food Service Program was begun in 1968 and supports meals for children as a part of summer programs. The Child and Adult Care Food program also was begun in 1968 and provides funding for meals in child care centers, day care homes, homeless shelters, and adult day care centers.

Children qualify for free and reduced-price meals in these programs based on family income; children from families with incomes at or below 130% of the poverty level are eligible for free meals. Those with family incomes between 130 and 185% pay a reduced price. A portion of the food served in these programs is provided directly by the U.S. Department of Agriculture (USDA) as commodities. All of the programs are administered at the federal level by the USDA through its Food and Nutrition Service (see www.fns.usda.gov/cnd). The professional organization for school foodservice managers is the School Nutrition Association (SNA) (formerly the American School Food Service Association). Information is available at SNA's Web site: www.schoolnutrition.org.

Colleges and Universities

College and university foodservice operations provide a variety of food options to students at more than 3,700 colleges and universities across the United States. Foodservice operations in this segment have grown from the traditional straight-line cafeteria in each dormitory to multiple retail venues including food courts, deli, kiosks, and convenience stores. Traditional board plans often are being replaced with declining-balance food accounts, which allow students to pay for only those foods eaten each day. Many college and university foodservice operations are offering extended hours and take-out or delivery service to better meet student demands for foods at nontraditional meal times and places. Portability of meal plans that allow students to eat in any of the on-campus food venues has been a driving force in many of the changes that have occurred in college and university foodservice operations.

The National Association of College and University Food Services (NACUFS) is the professional organization for college and university foodservice professionals. It is committed to becoming the organization that has insight on college and university foodservice, whether it is self-operated, outsourced, or cosourced. Additional information about NACUFS is available from the Web site www.nacufs.org.

Child Care

Increasing numbers of American children are enrolled in child care outside their homes as more mothers are working (American Dietetic Association, 1999). The position of The American Dietetic Association (ADA) is that all child care programs should achieve recommended standards for meeting children's nutrition and education needs in a safe, sanitary, supportive environment that promotes healthy growth and development. The standards recommended by the ADA are consistent with standards established by the U.S. Department of Health and Human Services for its Head Start Program, by the USDA for its Child and Adult Care Food Program, by the American Public Health Association and the American Academy of Pediatrics for out-of-home child care programs, and by the Society for Nutrition Education for child care settings.

Meals and snacks should be nutritionally adequate, consistent with the Dietary Guidelines for Americans, and follow recommended meal patterns for ages of the children, number of hours at the center, and cultural or ethnic differences in food habits. Plenty of fresh fruit, fresh or frozen vegetables, and whole-grain products should be used; the addition of fat, sugar, and sodium should be minimized. Foods should be provided in quantities that balance energy and nutrient intake with the children's small appetites.

More than half of American preschool children are in child care while their parents are at work (see the National Child Care Information Center online at www.nccic.org). Many organizations have child care centers at the worksite to facilitate child care for their employees. Support for managers working with child care operations is provided by the National Child Care Association. Training materials are available from the National Food Service Management Institute. Information can be found at the Web sites www.nccanet.org and www.nfsmi.org.

Senior Care

Many older people do not want to lose their independence by going to a nursing home, but they need assistance in preparing meals. The purpose of the Nutrition Services Program for Older Americans, as authorized by Title III of the Older Americans Act, is to provide nutritious, low-cost meals to homebound persons and congregate meals in senior centers. Many organizations also sponsor home-delivered meals, including the Visiting Nurse Service and the National Association of Meal Programs, which is subsidized partially by the USDA and the United Way. Typically, meals are prepared and packaged by outside contractors, hospitals, schools, or senior centers, and volunteers deliver them to people's homes at lunchtime five days a week. Recipients generally are charged what they can afford for the meal.

Many housing and meal options exist for seniors. The options differ based on the amount of care given:

- **Independent living.** For people who can take care of themselves in their own homes or apartments, a retirement community, or independent living apartment.
- **Congregate care.** A community environment with one or more meals a day served in a community dining room. Many services are provided, such as transportation, a pool, convenience store, bank, barber/beauty shop, laundry, housekeeping, and security.
- **Assisted living.** Apartment-style accommodations where assistance with daily living activities is provided. Fills the gap between independent living and nursing home care. Services include meals, housekeeping, medication assistance, laundry, and regular check-ins by staff.
- **Intermediate care.** Nursing home care for residents needing assistance with activities but not significant nursing requirements.
- **Skilled nursing.** Traditional state-licensed nursing facilities that provide 24-hour medical nursing care for people with serious illnesses or disabilities. Care provided by registered nurses, licensed practical nurses, and certified nurse aides.

Many elderly are signing contracts to live in **continuing care retirement communities (CCRCs).** These communities typically offer a variety of housing accommodations as well as a range of services, amenities, health and wellness programs, and security that allow for independent living, assisted living, or skilled nursing. Traditionally, such communities were sponsored by not-for-profit religious groups; however, corporate giants, such as Hyatt, have entered the industry. They are hoping their reputations as dependable, service-oriented leaders in the hospitality industry will give them a competitive edge. For independent and assisted living residents, a restaurant-style foodservice operation typically is offered allowing residents to choose their meals from restaurant-style menus. Many foodservice staffs include professional chefs and dietitians who provide not only daily meals but elaborate catering, room service, and elegant theme dinners for residents.

Several organizations provide support for foodservice managers working with seniors, including the Consultant Dietitians in Health Care Facilities (CD-HFC) and the Dietary Managers Association (DMA). Additional information can be found at the Web sites www.cdhcf.org and www.dmaonline.org.

Military

Military foodservice operations include dining hall and food court meal service for troops, hospital feeding for patients and employees, club dining for commissioned and noncommissioned officers, and mobile foodservice units for troops deployed to off-base locations. The Army & Air Force Exchange Service is adding more national chains, such as Taco Bell and Manhattan Bagel Co., to food courts on military bases. Food is being sold in on-base convenience stores or gas stations. The Army is taking advantage of the prime vendor program, which gives discounts from commercial suppliers and encourages the promotion of brands in dining halls. In many areas, enlisted men and women are being moved out of foodservice and these foodservice duties are being outsourced to contract companies (Matsumoto, 2002).

Over the last decade, military foodservice managers have responded to the same customer preference and trends that managers in other market segments have experienced. Soldiers, sailors, marines, and air force personnel are being offered a wide variety of menu choices and styles of service. Dining areas are decorated and colorful and have tables for four persons and booths instead of the long mess hall tables. Enlistees are bringing with them expectations formed by their exposure to commercial foodservice. Branding, the natural accompaniment to the privatization of military foodservice, is being incorporated into many of the operations; for example, McDonald's and Pizza Hut are on some bases.

Nutrition is a big trend in cafeterias and cash operations. The Department of Defense's Armed Forces Recipe Service is modifying recipes to reduce fat, salt, and cholesterol. The demand for more fresh fruits and vegetables, healthful entrées, and nutrition education and classes has increased on bases in all four service branches.

Correctional Facilities

Up until the early twentieth century, enforced silence was used to help maintain order in prisons, and food was characterized as "a meager and monotonous diet." Prison foodservice slowly improved over the years as the number of prisoners increased (Corrections, 1997). Up until 1960, prison foodservice offered only the basics. Lacking were standardization of procedures and menus, production equipment, computerization, and training. During this time, prison foodservice directors banded together to form the American Correctional Food Service Association (ACFSA; see online at www.acfsa.com). A major goal of ACFSA was to improve foodservice and reduce food riots started by prisoners unhappy with the quality of food served to them. At that time, foodservice had no voice in operations, budgeting, internal security, or anything else. Today, foodservice has a voice in the administration of most correctional facilities.

Correctional foodservice directors are challenged to produce meals on a budget of less than $3 per inmate per day. In the past, many prisons had their own farms and canneries, which allowed for production and preservation of food. Most farm operations were phased out in the early 1970s, although some prisons still work with local growers and are allowed to do gleaning of food items left in the fields postharvest. From the 1970s to the early 1990s, the U.S. government supplied prisons with commodities such as flour, dried milk, and dairy products, but in 1994, the USDA stopped supplying these items. Foodservice directors used "opportunity buying," or purchasing foods rejected by restaurants and others because they did not meet specifications. Many directors work with food processors to get food on short notice. Also, states with several prisons use group purchasing to control costs.

Today, many facilities have self-serve salad, pasta, and dessert bars and serve favorites like pizza, chicken nuggets, and homemade baked goods. Inmates are eating more nutritionally balanced meals. Foods served to inmates must meet standard dietary guidelines, and all menus must be approved by registered dietitians and are evaluated quarterly. Many prisons and jails are hiring full-time registered dietitians. Better production equipment, increased kitchen safety, and improved production methods are being used. Food safety is being emphasized and the old cook-and-hold methods are being replaced by cook-chill systems. Foodservice employees also are being trained.

One of the most obvious changes is the improvement of seating areas in prison cafeterias. Long tables are replaced with smaller ones, and the ambience has been improved. Many cafeterias today feature murals painted and signed by inmates.

Employee Feeding

Employee feeding has undergone many changes because of the rising cost of labor and the decrease in corporate subsidies. Managers are realizing that these operations must be self-supporting and revenue generating (Ninemeier & Perdue, 2005). Menus have been updated and facilities renovated to make the foodservice more like a commercial restaurant operation. Instead of a straight-line cafeteria with an employee serving the food, customers serve themselves from individually themed stations, such as salad, soup, pasta, grill, deli, desserts, and beverages.

Employee-feeding contractors are being used in many of these operations. Many are expanding wellness programs and encouraging healthful food in employee dining rooms. The objective of an employee-feeding program is to give employees food and service that exceeds the quality and value of local restaurants so they will not leave the building and take a longer lunch period.

Mangia, the dining operation for the nearly 10,000 employees of the Bellagio hotel and casino complex in Las Vegas, offers a rather unique approach to employee feeding. Mangia is open 24 hours a day, free to all employees, and access is unrestricted. Employees walk in when they would like and eat what they would like; there are no ID checks or limits on how much an employee can eat (Buzalka, 2005).

FOODSERVICE INDUSTRY OPERATING PRACTICES

Forecasts for the coming years present a number of interesting scenarios for foodservice operations and the industries supporting them. Changes in economics, technology, demographics, politics, and competition will have an impact on how foodservice operators manage their programs and meet their missions. The choice is whether to turn challenges into opportunities or to struggle as challenges mushroom into obstacles.

Customers are spending more than half of their food dollars on food prepared outside the home. Many foodservice operators have expanded their services to include take-out and delivery options for their customers.

Foodservice operators who deliver high-quality, ready-to-eat foods at reasonable prices will succeed. Many of them will seek partnerships with distributors, manufacturers, and brokers to provide new cost-effective and value-added services to meet their requirements.

- Distributors will expand their services in a cost-effective way to include more pre-preparation, menu development, signature products, customer tracking, and frequent delivery.
- Manufacturers will support foodservice organizations by developing packaging and presentation technology.
- Brokers will focus on meal solutions rather than ingredients and communicate product performance back to manufacturers.
- Onsite foodservice operators will serve a greater variety of customers in the community, maximize use of their facilities, lease off-hour time to other operations, and develop menu items that are nutritious, good-tasting, convenient, and inexpensive.

The foodservice industry includes a variety of operating practices, including self-operation, partnering, contracting, franchising, and multidepartment management.

Self-Operation

Self-operation means that the foodservice operation is managed by an employee of the company in which that foodservice operation is located. This manager has full responsibility and authority for all functions within the department and reports to an administrator employed by this same company (Puckett, 2002). Traditionally, all foodservice operations were self-operations.

Partnering

Partnering is a mutual commitment by two parties on how they will interact during a contract with the primary objective of improving performance through communications. It is primarily a relationship of teamwork, cooperation, and good faith performance through communications. Partnering

- Establishes mutual goals and objectives
- Builds trust and encourages open communication
- Helps eliminate surprises
- Enables the two parties to anticipate and resolve problems
- Avoids disputes through informal conflict management procedures
- Improves morale and promotes professionalism in the workforce
- Generates harmonious business relations
- Focuses on the mutual interests of the two parties

The term *partnering* is beginning to be used in the foodservice industry. A program that makes culinary education available to front-line cooks has been created in a partnership between the National Association of College and University Food Services (NACUFS) and The Culinary Institute of America (CIA) (Blake, 1998). Middlebury College in Vermont is partnering with the New England Culinary Institute on a year-long basis (King, 1998a).

Contracting

A **contract** is defined as an agreement between two or more persons to do or not to do something. A partnership between the two is necessary to make the contract work. The focus must be long term.

There are many companies who will contract with an organization to run the foodservice operations in that organization. The largest contracting companies are ARAMARK Corporation, Compass Group, and Sodexho, Inc. (Top 50, 2005).

Contracts that are signed by organizations with foodservice contract companies include many different conditions and terms and can vary from a single page to more than 50 pages in length. Contractors typically are expected to provide foodservice options that will satisfy the customers and provide revenue for the organization. In some organizations, the contract company is hired only to provide management services. The organization pays a fee for these management services. In such a contract, the management personnel will be employees of the contract company; however, all other employees remain employees of the contracting organization. In other cases, an organization will contract with a company to operate its foodservice, and in such contracts, all employees and managers are employees of the contracting company.

Franchising

Colonel Harlan Sanders perfected the secret recipe for his Kentucky Fried Chicken—now referred to as KFC—in 1939 after more than a decade of experimenting (Apfel, 1998a). Sanders was so convinced that his chicken was a winning business idea or concept that he wanted to feature it in restaurants around the nation. But how could one small business owner find the human resources and financial capital to realize his plans? The answer was simple—**franchising,** defined by *Webster's* as the right granted to an individual or group to market a company's concepts. The International Franchise Association (IFA) (www.franchise.org) suggests that the biggest advantage of becoming a franchisee is that it solves the two biggest expansion problems: people and money. It usually is a better solution than going public or taking a chance on finding capital in a particular territory. A **franchisee** is a person who is granted a franchise, and a **franchisor** is a person who grants a franchise.

In more recent times, this way of doing business includes arrangements known as franchises, licenses, dealerships, and distributors. It also is a potential method to foster more rapid growth. Owners can expand the concept more rapidly than they could using their own resources. A successful franchise must be built upon a solid concept that is unique but not necessarily unseen on the dining scene (Apfel, 1998a). A concept must be market tested and appeal to the many types of customers. The decision to begin franchising is a serious one because it has a complex organizational structure. Support services must be provided to the franchisee group in operations, finance, administration, marketing, human resource management, and research and development. Franchising might reduce risks because parent companies often provide advice and assistance. Franchisors also provide proven methods, training, financial support, and an established identity and image. Few McDonald's franchises fail.

Improving relationships with franchisees will increase business. Franchisors must give franchisees dignity and responsibilities for setting shared values, focusing on brands, fostering corporate culture, and marketing, with everyone working together toward the same goals. Computerization of an entire franchise system is required to compete. Training departments will develop programs to help franchisees learn technical methods of achieving a practical purpose. CD-ROMs will be available for training franchisees. Technology spending will increase in franchise budgets not only for hardware and software but also for staff to operate the system.

Multidepartment Management

Multidepartment management is coming back into the picture for many foodservice operations. In the early 1940s, foodservice management was just beginning to appear in hospitals, schools, restaurants, and any other operation that served food. In those days, a nurse in a hospital might have planned the menus, ordered the food, and served patients; a school principal might have assumed the same responsibilities. A cook in a restaurant might have scrubbed floors, baked bread, and gone to the grocery store in the morning on the way to work. This pattern was replaced by specialization. However, to control costs, management skills are again being emphasized and the area of expertise is secondary. Silverman et al. (2000) reported that more than half of hospital foodservice directors surveyed expected their management responsibilities to change in the future; many expected to manage more than just the foodservice department in the future. Schuster (1998) cites many reasons for becoming a multidepartment manager: professional development, value to the employer, and a higher position within the organization. A more compelling reason is job security and simple survival. Foodservice managers must be able to take advantage of such opportunities.

An excellent example of multidepartment management occurred in the gift shop located left of the front desk in the Jewish Hospital, Louisville, Kentucky (Convenience Retailing, 1999). A management change in early 1998 made gift shop operations the direct responsibility of the hospital foodservice director. In 16 months since the director assumed responsibility, the shop moved from breakeven to $300,000 in annual sales and a $100,000 profit. The foodservice department's cost controller revamped the shop's merchandise mix, accounting procedures, and promotional strategies. She took on responsibility for purchasing, selecting merchandise, and overseeing the general operations. She computerized the shop's daily reporting system with a spreadsheet and followed with a series of clearance sales.

Professional Organization Profile

American Dietetic Association

MISSION

Leading the future of dietetics

WHO BELONGS TO THE ORGANIZATION

Members include more than 65,000 registered dietitians, dietetic technicians, clinical and community dietetics professionals, consultants, foodservice managers, educators, researchers, and students.

ADVANTAGES OF MEMBERSHIP

Members have access to current scientific and professional information (website, *Journal of the American Dietetic Association,* ADA Times), are represented in Washington on policy issues, can participate in continuing education programs, have access to a network of professionals nationwide, and are provided optional benefits such as insurance and travel discounts. Student memberships are available.

WEBSITE

www.eatright.org

MEET THE PRESIDENT

Rebecca S. Reeves, DrPH, RD, FADA, president ADA 2005–2006, is a registered dietitian and researcher at the Baylor College of Medicine focusing on issues of weight loss and health. She began her dietetics career as a hospital clinical dietitian and has held a variety of public health and medical research positions. She holds distinction as a Fellow of the American Dietetic Association and was recognized with their Medallion Award. Dr. Reeves offers the following advice to students: "Always be positive about your abilities and skills and confident that you can accomplish challenges that are presented to you. A 'yes I can' attitude will take you farther in your career than an attitude of reluctance and doubt about taking on new tasks. Choosing to join the American Dietetic Association will offer you a wide array of services that will maintain your currency in the field of dietetics and provide unlimited networking opportunities."

CHAPTER SUMMARY

This summary is organized by the learning objectives.

1. There are several terms used to describe the characteristics of a system. *Interdependency* is the reciprocal relationship of the parts of a system; each part mutually affects the performance of the others. *Dynamic equilibrium,* or *steady state,* is the continuous response and adaptation of a system to its internal and external environment. *Equifinality* means that the same or similar output can be achieved by using different inputs or by varying the transformation process.
2. The foodservice systems model includes inputs which are transformed into outputs. Controls and memory impact the transformation process. Environmental factors can impact inputs, transformation, and outputs. Feedback provides a way to provide information about the quality of outputs to the inputs to effect change in the system. The operations in any foodservice operation can be described and analyzed using this systems model.
3. Inputs to the foodservice systems model include human (labor, skill), materials (food, supplies), facilities (space, equipment), and operational (money, time, utilities, information). Outputs include meals (quantity and quality), customer satisfaction, employee satisfaction, and financial accountability.
4. The transformation process involves changing inputs into outputs. This process includes the functional subsystems (procurement, production, distribution and service, and sanitation and maintenance), management functions (planning, organizing, staffing, leading, and controlling), and the linking processes (decision making, communication, and balance).
5. Factors in the environment can include things such as technological innovation, globalization, and changing demographics. Technological innovations could mean new equipment for preparing food that could reduce labor time or improve food quality. Globalization might mean that foods purchased will be coming from countries around the globe, which could have cost and food safety implications. Changing demographics could mean working with employees whose first language is not English.
6. The strategic management process involves analysis of the company and its environment, creation and implementation of strategies to move a company toward its goals, and evaluation of progress.
7. The foodservice industry is very large and diverse. Foodservice operations include schools, hospitals, universities, restaurants, country clubs, cruise lines, hotel food and beverage, correctional, military, and childcare.

TEST YOUR KNOWLEDGE

1. What is the difference between an open and closed system?
2. What are the inputs and outputs of the foodservice system?
3. Describe what occurs in the transformation process in the foodservice system.
4. Compare and contrast a foodservice operation using a cost strategy with one using differentiation as a strategy.
5. What is onsite foodservice? How does it differ from commercial foodservice? How is it the same?

CLASS PROJECTS

1. Discuss, using the foodservice systems model, how the following events might impact a foodservice operation:
   - Loss of the lettuce crop in California because of excessive rain
   - Delivery of 88-size oranges instead of the specified 113 size
   - Addition of too much salt to the vegetable soup
   - Absenteeism of the cook and dishwasher for the evening meal
   - Low final rinse temperature on the dishmachine
2. Divide into discussion groups. Each group should choose one of the following types of foodservice operations: casual theme restaurant, corrections foodservice, cruise ship foodservice, fine dining restaurant, hospital foodservice, hotel food and beverage, nursing home foodservice, quick-service restaurant, school foodservice, university foodservice. Describe how the following terms might be applied in your foodservice operation—interdependency, dynamic equilibrium, synergy, equifinality, and permeability of boundaries—as they specifically relate to that foodservice operation.
3. Individually keep a journal for a month on the source of each of your meals. Construct a frequency table showing each of the following: number of times meals are prepared at home, number of times food is consumed at (or purchased from) a commercial foodservice operation, number of times food is consumed at (or purchased from) an onsite foodservice operation. Prepare a list of all of the commercial and onsite foodservice operations you experienced food from during the month. Share results with other students.
4. Invite a local foodservice director to class and ask that he or she share the strategic plan in place for his or her operation. Discuss how different strategies discussed in this text might be used to help the organization meet its goals.

WEB SOURCES

Professional Association Websites

www.acfsa.org	American Correctional Food Service Association
www.eatright.org	American Dietetic Association
www.ahma.com	American Hotel and Lodging Association
www.ashfsa.org	American Society for Healthcare Food Service Administrators
www.cmaa.org	Club Managers of America Association
www.dmaonline.org	Dietary Managers Association
www.hfm.org	National Society for Healthcare Food Service Management
www.nacufs.org	National Association of College and University Foodservice
www.restaurant.org	National Restaurant Association
www.paii.org	Professional Association of Innkeepers International
www.schoolnutrition.org	School Nutrition Association
www.sfm-online.org	Society for Foodservice Management

Other Related Websites

www.fns.usda.gov/fns	USDA Food and Nutrition Services
www.nfsmi.org	National Food Service Management Institute

BIBLIOGRAPHY

Allen, R. L. (1999). Commerce's 'Year of the Restaurant' designation caps one prolific century. *Nation's Restaurant News, 33*(3), 31.

American Dietetic Association (1999). Position of the American Dietetic Association: Nutrition standards for child care programs. *Journal of the American Dietetic Association* 99:981–988.

Andorka, F. H. (1998a). Foodservice: Counting on food courts. [Online]. Available: http://www.hotel-online.com/Neo/SpecialReports1998.FoodCourts.html

Andorka, F. H. (1998b). Hotel Online Special Reports. Foodservice: Counting on Food Courts. [Online] Available: http://www.hotel-online.com

Apfel, I. (1998a). The art of franchising. *Restaurants USA, 18*(7), 22–26.

Apfel, I. (1998b). Getting into the swing: Country-club restaurants stay on par with current trends. *Restaurants USA, 18*(6), 35–38.

Banathy, B. H. (1996). *Designing social systems in a changing world.* New York: Plenum.

Best change practices. (1998). Industry specific insights: Health care, Expanding foodservice into a profit center. [Online]. Available: http://www.changecentral.com/bestindustryexpanding.html

Billings, L. (1998). Chefs of a different flavor. *Eating Smart, 11*(9), 46–49.

Blake, K. (1998). NACUFS and the CIA partner in education for university cooks. *Nation's Restaurant News, 32*(7), 50.

Brown, W. B. (1966). Systems, boundaries and information flow. *Academy of Management Journal, 9,* 318–327.

Buzalka, M. (2005). Inside Bellagio's secret servery. *Food Management,* September, 2005:58.

Checkland, P. (1989). *Systems thinking, Systems practice.* New York: John Wiley & Sons, Inc.

Children's Defense Fund. (2002). Child Care Basics. available at www.childrensdefense fund.org

Chon, K., & Sparrows, R.T. (2000) *Welcome to hospitality.* Albany, NY: Delmar Thompson Learning.

Convenience Retailing. A Supplement to *Food Management.* (1999). Gift shop management proves a profit opportunity, 1–8.

Corrections. (1997). *Food Management, 32*(10), 92, 94–96.

Coulter, M. (2005). *Strategic management in action.* 3rd ed. Upper Saddle River, NJ: Pearson/Prentice Hall.

Crecca, D. H. (1997). The new shape of airport restaurants. *Restaurants USA, 17*(6), 32–36.

Cruise With Us. (1998). Frequently asked cruise package questions. [Online]. Available: http://www.cruisewithus.com/faq.htm

David, B. D. (1972). A model for decision making. *Hospitals, 46*(15), 50–55.

David, F. R. (2005). *Strategic management concepts,* 10th ed. Upper Saddle River, NJ: Pearson/Prentice Hall.

Food Distributors International (2000). *Foodservice 2010: America's appetite matures.* Falls Church, VA: Author.

Freshwater, J. F. (1969). Future of food service systems. *Cornell Hotel and Restaurant Administration Quarterly, 10*(3), 28–31.

Georgia Dome. (1998). Attractions. [Online]. Available: http://www.stadianet.com/georgiadome/attractions/funstuff.html

Gue, R. L. (1969). An introduction to the systems approach in the dietary department. *Hospitals, 43*(17), 100–102.

Harrison, J. S., & Enz, C. A. (2005). *Hospitality strategic management.* Hoboken, NJ: John Wiley & Sons, Inc.

Jackson, R. (1997). *Nutrition and foodservice for integrated health care.* Gaithersburg. MD: Aspen Publishers, Inc.

Johnson, R. A., Kast, F. E., & Rosenzweig, J. E. (1973). *The theory and management of systems,* 3rd ed. New York: McGraw-Hill.

Kast, F. E., & Rosenzweig, J. E. (1985). *Organization and management: A systems and contingency approach,* 4th ed. New York: McGraw-Hill.

King, P. (1998a). Taking Harvard dining to next level with 'preferred futuring.' *Nation's Restaurant News, 32*(30), 37.

King, P. (1998b). Workshops are a start, but are they enough? *Nation's Restaurant News, 32*(7), 51.

Konnersman, P. M. (1969). The dietary department as a logistics system. *Hospitals, 43*(17), 102–105.

Kroll, D. (1992). The changing commercial and institutional foodservice industry. [Online]. Available: http://www.buscom.com/archive/GA078.html

Lanier, P., & Berman, J. (1993). Bed-and-breakfast inns come of age. *Cornell Hotel and Restaurant Administration Quarterly, 34*(2), 14–23.

Livingston, G. E. (1968). Design of a food service system. *Food Technology, 22*(1), 35–39.

Luchsinger, V. P., & Dock, V. T. (1976). *The systems approach: A primer.* Dubuque, IA: Kendall/Hunt.

Managed Care Terms. (1998). [Online]. Available: http://www.amso.com/terms.html.

Management has its privileges. (1998). *Restaurants USA, 18*(11), 6.

Martin J. (1999). Perspectives on managing child nutrition programs. In Martin, J., & Conklin, M. J. (eds.), *Managing child nutrition programs.* Gaithersburg, MD: Aspen Publishers, Inc., pp. 20–26.

Matsumoto, J. (1999). Nontraditional noncommercial. *Restaurants & Institutions, 109*(2), 44–45, 56, 58.

Matsumoto, J. (2002). Contract management companies look to military for new growth opportunities. *Restaurants and Institutions* (March 15). Online at www.rimag.com

Mintzberg, H. (1994). *The rise and fall of strategic planning.* New York: Prentice Hall.

NACS online. (2005). The convenience store industry: Fact sheet. Online at www.nacsonline.com.

National Restaurant Association. (1992). *The Americans with Disabilities Act: Answers for foodservice operators.* Washington, DC: National Restaurant Association and National Center for Access Unlimited.

National School Lunch Program. (2005). Nutrition program facts. [Online]. Available: http://www.fns.usda.gov/cnd/Lunch/default.htm

Ninemeier, J. D. and Perdue, J. (2005). *Hospitality operations.* Upper Saddle River, NJ: Pearson Education, Inc.

Nursing Home Services. Nursing Home INFO. (1998). [Online]. Available: http://www.nursinghomeinfo.com/nhserve.html

Parsons, T. (1960). *Structure and process in modern societies.* New York: Free Press.

Payne, K. D. (1998). Increasing food and beverage revenues in hotels. [Online]. Available: www.hotel-online.com/Trends/Payne/Articles/IncreasingFoodBeverageRevenues.html

Petit, T. A. (1967). A behavioral theory of management. *Academy of Management Journal,* 10, 341–350.

Porter, M. E. (1985). *Competitive advantage: Creating and sustaining superior performance.* New York: Free Press. (Republished with a new introduction, 1998.)

Porter, M. E. (1990). *The competitive advantage of nations.* New York: Free Press. (Republished with a new introduction, 1998.)

Powers, T. F. (1978). A systems perspective for hospitality management. *Cornell Hotel and Restaurant Administration Quarterly, 19*(1), 70–76.

Puckett, R. P. (2002). Alternate methods for managing a food service. Available online at www.dmaonline.org

Schuster, K. (1998). In the hot seat: the challenging role of the contract liaison. *Food Management, 33*(7), 52–54, 58, 60.

Silverman, M. R., Gregoire, M. B., Lafferty, L. J., and Dowling, R. A. (2000). Current and future practices in hospital foodservice. *Journal of the American Dietetic Association, 100:*76–80.

Slater, D. (1989). Coming to a truce: Ways to bridge the gap between the front and the back of the house. *Restaurants USA, 9*(10), 26, 28–29.

Sullivan, B. (1998). The second boom. *Restaurants and Institutions, 108*(26), 95–97.

The school breakfast program. (1998). [Online]. Available: http://www.usda.gov/fcs/ogapi/breakf~3.htm

Top 50 management companies. *Food Management.* September, 2005: 42–56.

U.S. Army Communications. (1998). Background & definition. [Online]. Available: http://acqnet.sarda.army_mil/acqinfo/bluprint/backg.htm

Walkup, C. (1998). Zoos upgrade foodservice to boost revenues. *Nation's Restaurant News, 32*(16), 16.

What Is a Convenience Store? 2005. [Online]. Available: http://www.nacsonline.com

Whyte, W. F. (1948). *Human relations in the restaurant industry.* New York: McGraw-Hill.

Wilkerson, J. (1998). Franchising trends for '98: Finding a balance between old and new business cycles. *Nation's Restaurant News, 32*(16), 46, 48.

Chapter 2

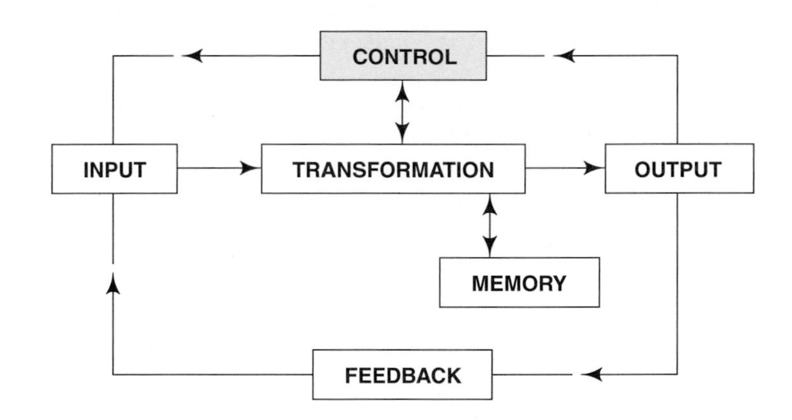

Managing Quality

Enduring Understanding

- Quality is defined by the customer through his or her satisfaction.
- Quality of foodservice operations needs to be improved on a continual basis.

Learning Objectives

After reading and studying this chapter, you should be able to:

1. Differentiate between quality assurance and process improvement programs.
2. Describe process improvement terms such as Six Sigma, reengineering, benchmarking, cause and effect diagrams, control charts, and Pareto analysis.
3. Discuss differences in Malcolm Baldrige Award criteria, ISO 9000 standards, and JCAHO standards.
4. Evaluate a foodservice operation using standards appropriate to that segment of the industry.

Quality is critical to the success of a foodservice operation. Achieving quality outputs requires attention to the quality of inputs and the quality of the transformation process. In this chapter, you will see how total quality management and process improvement are incorporated in the foodservice systems model. You will be introduced to the evolution of total quality management and process improvement in the United States and learn about individuals whose work has changed operational practices. We also will describe ways companies are recognized for their efforts to improve the quality of their operations.

QUALITY IN THE FOODSERVICE SYSTEM

Quality: Characteristics of a product or service that bears on its ability to satisfy stated or implied needs and a product or service that is free of defects.

Quality has intensely personal connotations and thus is a very difficult word to define. The American Society for Quality (ASQ), on its Web site (www.asq.org), provides two definitions for **quality**: "the characteristics of a product or service that bear on its ability to satisfy stated or implied needs, and a product or service that is free of defects." The ASQ suggests that

- Quality is not a program; it is an approach to business.
- Quality is defined by the customer through his or her satisfaction.
- Quality is aimed at performance excellence; anything less is an improvement opportunity.
- Quality increases customer satisfaction, reduces cycle times and costs, and eliminates errors and rework.

Figure 2.1 presents the foodservice systems model with emphasis on quality management/process improvement as a component of the control element. Organizational goals and objectives provide the beginning point for a total quality management (TQM) program. Quality customer service is the goal of both profit-oriented and not-for-profit organizations.

Goals and objectives provide the basis for defining quality standards, which in turn are used for developing policies and procedures for quality management/process improvement. The key to a TQM program is continuous monitoring and evaluation to determine if quality is being maintained in all aspects of operations. Feedback mechanisms are critical to providing information on the quality of both processes and products.

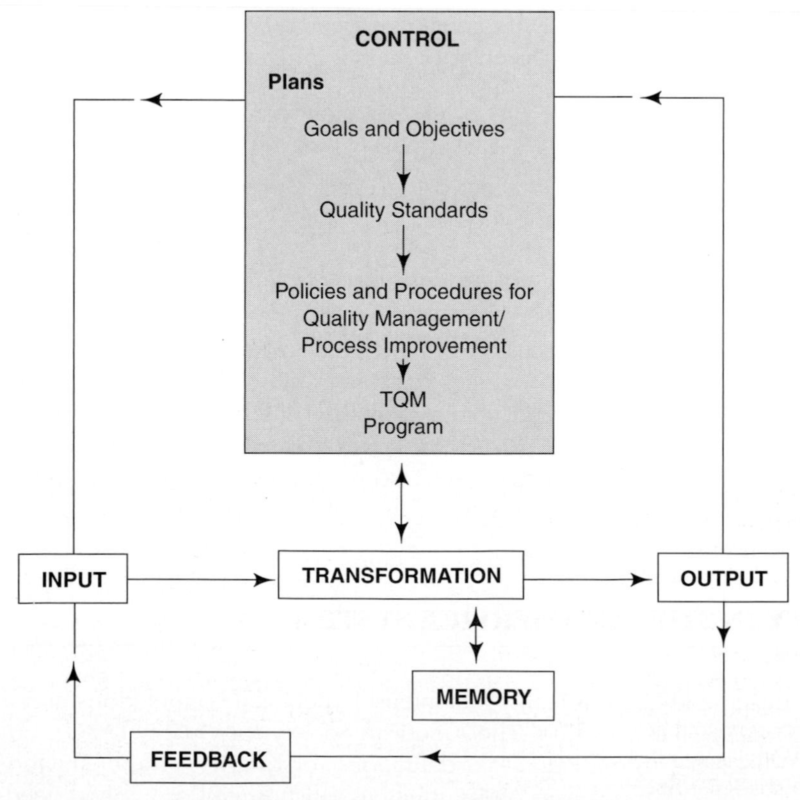

Figure 2.1. Total quality management in the foodservice system.

APPROACHES TO QUALITY IMPROVEMENT

Quality is important for the financial success of a business as well as the satisfaction of its customers. Buzzell and Gale (1987) were among the first to document empirically that the most important factor predicting success of a business is the quality of its products and services. A U.S. General Accounting Office study (1994) reported that companies using total quality management practices had higher profitability, greater customer satisfaction, better employee relations, and increased market share than companies that did not use TQM.

Concern about quality began in the 1940s with the advent of World War II. American statisticians W. Edwards Deming, Joseph Juran, and W. A. Shewhart developed new methods for managing quality in wartime industries that had to produce high-quality armaments with a largely unskilled labor force. The thought was that if inventories were kept low, good relationships with suppliers were made, and jobs were performed more efficiently, then better-quality products would be produced at lower cost.

Dr. W. Edwards Deming, often regarded as the intellectual father of total quality management, proposed a 14-point system to help companies increase their quality (Deming, 1986). His 14 points include the following:

1. Create consistency of purpose toward improvement of products and services.
2. Adopt the new philosophy of quality.
3. Cease dependence on inspection to achieve quality.
4. End the practice of choosing suppliers based solely on price.
5. Improve constantly and forever the production and service systems.
6. Institute extensive training on the job.
7. Shift focus from production numbers to quality.
8. Drive out fear.
9. Break down barriers among departments.
10. Eliminate slogans and targets for the workplace.
11. Eliminate numeric quotas for the workforce.
12. Remove barriers that rob employees of pride or workmanship and eliminate annual rating or merit systems.
13. Institute a vigorous program of education and self-improvement for everyone.
14. Make sure everyone in the company is put to work to accomplish the preceding 13 points.

Quality assurance: Procedure that defines and ensures maintenance of standards within prescribed tolerances.

Total quality management: Management philosophy in which processes are refined with goal of improving performance in response to customer needs and expectations.

After the war, the United States reverted to prewar manufacturing practices that depended on large inventories, labor-intensive production processes, and top-down management. The Japanese, however, were committed to rebuilding their country by following the teachings of Deming, Juran, and Shewhart. Their priority was to change customers' associations of "made in Japan" with poor quality. The Japanese learned to produce high-quality products that were cost effective.

An abundance of literature is available on dealing with quality issues in organizations. The terms **quality assurance**, **total quality management**, **continuous quality improvement, six sigma, reengineering, lean,** and the **theory of constraints** have been used to describe management approaches to improving performance. Definitions for each term are more precise:

- **Quality assurance (QA).** A procedure that defines and ensures maintenance of standards within prescribed tolerances for a product or service (Thorner & Manning, 1983).
- **Total quality management (TQM).** A management philosophy directed at improving customer satisfaction while promoting positive change and an effective cultural environment for continuous improvement of all organizational aspects (Gift, 1992).
- **Continuous quality improvement (CQI).** A focused management philosophy for providing leadership, structure, training, and an environment in which to improve continuously all organizational processes (Shands Hospital, 1992).
- **Six sigma.** A disciplined, data-driven approach for improving quality by removing defects and their causes (www.isixsigma.com).
- **Reengineering.** Radical redesign of business processes for dramatic improvement (Hammer, 1996).
- **Lean.** Using less human effort, less space, less capital, and less time to make products exactly as the customer wants with fewer defects than occur in mass production (Womack, Jones, & Roos, 1990).
- **Theory of constraints.** Concentration on exploiting and elevating constraints that slow production or service (Goldratt & Cox, 2004).

Quality Assurance

Quality assurance (QA) programs began in the 1970s as a way to improve quality. QA programs are output oriented and include the process of defining measurable quality standards and then putting controls in place to ensure that these standards are met. Measurement of performance is done to compare actual results to standards. Quality assurance is a reactive process and is predicated on follow-up and inspection and finding error after the fact.

Total Quality Management

Total quality management (TQM) is a management philosophy in which processes are refined with the goal of improving the performance of an organization in response to customer needs and expectations.

Zabel and Avery (1992) suggest that TQM has six components, as shown in Figure 2.2. Robbins and Coulter's (2002) definition of TQM includes

- Intense focus on the customer
- Concern for continual improvement
- Focus on process
- Improvement in quality of everything the company does
- Accurate measurement
- Empowerment of employees

TQM originated in manufacturing industries but currently is being applied to many organizations, including colleges and universities, governmental agencies, nonprofit organizations, restaurants, clubs, and schools.

Customers are people who use products or receive services, and TQM helps the organization focus on customers by identifying and satisfying their needs and expectations. A **customer** is anyone who is affected by a product or service and may

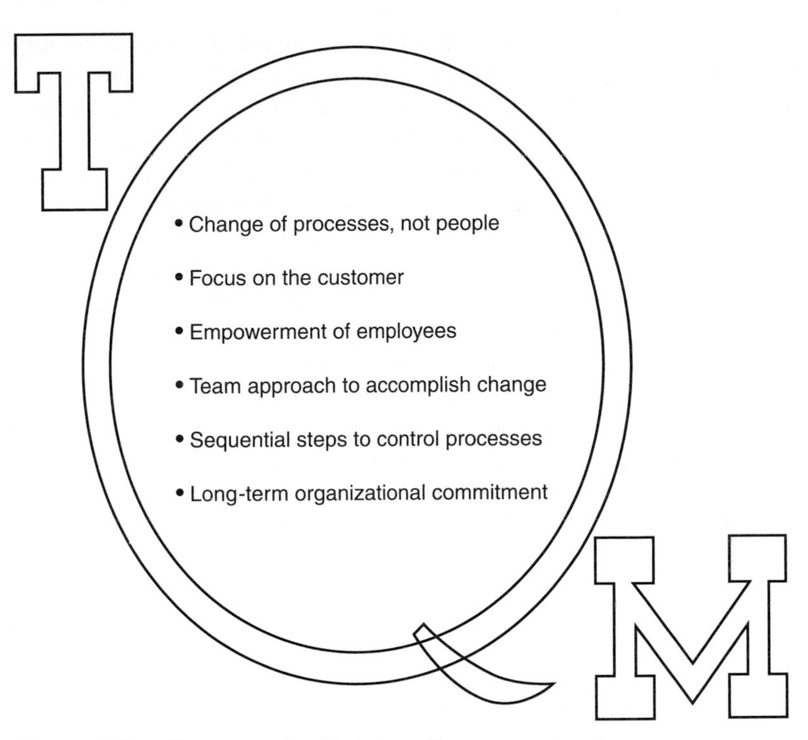

T
• Change of processes, not people
• Focus on the customer
• Empowerment of employees
• Team approach to accomplish change
• Sequential steps to control processes
• Long-term organizational commitment
Q
M

Figure 2.2. Components of total quality management.

be external or internal (Juran, 1992). **External customers,** like those who buy a hamburger or a pizza from a restaurant, are affected by the product but do not belong to the organization that produces it. **Internal customers** are both affected by the product and belong to the organization that produces it. Employees are internal customers, as are suppliers of food products. If a new supplier is added to the list, sales volume probably will decrease for the regular suppliers.

TQM encourages employees' participation in identifying problems and finding solutions for the improvement of the organization's overall performance. Management is responsible for facilitating and supporting employees' growth, thus empowering them to use their best abilities and improve their weaknesses (Ebel, 1991). **Empowerment** is the level or degree to which managers allow their employees to act independently within their job descriptions. It occurs when employees are permitted to make decisions in the areas where they work. Everyone in the organization requires training, retraining, and the opportunity to acquire and develop skills necessary to do their jobs better.

Empowerment: Level or degree to which managers allow employees to act independently within their job descriptions.

TQM usually involves teams of employees, a leader, and a facilitator for accomplishing change. Teams are assigned specific projects aimed at improving product or service quality while reducing costs. Facilitators assist the teams and usually are supervisors or specialists with special training for the role. Being a facilitator is a part-time assignment included in one's regular job. The facilitator helps trainees learn how to be team members by teaching them to communicate with one another, contribute information, challenge decisions, and share experiences. The facilitator also can serve as a consultant to the team chairperson.

The number of steps organizations use for implementing TQM processes varies depending upon the setting and the problem. TQM usually involves identifying a problem, determining causes, developing measurable improvement, selecting and implementing the best solution, collecting data to measure results statistically, refining the solution, and repeating the cycle.

TQM requires a long-term commitment from the organization and a more complete understanding of its operation. TQM cannot be implemented overnight; results take time to appear. Without upper-management commitment and participation, TQM initiatives are destined for failure. Free and open lines of communication among all departments must be established. Poor quality often can be traced to voids or overlaps of responsibility.

TQM programs have been implemented in many organizations. Quality has become increasingly important to organizations in part because of increased customer demand for high-quality products and services.

Taco Bell did a quality-led turnaround, which was used as a case study in the Harvard Business School. After a top-management change in the early 1980s, consumer studies were conducted in 1987 and 1989 to find out what customers really want. Before that, upper-level management thought that sales would increase if new items were added to the menu frequently. Surprisingly, customers were more interested in certain fundamentals than in menu changes. By analyzing and classifying customers' responses, management came up with the acronym FACT: **F**ast service and **A**ccurate orders with food served in a **C**lean restaurant and at the right **T**emperature. The company improved performance by redesigning the layout of its stores, preparing more menu items offsite to decrease preparation time, and scheduling employees more efficiently (Wood, 1993).

John Engstrom, director of Culinary Services at Virginia Polytechnic Institute and State University in Blacksburg, introduced TQM by leading a process he calls Triad of Excellence (TOE) for improving the department's quality, service, and cooperation. The second phase of the process at Virginia Polytechnic is a statistical analysis, called an environmental scan. Management and employees look at the department internally and externally, their customers, and the total market to evaluate their accomplishments. After Engstrom initiated TOE in 1992, the foodservice operation went from a deficit of $1.5 million to a $1 million profit (King, 1992).

Continuous Quality Improvement

Continuous quality improvement (CQI) involves reviewing operations on a routine basis with the goal of finding ways to continually improve the processes in and outcomes of that operation. CQI is commonly used within the medical community to describe a team-based approach to quality improvement efforts. The focus of CQI projects is on processes rather than people. A basic premise is to find ways to improve the process rather than blame individuals.

Sacred Heart Medical Center began its CQI efforts in 1990 (Weisberg, 2005). Howard Traver, the Medical Center's foodservice director, indicates that CQI has helped the department reduce full-time-equivalent employees by 18%, increase productivity by 70%, improve on-time delivery of late trays to 100%, and decrease costs by 40%.

Six Sigma

Six Sigma is a data-driven approach and technique for eliminating defects and reducing variations in any process. Started by the Motorola Corporation as a means for measuring quality, achieving Six Sigma means that a process cannot produce more than 3.4 defects per million opportunities.

Sigma is a Greek letter for a statistical unit of measurement used to define standard deviation. Six sigma is six standard deviations from the mean, meaning that very little variation in process occurs. As a point of comparison, average company performance in the United States is three sigma.

Six Sigma programs are designed using the DMAIC approach to quality improvement. This approach has managers **D**efine the project goals, **M**easure the current performance of the process, **A**nalyze and determine causes of defects, **I**mprove the process by eliminating defects, **C**ontrol future process performance, and **S**tandardize the process for the future.

The Manitowoc equipment company began incorporating Six Sigma in its operations in 2002 (see www.manitowocfsg.com). Company managers believe this process helps them better understand customer needs, produce more reliable products and services, make quicker decisions with data, and reduce operating costs.

Reengineering

The term reengineering was popularized in 1993 by authors Hammar and Champy in the first edition of their book *Reengineering the Corporation*. Reengineering was defined as a fundamental rethinking and radical redesign of business processes to achieve dramatic improvements in performance.

The concept of reengineering focuses on improving processes that are core to a business rather than on individual tasks completed by individual employees. A **process** is a complete end-to-end set of activities that together create value for a customer (Hammar & Champy, 2001). The foodservice system model depicts the core process for a foodservice operation, starting with inputs that are transformed into outputs.

Taco Bell serves as a model of positive results that can occur when reengineering principles are applied to an operation. Before reengineering, the average Taco Bell restaurant had only 30% of its floor space as customer seating. By reengineering their process for preparing food, Taco Bell stores increased their customer space to 70% of total space, decreased their labor needs by 15 hours/day, and increased the capacity of their kitchens from $400/hr to $1,500/hr. Automating company operations reduced manager time spent on paperwork and increased manager time on customer service activities. Taco Bell grew from a $500 million company to a $3 billion company in 10 years because it was able to redesign its business processes radically.

Lean

The concept of "lean" was first discussed in a benchmarking study of the automobile industry by Womack, Jones, and Roos (1990) in their book, *The Machine That Changed the World: The Story of Lean Production*. The book described how the

Toyota Company moved decision making to production employees, maintained a very low inventory, and grew to be a large and profitable company. The authors termed Toyota's production process as "lean production" because Toyota used less labor, less manufacturing space, and less equipment to develop a product in less time than its competition.

According to Nave (2002), there are five essential steps in lean:

- Identify which features create value for internal or external customers.
- Document the value stream (sequence of activities that create value).
- Improve flow (eliminate things that interrupt flow).
- Let the customer pull the product or service through the process (provide product or service only when customer wants it).
- Perfect the process.

The Food and Nutrition department at the University of Pittsburgh Medical Center Shadyside used lean principles to enhance its operations (www.lean.org). Patients in the hospital selected from preset menus for prescribed diets (such as low fat, low sodium, etc.). The department moved to a standard, restaurant-style menu for all patients. Menu icons highlighted good choices for specific conditions. The results were a reduction in food costs and an increase in patient and hospital staff satisfaction.

Theory of Constraints

Constraint: Something that limits an organization from reaching its goals.

The theory of constraints was introduced by Goldratt in 1984 in his book, *The Goal*. A **constraint** is anything in an organization that keeps the organization from reaching its goals. Constraints could be physical, like the capacity of a machine or individual worker, or could nonphysical, such as demand for a product or company policies.

Organizations applying the theory of constraints would first identify the constraints. Once identified, employees and managers would work to exploit and elevate (break) the constraint. Exploiting the constraint involves exploring ways to improve or support the process causing the constraint without major expense or upgrades. If this does not work, then major changes may need to be made to the constraint. This is termed elevating the constraint by taking whatever action is necessary to eliminate it.

TOOLS USED IN PROCESS IMPROVEMENT

Several tools have been used to help managers with their process improvement efforts: benchmarking, the plan-do-check-act cycle, cause and effect diagrams, control charts, and Pareto analysis. These tools assist in identifying problems and guiding decisions on changes that might be needed.

Benchmarking: Comparison against best performance in the field.

Benchmarking involves comparing one's performance with those believed to be "best in class." Such comparisons assist managers in identifying areas on which to focus improvement efforts. Many companies and some professional organizations compile data to assist organizations with this comparison. For example, hospital foodservice managers might work with Press Ganey Associates, Inc., a

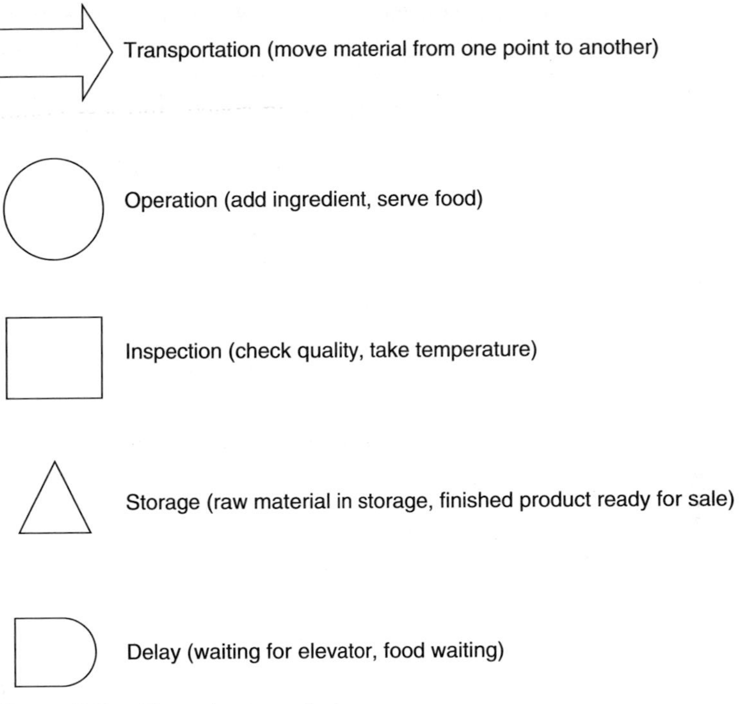

Figure 2.3. Flow chart symbols.

company that collects and prepares comparative analyses on patient satisfaction ratings in hospitals. The Press Ganey patient satisfaction survey includes several items related to patient foodservice.

Flow charts are a graphical representation of the steps in a process. A flow chart details all of the elements in a process and the sequence in which these elements occur. A unique set of symbols developed by the American Standard National Institute (ANSI) is used when drawing flow charts (Figure 2.3).

The **plan-do-check-act (PDCA)** cycle is a model for coordinating process improvement efforts. As shown in Figure 2.4, the cycle is depicted as a circle with four equal quadrants. The model was developed originally by Walter Shewhart, a statisti-

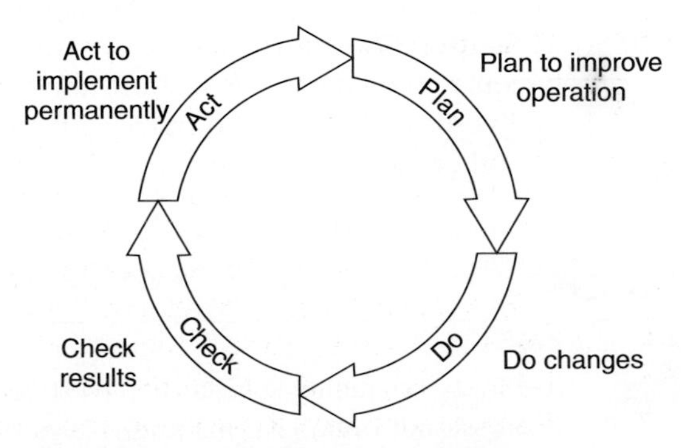

Figure 2.4. Plan-do-check-act cycle.

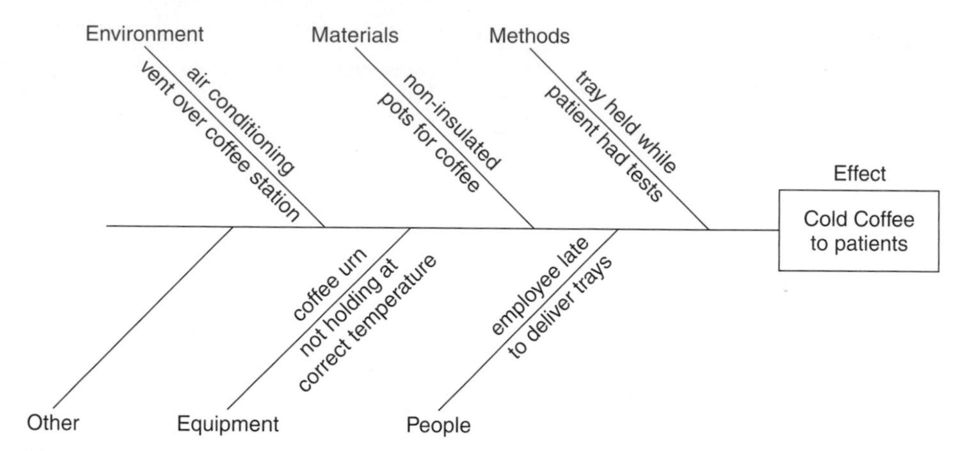

Figure 2.5. Cause and effect diagram.

cian working with process control for Bell Laboratories in the 1930s. The cycle gained national recognition through the work of W. Edwards Deming in the 1950s.

Cause and effect diagrams (also referred to as Ishikawa or fish-bone diagrams) illustrate the factors that may influence or cause a given outcome. Figure 2.5 provides an example of such a diagram for patient complaints of cold coffee. The diagram provides a way to identify and categorize factors that might be the cause of such a problem.

Histograms are bar graphs that are used to display graphically the frequency distribution of data. Using histograms provides a visual way to examine patterns in data that might not be evident when just looking at the numbers themselves.

Control charts are a graphical record of process performance over a period of time. Values of upper and lower control limits (often set as two or three standard deviations from the mean) are drawn to help identify potential problems. The control chart in Figure 2.6 tracks the number of requests for schedule change received by the foodservice manager for each day in a given month.

Pareto analysis, often called the 80–20 rule because 80% of a given outcome typically results from 20% of an input, was named after nineteenth-century economist Vilfredo Pareto, who suggested that most effects come from relatively few causes. Marketers, for example, have found that 80% of sales often come from 20% of customers. Juran (1992) described it as the "vital few and trivial many." Basically there are a vital few causes that if identified and corrected can have the greatest impact on improving quality. Managers should focus on these vital few causes.

Scatter diagrams provide a visual way to examine possible relationships between two variables. Data from the two variables are plotted on a horizontal (*X*) and vertical (*Y*) axis. The purpose is to display what happens to one variable when the other variable changes.

QUALITY STANDARDS

A variety of quality standards have been published. Some have been developed by professional organizations and others by industry groups. Some serve as guidelines for an operation; others are mandated standards to be met as part of accreditation.

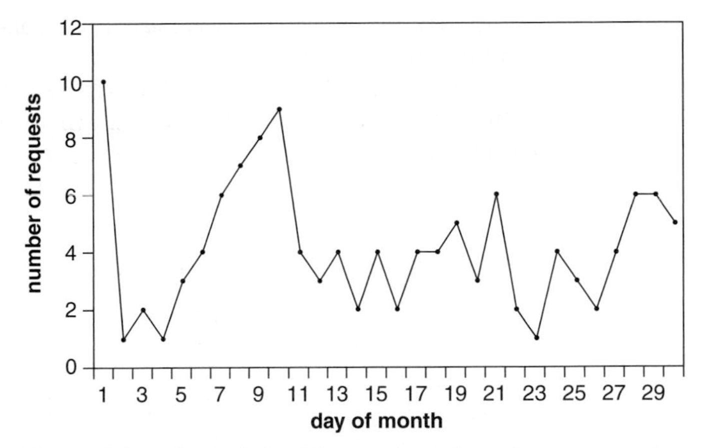

Figure 2.6. Control chart for number of employee requests to change work schedule.

ISO 9000 Standards

The ISO 9000 standards are a group of five individual but related international standards on quality management. These standards were developed by the International Organization for Standardization (ISO). The ISO is a worldwide, nongovernmental federation consisting of representatives from national standards groups in more than 140 countries (see www.iso.org).

Although the vast majority of the ISO standards are specific to a particular product, material, or process, the ISO 9000 series of standards are known as generic management standards. The standards are based on eight principles that can be used by management as a framework to guide an organization toward improved performance. The eight principles are

- Customer focus
- Leadership
- Involvement of people
- Process approach
- Systems approach to management
- Continual improvement
- Factual approach to decision making
- Mutually beneficial supplier relations

In 1998, Pulaski Community Hospital in Virginia was the first hospital in the United States to be registered as meeting the ISO 9000 standards; its Quality System registration was reviewed and renewed in 2002. The hospital's foodservice department offers room service to all patients as a way of providing patients what they want to eat, when they want it.

Keys to Excellence

The School Nutrition Association (formerly American School Foodservice Association) developed the *Keys to Excellence* to provide quality standards for Child Nutrition Programs (CNPs) (see www.schoolnutrition.org). The *Keys* contain indicators of quality in four areas: administration, communication and marketing, nutiriton and nutrition education, and operation.

CNP directors can do an online self-assessment of their program using the *Keys*. Schools demonstrating superior performance in providing nutritious food and a healthy educational environment for all students can earn designation as a District of Excellence in Child Nutrition.

Professional Practices in College and University Foodservice

The National Association of College and University Food Services (NACUFS) developed the *Professional Practices in College and University Food Services* manual as a tool for professional development, self-assessment, and continuous quality improvement (see www.nacufs.org). The manual contains objective principles and practices in 15 areas: professionalism, organization, planning, marketing, operational controls, human resources, menu management, purchasing/receiving/storage, food safety, service management, safety and security, facilities management, technology, capital improvement, and contract services.

Joint Commission on Accreditation of Healthcare Organizations

The healthcare segment of the foodservice industry was the first to give impetus to the establishment of formalized TQM programs. In 1979, the Joint Commission on Accreditation of Healthcare Organizations (JCAHO) published the original quality assurance standards for hospitals. JCAHO is an independent, not-for-profit organization that sets standards for and accredits healthcare organizations (see www.jcaho.org). The mission of JCAHO is to continuously improve the safety and quality of care provided to the public through the provision of healthcare accreditation and related services that support performance improvement in healthcare organizations. JCAHO evaluates and accredits more than 15,000 healthcare organizations and programs in the United States; approximately 80% of hospitals are accredited by JCAHO.

The joint commission initially began in 1910 when Ernest Codman, M.D., proposed the "end result system of hospital standardization." Under this system, a hospital tracks every patient treated long enough to determine whether the treatment was effective. If the treatment was not effective, the hospital would then attempt to determine why; thus, similar cases could be treated successfully in the future. In 1951, the joint commission developed state-of-the-art, professionally based standards and evaluated the compliance of healthcare organizations against these benchmarks. The joint commission has undergone numerous changes throughout its history. However, its basic philosophy has remained constant. The joint commission is the embodiment of professional leadership in stimulating the healthcare field to attain ever-higher standards for providing care to the American people.

Standard development is an ongoing dynamic process that continually incorporates the experience and perspective of healthcare professionals and other individuals and organizations throughout the country. The standards that result from this process focus on the essential elements of providing quality healthcare and are nationally recognized for the significant contributions they make to improving the care delivered to patients in the United States. JCAHO standards

- Address performance expectations in several areas (ethics, rights, and responsibilities; leadership; assessment of patients; management of the environment of care; care of patients; management of human resources; education; management of information; continuum of care; governance; provision of care, treatment, and services; management; medication management; medical staff; surveillance, prevention and control of infection; nursing; and improving organization performance)
- Focus on what an organization does not what is has
- Are developed in consultation with healthcare experts, providers, measurement experts, purchasers, and consumers

JCAHO launched a new initiative, ORYX®, in 1997 that integrates outcomes and other performance measurement data into the accreditation process. The initiative adds a quality improvement component to the accreditation process by having hospitals submit data in core measurement areas such as acute myocardial infarction; heart failure; community-acquired pneumonia; and pregnancy and related conditions.

In 2004, the JACHO survey process shifted to include a greater focus on tracing the care of specific patients as they moved through the healthcare system. This "tracer" process involves identifying specific patients and then tracing the care, treatment, and services they received as a way to measure the quality of care and services provided in that facility.

At one time, dietetic services had a separate set of standards which included standards related to the handling and preparation of food; the current JCAHO manual incorporates standards related to the management of the food and nutrition operation under many standards, including care of patients, environment of care, human resources, information management, and leadership.

JACHO accreditation is voluntary. Healthcare organizations desiring JCAHO accreditation complete a self-study and have their operations reviewed by a trained team of reviewers. Organizations that meet the standards receive accreditation status for a 3-year period. Organizations are subject to unannounced visits by reviewers at any point during their accreditation. Information about accredited hospitals is available at www.qualitycheck.org.

EXTERNAL RECOGNITION OF QUALITY

A variety of recognitions exist related to quality management and implementation in organizations. Individuals can receive certification and organizations can be recognized.

The American Society for Quality (ASQ) offers several options for certification: Quality Manager (CQM), Quality Engineer (CQE), Quality Auditor (CQA), HACCP Auditor (CHA), Biomedical Auditor (CBA), Reliability Engineer (CRE), Quality Technician (CQT), Mechanical Inspector (CMI), Software Quality Engineer (CSQE), Quality Improvement Associate (CQIA), Six Sigma Black Belt (SSBB), and Manager of Quality/Organizational Excellence (CMQ/OE). The iSixSigma organization offers certification as Champion, Master Black Belt, Black Belt, and Green Belt.

Many states also present quality awards each year. Organizations that include foodservice as a component of their operation that recently have received such awards include Charles R. Hadley Elementary School (received the Florida Governor's Sterling Award, 2004), St. Francis Hospital and Health Services (received the Missouri Quality Award, 2004), Swedish Medical Center (received the Washington State Quality Award in 2004), VA Medical Center (received the Vermont Governor's Award for Performance Excellence in 2003), Southern Ohio Medical Center (received the 2002 Ohio Governor's Award for Excellence), and San Jose Unified School District (received the California 2001 Eureka Award For Performance Excellence).

The Malcolm Baldrige National Quality Award

The Malcolm Baldrige National Quality Improvement Act of 1987, signed by President Reagan on August 20, 1987, established an annual U.S. national quality award, entitled the Malcolm Baldrige National Quality Award. The award was established to recognize companies for their achievements in quality and performance (see www.quality.nist.gov). The award promotes sharing of information on successful performance strategies. The award, managed by the U.S. Commerce Department's National Institute of Standards and Technology, is open to businesses headquartered in the United States or its territories. Up to three awards may be given each year in each of five categories:

- Manufacturing
- Service
- Small businesses
- Education
- Healthcare

Criteria for performance excellence provide organizations with an integrated, results-oriented framework for implementing and assessing processes for managing all operations. These criteria are also the basis for making awards and providing feedback to applicants. The criteria consist of seven categories:

- Leadership: The company's leadership system, values, expectations, and public responsibilities
- Strategic planning: The effectiveness of strategic and business planning and deployment of plans, with a strong focus on customer and operational performance requirements
- Customer and market focus: How the company determines customer and market requirements and expectations, enhances relationships with customers, and determines their satisfaction
- Measurement, analysis, and knowledge management: The effectiveness of information collection and analysis to support customer-driven performance excellence and marketplace success
- Human resource focus: The success of efforts to realize the full potential of the workforce to create a high-performance organization
- Process management: The effectiveness of systems and processes for assuring the quality of products and services
- Business results: Performance results, trends, and comparison to competitors in key business areas, including customer satisfaction, financial, marketplace, human resources, suppliers and partners, operations, governance, and social responsibility

Award winning organizations have included Ritz Carlton (1992, 1989—the only two time award winner), Sunny Fresh Foods (1999), Chugach and Pearl River School districts (2001), the University of Wisconsin–Stout (2001), Pal's Sudden Service (2002), Palatine Consolidated Community Schools (2003), Baptist Hospital (2003), St. Luke's Hospital (2003), and Robert Wood Johnson University Hospital (2004). Davis (1992) stated that perhaps the best lesson learned from Baldrige Award recipients is the vital importance of human resource practices that support quality work and encourage people at the middle and lower levels to create ideas for improvement.

CHAPTER SUMMARY

This summary is organized by the learning objectives.

1. Quality assurance involves setting standards, continuously monitoring and evaluating to determine if standards are being met, and modifying operations as needed. Process improvement focuses on analyzing current processes in an operation and developing ideas for improving these processes.
2. Six Sigma is a business method for improving quality by removing defects and their causes with a goal of having not more than 3.4 defects per million occurrences. Reengineering is the radical redesign of business processes to achieve dramatic improvements in performance. Benchmarking is the comparison of performance with others believed to be "best in class." Cause and effect diagrams (Ishikawa or fishbone diagrams) illustrate the main causes and subcauses leading to an effect. Control charts are a graphical record of process performance over time. Pareto analysis focuses on analyzing the 20% of processes that impact 80% of outcome.
3. The categories for the Malcolm Baldrige Award include leadership; strategic planning; customer and market focus; measurement, analysis, and knowledge management; human resource focus; process management; and business results. The ISO 9000 standards include leadership, customer focus, process approach, involvement of people, a systems approach to management, continual improvement, a factual approach to decision making, and mutually beneficial supplier relationships. JCAHO standards include assessment of patients, care of patients, an environment of care, human resources, information management, leadership, medical staff, and patient rights. All include standards related to leadership, customers, and employees.
4. General quality standards such as the ISO 9000 standards or the Malcolm Baldrige criteria or industry-specific standards such as the SNA *Keys to Excellence* or the NACUFS *Professional Practices* provide criteria that could be used to evaluate a foodservice operation.

TEST YOUR KNOWLEDGE

1. Write a definition for the term *quality*.
2. What is JCAHO and how has it affected quality management in hospitals?
3. What are Deming's 14 points?

Professional Organizational Profile

School Nutrition Association

MISSION

To advance good nutrition for all children

WHO BELONGS TO THE ORGANIZATION

Membership in this organization includes more than 54,000 school nutrition professionals from local school line workers and local directors to state personnel, and industry and college professionals who strive to provide healthful meals and nutrition education to all children.

ADVANTAGES OF MEMBERSHIP

Members have numerous opportunities for networking with other school foodservice professionals across the country and have access to a website that keeps them abreast of child nutrition issues, as well as organization news. Members receive the *School Foodservice and Nutrition* publication and have professional development opportunities available at the local, state, and national levels. Student memberships are available.

WEBSITE

http://www.schoolnutrition.org/

MEET THE PRESIDENT

Janey Thornton, SFNS, President 2006–2007, has worked in school child nutrition programs for more than 24 years. She graduated with a Masters and Rank 1 in vocational home economics and school administration and is currently pursuing a doctorate from Iowa State University. She worked in the Kentucky State Department of Education for ten years supervising home economics teachers and working extensively with curriculum development, directing several projects jointly with the Kentucky Restaurant Association. Janey is currently director of Child Nutrition Programs for Hardin County Schools in Elizabethtown, KY. Janey offers the following advice to students: "So many careers are out there that offer so many personal and professional opportunities; don't give up until you find your place to make a difference."

4. What are the components of TQM?
5. What is benchmarking? How is it related to quality management?
6. How are Six Sigma and reengineering similar?
7. What are the ISO 9000 standards?

CLASS PROJECTS

1. Ask a manager from a local foodservice operation that has an active quality improvement program to speak to the class about one of the operation's quality improvement projects. Have the manager include the objectives of the project, the type of data collected, how the data are shared with employees, and how results are used to improve operations.
2. Divide into groups of two to three students. Develop an Ishikawa (fish-bone) diagram to identify possible causes of being short of food on the school serving line (the effect). Share diagrams with other groups.

WEB SOURCES

www.asq.org	American Society for Quality
www.iso.org	International Organization for Standardization
www.jcaho.org	Joint Commission on Accreditation for Healthcare Organizations
www.nist.gov	National Institute of Standards and Technology
www.isixsigma.com	Online resource for Six Sigma information

BIBLIOGRAPHY

Beasley, M. A. (1991a). The story behind quality improvement. *Food Management, 26*(5), 52, 56–57, 60.

Beasley, M. A. (1991b). Quality improvement process—Part II. *Food Management, 26*(6), 21–26.

Blazeg, M. L. (2002). *Insights to performance excellence in health care! An inside look at the 2002 Baldrige Award criteria for health care.* Milwaukee, WI: ASQ Quality Press.

Buzzell, R. D. & Gale, B. T. (1987). *The PIMS Principles: Linking Strategy to Performance.* New York: The Free Press.

Carter, J. R., & Narasimhan, R. (1993). *Purchasing and materials management's side in total quality management and customer satisfaction.* Tempe. AZ: Center for Advanced Purchasing Studies.

Cox, D. R. (1990). Quality and reliability: Some recent developments and a historical perspective. *Journal of the Operational Research Society, 41*(2), 95–101.

Davis, L. E. (1980). Individuals in the organization. *California Management Review, 22*(3), 5–14.

Davis, T. (1992). Conference report: Baldrige winners link quality, strategy, and financial management. *Planning Review, 20*(6), 36–40.

Deming, W. E. (1986). *Out of crisis.* Cambridge, MA: MIT Center for Advanced Engineering Study.

Duncan, W. J., & Van Matre, J. G. (1990). The gospel according to Deming: Is it really new? *Business Horizons, 33*(4), 3–9.

Eacho, B. (1992). Quality service through strategic foodservice partnerships: A new trend. *Hosteur, 2*(2), 22–23, 28.

Ebel, K. E. (1991). *Achieving excellence in business: A practical guide to the total quality transformation process*. Milwaukee: American Society for Quality Control, and New York: Marcel Dekker.

Garwin, D. A. (1991). How the Baldrige Award really works. *Harvard Business Review, 69*(6), 80–93.

Gift, B. (1992). On the road to TQM. *Food Management, 27*(4), 88–89.

Goldratt, E. M., & Cox, J. (2004). *The goal: A process of ongoing improvement*. Great Barrington, MA: The North River Press.

Green, L. (1999). A simplified TQM diagnostic model. [Online]. Available: http://www.skyenet.net/~leg/tqm.htm

Hammer, M. (1996). *Beyond reengineering*. New York: Harper Business.

Hammer, M. (2001). *The agenda: What every business must do to dominate the decade*. New York: Crown Business.

Hammer, M., & Champy, J. A. (2001). *Reengineering the corporation: A manifesto for business revolution*. New York: Harper Business.

Hart, W. L., & Bogan, C. E. (1992). *The Baldrige: What it is, how it's won, how to use it to improve quality in your organization*. New York: McGraw-Hill.

Heymann. K. (1992). Quality management: A ten-point model. *Cornell Hotel and Restaurant Administration Quarterly, 33*(5), 51–60.

Joint Commission for Accreditation of Healthcare Organizations, Oakbrook Terrace. II. (2002). [Online]. Available: http://www.jcaho.org

Juran, J. M. (1992). *Juran on quality by design: The new steps for planning quality into goods and services*. New York: Free Press.

Juran, J. M., & Godf, A. B. (2001). *Total quality management*. New York: McGraw-Hill Professional Publishing.

Kanter, R. M. (1991). Transcending business boundaries: 12,000 world managers view change. *Harvard Business Review, 69*(3), 151–164.

Khoong, C. M. (ed.). (1998). *Reengineering in action: The quest for world-class excellence*. London: Imperial College Press.

King, P. (1992). A total quality makeover. *Food Management, 27*(4), 96–98, 102, 107.

Learning the language of quality care. (1993). *Journal of The American Dietetic Association, 93,* 531–532.

Malcolm Baldrige National Quality Award. (2002). [Online]. Available: http://quality.nist.gov

Management Practices: U.S. Companies Improve Performance through Quality. (1994). Washington, DC: US General Accounting Office (GAO/NSIA 0-91-190).

National Association of College and University Food Services. (2004). *Professional practices in college and university food services,* 4th ed. East Lansing: Michigan State University.

Nave, D. (2002). How to compare Six Sigma, lean, and the theroy of constraints. *Quality Progress*. March: 73–78. [Online]. Available: www.asq.org.

Niven, D. (1993). When times get tough, what happens to TQM? *Harvard Business Review, 71*(3), 20–22, 24–26, 28–29, 32–34.

Robbins, S. P., & Coulter, M. (2002). *Management,* 7th ed. Upper Saddle River, NJ: Prentice Hall.

School Nutrition Association. (Shands Hospital at the University of Florida. (1992). *Management initiative for continuous quality improvement*. Gainesville, FL: Author.

Shands Hospital at the University of Florida (1992). *Management Initiative for Continuous quality improvement*. Gainesville, FL: Author.

Thorner, M. E. & Manning, P. B. (1983). *Quality Control in Foodservice* (rev. ed.). Westport, CT: AVI.

Walton, M. (1991). *Deming management at work*. New York: Putnam Sons.

Weisberg, K. (2005). Continuing the journey. *Foodservice Director, 13*(3), 22–23.

Whitley, R. (1993). *The customer driven company: Moving from talk to action*. Reading, MA: Addison-Wesley.

Womack, J. P., Jones, D. T., & Roos, D. (1990). *The machine that changed the world: The story of lean production*. New York: Rawson and Associates.

Wood, T. (1993). Total quality management: Learning from other industries. *Restaurants USA, 13*(2), 16–19.

Zabel, D., & Avery, C. (1992). Total quality management: A primer. *RQ,* 32(2), 206–216.

CHAPTER 3

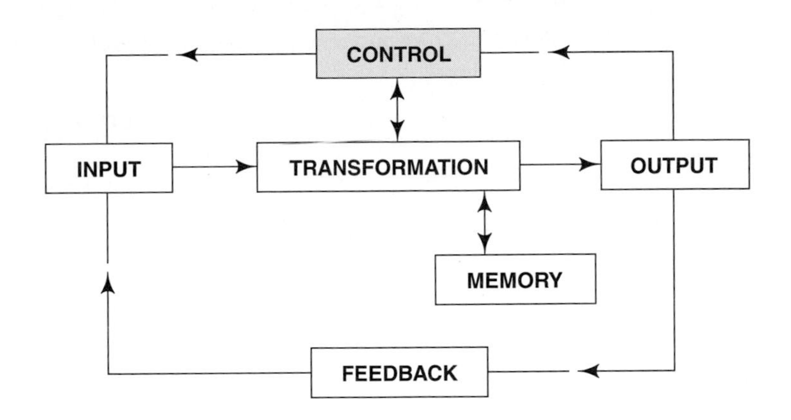

The Menu

Enduring Understanding

- The menu is the primary control for the foodservice system and impacts all components of the system.

Learning Objectives

After reading and studying this chapter, you should be able to:

1. Differentiate menu-related terms such as static, cycle, and single-use menus and à la carte versus table d'hôte.
2. Evaluate the aesthetic characteristics of a menu.
3. Plan a static or cycle menu for a foodservice operation.

Menu: List of items available for selection by a customer and the most important internal control of the foodservice system.

The **menu** is a key component of a foodservice operation. It establishes the inputs needed by the system. A change in menu can impact all aspects of the foodservice system. In this chapter, you will learn about different types of menus and how to plan and evaluate menus. You will come to see why the menu is considered the primary control of the foodservice system.

THE MENU

The menu, a list of food items, serves as the primary control of the foodservice operation and is the core common to all functions of the system (Figure 3.1). The menu controls each subsystem, is a major determinant for the budget, and reflects the "personality" of the foodservice operation. The menu impacts the layout of the operation and equipment needed to produce it. To the production employee, the menu indicates work to be done; to the waitstaff, the foods to be served; and to the dishroom staff, the number and types of dishes, glasses, and flatware requiring washing and sanitizing. According to Panitz (2000), the menu gives customers a sense of who you are as an operation; it is part of an organization's brand identity. Different foodservice operations will have different menus because organizational objectives and customers are different. The menu, therefore, expresses the character of a foodservice operation and is largely responsible for its reputation, good or bad.

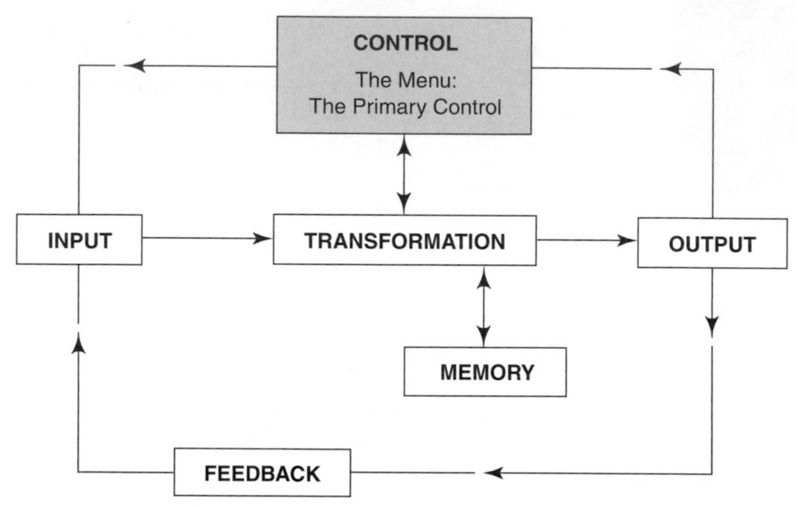

Figure 3.1. The menu: The primary control of the foodservice system.

MENU TRENDS

As Americans continue to travel and become more exposed to foods of different cultures and various regions of countries, they will continue to crave foods from these countries. Most popular are foods with Italian, Cajun, Spanish, French, Japanese, Greek, Middle Eastern, Cuban, Thai, Tex-Mex, and Caribbean influence (Davis, 2003). Meats, poultry, and fish are regularly prepared with wine or spirits or layered with trendy crusts and savory sauces (Sloan, 2004). Milk, especially flavored milk, and bottled waters are much more popular beverages than in past years. Growing concerns about obesity have resulted in increased consumer interest in products considered healthy, such as fruits, vegetables, grains, nuts, and yogurt (Sloan, 2005). Restaurants have responded to this interest in a variety of ways: McDonald's added options of milk, juice, and apple wedges as options in its children's Happy Meals; Wendy's offers choices of salad, fruit, or baked potato in place of its french fries in its combo meals; Applebee's Neighborhood Bar & Grill linked with Weight Watchers and provides Weight Watcher options and points on its menus. More entrées are meatless and different varieties of vegetables will be on the plate, including red and gold beets, white and gold turnips, and red and white carrots. Asian vegetables such as choy sum, long beans, and Chinese chives are becoming more popular as is the use of various peppers such as ancho, poblano, guajillo, and piquillo. Coffee will become mellower and tea, especially white, will be rediscovered. Self-contained food choices like wraps and calzones, which can be eaten easily at a desk or in the car, are becoming increasingly popular. In 2003, 40% of consumers reported eating to-go foods as part of breakfast, 55% ate to-go foods for lunch, and 37% had to-go foods as part of their dinner (Information Resource Inc., 2003). Curbside to-go, which offers consumers the option of phoning in an order and then being able to park in a designated area for pick-up, is becoming available at many restaurants. Focaccia and breads with herbs, olives, or anything else that has flavor will live long lives. "Food with a face" campaigns have increased purchase of locally grown foods; the grower is known and thus becomes to the consumer a real person with a face (Yee, 2004). Demand is increasing for organic, sustainably grown, free-range and grass-fed, and grown-from-heirloom-seeds foods. Fresh fruits are becoming more common on dessert menus, and salsas made with exotic fruits such as persimmon, guava, and cherimoya often accompany entrées (Yee, 2005). Raw dishes such as sushi, sashimi, ceviche, and seafood tartare are gaining consumer acceptance in many areas of the country. The distinctions between commercial and onsite foodservice are becoming fainter, and more customers will order food from the Internet.

MENU PRESENTATION

The menu provides a list of food items available for selection by a customer, and thus it is a sales tool for foodservice operations. An additional function of the menu is customer education. For example, the menu is a valuable tool for teaching children about good nutrition through the school lunch program and for instructing hospital patients on modified diets.

Hospitals are undergoing some unique changes in menu presentation which have improved patient satisfaction. In some hospitals an employee, often a dietetic technician, visit patients prior to the next meal to read the menu and record patient choices on a printed menu. Patient satisfaction often is increased because patients are able to interact in the menu selection process and can select their menu closer to the time of service. This type of menu is being referred to as a **spoken menu** because it is presented orally to the patient. Another trend is the implementation of **room service** menus, similar to that used in hotels. Patients call the foodservice department when they are hungry and ready to eat and order from a room service menu. Food items are delivered to the patient usually within 30–45 minutes.

Spoken menu: Menu that is presented by the technician orally to the patient.

Restuarants use a variety or menu presentation techniques to share menus with customers, including printing the menus in large font, Braille, and multiple languages. The use of a menu board provides a way to change menus daily and advertise daily specials. Menus can be handwritten on chalkboards or on fluorescent illuminated blackboards. Letters and numbers are commercially available in a wide selection of colors for mounting on grooved felt boards. Many operations post their menus on their Web site as well.

The video menu board, a recent innovation in menu merchandising, has the advantage of being changed readily. Current menus are displayed on video monitors that are placed in locations easily seen by customers. The video menu system has been used in plant and corporate dining facilities and has application for hospitals (to inform patients of daily menu offerings). Video menu software is available and may be obtained in a range from black and white to complex color graphics. Ordering by computer or cell phone text messaging is being used in some operations. Facsimile machines electronically transmit printed or photographed material over telephone lines for reception and reproduction by the receiving machine. Many restaurants are providing a fax number to customers to place orders for take-out, delivery, or catered meals.

Often, when table d'hôte meals are served, a verbal menu is recited by the waitstaff. **Table d'hôte**, a French term for the host's table, is a complete meal consisting of several courses at a fixed price. This differs from **à la carte**, in which food items are priced individually.

Table d'hôte: Several food items grouped together and sold for one price.

À la carte: Food items priced individually.

Menu psychology: Designing and layout a menu in such a way as to influence the sale of foods served on that menu.

Techniques used in the graphic design and layout of a menu to influence menu selections by customers often are referred to as **menu psychology**. These techniques include print style and size, paper type and color, ink color, graphic illustrations and designs, and placement on a page. These techniques are not designed to make customers buy things they do not want but rather to showcase items in such a way as to encourage customers to give thought to items they might not have otherwise considered.

The printed menu is an important communication tool for a foodservice operation. The menu cover should project an accurate image of the operation and suggest to the customer the formality, price range, and even theme of foods served. According to Pavesic and Magnant (2005), key elements used in menu psychology include

- **Eye gaze motion:** The eye will travel in a set pattern when viewing a menu (Figure 3.2). Thus the center of a three-fold menu is considered the prime menu sales area.
- **Primacy and recency:** Position menu items you want to sell more of in the first and last positions within a category as the first and last things a customer reads. These are the items more likely than others to be chosen.
- **Font size and style:** Increase the size of font to attract the customer's attention to an item; decrease the size to deflect attention from an item. Avoid use of fonts that are difficult to read, especially in dim lighting.
- **Color and brightness:** Increase the brightness, color, or shading of visual elements to attract customer attention.
- **Spacing and grouping:** Use borders around or placement of items together within a space to draw attention to items.

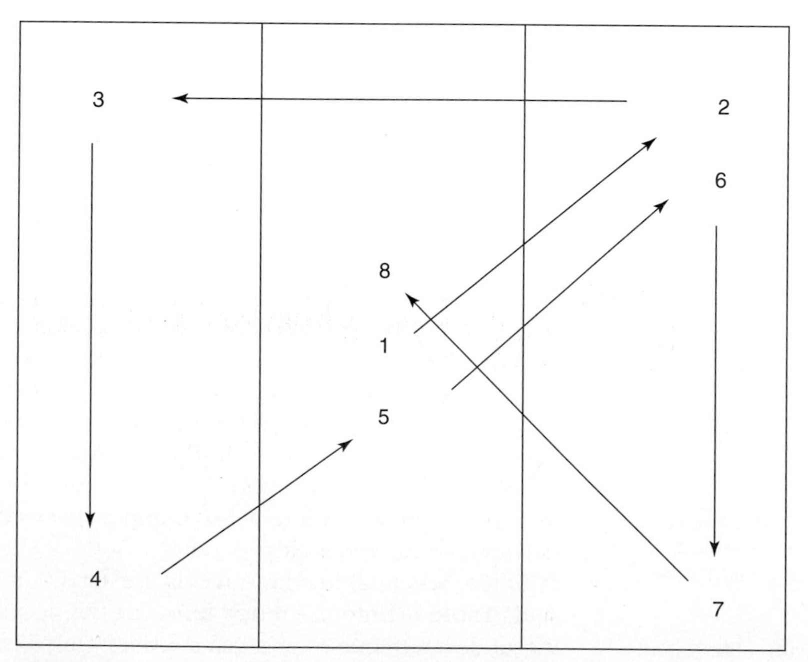

Figure 3.2. Eye movement across a three-fold menu.

MENU PATTERN

A menu pattern is an outline of the menu item categories for each meal, such as appetizers, entrées, and desserts on the dinner menu. One of three basic types of menus generally is used, and the number of menu item choices in each of these can vary according to the goals of the foodservice operation.

Types of Menus

The three types of menus are static, cycle, and single use. Deciding which type to use depends upon expectations of the customer when eating away from home.

Static

Static menu: Same menu items are offered every day; that is, a restaurant-type menu.

A **static menu** is one in which the same menu items are offered every day. Traditionally, the static menu has been characteristic of many restaurants; however, many hospitals are using restaurant-type menus as well.

The entire concept for a restaurant frequently is built around the menu, including the décor, advertising campaign, and market segment identified as the target audience. Examples include Red Lobster restaurants (www.redlobster.com), whose décor includes fishnets, sea shells, and other ocean-related items and whose menu features a variety of seafood items; or Olive Garden restaurants (www.olivegarden.com), which create an Italian environment with both the décor and the menu offerings.

Cycle

Cycle menu: Series of menus offering different items daily on a weekly, biweekly, or some other basis, after which the menus are repeated.

A **cycle menu** is a series of menus offering different items each day on a weekly, biweekly, or some other basis, after which the cycle is repeated. In many onsite foodservice operations, seasonal cycle menus are common. Cycle menus typically are used in healthcare institutions and schools and offer variety with some degree of control over purchasing, production, and cost.

Single Use

Single-use menu: Menu that is planned for service on a particular day and not used in the exact form a second time.

The last of the three basic menu types, the **single-use menu**, is planned for service on a particular day and is not used in the exact form a second time. This type of menu is used most frequently for special events. Many college and university foodservices use a single-use menu as a "monotony breaker"; the menu in Figure 3.3 is an example of a menu for an event at Kansas State University.

Degree of Choice

Menus also may be categorized by the degree of choice. The number of choices in each menu item category is determined by the goals of each foodservice operation. Choices may vary from many to few in static, cycle, and single-use menus. A static menu might have few choices in a limited-menu restaurant. A cycle or single-use menu may provide selections for some menu items but not for others. The menu of a catered business lunch, for example, may include no choice for the

"Planet on a Platter – College, Culture, Cuisine"
Spring Dinner
Kansas State University Dining

Appetizers
Crudités with Olive Oil & Salt
Specialty Cheeses & Olives
Sangria Punch

Entrees
Sautéed Trout with Macadamia Nuts
Thai-Style Chicken Breast
Broccoli Gruyere Gratin

Carved in Dining Room
Leg of Lamb with Tomato-Olive Ragout & Apricot Chutney
Steamship Round

Accompaniments
Patatas Bravas
Tropical Coconut Rice
Spicy Green Beans
Mediterranean Baked Tomato Halves

Breads
Portuguese Sweet Bread
Fresh Garlic & Rosemary Focaccia

Salads
Mixed Greens with Avocado & Mango
Roasted Almonds with Coriander, Red Pepper Flakes & Olive Oil
Panzanella
Ancho Chile and Garlic Marinated Shrimp
Tabouli
Smoked Eggs

Desserts
Australian Torte
European-Style Chocolate Hazelnut Layer Cake
Amaretto Cheesecake
Fresh Strawberries & Whipped Cream

Figure 3.3. Example of a single-use menu.
Source: From Kansas State University. Used by permission.

center-of-the-plate item and salad but offer the customer a choice of bread, beverage, and dessert.

MENU STRUCTURE

The menu always has been the heart of every foodservice operation, but today it is more crucial to the operation than ever. The menu is the sales tool that leads customers to the operation. Bartlett (1991) compared a well-executed menu to a television series. It has captivating lead characters, a strong supporting cast, good production values, and plot twists that keep viewers coming back for more. As a bonus, it produces profitable spin-offs. Foodservice managers must check their menus often to be sure they satisfy cost-conscious but value-greedy customers.

"Making a menu is a balancing act," says Twyla Fultz, executive chef at the Quality Hotel Capitol Hill, in Washington, D.C. (Ryan, 1993). The menu can be approached from many angles: location, the name of the restaurant and what it says to potential customers, primary target market, menu priorities, balancing traditional and hearty with light and healthful menu items, and the capabilities of the staff. According to Steve Miller, the menu "should tell a customer what you want them to buy." The placement of menu items, the graphics, and the descriptions all send messages about what you want a customer to order (Panitz, 2002).

Managers of various types of foodservice operations spend many hours in planning menus that will meet customer expectations. Multiunit restaurant chains need menus that have broad appeal but are short because customers want fast service and do not want to waste time making decisions. Menus must have regional appeal. The Olive Garden chain, for example, has approximately 18 different lunch menus and 24 different dinner menus for 472 restaurants in the United States (Olive Garden Restaurants, 2001). Hospital menus not only serve patients with special nutritional needs, but also healthy, active customers, such as patients' families, doctors, nurses, and employees. Because most foodservice operations serve a minimum of two, and often three, meals a day, menus must be geared to breakfast, lunch, and dinner and to style of service and menu price range. Balancing labor and food costs also is a challenge. Some other concerns of the menu planner include adding variety to seasonal menus, keeping cycle menus exciting by including new items along with old favorites, and offsetting high-priced items with those that are low priced (Ryan, 1993).

Typical American meals used to be a quick breakfast, a light lunch, and a heavy dinner in the evening, particularly for those living alone or in dual-career families. Three meals a day, popular in the 1940s and 1950s, is atypical for many people today. "Grazing," eating small amounts of food throughout the day, has become habitual for many people. Adams (1987) succinctly described grazing in mathematical terms:

grazing = flexibility + frequency + food

Customer emphasis on quality, convenience, nutrition, and value, as well as changing lifestyles, affects meal patterns. Whatever the meal of the day may be, it is the primary factor influencing menu planning.

Breakfast and Brunch

Breakfast is no longer just a bacon-and-egg affair. Even blueberry pancakes are passé these days. Consumers of all ages want egg dishes that set their tongues on fire and cakes as sweet as honey. They want peppered meats and French toast spiked with cinnamon and vanilla (Waters, 1998). Even cold cereal must be more than just milk and flakes.

Fewer people eat breakfast than eat lunch or dinner, but the morning victuals account for approximately 20% of daily restaurant traffic. Breakfast can be so lucrative that places you'd never think of as breakfast spots are serving early morning meals. (Grocers, convenience stores, and even gas stations are selling breakfast items.)

Breakfast items range from those that are considered light to traditional items. Bob Evans Farm Restaurants, which emphasize breakfast menu items all day, offer an oatmeal breakfast on their light menu consisting of a bowl of oatmeal, a blueberry muffin, and orange juice; the traditional favorite is sausage gravy and biscuits with home fries. To be responsive to customers, the Bob Evans restaurant chain has added new items, such as the Border Skillet, which draws on the popularity of Southwestern cuisine.

Both commercial and onsite foodservice operations usually offer traditional breakfast items and light and healthful options. Hearty breakfast menu items—eggs, sausage, bacon, hash browns, pancakes, waffles, breakfast breads, and pastries—are available for customers who want them. Health and fitness awareness and the high profile of carbohydrates and fiber have increased consumption of cold and hot cereals. McDonald's Egg McMuffin is the grandfather of all fast-food breakfasts. The chain's overcoddled egg and Canadian bacon on English muffin brought new meaning to breakfast-on-the-go when it was introduced in the 1970s (Waters, 1998). Since then, quick-service chains have had varying degrees of success by reinventing breakfast items as finger foods, French toast sticks, mini cinnamon rolls, muffins, and knock-offs of the Egg McMuffin. Casual dining chains also are getting in on the breakfast-to-go concept with special call-ahead programs.

Brunches are designed as a mid- to late-morning meal that combines breakfast- and lunch-type items. An example of a brunch menu is shown in Figure 3.4. In many types of foodservice operations, brunch is served on weekends or for catered events. Many restaurants have turned Saturday and Sunday prelunch hours into a profitable brunch business by creating signature brunches. Onsite foodservice operations, particularly retirement centers and colleges and universities, may serve brunch in lieu of both breakfast and lunch on weekends and for special occasions, such as a birthday or commencement.

Brunch Menu
(Prix Fixe: $18.95)

Beverages—Choice of
Wine, Mimosa, Bloody Mary, Orange Juice, Cranberry Juice

Appetizer—Choice of
Cream of Asparagus Soup
Shrimp and Mango Skewers with Guava Lime
Glaze Mini Bagels with Brie Cheese, Preserves, and Fruit Compote
Caesar Salad

Main Course—Choice of
Eggs Benedict
Cheddar Biscuits with Fried Eggs, Ham, and Brown Butter
Linguine with Chicken and Sun Dried Tomatoes
Smoked Salmon and Sweet Potato Hash
Pork Tenderloin with Rhubarb Chutney

Figure 3.4. Example of a brunch menu.

The basic brunch menu pattern generally begins with fruit and juice; when alcoholic beverages are desired, the Bloody Mary and Mimosa are popular. The champagne brunch, often associated with wedding parties, usually is held in elegant restaurants or hotels or in country clubs. Menu selections vary depending on type of service; the menu for a buffet generally includes a greater variety of selections than that planned for a sit-down brunch. Entrée brunch offerings usually include egg dishes and breakfast meats, in addition to such typical lunch and dinner entrées as steamship round of beef, chicken breast in mushroom wine sauce, or lobster Newburg. A variety of hot breads is included along with an assortment of cheeses, fruits, and salads.

Lunch

The restaurant lunch menu is hard to balance between trends and tradition, healthy and hearty foods, variety and the just-right menu mix. Lunch is a difficult meal to deliver to the customer because menu items are more complicated than those served at breakfast and must be produced faster than dinner items (Ryan, 1993). Lunch is the meal most often eaten away from home; approximately 6 in 10 individuals consume a commercially prepared lunch at least once a week.

Lunch-to-go is a quickly growing trend for workers who run errands at noon or who want to relax on a park bench while enjoying a salad and a change of scenery. These customers want a good meal that travels well but do not want to go into debt paying for it, and they also want it fast. Sandwiches are the quintessential fare for people on the go, with fillings being enclosed in bread, rolls, pita, and tortilla. Sandwich offerings are more common on lunch menus rather than on dinner menus, although most casual dining restaurants offer sandwiches at any time of the day. The secret to a successful lunch-to-go program is the packaging required to make eating the lunch easy for customers; packaging, however, adds to the cost of the lunch.

Catering lunches remains one of the big profit makers in the foodservice industry. In no other operation is time more important than in catering. The customer waiting for a catered brown bag or boxed lunch watches the clock because timing is part of the contract. Many restaurants that offer sandwich and salad menus, such as Au Bon Pain, do a very large amount of lunch catering.

In onsite foodservices, lunch often is the busiest meal served. Lunch menu items in onsite operations are becoming more like those served in commercial operations, especially limited-menu restaurants. The idea that people should eat whatever is put before them is gone. Managers of hospital, nursing home, college and university, and school foodservices are as eager to satisfy customers as are restaurateurs. Thus pizza, fajitas, chili, sandwiches, soup, salads, chicken nuggets, ethnic foods, and pasta appear regularly on menus. An example of a lunch menu served as part of the National School Lunch Program at Hammond, Indiana High School is shown in Figure 3.5.

Dinner

Dinner does not follow the distinctive pattern of food choices that characterizes breakfast and lunch menus. The traditional dinner menu includes an entrée of meat, fish, or poultry; potato or substitute; vegetable; and salad. For lighter or late evening meals, often referred to as supper, menus may be similar to those served at breakfast, brunch, or lunch.

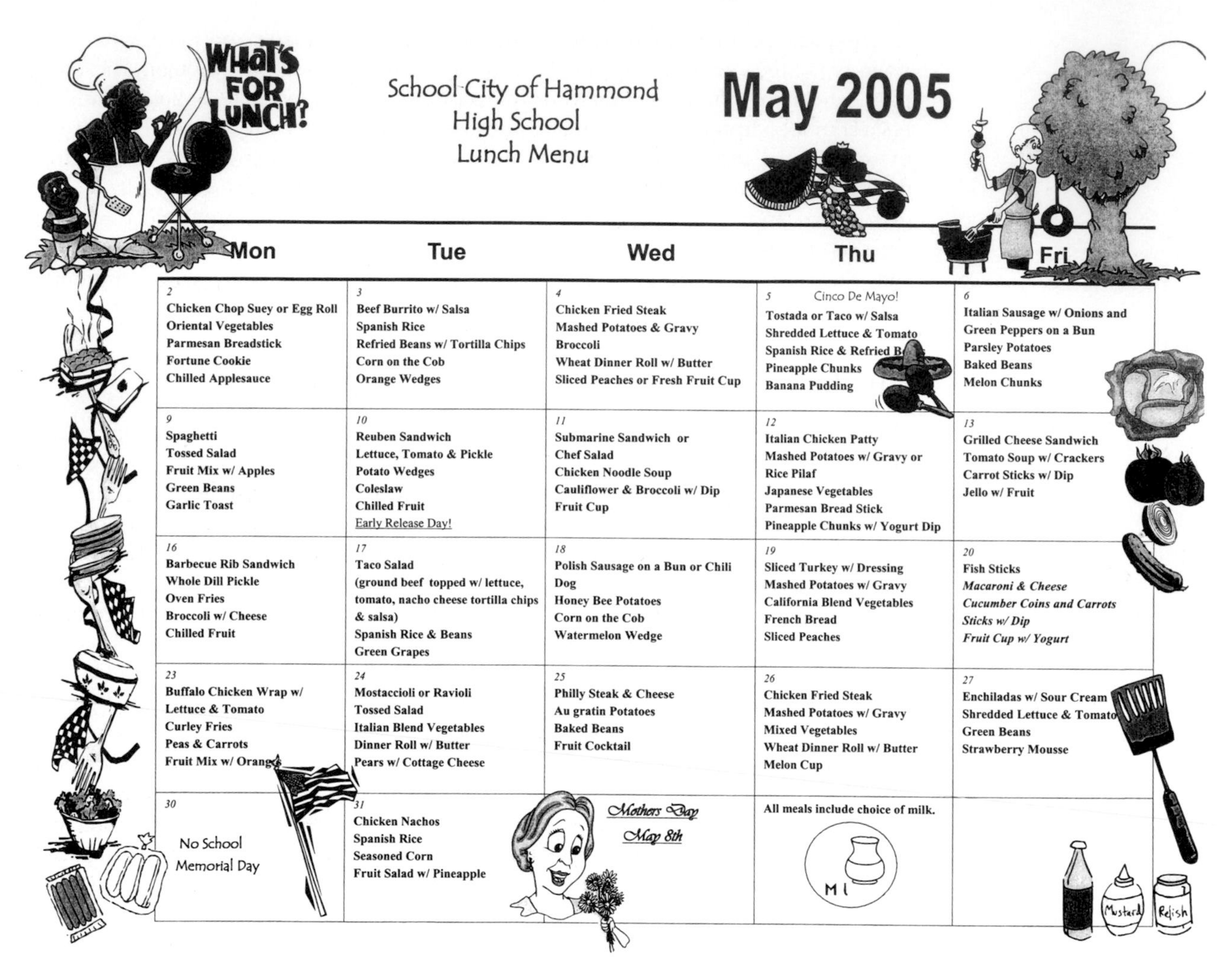

WHAT'S FOR LUNCH?

School City of Hammond
High School
Lunch Menu

May 2005

Mon	Tue	Wed	Thu	Fri
2 Chicken Chop Suey or Egg Roll Oriental Vegetables Parmesan Breadstick Fortune Cookie Chilled Applesauce	3 Beef Burrito w/ Salsa Spanish Rice Refried Beans w/ Tortilla Chips Corn on the Cob Orange Wedges	4 Chicken Fried Steak Mashed Potatoes & Gravy Broccoli Wheat Dinner Roll w/ Butter Sliced Peaches or Fresh Fruit Cup	5 Cinco De Mayo! Tostada or Taco w/ Salsa Shredded Lettuce & Tomato Spanish Rice & Refried B Pineapple Chunks Banana Pudding	6 Italian Sausage w/ Onions and Green Peppers on a Bun Parsley Potatoes Baked Beans Melon Chunks
9 Spaghetti Tossed Salad Fruit Mix w/ Apples Green Beans Garlic Toast	10 Reuben Sandwich Lettuce, Tomato & Pickle Potato Wedges Coleslaw Chilled Fruit Early Release Day!	11 Submarine Sandwich or Chef Salad Chicken Noodle Soup Cauliflower & Broccoli w/ Dip Fruit Cup	12 Italian Chicken Patty Mashed Potatoes w/ Gravy or Rice Pilaf Japanese Vegetables Parmesan Bread Stick Pineapple Chunks w/ Yogurt Dip	13 Grilled Cheese Sandwich Tomato Soup w/ Crackers Carrot Sticks w/ Dip Jello w/ Fruit
16 Barbecue Rib Sandwich Whole Dill Pickle Oven Fries Broccoli w/ Cheese Chilled Fruit	17 Taco Salad (ground beef topped w/ lettuce, tomato, nacho cheese tortilla chips & salsa) Spanish Rice & Beans Green Grapes	18 Polish Sausage on a Bun or Chili Dog Honey Bee Potatoes Corn on the Cob Watermelon Wedge	19 Sliced Turkey w/ Dressing Mashed Potatoes w/ Gravy California Blend Vegetables French Bread Sliced Peaches	20 Fish Sticks *Macaroni & Cheese* *Cucumber Coins and Carrots Sticks w/ Dip* *Fruit Cup w/ Yogurt*
23 Buffalo Chicken Wrap w/ Lettuce & Tomato Curley Fries Peas & Carrots Fruit Mix w/ Orang	24 Mostaccioli or Ravioli Tossed Salad Italian Blend Vegetables Dinner Roll w/ Butter Pears w/ Cottage Cheese	25 Philly Steak & Cheese Au gratin Potatoes Baked Beans Fruit Cocktail	26 Chicken Fried Steak Mashed Potatoes w/ Gravy Mixed Vegetables Wheat Dinner Roll w/ Butter Melon Cup	27 Enchiladas w/ Sour Cream Shredded Lettuce & Tomato Green Beans Strawberry Mousse
30 No School Memorial Day	31 Chicken Nachos Spanish Rice Seasoned Corn Fruit Salad w/ Pineapple	Mothers Day May 8th	All meals include choice of milk.	

Figure 3.5. School lunch menu.
Source: From School City of Hammond. Used by permission.

In many restaurants, menus are getting shorter, but the appetizer section is getting longer. On many menus, appetizers equal or outnumber entrées. Customers are sharing appetizers or ordering two appetizers instead of an entrée. Tapas, small plates of food originally eaten with a drink at the bar, and mezzes, little dishes originating in the Middle East, have migrated to the dinner table as the idea of grazing attracts many customers. Evidence of tapas intertwined with foods from other parts of the world can be seen when such items as Chinese dumplings and Thai or Vietnamese spring rolls begin to appear on casual dining chain menus (Menu Engineering, 1998). Another example is the Chinese tradition of an all-appetizer feast, the dim sum, which is shared by many people. The appetizer in a hospital or nursing home usually is soup, seafood cocktail, fruit cup, or fruit or tomato juice.

Ethnic cuisines continue to impact menus. They include Thai, Mediterranean, Vietnamese, Middle Eastern, Caribbean, and certain regions in France, Italy, and Spain. The Italian and Asian influence and the move toward spicy and strong flavors are evident. Wise foodservice operators include menu items to satisfy newer and younger as well as regular and older customers. The Food Marketing Institute (FMI, 2004) reported that one-third of young adults aged 15–24 eat their evening meal in a restaurant three or more times per week.

Desserts commonly are included as a menu component in foodservice operations. Enticing customers with rich desserts is easy, but the challenge is to create a light dessert that still dazzles. Examples of elegant fruit desserts are fresh pears poached in raspberry sauce that turn a brilliant pink color after being refrigerated overnight or baked apples with raisins and a little brown sugar baked in phyllo dough.

FACTORS AFFECTING MENU PLANNING

The crux of menu planning is that the menu is customer driven. The overriding concern in all facets of planning should be the satisfaction of customer desires. The concept of value cannot be ignored in menu planning; value prompts the clientele to select a particular item from the menu. Although satisfying customers is a primary concern for all foodservice managers, producing menu items at an acceptable cost takes priority. Customer satisfaction, government regulations, and management decisions must be involved in menu planning.

Customer Satisfaction

Sociocultural factors should be considered in planning menus to satisfy and give value to the customer. Nutritional needs provide a framework for the menu and add to customer satisfaction. Probably the most important aspects for satisfying customers are the aesthetic factors of taste and appearance of the menu items. Will the customer be satisfied with the meal and want to return?

Sociocultural Factors

Sociocultural factors include the customs, mores, values, and demographic characteristics of the society in which the organization functions (Griffin, 1999). Sociocultural processes are important because they determine the products and services people desire. Customers have food preferences that influence the popularity of menu items.

Food Habits and Preferences

Consideration of food habits and preferences should be a priority in planning menus for a particular population. Cultural food patterns, regional food preferences, and age are related considerations. Too often, menu planners are influenced by their own likes and dislikes of foods and food combinations rather than those of the customer.

Food habits are the practices and associated attitudes that predetermine what, when, why, and how a person will eat. Regional and cultural food habits still exist in the United States, but the mobility of the population and the sophistication of food marketing and distribution have lessened these distinctions. **Food preferences** express the degree of liking for a food item.

Analysis of food habits and preferences should be conducted to provide data for menu planning. Formal and informal methods may be used to examine customer reactions to various menu items. Highly sophisticated market research studies are conducted by large national multiunit foodservice corporations before a major menu change or even the introduction of a new item.

Small-scale surveys, formal and informal interviews with customers, observations of plate waste, customer comment cards, and tallying of menu selections are methods used to collect data on food preferences and menu item popularity. Too often, menu planners offer variety for variety's sake.

Food preference surveys usually employ a hedonic scale in which foods are rated by an individual on a continuum from "like extremely" to "dislike extremely" (Hegenbart, 1992). For measurement of food preferences among children, facial hedonic scales have been widely used and are sometimes called the "smiley face" rating scale (Comstock, St. Pierre, & Mackiernan, 1981; Lachance, 1976). An example of a facial hedonic scale is shown in Figure 3.6. Wells (1965) found that the facial method is easier to use with children than words or numbers because it allows good communication and understanding regardless of age, intelligence, education, or even the ability to speak English.

Sensory evaluation has been used to measure reactions to food by asking individuals to rate menu items on various dimensions such as flavor, appearance, temperature, and portion size. Figure 3.7 shows a sensory evaluation scorecard for evaluation of elementary students' reactions to menu items.

Plate waste, or the amount of food left on a plate, is a method used as a measure of food acceptability. Plate waste often is weighed to provide numerical results that can be used in many studies, particularly in the school lunch program. Plate waste also is being used in environmental control studies. It can be used to weigh uneaten menu items on an individual or group basis or the total waste for a meal.

Observation is a method that requires trained observers to estimate visually the amount of plate waste. Results of several studies indicate that the visual estimation of plate waste is a sufficiently accurate and simple method for assessing food acceptability. **Self-reported consumption** is another technique for measuring plate waste in which individuals are asked to estimate their plate waste using a scale

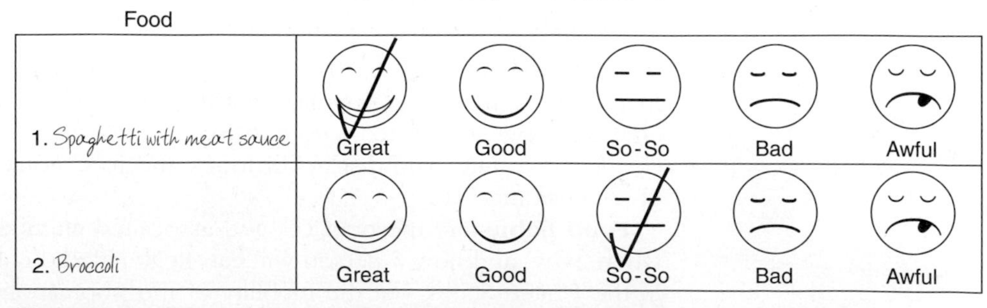

Figure 3.6. Facial hedonic scale used for measuring children's food preferences.

***Circle* the one word that *best* describes how you feel about the food today.**

Food					
1. Spaghetti with meat sauce	TASTES	Great	Good	So-So	Awful
	LOOKS	Great	Good	So-So	Awful
	TEMPERATURE	Just Right	OK	Too Cool	
	AMOUNT	Too Much	Right Amount	Not Enough	
2. Broccoli	TASTES	Great	Good	So-So	Awful
	LOOKS	Great	Good	So-So	Awful
	TEMPERATURE	Just Right	OK	Too Cool	
	AMOUNT	Too Much	Right Amount	Not Enough	

Figure 3.7. Example of a sensory evaluation scorecard for menu items.

similar to one used by trained observers. An example of this type of scale is shown in Figure 3.8.

Menu planners must be cognizant of the food habits and preferences of the target population to plan menus that are acceptable and will generate sales and overall customer satisfaction. Age, cultural, and regional food patterns are important to consider, along with changing food patterns over time.

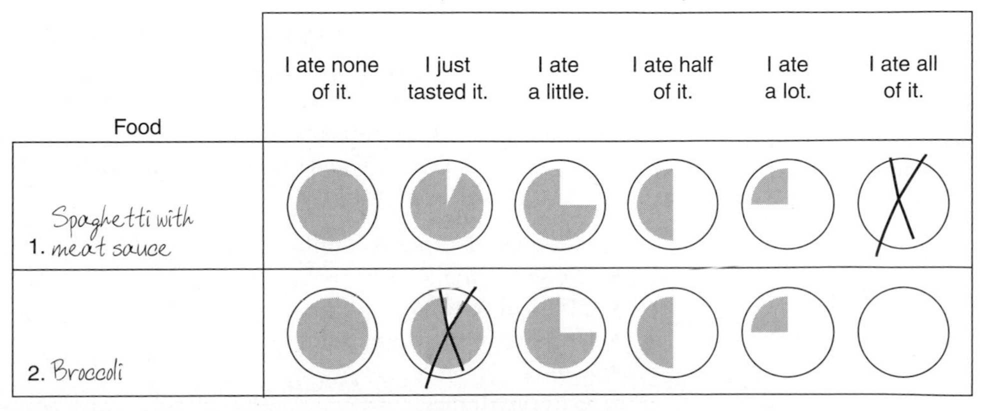

How much did you eat?

For each food, Please put an "X" on the amount you ate.

Food	I ate none of it.	I just tasted it.	I ate a little.	I ate half of it.	I ate a lot.	I ate all of it.
1. Spaghetti with meat sauce						X
2. Broccoli		X				

Figure 3.8. Scale for measuring self-reported consumption.

Nutritional Influence

Nutritional needs of the customer should be a primary concern for planning menus for all foodservice operations, but they are a special concern when living conditions constrain persons to eat most of their meals in one place. In healthcare facilities, colleges and universities, and schools, for example, most of the nutritional needs of the customer are provided by the foodservice.

Increasing public awareness of the importance of nutrition to health and wellness also has motivated commercial foodservice operators to consider the nutritional quality of menu selections. Interest in nutrition increased after the release of two reports: *The Surgeon General's Report on Nutrition and Health* (1988) and the report of the National Research Council (National Research Council, 1989), *Diet and Health: Implications for Reducing Chronic Disease Risk.* The U.S. Department of Health and Human Services publications, *Healthy People 2000* (1990) and *Healthy People 2010* (2000), detail health objectives for the nation, many of which are related to food consumption. These publications increased customer awareness of the relationship of diet to such chronic diseases as heart disease, stroke, cancer, and diabetes. As a result, foodservice managers cannot afford to ignore customer demand for nutritionally adequate menu offerings.

As many as 60% of Americans don't know at 4 P.M. what they will be eating for dinner that night (American Meat Institute, 1998). It is no wonder that consumers are buying and preparing their meals differently than in the past. Behind these changes are major shifts affecting not only how Americans eat but where their meals are coming from. The National Restaurant Association reported that the typical person (age 8 and older) consumes an average of 4.2 meals prepared away from home each week (National Restaurant Association, 2002).

Although convenience is a major driving force to this change, another trend behind consumer purchasing decisions is consumers' desire for better nutrition. A total of 75% of consumers in a 2002 survey sponsored by the American Dietetic Association reported that "they carefully select foods in order to achieve balanced nutrition and a healthful diet" (American Dietetic Association, 2002).

A market survey of current or potential customers may help to determine their needs, desires, and attitudes. An informal method of determining the desires of customers is to review questions asked of the service staff concerning nutritional values and preparation of menu items.

Dietary Guidelines for Americans: Recommendations for good health developed by the USDA and the U.S. Department of Health and Human Services.

The sixth edition of the ***Dietary Guidelines for Americans*** (Figure 3.9) and the companion consumer brochure, *Finding Your Way to a Healthier You* (Figure 3.10), issued in 2005 by the USDA and U.S. Department of Health and Human Services, suggest now Americans should eat to stay healthy (see www.healthierus.gov/dietaryguidelines). The 41 key recommendations in the *Dietary Guidelines* are grouped into nine categories:

- Adequate Nutrients within Calorie Needs
- Weight Management
- Physical Activity
- Food Groups to Encourage
- Fats
- Carbohydrates
- Sodium and Potassium
- Alcoholic Beverages
- Food Safety

Dietary Guidelines
for Americans
2005

U.S. Department of Health and Human Services
U.S. Department of Agriculture
www.healthierus.gov/dietaryguidelines

Figure 3.9. Dietary Guidelines for Americans.
Source: U.S. Departments of Agriculture and Health and Human Services, available at http://www.healthierus.gov/dietaryguidelines/

MyPyramid: Illustration of nutrition and physical activity recommendations.

A companion to the Dietary Guidelines is **MyPyramid** (Figure 3.11) (see http://www.mypyramid.gov/). MyPyramid is a complex illustration that can be used interactively at the MyPyramid.gov Web site. It includes nutrition and physical activity messages from the Dietary Guidelines:

- Make half your grains whole.
- Vary your veggies.
- Focus on fruits.

Figure 3.10. Finding Your Way to a Healthier You.
Source: U.S. Departments of Agriculture and Health and Human Services, available at http://www.healthierus.gov/dietaryguidelines/

- Get your calcium-rich foods.
- Go lean on protein.
- Know the limits on fats, sugars, and salts.
- Find your balance between food and physical activity.

This national focus on good nutrition and physical activity has resulted in related campaigns, such as The Produce for Better Health Foundation, a nonprofit organization incorporated in 1991; and the National Cancer Institute's **5 A Day** program, launched to encourage increased consumption of fruits and vegetables to at least five servings daily (see www.5aday.gov). Two campaigns were started to improve consumption of milk and other calcium-rich foods. The National Institute of Child Health and Human Development began the **Milk Matters** campaign to increase milk consumption among America's children and teens (see www.nichd.nih

Figure 3.11. MyPyramid.
Source: U.S. Department of Agriculture, available at http://www.mypyramid.gov/

.gov/milk), and the American Dairy Association/National Dairy Council are supporting the **3-A-Day** campaign to increase consumption of dairy products to three servings per day (see www.3aday.org). The Centers for Disease Control and Prevention started the ***VERB***™ campaign to increase the physical activity of 9–13-year-old youths (see www.cdc.gov/youthcampaign).

For some food programs, specific nutritional guidelines have been made mandatory. For example, the nutritional goal of the National School Lunch Program is to provide one-third of the Recommended Dietary Allowances (RDAs) for specified nutrients. According to the Menu Planning Guide for School Food Service (National School Lunch Program, 1998), each lunch is not expected to provide one-third of the RDAs for all nutrients, but the average over a period of time should meet the goal.

Many onsite foodservice operations have either a registered dietitian on staff or as a consultant on nutritional aspects of menu planning. Commercial foodservice operators also are positioning and marketing nutrition to respond to consumer demand and to gain a competitive edge.

Because of consumer interest in nutrition and consumer requests for more healthful items, chefs are being trained to meet these needs. The Culinary Institute of America (CIA) curriculum includes nutrition preparation for its students. The General Foods Nutrition Center, the first of its kind in the country, was opened on the CIA campus in 1989. It offers valuable training in nutritional concepts and principles to students and foodservice professionals.

Aesthetic Factors

Flavor, texture, color, shape, and method of preparation are other factors to consider in planning menus. *Flavor* is the taste that occurs from a product in the mouth and often is categorized as salty, sour, sweet, or bitter. A balance should be maintained among flavors, such as tart and sweet, mild and highly seasoned, light and heavy. Certain combinations have become traditional, such as turkey and cranberry sauce and roast beef and horseradish sauce. The flavors are complementary, and customers tend to expect these combinations to be served together. Foods of the same or similar flavors generally should not be repeated in a meal. A variety of flavors within a meal is more enjoyable than duplications, although there are exceptions to this rule. For instance, tomato in a tossed salad is an acceptable accompaniment to spaghetti and meatballs in a tomato sauce.

Texture refers to the structure of foods and is detected by the feel of foods in the mouth. Crisp, soft, grainy, smooth, hard, and chewy are among the descriptors of food texture, which should be varied in a meal. For example, a crisp salad served with soup on a luncheon menu is a more pleasing textural combination than a gelatin salad with soup.

Consistency of foods is the degree of firmness, density, or viscosity. Runny, gelatinous, and firm describe the characteristics of consistency, as do thin, medium, and thick when referring to sauces.

Color on the plate, tray, or cafeteria counter has eye appeal and helps to merchandise the food. The combination of colors of foods always should be considered in selecting menu items. The orange-red of tomatoes and the purple-red of beets, for example, is an unappealing combination, and a menu with several white foods is unimaginative. A grilled steak with a baked potato and sauteed mushrooms is a wonderful meal but lacks eye appeal. The presentation could be improved by the addition of a broiled tomato half or small wedge of watermelon with the green rind.

The *shape* of food also can be used to create interest in a menu through the variety of forms in which foods can be presented. Food processors, slicers, and mixers with attachments can produce a wide variety of shapes. As an example, the ever-popular French fried potato can be served as regular, curly, waffle, or steak fries.

Combinations of foods using different methods of preparation can add variety to the menu. Foods prepared in the same manner generally should not be served in the same meal, barring some common exceptions such as fish and chips, both of which are fried. Southern fried chicken with cream gravy and steamed asparagus would be appealing, but if the asparagus were dressed with hollandaise sauce, the appeal would be lost. Other ways to introduce variety and texture are to serve both cold and hot foods or both raw and cooked foods together.

Government Regulations

Menu planning in some foodservice organizations will be impacted by local, state, and/or federal regulations governing the types and quantities of food items to be served at a meal. Schools and long-term care facilities that receive state and/or federal funding are required to meet menu planning guidelines.

The USDA School Meals Initiative for Healthy Children (SMI), which began in 1995, requires Child Nutrition Programs to plan menus to meet specific minimum standards for key nutrients and calories. The SMI stipulates that meals must

Recommended Dietary Allowance: Recommendations for dietary intake of nutrients for healthy growth.

- Meet the **Recommended Dietary Allowances**
  - 1/4 RDA for age/grade group for breakfast
  - 1/3 RDA for age/grade group for lunch
- Meet calorie goals
  - Appropriate for age/grade
- Be consistent with the *Dietary Guidelines for Americans*

The menu patterns for the USDA National School Breakfast and Lunch Programs are defined in federal regulations. Schools have the option to choose one of four systems for their menu planning:

- Enhanced food-based menus
- NuMenus (Nutrient Standard Menu Planning)
- Assisted NuMenus (Assisted Nutrient Standard Menu Planning)
- Traditional food-based menus

In Figure 3.12, the four menu planning options are compared. Meeting dietary guidelines, advantages, disadvantages, computer and recordkeeping needs, and reimbursable lunch requirements under offer-versus-serve are listed for all four options (National School Lunch Program, 1998). Both NuMenus and Assisted NuMenus planning systems base their planning on a computerized nutritional analysis of the week's menu. The traditional and enhanced food-based meal pattern options base their menu planning on minimum component quantities of meat or meat alternate, vegetables and fruits, grains and breads, and milk.

The USDA has made a commitment to improve the nutritional quality of all school meals. To help implement the SMI, the Food and Nutrition Service of USDA developed Team Nutrition, an integrated, behavior-based, comprehensive plan for promoting the health of school children. Team Nutrition includes strategies for training child nutrition food service professionals, providing nutrition education for students and their parents, and providing support for healthy eating and physical activity. (Additional information is available at www.fns.usda.gov/tn.)

Long-term care facilities receiving Medicare and Medicaid funding also have mandated guidelines for meals served to residents. These guidelines are established by the Centers for Medicare and Medicaid Services (see www.cms.hhs.gov). Meal requirements are specified on a daily basis for the following groups: milk/milk products; meat; vegetables and fruits; bread, cereal, rice, and pasta; and butter or margarine. Quantities of food items considered equivalent for meeting the meal requirements are specified. For example, 1 serving from the meat group would be equivalent to 3 ounces of cooked meat, 3 eggs, or 1 1/2 cups of cooked dried beans. Additional information about these menu planning guidelines is available online at www.hcfa.gov.

Figure 3.12. Comparison of four menu planning options.

	Option 1 NuMenus	Option 2 Assisted NuMenus	Option 3 Enhanced Food-Based	Option 4 Traditional Food-Based
Meeting dietary guidelines	Schools must meet dietary guidelines and are required to do nutrient analysis at school level. Nutrient analysis is reviewed to determine if dietary guidelines are met.	Schools must meet guidelines and are required to provide nutrient analysis at school level. Schools must document that they have served the recipes and menus used in the nutrient analysis. Nutrient analysis is reviewed to determine if dietary guidelines are met.		Schools must meet dietary guidelines. Nutrient analysis conducted by the school is optional. Nutrient analysis is conducted during normal review cycle to determine if dietary guidelines are met.
Advantages	Local district retains flexibility and control. Menus will comply with the DGA.* After initial setup and planning are completed, menu planning process will be faster and easier. Comprehensive nutritional data will quickly and readily be available. There is greater flexibility in menu planning. Food costs may be lowered by reducing waste.	Requires minimal training and change for local personnel. Menu will comply with the DGA. After initial setup and planning are completed, menu planning process will be faster and easier. Comprehensive nutritional data will be quickly and readily available. There is greater flexibility in menu planning. Food costs may be lowered by reducing waste.		Requires minimal training and change for local personnel.
Disadvantages	Costs will be incurred for hardware and software. Personnel will need to spend time learning the software and setting up the system, i.e., entering local recipe and product data.	Costs may be incurred to have an outside party perform nutrient analysis. If a food vendor performs nutrient analysis, there may be financial implications. For example, schools might be required to use specified products.		School will not know if it is meeting the DGA until it is reviewed. Length of review will increase substantially to allow time to perform nutrient analysis.

		Schools will lose some control and flexibility.		Due to length of time between reviews, problems could go uncorrected for long periods.
Computer needs	Required—District must have computer hardware and USDA approved nutrient analysis software.	Not required because nutrient analysis may be done by another school, a consultant, or a school food co-op.	Not required	Not required
Record keeping	Production records document quantities planned and served. Nutrient analysis required at school level.	Production records document quantities planned and served. Nutrient analysis required at school level.	Production records document quantities planned and served. CN label or product analysis required for pre-prepared items. Standardized recipes for all prepared foods. Nutritional analysis of all prepared items.	Production records document quantities planned and served. CN label or product analysis required for pre-prepared items. Standardized recipes for all prepared foods. Nutritional analysis of all prepared items.
Reimbursable lunch requirements under offer-vs.-serve	Offer a minimum of 3 menu items. Entrée and milk must be offered. Students must accept 2 menu items, one of which must be the entrée. If more than 3 menu items are offered, no more than 2 can be declined.	Offer a minimum of 3 menu items Entrée and milk must be offered. Students must accept 2 menu items, one of which must be the entrée. If more than 3 menu items are offered, no more than 2 can be declined.	Offer a minimum of 5 food items - 1 meat/meat alt. - 2 vegetables/fruits - 1 grain/bread - 1 milk Senior high students must accept a minimum of 3 food items. Students below senior high must accept 3 or 4 food items at the discretion of the school food authority.	Offer a minimum of 5 food items: - 1 meat/meat alt. - 2 vegetables/fruits - 1 bread/bread alt. - 1 milk Senior high students must accept a minimum of 3 food items. Students below senior high must accept 3 or 4 food items at the discretion of the school food authority.

*DGA: Dietary Guidelines for Americans
Source: USDA.

Schools and long-term care facilities are required to plan menus in advance and keep copies of their menus for periodic review by state or federal reviewers. Documentation of serving size, recipe/product information, and production sheets is also needed to facilitate nutrient analysis of menus served.

Management Decisions

The menu should be viewed as a managerial tool for controlling cost and production. A number of management-related factors must be considered in the menu's design: food cost, production capability, type of service, and availability of foods.

Food Cost

Food cost is the cost of food as purchased. Foodservice managers in a competitive situation must be cost conscious in all areas of operations. Because the menu is a major determinant of pricing for food items, the manager must be particularly aware of both raw and prepared food costs for each menu item.

A very important part of most food-cost accounting systems is the determination of the food-cost percentage figure, the ratio of the cost of food sold over the dollars received from selling the food (Keiser & DeMicco, 2000). For example, a 40% raw-food cost in relation to sales revenues has been a rule of thumb for many commercial operations. This objective does not necessarily apply to each menu item, but to the overall sales. Some menu items, such as beverages, may have a much lower food cost in relation to sales, and other items, especially entrées, may be higher. What the customer is willing to pay also must be considered in menu planning.

In most hospitals and college and university foodservices, a daily food cost per customer is provided in the budget. In school foodservice, the amount of federal and state reimbursement is an important factor in determining the budget available for food. Food cost control is discussed further in Chapter 13.

Production Capability

To produce a given menu, several resources must be considered, of which a primary one is labor. The number of labor hours and the number and skill of personnel at a given time determine the complexity of menu items. Some menu items may be produced or their preparation completed during slack periods to ease the production load during peak service times; however, the effect on food quality may limit the amount of production in advance of service that could be completed. Employees' days off may need to be considered in menu planning because relief personnel may not have equal skill or efficiency. In quick-service operations, however, variations in employees' skill are less important because of product standardization. Planning less complicated menu items or using convenience items may be alternatives.

Production capability also is affected by the layout of the food production facility and the availability of large and small equipment. The menu should be planned to balance the use and capacity of ovens, steamers, fryers, grills, and other equipment. Refrigeration and freezer capacity must also be considered. Many novice menu planners have planned menus that overtax the oven capacity in a foodservice facility. In a small hospital foodservice operation, for example, a menu including meat loaf, baked potato, rolls, and brownies may present a production problem because all the items require use of the oven.

Type of Service

Type of service is a major influence on the food items that can be included on a menu. A restaurant with table service will have a different menu from that of a school foodservice.

Food items with longer holding capability should be selected for menus in establishments where last-minute preparation may present a problem, like healthcare facilities with patient tray service. In some hospitals and other facilities, galley kitchens on patient floors may be available for last-minute preparation of menu items that do not hold well, such as toast and coffee.

Equipment for holding and serving will affect the menu selections that can be offered. Hot foods may be held in either stationary or portable heated equipment. Required temperatures are relatively low, and humidification is necessary for some foods. Cold foods may be held in refrigerated units or iced counters. Central commissaries and satellite units require insulated transport units for both hot and cold foods. The availability of sufficient china, flatware, and glassware is another problem. Certain menu items may require special serving equipment, such as chafing pans for flamed desserts or small forks and iced bowls for seafood cocktail appetizers. This impact of menu items and combinations on dishroom capability also should be considered.

The temperature, color, or texture of some menu items deteriorate during the time between food production and service. Items that present a particular problem should be eliminated or modified. For example, school children may like grilled cheese sandwiches, but that sandwich does not lend itself to a menu for a satellite service center unless a grill and enough employees are available for last-minute preparation.

Availability of Foods

Improvement in transporting food both nationally and internationally and in food preservation makes many foods that were once considered seasonal available during most of the year. Strawberries are imported into the United States from Mexico in January, and frozen foods are available all year. During the growing season, however, local food products often are of better quality and are less expensive than those shipped from distant markets, and they should be added to the menu at that time. Foodservice managers in small communities also may need to consider frequency of delivery from various food distributors in planning menus.

MENU PLANNING

Thus far, this chapter has been devoted to the menu planner. In many foodservice operations, however, menu planning is often the responsibility of a team rather than an individual. This is especially true in large organizations in which the viewpoints of managerial personnel in both production and service are important in menu planning. In addition, personnel responsible for procurement have valuable input on availability of food, comparative cost, and new products on the market. In a healthcare organization, the clinical dietitian should be included on the menu planning team to ensure that patients' needs and food preferences are given appropriate consideration.

General Considerations

Computer-assisted menu planning is not widely used today because of the difficulty in quantifying the many variables involved in menu planning, such as flavor, color, and texture. Instead, the computer is used more often for analyzing menus for cost and nutrient composition.

Menu planning should proceed from the premise that the primary purpose of any foodservice is to prepare and serve acceptable food at a cost consistent with the objectives of the operation. Certain decisions must be made in advance and policies and procedures established for planning menus in a systematic manner. In addition to decisions on the design and format of the menu, other considerations are the number of choices to be offered, type of menu, and frequency of revision.

This planning must be far enough in advance of actual production to allow delivery of food and supplies and permit labor to be scheduled. Many operations also need time for printing or other reproduction of menus.

The initial decision is the design of the menu pattern or the outline of menu items to be served at each meal. The meal of the day is the key influence on the menu pattern, although it may vary from commercial to onsite operations. Menus for either type of operation, however, should be designed to inform the customer of what items are available and, in many instances, what they cost. Simple descriptions of menu items should be used, and confusing or overstated terms should be avoided. A pitfall for many menu planners is the use of interesting names to enliven a menu when these flowery terms often only confuse the patron. Any special names or menu items should be readily understood in the region where they are served. For example, Hopping John, a traditional southern New Year's Day dish consisting of black-eyed peas, rice, salt pork, and onions, would be puzzling to most New Yorkers.

Legislation in a number of states requiring truth in menus has had an impact on names of menu offerings. The California Restaurant Association, for example, has distributed to its membership a comprehensive special report explaining the labeling and advertising requirements of the California Business and Professional Code for food items served in restaurants. The National Restaurant Association and committees of restaurant operators defined menu terms and have distributed them in printed form, called Accuracy in Menus. The following are areas of potential misrepresentation with a short explanation of each (Miller, 1987):

Quantity. Proper procedures should preclude any concerns with misinformation of quantities (e.g., extra-large salad).

Quality. Federal and state standards of quality grades exist for many products, including meat, poultry, eggs, dairy products, fruits, and vegetables (e.g., choice sirloin of beef).

Price. Extra charges for service or special requests for food items should be brought to the customer's attention.

Brand names. Any product brand that is advertised must be the one served.

Product identification. Substitutions for products must be on the menu, such as blue cheese for Roquefort cheese.

Points of origin. Claims of origin should be documented (e.g., Maine lobster). Geographic names used in a generic sense, such as New England clam chowder, are permitted.

Merchandising terms. Terms for specific products need to be qualified (e.g., flown in daily).

Means of preservation. For preservation, food may be canned, chilled, bottled, frozen, or dehydrated. If a method is identified on the menu, it should be correct (e.g., frozen orange juice is not fresh).

Food preparation. Absolute accuracy is a must (e.g., charcoal broiled).

Verbal and visual presentation. If a picture of a meal is shown, the actual meal must be identical (e.g., if six shrimp are shown, six—not five—must be served).

Dietary or nutritional claims. Misrepresentation of nutritional content of food is not permitted (e.g., "low calorie" must be supportable by specific data).

Customers who are allergic to foods such as eggs, wheat, or shellfish want ingredients to be identified. In July 1996, the FDA set standards for nutritional claims made in restaurant menus to be sure that customers get what they order. If claims like "low fat" or "heart healthy" are made on a restaurant menu, the restaurant owner must be able to demonstrate that the food is qualified to bear this claim (FDA Talk Paper, 1996). Restaurant menu selections are not required to supply complete nutrition information. Nutrition information can be provided to customers by various means. It need not be presented in the "Nutrition Facts" format seen on packaged food labels, nor does it have to appear on the menu. The new menu rules are identical to the standards that have been in effect since May 1994 for nutrient content on placards and signs in large and medium-sized restaurants and since May 1995 for smaller restaurants.

Keeping various types of records can assist in menu planning. Data on past acceptance of items, weather, day of the week, season, and special events that may have influenced patronage are essential for menu planning. Other resources should include files of standardized recipes with portion size and cost, market quotations, suggestions from clientele, lists of food items classified by category (vegetables, entrées, desserts, etc.), trade publications, and cookbooks.

Planning Process

The general principles of menu planning are applicable to both onsite and commercial foodservice operations. Even though planning principles are similar in both operations, more variety may need to be incorporated into menus in onsite foodservice because the clientele remains very similar each day.

Onsite Foodservice Operations

Most onsite foodservice menus, with the exception of school foodservice, are designed on a three-meal-a-day plan. Some foodservice operations use a four- or five-meal plan built around brunch and an early dinner with some light, nutritious snack meals at other times of the day.

Cycle menus are used widely in onsite foodservices, with the length of the cycle varying from 1 to 3 weeks or longer. Also, cycles may change according to the season of the year to take advantage of plentiful foods on the market and to satisfy clientele expectations. The average length of stay is an important consideration in determining the length of the menu cycle in healthcare institutions. Cycles of 1 or 2 weeks have been used successfully in hospitals with a 2- to 4-day patient stay. Many larger hospitals, in recognition of the relatively short-term patient stay, use a restaurant-type static menu. In long-term care facilities, a 3- or 4-week cycle menu may be used. The general or regular menu provides the basis for planning menus for the various modified diets in healthcare facilities.

A step-by-step procedure for onsite foodservice menu planners for a three-meal-a-day pattern follows. Note that the entrée is the main item around which the meal is planned and must therefore be selected before any complementary foods.

1. *Plan the dinner meats or other entrées for the entire cycle.* Because entrées are the most expensive foods on the menu, total food cost can be controlled to a great extent through careful planning at this stage. A balance between high- and low-priced items will average out the cost over the week or period covered by the cycle. If choices of entrée are offered, the alternatives should include meat, chicken or other poultry, fish, a vegetarian entrée, and a meat extender, such as meat loaf or stew. Choices should be available for persons who have religious or medical dietary restrictions. Menus for preceding and subsequent days should be considered to preclude repetition.
2. *Select the luncheon entrées or main dishes, avoiding those used on the dinner menu.* Provide variety in method of preparation. A desired meal cost per day can be attained by serving a less expensive item at one meal of the day when a more expensive food has been planned for the other meal. Soups, sandwiches, main-dish salads, and casseroles are commonly served as luncheon entrées.
3. *Decide on the starch item appropriate to serve with the entrée.* Usually, if the meat is served with gravy, a mashed, steamed, or baked potato would be on the menu. Scalloped, creamed, or au gratin potatoes are most appropriate with meats having no gravy or sauce. Rice or pasta are common substitutes for potatoes. Variations in nonstarchy vegetables are obtained by serving them raw, cooked, peeled, or unpeeled; cutting vegetables in different sizes and shapes is another alternative. Methods of preparation can add variety to a vegetable, as can seasonings and sauces.
4. *Select salads, accompaniments, and appetizers next.* Work back and forth between the lunch and dinner meals to avoid repetition, introduce texture and color contrast into the meal, and provide interesting flavor combinations.
5. *Plan desserts for both lunch and dinner.* Desserts may be selected from the following main groups: fruit, pudding, ice cream or other frozen desserts, gelatin, cake, pie, and cookies.
6. *After the luncheon and dinner meals have been planned, add breakfast and any others.*
7. *Review the entire day as a unit and evaluate if clientele, governmental regulations, and managerial considerations have been met.* Check the menu for duplication and repetition from day to day. The use of a checklist aids in making certain that all factors of good menu planning have been met.

Commercial Foodservice Operations

Merchandising is the primary consideration in planning menus for commercial operations. Because of the varied types of operations, the menu takes many forms. The static-choice menu is the predominant type used in commercial foodservices, including upscale restaurants, limited-menu operations, and coffee shops. Menus are revised infrequently. Either all meals are included on one menu or separate printed menus are available for each meal with clip-ons for daily specials. Usually, entrées and main dishes are planned first, as in onsite foodservice operations. *Signature food items* (food items unique to that operation) are common in commercial foodservice operations.

A restaurant's menu is a powerful merchandising and marketing tool. The outcome of this planning process, according to Gray (1986), should be a menu that is efficiently and consistently produced in the kitchen and is pleasing to guests.

CHAPTER SUMMARY

This summary is organized by the learning objectives.

1. There are several terms used to describe characteristics of a menu. A static menu is one that offers the same items each day; a cycle menu has different menu items each day of the cycle and then repeats the entire menu; and a single-use menu is used for only one event. À la carte is pricing each menu item separately; table d'hôte is one price for the entire meal.
2. Evaluating the aesthetic qualities of a menu involves an assessment of the flavor, color, shape, texture, and appearance of the menu items. Variety and balance are important components of this assessment.
3. Techniques discussed in the chapter can be used to plan static or cyle menus.

TEST YOUR KNOWLEDGE

1. Why is the menu considered to be the primary control of the foodservice system?
2. What are important customer and management factors that affect menu planning?
3. Describe the differences in commercial and onsite foodservice menu planning.
4. What is menu psychology and how is it used in developing a printed menu?

CLASS PROJECTS

1. Divide into groups of two to three people. Find examples of the different kinds of menus either from a Web site or from an actual operation. (Don't steal the menus; ask if you could have one, or take pictures, then label the menus.)
2. Talk to managers of at least two different foodservice operations; find out how they plan their menus and how often their menus change.
3. Invite a manager of a chain operation to class to discuss the pros and cons of having system-wide menus versus locally developed menus.

Professional Organization Profile

National Association of College and University Foodservice

MISSION

To assist members and advance the collegiate foodservice industry by providing insight, education, services, and knowledge exchange

WHO BELONGS TO THE ORGANIZATION

NACUFS membership includes 600+ institutional members, professionals (directors, managers, buyers, chefs) at schools, colleges, and/or universities who are involved in providing a food program (either directly or through a campus auxiliary association) and whose sole purpose is to provide education-related services for the benefit of the campus community, including faculty, staff, and students in harmony with the educational mission and goals of the school, college, or university, and 400+ industry members including representatives of food and equipment manufacturers, distributors, brokers, foodservice support companies, councils, boards, trade associations, advisory commissions, and other related professional groups.

ADVANTAGES OF MEMBERSHIP

Members have access to a full range of educational programs, publications, training, technical assistance, industry information, scholarships, research support, networking, and volunteer leadership. The organization operates a website, publishes a monthly newsletter, and sponsors seminars and professional development courses. Student memberships are not available.

WEBSITE

www.nacufs.org

MEET THE PRESIDENT

Sharon Coulson, president 2005–2006, is foodservice director for the associated students at the University of California–Davis, a position she has held for more than 20 years. She is a registered dietitian and started her career as dietary manager in a convalescent hospital. Sharon offers the following advice to students: "Consider the college foodservice profession a viable career choice. It's much more than serving food out of a steam table. This profession is rich with diversity—you're a teacher, a motivator, an executive chef, a mentor, a culinary trend-tracker, a coordinator, a team leader, a business person, a psychologist, a kitchen designer, and a vital component on any campus. It's exciting, dynamic, challenging, and it's based around one of the great pleasures in life—eating!"

WEB SOURCES

www.fns.usda.gov/fns	USDA Food and Nutrition Service
www.fns.usda.gov/tn	USDA Team Nutrition
http://schoolmeals.nal.usda.gov/ Recipes/menuplan/menuplan.html	USDA Child Nutrition Program menu plans
www.healthypeople.gov	Healthy People 2010
www.eatright.org	American Dietetic Association
www.healthierus.gov/dietaryguidelines	Dietary Guidelines for Americans
www.mypyramid.gov	My Pyramid
www.5aday.gov	5 A Day Program
www.3aday.org	3 A Day Program
www.cms.hhs.gov	Centers for Medicare and Medicaid Services
www.nichd.nih.gov/milk	Milk Matters campaign
www.cdc.gov/youthcampaign	***VERB***™ campaign

BIBLIOGRAPHY

Adams, A. J. (1987). Consumers, what are they telling us? In *Agriculture research for a better tomorrow: Commemorating the Hatch Act Centennial (1887–1987).* Washington DC: U.S. Department of Agriculture.

Allen, R. L. (1993). Think big! Kids' menus help create future customers. *Nation's Restaurant News, 27*(3), 12.

American Dietetic Association. (2002). American Dietetic Association survey finds consumers know importance of diet and activity, but misconceptions persist. Available online at www.eatright.org

AMI: Just the facts. (1998). New trends in food preparation and eating. [Online]. Available: http://www.meatami.org/trndpg02.htm

Bayou, M. E., & Bennett, L. B. (1992). Profitability analysis for table-service restaurants. *Cornell Hotel and Restaurant Administration Quarterly, 33*(2), 49–55.

Callaway, C. W. (1992). The marriage of taste and health: A union whose time has come. *Nutrition Today, 27*(3), 37–42.

Carlson, B. L. (1987). Promoting nutrition on your menu: Three myths, eight tarnished rules, and five hot tips. *Cornell Hotel and Restaurant Administration Quarterly, 27*(4), 18–21.

Comstock, E. M., St. Pierre, R. G., & Mackiernan, Y. O. (1981). Measuring individual plate-waste in school lunches. *Journal of the American Dietetic Association* 79: 290–296.

Cummings, L. E., & Kotschevar, L. H. (1989). *Nutrition management for foodservices.* New York: Delmar Publishers.

Davis, J. L. (2003). America's food trends. [Online]. Available: http://www.webmd.com

DiDomenico, P. (1992). Adding a profit center that promotes: Selling signature items. *Restaurants USA, 12*(10), 18–21.

Donovan, M. D. (ed). (1993). *The professional chef's® techniques of healthy cooking.* New York: Van Nostrand Reinhold.

Drysdale, J. A., & Aldrich, J. A. (2002). *Profitable menu planning.* Upper Saddle River, NJ: Prentice Hall.

Dulen, J., & Sheridan, M. (1997). Star Grazing: The R & I menu census. *Restaurants and Institutions, 107*(25).

FDA Talk Paper. (1996). [Online]. Available: http://www.fda.gov

FMI 2004: Trends in the United States: Consumer attitudes and the supermaker. Food Marketing Institute, Washington, DC. [Online]. Available: http://www.fmi.org

Finn, S. (1992). *The real life nutrition book.* New York: Penguin.

5 a day—word is getting out! (1993). *5 A Day News, 2*(2), 1–5.

5 a Day for Better Health. 2002. [Online]. Available: http://www.sadey.com

Ganem, B. C. (1990). *Nutritional menu concepts for the hospitality industry.* New York: Van Nostrand Reinhold.

Gordon, E. (1992). Taking a look at lunch. *Restaurants USA, 12*(5), 36–38.

Gray, N. J. (1986). 17 steps to developing a winning menu. *NRA News, 6*(2), 16–20.

Griffin, R. W. (1999). *Management,* 6th ed. Boston: Houghton Mifflin.

Hegenbart, S. (1992). An Inventory of the Toolbox. Food product design. [Online]. Available: www.foodproductdesign.com

Howat, M. (1991). Menu labeling: the pros and cons. *Nation's Restaurant News, 25*(4), 42.

Information Resource Inc. (2003). Snacks arise meals demise. IRI update on snacking behavior. [Online]. Available: http://www.infores.com

Kasavana, M. L., & Smith, D. L. (1990). *Menu engineering: A practical guide to menu analysis.* Lansing, MI: Hospitality Publications.

Kass, M. (1990). Build a better menu. *Restaurants & Institutions, 100*(17), 36–39, 44, 48, 56, 58.

Keiser, J., & DeMicco, F. J. (2000). *Controlling and analyzing costs in foodservice operations,* 4th ed. New York: Macmillan.

Kotschevar, L. H. (2002). *Management by menu,* 3rd ed. Dubuque, IA: Wm. C. Brown in cooperation with the Educational Foundation of the National Restaurant Association.

Kreul, L. M., & Scott, A. M. (1982). Magic numbers: Psychological aspects of menu pricing. *Cornell Hotel and Restaurant Administration Quarterly, 23*(2), 70–75.

Kroll, B. J. (1990). Evaluating rating scales for sensory testing with children. *Food Technology, 44*(11), 78–86.

Lachance, P. A. (1976). Simple research techniques for school foodservice. Part I: Acceptance testing. *School Food Service Journal* 30: 54–56, 58, 61.

Lorenzini, B. (1992). Lighthouse offers braille menus. *Restaurants & Institutions, 102*(22), 22.

Lunch at court. (1993). *School Food Service Journal, 47*(1), 34–37.

Lydecker, T. (1991). Money-making menus. *Restaurants & Institutions, 101*(8), 96–97, 100, 102, 104, 109–110, 112.

McCarthy, B. (1992). Lunch to go. *Restaurants & Institutions, 102*(20), 61, 64, 66, 67.

McCarthy, B. (1993). Breakfast brings home the bacon. *Restaurants & Institutions,* 103(6), 26–27, 34, 39, 42, 46.

McCarthy, B., & Straus, K. (1992). Tastes of America 1992. *Restaurants & Institutions, 102*(29), 24–25, 28–29, 34, 36, 38, 44.

Menu Engineering. (1998). [Online]. Available: http://orbit.net.mt/magro/FoodandBev/FoodandBev1.htm

Mermelstein, N. H. (1993). Nutrition labeling in foodservice. *Food Technology, 47*(4), 65–68.

Miller, J. E. (1987). *Menu pricing and strategy,* 2nd ed. New York: Van Nostrand Reinhold.

Miller, S. G. (1992). The simplified menu-cost spreadsheet. *Cornell Hotel and Restaurant Administration Quarterly, 33*(3), 85–88.

National Research Council. (1989). *Diet and health: Implications for reducing chronic disease risk. Executive summary.* Washington, DC: National Academy Press.

National Restaurant Association. (1987). *Guidelines for providing facts to foodservice patrons: Ingredient and nutrient information.* Washington, DC: Author.

National Restaurant Association. (January 1990). *Nutrition awareness and the foodservice industry. Current issues report.* Washington, DC: Author.

National Restaurant Association. (1993). FDA releases new proposal on menu claims, asks for public comment. *Washington Weekly, 13*(25), 1, 4.

National Restaurant Association Research Department. (1998). *Menu analysis 1997.*

National Restaurant Association. (2002). Industry at a Glance. Available online at www.restaurant.org

Nation's Restaurant News. (2002). *What's for Lunch. On-site Restaurants Supplement, 5*(7):9. Washington, DC: Author.

National School Lunch Program: Nutrition program facts. (1998). [Online]. Available: http://www.usda.gov

Olive Garden Restaurants. (2001). SEC 10-K report online at www.sec.gov

Panitz, B. (2000). Reading between the lines: The psychology of menu design. [Online]. Available: www.restaurant.org

Pavesic, D. (1988). Taking the anxiety out of menu pricing. *Restaurant Management, 2*(2), 56–57.

Pavesic, D. V., & Magnant, P. F. (2005). *Fundamental principles of restaurant cost control.* Upper Saddle River, NJ: Prentice Hall.

Pettus, E. M. (1989). Menu engineering. *Club Management, 68*(5), 24–25.

Position of The American Dietetic Association: Nutrition in foodservice establishments. (1991). *Journal of the American Dietetic Association, 91,* 480–482.

Restaurants and Institutions 2001 menu census. (2002). Available online at www.rimag.com

Rose, J. C. (1988). Pricing I: Three menu pricing systems. *Food Management, 23*(1), 40.

Ryan, N. R. (1993). Good 'mixes' break the lunch routine. *Restaurants & Institutions, 103*(6), 50–51, 56, 62, 64, 68.

Schultz, H. G. (1965). A good action rating scale for measuring food acceptance. *Journal of Food Science, 30*(2), 365–374.

Sensory evaluation and the consumer—points of interaction. (1990). *Food Technology, 44*(11), 153–172.

Sloan, A. E. (2004). Taking off & piling on: Restaurants take new approaches. *Food Technology, 58*(10), 20–27.

Sloan, A. E. (2005). Top 10 global food trends. *Food Technology,* 59(4), 20–32.

Sneed, J., & Burkhalter, J. P. (1991). Marketing nutrition in restaurants: A survey of current practices and attitudes. *Journal of The American Dietetic Association, 91,* 459–462.

Solomon, J. (1992). A guide to good menu writing. *Restaurants USA, 12*(5), 27–30.

Spears, M. C. (1999). *Foodservice procurement: Purchasing for profit.* Upper Saddle River, NJ: Prentice Hall.

Straus, K. (1992b). New-size entrees multiply sales. *Restaurants & Institutions, 102*(6), 62, 66, 68.

Tableservice restaurants offer fax service and other conveniences. (1993). *Restaurants USA, 13*(3), 36.

Thomas, P. R. (ed.). (1991). *Improving America's diet and health: From recommendations to action.* Washington, DC: National Academy Press.

The Surgeon General's report on nutrition and health. (1988). DHHS (PHS) Publication Number 88-50210. Washington, DC: U.S. Government Printing Office.

USDA pyramid stacks up nutritional advice. (July/August 1992). *Food Insight, 8.*

U.S. Department of Agriculture, Food and Nutrition Service. (1983). *Menu planning guide for school food service.* Washington, DC: Author.

U.S. Department of Agriculture, Food and Nutrition Service. (2001). *SMI Implementation Study: Second Year Report.* Available online at www.fns.usda.gov

U.S. Department of Agriculture, Human Nutrition Information Service. (1989). Eating better when eating out: Using the dietary guidelines. *House and Garden Bulletin No. 232-11.* Washington, DC: Author.

U.S. Department of Health and Human Services. (1990). *Healthy People 2000.* Washington, DC: U.S. Government Printing Office, November 1990. Available online at www.health.gov/healthy people

U.S. Department of Health and Human Services. (2000). *Healthy People 2010.* Washington, DC: U.S. Government Printing, November 2000. Available online at www.health.gov/healthy people

U.S. Departments of Agriculture and Health and Human Services. (1990). *Nutrition and your health: Dietary guidelines for Americans,* 3rd ed. Washington, DC: Author.

Wallace, J. (1991). The menu challenge. *Restaurants & Institutions, 101*(8), 5.

Warshaw, H. S. (1993). America eats out: Nutrition in the chain and family restaurant industry. *Journal of The American Dietetic Association, 93,* 17, 19–20.

Waters, J. (1998). What's for breakfast? *Restaurants & Institutions, 108*(14), 63–64, 72–73.

Wells, W. D. (1965). Communicating with children. *Journal of Advertising Research, 5*(2), 2–14.

Welsh, S., Davis, C., & Shaw, A. (1992). A brief history of food guides in the United States. *Nutrition Today, 27*(6), 6–11.

Welsh, S., Davis, C., & Shaw, A. (1992). Development of the food guide pyramid. *Nutrition Today, 27*(6), 12–23.

What's to be in '93. (1993). *Sysco's Menus Today* (January), 20–24, 26, 28, 30, 32.

What will America be eating? (1998). [Online]. Available: http://www.foodexplorer.com/BUSINESS/Products/FoodServices/

Yee, L. (1999). Outlook '99 fare forecast. *Restaurants & Institutions, 109*(1), 34–36, 38, 40, 46, 49–50.

Yee, L. (2004). Taste 2004. *Restaurants & Institutions, 115*(Jan. 1). [Online]. Available: http//www.foodservice411.com/rimag

Yee, L. (2005). Custom fit. *Restaurants & Institutions, 115*(Jan. 1). [Online]. Available: http//www.foodservice411.com/rimag

CHAPTER 4

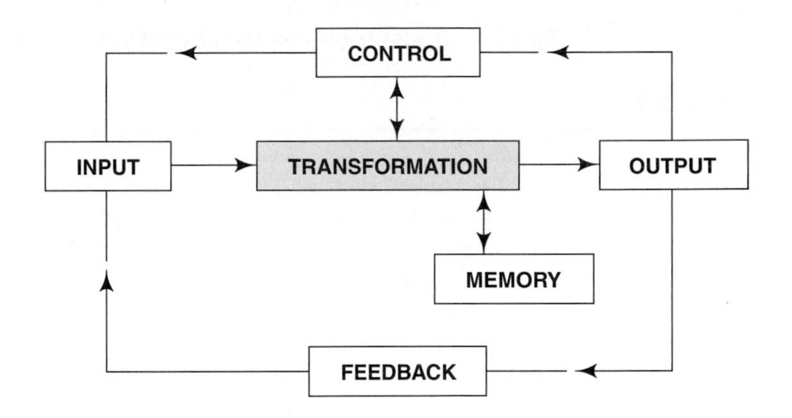

Food Product Flow and Kitchen Design

Enduring Understanding

- Food product flow differs in each foodservice operation
- The most desirable flow plan for a kitchen is a straight line

Learning Objectives

After reading and studying this chapter, you should be able to:

1. Differentiate among conventional, ready prepared, commissary, and assembly/serve foodservice operations.
2. Describe differences in food product flow in conventional and ready prepared foodservice operations.
3. Define terms such as cook-chill, cook-freeze, sous vide, centralized service, and decent decentralized service.
4. Observe a foodservice operation and draw a diagram showing the flow of food in that operation.
5. Differentiate the concepts of design, layout, and flow.
6. Describe the planning process for designing a foodservice operation.

Food can travel along many different paths from the time it enters a foodservice operation until it is served to the public. The production process in a foodservice operation often is categorized into one of four types. In this chapter, you will learn about these different types of foodservice and about the different patterns for the flow of food in a foodservice operation. Information in the chapter expands on the transformation phase of the foodservice system model and shows how this process can differ in many foodservice operations. You also will learn about designing space for a new or renovated foodservice operation.

FLOW OF FOOD

Food is the primary resource in the foodservice system. In the early 1970s, developments in food technology influenced changes in food product flow through the subsystems: procurement, production, distribution, and service.

Food product flow: Alternative paths within foodservice operations that food and menu items may follow, beginning with receiving and ending with service to the customer.

Food product flow refers to the alternative paths within foodservice operations that food and menu items may follow, initiating with receiving and ending with service to the customer (Unklesbay, Maxcy, Knickrehm, Stevenson, Cremer, & Matthews, 1977). Several authors (Bobeng, 1982; Matthews, 1982; Escueta, Fiedler, & Reisman, 1986; Gregoire & Nettles, 1997; Gregoire & Bender, 1999) have proposed models to show the flow of product through a foodservice operation. Alternative paths have been aimed primarily at increasing productivity, decreasing cost, or strengthening control of operations. Physical, chemical, and microbiological changes occurring in food throughout all stages of procurement, production, and service must be controlled to ensure the quality and safety of the finished products. Figure 4.1 builds on the work of previous authors and shows the multitude of possible product flow options in a foodservice operation. This figure will be used to describe food product flow in specific operations throughout this chapter.

TYPES OF FOODSERVICES

Faced with both increasing labor costs and a shortage of highly skilled employees, foodservice managers have been receptive to using new forms of food with built-in convenience or labor-saving features. New food products, available in various forms and stages of preparation, have appeared on the market in increasing numbers each year. Many require specialized equipment for final production, delivery, and service.

In what is now considered a landmark publication, Unklesbay et al. (1977) identified four types of foodservice operations:

- Conventional or traditional
- Ready prepared
- Commissary
- Assembly/serve

These types of foodservice operations were identified as foodservice systems in the original research. Because the conceptual framework of this text is the managerial foodservice system, these operations will be referred to as types of foodservice.

Coincidental with the evolution of the various types of foodservice operations, the interdependence between food processing and foodservice industries has become more evident, requiring coordination of functions. Food processing is a commercial industry in which food is processed, prepared, packaged, or distributed for consumption in the foodservice operation. The foodservice industry includes restaurants, hospitals, schools, and other specialized operations.

Conceptual diagrams developed by Unklesbay et al. (1977) (Figure 4.2) to illustrate food product flow within various foodservice operations will be used to show the relationship between food processing and the type of foodservice operation.

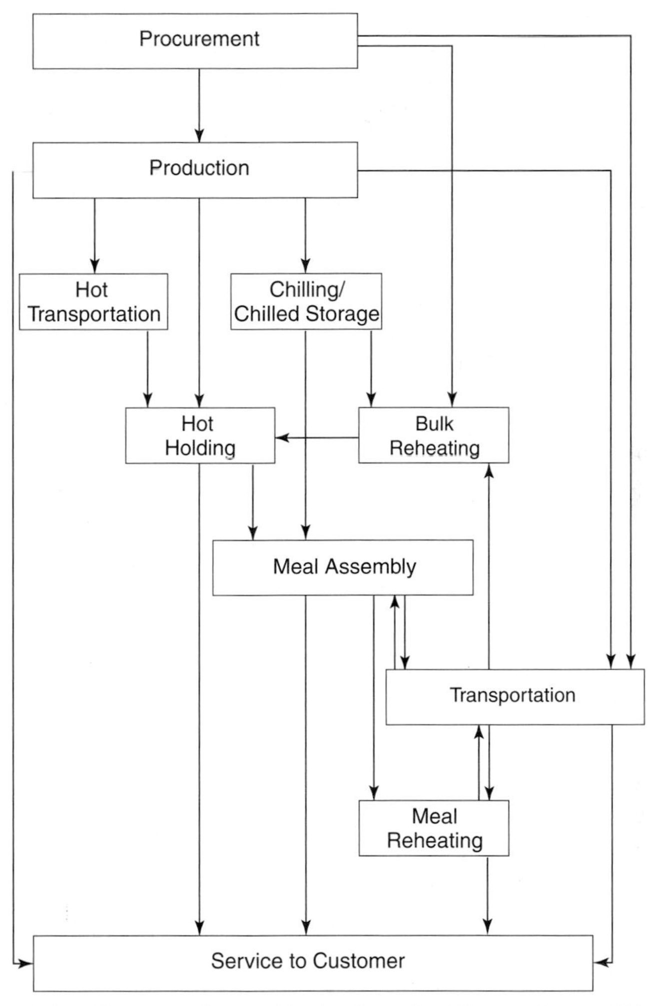

Figure 4.1. Diagram of possible food product flow options within a foodservice operation of Missouri–Columbia Agricultural Experiment Station.

The distinguishing feature of the diagram in Figure 4.2 is the food processing continuum and its interface with a particular foodservice type and the consumer. At the left side of the continuum, the point of origin of the elongated black triangle indicates no processing of the food entering the food product flow. The increasing altitude of the triangle represents the amount of food processing, reaching a maximum at the far right, required for complete processing. For example, fresh apples, sugar, flour, and shortening used in baking an apple pie are foods with almost no initial processing, and a frozen baked apple pie is purchased completely processed. The continuum is related to food procurement alternatives, depending upon the needs of a particular type of foodservice operation.

The central block of the diagram, labeled Foodservice Type, covers the production and delivery phase of a food product flow path. The requisite interface with the Food Processing Continuum is shown by lines from the appropriate point on

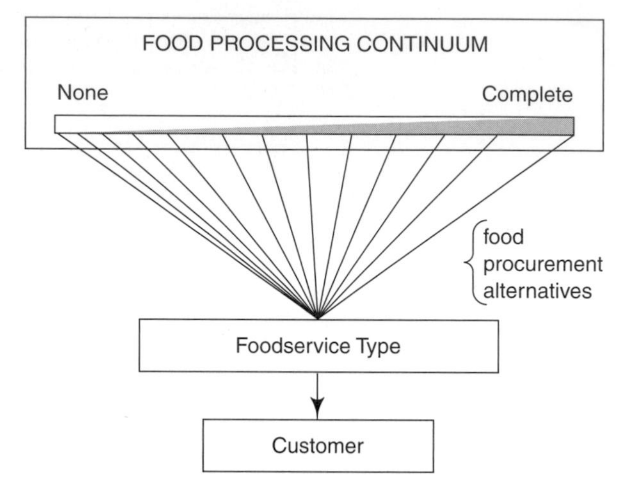

Figure 4.2. Food processing/foodservice operation interface.
Source: Foodservice Systems: Product Flow and Microbial Quality and Safety of Foods. North Central Research Public No. 245, 1977, Columbia, MO: University of Missouri—Columbia Agriculture Experiment Station.

the continuum to the Foodservice Type block. Completed menu items are served to customers as shown in the diagram.

Conventional Foodservice

Conventional foodservice: Foods are purchased in different stages of preparation for an individual operation, and production, distribution, and service are completed on the same premises.

Conventional foodservice traditionally has been used in most foodservice operations. Foods are purchased in various stages of preparation for an individual operation, and production, distribution, and service are completed on the same premises. Following production, foods are held hot or refrigerated, as appropriate for the menu item, and served as soon as possible. In the past, conventional foodservices often included a butcher shop, bake shop, and vegetable preparation unit. Currently, many conventional foodservice operations purchase preportioned meats, baked goods, and fresh, canned, frozen, or preprocessed fruits and vegetables instead of completely processing raw foods on premises. Figure 4.3 illustrates the simplest flow for a conventional foodservice operation, in which food goes from procurement through production and then is served directly to the customer, as shown by the gray highlighted portions of the diagram. This flow pattern is seen most often in restaurants where food is prepared to order. Flow of food in a conventional foodservice operation in a hospital characteristically appears as in Figure 4.4. In this diagram food is purchased, and products are prepared in production and then held hot until patient meal trays are assembled. Assembled trays are transported to the patient areas and then served to customers (patient). Processes such as hot holding, meal assembly, and transportation increase the time between production and service. Although alternative food product flow paths have evolved, the conventional foodservice remains the dominant one in the United States. Managers of conventional foodservices frequently have made changes in purchased ingredients and menu items to reduce labor costs in production. Types of food procured for conventional operations vary from no processing to complete

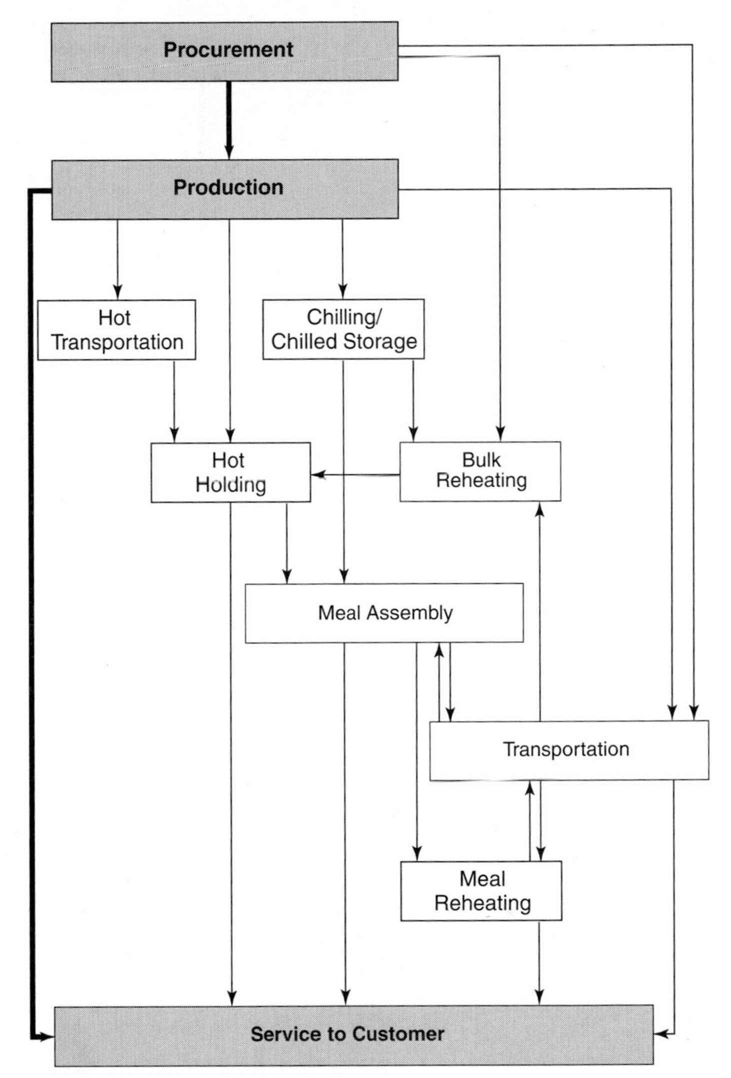

Figure 4.3. Food product flow in a conventional restaurant foodservice.

processing. Some foods are merely purchased and held chilled before service, such as milk or butter pats. Other menu items are produced from raw foods and may be held either heated or chilled until time of service. Items such as cereal, bread, and condiments are placed in dry storage for withdrawal as needed. Thus, foods for the conventional system come from the entire food processing continuum shown in Figure 4.2.

Following receipt and appropriate storage of food items and ingredients, menu items are prepared as near to service time as possible to assure quality. Because food subjected to hot-holding conditions is affected by temperature, humidity, and length of holding time, however, nutritional and sensory quality can be affected adversely.

Foods prepared in the conventional foodservice may be distributed for service directly to an adjacent or nearby serving area, such as a cafeteria or dining room.

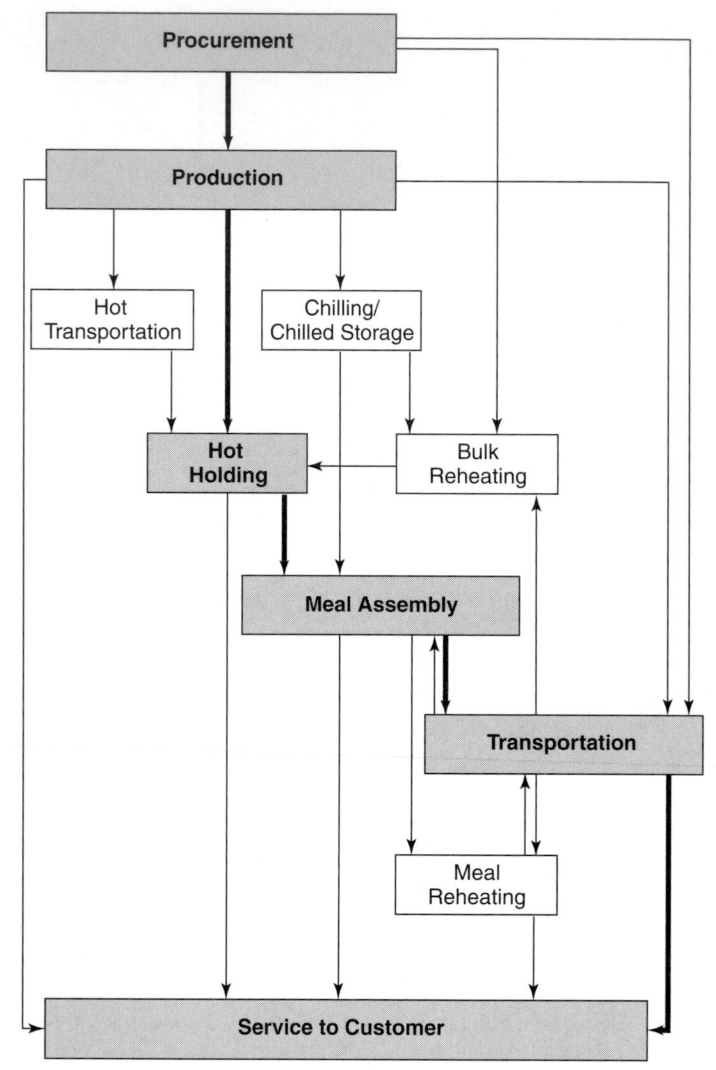

Figure 4.4. Food product flow in a conventional hospital foodservice operation.

Centralized service: Plates or trays are assembled in area close to production.

Decentralized service: Food is transported in bulk to a location separate from production and plates or trays of food are assembled in that location.

In hospitals and other healthcare facilities, food may be served on trays, using centralized or decentralized service. In **centralized service**, individual patient trays are assembled in or close to the production area. Trays then are distributed by carts or conveyors to patient units for delivery to patients' rooms. In **decentralized service**, food is distributed in bulk quantities for tray assembly in an area close to patients' rooms, such as a galley located in a hospital wing. In some facilities, a combination of these two approaches is used.

The systems model discussed in Chapter 1 is directly applicable to the conventional foodservice. Foods are brought directly into the operation, menu items are produced using conventional methods, and the meals then are served without extensive holding. In discussions of the other three types of foodservice operations, modifications of the functional subsystems of the systems model are indicated.

Traditionally, a skilled labor force for food production has been used in conventional foodservices for long periods each day. As the cost of labor increases within the foodservice industry, foodservice managers often explore procuring more extensively processed foods.

Ready Prepared Foodservice

Ready prepared foodservice: Menu items are produced and held chilled or frozen until heated for serving.

Ready prepared foodservices have evolved because of increased labor costs and a critical shortage of skilled food production personnel. In ready prepared foodservices, menu items are produced and held chilled or frozen until heated for service later. A significant difference between ready prepared and conventional foodservices is that menu items are not produced for immediate service but for inventory and subsequent withdrawal. Food items, upon receipt, are stored and recorded in a storage inventory. When needed for production, the food item is withdrawn. After production, menu items are stored in refrigerators or freezers and entered in the distribution inventory.

Purchasing foods for the ready prepared foodservice operation is similar to conventional foodservice, in that foods from the entire spectrum of the food processing continuum are used (Figure 4.2). Completely processed foods brought into the operation are stored either chilled or frozen, as appropriate to the food item. Items such as cereal and bread do not go through production but are put into dry storage for withdrawal as needed. Foods procured with little or no processing are used to produce menu items that are stored either chilled or frozen. A distinct feature of these foodservices is that prepared menu items are readily available at any time for final assembly and heating for service.

Center-of-the-plate menu items and hot vegetables require two phases of heat processing in ready prepared foodservice. The first occurs during the production of menu items, and the second occurs after storage, when items are brought to the appropriate temperature for service to the consumer. In some operations, chilled or frozen food items are reheated in bulk prior to the assembly of customer meals (Figure 4.5). As shown in the grayed boxes, the food product flow would be procurement, production, chilling/chilled storage, bulk reheating, hot holding, meal assembly, and service to customers. Many hospitals using the **cook-chill** technology will plate the food for patients in a chilled state. These plates of food are then reheated just prior to service using either special carts or convection or microwave ovens (Figure 4.6). The flow in these operations would be procurement, production, chilling/chilled storage, meal assembly, transportation, meal reheating, and service to customer.

Cook-chill: Method in which menu items are partially cooked, rapidly chilled, held in chilled storage, and reheated just prior to service.

Cook-freeze: Method in which menu items are partially cooked, rapidly frozen, held in freezer storage, and reheated just prior to service.

Sous vide: A process of sealing raw, fresh food items in plastic pouches to allow chilled storage and then cooking in boiling water prior to service.

Cook-chill and **cook-freeze** are two methods used in ready prepared foodservices. Cook-chill technology is moving into all onsite and some commercial market sectors. Although industry trade journals often contain articles detailing the implementation of cook-chill operations, their use remains limited throughout the industry; less than 10% of hospitals (Greathouse & Gregoire, 1988; Nettles & Gregoire, 1993; Nettles, Gregoire, & Cander, 1997) and schools (Mann, Shanklin, & Cross, 1993; Nettles & Gregoire, 2000; Brown, 2005) report use of cook-chill production. In the cook-freeze method, menu items are stored in the frozen state for periods generally ranging from 2 weeks to 3 months.

Sous vide, from the French term for "under vacuum," is classified as ready serve because it involves chilling and sometimes freezing menu items. Sous vide, which

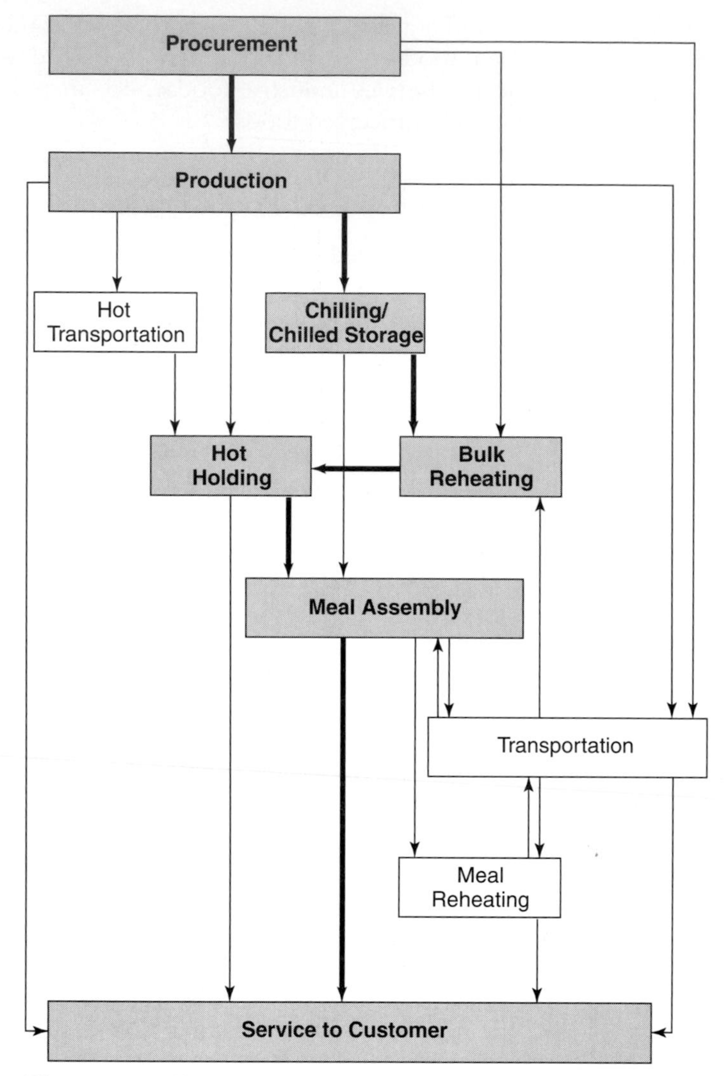

Figure 4.5. Food product flow in a ready prepared foodservice operation using bulk reheating.

originated in Europe, entails sealing raw, fresh food in impermeable plastic pouches with special equipment. Air is forced out of the pouches, pouches are sealed, and a vacuum is created. The foods then are partially or completely cooked slowly in low-temperature circulating water, rapidly cooled, and stored in temperature-controlled refrigerators (32° to 38°F), extending shelf life to about 21 days (National Restaurant Association Educational Foundation, 1992; Snyder, 1993; Ghazala, 1998). Menu items then are heated for service typically by placing the bag of food in simmering water. The Food and Drug Administration (FDA) states that only licensed food processors can perform the sous vide process. Foodservice operations are allowed to purchase sous vide products only from a reliable supplier that has an HACCP (Hazard Analysis Critical Control Points) program, as improper handling of the products can cause microbiological health hazards (Birmingham, 1992; National Restaurant Association Educational Foundation, 1992).

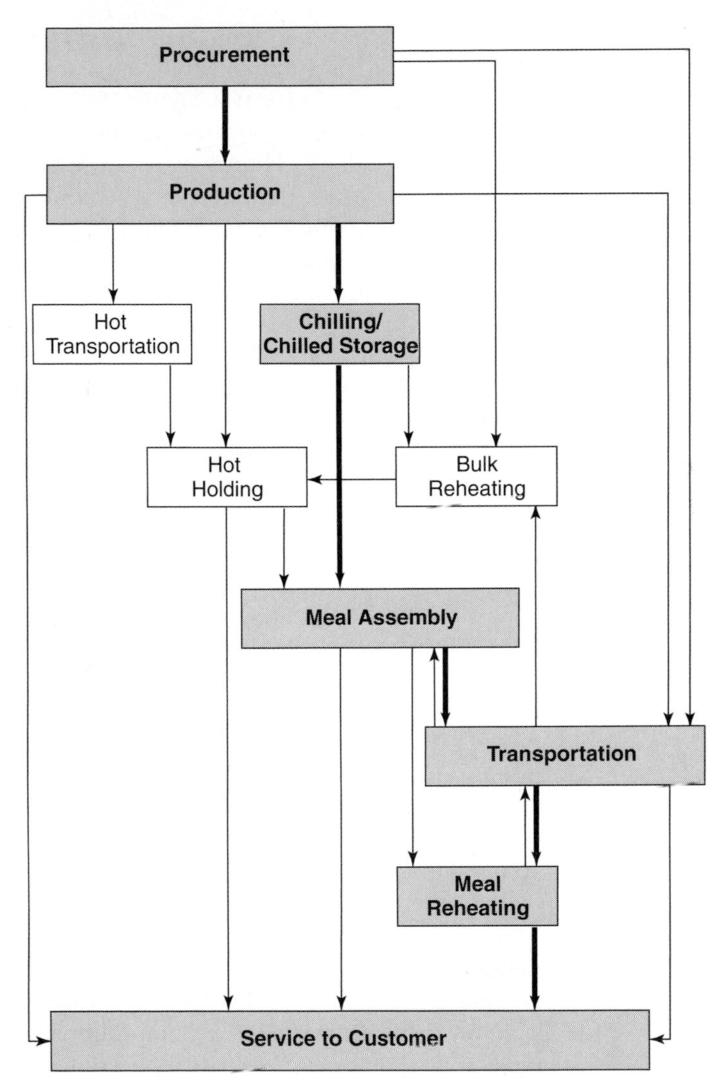

Figure 4.6. Food product flow in a ready prepared foodservice operation in a hospital with centralized tray assembly and decentralized reheating.

Accurate forecasting and careful production scheduling are needed to maintain quality of foods and to avoid prolonged holding, especially in the chilled state. In the cook-chill method, the cooked food is chilled in bulk and then portioned for service several hours or as much as a day in advance of the serving period. In the cook-freeze method, however, the food is portioned immediately after cooking and held in the frozen state until ready for service. In both cook-chill and cook-freeze operations, final heating occurs just before service, often in units near the serving areas. For example, some hospitals have galleys on patients' floors. Specialized heating equipment, generally microwave ovens or special induction heat carts, is used to finish the preparation of menu items and bring them to the correct serving temperature. Minor preparation, such as toasting bread or making coffee, also may occur in these galleys. Menu items usually are slightly undercooked during the initial production process to avoid overcooking and loss of sensory quality in the final heating for service (Castagna, 1997). Special recipe formulations are needed for many menu items because of the changes that occur during the chilling or freezing of food. Development of off-flavors may be a problem with some food items. Some of these changes may be controlled by substituting more stable ingredients; by exercising greater control of storage time, temperature, and packaging; or by adding stabilizers. To thicken sauces, for example, waxy maize starch, which has much greater thickening ability than flour, may be used to prevent product breakdown during thawing and heating (McWilliams, 1993).

Challenges in ready prepared foodservices are retention of microbiological, nutritional, and sensory qualities of food; the critical control points for cooling and reheating are extremely important in this type of foodservice (Gregoire & Nettles, 1997). Note that if the time and temperature standards are not adhered to in the ready prepared system, the corrective action is to discard the product. If thawing of frozen products is required prior to heating, thawing must be done under appropriate conditions in tempering boxes held at a temperature of less than 40°F and not scheduled too long before service. Prolonged holding is avoided, and careful control in the final heating stage before service is critical.

The foodservice systems model discussed in Chapter 1 has direct application to the ready prepared foodservice. Procurement of food items in various stages of processing greatly affects the activities of the subsystems in the transformation process. Food with little processing passes through production before being placed in either chilled or frozen storage; processed food items bypass production and go directly to storage. The ready prepared foodservice is unique in that it requires separate inventories for purchased and for ready-to-serve food. In the foodservice systems model, the subsystems of distribution and service are emphasized for the ready prepared foodservice.

Ready prepared foodservices, adopted in many operations to reduce labor expenditures and use staff more effectively, are expected to increase (World of Noncommercial, 1992). Harrah's Casino in Lake Tahoe, Nevada, reduced its production staff by half and reduced overtime labor by 25% with the implementation of a cook-chill operation (Sheridan, 1998). John Benke, vice president of operations for Sodexho, described cook-chill as a solution to rising food costs and the declining pool of skilled labor (Sheridan, 2004). Peak demands for labor are removed because production is designed to meet future rather than immediate needs. Production personnel can be scheduled for regular working hours rather than during early morning, late evening, and weekend shifts required in conventional foodservices. The heating and service of menu items do not require highly skilled employees; however, since this process is often decentralized, additional labor may be needed. Thus, an overall savings in labor cost does not often result from a shift from conventional to a ready food system. Food procurement in volume also may decrease food costs in these foodservices. Integration of computer controls with large-scale production allows management to react quickly and accurately to changes in the production environment and service needs of units (Sheridan, 2004). Other foodservice directors believe that product quality and food safety are important contributions of cook-chill systems. Andy Allen, executive chef of Harvard University Dining Services, indicated that implementation of a cook-chill system provided a much better consistency in the quality of products produced (Doty, 2005).

Any food production operation using cook-chill or cook-freeze methods requires special equipment to chill or freeze food quickly after the initial production phase. After production, food is portioned into either bulk or individual containers, wrapped, and chilled or frozen. Rapid chilling can occur in either blast chillers or freezers or in tumble chillers. Covered or wrapped pans of bulk food or plates of portioned food items are placed in a blast chiller or blast freezer, depending on the type of ready prepared foodservice. In both kinds of blast equipment, very rapid circulation of cold air reduces the temperature quickly, thus preserving the microbiological, nutrient, and sensory qualities of the food. Food items prepared for the cook-freeze method are completely frozen in a blast freezer before being moved to a freezer for holding at 0°F or below. Foods frozen in bulk need to be tempered in a rapid thawing refrigerator, which uses forced air or a slightly higher than normal temperature to defrost food safely and quickly. Plated food begins to defrost quickly after removal from the freezer and generally can be held in a regular refrigerator.

Machines may be used for portioning some food items, such as mashed potatoes on a plate, and may be part of a sequential service on a conveyor terminating at a shrink-wrap machine. The plate with the portion of food is covered with a special plastic wrap and passed through a shrink tunnel that, by removing air, shrinks the wrap to the shape of the food. The wrapped portions are then put through the blast freezer and finally stored in the inventory freezer.

Commissary Foodservice

Commissary foodservice: Centralized procurement and production facilities with distribution of prepared menu items to several remote areas for final preparation and service.

Technological innovations and the design of sophisticated foodservice equipment have led to the evolution and development of **commissary foodservices**. Unklesbay et al. (1977) described commissary foodservices as centralized procurement and production facilities with distribution of prepared menu items to several remote areas for final preparation and service. Centralized production facilities often are referred to as central commissaries, commissariats, or food factories, and the service units are known as satellite service centers. The menu items usually are delivered hot or cold/hold and served (Figure 4.7), although some commissaries chill or freeze the items and then reheat them at the remote site (Figure 4.8). The potential for economies from large-scale purchasing and production in a central facility is a common justification for design and construction of these facilities. In addition, expensive automated equipment for production of foods from unprocessed states is required.

In commissary foodservices, foods purchased typically have received little or no processing, as indicated in the left side of the continuum at the top of the diagram in Figure 4.2. These foods generally are purchased in large quantities and held after delivery under appropriate environmental conditions in dry, refrigerated, or frozen storage. Among the advantages of large-scale purchasing are increased supplier competition and cooperation and volume discounts. Operational advantages include centralized receiving, storage, and inventory control.

Most menu items in commissary foodservices are processed completely in the central facility. Because of the large quantities produced, the equipment for preprocessing and production often is different from that used in conventional foodservices. These large central production units may be designed using equipment

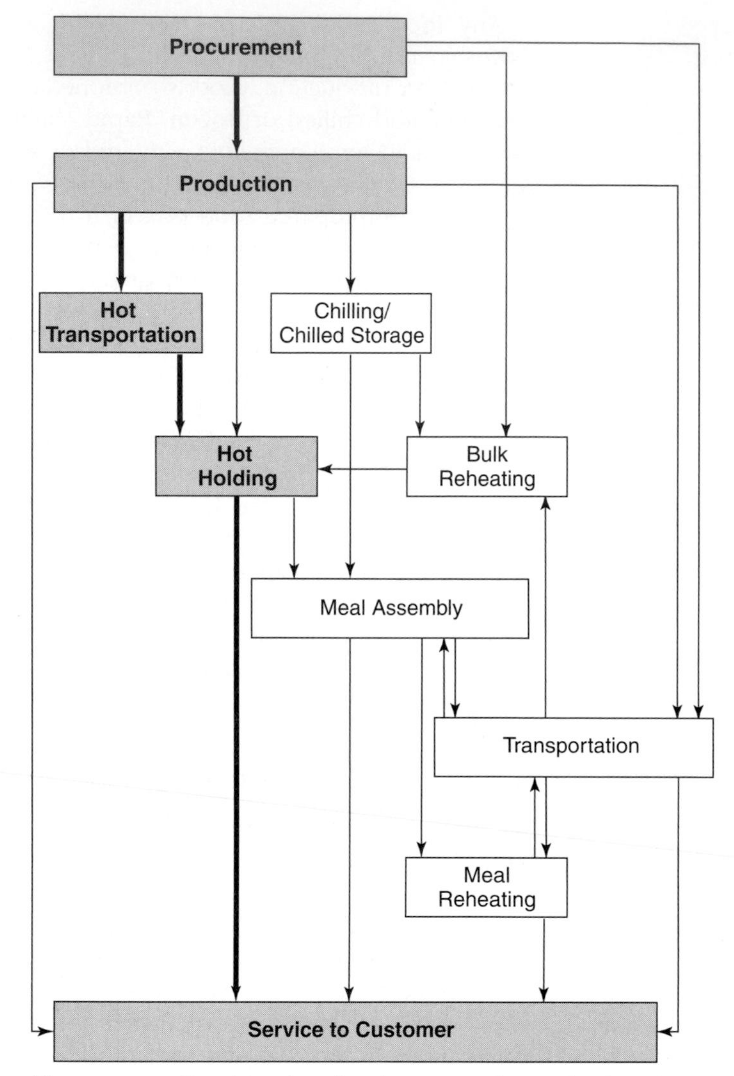

Figure 4.7. Food product flow in a commissary foodservice operation using hot transport of food.

frequently seen in food industry operations, such as canneries or frozen food processing plants. Large-scale production quantities also require major modifications of recipes and food production techniques. For example, cooking actually is done in two stages: first to a satisfactory and safe condition for transport and second for complete doneness and acceptable temperature.

Menu items prepared in a commissary foodservice may be stored in bulk or in individual portions. The type of storage may depend on the time necessary between production and service. In many instances, however, the type of storage for prepared menu items guides the design of the foodservice. For example, a decision to use frozen storage for menu items would be made before proceeding with the design. Many menu items that are held chilled or frozen require additional heating to reach desirable service temperatures. Highly specialized distribution equipment may be needed, depending on the type and location of satellite service centers.

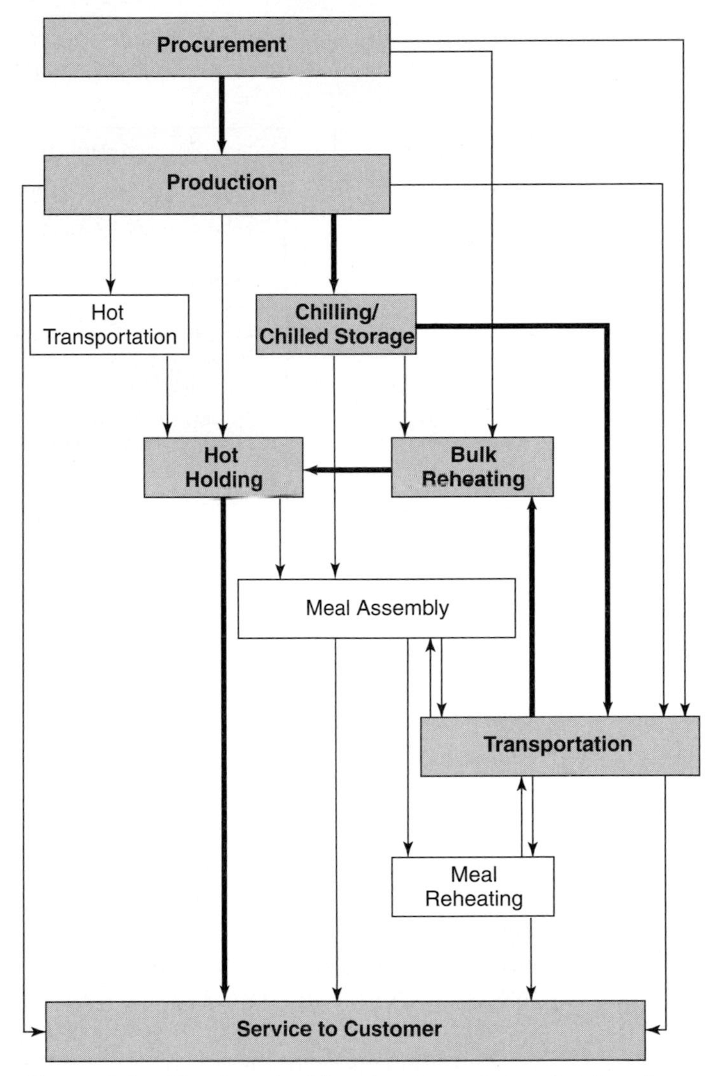

Figure 4.8. Food product flow in a commissary foodservice operation using cook-chill and reheating food on site after transporting.

Chilling and freezing of food are discussed in detail in the section titled "Ready Prepared Foodservices."

In Figure 4.9, a modification of the foodservice systems model discussed in Chapter 1 illustrates the uniqueness of commissary foodservices. As indicated, the major changes affecting inputs to the system are the type of food and facilities used. In the transformation element, the nature of the functional subsystems differs greatly from the conventional foodservice, primarily due to larger production capacity, storage for prepared menu items, and distribution capabilities for transporting prepared menu items to many satellite centers.

The packaging and storage of prepared menu items present challenges for control in commissary foodservices. Various packaging materials and techniques are now being used, ranging from individual pouches or serving dishes designed for

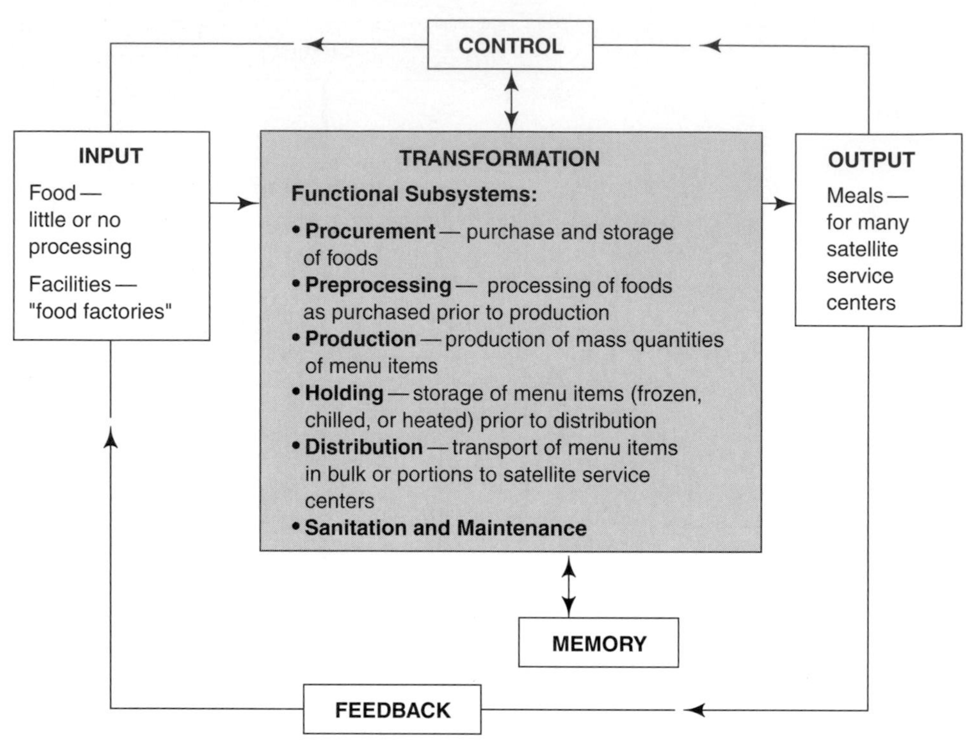

Figure 4.9. Uniqueness of a commissary foodservice.

chilled or frozen holding to disposable or reusable metal pans for different types of distribution and transportation equipment. In addition, specialized equipment is required for packaging, storing, and distributing products. Preserving the microbiological, nutritional, and sensory qualities of foods during holding and heating at point of service can present problems. The importance of establishing critical control points for hot holding and service or for cooling and reheating needs to be emphasized. In large operations, a food technologist or microbiologist is frequently on staff and is responsible for quality control.

Commissary foodservice principles have been adopted in operations in which service centers are remote from, yet accessible to, the production unit. Reducing the duplication of production, labor, and equipment that occurs if production is done at each service center has been the objective. Space requirements at the service centers also can be minimized because of limited production equipment. The high initial costs of construction of commissaries and purchase of transportation equipment and the operating costs of producing and transporting food are current concerns in evaluating the cost-effectiveness of these foodservices.

Commissary foodservice has long been used in schools, although many districts have combined it with conventional foodservice. In recent years, centralized production facilities frequently have been constructed in urban districts with a large number of schools. More often, however, large secondary school kitchens serve as commissaries, or base kitchens as they are frequently called, producing meals for transport either in bulk or individual portions to elementary schools. In this case, the operation is a combination of commissary and conventional foodservices because secondary students also are served in an adjacent cafeteria. In 1996, Nettles reported that 35% of high schools serve as a base kitchen for other schools, and 12% of school districts use a central commissary. The numbers of each are increasing; Brown (2005) reported that 40% of schools use one school as a base kitchen to provide food to other kitchens and 14% of schools use a commissary kitchen. Ready prepared food operation concepts have also been combined with the commissary foodservice. Large commissary kitchens have been developed in which food is prepared using cook-chill technology and then transported to remote sites for reheating prior to service. School districts such as Montgomery County, Maryland (Sheridan, 2004) have built large central commissaries in which food products are prepared, chilled in bulk, and then transported to schools throughout the district. Foods are reheated prior to service at these schools.

The use of a commissary is not restricted to a metropolitan area. Several of the large national restaurant chains operate a mass production commissary at one location and transport their products all over the United States. Taco Bell greatly reduced the amount of space and kitchen equipment needed in its restaurants by preparing its food in central commissaries.

Many organizations have used their commissary/ready food operation as a means for generating revenue. Food prepared in the commissary kitchen is sold to other operations in the same geographic area (Healthcare Re-visits, 1997).

One of the most appealing advantages of a commissary is that the equipment and personnel operate at a high efficiency rate during the day with no idle periods. Also, a commissary often operates 8 hours a day for 5 days a week. With a greater number of dependent service centers, the operation may be extended to multiple shifts, 24 hours a day, 7 days a week. The commissary cannot be operated efficiently unless highly skilled personnel are employed.

Assembly/Serve Foodservice

Assembly/serve: Menu items are purchased preprepared and require minimal cooking before service.

The development of **assembly/serve** foodservices—called convenience-food foodservices or the minimal cooking concept—occurred primarily because of the market availability of foods that are ready to serve or require minimum cooking. Another factor has been the chronic shortage of skilled personnel in food production and the increasing cost of labor.

Food products are brought into the operation with a maximum degree of processing. In Figure 4.2, foods would come from the far right side of the food processing continuum. Only storage, assembly, heating, and service functions are commonly performed in these foodservices, thus reducing labor and equipment costs. Little if any preprocessing is done onsite, and production is very limited. Fresh, frozen, and dried items like shredded lettuce, sliced carrots, beef stew, preportioned roast beef, and lasagna are purchased in an assembly/serve foodservice.

The three market forms of foods used predominantly in these foodservice operations are bulk, preportioned, and preplated. The bulk form requires portioning before or after heating within the foodservice operation, whereas the preportioned market form requires only assembly and heating. The preplated products require only heating for distribution and service and thus are the most easily handled of the three forms.

Figures 4.10 and 4.11 show possible flow of food in assembly/serve operations. Figure 4.10 represents the flow in operations such as Cedars Sinai Medical Center (Berkman & Schechter, 1991), where frozen meals and preportioned salads and

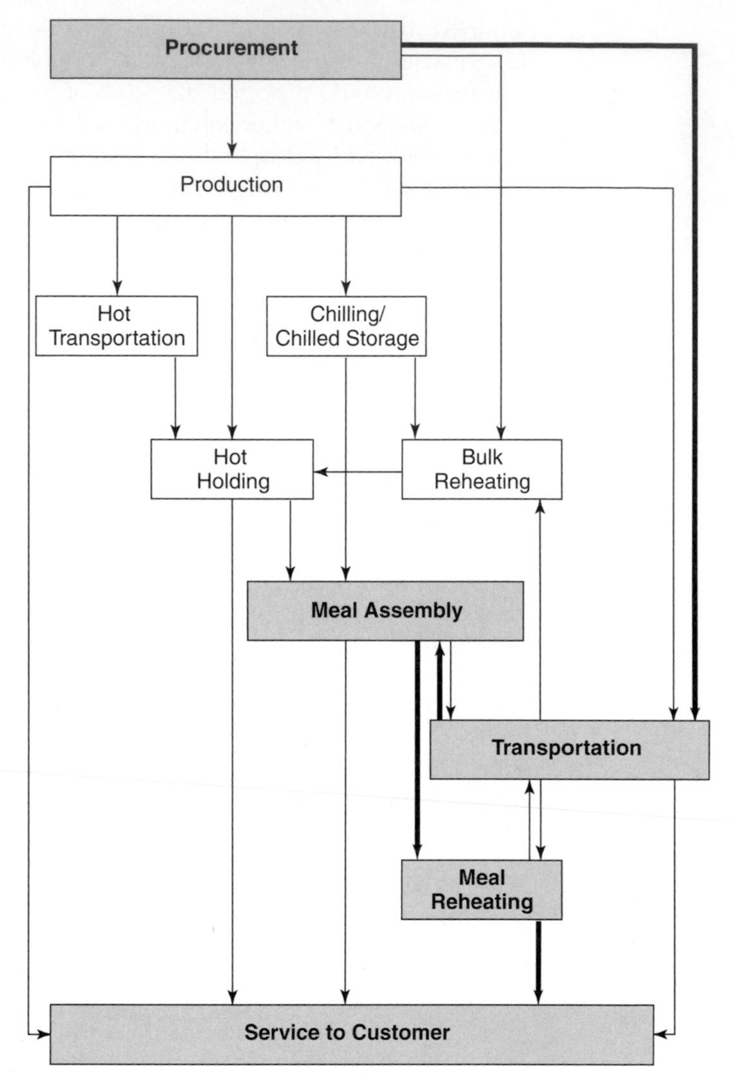

Figure 4.10. Food product flow in an assembly/serve foodservice operation using preportioned meals.

desserts are purchased. Food products are transported to floor galleys for storage. Patient trays are assembled and food items reheated in the galleys before service to patients. Such a concept eliminates the need for a central production kitchen. A more common form of assembly/serve is shown in Figure 4.11. Food is procured, heated in bulk centrally, and held hot before service to customers.

In many assembly/serve operations, a combination of foods is used, some requiring a limited degree of processing in the foodservice operation and others requiring none. Often, partially prepared foods are purchased to be combined with other ingredients before heating or chilling. In many operations, completely

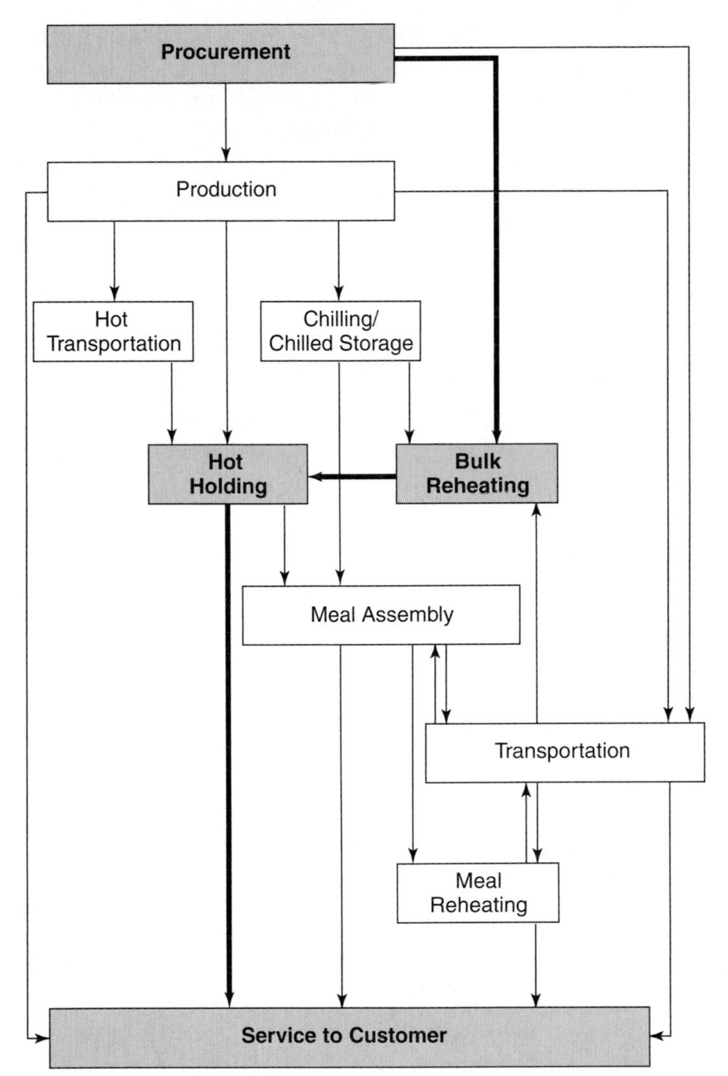

Figure 4.11. Food product flow in an assembly/serve foodservice operation using bulk reheating.

processed foods may be enhanced in assembly/serve foodservices as a way of individualizing menu items; for example, a sauce may be added to an entrée.

Following procurement, food items are held in dry, refrigerated, or frozen storage. When menu items are heated in either bulk or preportioned form, quality control is a definite concern. Thawing of frozen food must be done with rigorous control of time and temperature. The same technology that made ready prepared foodservices feasible also is used in assembly/serve operations.

The assembly/serve foodservice has gained some degree of acceptance because it appears to offer an easy solution to labor and production problems. Nevertheless, a readily available supply of highly processed, high-quality food products is a prerequisite for a successful assembly/serve operation. Availability of such items is often a problem, especially for modified diets. Although modified dietary products have been developed in recent years, these items are not always readily available, particularly in rural and small communities. Therefore, if an assembly/serve foodservice is used for hospital patients or nursing home residents, menu items may need to be prepared for those on modified diets.

Another common complaint about assembly/serve operations is the lack of individuality of food products. These complaints duplicate the comments frequently heard about the "sameness" of convenience foods available on the retail market. As discussed in the section on conventional foodservices, a trend toward the use of foods with some degree of processing is evident. This trend appears to be more predominant than total adoption of the assembly/serve foodservice. In some instances, however, an assembly/serve foodservice meets the needs of particular operations in which space is very limited for production facilities or competent labor is not available.

KITCHEN DESIGN AND LAYOUT

Planning a kitchen and arranging work areas to minimize operating costs and maximize productivity is an important activity for a foodservice manager. Planning for a new kitchen can take months or years to complete; most hospital and school foodservice directors who had been involved in the process to plan for a new kitchen indicated the planning process took more than 7 months to complete (Nettles, Gregoire, & Canter, 1997; Nettles & Gregoire, 2000).

Design: Defining the size, shape, style, and decoration of a space.

Layout: Detailed arrangement of equipment, floor space, and counter space.

Flow: Movement of product or people through an operation.

Planning a kitchen involves both design and layout components and consideration of flow principles. **Design** focuses on the overall space planning and includes defining the size, shape, style, and decoration of a space. **Layout** refers to the detailed arrangement of the equipment, floor space, and counter space.

Flow, the movement of product or people in an operation, should be an important topic in the planning process. Two types of flow are usually discussed, product and traffic. **Product flow** is the movement of food from receiving through production to the customer and then through warewashing and trash removal. **Traffic flow** is the movement of employees through the operation as they complete their work. One goal in planning for a new kitchen and serving space is to have a straight-line flow from receiving through warewashing. A straight line flow minimizes backtracking and cross-over movement of food and people. Almanza, Kotschevar, and Terrell (2000) recommend the following materials-handling rules in relation to flow.

- Store at point of first use.
  Have dry, refrigerated, and frozen storage close to the work areas that will use these products.
- Allow for economy of motion.
  Store products based on use, with those used most often within normal reach.
  Store heavier products lower and lighter products higher.

- Use space economically by providing for specific sizes.
  Plan depth and distance between shelves based on products to be placed on the shelves.
- Minimize handling and storage.
  Locate storage close to the receiving area.
  Use carts and pallet movers to move product from receiving to storage.
- Systemize.
  Organize storage area.
  Group like products in close proximity to each other.
- Use good handling practices.
  Use lift trucks and other mechanized devices for lifting and moving product.
  Keep aisles clean and cleared of product.
  Establish a designated traffic flow.
- Communicate.
  Use signal lights, mirrors, buzzers, and electronic transmission of messages to facilitate communication of product movement.

Planning Team

Planning for a new or remodeled foodservice operation typically involves a team of five to ten individuals and includes members of the foodservice department, facility administration, and often a foodservice consultant and/or architect. Each will bring different, yet important, perspectives to the planning process.

The planning team is responsible for articulating the vision for the new space, establishing goals for the planning process, and describing expectations for desired flow, operations, and service. The team will work with the architect to plan the new or remodeled space.

Planning Process

Charrette: Collaborative planning session.

The planning process often starts with gathering input from various constituents about the potential needs and use of the space. Often the architects or designers will hold **charrettes**, collaborative planning sessions, to gain input from all who might be impacted by the new kitchen such as foodservice employees, other employees of the facility (teachers, doctors, nurses, administrators), and clients (patients, students, visitors). Another common technique used to gather information for kitchen design is the development of bubble diagrams. A **bubble diagram** defines spaces in the planned kitchen and the relationship of these spaces to each other. As shown in the bubble diagram in Figure 4.12, the dry and refrigerated/freezer storage areas of this planned university dining center would be adjacent to receiving and to the cold and hot food preparation areas. Issues considered at this stage of planning include the following:

- What quantity of food and supply items will be purchased in this new facility (which influences type and amount of storage needed)?
- What amount of production will be done onsite in the new facility (which influences type and size of production equipment needed)?

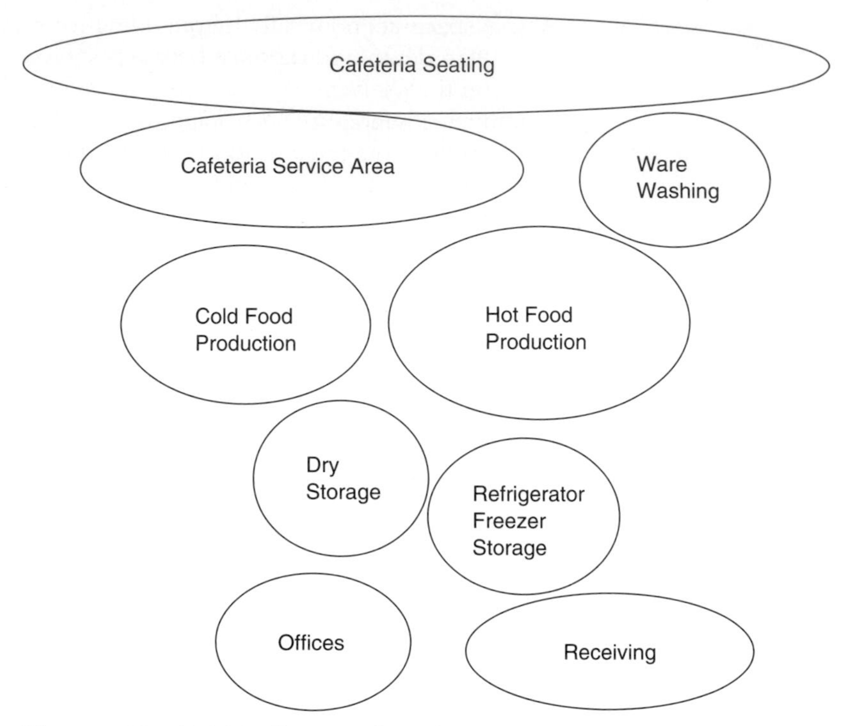

Figure 4.12. Bubble diagram of a university dining center.

- What type of production will be done in the new facility (conventional, ready foods, and assembly/serve have differing equipment needs)?
- How will food be served to customers from or in this new facility (which influences holding, delivery, and service equipment needed)?
- What funds are available for the renovation (which influences project scope and quantity and quality of supplies and equipment)?
- What will be the operating costs of the new facility?
- What food safety issues will need to be addressed (which influences work space layout and food holding and delivery)?
- How many employees will be working in this space (which influences size and number of needed work spaces and amount of locker room and restroom facilities needed)?
- What future changes in capacity are anticipated; what flexibility is needed (which influences equipment purchased and space needed)?

Foodservice directors use a variety of sources to help them gather information and ideas in this planning process. Discussions with other foodservice directors of similar operations; visits to other facilities to see equipment, workflow, and service; attendance at seminars and trade shows such as the National Association of Foodservice Equipment Manufacturers (NAFEM); meetings with equipment manufacturers; and reading of trade journal articles all provide valuable information for those planning a new or renovated facility.

After gathering information from the planning group and various constituent groups, the architect and/or designer will develop what is known as a **program.** The program typically states the goals for the project, gives an overview of the occupants of the planned space, presents a projected timeline for the project, proposes allocation of space to the desired units/activities within the planned space, and details guidelines to be followed in the actual design phase. These guidelines address topics such as access, acoustics, ceilings, doors, environmental conditions, fire protection, floors, height, insulation, lighting, mechanical needs, power, security, walls, and, in some cases, elevators, landscaping, and parking.

Developing a **schematic design** is the next step in the planning process. The schematic design includes the preliminary plans for the project, space drawings, proposed materials, electrical and mechanical issues, and projected costs. The design is done to give building occupants a general idea of the architect/designer's vision for the space.

Meetings with the planning committee and the architect/designer continue as the schematic design moves to the more detailed **blueprint** or, as it is sometimes called, a layout design. The blueprint shows specifically the layout of the planned space; the mechanical, electrical, and plumbing details; and the proposed finish materials. The blueprint is drawn to scale, often 1/4 inch equals 1 foot. Many of the items shown on a blueprint will have a number and/or letter on this. Refer to the equipment or specifications key for additional details related to this item. The blueprint is accompanied by a **specification book,** which details specifications for all of the equipment planned for the space.

When the blueprint is approved by the planning committee, a more detailed set of **construction drawings** is prepared. These drawings, which can number 50 pages or more, provide not only the blueprint layout of the space but also more detailed elevations and graphic drawings to help clearly convey to the contractor the proposed plans for the facility. These construction drawings are typically then put out for bid by local contractors interested in working on the project.

Components of a Foodservice Design

Lighting

Direct lighting: Lighting aimed at a certain place.

Indirect lighting: Lighting shining over a space rather than at a certain place.

Lumen: Amount of light generated when 1 foot-candle of light shines from a source.

Foot-candle: Measurement of illumination equal to 1 lumen of light on 1 square foot of space.

Lighting decisions focus on what to light and how to light it and can be the most important environmental decisions made. Katsigris and Thomas (1999) indicate that a restaurant's lighting system can account for approximately one-third of its energy costs, so the lighting decisions have cost implications as well for a foodservice operation.

Lighting can be used in many different ways. The most common are direct and indirect lighting. **Direct lighting** aims a certain light at a certain place. Fluorescent light fixtures in a kitchen area over a workstation would provide direct lighting. **Indirect lighting** shines light over a space instead of aiming it at a particular spot. Wall sconces in a dining room would provide indirect lighting.

Light and illumination (the effect achieved when light strikes a surface) are measured in many different ways. A **lumen** (lm) is the amount of light generated when 1 foot-candle of light shines from a single source. One **foot-candle** is the light level of 1 lumen on 1 square foot of space. The **color rendering index** (CRI) indicates the effect of a light source on the color of objects based on a 0 to 100 scale.

A higher CRI means that lighting is more likely to show things in their natural color. The **correlated color temperature** (CCT) is a measure of the coolness (red, orange) to warmness (blue-white) appearance of a light. CCT is measured in kelvins (K). The **power factor** (PF) is a measure of how efficiently a light source uses power. A light source that converts all of the power supplied to it into watts for lighting would have a PF of 1.

Artificial lighting can be from two sources; incandescent lamps or electric discharge lamps (fluorescent, mercury vapor, halogen, high- and low-pressure sodium). **Incandescent lights** are like the common light bulb in which a tungsten filament is encased in a sealed glass bulb. These bulbs are screwed into a socket connected to an electrical source, and when electricity is provided to the bulb, it flows through the filament and creates light that is diffused through the glass bulb. **Electric discharge lights** generate light by passing an electric arc through a tube filled with a special mixture of gasses. These lights require an electronic or magnetic **ballast** to control the electric current and "turn on" the light. Fluorescent lights are commonly used in kitchen workspaces because although they are more expensive to purchase than incandescent lights, they will last much longer and give off much more heat.

Lighting planned for the kitchen work area typically will be very different from that planned for the dining area. Factors impacting lighting include ceiling height and color, wall color, amount of texture in surfaces, and floor finish and color. Lighter colored, smooth, and shinny surfaces (walls, ceilings, floors, equipment) will reflect more light than darker colored and textured surfaces. Lighting should be placed over work areas and traffic aisles in such a way as to prevent shadows and glare.

The Food Code (2001) requires lighting levels of at least 10 foot-candles in storage areas, 20 foot-candles in handwashing and warewashing areas, and at least 50 foot-candles where an employee is working with food. Almanza, Kotschevar, and Terrell (2000) recommend 70 to 100 foot-candles in work areas which involve detail work. The amount of lighting in the dining area could range from 5 to 100 foot-candles depending on the mood desired in the dining area. Many areas will have minimum lighting levels specified in local building codes.

Several special features are required for lighting in the kitchen area. Guards are needed to prevent glass from falling into food should a bulb break. Vapor-proof/moisture-resistant lights are needed in ventilation hoods and dishwashing areas because of the moisture generated in those areas.

Heating and Air Conditioning

The quality of the air in the workspace is controlled by the heating, ventilation, and air conditioning (HVAC) system. Katsigris and Thomas (1999) indicate that the key environmental factors related to HVAC are indoor temperature, humidity, air movement, room surface temperature, and air quality. They state that to control and modify these factors, HVAC systems will typically include "furnaces or boilers (to produce hot air), air conditioners or chillers (to produce cold air), fans (to circulate air), duct work (to move air) and filters (to clean air)."

HVAC systems work with air in a variety of ways to achieve a comfortable work environment. As air passes through the HVAC system it may be referred to as

- Supply air—air coming to the workspace from the HVAC system
- Return air—air returning to the HVAC system from the workspace
- Exhaust air—air that is being removed from the workspace and building
- Makeup air—air that is being brought in to replace exhaust air

Negative air pressure occurs when more air is removed from a space than is brought into the space. **Positive air pressure** occurs when more air is brought into a space than is removed from the space. Creating a slight negative pressure in the kitchen area will prevent kitchen smells and odors from drifting into the dining area.

Ventilation: The circulation of fresh air in a space.

Ventilation, the circulation of fresh air in a space, is measured in cubic feet per minute (cfm). The American Society of Heating, Refrigerating, and Air Conditioning Engineers recommends 35 cfm of fresh air per person for kitchens, 15–20 cfm per person for dining rooms, and 15–25 cfm per person for general offices. Ventilation in the kitchen area is particularly important because of the heat, odors, grease, steam, and moisture generated in this space. Ventilation hoods placed over the cooking equipment are used to draw this air out of the kitchen area. Katsigris and Thomas (1999) recommend that a kitchen ventilation system provide 20 to 30 total air changes in an hour. The ventilation hoods are equipped with filters designed to filter grease from the air. These filters must be cleaned regularly to prevent fires in the hoods. The hoods also are equipped with automatic fire protection systems with nozzles that shoot water or liquid fire retardant at the cooking surface when activated.

Materials

Metal. One of the most common materials found in a foodservice kitchen is stainless steel. Stainless steel is an alloy of many different metals. Almanza, Kotschevar, and Terrell (2000) indicate that stainless steel No. 304 is the most commonly used in foodservice. Often called 18-8 stainless steel because of its chromium (18%) and nickel (8%) content, it also includes carbon, manganese, phosphorus, sulfur, and silicon. Stainless steel is highly desirable for use in foodservice operations because it is inert chemically and is stainproof, strong, and durable.

Gauge: The weight of the material per square foot.

Metal is specified in a standard **gauge** that indicates the weight of the material per square foot. A common gauge of stainless steel used in foodservice operations is 20 gauge, which weighs 1.5 pounds per square foot and is 1/32-inch thick. The larger the number, the thinner the metal. Metals also will have a finish or polish. A No. 4 finish, a bright satin finish, is common on foodservice equipment.

Plastics. Plastics are synthetically produced, nonmetallic, polymeric compounds that can be molded into various forms and then hardened. Plastics are being used in foodservice operations for storage containers, shelving, trays, equipment paneling, and so on. In many cases they provide a lighter weight, less expensive option when compared to equivalent metal items. Plastics used for storing food should meet local and national standards for odor and taste transfer.

Equipment

Criteria for selecting equipment for a foodservice operation include functionality, durability, ease of cleaning, capacity, size, energy use, and availability of parts and service. Almanza, Kotschevar, and Terrell (2000) recommend the following when designing a kitchen and the equipment for that kitchen:

- Angles and corners on equipment and work tables should be rounded.
- Edges that are turned down should be closed against the body.
- All tubing should be seamless and easily cleaned.
- All joints and seams in food contact areas and areas where seepage or condensation might occur should be sealed.
- Doors and covers should be properly sized and should fit tightly.
- Drawers and bins should be readily removable and easily cleaned.
- Shelving should be removable and adjustable.
- Thermostats on equipment should be easily cleaned.

Additional information about specific pieces of foodservice equipment is found in Chapter 6, "Food Production."

Floors

Covering materials for floors, walls, and ceilings should be selected for ease of cleaning and maintenance and for appearance. In back-of-the-house areas, covering materials are generally very sturdy and are selected primarily on the basis of cleanability and durability. For the front of the house, appearance will generally have greater priority; as a result, dining areas may present some unique cleaning problems. Customers are sensitive to dusty drapes, streaked or grimy windows, smudged or cracked walls, cobwebbed ceilings, loose floor tiles, and a heavily soiled carpet. Dining room floors need continuous care and in many establishments need to be swept, vacuumed, or mopped when customers have departed after a meal.

Several types of flooring materials are used in construction of foodservice facilities, and each type presents a different cleaning challenge. The material descriptions, advantages, and disadvantages are outlined in Table 4.1. Wood floors and carpeting, often used in dining rooms, are not acceptable in production areas because of problems with cleaning and sanitation.

Principles of Work Design

Konz and Johnson (1999) emphasized the importance of the concept of "work smart and not hard" when designing work areas. To *work smart* means to work more efficiently; to *work hard* is to exert more effort.

Work design refers to a program of continuing effort to increase the effectiveness of work systems (Kazarian, 1979). Industrial engineers, for example, have applied work analysis and design techniques in the manufacturing industries for many years. More recently, these principles have been applied to the service industries.

Table 4.1 Description, Advantages, and Disadvantages of Common Flooring Materials

Material	Description	Advantages	Disadvantages
Asphalt	Mixture of asbestos, lime rock, fillers, and pigments, with an asphalt or resin binder	Resilient; inexpensive; resistant to water and acids	Buckles under heavy weight; does not wear well when exposed to grease or soap
Carpeting	Fabric covering	Resilient, absorbs sound, shock; good appearance	Not to be used in food preparation areas. Sometimes problems with maintenance elsewhere
Ceramic tiles	Clay mixed with water and fire	Nonresilient; useful for walls; nonabsorbent	Too slippery for use on floors
Concrete	Mixture of Portland cement, sand, and gravel	Nonresilient; inexpensive	Porous; not recommended for use in food preparation areas
Linoleum	Mixture of linseed oil, resins, and cork pressed on burlap	Resilient; nonabsorbent	Cannot withstand concentrated weight
Marble	Nature, polished stone	Nonresilient; non-absorbent; good appearance	Expensive; slippery
Plastic	Synthetics with epoxy resins, polyester, polyurethane, and silicone	Most resilient; nonabsorbent	Should not be exposed to solvents or alkalies
Rubber	Rubber, possibly with asbestos fibers. Comes in rolls, sheets, and tiles	Antislip; resilient	Affected by oil, solvents, strong soaps, and alkalies
Quarry tiles	Natural stone	Nonresilient; nonabsorbent	Slippery when wet, unless an abrasive is added
Terrazzo	Mixture of marble chips and Portland cement	Nonresilient; nonabsorbent. If sealed properly, good appearance	Slippery when wet
Commercial-grade vinyl tiles	Compound or resins and filler and stabilized	Resilient; resistant to water, grease, oil	Water seepage between tiles can lift them, making floor dangerous and providing crevices for soil and pests
Wood	Maple, oak	Absorbent; sometimes inexpensive; good appearance	Provides pockets for dust, insects. Unacceptable for use in food preparation areas. Sealed wood may be used in serving areas

Source: From *Applied Foodservice Sanitation,* 4th ed. (pp. 161–162) by the Educational Foundation of the National Restaurant Association, 1992, Chicago: Author. Used by permission.

Principles of work analysis and design have been developed from several different fields of study over time. Among the major contributions to the study of scientific management is the classic work of Frederick Taylor and the Gilbreths. Their principles of materials handling have impacted the design of work areas in a foodservice operation.

Materials handling refers to the movement and storage of materials and products as they proceed through the foodservice system. Good design of materials

1. Minimize all material movements and storages.
2. Use the shortest and straightest routes for the movement of materials across the workplace.
3. Store materials as close to the point of first use as possible.
4. Minimize handling of materials by workers unless absolutely necessary.
5. Preposition all materials at the workplace as much as possible to reduce handling effort.
6. Handle materials in bulk if at all possible.
7. Make provisions to remove scrap, trash, and other wastes at the point of creation.
8. Take advantage of gravity to move materials when feasible.
9. Use mechanical aids to lift heavy materials that are frequently used at workplaces.
10. Use built-in leveling devices to keep materials at a convenient working height.
11. Use mechanized conveyors to move materials that follow a fixed route across the workplace if they do not interfere with the work.
12. Use well-designed containers and tote pans that are easy to pick up and move.
13. Consider the use of interlocking containers for moving greater loads with ease and safety.
14. Consider changing design of products involved to improve their materials-handling characteristics.

Figure 4.13. Principles of materials handling.

movement will lead to increased efficiency and decreased activities that do not add appreciable value to the end product. The amount of materials handling often depends on the location and arrangement of storage areas, preparation and production areas, and equipment. The key principles of materials handling, outlined in Figure 4.13, should be used to develop a system for moving materials efficiently within the foodservice facility.

Principles of motion economy, primarily from the early work of the Gilbreths, relate to the design of work methods, of the workplace, and of tools and equipment. These principles, summarized in Figure 4.14, specify that movement should be

- simultaneous
- symmetrical
- natural
- rhythmic
- habitual

Principles of motion economy that pertain to the human body—the use of both hands, coordination of hands and eyes, and continuous motion—are aimed specifically at reducing the effort and energy required to do a job. Principles related to the design of the workplace, tools, and equipment identify situations that lead to easy body motions, such as locating tools within easy reach and placing objects in fixed positions. The graphics in Figure 4.15, which have been adapted from employee training materials on work simplification, illustrate several of the key principles. Figures 4.16a and 4.16b show how improper design can create situations of back strain and fatigue because employees must reach above the head with heavy loads or bend for extended periods of time.

A. Use of the human body

1. The number of motions required to complete a task should be minimized.
2. The length of necessary motions should be minimized.
3. Both hands should be used for work, and they should begin and end their activities simultaneously.
4. Motions of hands and arms should be in symmetrical and opposite directions.
5. Both hands should not be idle at the same time except for rest.
6. Motions should be confined to the lowest possible classifications needed to perform the task satisfactorily.
7. Smooth curved motions should be developed in preference to straight-line or angular motions.
8. Motion patterns should be developed for rhythmic and habitual performance.
9. The motions should be arranged to take advantage of momentum.
10. The number of eye fixations required for the task should be minimized.
11. Intermittent use of the different classifications of movements should be provided to combat fatigue.

B. Design and layout of the workplace

1. Materials, tools, and controls should be located within the normal working area.
2. Materials and tools should have a fixed location.
3. Work requiring the use of eyes should be done within the normal field of vision.
4. Tools and materials should be prepositioned to facilitate picking up.
5. Gravity feed bins or containers should be used to deliver incoming materials close to the point of use.
6. Gravity should be used to deliver outgoing materials.
7. The height of the working surface should be designed to allow either a standing or sitting position.
8. The environment of the workplace should be conducive to productive motions.

C. Design of tools and equipment

1. Tools, hand equipment, and controls should be designed for easy grasp.
2. Two or more tools should be combined if possible.
3. Jigs, fixtures, or foot-operated devices should be used to relieve the work of the hands.
4. Equipment should be designed so the inherent capabilities of the body members are fully utilized.
5. Levers and controls should be designed to make maximum contact with the body member.

Figure 4.14. Principles of motion economy.

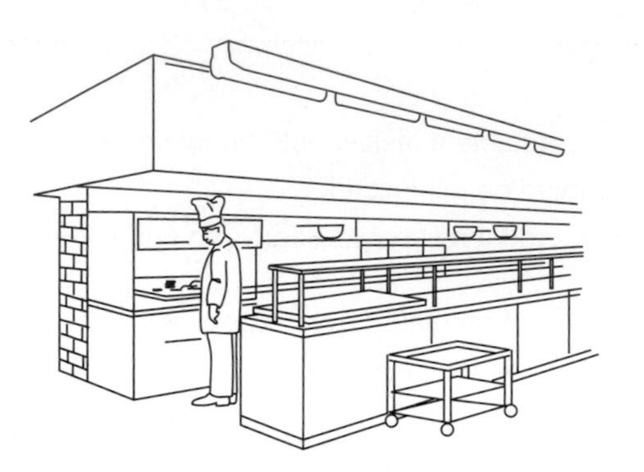

Proper equipment arrangement helps you "work smart."

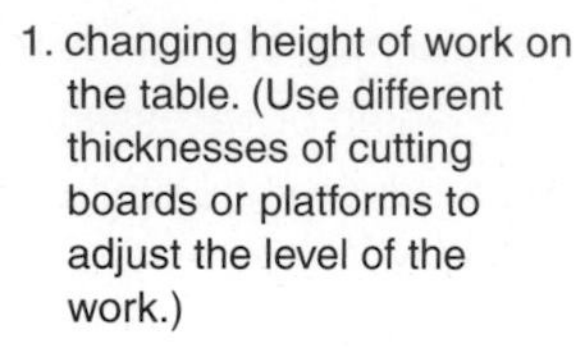

Adjust work heights to your elbow by:

1. changing height of work on the table. (Use different thicknesses of cutting boards or platforms to adjust the level of the work.)

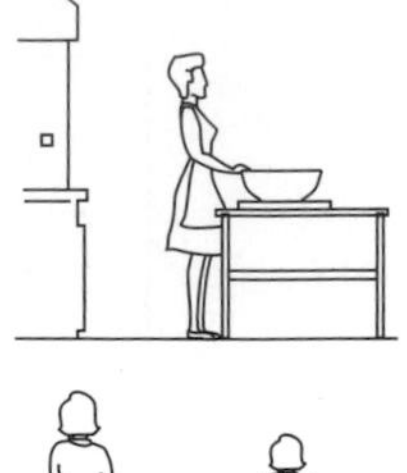

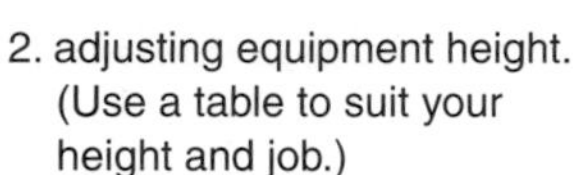

2. adjusting equipment height. (Use a table to suit your height and job.)

3. adjusting the chair. (Use adjustable-height chairs.)

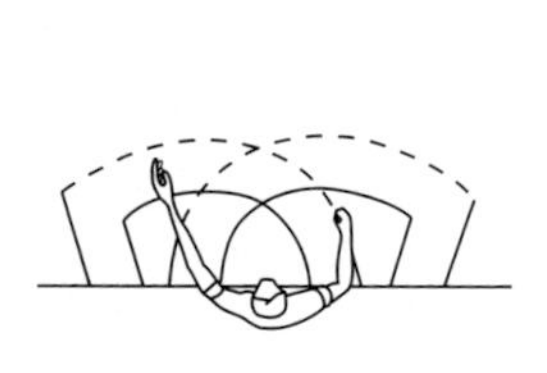

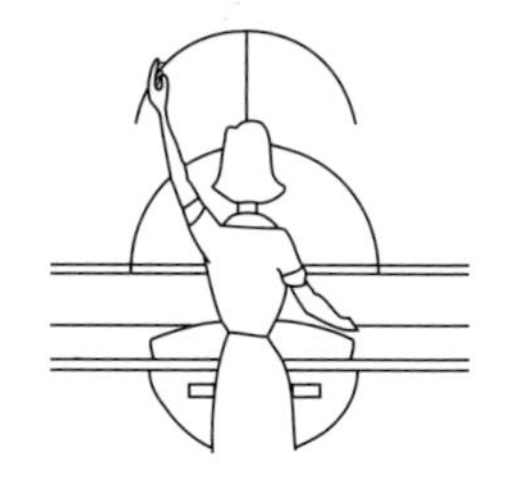

Maximum work areas, based on your reach distance, affect tiring.

Arranging work and using both hands in rhythm and order improves your methods.

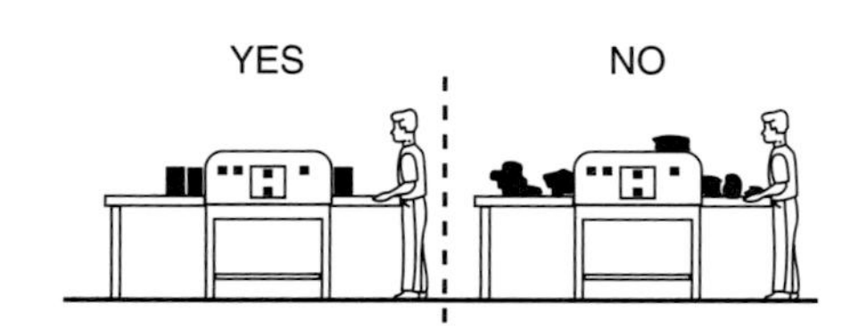

Planned arrangements save time, motion, and effort.

Drop delivery using proper heights of equipment helps you "work smart."

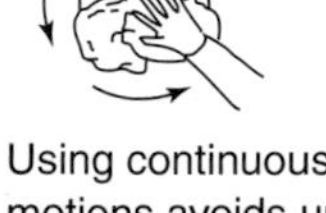

Using continuous curved motions avoids unneeded starts and stops.

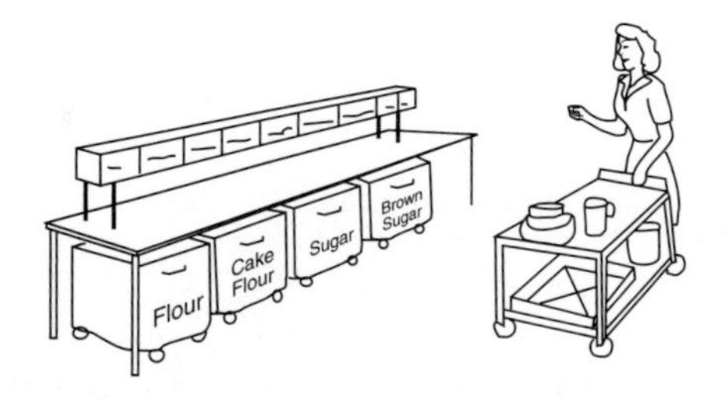

Equipment on wheels can help organize work.

Using muscles smoothly when lifting prevents strain.

Figure 4.15. Illustrations of work design principles in foodservice operations.
Source: The Basic Four of Work, by S. Konz and J. Maxwell, 1980. Unpublished manuscript. Manhattan, KS: Kansas State University.

(a)

(b)

Figure 4.16. **(a)** Cook reaching to upper oven. **(b)** Foodservice employee bending over pot and pan sink.

CHAPTER SUMMARY

This summary is organized by the learning objectives.

1. In a conventional foodservice operation, food is prepared, held hot or cold as appropriate, and served at the same site. Ready prepared foodservice operations include cook-chill and cook-freeze operations. In each of these operations, food is prepared in advance of service, chilled or frozen, held in inventory until needed, and then reheated prior to service. Commissary foodservice operations are those in which food production and service are separated. Food is prepared at one location and then transported for service at another location. In assembly serve foodservice operations, food items are brought into the operation ready prepared and require only heating before service.
2. In conventional foodservice operations, the flow of food is from production to service, whereas in ready prepared systems, food flows from production to chilled or frozen storage. The food is reheated prior to service.
3. Cook-chill is a process where foods are cooked to a partial state of doneness and then chilled rapidly. They are held chilled until they are reheated just prior to service. Cook-freeze is a similar process, although the foods are frozen rapidly after cooking and are held frozen until reheating just before service. Sous vide is a process of sealing raw, fresh foods in impermeable plastic pouches. The pouches are placed in heated water to cook the food items. Centralized service is one in which meals are plated for service in or near the

Professional Organization Profile

Foodservice Consultants Society International

MISSION

To promote professionalism in foodservice and hospitality consulting while returning maximum benefits to all members

WHO BELONGS TO THE ORGANIZATION

Members include more than 1000 foodservice consultants, consultant firms, and allied organizations in 35 countries. Student memberships are available.

ADVANTAGES OF MEMBERSHIP

Members have access to the society's website, including an online network and listserve for interacting with other foodservice consultants around the world; receive organization publications including a monthly newsletter, quarterly journal *The Consultant,* and Membership Directory; attend association-sponsored conferences and seminars; become part of an international consultant referral service; and are eligible for awards recognizing their consulting expertise.

WEBSITE

www.fcsi.org

MEET THE PRESIDENT

Kenneth Winch, president 2004–2006, is director of kenwinchdesign, a foodservice design firm in the United Kingdom. He worked as an engineer for a manufacturing company and in sales and branding for a fabrication company prior to starting his consulting practice 35 years ago. He has been recognized with more than 23 design awards for his foodservice work.

production area. Decentralized service involves transporting food items in bulk to a distant location and then plating the food for service at this site.

4. When observing a foodservice operation, watch for what happens to the food between when it is produced and then served to the customer. Draw a flow chart to show the steps that occur.
5. Design focuses on the overall space planning and includes defining the size, shape, style, and decoration of a space. Layout refers to the detailed arrange-

ment of the equipment, floor space, and counter space. Flow is the movement of product or people in an operation.

6. The planning process typically involves a planning committee that works with an architect and/or designer to develop plans for a new or renovated step. The planning process usually involves development of bubble diagrams, a program, and schematic designs prior to completion of the final blueprints and construction documents.

TEST YOUR KNOWLEDGE

1. How does sous vide differ from cook-chill?
2. Why are cook-chill and cook-freeze foodservice operations referred to as ready prepared?
3. Food in the local school district is prepared at the middle school kitchen, placed in steam table pans, and then transported to several elementary schools to be served to children there. Is this an example of a centralized or a decentralized service? Why?
4. At the University of Kansas Medical Center, food is prepared in the kitchen and dished onto individual plates, and then the plates of food are frozen. On the day of service, plates of food are placed on trays on an assembly line in the kitchen. The trays are transported in a cart to the patient care area. Plates of food are reheated in microwave ovens in galley kitchens on each floor, and then the tray is served to the patient. Draw a flow chart of the food product flow in this operation.
5. How and why would lighting and ventilation needs differ between the dining area and the kitchen used to prepare food in a university dining center?

CLASS PROJECTS

1. Divide into groups of two to three people. Visit two different foodservice operations in your area. Draw a diagram showing the flow of food in those operations.
2. Find an example of each type of foodservice operation in your area. Invite the manager to come to speak with the class about the strengths and weakness of that type of foodservice.
3. Design a new kitchen for a local foodservice operation. Use the existing space in that operation but design a better layout for the operation and specify the type and location of equipment and workspaces for that operation.

WEB SOURCES

www.aia.org	American Institute of Architects
www.cfesa.com	Commercial Food Equipment Service Association
www.nafem.org	North American Association of Food Equipment Manufacturers
www.nfs.org	NFS, International

BIBLIOGRAPHY

Alexander, J. (1991). Cook-chill automation. *Hospital Food Service, 24*(3), 7–8.

Almanza, B. A., Kotschevar, L. H., & Terrell, M. E. (2000). *Foodservice layout, design, and equipment planning,* 4th ed. Upper Saddle River, NJ: Prentice Hall.

American Society of Heating, Refrigerating, and Air Conditioning Engineers. (1997). *ASHREA handbook—Fundamentals.* Atlanta, GA: Author.

Berkman, J., & Schechter, M. (1991). Today, I closed my kitchen. *Food Management, 26*(11), 110–114, 118, 122.

Birmingham, J. (1992). Whatever happened to sous vide? *Restaurant Business, 91*(6), 64–65, 68, 72.

Bobeng, B. J. (1982). Alternatives for menu flow in hospital patient feeding systems. In *Hospital patient feeding systems* (pp. 113–117). Washington, DC: National Academy Press.

Boss, D. (1991). A kitchenless future? *Food Management, 26*(11), 16.

Brown, D. (2005). Prevalence of food production systems in school foodservice. *Journal of the American Dietetic Association. 105,* 1261–1265.

Castagna, N. G. (1997). Know your cook-chill. *Restaurant & Institutions, 107*(26), 84, 86.

Doty, L. (2005). Whole lotta food: Bulk preparation and production. *Restaurants & Institutions.* [Online]. Available: http://www.foodservice411.com/rimag/default.asp

Eck, L. S., & Ponce, H. (1993). HACCP: A food safety model. *School Food Service Journal, 47*(2), 50–52.

Escueta, E., Fiedler, K., & Reisman, A. (1986). A new hospital foodservice classification system. *Journal of Foodservice Systems, 4,* 107–116.

Food Code (2001). [Online]. Available: http://www.cfsan.fda.gov/~dms/fc01-toc.html

Fox, M. (1993). Quality assurance in foodservice. In Kahn, M., Olsen, M., & Var, T. (eds.), *VNR's encyclopedia of hospitality and tourism* (pp. 148–155). New York: Van Nostrand Reinhold.

Friedland, A. (1993). 3 districts, 1 kitchen. *Food Management, 47*(5), 42–43.

Ghazala, S. (1998). *Sous vide and cook-chill processing for the food industry.* Rockville, MD: Aspen Publishing.

Greathouse, K. R., & Gregoire, M. B. (1988). Variables related to selection of conventional, cook-chill, and cook-freeze systems. *Journal of the American Dietetic Association, 88,* 476–478.

Greathouse, K. R., Gregoire, M. B., Spears, M. C., Richards, V., & Nassar, R. F. (1989). Comparison of conventional, cook-chill, and cook-freeze foodservice systems. *Journal of the American Dietetic Association,* 89, 1606–1611.

Gregoire, M. B., & Bender, B. (1999). Food production systems for the future. In Martin, J. (ed.), *Managing child nutrition programs: Leadership for excellence* (pp. 429–461). Gaithersburg, MD: Aspen Publishing.

Gregoire, M. B., & Nettles, M. F. (1997). Alternative food production systems. In Jackson, R. (ed.), *Nutrition and food services for integrated health care* (p. 331). Gaithersburg, MD: Aspen Publishing.

Healthcare re-visits the commissary concept. (1997). *Food Management, 32*(6), 42–44, 46, 48.

Jones, P. (1993). Foodservice operations management. In Khan, M., Olsen, M., & Var, T. (eds.), *VNR's encyclopedia of hospitality and tourism* (pp. 27–36). New York: Van Nostrand Reinhold.

Katsigris, C., & Thomas, C. (1999). *Design and equipment for restaurants and foodservice.* New York: John Wiley & Sons, Inc.

Kazarian, E. A. (1979). *Work analysis and design for hotels, restaurants and institutions,* 2nd ed. Westport, CT: AVI.

King, P. (1992d). Sky Chefs enters new era with central kitchen. *Food Management, 27*(1), 48, 50.

Konz, S., & Johnson, S. (1999). *Work design: Industrial ergonomics,* 6th ed. Scottsdale, AZ: Holcomb Hathaway.

Kujava, G. (1991). Cook-chill technology: Serving savings on every tray. *The Consultant, 24*(2), 22–24.

Longrée, K., & Armbruster, G. (1996). *Quantity food sanitation,* 5th ed. New York: John Wiley & Sons, Inc.

Lundberg, D. E. (1984). A look at restaurant commissaries. *FIU Hospitality Review, 2*(1), 7–13.

Mann, N. L., Shanklin, C. W., & Cross, E. W. (1993). An assessment of solid waste management practices used in school food service operations. *School Food Service Research Review, 17,* 109–114.

Matthews, L. (1991). Gearing up for cook-chill. *Food Management, 26*(3), 39.

Matthews, M. E. (1982). Foodservice in healthcare facilities. *Food Technology, 36,* 53–64, 71.

McWilliams, M. (1993). *Foods: Experimental perspectives,* 2nd ed. New York: Macmillan.

National Restaurant Association. (1991). *Make a S.A.F.E. choice: Sanitary assessment of food environment.* Washington, DC: Author.

National Restaurant Association, Educational Foundation. (1992). *Applied foodservice sanitation,* 4th ed. Chicago: Author.

Nettles, M. F. (1996). *Issues related to equipment and the dietary guidelines for Americans.* University, MS: National Food Service Management Institute.

Nettles, M. F., & Gregoire, M. B. (1993). Operational characteristics of hospital foodservice departments with conventional, cook-chill, and cook-freeze systems. *Journal of the American Dietetic Association, 93,* 1161–1163.

Nettles, M. F., & Gregoire, M. B. (2000). Analysis of the process used to select a food production system. *Journal of Child Nutrition and Management, 24,* 84–91.

Nettles, M. F., Gregoire, M. B., & Canter, D. D. (1997). Analysis of the decision to select a conventional or cook-chill system for hospital food service. *Journal of the American Dietetic Association, 92,* 626–631.

Norton, C. (1991). What is cook-chill? *Hospital Food Service, 24*(2), 5–7.

Rethermalization—A hot topic! (1991). *Hospital Food Service, 24*(4), 6–7.

Rhodehamel, E. J. (1992). FDA's concerns with sous vide processing. *Food Technology, 46*(12), 73–76.

Sheridan, M. (1998). Cook-chill: From colleges to casinos. *Restaurants & Institutions, 108*(21), 135.

Sheridan, M. (2004). Critical mass. *Restaurants & Institutions.* [Online]. Available: http://www.foodservice411.com/rimag/default.asp

Snyder, O. P. (1993). Hazard analysis and critical control point in foodservice. In Khan, M., Olsen, M., & Var, T. (eds.). *VNR's encyclopedia of hospitality and tourism* (pp. 185–219). New York: Van Nostrand Reinhold.

Snyder, O. P., Gold, J. I., & Olson, K. A. (1987). Quantifying design parameters for foodservice systems in American hospitals. *Journal of Foodservice Systems, 4*(3), 171–186.

Surveying the scene. (1994). *School Food Service Journal, 48*(2), 40–45.

Unklesbay, N. (1977). Monitoring for quality control in alternate foodservice systems. *Journal of the American Dietetic Association, 71,* 423–428.

Unklesbay, N., Maxcy, R. B., Knickrehm, M., Stevenson, K., Cremer, M., & Matthews, M. E. (1977). *Foodservice systems: Product flow and microbial quality and safety of foods.* North Central Regional Research Public. No. 245. Columbia: University of Missouri–Columbia Agricultural Experiment Station.

Walkup, C. (1994). NRA Educational Foundation: Industry's classroom. *Nation's Restaurant News* [The National Restaurant Association 1919–1994, special commemorative issue], 84, 89, 96.

World of noncommercial foodservice. (1992). *Food Management, 27*(1), 121–123, 128.

CHAPTER

5

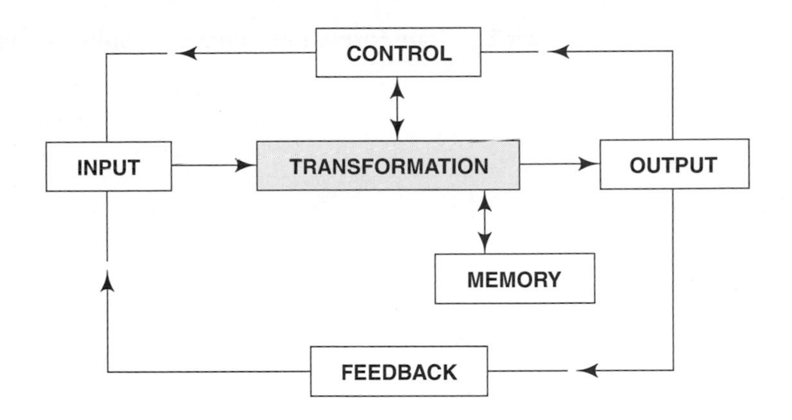

Procurement

Enduring Understanding

- The procurement process is critical to the financial success of a foodservice operation and includes the activities of purchasing, receiving, storage, and inventory control.
- Quality, quantity, and price are equally important in a purchase decision.

Learning Objectives

After reading and studying this chapter, you should be able to:

1. Describe what is involved in purchasing, receiving, storage, and inventory control.
2. Write a specification.
3. Differentiate quality grades and yield grades.
4. Describe different purchasing methods.
5. Identify controls that are needed in purchasing, receiving, and storage.

The goal of any foodservice operation is to serve quality meals while maximizing value for both the operation and the customer. Before this foodservice goal can be achieved, however, the necessary food items must be procured. In this chapter, you will learn about the procurement subsystem of the transformation process of the foodservice systems model. Purchasing, receiving, and storage are all activities of the procurement subsystem.

PROCUREMENT

According to the foodservice model presented in Chapter 1, procurement is the first functional subsystem of the transformation element, as highlighted in Figure 5.1. Several important activities exist within this subsystem: purchasing, receiving, storage, and inventory control.

In common with other subsystems of the transformation element, the management of procurement involves planning, organizing, staffing, leading, and controlling. Because procurement is considered an important profit generator, those

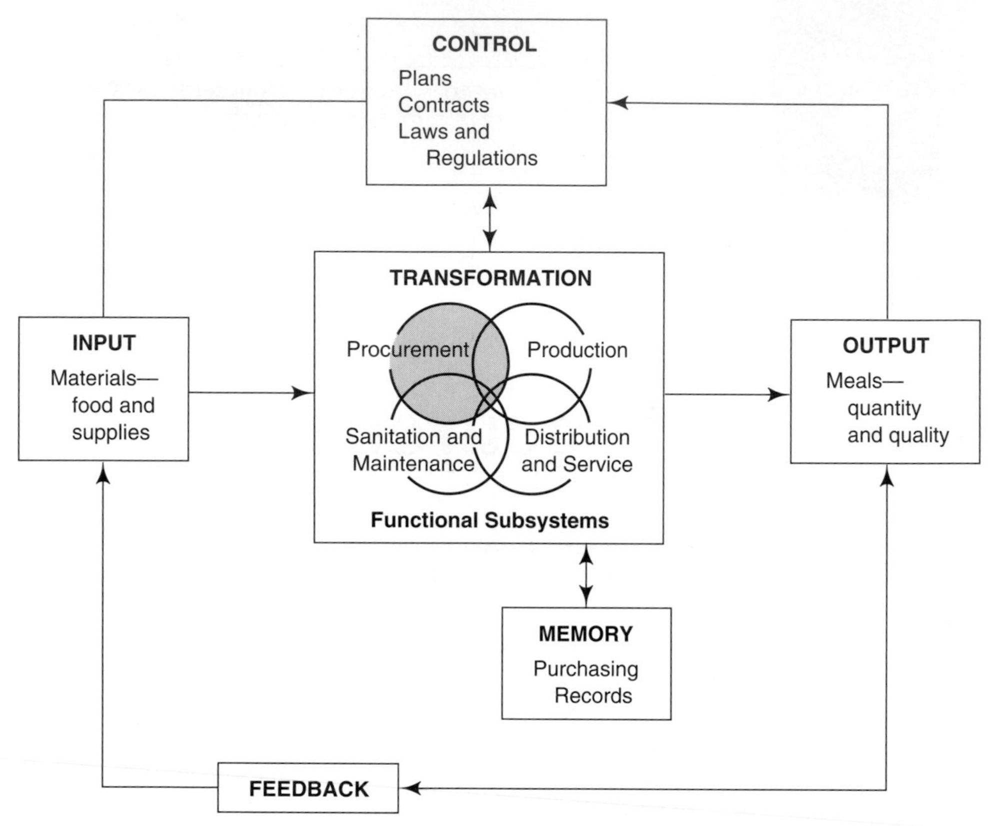

Figure 5.1. Foodservice systems model with procurement subsystem highlighted.

responsible for it, often termed the Director of Procurement or Purchasing Director, should be members of the top management team and involved in all operational decision making.

PURCHASING

Purchasing: Activity concerned with the acquisition of products.

Purchasing is an activity concerned with the acquisition of products, both food and nonfood. It is often described as obtaining the right product, in the right amount, at the right time, and at the right price. To do this, buyers must know the market and the products and must have general business acumen. They also rely on sales representatives to give advice on purchasing decisions and to relay valuable information about available food items and new products.

Purchasing Managers

Today's buyer has a powerful influence on the world food distribution system. The buyer is the one who listens to the desires of customers, who determines what is grown and packaged, and who understands how items are processed or manufactured, shipped, sold, and consumed. If an item cannot be sold, it will not be grown or manufactured or marketed (Spears, 1999).

Purchasing executives across the United States were asked to rank skills of buyers that would be valuable in the year 2000 (Kolchin & Giunipero, 1993). The following 10 skills were rated in descending order of importance:

- Interpersonal communication
- Customer focus
- Ability to make decisions
- Negotiation
- Analytical ability
- Managing change
- Conflict resolution
- Problem solving
- Influence and persuasion
- Computer literacy

According to the National Restaurant Association's Foodservice Purchasing Managers (FPM) executive study group, the role of purchasing managers is changing (Patterson, 1993). Change is especially evident in three areas: writing food and equipment specifications, ethics in purchasing, and the food industry in cyberspace.

Managerial productivity once was based primarily on dollars saved, but today total quality management (TQM), partnering with suppliers, and ultimate customer satisfaction with products purchased also must be measured (Patterson, 1993). Purchasing has a role in TQM and, therefore, a significant impact on quality, customer satisfaction, profitability, and market share (Carter & Narasimhan, 1993). In the Kolchin and Giunipero (1993) study, TQM topped the list for the common body of knowledge needed for a purchasing manager in the year 2000.

In many foodservice operations, purchasing is a profit center (Patterson, 1993). A large percentage of the sales revenue in a foodservice operation is spent for purchases. Better purchasing saves dollars paid to suppliers for needed products, supplies, and services. These savings go directly to the bottom line, before taxes, on the profit and loss statement (Leenders & Fearon, 1992). Purchase dollars are high-powered dollars! Purchasing, therefore, can contribute to the profit of an operation. Until recently, purchasing in a large organization, such as a hospital, was considered a service to other departments, and top management thought of it as a cost center rather than a profit generator.

Reck and Long (1983) discussed the procedures for organizing the purchasing department as a profit center. A **profit center** is any department that is assigned both revenue and expense responsibilities. The purchasing manager is expected to manage expenses while creating profit for the organization. A **cost center**, however, is a department that is expected to manage expenses but not generate profits for the organization; it is expected to help other departments contribute to the creation of profit.

Profit center: Department that generates revenues greater than its expenses and creates profit for the organization.

Cost center: Department that has expenses but does not generate profits to cover those expenses.

Purchasing managers use computers for pricing, inventory monitoring, product usage, tracking product movement, paying bills, and communicating with other departments and suppliers. Expert systems have been developed to interact with Universal Product Code (UPC) readers and facilitate taking physical inventories. Big central warehouses with large food and supply inventories will be phased out as foodservice operators reduce the capital in inventories (Spears, 1999).

Marketing Channel

The **market** is the medium through which a change in ownership moves commodities from producer to consumer. Foodservice markets may be meat or fresh produce establishments or locations in Chicago or Mexico.

Knowledge of the food market involves finding sources of supply and determining which food items can be obtained from which supplier. It also requires understanding the flow of supplies through the marketing channel and the effect of market regulations on the distribution process.

Exchange of ownership of a product occurs in the marketing channel, sometimes identified as the channel of distribution, as shown in Figure 5.2. The marketing channel indicates the exchange of ownership of a product from the producer through the processor or manufacturer and the distributor to the customer. Products are distributed through the channel from the producer to the customer and procured by reversing the path of the product from the customer to the producer.

The marketing channel has five major components: producers, processors or manufacturers, distributors, suppliers, and customers. Value and cost are added to the product in each of these components and are reflected in the final price paid by the customer.

Producers

Producer: Someone who produces raw food to sell.

Producers, generally farmers or ranchers, produce raw food to sell to processors or manufacturers, who sell to distributors or directly to the foodservice operation. The products then are sold to customers. The abundance of food in the United States is not caused by more people, animals, and land on today's farms but is rather the result of increased efficiency in agricultural operations resulting from application of advances in science and technology. The amount of land used for producing food

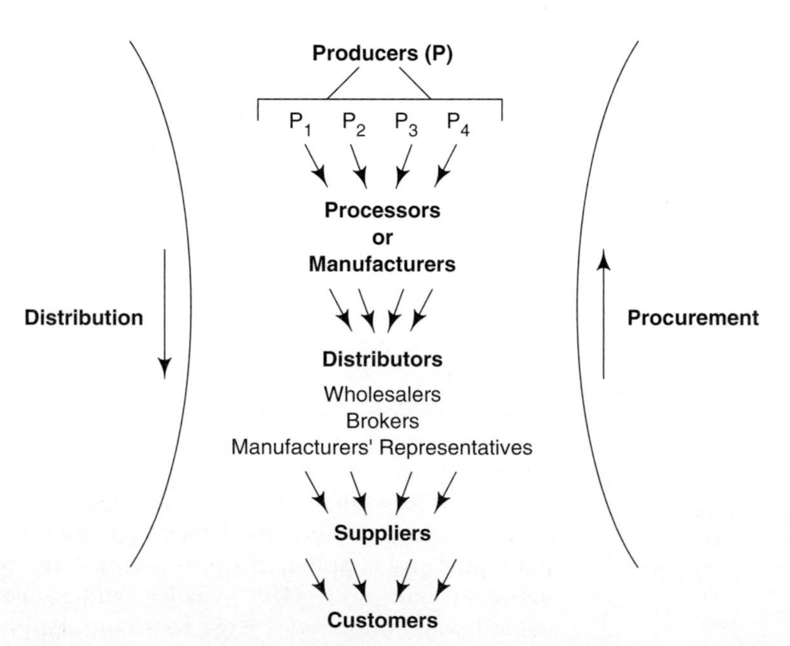

Figure 5.2. The marketing channel.

has not expanded, but the amount of food produced per acre has increased tremendously. This increase results from improvements in production methods, animal and plant genetics, and farm mechanization. Each year more food for more people is produced by fewer farmers. In 1940, one farmer produced food for about 12 people. Today one farmer can produce food for more than 120 people.

Processors or Manufacturers

Processor: Organization that takes raw food items and transforms them into packaged products for sale to consumers or foodservice operations.

The food supply in the United States is marketed not only by quantity but also by quality, variety, and convenience. In the early 1900s, food spoiled quickly and variety was limited because most products were seasonal. In contrast, today supermarkets stock more than 12,000 items, and distributors vie for shelf space in heavy-traffic aisles. The **processor** is responsible for the many forms of a food available to the customer.

For example, ranchers raise calves to weights of 500 to 600 pounds before transferring them to a feed lot, which functions as a feed growing area, where the cows' weights are brought up to 1,000 to 1,200 pounds. The cattle then are sold directly to a meat packing plant near the feed lot in which cattle are slaughtered and the meat is processed, portioned, and packaged in plastic film before boxing it. The boxes are shipped to all parts of the country in refrigerated trucks in 50,000-pound loads to warehouses. Distributors store the meat temporarily under refrigeration before selling it to the supplier, who sells it to a commercial or onsite foodservice operation, which converts it to menu items and sells it to the customer. Another example is an equipment manufacturer who buys stainless steel from a producer to manufacture dish machines to sell to a foodservice operation through a distributor or supplier.

Buyers for large foodservice operations—quick-service restaurant chains, for example—often bypass distributors by purchasing directly from the producer or processor; they act as their own wholesalers.

Distributors

Distributor: Wholesaler responsible for transferring products from the processor or manufacturer to the supplier.

Distributors, the third component in the marketing channel, are responsible for transferring products from the processor or manufacturer to the supplier. They are classified as wholesalers, brokers, and manufacturers' representatives. Some small manufacturing or processing plants sell products directly to the foodservice operation, but the large size of most plants precludes this type of selling. These operations must rely on distributors, especially if they do not have their own salespeople.

Wholesaler: Distributor that purchases from various manufacturers or processors, provides storage, sells, and delivers products to suppliers.

Wholesalers. **Wholesalers** are distributors who purchase from various manufacturers or processors, provide storage, sell, and deliver products to suppliers. The wholesaler protects the quality of a product from the time it is purchased to the time it is delivered to the foodservice operation. For example, the wholesaler learns from agents in the field the best time to purchase products and the grade, packaging, and handling of products before delivery and storage in the wholesale warehouse. Wholesalers can be either full or broadline, speciality, or special breed distributors (Gunn, 1995). **Full or broadline wholesalers** generally carry large amounts of stock, permitting the buyer to purchase everything from frozen and canned products to kitchen equipment and furniture. Specialty wholesalers deal in a particular product category, such as meat, produce, dairy, paper, or detergent. Milk and bread companies are specialty manufacturers and distributors.

Special breed distributors are purchasing and product movement specialists whose customers are restaurant chains that purchase food directly from processors and hire a distributor to deliver the products. The distribution companies include such industry giants as Martin-Brower Company, with 26 warehouses delivering more than $3 billion worth of food a year, and Golden State Foods, the second largest, delivering $990 million a year. More than 30 of these companies deliver at least 13% of all food sold wholesale.

Another type of distribution company which has become very popular in the United States is the **wholesale club or supermarket.** These companies, such as Sam's Club, started in big cities but have extended to smaller communities. Not only do they carry food products in household sizes, but they also have large sizes, such as number 10 cans of vegetables and fruits, 10-pound bags of chicken breasts, and 25-pound bags of detergent. Many small foodservice operators purchase most of their supplies from these wholesale clubs because the large distributors could not justify the cost effectiveness of delivering to them. The one disadvantage is the purchases are cash and carry, but the advantage is that the price often is competitive with the large distribution companies.

Broker: Independent sales and marketing representative who contracts with manufacturers, processors, or prime source producers to sell and conduct local marketing programs with wholesalers, suppliers, or foodservice operations.

Brokers. **Brokers** are independent sales and marketing representatives who contract with manufacturers, processors, or prime source producers to sell and conduct local marketing programs with wholesalers, suppliers, or foodservice operators. They represent a variety of products. Buyers often seek out brokers when they are "sourcing out" a new or existing product. Brokers do not take title to the products they sell. They are retained by manufacturers, processors, or producers, who contract and pay a commission for their services. The broker is not a distributor but rather an aid to sales and marketing in an operation. Brokers are not employed by or affiliated with the buyer to whom they sell and are not subject to the buyer's direct or indirect control. In the marketing channel, however, they are classified as distributors.

Today's brokerage firm is a sophisticated business employing staff with a variety of responsibilities, including account management, marketing, retail service, menu planning, foodservice promotion concepts, and technical support. Brokers vary in size from small firms with fewer than 10 employees to large regional companies with hundreds of employees. Processors of products needed by foodservice operations can go through regional managers who manage a number of brokers in a geographic area. Brokers often have test kitchens in which products for the foodservice market are prepared and tested. Foodservice brokers must be aware of their customers' needs and must know their products to find the right one for restaurant owners, dietitians, and chefs.

For example, potato producers have been using the services of specialty brokers for many years. The broker seeks out growers with 1-pound baking potatoes to sell to a wholesaler or directly to a restaurant with a potato bar. A different kind of potato is needed for a french fry processing plant or a quick-service restaurant chain, in which case the broker would have to test different kinds of potatoes to find the best ones for frying.

Manufacturer's representative: Distributor who represents a manufacturing company and informs suppliers of products by this manufacturer.

Manufacturers' Representatives. Manufacturers' representatives do not take title, bill, or set prices. They usually represent small manufacturing companies, often including foodservice equipment manufacturers. The companies pay a flat commission on sales volume. This method of selling is probably the most economical because the company does not have to maintain sales offices in every area where it has customers. Manufacturers' representatives are not empowered to enter into contracts on behalf of the companies they represent and can only make small decisions. They have fewer and more specialized lines than a broker and a minimum number of manufacturers. They have greater product expertise than brokers. For example, they might represent a manufacturer of ranges and ovens, refrigerators and freezers, transport carts, or dish machines. Often they represent two or more of these products.

Suppliers

Supplier: Organization that offers products for sale.

Suppliers are the fourth component in the marketing channel. They sell products to the ultimate buyer, the customer. The foodservice manager generally will buy more often from the supplier than from a wholesaler, broker, or manufacturer's representative because of the convenience of doing one-stop shopping by dealing with a single supplier, often referred to as a **prime vendor** or single-source supplier. This type of supplier is one who is used for most purchases. Many foodservice managers favor the "don't put all your eggs in one basket" approach, preferring instead to bid out individual line items. By doing so, they believe, they get the most competitive prices and avoid being at the mercy of one particular supplier (Savidge, 1996). Some managers, especially in small operations, like the more personalized service a prime vendor gives them. They agree that using this type of supplier is not necessarily the cheapest way to buy, but it certainly makes purchasing a simpler process. In the long run, it pays off.

The co-op/prime supplier arrangement often used in school foodservices offers advantages to both the supplier and buyer. Suppliers enjoy a guaranteed portion of a big school foodservice pie, while school districts enjoy the lower costs associated with volume purchasing along with the ease of dealing with one supplier. Savidge (1996) warned that despite the pluses of relationships with one supplier, risks exist: When the bottom falls out, as in a market crash, the destruction of crops by bad weather, or the supplier going bankrupt, the foodservice buyer would need to find new sources for products.

Customers

The customer, anyone who is affected by a product or service, is a fifth component of the marketing channel. According to the National Restaurant Association (2002), the typical person (age 8 and older) consumes an average of 4.2 meals prepared away from home per week. Customer satisfaction becomes the goal of the foodservice industry, and purchasing quality food and related products should be the first objective for the foodservice operation.

Local Sourcing

In many areas, foodservice buyers are doing more local sourcing of food products. Such sourcing often eliminates many of the steps in the market channel and allows the grower/producer to interact directly with the foodservice buyer. According to

Gregoire and Strohbehn (2002), foodservice managers indicate that the greatest benefits to local purchasing include aiding the local economy, maintaining good public relations, buying food in smaller quantities, and having fresher food. Buying locally decreases the distance food needs to travel and provides "food with a face," allowing customers to know the source of the food they are eating. The greatest obstacles of purchasing local foods have been the ability to purchase these foods the entire year, having access to a sufficient quantity to meet the operations needs, and often having to work with multiple growers to obtain the quantities needed (Gregoire & Strohbehn, 2002).

In 1997, the USDA started the Farm/School Meals Initiative to link small farms and Child Nutrition Programs (see www.usda.gov). Occidental College has helped develop several successful farm-to-school initiatives in California and has become a recognized source of valuable information for programs wanting to include more locally grown foods on their menus (see www.farmtoschool.org). The Santa Monica school district was one of the first in California to increase its use of locally grown foods. Foodservice employees there work with local farmers' markets and parent volunteers to provide salad bars in each school featuring only locally grown produce. One of the more well-known projects is the Berkley United School District project and its edible school yard developed by well-known restaurateur Alice Waters.

Value Added

The cost of taking food products through the marketing channel often equals or even exceeds the initial cost of the products. Value added is the increase in value caused by both processing or manufacturing and marketing or distribution, exclusive of the cost of materials, packaging, or overhead. The objective of value added is to increase the marketing value of raw and semiprocessed products.

Processing flour to adapt it for specific baking purposes is an example of the value-added concept. Research has proven that wheat flour after original milling is not satisfactory for products such as bread, rolls, biscuits, cakes, doughnuts, pie crust, crackers, pasta, and thickening agents for sauces and puddings. An example of modification of wheat flour is cake flour in which the amount of gluten is reduced to produce a fine-textured cake. The value-added cost of a special-purpose flour includes the miller, research scientist, necessary production facilities, and distribution of the product.

Produce distributors have been increasing the market value of fresh fruits and vegetables for some time and are constantly coming up with new ideas for the foodservice manager to control labor costs while improving quality. Peeled oranges are available, as are grapefruit sections and fresh pineapple chunks. Peeled potatoes have virtually eliminated potato peeler equipment, which was a necessity in every foodservice operation not too many years ago. Not only are peeled potatoes available, but potatoes can be purchased cut into any shape the operation needs. Lettuce heads still can be purchased, but many operators now buy lettuce torn or chopped. A tossed salad mix currently is a popular product. Caterers are putting more of their labor dollars into service rather than production. For example, watermelons are now carved into baskets that can be filled with a large assortment of fruits cut into various shapes.

Market Regulation

The food industry is the most controlled industry in the United States today, covered as it is by comprehensive and complex federal regulations. The purpose of market regulation through federal legislation is to protect the consumer without stifling economic growth. Government, industry, and consumers need to interact to accomplish this purpose; each is interdependent and mutually affects the performance of the other two. Government is responsible for enacting legislation that safeguards consumers and at the same time promotes competition among industries. Industry, with the ultimate goal of satisfying the consumer while making a profit, is responsible for complying with this legislation. Consumers, who are becoming increasingly vocal about such issues as food safety, nutrition, and the environment, alert government about their concerns and expect industry to produce appropriate products. Although many federal agencies are responsible for regulations that directly affect the industry, the U.S. Department of Health and Human Services (HHS) and the U.S. Department of Agriculture (USDA) are the most often involved. The U.S. Department of Commerce also is becoming involved because of the great concern for the microbiological safety of seafood for human consumption.

U.S. Department of Health and Human Services

Under the umbrella of the U.S. Department of Health and Human Services (HHS) are the Food and Drug Administration (FDA) and the U.S. Public Health Service (PHS). HHS units cooperate closely with the U.S. Department of Agriculture (USDA).

Food and Drug Administration. The Food, Drug, and Cosmetic Act, passed in 1938, is enforced by the FDA. The purposes of this act are to ensure that foods other than meat, poultry, and fish are pure and wholesome, safe to eat, and produced under sanitary conditions; the Act also requires that packaging and labeling are in agreement with the contents. The FDA is one of our nation's oldest consumer protection agencies. It has approximately 9,000 employees who monitor the manufacture, import, transport, storage, and sale of about $1 trillion worth of products each year at a per capita cost of only $.20 per day (see www.fda.gov).

Standards of identity: Federal standards that identify what a given food product contains.

The Act provides three mandatory standards for products being shipped across state lines: identity, quality, and fill of container. **Standards of identity** establish what a given food is named and what the product contains. Certain ingredients must be present in a specific percentage before the standard name may be used. For example, the consumer is assured that any product labeled mayonnaise, regardless of its manufacturer, contains 65% by weight of vegetable oil, along with vinegar or lemon juice and egg yolk. Products labeled mayonnaise that do not contain these ingredients could not use the term *mayonnaise* in their title as this would violate FDA regulations. The FDA periodically reviews standards of identity for food products. The reviews include public hearings, and the results could result in a change in regulations or prosecution of current violators.

Standards of quality: Federal standards for product quality attributes such as tenderness, color, and freedom from defects.

Standards of quality, or minimum regulatory standards for tenderness, color, and freedom from defects, have been set for a number of canned fruits and vegetables to supplement standards of identity. If a food does not meet the FDA quality standards, it must be labeled "Below Standard of Quality" and can bear an explanation such as "Good Food—Not High Graded," "Excessively Broken," or "Excessive

Peel." Standards of fill of container tell the packer how full a container must be to avoid the charge of deception.

Two major amendments to the Federal Food, Drug, and Cosmetic Act are the Food Additive and Color Additive amendments of 1958 and 1960, respectively, both of which safeguard the consumer against adulteration and misbranding of foods. Food is considered adulterated if it contains substances injurious to health, if it was prepared or held under unsanitary conditions, or if any part is filthy, decomposed, or contains portions of diseased animals. Food also is considered to be adulterated if damage or inferiority is concealed, if its label or container is misleading, or if a valuable substance is omitted. Another addition was the Miller Pesticide Amendment of 1954, which provides procedures for establishing tolerances for residues of insecticides used with both domestic and imported agricultural products. Tolerances of permitted residues are established by the Environmental Protection Agency (EPA), but the FDA is responsible for their enforcement. The FDA samples dairy products, fish, fresh and processed fruits and vegetables, and animal foods for unsafe residue levels.

The FDA also regulates food irradiation, genetically engineered foods, and nutrition labeling. Food irradiation is classified as a food additive and is regulated by the FDA. **Irradiation** refers to exposure of substances to gamma rays or radiant energy. The process for some foods has been approved since the 1960s; followed in 1985 by approval of irradiation to control trichinella in pork; in 1992 to control foodborne pathogens and other bacteria in frozen poultry, including ground poultry products; and in 2000 to control refrigerated or frozen uncooked meat, meat byproducts, and other meat food products (see www.usda.gov). Ionizing radiation has energy high enough to change atoms by knocking an electron from them to form an ion, but not high enough to split atoms and cause foods to become radioactive. The amount of radiation energy absorbed by a food is measured in units called kilograys (kGy) (Paparella, 1998). Less than 1 kGy inhibits the sprouting of tubers in potatoes, delays the ripening of some fruits and vegetables, controls insects in fruits and stored grains, and reduces parasites in foods of animal origin. One to 10 kGy control microbes responsible for foodborne illness and extend the shelf life of refrigerated foods. More than 10 kGy are used only on spices and dried vegetable seasonings.

Irradiation: Exposure of foods to gamma rays or radiant energy to reduce harmful bacteria.

Irradiation does not make foods radioactive (International Food Information Council [IFIC], 1997). The process moves food through a radiant energy field, but the food never touches the energy source. Irradiation does not change the nutrient content or flavor of foods. It produces virtually no heat within food and does not "cook" foods. Foods processed with irradiation are just as nutritious and flavorful as other foods in the marketplace. Just as any spoilage or potentially harmful bacteria in milk are killed through pasteurization using heat, irradiation kills most harmful bacteria in foods other than milk. Irradiated food, however, can spoil or foster growth of bacteria that it may come into contact with during food handling and preparation. Safe food handling measures are required. These include washing hands and preparation surfaces often, separating foods to avoid cross-contamination, cooking foods thoroughly, and refrigerating and storing food properly. In 2000, the American Dietetic Association issued its position on food irradiation. The position states that "food irradiation enhances the safety and quality of the food supply and helps protect consumers from food borne illness" (see www.eatright.org).

Figure 5.3. International logo for irradiated food.

Federal law requires that all irradiated foods must be labeled with the international symbol for irradiation identified as the **radura** (Paparella, 1998). This symbol, shown in Figure 5.3, must be accompanied by the words, "Treated by Irradiation" or "Treated with Radiation" (IFIC, 1997).

Genetically engineered foods: Food whose genetic structure has been altered by adding or eliminated genes to enhance qualities of the product.

In May 1992, the FDA initiated a policy for **genetically engineered foods** or, as they are more commonly referred to, **genetically modified organisms** (GMOs) that would be regulated no differently than foods created by conventional means (see www.usda.gov). The FDA determined that a special review of genetically engineered foods was needed only when specific safety issues were raised. In general, only genetically engineered food crops that contain substances significantly different from those in the consumer's diet need to seek FDA approval. Currently, the FDA does not require products to be labeled regarding the content as it relates to genetic modification unless the modification significantly changes the product's composition. Genetically modified foods that include proteins from commonly known allergenic foods such as peanuts and fish would need to be labeled unless it has been shown that persons allergic to the allergen used are not allergic to the GMO product. Since most genetically engineered foods will be indistinguishable in appearance from nonengineered foods, consumers will generally not know what they are buying.

Consumer concerns related to source of food ingredients may prompt legislation to change current practices. The European Union requires labeling of genetically engineered foods and feeds (see www. eurunion.org)

Requirements for **nutrition labeling** are spelled out in regulations issued in late 1992 by the FDA and by the USDA's **Food Safety and Inspection Service (FSIS).** The FSIS is the public health agency in the USDA responsible for ensuring that the nation's commercial supply of meat, poultry, and egg products is safe, wholesome, and correctly labeled and packaged as required by the Federal Meat Inspection Act, the Poultry Products Inspection Act, and the Egg Products Inspection Act (USDA Food Safety and Inspection Service, 1998). FDA regulations meet the provisions of the Nutrition Labeling and Education Act (NLEA) of 1990, which requires nutrition labeling for most foods, except meat and poultry, and authorizes the use of nutrient content claims and appropriate FDA-approved health claims. Meat and poultry products regulated by USDA are not covered by NLEA; however, USDA's regulations closely parallel the FDA's rules (FDA, 1995). The current label, identified as "Nutrition Facts," is shown in Figure 5.4. Macronutrients (fat, cholesterol, sodium, carbohydrate, and protein) are shown as a percent of the recommended daily value. The amount, in grams or milligrams per serving, of these nutrients is listed to their immediate right, and a column headed "% of Daily Value," based on a 2,000-calorie diet, is on the right side. In products such as macaroni and cheese or stuffing mix, the percent of daily value is as "packaged" or "in the box." Most nutrition labels list daily values for a 2,000- and 2,500-calorie diet and the number of grams of fat, carbohydrate, and protein. The content of micronutrients, vitamins and minerals, also is expressed as a percentage.

According to food labeling regulations, specific definitions for all descriptors are now spelled out (National Center for Nutrition and Dietetics, 1993). A product described as "free" of fat, saturated fat, cholesterol, sodium, sugar, or calories must contain "no amount" or only a "trivial amount" of that component, according to the labeling regulations. For example, the regulations state that "calorie free = less than 5 calories per serving" and that "sugar or fat free = less than .5 grams per serving." The term *low* can be used with fat, saturated fat, cholesterol, sodium, and calories only when the food could be eaten frequently without exceeding dietary guidelines for these nutrients, such as 3 grams or less of fat, less than 140 milligrams of sodium, or 40 or less calories per serving. The terms *lean* and *extra lean* can be used to describe the fat content of meat, poultry, and seafood. Lean is less than 10 grams of fat, 4.5 grams of saturated fat, and 95 milligrams of cholesterol per serving; extra lean is less than 5 grams of fat, 2 grams of saturated fat, and 95 mg cholesterol per serving.

The FDA amended its food-labeling regulations in 1994 to remove the provisions that exempted restaurant menus from conforming to the standards for nutrient and health claims that appear on menus. The FDA has decided that all restaurant menus are subject to the same rules regarding claims; if asked, restaurateurs must provide customers with backup nutritional information for claims made on menus. Any restaurant that uses nutrient-content claims or health claims on its menu must comply with the regulations. Establishments that do not make those types of claims about their menu items need not supply nutrient-content information.

U.S. Public Health Service. Under the Public Health Service Act, the FDA advises state and local governments on sanitation standards for prevention of infectious diseases. The most widely adopted standards deal with production, processing, and distribution of Grade A milk. In contrast to USDA quality grade standards for food, the Public Health Service standard for Grade A milk is largely a standard of wholesomeness. The Grade A designation on fresh milk means that it has met state or local requirements that equal or exceed federal requirements.

U.S. Department of Agriculture

The USDA facilitates strategic marketing of agricultural products in domestic and international markets by grading, inspecting, and certifying the quality of these products in accordance with official USDA standards or contact specification. U.S. grade standards are quality driven and provide a foundation for uniform grading of agricultural commodities nationwide. The USDA's food safety organization assures that the nation's meat and poultry supply is safe, wholesome, unadulterated, and properly labeled and packaged (USDA, 1998).

Nutrition Facts

Serving Size ½ cup (114g)
Servings Per Container 4

Amount Per Serving

Calories 260 Calories from Fat 120

	% Daily Value*
Total Fat 13g	**20%**
Saturated Fat 5g	**25%**
Cholesterol 30mg	**10%**
Sodium 660mg	**28%**
Total Carbohydrate 31g	**10%**
Dietary Fiber 0g	**0%**
Sugars 5g	
Protein 5g	

Vitamin A 4% • Vitamin C 2%

Calcium 15% • Iron 4%

* Percent Daily Values are based on a 2,000 calorie diet. Your daily values may be higher or lower depending on your calorie needs:

	Calories:	2,000	2,500
Total Fat	Less than	65g	80g
Sat Fat	Less than	20g	25g
Cholesterol	Less than	300mg	300mg
Sodium	Less than	2,400mg	2,400mg
Total Carbohydrate		300g	375g
Dietary Fiber		25g	30g

Calories per gram:
Fat 9 • Carbohydrate 4 • Protein 4

Figure 5.4. Example of the current nutritional label.
Source: From the FDA and USDA Food Safety and Inspection Services (1992).

The USDA has established grade standards for several categories of food: fresh fruits, vegetables, specialty crops, processed fruits and vegetables, milk and other dairy products, livestock and meat, poultry, eggs, cotton, and tobacco (see www.usda.gov). Grading services are often operated cooperatively with state departments of agriculture. Quality grades provide a common language among buyers and sellers, which in turn assures consistent quality for consumers. A summary of the grades in use today is shown in Table 5.1.

The USDA also has the responsibility for enforcing federal meat, poultry, and egg products inspection programs. Meat and poultry processing plants are required to follow HACCP regulations. Plants must install their own facility's preventive measures to reduce *E. coli* and salmonella bacteria and improve sanitation. HACCP systems involve identifying points in a processing plant where contamination is most likely to occur and finding methods to combat it. Processing plants can design their own HACCP system but must meet certain standards (Anderson, 1998).

FSIS established requirements for poultry plants to improve food safety (USDA, 1996). The four major elements require that all slaughter and processing plants:

- Have an HACCP plan.
- Conduct microbial testing for generic *E. coli* to verify that their control systems are working as intended to prevent fecal contamination, the primary avenue of contamination for harmful bacteria.
- Meet pathogen reduction performance standards for salmonella for raw products.
- Adopt and carry out a written plan (sanitation standard operating procedures) for meeting its basic sanitation responsibilities.

Under the Egg Products Inspection Act (EPIA) of 1970, EPIA provided mandatory continuous inspection of the processing of liquid, frozen, and dried egg products. On May 28, 1995, USDA's Food Safety and Inspection Service became responsible for the inspection of egg products when elements of various USDA agencies were combined into one food safety agency. The FSIS inspects all egg products, with the exception of those products exempted under the Act, that are used by food manufacturers, foodservice, institutions, and retail markets. The FDA is responsible for the inspection of egg substitutes, imitation eggs, and similar products that are exempt from continuous inspection under EPIA (FSIS, 1995).

U.S. Department of Commerce

The FDA has set seafood inspection requirements for processors and importers. The regulations require HACCP plans be followed. (Seafood Safety Regulations, 1995).

The National Marine Fisheries Service (NMFS) is a part of the National Oceanic and Atmospheric Administration (NOAA), U.S. Department of Commerce (see www.nmfs.noaa.gov). NMFS administers NOAA's programs which support the domestic and international conservation and management of living marine resources. NMFS provides services and products to support operations, fisheries development, trade and industry assistance activities, enforcement, protected species and habitat conservation operations, and the scientific and technical aspects of NOAA's marine fisheries programs. Five fisheries' science centers provide scientific and technical

Table 5.1 Summary of major grades by food categories

Quality Grades

Product	Grading Criteria	Highest ⟶ Lowest
Beef	Tenderness Juiciness Flavor	Prime, Choice, Select, Standard, Commercial, Utility, Cutter, Canner
Veal	Tenderness Juiciness Flavor	Prime Choice Good Standard Utility
Lamb	Tenderness Juiciness Flavor	Prime Choice Good Utility Cull
Pork		No quality grades
Poultry	Confirmation Fleshing Fat covering	Grade A Grade B Grade C
Eggs	Interior quality of the egg Appearance and condition of the shell	Grade AA Grade A Grade B
Fish	Appearance Uniformity Absence of defects Texture Flavor and odor	Grade A Grade B Grade C Substandard
Milk, fluid	Bacterial count Sanitary conditions	Grade A
Milk, nonfat dry	Flavor and odor Bacterial count Scorched particle content Lumpiness Solubility	U.S. Extra Grade
Butter	Flavor and odor Freshness Plasticity Texture	Grade AA Grade A
Cheese Cheddar Swiss Colby Monterey Jack	Flavor and odor Texture Body Appearance Finish Color	Grade AA Grade A Grade B Grade C
Vegetables, Fruits	Maturity Shape Color Size Uniformity Texture Presence of defects	U.S. Fancy U.S. No. 1 U.S. No. 2 U.S. No. 3

support to each of the six regions: Southeast, Northeast, Southwest, Northwest, Alaska, and Pacific Islands.

Imported Food Regulations

Meat and poultry imported into the United States must be produced under standards equivalent to those of the United States for safety, wholesomeness, and labeling accuracy (FSIS, 1998). USDA's Food Safety and Inspection Service is responsible for assuring that those standards are met. The FDA has jurisdiction over all foods not covered by federal meat and poultry inspection laws.

Evaluation of a country's inspection system to determine eligibility involves two steps: a document review and an onsite review. The document review is done by technical experts who evaluate the information to assure that critical points in the five risk areas are being covered satisfactorily with respect to standards, activities, and resource allocations of contamination, disease, processing, residues, and compliance and economic fraud. If the review process is satisfactory, a technical team visits the country to evaluate the five risk areas and inspect the plant facilities and equipment, laboratories, training programs, and in-plant inspection operations. If FSIS finds that the operation is equivalent to that in the United States, the country becomes eligible to export meat or poultry to the United States. FSIS periodically reviews the operation to assure it continues to meet U.S. requirements. The United States imports from 2.4 to 2.8 billion pounds of meat per year, always less than 10% of the domestic meat supply. Most is fresh and processed meat; only a small amount of poultry is imported.

FSIS checks shipping container labels during reinspection at the port of entry. Importers of any merchandise into the United States must file a customs entry form within 5 working days after the shipment arrives at a port. The Customs Service also requires the importer to post a bond until FSIS notifies the service of the results of the reinspection. After the Customs Service and Animal and Plant Health Inspection Service (APHIS) requirements are met, the shipment must be inspected by FSIS at an approved import inspection facility (FSIS, 1998). FSIS enters information about the shipment into a centralized computer system called the Automated Import Information System (AIIS). The AIIS scans its memory bank to determine if the country, plant, and product are eligible for export to the United States, where the product will be inspected again. When a shipment passes inspection, each shipping case is stamped with the official mark of inspection and released into U.S. commerce. The shipment then is treated as a domestic product.

PRODUCT SELECTION

Purchasing for a foodservice operation is a highly specialized job function. Buyers must know not only the products to be procured but also the market, buying procedures, and market trends. They must know how the materials are produced, processed, and moved to market. In addition, they must be able to forecast, plan, organize, control, and perform other management-level functions.

The primary function of the buyer is to procure the required products for the desired use at minimum cost. The accomplishment of this function frequently involves some research by the buyer to aid in decision making. For this purpose, techniques have been adapted from industry, notably value analysis and make-or-buy decisions.

Value Analysis

Value analysis: Methodical investigation of all components of an existing product or service with the goal of discovering and eliminating unnecessary costs without interfering with the effectiveness of the product or service.

In the most liberal sense, value analysis is virtually any organized technique applicable to cost reduction. More precisely, however, **value analysis** is the methodical investigation of all components of an existing product or service with the goal of discovering and eliminating unnecessary costs without interfering with the effectiveness of the product or service (Miles, 1972). This definition of value analysis is related primarily to industry, but it contains broad implications for foodservice as well. For example, a value analysis of a menu item may reveal that some quality features may be eliminated without detracting from the utility of the final product.

The essence of quality is suitability (Heinritz, Farrell, Giunipero, & Kolchin, 1991). The supplier gives the buyer the quality of the product identified in the specification. Part of the price may be for quality features that do not contribute substantially to the suitability of the product for the foodservice operation and thus make the expenditure wasteful. A foodservice, for example, may purchase an expensive, sophisticated software package although the manager does not know what is needed and no employee has been trained in computer usage. Probably, many of the software features would never be used. A simple package would probably have been adequate in the beginning, and it certainly would have been less expensive.

Value analysis is an important element of scientific purchasing and has brought about the realization that purchasing is a profit-making activity. The concept of value analysis is important in enhancing efficiency and profitability (Williams, Lacy, & Smith, 1992). Reductions in purchasing costs resulting from making the process more efficient have a strong impact on profitability, which is quite different from increasing sales by selling more products.

Value Analysis in Foodservice

Value is the result of the relationship between the price paid for a particular item and its utility in the function it fulfills. Value analysis permits the foodservice manager to look at a problem from a new perspective (Williams, Lacy, & Smith, 1992). The tendency to become locked into a pattern can be reduced.

Although value analysis often has been used in the development of new products in industry, its effectiveness is not confined to such development; it is used much more frequently in evaluating existing product specifications and thus is readily applicable in foodservice. For example, a major university residence hall foodservice applied value analysis to a lasagna entrée after students complained that its flavor was not like that at a local Italian restaurant. An associated problem was the foodservice director's concern that the product was extremely labor intensive. In conference with the director, the foodservice manager decided to compare lasagna made by the current, standardized recipe that called for American cheese with a revised recipe substituting mozzarella cheese for American. For the secondary problem of labor intensity, one well-known brand of frozen lasagna was evaluated.

Reporting on Value Analysis

A brief value analysis report prepared for the foodservice director, based on the residence hall foodservice lasagna example, is shown in Figure 5.5. This example illustrates a primary requirement for reporting to management—brevity. Complete data on the methodology and the test procedures should be prepared as a reference and excerpted only for the report to administration. Value analysis results for quality and cost and recommendations are of major interest to the foodservice director. This portion of the report should be concise and definite.

This lasagna example illustrates a good approach to a secondary problem. Properly, the evaluator gave a recommendation on the primary problem, that of customer dissatisfaction, and suggested that a separate analysis should be made for several brands of frozen lasagna to examine the secondary problem, labor intensity of the onsite product.

Make-or-Buy Decisions

The procedure of deciding whether to purchase from oneself (make) or purchase from suppliers (buy) is continuous, and reviews of previous make-or-buy decisions should be conducted periodically. A foodservice manager has three basic choices for production of a menu item:

- Produce the item completely, starting with basic raw ingredients.
- Purchase some of the ingredients preprepared and assemble them.
- Purchase the item in its final form from a wholesaler.

Few foodservice managers consider the first alternative, primarily because of the labor time and skill level required. The second two alternatives were compared in the value analysis report on lasagna (Figure 5.5). A product prepared onsite from a recipe using many purchased items (such as lasagna noodles, canned tomatoes, and cheese, which were assembled for baking in the foodservice kitchen) was compared with a commercial frozen product. Based on cost and quality criteria, the alternative of making the lasagna on the premises was recommended rather than that of purchasing a frozen product.

Decision factors in foodservice are quality, quantity, service, and cost. The serving of quality food is a prime consideration in all foodservices, whether menu items are made on the premises or purchased prepared from the processor. Quality standards have become well established among most processors of ready-to-serve foods, although variations are evident.

In the lasagna analysis, the testers evaluated **quality** and said the noodles in the convenience product were tender but firm enough to retain their shape, and the product held its square form when plated, as specified in the standard. The top was covered with meat sauce and cheese with no juices on the plate. The problem was the flavor of the product, which was not acceptable to the students. The standard was that the flavor should be a blend of cheese, meat, and pasta with evidence of Italian seasoning, but should not be overpowering. The taste panel indicated that the frozen product had a strong tomato flavor and was too highly seasoned; however, the made-on-premises product with the mozzarella cheese did satisfy the flavor standard.

Quantity enters into the decision process when the ability to produce in the desired amount is considered. The quantity needed may be too large for a satisfactory

VALUE ANALYSIS
Report to Administration

Product Evaluated: Lasagna

Evaluator: Mary Smith
Date: November 2, 2002

Statement of Problem

- Students criticize flavor of lasagna prepared in Residence Hall foodservice.
- Management is concerned about labor intensity of the product.

Products Evaluated

- Residence Hall recipe with American cheese
- Residence Hall recipe with mozzarella cheese
- Frozen prepared commercial brand

Methodology

- *Preparation of product*
  Two one-half-size counter pans of the Residence Hall recipe were prepared at the same time from identical ingredients and the same recipe except one contained American cheese and one mozzarella cheese. The frozen product was heated according to manufacturer's directions.
- *Test procedure*
  The three products, each identified by a three-digit number, were cut into sample size portions and displayed side by side. Each product was rated on flavor, texture, and appearance by a taste panel consisting of 16 students, 6 foodservice employees, and 4 staff members.

Value Analysis Results

- *Cost Evaluation*

Cost Items	Residence Hall Recipe (American cheese)	Residence Hall Recipe (mozzarella cheese)	Commercial Frozen
		cost per portion	
Food	$ .6216	$ .6256	$ 1.165
Labor	.0676	.0676	—
Total	$ .6892	$ .6932	$ 1.165

- *Quality Evaluation*
  The lasagna prepared from the Residence Hall recipe with mozzarella cheese ranked highest in flavor. Both Residence Hall products ranked similar in texture and appearance. The frozen commercial product ranked lowest in all three quality criteria.

Recommendation

- The recommendation based on taste panel evaluations is that the Residence Hall lasagna recipe with mozzarella cheese be used. Also, the made-on-premises product is much less expensive than the purchased frozen. If a decision to serve a frozen prepared product is made because of labor intensity, other brands should be evaluated.

Figure 5.5. Example of a value analysis report prepared for the Kansas State University dining services.

"make" decision or too small for the processor to consider. As a result of past make-or-buy decisions, many foodservices have closed bake shops or omitted them from new facilities because acceptable bakery products are commercially available.

Service includes a wide variety of intangible factors influencing the satisfaction of the buyer. Two important factors in the decision process are reliable delivery and predictable service. Foodservices must operate on a rigid time schedule; a late delivery of a menu item cannot be tolerated. The dependability of a supplier in all circumstances must be assured. Clues to suppliers' reliability may be their record of labor relations and reputation in the industry, including the opinions of other customers. The geographic proximity of a supplier also should be considered.

When quality, quantity, and service factors are equal, a make-or-buy decision will be made by comparing the known cost from the supplier with the estimated cost of making the product. In this process, the cost of raw food is easily determined, but the costs of labor, energy, equipment depreciation, and overhead may be difficult to calculate.

Set policies for reaching make-or-buy decisions should be avoided because these and other purchasing decisions are influenced by a number of interrelated factors that are extremely variable. The critical factor on one occasion may be not important on the next. The buyer must strike a fine balance among all factors and make a decision on the basis of what is best for the foodservice operation.

SPECIFICATIONS

Quality has become the watchword in foodservices. In addition to the assurance of quality in product selection, critical elements in producing quality food and foodservice are the development of and strict adherence to rigorous purchasing specifications.

The root of the word *specification* is "specific," which means that some condition or status must be met, definitely and without equivocation. Thus the primary safeguard of foodservice quality is adherence to specifications.

Specification: Statement understood by buyers and suppliers of the required quality of products, including allowable limits of tolerance.

A **specification** has been defined in many different ways, but it is essentially a statement, readily understood by both buyers and suppliers, of the required quality of products, including the allowable limits of tolerance. In the simplest terms, a specification, or spec, may be described as a list of detailed characteristics desired in a product for a specific use.

Types

The three types of specifications applicable to foodservices are technical, approved brand, and performance. Selection of one, or a combination of two or all three, of these types is based on the product being purchased, whether it is food, supplies, or equipment.

Technical specification: Specification that indicates quality by objective and impartial test results.

Technical specifications are applicable to products for which quality may be measured objectively and impartially by testing instruments. These are particularly applicable to graded food items for which a nationally recognized standard exists. Other examples are parts and metals in fabricated equipment, such as stainless steel and aluminum, which are subject to a thickness requirement that can be measured by gauges. Technical specifications also can be written for detergents and cleaning compounds that can be chemically analyzed.

Approved brand specification: Specification that indicates quality by specifying a brand name or label.

Approved brand specifications indicate quality by designating a product of known desirable characteristics, such as a manufacturer's specification for a Combi-Oven or a processor's formula for frozen pizza (Spears, 1999). Paper and plastic disposable items also could be identified by brand name. The brand name specification often is used in smaller foodservice operations in which a cook or chef might be responsible for purchasing in addition to operational duties. Nothing captures customers' attention and enthusiasm more than branded products that meet their expectation as a status symbol. Examples include packets of Heinz ketchup, Grey Poupon mustard, or Mrs. Dash no-salt seasonings. Pepsi and Coca Cola compete fiercely to keep their brand names in front of the public. Alcoholic beverages are purchased almost exclusively by brand names. Retail brand names on menus have customer appeal.

Performance specification: Specification that indicates quality by functioning characteristics of the product.

In **performance specifications**, quality is measured by the effective functioning of large or small equipment, disposable paper and plastic items, or detergents (Spears, 1999). Examples include the minimum and maximum number of pounds to be weighed on a scale, the number of dishes to be washed per minute, the minutes coffee will remain at serving temperature in a styrofoam cup, or the pH level of detergents.

Of the three types of specifications, the technical is used most often for food products purchased by schools, colleges and universities, large healthcare facilities, and multiunit commercial operations. A specification for a particular product also might be a combination including both technical and performance criteria.

Writing Criteria

Written specifications are necessary for an efficient foodservice operation. When specifications are written, suppliers, the receiving clerk, and the foodservice manager all can determine if the products received are what were ordered. Specification writing can be time consuming and labor intensive, especially for small operations. However, if specifications for high-priced products, such as meat and seafood, are written first, then writing them for other products will be easier. Specification writing requires a team approach and generally includes the foodservice manager, dietitian, procurement and production unit heads, buyer, cook or chef, and often the financial manager. A specification can be simple or complex, depending on the type used; the brand name type is the simplest, and the technical type is the most complex.

Good specifications are

- Clear, simple, and sufficiently specific so both buyer and supplier can readily identify all provisions required
- Consistent with products or grades currently on the market
- Verifiable by label statements, USDA grades, weight determination, and so on
- Fair to the supplier and protective to the buyer
- Realistic quality standards that would find at least some products acceptable
- Capable of being met by several bidders to enable competition

Specific Information

All specifications for food products should include the following information:

- Name of product (trade or brand) or standard
- Federal grade, brand, or other quality designation
- Size of container (e.g., weight, can size)

- Count per container or approximate number per pound (number of pieces per container if applicable)
- Unit on which price will be based

Information that would describe the product in more detail might include

- Product use (e.g., for salads, soups)
- Product test procedures used by the foodservice operation to determine quality compliance (e.g., degree of ripeness, flavor characteristics)
- Quality tolerance limits (number of substandard products in a container of produce, for example)
- Weight tolerance limits (range of acceptable weight, such as for meat, poultry, or seafood)

Any other information that helps to describe the condition of the product should be included, as in the following:

- Canned goods: Type or style, pack, syrup density, size, specific gravity
- Meat and meat products: Age, exact cutting instructions, weight tolerance limit, composition, condition upon receipt of product, fat content, cut of meat to be used, market class
- Fresh fruits and vegetables: Variety, weight, degree of ripeness or maturity, quality tolerance limit, geographical origin
- Frozen foods: Temperature during delivery and upon receipt, variety, sugar ratio
- Daily products: Temperature during delivery and upon receipt, milk fat content, milk solids, bacteria content

Were it not for the availability of nationally accepted grades and other criteria developed by the USDA and other organizations, writing a specification would be an almost insurmountable task for the buyer and difficult for the supplier to interpret. Buyers can now write definite specifications by citing a known standard, which is in itself a rigorous specification. These referenced standards constitute a common technical language for buyers and suppliers. Examples of specifications for some food products are found in Appendix A.

Additional Information

The USDA has published a greater volume of specification material pertaining to foodservice than any other agency. The USDA has established grading standards for fruits, vegetables, eggs, poultry, dairy, and meat products.

The USDA Institutional Meat Purchase Specifications, commonly referred to as IMPS, simplify specification writing for large-volume users of meat (available at www.ams.usda.gov/lsg/stand/imps.htm). Use of these specifications is related to the food acceptance service of the USDA, a part of the total service of federal inspection and grading.

An extremely valuable feature of IMPS is the numbering system for the identification of carcass cuts and various cuts or types of meat products. In addition to the benefits of the numbering system, IMPS information on ordering includes USDA quality and yield grades, weight ranges, portion cut tolerances, fat limitations, and refrigeration requirements for beef, pork, veal, and lamb.

One of the most useful guides for writers of meat purchase specifications is the *Meat Buyers Guide,* published by the North American Meat Processors Association.

This organization, commonly referred to as NAMP, coordinated the publication of the guide with the USDA. The numbering system and the general purchasing information are identical to those of IMPS. The advantage of the NAMP *Meat Buyers Guide* is the arrangement of the text material and the excellent colored illustrations of each numbered item. Most wholesalers of meat products are members of NAMP, so the guide provides an excellent adjunct to communication between buyer and supplier. Also included in the guide is a description of USDA identification marks for quality and yield grades and federal inspection stamps for beef.

Grades and Inspection

The USDA **quality grades** for beef are U.S. Prime, U.S. Choice, U.S. Select, U.S. Standard, U.S. Commercial, U.S. Utility, U.S. Cutter, and U.S. Canner. These grades pertain to the palatability qualities of beef, namely tenderness, juiciness, and flavor. The USDA conducts a voluntary meat grading service to identify beef, and when this service is used, a USDA Quality stamp will appear on the carcass (Figure 5.6). Resource material for writing specifications for other food products is included in the references in Appendix B.

Yield grades for beef provide a nationwide uniform method of identifying "cutability" differences among beef carcasses. Specifically, these grades are based on the predicted percentage of carcass weight in closely trimmed, boneless, retail cuts from the chuck, rib, loin, and round. The Yield Grade stamp is shown in Figure 5.6. Because of trimming done by the purveying industry, however, Yield Grade stamps will rarely appear on cuts purchased by the foodservice industry. Beef carcasses, as of April 1989, no longer need to be graded for both quality and yield; they may be graded for quality or yield, or both. Because quality denotes palatability and yield cutability, each quality grade could have a yield from 1, increased muscle and decreased fat, to 5, decreased muscle and increased fat.

The mark of federal inspection is a round stamp (Figure 5.6) identifying the slaughterhouse of origin for carcasses and the meat fabricating house for further cuts. This stamp indicates that all processing was done under government supervision and that the product is wholesome.

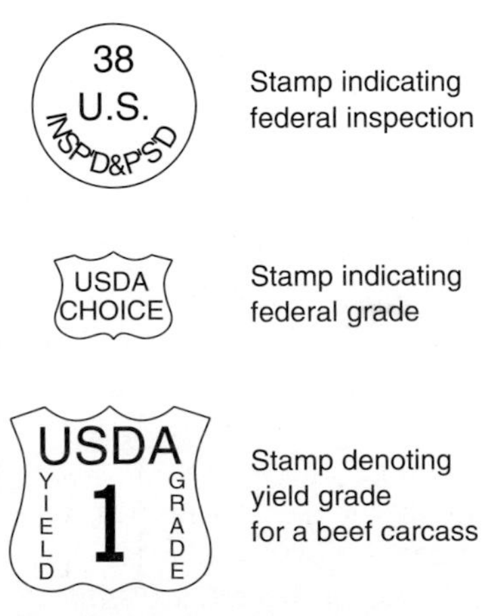

Figure 5.6. USDA meat stamps.

Kosher Inspection

Some foodservice operations, such as restaurants or retirement centers with a predominantly Jewish population, purchase meat that is kosher, a term meaning "fit and proper" or "properly prepared" (National Livestock and Meat Board, 1991). Kosher meat must meet the standards of Mosaic and Talmudic laws. It must come from an animal that has split hooves and chews its cud. Because hogs do not meet this requirement, pork, bacon, and ham cannot be kosher. Kosher meat is processed under the supervision of a specially trained rabbi who performs the ritual slaughter and drains the blood from the animal. The fresh meat is soaked and salted on all sides and in all cuts and folds to draw out the remaining blood. Biblical law (Leviticus 17:14) states that "You shall not eat the blood of any creature, for the life of every creature is its blood; whoever eats it shall be cut off." Hindquarters cannot be used for kosher meat because the blood cannot be removed completely from them. The rabbi stamps the meat with the kosher symbol, as shown in Figure 5.7, to show that the meat meets the Jewish dietary laws. Kosher meat must also meet all federal or state meat inspection laws before it can be sold.

Packers' Brands

Many food producers use their own brand names and define their own quality standards. The grading system closely follows federal grades and standards but with more flexibility than is allowed in the USDA system. For example, one brand or label may mean the highest quality, while another is used to indicate a lower quality. Canners, packers, and distributors usually put their top-ranking label on the highest-quality products. For example, many packers of canned and frozen fruits and vegetables choose their own quality labels rather than those established by the USDA; some packers also use their own color code for various quality levels. The buyer must realize that the word *Fancy* as a brand name does not mean that the product meets the requirements of the U.S. Fancy grade. Because not all canned fruits and vegetables are graded by the USDA, the careful buyer will open cans of different brands for comparison before placing major orders. This process is commonly called **can cutting** and constitutes the only means of acquiring first-hand information about drained weight, appearance, texture, flavor, or any scoring factors of particular interest to the buyer. Following selection of a product by this or some other process, the buyer of larger-volume quantities might use the USDA acceptance service for inspection at the cannery.

Figure 5.7. Kosher meat symbol.

METHODS OF PURCHASING

The buyer who has forecast a need for specific products and written specifications must then decide how to make the purchase. Whatever method is chosen, it will belong to one of two general categories of procedures, identified as either informal or formal, both applicable to independent and organizational buying.

Informal

Informal purchasing, in which price quotations are given and orders made by telephone or personally with a salesperson, is often used when time is an important factor. It is usually done under the following circumstances:

- Amount of purchase is so small that time required for formal purchasing practices cannot be justified.
- An item can be obtained only from one or two sources of supply.
- Need is urgent and immediate delivery is required.
- Stability of market (and prices) is uncertain.
- Size of operation may be too small to justify more formal procedures.

If possible, at least two prices for each item should be obtained because informal quotations, being oral, have little legal force. Any prices quoted by telephone should be recorded by the buyer and then checked on the invoice at the time of delivery. Because of the lack of formal records, federal, state, or local laws often determine conditions under which informal purchasing can be used for tax-supported institutions.

Formal

Tax-supported institutions usually are required to use competitive bidding, which is optional for private institutions and commercial foodservices. Competitive bidding usually culminates in a formal contract between buyer and supplier. Understanding legal implications of contract buying is important for both parties. These legal considerations apply equally to buying for a single independent unit, a department within an organization, several departments of an organization, or a group of organizations.

Bid Buying

Bid buying: Buyer decides which supplier will be chosen for the order based on bids submitted from the seller.

Bid: A price quotation from seller.

Bid buying occurs when a number of suppliers are willing to compete (**bid**) over price quotations on buyers' specifications. Buyers generally use the fixed bid or the daily bid.

The fixed bid often is chosen for large quantities, particularly for nonperishable items purchased over a long period of time. Suppliers avoid committing themselves to that much time because of potential price fluctuations. The buyer selects a group of suppliers and gives them specifications and bidding forms for desired products, which they complete and return. Figure 5.8 is an example of a bid request form.

The **daily bid** often is used for perishable products that last only a few days. Sometimes this bid is referred to as daily quotation buying. Daily bids usually are made by telephone followed by a written confirmation before the buyer accepts the bid. Bids quoted by telephone are recorded by the buyer on a quotation record form similar to that in Figure 5.9.

Whether fixed bids or daily bids are used, only two basic methods for awarding bids are available.

- **Line-item bidding.** Each supplier bids on each product on the buyer's list, and the one offering the lowest price receives the order for that product. Foodservice buyers must consider the size of orders wholesalers are required to deliver. According to Gunn (1992), the typical wholesaler needs a guarantee of a $600 minimum order to realize a return on investment. The line-item bid is much more time consuming than the all-or-nothing bid and costs more for both the buyer and supplier.
- **All-or-nothing bidding.** This type of bidding, often referred to as the bottom-line approach, requires suppliers to bid the best price on a complete list of items (Gunn, 1992). Milk bids in school foodservice operations have been awarded on a bottom-line basis for a long time. Gunn (1992) offers this scenario to explain why: Imagine having one dairy deliver 2% unflavored milk while another delivers whole and skim milk, and maybe a third dairy delivers cottage cheese and yogurt. The foodservice manager would have to receive three deliveries and process three invoices if a single source wholesaler is not used.

Buyers for federal, state, and local institutions must allow bidding by all qualified distributors, but buyers for private institutions and commercial foodservices may select any supplier they want to submit bids. A bid request form, on which the specification and date for closing bids are indicated, is used to solicit bids from prospective suppliers. Bids usually are required to be submitted in an unmarked, sealed envelope and are opened at a specified time. Government purchasing policies usually require that this be done publicly, and the award should be made to the lowest bidder. No such requirement is dictated to the foodservice industry; however, if buyers select qualified suppliers and bids are responsive to the specifications, the lowest bid should be the one accepted. Computer software programs that allow managers to compare bid prices by food category and suppliers are available to assist with the bid analysis process.

Legal Considerations

Purchasing practices and the buyer/supplier relationship usually are predicated upon good faith and not dependent on legal considerations. The purchase/sale interchange, however, is a legal and binding commitment covered in the Uniform Commercial Code (UCC), which has been passed by all state legislatures except Louisiana's. The Code provides uniformity of law pertaining to business transactions in eight areas, which are identified as articles. Article 2, which governs purchase/sales transactions, is of prime concern to buyers because the buyer is

BID REQUEST

Bids will be received until November 22, 2002

Issued by: Community Hospital

Address: 100 North Street
Sunnyvale, OK

Date issued: September 27, 1999

Date to be delivered: Weekly as ordered between 01/02/03 to 6/30/2003

Increases in quantity up to 20% will be binding at the discretion of the buyer. All items are to be officially certified by the U.S. Department of Agriculture for acceptance no earlier than 2 days before delivery; costs of such service to be borne by the supplier.

Item No.	Description	Quantity	Unit	Unit Price	Amount
1	Tomatoes, whole or in pieces, U.S. Grade B, #10 cans, 6/cs.	100	cs.		
2	Sweet Potatoes, vacuum pack, U.S. Grade B, #3 vacuum cans (enamel lined), 24/cs.	50	cs.		
3	Asparagus, all green, cuts and tips, U.S. Grade A, #10 cans, 6/cs.	50	cs.		
4	Corn, Cream Style, golden, U.S. Grade A, #10 cans (enamel lined), 6/cs.	50	cs.		
5	Blueberries, light syrup, U.S. Grade B, #10 cans (enamel lined), 6/cs.	20	cs.		

Supplier______________________

Figure 5.8. Example of a bid request form.

COMMUNITY HOSPITAL
QUOTATION RECORD FORM

Date: Monday, 8/06/02 *Delivery Date:* Wednesday, 8/08/02

Circle accepted price quotation.

Item	Specifications	Amount Needed	Amount on Hand	Amount to Order	Price Quotations: Supplier Jones per unit	Supplier Jones total	Supplier L. & M. per unit	Supplier L. & M. total
Tomatoes	U.S. #1, 20# lug, 5 × 6	4 lugs	1	3	$10.50	$31.50	$10.75	$32.25
Lettuce	U.S. #1, Untrimmed Iceberg, 40% hard head, 24 heads, 1 crate	4	1	3	12.00	36.00	11.70	35.10
Potatoes	U.S. #1, Long Russet Bakers 100 count (50# box)	6	2	4	8.82	35.28	9.07	36.28
Onions	U.S. #1, Yellow, med. size, 50# sack	1	0	1	10.25	10.25	10.00	10.00
Watermelon	U.S. #1, Red flesh	400#	0	400#	.18/#	72.00	.155/#	62.00
						$185.03		($175.63)

Figure 5.9. Example of a form for telephone bids.

protected legally against trickery by the supplier. The major legal areas involve the laws of agency, warranty, and contracts. Buyers should understand the basic principles of each to avoid litigation.

Law of agency: Buyer's authority to act for the organization, the obligation each owes the other, and the extent to which each may be held liable for the other's actions.

Law of Agency. Every purchasing transaction is governed by at least one law, the **law of agency**; in most purchases, however, many laws are involved. Buyers may involve the foodservice operation and themselves in expensive legal disputes if they do not understand how these laws affect the purchasing function. The law of agency defines the buyer's authority to act for the organization, the obligation each owes the other, and the extent to which each may be held liable for the other's actions.

Some terms need to be defined before the law can be understood. An **agent** is an individual who has been authorized to act on behalf of another party, known as the **principal.** The business relationship between the agent and the principal is the **agency.** Agents have the power to commit their principals in a purchase contract with suppliers. In large organizations, the power of the agency is created by actions of the board of directors and delegated to the buyer, but in small organizations, this authority may have developed over years without ever having been recorded in a written agreement.

Often the buyer will deal with a salesperson and not the supplier. Generally, salespeople are considered to be special agents empowered to solicit business for their principal or company, and nothing else. The primary interest of most salespeople, quite rightly, is to make as many sales as possible. Most companies protect themselves and their customers by specifying in sales agreements that they are not bound by a salesperson's promises unless they appear in writing in a contract approved by an authorized person in the supplier's office.

Law of warranty: Guarantee by the supplier that an item will perform in a specified way.

Law of Warranty. The **law of warranty** defines *warranty* as a supplier's guarantee that an item will perform in a specified way (Kotschevar & Donnelly, 1998). The UCC recognizes three types of warranties (Leenders & Fearon, 1992):

- **Express warranty.** Promises, specifications, samples, and descriptions of goods that are under negotiation.
- **Implied warranty of merchantability.** Suppliers "puff" the virtues of their products for the purpose of making a sale.
- **Implied warranty of fitness for a particular purpose.** Buyer relies on the supplier's skill or judgment to select or furnish suitable goods.

The UCC provides that if the supplier knows the particular purpose for which products are required and the buyer is relying on the supplier's skill or judgment to select or furnish suitable products, the implied warranty is that the goods shall be fit for such a purpose. If, however, the buyer provides detailed specifications for the product, the supplier is relieved of any warranty of fitness for a particular purpose.

Law of contract: Signed agreement between two or more parties related to the purchase of a product or service.

Law of Contract. Under the **law of contract,** a **contract** is an agreement between two or more parties. Because contracts constitute an agent's primary source of liability, buyers should always be certain that each contract bearing their signature is legally sound. A contract must fulfill the following five basic requirements to be considered valid and enforceable: the offer, acceptance of the offer, consideration, competent parties, and legality of subject matter. An example of a contract award form for purchasing by a school district is shown in Figure 5.10.

- **The offer.** The first step in entering into a contract is making an offer, usually by the buyer. A purchase order becomes a buyer's formal offer to do business. When the supplier agrees to terms, a contractual relationship will be in force. A contract does not exist until both parties agree to the terms stipulated therein.
- **The acceptance.** The next step is the acceptance of the offer by the supplier. Generally, a clause is included that requires the supplier to indicate acceptance of the terms in the purchase order. Two copies of the order are sent to the supplier, one for the supplier and the other to be signed and returned to the buyer.
- **The consideration.** Consideration is the value, in money and materials, that each party pays in return for fulfillment of the other party's contract promise. Failure to do this usually means that the contract will not be legally enforceable. Agreement must be made on quantity, price, and time of delivery between the buyer and the supplier.

Contract Award

Board of Education or School ______________________ Contract Award No. __________

______________________ Date Awarded __________
Address

______________________ Date Bid Opened __________

This is a notice of the acceptance of Bid #________ for the period of ____________ 19 ____ to ____________ 19 ____.

Delivery
Delivery is to be made in two shipments: Week of ____________ and ____________ between ______ a.m. and ______ p.m.

Notice to Contractors:
This notice of award is an order to ship. Orders against contract are listed by ________ and invoices shall be rendered direct to the ________. The price basis, unless otherwise noted, ________ includes delivery and transportation charges fully prepaid F.O.B. agency. No extra charge to be made for packing or packages.

Names and Addresses of Successful Bidders

Offer
In compliance with the above award, and subject to all terms and conditions listed on the Bid Request, the undersigned offers and agrees to sell to ____________ the items listed on the attached schedule.

Bidder ______________________

____
Address

By ______________________
Signature of person authorized to sign this contract

Title ______________________

Accepted as to items numbered ______________________ Accepted by ______________

By ______________________ Date ______________

Title ______________________

Figure 5.10. Example of a contract award form.
Source: Food Purchasing Pointers for School Food Service (p. 17) by U.S. Department of Agriculture, Food and Nutrition Service, August 1977, Washington, DC: U.S. Government Printing Office.

- **Competent parties and legality.** The final two requirements of a valid contract are competency and legality. **Competency** means that the agreement was reached by persons having full capacity and authority to enter into a contract. **Legality** means that a valid contract cannot conflict with any existing federal, state, or local regulations or laws.

Independent and Organizational Buying

Buying methods and legal concerns are just as applicable to an independent unit as they are to central and group purchasing. As mentioned previously, buyers must be legally authorized to act as agents of those for whom they purchase.

Independent Purchasing

Independent purchasing is done by a unit or department of an organization that has been authorized to purchase. In the simplest situation, the owner of a restaurant might be the manager or the buyer, who would have full legal authority to execute binding contracts. In small hospitals, the head of the Food and Nutrition Department may have authority to do the purchasing and thus becomes the agent for the principal, which is the hospital.

Centralized Purchasing

Centralized purchasing is based on the principle that the purchasing activity is done by one person or department. In operations that have centralized purchasing, the head of the department usually reports directly to top management, which has the overall responsibility for making a profit. Advantages of centralized purchasing are many and include the following:

- Better control with one department responsible for purchasing and one complete set of records for purchase transactions and expenditures
- Development of personnel with specialized knowledge, skills, and procedures that result in more efficient and economical purchasing
- Better performance in other departments when managers are relieved of purchasing and of the interruptions and interviews incidental to purchasing
- Economic and profit potentials of purchasing, making it a profit rather than a cost center (Heinritz et al., 1991)

Most hospitals have used centralized purchasing for many years, even though they are part of a group purchasing organization. The foodservice department purchases canned and frozen products, staples, paper goods, and detergents from the central storeroom. Foodservice managers, because they know the products that will satisfy the customer, often are responsible for ordering perishable foods, such as fresh fruits, vegetables, and meat, and storing them in department refrigerators or freezers.

Widely dispersed units under one central management also may use centralized purchasing. A classic example is a school foodservice operation in which a central purchasing office, usually with warehouse facilities, meets the requirements for all schools in a district. Warehousing and delivery costs add to the purchasing budget, but these costs may be somewhat offset by lower prices from suppliers who do not have to make individual deliveries to all schools in the district.

The profit center concept is most successful in a centralized purchasing department (Reck & Long, 1983). The purchasing department can make a profit by adding a percentage that includes overhead, labor, and profit to the price of the product. As a check, individual departments need to be given the authority to purchase products and services directly from suppliers if significant cost savings can be achieved. A buyer in the centralized purchasing department, however, can negotiate for lower prices, consolidate orders from several departments into one large order, develop long-term supply agreements with suppliers in return for larger discounts, and search for low-cost suppliers and substitute materials. In most cases, the ultimate cost of products is lower than individual departments could achieve.

Group Purchasing

Group purchasing, often erroneously called centralized purchasing, involves bringing together foodservice managers from different operations and schools, most often onsite foodservices, for joint purchasing. Many healthcare organizations participate in group purchasing, which is external, but have internal centralized purchasing. The economic advantage of group purchasing is that the volume of purchases usually is large enough to warrant volume discounts by suppliers.

A site is selected for the group purchasing office, and purchasing personnel are hired by the group of managers and paid from group funds. Warehouse space is not required because products are delivered directly to the foodservice operation by a distributor on a cost-plus basis. The participating managers serve as an advisory committee to the purchasing personnel and also have some decision-making power. An example is their agreement on specifications for each item to be purchased. Obviously, wide variations in specifications defeat the purpose of group purchasing because the quantity of a specific product ordered by one foodservice operation would be low, thus decreasing high-volume savings.

The earliest group purchasing program was started in 1918 by the Greater Cleveland Hospital Association and was supported by hospitals in a 100-mile radius of Cleveland. Currently, many cities, such as Cincinnati, Chicago, Los Angeles, and Philadelphia, and states, such as Connecticut, Idaho, Mississippi, and Missouri, have group purchasing organizations. Both large medical centers and small hospitals as well as large and small school districts can profit from group purchasing. These group purchasing organizations have contracts for products or services with various companies.

Warehouse Club Purchasing

Many small independent foodservice operations have used self-service, cash and carry, and wholesale warehouses, often saving money in the process. These wholesale units, such as Sam's Club, operate on a no-frills approach to purchasing that enables prices to be kept exceptionally low. Products remain in original cartons stacked on pallets or metal shelves separated by wide aisles. A wide variety of brand-name items, including food, cleaning supplies, furniture, electronic equipment, and many others needed for foodservice operations, are stocked in plain and unfinished warehouses. Personnel often are not available to assist customers. Markups on most articles average 10% over wholesale cost, and food products have an even lower percentage markup.

Just-in-Time Purchasing

The objective of just-in-time (JIT) purchasing in foodservice is to purchase products as needed for production and immediate consumption by the customer without having to store and record products in inventory. The greatest challenge for most manufacturers and processors in achieving JIT is the purchase of raw materials (University of Technology Perth Western Australia, 1995). JIT purchasing can only be successful when the material is of sufficient quality and delivered in time. The goals of JIT should be to

- Secure a steady flow of raw materials
- Reduce the lead time required for ordering a product
- Reduce the amount of inventory in the storage and production areas
- Reduce the cost of purchased materials

The cost of the products is included in daily food costs, and capital is not tied up in inventory. Receiving clerks check the food in and then immediately take it to the area in which it will be used. For example, chopped lettuce, shredded cabbage, and diced fruit go directly to the unit in which employees assemble salads or to the production unit where chefs use fresh-cut vegetables in stir-fry and other menu items. The need for large refrigerated areas for storing fresh produce is eliminated. The ideal situation would be to have reach-in refrigerators in the units using fresh-cut produce for temporary storage until the product is used. More and more suppliers of value-added products are offering daily deliveries or a minimum of two or three deliveries a week to customers (Spears, 1999).

A comparison of JIT to non-JIT buying firms was made on context, organizational, and performance dimensions (Germain & Droge, 1998). Many differences were found in the organizational design variable. JIT buyers engage in far more extensive performance measurement on internal, benchmark, and supplier dimensions. JIT buyers probably will write purchasing mission statements and strategic plans. Organizations using JIT are more likely to rely on cross-functional teams to formulate purchasing strategy and to decentralize line-operating decision making. Material handling, market research, and transportation scheduling specialists are more often employed by JIT buying firms than by non-JIT buying firms.

SUPPLIER SELECTION AND EVALUATION

Supplier selection may be the single most important decision made in purchasing. The trends to fewer suppliers, buying rather than making products, electronic data interchange, and continuing improvement in quality, price, and service require greater communication between buyers and suppliers than ever before. Improving buyer and supplier relationships, therefore, is one of the key concerns in the overall area of supplier selection (Leenders & Fearon, 1992).

The ideal method of purchasing would be based on price, quality, and delivery from the supplier offering the best possible combination of all three. This ideal method is not always an easy task because salespeople might have their own ideas of right and wrong and buyers might be prejudiced against certain suppliers. A supplier, often identified as a seller or vendor, is a person who offers products for sale. *Supplier* is the term used by the National Association of Purchasing Management in the *International Journal of Purchasing and Materials Management* and, therefore, will be used in this text. A buyer, often called a purchaser, has charge of the selection and purchasing of products.

Traditionally, the supplier seeks out the buyer, but for the purchasing professional, this leaves too much to chance. Few foodservice managers need to be convinced of the advantages of consolidating most purchases with a primary distributor. Gone are the days when buyers had many suppliers or distributors to supply their needs. According to Goebeler (1989), because of mounting competitive pressures, every manager should pay careful attention not only to selecting a distributor but also to maintaining the relationship that ensues. The supplier selection process consists of the survey and inquiry stages followed by an evaluation.

Survey Stage

The purpose of a supplier survey is to explore all possible sources for a product. The buyer's experience and personal contacts with various suppliers provide the most valuable and reliable information and should be kept in complete and up-to-date supplier files in the purchasing office. These files should contain the name and address of every supplier with whom a foodservice organization has transacted business, plus information on products purchased from the supplier. Additional helpful information would be reliability in meeting commitment dates, willingness to handle emergency orders, and number of times orders were incorrect. The supplier file can be cross-referenced with a file of products listing from whom they were purchased, the prices paid, and the points of shipment.

Buyers should learn to question salespeople because they have a lot of information even though they could be biased about their own products. If they do not have a product a buyer wants, they often will suggest other companies that carry it.

Inquiry Stage

In the inquiry stage of the supplier selection process, the field must be narrowed from possible to acceptable sources. In general, this process involves comparing potential suppliers to provide the right quality and needed quantity at the right time, all at the right price with the desired degree of service. Quality, quantity, and price should be compared and balanced against one another. The common factor of service is applicable to all purchases, and although it cannot be quantified, it is an important subjective consideration.

Geographic location is a major concern in evaluating a supplier's service. Certainly, shorter delivery distances offer better opportunities for satisfactory service. The supplier's inventory also is an important consideration, as are the kinds and quantities of products available to buyers. The selected supplier should be one who keeps current with technological developments, is capable of providing new and improved products as they become available, and whose quality control standards ensure that inspection and storage methods adhere to FDA and USDA standards. Warranty and service offered by an equipment supplier also are a vital concern for the buyer.

Another factor to consider is that the financial condition of a supplier is crucial for maintaining a satisfactory business relationship. In general, the controller or credit manager of the buyer's organization can provide a credit report. When feasible, a visit to the prospective supplier's warehouse facilities will give an indication of the efficiency and general cleanliness associated with a good business. The visit gives the buyer an opportunity to observe employees, including union relations, and order-handling procedures.

After the survey and inquiry stages of selection, the buyer should have a few suppliers from which to choose one or more. The selection process is relatively subjective; more objective evaluations can be made after an order is placed and the product delivered. An evaluation process needs to be developed that can be used for newly selected suppliers and for those who have been servicing the operation for some time.

Supplier Performance Evaluation

The evaluation of suppliers is a continuing purchasing task (Leenders & Fearon, 1992). Current suppliers need to be monitored to be sure they are meeting performance expectations, and new ones need to be screened to determine if they should be seriously considered in the future. Buyers, however, tend to use suppliers who have proven to be reliable. The new supplier should be given the opportunity to make a few deliveries to the foodservice operation before a performance evaluation is conducted. It is difficult to select a supplier performance evaluation instrument that is fair but truly evaluates performance. Many foodservice managers develop their own forms. Figure 5.11 is an example of an evaluation form that can be modified for a specific operation.

Evaluations should be conducted periodically to keep the supplier's record file up to date. Suppliers who have good evaluations will appreciate knowing the results. Suppliers who do not rate well should be told why before severing the business relationship.

PURCHASING PROCESS

Procurement in any foodservice operation requires procedures to accomplish the routine purchasing transaction as quickly and accurately as possible. Undoubtedly, each foodservice operation has its own procedures unique to its needs, but very likely these procedures conform to the basic pattern followed by many other foodservices. The adoption of definite purchasing procedures implies use of appropriate records for each phase in the purchasing process.

Purchasing Procedures

Several basic procedures appear in some form in every purchasing unit (Figure 5.12). These procedures are simple and should be adapted to the particular needs of the various departments of an organization. Buyers can add, delete, or modify these procedures to fit their operation.

Recognition of Need

The most obvious place for a need to occur is the production unit. The second location is the storage area, in which the objective is to have on hand the right product, in the right quantity, at the right time. A third location exists in larger

SUPPLIER PERFORMANCE EVALUATION

Supplier: ______________________ *Date:* ______________

Company	**Excellent**	**Good**	**Fair**	**Poor**
Size and/or capacity				
Financial strength				
Technical service				
Geographical locations				
Management				
Labor relations				
Trade relations				
Products				
Quality				
Price				
Packaging				
Uniformity				
Service				
Delivers on time				
Condition on arrival				
Follows instructions				
Number of rejections				
Handling of complaints				
Technical assistance				
Emergency deliveries				
Supplies price changes promptly				

Figure 5.11a. Example of a supplier evaluation form.

organizations in which the centralized purchasing department has inventory responsibility. Recognition of a need should be followed by action to remedy the deficiency by preparing a requisition.

Description of the Needed Item

In most organizations, the production unit cooks, having recognized a need, initiate a requisition, preferably written, to the storeroom for the required amount of the product. If the storeroom has an adequate supply of the item, the requisition can be honored. If, however, honoring this requisition brings the inventory stock below the acceptable minimum, storeroom personnel initiate another requisition to purchasing for replenishment of the product to the desired level.

Sales Personnel	Excellent	Good	Fair	Poor
• Knowledge:				
Of company				
Of products				
Of foodservice industry				
• Sales calls:				
Properly spaced				
By appointment				
Planned and prepared				
Mutually productive				
• Sales service:				
Obtains information				
Furnishes quotations promptly				
Follows through				
Expedites delivery				
Handles complaints				
Accounting				
Invoices correctly				
Issues credit promptly				

Figure 5.11b. Continued.

Computer software programs are available that will generate a list of the quantities of ingredients needed to prepare menu items. The caveat is that many of these programs round quantities to the nearest purchase unit; for example, a teaspoon of salt might become a case of salt.

Authorization of the Purchase Requisition

The third procedure in the purchasing process is authorizing the purchasing requisition. In every foodservice organization, a policy should be established to indicate who has the authority to requisition food, supplies, and equipment. No requisitions should be honored unless the person submitting them has the authority to do so. Furthermore, suppliers should know the names of persons authorized to issue purchase orders.

Negotiation with Potential Suppliers

Negotiation in purchasing is the process of working out both a purchasing and sales agreement, mutually satisfactory to both buyer and supplier, and reaching a common understanding of the essential elements of a contract. It is one of the most important parts of the procurement function, if not the most important. Negotiation

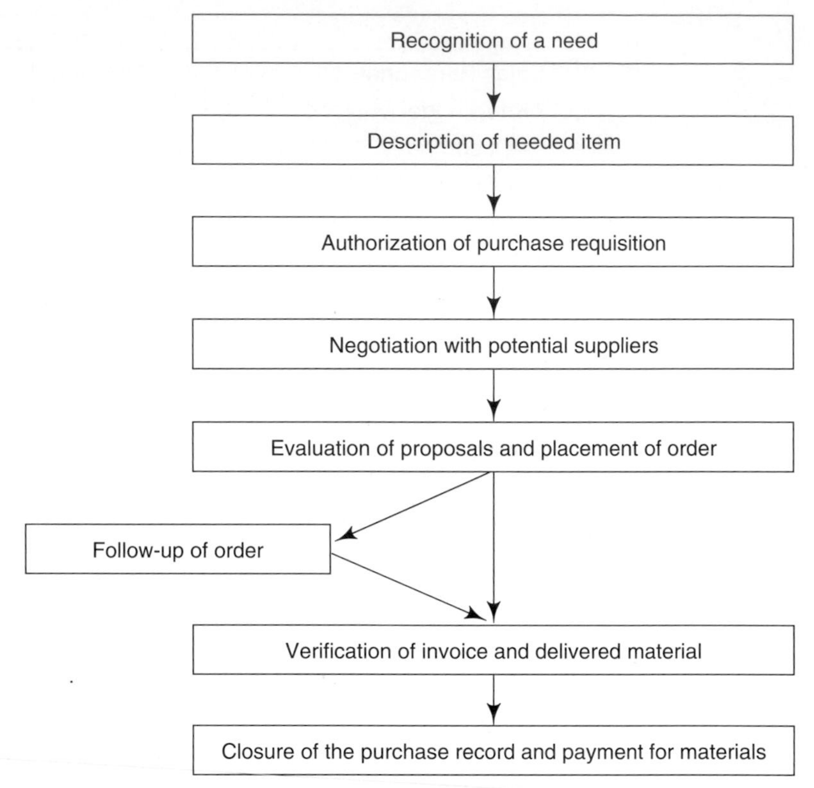

Figure 5.12. Fundamental steps in purchasing.

usually is responsible for vital details such as establishing the qualifications of a particular supplier, determining fair and reasonable prices to be paid for needed products, setting delivery dates agreeable to both the buyer and supplier, and renegotiating contract terms when conditions change.

Evaluation of Proposals and Placement of Order

All supplier proposals are evaluated for compliance with the preceding four fundamental steps in purchasing (Figure 5.12). The actual placement of an order follows the evaluation of proposals. Ideally, all orders should be in writing, but if an order is placed by telephone, confirmation in writing should be made promptly. A written record of every purchase should always be on file.

Many software programs will generate a purchase order for the supplier. These purchase orders are given to the supplier in a variety of ways. The foodservice buyer may meet in person with a salesperson that represents a supply company and give the order to that salesperson. Or the foodservice buyer may place the order over the telephone or via fax transmission to the supplier. Some suppliers link their computers with those of buyers. This process, termed electronic data interchange (EDI), allows for fast, accurate, and efficient transmission of the order (Keiser, DeMicco, & Grimes, 2000).

Follow-Up of Order

Theoretically, buyers should not have to follow up an order after it has been placed and accepted by a supplier. A follow-up in a foodservice operation is justified, however, when a specific delivery time of certain products is critical to an occasion, such as a major banquet.

Verification of Invoice and Delivered Materials

Invoice: Document prepared by the supplier that includes product name, quantity, and price.

The **invoice** is the supplier's statement of what is being shipped to the buyer and the expected payment. The invoice should be checked against the purchase order for quality, quantity, and price. Delivery and condition of materials should be in agreement with the purchase order and the invoice. Any differences require immediate action by the buyer. For example, the condition of frozen food should be verified, and any indication of exposure to higher temperatures would be a cause for rejection and immediate communication with the supplier.

Closure of Purchase Record

Closing the purchase record consists of the clerical process of assembling the written records of the purchase process, filing them in appropriate places, and authorizing payment for the goods delivered. The filing system need not be complex; its only purpose is to provide an adequate historical record of these business transactions.

Purchasing Records

The essential records for the purchasing process are the requisition and purchase order, originating with the buyer, and the invoice, prepared by the supplier. These records differ among foodservice operations but all have the same essential information.

Purchase Requisition

Requisition: Form used by foodservice manager to request items from purchasing manager or department.

The **requisition** is the first document in the purchasing process and may have originated in any one of a number of units in the foodservice operation. A requisition is used by foodservice managers to request food items from a purchasing manager or department. An example is shown in Figure 5.13. The following five basic items of information generally are included on all requisitions:

- **Requisition number.** This number is necessary for identification and control purposes and is generally accompanied by a code for the originating department.
- **Delivery date.** This date, on which the product should be in the storeroom for use by the cooks, should always allow sufficient time to secure competitive bids and completion of the full purchase transaction, if possible.
- **Budget account number.** This number indicates the account to which the purchase cost will be charged.
- **Quantity needed.** Quantities should be expressed in a common shipping unit, such as cases, along with the number of items in a unit. For example,

COMMUNITY HOSPITAL
PURCHASE REQUISITION

To: Purchasing Office *Requisition No.:* FS1201

Date: August 17, 2002 *Purchase Order No.:* 1842

From: Foodservice *Date Required:* September 10, 2002

Budget Account No.: FS 1101

Total Quantity	Unit	Description	Supplier	Unit Cost	Total Cost
20	cases	Tomatoes, diced, U.S. Grade B (Extra Standard) #10 cans, 6/case	L. & M. Wholesale Grocers	$13.86	$277.20

Requested by ________ Approved by ________ Date Ordered 8/24/02

Figure 5.13. Example of a purchase requisition.

the entry on the requisition is for 20 cases of diced tomatoes, with six No. 10 cans to a case.

- **Description of the item.** The description and quantity needed are the two most important pieces of information on the requisition form and, therefore, occupy the central space. The information may include the product specification, brand, or catalog number.

In addition to the preceding information, which is provided by the person responsible for the requisition, the buyer adds the name and address of the supplier and the details of the purchase, including the purchase order number and the price and date ordered.

Purchase Order

Purchase order: Document completed by the buyer and given to the supplier listing items to be purchased.

A **purchase order** is the document, based on the information in the requisition, completed by the buyer, who gives it to the supplier. It states in specific terms the purchase and sales agreement between the buyer and the supplier. Before acceptance, the purchase order represents an offer to do business under certain terms and, once accepted by a supplier, becomes a legal contract. An example is shown in Figure 5.14.

COMMUNITY HOSPITAL
1010 Main Street
Sioux City, Oklahoma

To: L. & M. Wholesale Grocers
200 South Street
Sunnyvale, OK 54123

Purchase Order No.: 1842

Please refer to the above number on all invoices, two copies required.

Date: August 23, 2002

Requisition No.: FS1201

Dept.: Foodservice

Date Required: September 10, 2002

Ship to: ________ F.O.B.: ________ Via: ________ Terms: ________

Total Quantity	Unit	Description	Supplier	Unit Cost	Total Cost
20	cases	Tomatoes, diced, U.S. Grade B (Extra Standard) #10 cans, 6/case	L. & M. Wholesale Grocers	$13.86	$277.20

Approved by ______________________
Title: Director of Purchasing
Community Hospital

Figure 5.14. Example of a purchase order form.

Format. The purchase order, like the requisition, exists in a wide variety of formats that have been developed to meet the needs of individual foodservice operations. The principal reason for this variety is that the purchase order is a legal document; the terms presented in it are intended to protect the buyer's interests, which differ from one operation to another. Almost every purchase order, however, includes the following items:

- Name and address of the foodservice organization
- Name and address of the supplier

- Identification numbers (purchase order and purchase requisition)
- General instructions to supplier
- Complete description of purchase item
- Price data
- Buyer's signature

Required Number of Copies. Three copies of the purchase order are sufficient for basic ordering: the original, which is sent to the supplier; an acknowledgment copy, which is actually a formal acceptance that the supplier returns to the buyer; and a file copy for the buyer's own record. This simplified plan offers only a minimum of control and, therefore, should not be used in large operations.

Ordering for large foodservice organizations requires six copies of the purchase order. The additional three copies are given to the following individuals or departments: receiving copy, which informs receiving personnel that a delivery is scheduled; accounting copy, which is sent to the accounting office; and requisitioner's copy, which notifies the requisitioner that the order has been placed.

F.O.B.: Free on board.

Shipping Terms. **F.O.B.** (free on board) means that products are delivered to a specified place with all transport charges paid. Once the point is specified, the following questions can be answered:

- Who pays the carrier?
- When is the legal title of the products being shipped given to the buyer?
- Who is responsible for claims with the carrier if products are lost or damaged during shipment?

F.O.B. origin: Buyer selects freight company, pays freight charges, and owns product during transit.

F.O.B. destination: Buyer pays freight charges, but seller owns product during transit.

F.O.B. origin refers to the place from which the product is originally transported; **F.O.B. destination** refers to the place the product is going. If F.O.B. origin is stipulated, the buyer selects the transport company and pays the freight charges, owns the title, and files claims for loss or damage to the products. The title and control of products come to the buyer when the carrier signs for the products at the point of origin. If F.O.B. destination is stipulated, the buyer pays the freight charges, but the supplier owns the products in transit and files claims. The title remains with the supplier until the products are delivered (Spears, 1999).

Invoice

The invoice, prepared by the supplier, contains the same essential information as the purchase order; that is, quantities, description of items, and price. When products are delivered, the supplier's invoice must be compared with the purchase order and the quantity received. Any discrepancies or rejections at the receiving point should be noted immediately. In some operations, these problems are handled by the receiving clerk, who notes the discrepancy on the invoice and has the supplier's delivery person initial it.

If the invoice and the purchase order are in agreement, the invoice is forwarded to the accounting office for payment. In small operations using only three copies of the purchase order, the receiving clerk only verifies the arrival of the delivered items and sends the invoice to the accounting office for comparison with the purchase order and subsequent payment.

RECEIVING

Receiving: Activity for ensuring that products delivered by suppliers are those that were ordered.

Receiving can be defined as an activity for ensuring that products delivered by suppliers are those that were ordered in the purchasing activity. Receiving has become known as the missing link in the procurement of materials because quite often food quality problems are caused by breakdowns in receiving procedures. One of the problems is related to a different product arriving on the receiving dock than what was ordered (Gunn, 1992). The receiving process involves more than just acceptance of and signing for delivered products. It also includes verifying that quality, size, and quantity meet specifications; that the price on the invoice agrees with the purchase order; and that perishable goods are tagged or marked with the date received. Furthermore, the products received should be recorded accurately on a daily receiving record, then transferred promptly to the appropriate storage or production areas to prevent loss or deterioration. Consistent and routine procedures are essential to the receiving process, along with adequate controls to preserve quality of products and prevent loss during the delivery and receipt of products.

In a foodservice establishment, between 30 and 50% of the revenues are spent on food purchases. In far too many operations, however, the receiving of these valuable purchases is entrusted to any employee who happens to be near the unloading or storage area when shipments from suppliers arrive. When responsibility for receiving is not assigned and procedures are not systematized, a number of problems may arise, such as careless losses, failure to assure quality and quantity of goods delivered, and pilferage. These potential losses can cost a foodservice operation more than its net profit each year.

The economic advantages gained by competitive purchasing based on complete and thorough specifications can be lost by poor receiving practices. Properly designed and enforced controls in receiving will ensure management that a dollar value in quantity and quality is received for every dollar spent.

Elements of the Receiving Activity

Elements of good receiving practices include competent personnel with specified responsibilities, proper facilities and equipment, well-written specifications, procedures for critical control, good sanitary practices, adequate supervision, scheduled receiving hours, and procedures to ensure security. Receiving is generated in purchasing through development of specifications and issuance of purchase orders.

Competent Personnel

Responsibility for receiving should be assigned to a specific member of the foodservice staff. In a small operation, this individual may have additional duties, but, if possible, the person responsible for the receiving should not be involved in food purchasing or production. Separating the duties of purchasing and receiving is basic to a check-and-balance system for ensuring adequate control. The owner or foodservice manager should assume the responsibility for either purchasing or receiving if sufficient employees are not available.

In addition to being familiar with food products and quality, receiving personnel must be able to detect old products, excess shrinkage, short weights, and products that do not meet specifications (Keiser & DeMicco, 1993). The receiving clerk should be provided with the specifications and purchase order sent to the supplier as a basis for checking all products delivered.

Employees need to be well trained in the skills and knowledge needed to perform receiving tasks competently. Many of the receiving tasks can be explained and demonstrated through on-the-job training; however, learning how to use specifications for evaluating products is more difficult and may require specific training sessions.

Facilities and Equipment

Adequate space and equipment are necessary to perform receiving tasks properly. In many foodservice operations, the receiving area may serve as an entrance for employees and salespeople, a place for general storage, and a passage to where trash is stored, all of which suggest a need for monitoring and good security procedures.

Ideally, the receiving area should be located near the delivery door, storeroom, refrigerators, and freezers to minimize the time and effort in movement of food into appropriate storage. In small operations, a wide hallway may be used as the receiving area; in larger operations, additional space is needed. In either case, the receiving area should be located near the delivery door for two reasons: Union contracts of delivery persons may stipulate "inside-door" delivery only, and security and sanitary concerns arise when persons who are not foodservice employees are permitted in back-of-house areas (Dietary Managers Association, 1991).

In large facilities, a receiving office generally is located near the delivery entrance. Enough space should be available to permit all incoming products to be inspected and checked at one time. In smaller facilities, a desk at the receiving entrance facilitates the check-in of products.

Many products require such minimal inspection as checking the package, label, and quantity on boxes of cereal and bags of flour. Deliveries of other products, such as meat, may require opening packages to inspect quality, count, and weight. To eliminate confusion for the receiving clerk, storage should not begin until the delivery personnel have left. Time and money can be saved by providing facilities that require a minimum handling of products and permit direct transfer from receiving to storage and areas of use.

The size of the receiving area for a specific food facility is influenced by the nature and volume of materials received or being transferred out at any one time. For example, a hand truck may be sufficient in some operations, but a forklift may be required for pallet deliveries. In commissary operations, the space must be sufficient for the various kinds of carts used for transporting foods to satellite service centers. A typical cart for transporting food from a commissary or base kitchen to a satellite school foodservice operation is shown in Figure 5.15. The receiving area in the satellite service center also must be designed to accommodate delivery carts.

The receiving department requires certain equipment, including scales in good working order. Both platform and counter scales should be available (Figures

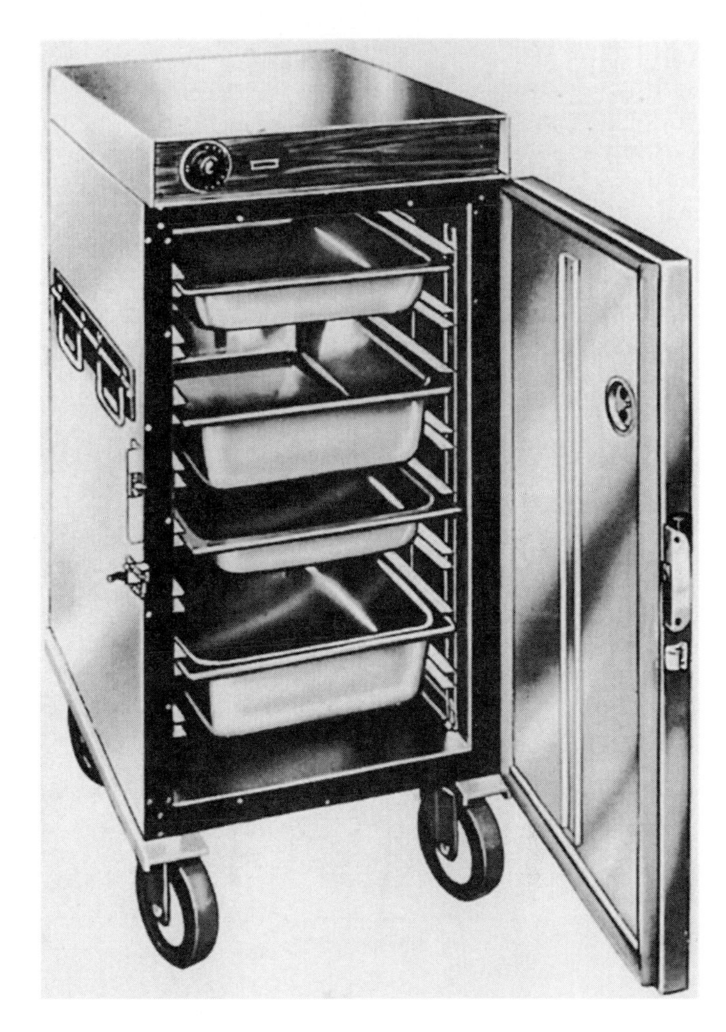

Figure 5.15. Bulk food transporter insulated to keep food hot or cold.
Source: Crescent Metal Products, Inc., Cleveland, OH. Used by permission.

5.16 and 5.17); portion scales are useful for checking portion cuts of meat. In large operations, the scales print the weight of the product on the reverse side of the invoice or packing slip or print a tape that can be attached to eliminate any doubt in the weighing-in process. Accuracy of all scales should be checked periodically.

An unloading platform of a convenient height for delivery trucks is needed, along with a ramp to facilitate the unloading of trucks that do not match the platform height. Dollies and hand trucks are important to expedite the movement of products to storage with the least amount of effort.

Other equipment in the receiving area should include a table for inspection of deliveries, and tools, such as a can opener, crowbar, claw hammer, and short-blade knife for opening containers and packages. A thermometer is needed to check if chilled or frozen products are delivered at appropriate temperatures according to specification. Clipboards, pencils, and marking and tagging equipment are also necessary. A file cabinet should be available for storing records and reports, and a calculator is needed to verify the computations on the invoice.

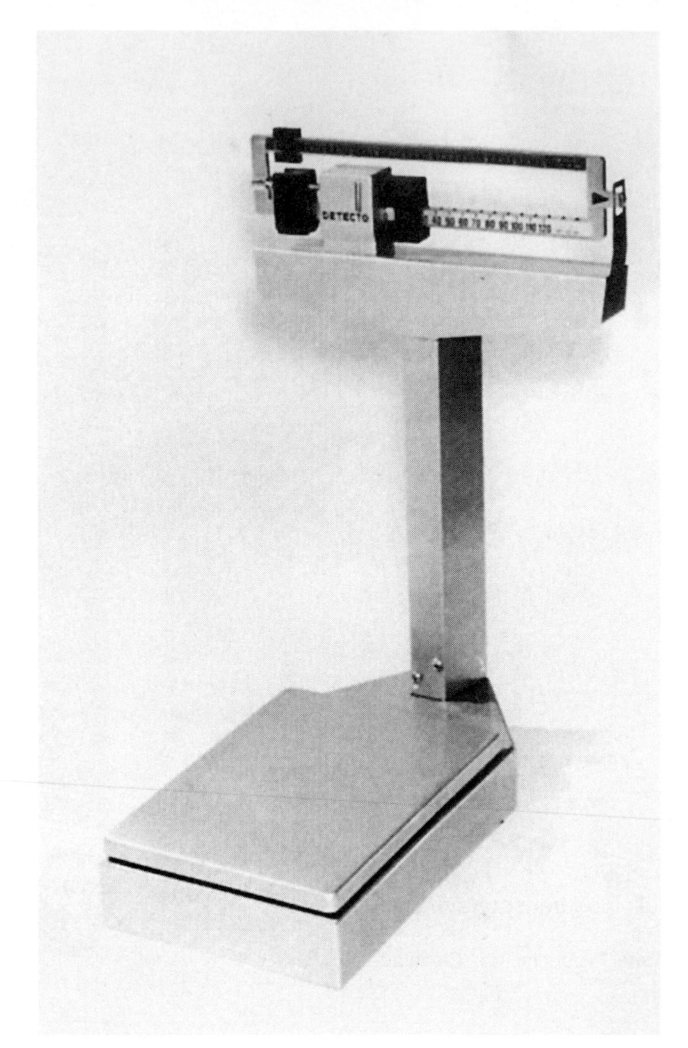

Figure 5.16. Platform scale.
Source: Detecto Scale, a division of Cardinal Scale Manufacturing Co., Webb City, MO. Used by permission.

Specifications

The employee who receives orders should know the standards the suppliers must meet and have a notebook or file box of specifications available for reference. All deliveries should be checked against these specifications and nothing below standard should be accepted.

A copy of all purchase orders should be provided to the receiving personnel. The purchase order includes a brief specification, data on quantity ordered, pricing information, and general instructions to suppliers such as delivery date and shipping instructions. With this information, receiving personnel are alerted that a delivery is scheduled on or before a certain date and that they must be prepared to count and inspect anticipated shipments.

Figure 5.17. Counter scale.
Source: Detecto Scale, a division of Cardinal Scale Manufacturing Co., Webb City, MO. Used by permission.

Critical Control in Receiving

Receiving is a critical control point for many foods, especially perishable ones such as raw meat and vegetables. Procedures for inspection and standards for acceptance are necessary to prevent foodborne illness. An HACCP flow chart for safe receiving procedures should be included in the procedures manual of a foodservice operation for all potentially hazardous foods.

Sanitation

The receiving area should be designed for easy cleaning. The floor should be of material that can be easily scrubbed and rinsed and has adequate drains and a water connection nearby to permit hosing down of the area. Storage for cleaning supplies should be conveniently located.

Because insects tend to congregate near loading docks, adequate screening must be provided. Electrical or chemical devices for destroying insects often are mounted near the outside doors in the delivery area.

Adequate Supervision

The management of a foodservice operation should monitor the receiving area at irregular intervals to check security and ensure that established receiving procedures are being followed. A member of the management staff should recheck weights, quantities, and quality of products received at various times as part of the control system for the foodservice system.

Scheduled Hours

Suppliers should be directed to make deliveries at specified times. This policy reduces the confusion of too many deliveries arriving at one time and ensures that deliveries will not arrive during meals or after receiving personnel are off duty.

In small operations, receiving personnel are frequently assigned other duties, especially during periods around meal service. In many hospitals, for example, a porter may have the responsibility for receiving, for transporting meal carts to patient floors, and for maintenance duties after meal service. When deliveries arrive at inappropriate times, the receiving clerk may be pressured and thus not check in goods properly. Therefore, midmorning and midafternoon deliveries are best in such operations. Another consideration in scheduling deliveries is the nature of the shipment. The receiving schedule in some operations requires that perishable foods be delivered during the morning hours and staple goods during the afternoon.

Security

An owner of an investigation company stated that foodservice managers can reduce internal theft by 60% if employees know that management is watching (Lorenzini, 1992). One practice to prevent theft at the time products are received is not to have the same person responsible for purchasing and receiving. Other elements of good receiving, previously discussed, also serve as components of a receiving security system; scheduled hours for receiving and adequate facilities and equipment for performing receiving tasks will prevent many problems. Another important practice is to move products immediately from receiving to storage. In addition, delivery persons generally should not be permitted in storage areas, further protecting the security of the area. Salespeople also should be excluded from the area, but usually for a different reason. Barring them from the storeroom prevents them from checking stock and influencing the purchaser.

Receiving Process

Detailed procedures are important to assure that incoming products are received properly. The steps in the receiving process are outlined in Figure 5.18.

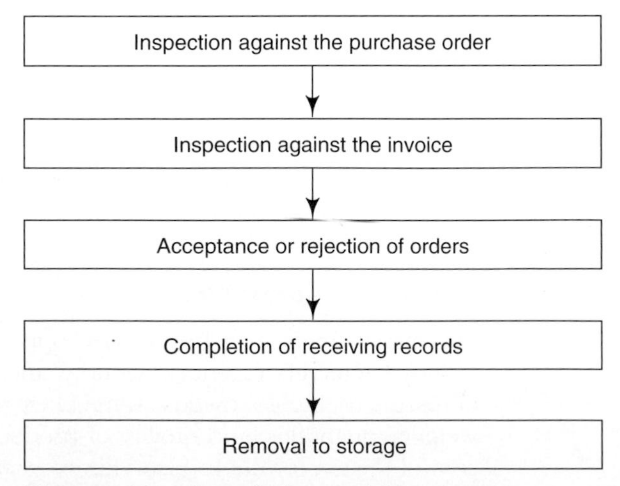

Figure 5.18. Steps in the receiving process.

Inspection against the Purchase Order

A written record of all orders must be kept to provide a basis for checking deliveries. This record, the purchasing order, as was shown in Figure 5.14, becomes the first control in the receiving process by including a brief description of the product, quantity, price, and supplier. In small foodservice operations using informal purchasing procedures, this record may be as simple as a notebook for listing this basic ordering information. In large operations with more formal procedures, however, one copy of the purchase order generally is transmitted to the receiving department as a record of scheduled deliveries.

Incoming shipments should be compared with the purchase order to ensure that the products accepted were in fact ordered. The purchase order also will permit the receiving personnel to determine partial deliveries or omission of ordered products. If a comparison of incoming deliveries with purchase orders indicates that the appropriate products have been delivered, then quality should be checked according to the receiving HACCP flow chart in the procedure manual. In reality, receiving is the initial phase of total quality control in the procurement subsystem.

Specialized training of receiving personnel is important to ensure that quality standards are known and can be recognized. All products should be compared with the characteristics in the established specifications. Without exception, the count or weight and quality tolerance levels of all products must be verified before the delivery invoice is signed. When too many products are below the tolerance levels for weight or quality, the receiving clerk should reject the shipment. A printed credit memorandum form may be used; one copy should be attached to the invoice with the discrepancy noted, initialed by the delivery person, and sent to the controller (Keiser & DeMicco, 1993). Another copy should be given to the delivery person to transmit to the supplier, and the receiving department should keep the third copy.

For some products, a foodservice operation may have a standing order and, therefore, not maintain a specific purchase order or other record for each delivery. In some operations, an inventory quantity level may have been established for such products as bread, dairy products, coffee, and eggs. Some foodservice operations permit delivery personnel employed by the supplier to check the quantity on hand and deliver enough to restore the inventory level. This practice cannot be recommended for reasons of security and quality control.

Inspection against the Invoice

After products have been checked against the purchase order and specifications, the delivery should be compared to the invoice prepared by the supplier. The invoice is the supplier's statement of what is being shipped to the foodservice operation and the expected payment.

Obviously, checking the quantities and prices recorded on the invoice is a critical step in the receiving process. Three receiving methods are used in foodservice operations: invoice, blind, and electronic receiving.

- **Invoice receiving.** The receiving clerk checks the quantity of each product against the purchase order. Any discrepancies are noted on both the purchase order and the invoice. This method is quick but can be unreliable if the receiving clerk does not compare the two records and only looks at the delivery invoice.

- **Blind receiving.** The receiving clerk uses an invoice or purchase order with the quantity column blanked out and records on it the quantity of each product received. This method requires that each product be checked because the amount ordered is unknown; it is time consuming for both the receiving clerk and the deliverer and, therefore, expensive in labor costs.
- **Electronic receiving.** Technology is speeding up the receiving process although it is still too expensive for small foodservice operations. Tabulator scales, which weigh and automatically print the weight on paper, are being used in large operations as is the Universal Product Code (UPC). Bar codes are appearing on cartons and packages in foodservice operations. Handheld scanners that can read a UPC bar speed up the receiving process (Stefanelli, 1997).

If the quantities and prices are correct and the receiving clerk has checked the quality of the products, the invoice should be signed. Generally, two copies of the invoice are required, one for foodservice records and the other for the accounting office. In small operations, only one copy may be required. If errors have been made in the delivery or pricing, corrections must be reported on the invoice before it is signed. The delivery person also should initial any correction of errors.

Acceptance or Rejection of Orders

Delivered products become the property of the foodservice operation when the purchase order, specifications, and supplier's invoice are in agreement. Payment then will be due at the agreed-upon time for products charged on the invoice.

Rejection at the time of delivery is much easier than returning products after they are accepted. If, however, errors are discovered after a delivery has been accepted, the supplier should be contacted immediately. Reputable suppliers generally are willing to correct problems. The foodservice manager should find out why the problem was not detected at the time of delivery and should make changes in receiving procedures. Whether products are returned after acceptance or rejected at the time of delivery, accounting personnel must be sure credit is given by the supplier.

Occasionally, a substandard product may be accepted because it is needed immediately and time does not permit exchanging it or finding an alternate source. The buyer might try to negotiate a price reduction with the supplier. If a product is not available or only a partial amount is delivered, the buyer needs to decide if it should be back ordered.

Completion of Receiving Records

The receiving record provides an accurate list of all deliveries of food and supplies, date of delivery, supplier's name, quantity, and price data. This information is helpful in verifying and paying invoices and provides an important record for cost control of all foods delivered to the kitchen and storeroom.

An example of a receiving record used in a community hospital is shown in Figure 5.19. The columns on the form that show distribution of products to the kitchen or storage are useful in some food cost-control systems. Products delivered directly to the kitchen will be included in that day's food cost, whereas products sent to the storeroom will be charged by requisition when removed from stores for production and service.

This kind of receiving record documents the transfer of products to storage in facilities in which receiving and storage tasks are performed by different employ-

COMMUNITY HOSPITAL
RECEIVING RECORD

Date: August 23, 2002

Quantity	Unit	Description of Item	Name of Supplier	Inspected and Quantity Verified by	Unit Price	Total Cost	Distribution To Kitchen	Distribution To Storage
3	lugs	Tomatoes	L. & M.	JH	$10.75	$32.25		X
3	crates	Lettuce	L. & M.	JH	11.70	35.10		X
4	boxes	Bakers	L. & M.	JH	9.07	36.28	X	
1	sack	Onions	L. & M.	JH	10.00	10.00		X
400	lbs.	Watermelon	L. & M.	JH	.155	62.00	X	

Figure 5.19. Example of a receiving record.

ees. The receiving record provides a checkpoint in the control system. This record usually is prepared in duplicate, with one copy sent with the invoices to the accounting department and the other retained in the receiving department. If the receiving clerk is responsible for verifying price extensions on invoices, this should be done before the invoice is forwarded to the accounting office. Accounting personnel, however, also should verify the arithmetic extensions on invoices.

Removal to Storage

Products should be transferred immediately from receiving to the secure storage area; personnel should not be permitted to wait until they have time to move food and supplies to storage. Because the products are now the property of the foodservice operation, security measures are important to prevent theft and pilferage. Also, spoilage and deterioration may occur if refrigerated and frozen products are held at room temperature for any period of time.

Foodservice operations may indicate on the receiving record various procedures for marking or tagging products for storage. Marking consists of writing information about delivery date and price directly on the case, can, or bottle before it is placed in storage. Daily food cost calculations can be done quickly because prices do not have to be looked up, and fewer products will spoil on the shelves because the ones with the oldest dates will be used first. Tagging products also facilitates stock rotation to ensure that older products are used first, which is particularly important with perishable food products such as expensive meat and fish. Data such as date of receipt, name of supplier, brief description of product, weight or count upon receipt, and place of storage may be written on the tag.

STORAGE

Storage: Holding of products under proper conditions to ensure quality until time of use.

After food and supplies have been received properly, they must be placed in appropriate **storage**, which is the holding of goods under proper conditions to ensure quality until time of use. Storage is important to the overall operation of a foodservice because it links receiving and production. Dry and low-temperature storage facilities should be accessible to both receiving and food production areas to reduce transport time and corresponding labor costs. Proper storage maintenance, temperature control, and cleaning and sanitation are major considerations in ensuring quality of stored foods. Competent employees are as essential for storage positions as they are for all other positions in a foodservice operation. Storage employees check in products from the receiving unit, provide security for products, and establish good material-handling procedures. Only those personnel authorized to store goods, issue foods, check inventory levels, or clean the areas should have access to storage facilities.

Prevention of theft and pilferage in storage facilities is a major concern. The sad commentary is that, once again, the customer pays for the problem. Stefanelli (1992) defines **theft** as "premeditated burglary" and **pilferage** as "inventory shrinkage." Theft occurs when someone drives a truck up to the back door and steals expensive products and equipment. Pilferage focuses on the employee who steals a couple of steaks before checking out. In some foodservice operations, locks for storage areas are replaced periodically to prevent entry with unauthorized duplicate keys; only authorized persons should have access to keys.

Ideally, storage should be located on the same floor as the production area and be visible to the foodservice manager, but this is not always possible. In large operations, the main storage areas often are located on a lower level or in another area of the building, making surveillance by the manager difficult. In institutions such as a large hospital, dry storage for food and supplies may be in the central storeroom for the entire hospital, and security responsibility is assumed by the centralized purchasing department. In multiunit restaurant operations, a central warehouse usually is kept for major storage, and a limited storeroom and small refrigerators and freezers for a few days' supply are provided in the individual restaurant operations.

The type of storage facilities will vary greatly, depending on the type of foodservice (refer to Chapter 4 for discussion of types of foodservices). In an assembly/serve foodservice, in which many fully processed frozen food products are used, the need for frozen storage may be much greater than in a conventional operation. In a commissary foodservice, the central production area demands large-storage capability.

Regardless of the type of foodservice, all foods should be placed in storage as soon as possible after delivery, unless they are to be processed immediately. Dry groceries, canned foods, and staples should be placed in dry stores. Perishable foods must be placed in refrigerated or frozen storage promptly. In addition, storage facilities must be available for china, flatware, trays, utensils, and nonfood products, such as paper supplies, detergent, and cleaning products.

In some storage areas, fast-moving products are placed near the entrance and slower-moving ones are stored in less accessible locations. Generally, however, products are categorized into groups, then arranged either alphabetically or according to frequency of use within the groups. Foods that give off odors should be stored separately. New stock should be placed in back of older stock to prevent loss from deterioration. Foods should be checked periodically for evidence of spoilage, such as bulging or leaking cans. Ideally, canned foods should not be kept more than 6 months even when stored under proper conditions (Thorner & Manning, 1983), because it is difficult to determine the length of time foods have been canned prior to delivery.

The advantage of purchasing large quantities to save money must be weighed against the possibility of storage loss and the costs of maintaining inventories. Inventory costs are discussed in more detail later in this chapter.

Dry Storage

The dry storage area for food provides orderly storage for foods not requiring refrigeration or freezing and should provide protection of foods from the elements, insects, rodents, and theft. Cleaning supplies and pesticides should not be kept in the food storeroom. Instead, a separate locked room should be provided to prevent such items from contaminating food products.

Facilities

The floors in the dry storage area should be slip resistant and easy to clean. Walls and ceilings should be painted light colors and have smooth surfaces that are impervious to moisture and are easy to wash and repair. The number of doors allowing access to the storeroom should be limited. The main door should be heavy-duty and of sufficient width for passage of the equipment used to transport foods from receiving to dry storage or from dry storage to production areas. This door should lock from the outside; however, a turnbolt lock or crash bar should be provided to permit opening from the inside without a key in case people are inadvertently locked in. A Dutch door often is used as the main entrance; the lower half is locked at all times except when accepting large shipments, and the upper door is open during issuing hours.

In large operations, a recording time lock may be used on the door to the storeroom. When the storeroom is opened at an unscheduled time, it is recorded by this type of locking system; the tape from the time lock should go to the controller each day for review.

Storeroom windows should be opaque to protect foods from direct sunlight. For security, most storerooms are designed without windows or have windows protected with grates or security bars. Adequate lighting is needed in storage areas, not only for sanitation purposes but also to prevent theft. Closed circuit television systems are being used more and more in storage areas.

Good ventilation in the dry stores area is essential to assist in controlling temperature and humidity and preventing musty odors. A thermometer should be mounted in an open area for easy reading. Dry storage temperature should be cool, within a range of 50 to 70°F. Spices, nuts, and raisins, which should be stored in temperatures not over 52°F, often are kept in a refrigerator. Humidity often is overlooked in storerooms. For most food products, a relative humidity of 50 to 60% is considered satisfactory. Humidity above this level may result in rusting cans, caking of dry and dehydrated products, growth of bacteria and mold, and infestation of insects in the storeroom (Longrée & Armbruster, 1996).

Dry storerooms frequently are located in a basement and, therefore, have all sorts of pipes along the ceiling, such as water, heating, and sewer pipes. Leakage from these pipes is a common source of trouble in basement storerooms, especially leakage from sewer pipes that causes contamination of food. Hot water and steam pipes may create high temperatures in the storeroom and, therefore, should be well insulated (Longrée & Armbruster, 1996).

Equipment

Sectional slatted platforms, delivery pallets, and metal platforms with wheels provide a useful type of storage for case lots of canned goods or for products in bags. Their distance from the floor must be in accordance with local health department requirements. Shelving is required whenever less-than-case lots must be stored or when management prefers that canned goods be removed from cases for storage. Adjustable metal shelving is desirable because it allows for various shelf heights and is vermin-proof (Figure 5.20). Shelving must be sturdy enough to support heavy loads without sagging or collapsing, and it should be located at least 2 inches from walls to provide ventilation. If the size and shape of the room permit, shelving should be arranged for accessibility from both sides.

AMCO® has introduced Ultra Density™ Storage shelving, as shown in Figure 5.21, as an alternative to traditional track shelving, which is very expensive to install. Ultra Density™ shelving consists of individual units that fit together in the storage area. These units roll forward or backward. One cart can be rolled to the receiving area and then rolled back to the storeroom. In addition, several people can have access to stored goods instead of just one person in a track shelving area.

Metal or plastic containers with tight-fitting covers should be used for storing cereal products, flour, sugar, dried foods, and broken lots of bulk foods. These containers should either be placed on dollies or have built-in wheels for ease of movement from one place to another (Figure 5.22).

Aisles between shelves and platforms should be wide enough for equipment with wheels. Forklift trucks for moving food and supplies or pallets require much wider aisles than do handcarts.

Products should be arranged in the storeroom according to a plan, and every product should be assigned a definite place. Time can be saved if forms for checking inventory are designed to match the arrangement of products on the shelves.

Low-Temperature Storage

Perishable foods should be held in refrigerated or frozen storage for preservation of quality and nutritive value immediately after delivery and until use. The type and amount of low-temperature storage space required in a foodservice operation

Figure 5.20. Metal shelving for storage area.
Source: InterMetro Industries Corporation. Used by permission.

will vary with the menu and purchasing policies. An excessive amount of refrigeration and frozen storage increases capital costs and operating expenses. Too much also encourages a tendency to allow leftovers to accumulate and spoil. Astute foodservice managers have found that limiting the amount of low-temperature storage discourages excess purchases and overproduction.

Facilities

In very large organizations, especially commissaries, separate refrigerated units are available for meats and poultry, fish and shellfish, dairy products and eggs, and fruits and vegetables. Freezers are available for frozen foods and ice cream; also, tempering boxes are used for thawing frozen products. Ideal storage temperatures vary among food groups, and the more precise the temperature, the better the quality of the products. Table 5.2 provides a detailed list of recommended storage temperatures and maximum storage times according to type of food, most of which has little processing.

Low-temperature storage units can be categorized into the following three types:

- **Refrigerators.** Storage units designed to hold the internal temperature of food products below 41°F

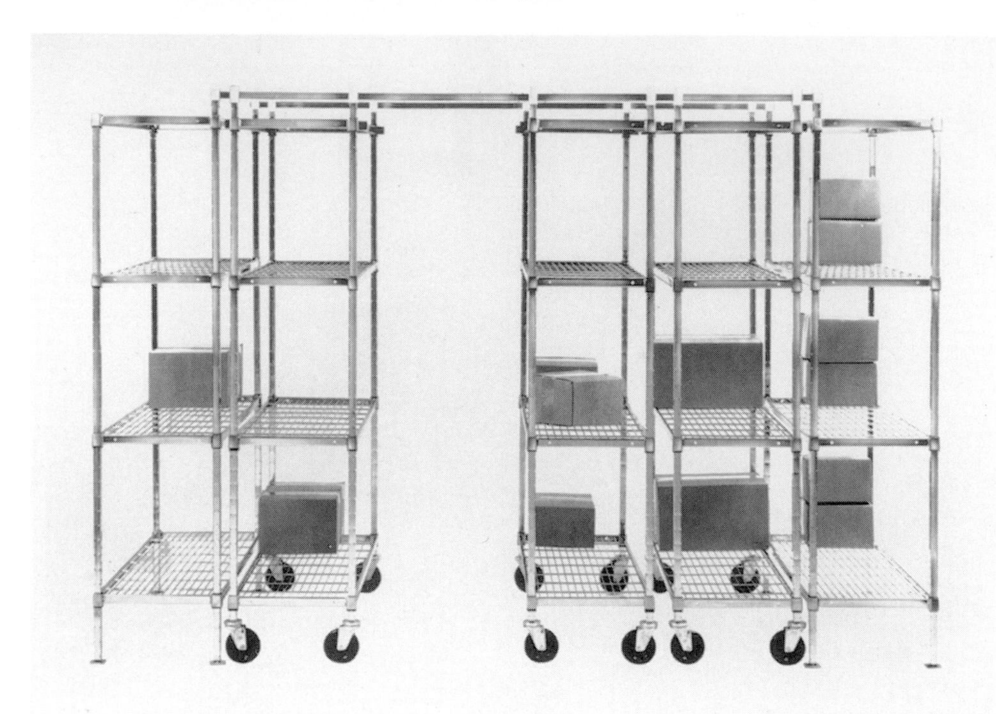

Figure 5.21. High-density storage shelves. *Source:* AMCO®. Used by permission.

Figure 5.22. Dollies and containers for use in storage facilities. *Source:* From *Food Storage Guide for Schools and Institutions* (p. 11) by U.S. Department of Agriculture, Food and Nutrition Service, 1975, Washington, DC: U.S. Government Printing Office.

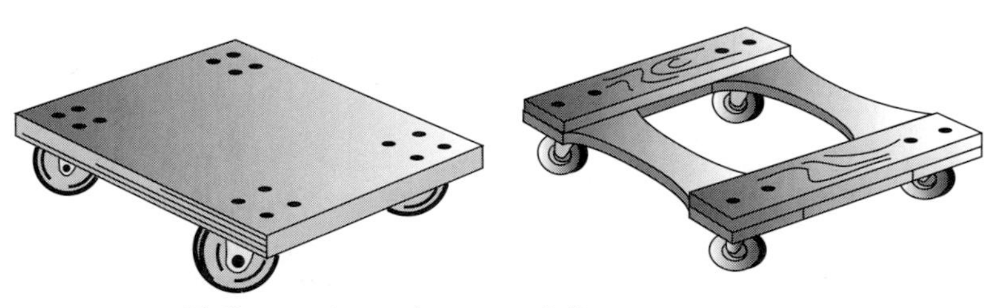

Platform and open-frame can dollies

Container on dolly

Container equipped with casters

Table 5.2 Recommended Storage Temperatures and Times

Food	Refrigerated Storage* < 41°F	Freezer Storage* < 0°F	Dry Storage* < 70°F
Fresh chops, steaks, roasts	3–5 days	4–9 months	Never
Ground meat, stew meat	1–2 days	3–4 months	Never
Cooked meat dishes	3–4 days	2–3 months	Never
Bacon	7 days	1 month	Never
Fresh chicken, turkey	1–2 days	9 months	Never
Cooked poultry dishes	3–4 days	4–6 months	Never
Processed meats (opened)	3–5 days	1–2 months	Never
Cheese (opened)	2–3 weeks	6 months	Never
Milk	7 days	3 months	Never
Eggs, in shell	2–5 weeks	Not recommended	Never
Fruit (berries)	1–2 days (fresh)	8–12 months (frozen) 12 months (canned)	6 months (dried)
Fruit (citrus and apples)	1–3 weeks (fresh)	4–6 months (frozen)	9 months (canned)
Vegetables	3–7 days (fresh)	8–12 months (frozen)	12 months (canned or dried)
Flour			6–8 months
Cereals (unopened)			6–12 months
Mayonnaise (unopened)			2–3 months
Salad oil (unopened)			6 months
Pasta			24 months
Mixes (cake, muffin, etc.)			9 months
Spices/herbs			6 months

*Amount of time from time purchased; use product manufacturer expiration date if given.

Source: Data obtained from Kansas Extension Service (www.oznet.ksu.edu); North Dakota Extension Service (www.ext.nodak.edu); and USDA (www.foodsafety.gov).

- **Tempering boxes.** Separate units for thawing frozen foods, specially designed to maintain a steady temperature of 40°F regardless of room temperature or product load
- **Storage freezers.** Low-temperature units for frozen foods that maintain a constant temperature in the range of −10° to 0°F

Humidity control also is important for maintaining food quality in low-temperature storage. A humidity range between 75 and 95% is recommended for most foods. Perishable foods, however, contain a great deal of moisture; therefore, a relative humidity of 85 to 95% is recommended. If humidity is not sufficient, evaporation will cause deterioration such as wilting, discoloration, and shrinking.

Low-temperature storage units are designed as walk-in or reach-in refrigerators or freezers. In large operations, walk-in units generally are located in the storage area with separate reach-in units in the production and service areas for dairy products, salad ingredients, or desserts. In small operations, the trend is away from walk-ins and toward reach-ins because less floor space is required and the capital investment generally is less.

Hard-surface, easily cleaned floors, walls, and shelves should be made of smooth, nonabsorbent material. The floor level should be the same as that of the area in which the walk-in is located to permit carts to be rolled in. As with dry storage, cartons of food products should not be placed on the floor. Drains are needed for removal of scrubbing water and condensate, and they should be located inside the low-temperature storage units. Uniform ventilation and adequate lighting are essential in the unit to maintain sanitary conditions. All low-temperature storage units should be cleaned on a regularly scheduled basis according to manufacturers' instructions. Most refrigerators and freezers in foodservice operations today are self-defrosting; if not, periodic defrosting of these units must be scheduled.

All refrigerators and freezers should be provided with one or more of the following kinds of thermometers:

- Remote reading thermometer. Placed outside the unit to permit reading the temperature without opening the door
- Recording thermometer. Mounted outside the unit, continuously records temperatures in the unit
- Refrigerator/freezer thermometer. Mounted or hung on a shelf in the warmest area inside the unit.

Temperatures in all units should be checked at least twice a day. An employee should be assigned to check and write down temperatures at specified times as a control measure. Many newer refrigeration units are equipped with temperature recording devices to facilitate the recording refrigerator temperatures. Some foodservice operations have an alarm or buzzer that is activated when temperatures rise above a certain level. Employees should be trained to open refrigerator doors as infrequently as possible; obtaining all foods needed at one time keeps temperatures down while conserving energy. If temporary power failures occur, refrigerators and freezers should be opened as seldom as possible.

In both low-temperature and dry storage, foods that absorb odors must be stored away from those that give off odors. Typical foods that emit and absorb odors are listed in Table 5.3.

Frozen food should be wrapped in moisture-proof or vapor-proof material to prevent freezer burn and loss of moisture. The original packages of most frozen foods

Table 5.3 Foods That Give Off and/or Absorb Odors

Food	Gives Off Odors	Absorbs Odors
Apples, fresh	Yes	Yes
Butter	No	Yes
Cabbage	Yes	No
Cheese	Yes	Yes
Cornmeal	No	Yes
Eggs, dried	No	Yes
Eggs, fresh shell	No	Yes
Flour	No	Yes
Milk, nonfat dry	No	Yes
Onions	Yes	No
Peaches, fresh	Yes	No
Potatoes	Yes	No
Rice	No	Yes

Source: From *Food Storage Guide for Schools and Institutions* (p. 27) by U.S. Department of Agriculture, Food and Nutrition Service, 1975. Washington, DC: U.S. Government Printing Office.

are designed with this protection in mind. For fresh foods that are to be held in storage, specifications may include special instructions for frozen storage wrapping.

Precooked frozen foods are being used more frequently in foodservice operations. Because the shelf life varies widely, the quality and stability of these foods require special attention.

Critical Control in Storage

Storage can be a critical control point for food items. The microbiological safety of raw food products while in storage before production is critical. The HACCP flow chart should identify storage procedures.

INVENTORY

As previously defined, inventory is a record of material assets owned by an organization. Inventory is supported by the actual presence of products in the storage areas. Materials held in storage represent a significant investment of the organization's assets. Although the monetary value of food and supplies in storage will be clear to management, employees may not always grasp this concept. Foodservice employees who would not steal money from the cash register may see nothing wrong in taking products from storage now and then. They may also not see as money down the drain the losses that result from failure to store foods promptly or rotate stock.

Foods, beverages, and supplies in storage areas must be considered as valuable resources of the operation and treated accordingly. For inventory control to be effective, access to storage areas should be restricted, authorized requisitions for removing goods from storage required, and inventory levels monitored. The inventory control system also requires maintenance of accurate records. Improvements in management effectiveness for inventory control require a plan and timely measures of performance. Issuing products, conducting inventories, and controlling methods all need some type of recordkeeping.

Issuing Products

Issuing is the process used to supply food to production units after it has been received. Products may be issued directly from the receiving area, especially if they are planned for that day's menu, but, more often, food and supplies are issued from dry or low-temperature storage. The issuing process entails control of food and supplies removed from storage and provides information for food cost accounting and, in some cases, information for a perpetual inventory system (Keiser & DeMicco, 1993).

Direct Issues

Products sent directly from receiving to production without going through storage are usually referred to as direct purchases or direct issues. To maintain accurate food cost information and better control, direct issues should be limited to food that will be used on the day it is delivered (Keiser & DeMicco, 1993). If the food is not used the same day, the recorded food cost will be unrealistically high on the day of delivery, and the food cost on the day of actual use of the product will be unrealistically low.

Storeroom Issues

Foods that are received but not used the day they are purchased are identified as storeroom issues; these products are issued from a storage area when needed for production or service. Control of issuing from storage has two important aspects. First, goods should not be removed from the storeroom without proper authorization and, second, only the required quantity for production and service should be issued. A requisition is needed to provide these two controls. The cook or ingredient room supervisor requisitions from the storeroom the desired products and quantity and size for each according to a recipe. The ingredient room concept is discussed in detail in Chapter 6. At the time of issue, the storeroom employee completes a daily issue record form, shown in Figure 5.23, which contains the daily issue number, issuing storeroom identification, and date of issue. The columns consist of the requisition number, quantity, unit, and product description. In addition, columns for unit receiving the issue, unit price and total price of each product, and identification of issuing storeroom employee are included on the daily record form.

In foodservice operations that employ an inventory control person, the production or ingredient room requisition would be sent to that person, who would send

Issue Number: 92

Issuing Storeroom: Dry Stores #1 *Date of Issue:* 9/13//02

Req. No.	Quantity	Unit	Description of Item	Issued to	Unit Price	Total Price	Issued by
823	10	#10 cans	Tomatoes, diced	Cook's unit	$ 2.31	$23.10	AV
	1	1½# box	Oregano, dried leaf		10.95	10.95	
	1	3# box	Dried, minced onion		4.93	4.93	
	1	1# box	Dried, diced green pepper		7.24	7.24	
	1	20# box	Spaghetti, long thin		10.75	10.75	
	1	1-gal. bag	Catsup		3.18	3.18	
	8	2-oz. can	Bay leaves		3.59	26.32	
	1	11-oz. can	Thyme, grd.		4.99	4.99	
	1	1 gal.	Worcestershire sauce		4.38	4.38	
	1	26-oz. box	Salt, iodized		.26	.26	
	3	#10 can	Tomato puree		2.07	6.21	

Figure 5.23. Daily issue record.

it to the storeroom employee after checking availability of the product in inventory. If a computer-assisted inventory system is used, an inventory number for each product is required. Costs would not have to be entered, because these data are available from the stored information in the computer. The storeroom requisition may be generated by the computer. The person responsible for ordering food and supplies would only need to review the computerized list of issues and make needed adjustments.

If an ingredient control room has been established, the employees in this unit are responsible for requisitioning supplies from various storage and prepreparation areas, weighing and measuring ingredients for menu items, and providing the needed products to production employees on a scheduled basis. A limited inventory may be maintained in the ingredient room, usually the balance of a product not needed that day, such as the remainder of a bag of flour or sugar.

Inventory Records

Placing products in storage, taking them out when needed, and ordering more when necessary are inadequate for control of valuable resources. Inventory control records must include adequate procedures to provide the foodservice manager with up-to-date and reliable data on costs of operation. Inventory records have four basic objectives:

- provision of accurate information of food and supplies in stock
- determination of purchasing needs
- provision of data for food cost control
- prevention of theft and pilferage

Issuing procedures are only one component of inventory records. A periodic physical count of food and supplies in storage is another requisite element of any inventory control system. More sophisticated records, such as a perpetual inventory, are maintained in many operations to assist in achieving the objectives.

Physical Inventory

Physical inventory: Periodic actual counting and recording of products in stock in all storage areas.

A **physical inventory** is the periodic actual counting and recording of products in stock in all storage areas. Usually, inventories are taken at the end of each month. In large operations, a complete inventory rarely is taken at one time. Instead, inventories often are taken by storage areas, or a section of one, each week with all areas covered by the end of a month.

The process involves two people, one of whom, as a control measure, is not directly involved with storeroom operations. One person counts the products, which should be arranged systematically, and the other person records the data on a physical inventory form, an example of which is shown in Figure 5.24. This form is designed to match the physical arrangement of products on the shelves, thereby greatly facilitating the physical count. Space should be included to record quantity in stock, unit size, name of food item, item description, unit cost, and total value of the inventory on hand.

As discussed in the section on receiving, in some operations, the price is marked on a good before it is stored. The unit cost is recorded at the time of the physical count; otherwise, the pricing of the inventory will be completed by the bookkeeper. If the physical inventory is computerized, employees taking the

Quantity on Hand	Unit Size	Food Item	Item Description	Unit Cost	Total Inventory Value
	#10	Asparagus	All green cuts and tips, 6/#10/case		
	#10	Beans, green	Cut, 6/#10/case		
	#10	Beans, lima	Fresh green, small, 6/#10/case		
	#10	Beets	Whole, 6/#10/case		
	#10	Carrots	Sliced, medium, 6/#10/case		
	#10	Corn	Whole kernel, 6/#10/case		
	#10	Peas	Sweet peas, 4 sv., 6/#10/case		
	#10	Potatoes, sweet	Whole, 6/#10/case		
	#10	Tomatoes	Whole peeled, juice packed, 6/#10/case		
	46 oz.	Tomato juice	Fancy, 12/46 oz./case		

PHYSICAL INVENTORY

Date ______________ Taken by ______________ Beginning Inventory $_________

Figure 5.24. Physical inventory form.

inventory only need to record the amount in stock and enter it into the computer for calculating the beginning inventory value.

In operations in which a monthly inventory is taken, food costs can be determined in one of two ways. The simplest method is to calculate cost of food in the following manner.

Beginning inventory
+ Purchases
= Cost of food available
− Ending inventory
= Cost of food used

For inventories conducted as infrequently as two or three months, daily food cost is determined by computing the cost of the direct and requisition issues, as shown in Figure 5.25. Direct issues for milk, bread, and produce often are

Date ______________

Direct Issues ($)					R equisition Issues ($)			Nonfood ($)	
Dairy		Bakery							
Milk	Other	Bread	Other	Produce	Meat/Fish & Poultry	Frozen	Groceries	Linen	Supplies

TOTAL DIRECT ISSUES ______________

TOTAL REQUISITION ISSUES ______________

TOTAL FOOD TODAY ______________

TOTAL FOOD TO DATE ______________

TOTAL NONFOOD TODAY ______________

TOTAL NONFOOD TO DATE ______________

Figure 5.25. Example of a daily expense worksheet.

delivered by the supplier directly to the using units. Requisition issues are made only on requisitions from using units and are taken from inventory. The monthly food cost is then calculated by adding the daily food costs for the entire month. The lower part of the daily expense worksheet has space for entry of totals of the direct and requisition issues, total food today, and total food to date. Totals also are recorded for nonfood today and to date. All entries on this form are in dollars.

As a check, the cost of food used may be found by adding the cost of purchases to the beginning inventory cost that yields the cost of food available. Subtracting the cost of ending inventory yields the cost of food used. An example of reconciliation of food cost records is shown in Figure 5.26, in which total direct and requisition issues are subtracted from the cost of food used. A zero difference would be quite unlikely. If the difference appears to be significantly large, the foodservice manager will need to analyze the reasons for the discrepancy and implement tighter controls over products in storage.

Perpetual Inventory

Perpetual inventory: Purchases and issues continuously are recorded for each product in storage, making the balance in stock available at all times.

A **perpetual inventory** involves keeping records of purchases and issues for each product in storage so that the balance in stock is known at any point in time as is the value of the inventory at that point in time. An example of a perpetual inventory record is shown in Figure 5.27.

If a perpetual inventory record is used, it is generally restricted to products in dry and frozen storage. Produce, milk, bread, and other fast-moving products usually are not kept on perpetual inventory but are considered to be direct issues. Fresh meats, fish, and poultry delivered on the day of use or 1 or 2 days in advance are not recorded on this record but are charged to the food cost for the day on which they are used. If large quantities of frozen meat are purchased at one time and stored until needed, however, these products will be included on perpetual inventory records. This method of purchasing is a modification of the just-in-time (JIT) manufacturing philosophy introduced into the United States by the Japanese in the late 1970s.

JIT actually is a philosophy and strategy that has effects on inventory control, purchasing, and suppliers (Heinritz, Farrell, Guinipero, & Kolchin, 1991). The ob-

	Beginning inventory		$ 2,000
+	Purchases	+	8,000
=	Cost of food available	=	$10,000
−	Ending inventory	−	1,925
=	Cost of food used	=	$ 8,075
−	Total direct and requisition issues	−	7,850
=	Difference	=	$ 225

Figure 5.26. Reconciliation of food cost records.

Item: Tomatoes, diced *Purchase Unit:* 6/10 cs *Issue Unit:* #10 can

Issuing Storeroom: Dry Stores #1

Date	Order No. or Requisition No.	Quantity In (purchase unit)	Quantity Out (issue unit)	Quantity on Hand (issue unit)
8/23/99	PO 1842	20 cs		120 cans
8/27/99	R 823		10 cans	110 cans

Figure 5.27. Perpetual inventory record.

jective is "to draw material through the production and distribution system on an 'as needed' basis rather than on a forced feed flow driven by an order quantity" (Dion, Banting, Picard, & Blenkhorn, 1992, p. 33). In foodservice operations, the objective of **just-in-time (JIT) purchasing** is to purchase products just in time for production and immediate consumption by customers without having to record them in inventory. Once a product is put into inventory, capital is tied up. If, however, the product is considered part of daily food cost, the money is not sitting on storeroom shelves making no interest. Buyers must have a good relationship with local suppliers to keep transportation costs low and delivery time short. Serving high-quality perishable products at the lowest possible cost is a priority for foodservice operators, and using JIT concepts might help achieve that end.

Perpetual inventories require considerable labor to maintain and are used only in large operations that keep a large quantity of products in stock. The increasing use of computers, however, makes maintaining a perpetual inventory record much easier. After the computerized inventory control system has been established, the perpetual inventory record can be kept up to date very simply by recording issues from the storeroom on a daily basis. However it is done, a perpetual inventory record is not sufficient for accurate accounting and control of food and supplies. A physical inventory should be conducted on a periodic basis for verification.

Inventory Files

Inventory control is a requirement for an efficient operation. Inventory control is the technique of maintaining assets at desired quantity levels. Because public funds are involved, managers of publicly owned onsite foodservice operations such as hospitals or schools are required by the government to have records indicating the amount of food on hand at the beginning of a month, plus the amount ordered, minus the amount served and wasted, to equal the amount left at the end of the period. The inventory record then is checked against actual supplies. A small percentage of errors is unavoidable in the inventory record, but it should remain fairly constant. If a major variation occurs, theft or poor recordkeeping may be a problem. Large commercial operations, especially chains, have similar requirements.

Computerization of the inventory can alleviate many problems by letting the manager know how much money is tied up in storage. An accurate inventory in many operations is difficult, however, because food products probably are in many different areas, including dry, refrigerated, freezer, and point-of-use storerooms. A computer can simplify the task by generating an inventory form by location.

Several software packages are available for inventory management control. Inventory management software maintains inventory records electronically, creates and uses inventory templates, analyzes inventory costs and turnover rate, and totals physical inventory automatically.

Universal Product Code (UPC): System for uniquely identifying products that consists of a rectangular box with vertical lines of various widths.

Bar codes and scanning technology permit managers to conduct a physical inventory and calculate the value of stock more quickly than a manual operation. The adoption of the **Universal Product Code (UPC)** in 1973 transformed a technological curiosity into a standardized system for designating products (Seideman, 1993). The UPC is a system to identify uniquely the thousands of different suppliers and millions of different products that are warehoused, sold, delivered, and billed throughout retail and commercial channels of distribution. It provides an accurate, efficient, and economical means of controlling the flow of products through the use of an all-numeric product identification system. Up until 1973, some companies used letters, some used numbers, some used both, and a few had no codes at all. When the UPC was established, these companies had to give up their individual methods and register with a new Uniform Code Council (UCC).

The UPC is a rectangular box with black vertical lines of various widths identifying the contents of the package, carton, and case (Figure 5.28). There are several versions of the UPC. In Version A, the basic code, the code is split into two halves of six digits each. The first digit (number system digit) identifies the product type. It is zero (for groceries) for most food products except for products like meat and produce that have variable weight. The next five are the manufacturers' or processors' code, followed by a five-digit product code. The final number is a check digit used to verify that the preceding digits have been scanned properly. Manufacturers register with the UCC to get an identifier code for their company and then register each of their products. Thus each package has its own unique identification number. After many years of research, scanning is currently done by a microchip. UPC has been used in retail grocery stores for some time and currently is being introduced into wholesale food operations serving foodservice operations. Manufacturers, processors, and distributors must contend with multiple distribution channels and conflicting industry standards. Adoption of UCC standards by trade associations has helped minimize industry-specific standards. Use of UPC numbers simpli-

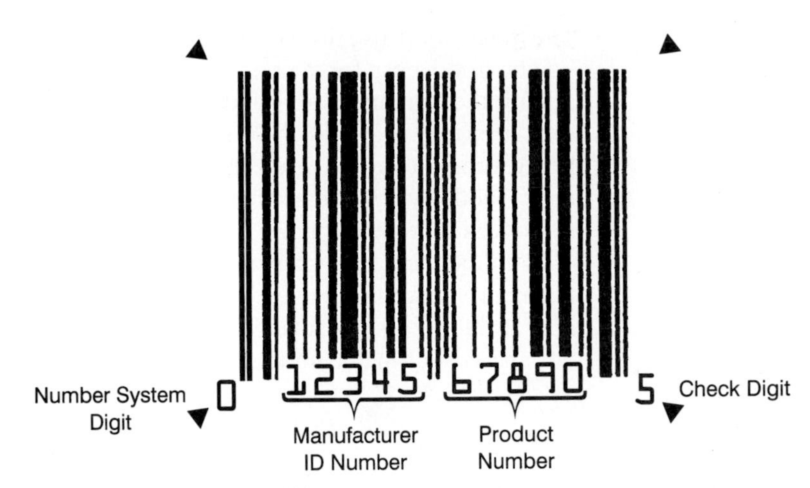

Figure 5.28. Universal Product Code (UPC) and symbol.

fies the ordering procedure by making the buyer's file interface with that of the suppliers. Use of this technology helps buyers achieve more accurate ordering, receiving, and inventory control. In large operations, electronic scanners are used, but in small ones, hand scanners are adequate (Spears, 1999).

Inventory Control Tools

With increased pressures for cost containment in all foodservice operations, the need for inventory control has become more important. The major functions of a control system are to coordinate activities, influence decisions and actions, and assure that objectives are being met. In foodservice operations, the activities to be coordinated and integrated are those in the procurement, production, distribution, and service subsystems. For example, if chili is a menu item and canned tomatoes are not available in the inventory, production will be disrupted and, in turn, service will be affected.

The second major function of inventory control, decision making, can be made somewhat routine through the use of computers. However, the role of the manager as the one responsible for establishing policies and procedures and for monitoring operations to ensure that plans are being implemented appropriately remains paramount.

Using tomatoes as an example, the computer may provide information on when and how many cases to order. The established policies and procedures may detail the processes for ordering, receiving, and storing products. Management, however, still must decide issues such as whether or not to take advantage of forward buying if a tomato crop failure is anticipated. An essential factor in this control system is feedback from the production and service units to procurement, the unit responsible for inventory control.

One of the indicators used to monitor the effectiveness of inventory control is **inventory turnover.** This calculation provides an indicator of the ability of the operation to control the amount of product held in inventory. The formula for inventory turnover is

$$\frac{\text{Food cost}}{\text{Average food inventory}}$$

The calculation usually is done on a monthly basis so the food cost figure would be the food cost for that month. The average food inventory is determined by adding the beginning and ending food inventory figures and dividing by 2. Inventory turnover provides the manager with an estimate of how rapidly product is being brought in and used. A turnover of 2 to 4 times per month is desirable in most foodservice operations. Those using just-in-time purchasing should see a turnover figure even higher.

With the increasing size and complexity of foodservice operations, inventory management and control have become more complicated and critical. A variety of tools is available to assist managers in determining quantities for purchase, inventory levels, and cost of maintaining inventories. Several of these techniques are discussed.

ABC Method

In most foodservice operations, a small number of products account for the major portion of the inventory value; therefore, a method for classifying products according to value is needed. This method is generally referred to as the ABC analysis and is shown in Figure 5.29.

The principle of the **ABC inventory method** is that effort, time, and money for inventory control should be allocated among products according to their value. Products should be divided into three groups, as shown in Figure 5.29, with the high-value, A, and medium-value, B, products given priority.

The high-value A products represent only 15 to 20% of the inventory but typically account for 75 to 80% of the value of total inventory. The inventory level of these expensive products, such as frozen lobster tails, should be maintained at an absolute minimum.

The medium-value B products represent between 10 and 15% of total inventory value and 20 to 25% of the products in inventory. C products are those whose dollar value accounts for 5 to 10% of the inventory value but make up 60 to 65% of

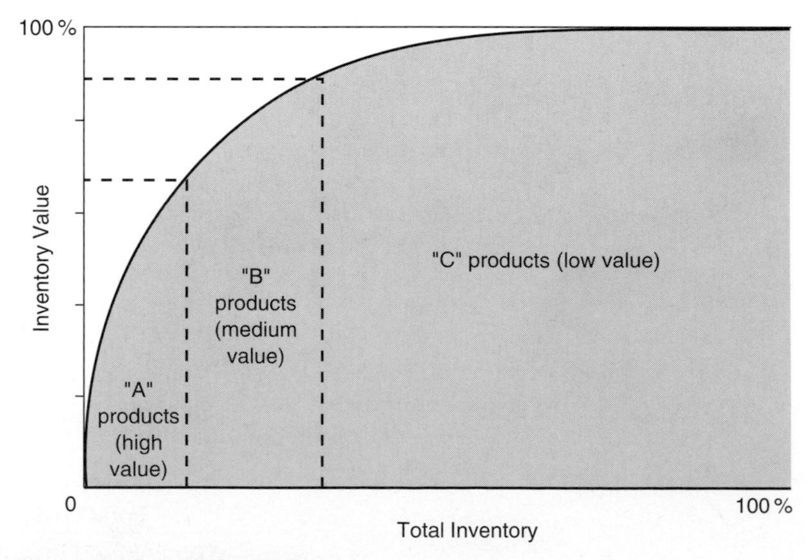

Figure 5.29. ABC analysis.

the inventory. Less concern, obviously, should be directed to the proportionally lower value products (Zenz, 1993).

In applying the ABC concept to a foodservice operation, an analysis of inventory products and classification of them into the three categories would assist the foodservice manager in deciding the amount of time and effort that should be spent in controlling inventory. In a restaurant with both food and beverage operations, liquor would be classified as an A product and controlled very closely; sugar packets would be classified as C products and monitored less closely.

Minimum-Maximum Method

An often-used method for controlling inventory involves the establishment of minimum and maximum inventory levels, commonly called the **mini-max method.** Figure 5.30 is a graphic representation of the mini-max principle.Theoretically, the minimum inventory level could be zero if the last product were used as a new shipment arrived. The maximum inventory then would be the correct ordering quantity.

In reality, however, this extreme policy is not practical because it would likely involve running out of products at a critical time. In the mini-max method, a **safety stock,** which is maintained at a constant level both on the inventory record and in the storerooms, is established. The safety stock is a backup supply to ensure against sudden increases in usage rate, failure to receive ordered products on schedule, receipt of products not meeting specifications, and clerical errors in inventory records. The size of safety stock depends on the importance of the products, value of the investment, and availability of substitutes on short notice. Size is dependent upon lead time and usage rate. Safety stock of a product is part of the quantity on the shelves in the storeroom and must be rotated; it is noted only on inventory records. Stock rotation, in which the oldest products are used first, should be a policy in the storeroom.

The **maximum inventory** level is the highest quantity desired to be in inventory. **Lead time** is the interval between the time that a requisition is initiated and

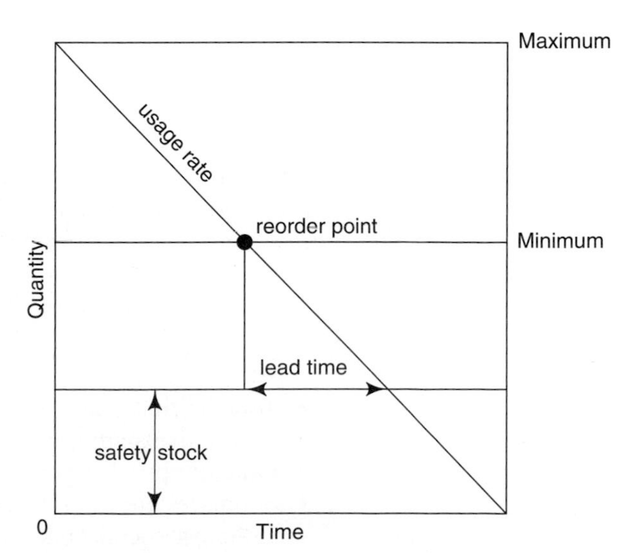

Figure 5.30. Graphic representation of mini-max principle.

receipt of the product. The shortest lead times occur when material is purchased from a local supplier that carries it in stock, and the longest when products are ordered from suppliers who travel some distance to deliver the product and/or do not carry it in stock. The **usage rate** of a product is determined by experience and forecasts. The **reorder point** is established from the lead time and usage rate. Reordering occurs when the quantity in inventory reaches the **minimum inventory** level, the lowest stock level that safely can be maintained to avoid a stock-out or emergency purchasing. The quantity ordered is the difference between the safety stock and the maximum inventory.

Economic Order Quantity Method

The total annual cost of restocking an inventory product depends directly on the number of times it is ordered in a year. To decrease these costs, orders should be placed as seldom as possible by ordering larger quantities. The holding cost of an inventory, however, is directly opposed to the concept of large orders.

The **economic order quantity** (EOQ) concept is derived from a sensible balance of ordering cost and inventory holding cost. Ordering cost diminishes rapidly as the size of the orders is increased, and holding cost of the inventory increases directly with the size of the order. In Figure 5.31, the ordering cost is a curve diminishing in ordinates as the abscissae, the order quantities, increase. The holding cost varies directly with the order quantity and, therefore, shows as a straight line. The objective of EOQ is to determine the relationship between the ordering cost and the holding cost that yields the minimum total cost, which is the point at which both costs are equal and the two lines intersect. This relationship expressed mathematically yields the formula for EOQ:

$$\text{EOQ} = \sqrt{\frac{2(\text{Annual usage in units})(\text{Order cost})}{(\text{Annual carrying cost per unit})}}$$

In this model, the ordering cost includes the total operating expenses of the purchasing and receiving departments, expenses of purchase orders and invoice payment, and data processing costs pertinent to purchasing and inventory. Total annual usage is the number of units to be used annually. Holding cost is the total of all expenses in maintaining an inventory and includes the cost of capital tied up in inventory, obsolescence of products, storage, insurance, handling, taxes, depreciation, deterioration, and breakage. The EOQ may be determined by solving a mathematical formula or by using tables that are the result of formula calculations.

In the development of the basic EOQ formula, the following several assumptions were made (Montag & Hullander, 1971):

- Total annual usage is known and constant.
- Withdrawals are continuous at a constant rate.
- Quantity purchases are available instantly.
- Shortages are not tolerated.
- Unit cost is constant.

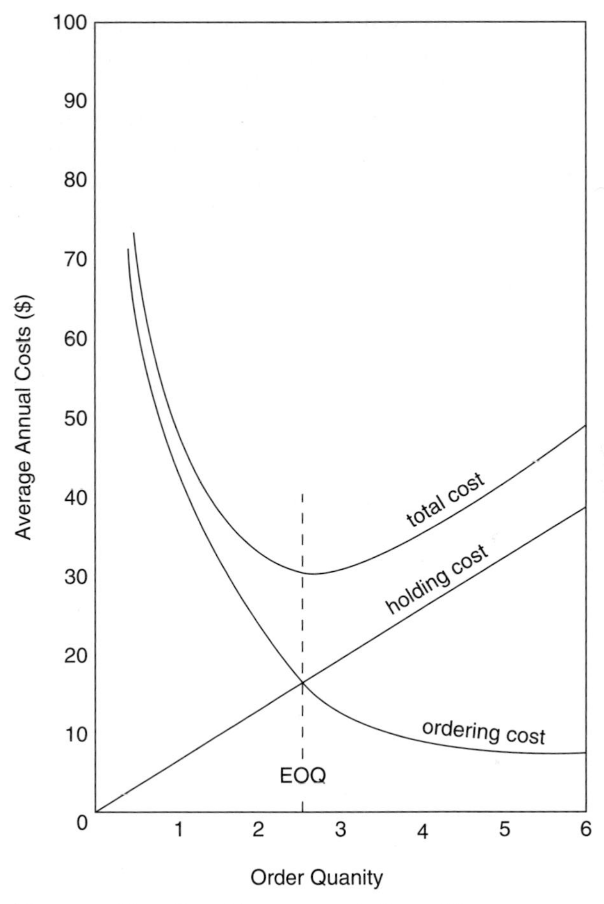

Figure 5.31. Graphic representation of the economic order quantity concept.

- Ordering cost is constant.
- Unit cost of inventory products is constant.

The disadvantages of the EOQ method include its unreliability when the demand rate is highly variable, the requirement of forecasting all products, and the necessity of past demand data. In foodservice operations, use of the EOQ concept is contraindicated when the following conditions exist (Montag & Hullander, 1971):

- Total annual usage is uncertain.
- Withdrawals are discontinuous or variable.
- Unit cost is dependent on price breaks.
- Lead time is not constant.
- Ordering cost is variable.
- Stock holding cost is variable.

The EOQ procedure is impractical unless computer assistance is available. Furthermore, the expenditure of large amounts of time and money required by this procedure is not justified on inventory analysis of products that account for very little of the total inventory costs. The EOQ method is not suitable for many foodservice operations because of a variable demand for certain food products, seasonal menu changes, and indefinite lead times. EOQ is acceptable, however, for large-scale food processing operations such as preparation of commercial frozen meals and wholesale bakeries. The EOQ method should not be considered a panacea for all inventory problems. The quantitative approach to inventory problems must be augmented by the knowledge, skills, and judgment of the foodservice manager.

Inventory Valuation Methods

Inventories, which are subject to change because usage and product replacement are continual, represent a significant portion of current assets owned by the foodservice operation. In every accounting period of generally 1 year, an accounting cycle usually begins and ends each month. Hence, inventories are taken monthly and the dollar value is included as an asset, or a debit, on the monthly balance sheet. A beginning and ending inventory occurs each month; the ending inventory becomes the beginning inventory for the next month.

Rotation of products, using the oldest first, is imperative if food quality is important. Pricing, or valuing, products does not have to follow this pattern in keeping inventory records. The important factor is that the method of valuing be consistent within an accounting period. The five principal methods of inventory valuation are actual purchase price, weighted average, FIFO (first in, first out), LIFO (last in, first out), and latest purchase price (Kim, Finley, Fanslow, & Hsu, 1992).

Actual purchase price involves pricing the inventory at the exact price of each individual product. Marking each product with the purchase price, as it is received, is necessary. This method requires detailed recordkeeping and, therefore, is used only in very small foodservice operations.

The second method is **weighted average,** in which a weighted unit cost is calculated, based on both the unit purchase price and the number of units in each purchase. This weighted unit cost is then used to determine the value of the inventory.

FIFO: First in, first out.

Inventory valuation using the **FIFO** method means that pricing closely follows the physical flow of products through the operation. The oldest products in the storeroom are used before the newest ones. The ending inventory reflects the current cost of products because inventory is valued at the prices for the most recent purchases.

LIFO: Last in, first out.

LIFO is based on the assumption that current purchases are largely, if not completely, made for the purpose of meeting current demands of production. The purchase price of the oldest stock, therefore, should be charged out first. Generally, the value of the inventory will be lowest using LIFO and highest using FIFO based on the assumption that current prices will be higher than older ones. Foodservice managers choose LIFO when determining the value of inventory to reduce profit on financial statements in order to decrease income taxes, particularly in a high-inflation period. LIFO minimizes the value of the closing inventory, which maximizes food cost for the time period.

The **latest purchase price** method uses the latest purchase price in valuing the ending inventory. This method often is used in foodservice operations because it is simple and fast (Kim et al., 1992).

The method chosen for valuing inventory is important because it will affect the determination of price of menu items sold, which in turn will affect the profit or loss figure. The balance sheet or statement of financial condition of a foodservice organization will also be affected because inventories are a current asset. These concepts are discussed in more detail in Chapter 13, "Management of Financial Resources."

ETHICAL CONSIDERATIONS

Ever since the Watergate scandal resulted in the resignation of a U.S. president, the public, businesses, and professions have become sensitive to what is and what is not ethical and to the standards by which people are measured. The news media have kept the word *ethics* in public view by reports of business practices of company executives such as those with Enron and Tyco. Congressional hearings usually include discussion on ethical conduct of appointees to important positions. With all of this emphasis on ethics, politicians, appearing almost daily in newscasts, occasionally are accused of lying and businesspeople of accepting favors under the table.

Ethics: Principles of conduct governing an individual or business.

Business and professional organizations are updating codes of ethics or standards of practice and are emphasizing them to their members. **Ethics** are defined as the principles of conduct governing an individual or a business. **Personal ethics** should be distinguished from business ethics. The source of personal ethics lies in a person's religion or philosophy of life and is derived from definite moral standards. **Business ethics** may be defined as self-generating principles of moral standards to which a substantial majority of business executives gives voluntary assent. A force within business leads to industrywide acceptance of certain standards of practical conduct (Zenz & Thompson, 1993). Ethics are created by society; morals are an individual's personal belief about what is right or wrong (Kapoor, 1991).

Code of Ethics

Code of ethics: Set of rules for standards of professional practice or behavior established by a group.

Organizations, to be professional, must have a **code of ethics** to which their members subscribe. Many business also have a code of ethics. A code is a set of rules for standards of professional practice or behavior established by a group. A code of ethics is influenced by codes of individuals, but the major emphasis is on the relationships within businesses and professional organizations. A standard is the result of the managerial process of planning and is defined as the measurement of what is expected to happen. Many businesses and professional organizations are changing from a code of ethics to standards of practice. The Institute for Supply Management (ISM) uses Principles and Standards of Ethical Supply Management Conduct as a guide for ethical behavior by members (Figure 5.32).

The ISM principles and standards of ethical conduct touch on many areas in which the buyer might err ethically. In addition to conforming to the standards, buyers are subject to the law of agency, in which they are the agent and the principal is the employing agency.

A prime example is the code of ethics for the Foodservice Purchasing Managers of the National Restaurant Association, as shown in Figure 5.33. This code was developed to promote and encourage ethical practices in the industry. To give further

Institute for Supply Management™
Principles and Standards of Ethical Supply Management Conduct

Loyalty to Your Organization
Justice to Those With Whom You Deal
Faith in Your Profession

From these principles are derived the ISM standards of supply management conduct. (Global)

1. Avoid the intent and appearance of unethical or compromising practice in relationships, actions and communications.
2. Demonstrate loyalty to the employer by diligently following the lawful instructions of the employer, using reasonable care and granted authority.
3. Avoid any personal business or professional activity that would create a conflict between personal interests and the interests of the employer.
4. Avoid soliciting or accepting money, loans, credits or preferential discounts and the acceptance of gifts, entertainment, favors or services from present or potential suppliers that might influence, or appear to influence, supply management decisions.
5. Handle confidential or proprietary information with due care and proper consideration of ethical and legal ramifications and governmental regulations.
6. Promote positive supplier relationships through courtesy and impartiality.
7. Avoid improper reciprocal agreements.
8. Know and obey the letter and spirit of laws applicable to supply management.
9. Encourage support for socially diverse practices.
10. Conduct supply management activities in accordance with national and international laws, customs and practices, your organization's policies and these ethical principles and standards of conduct.
11. Develop and maintain professional competence.
12. Enhance the stature of the supply management profession.

Figure 5.32. Principles and standards of purchasing practice.
Source: Institute for Supply Management; available at www.ims.ws.

credence to the code, the purchasing group agreed that each member should indicate support for the principles by signing the code of ethics document.

Ethical Issues

Buyers face many ethically critical situations in the performance of their duties. Reid and Riegel (1989) refer to such situations as ethical dilemmas, and they group them in the following three categories:

- Efforts to gain inside information about competitors that will benefit competition (for example, receiving information about competitors through shared suppliers)
- Activities that allow buyers to gain personal benefits from suppliers (free lunches, dinners, entertainment, trips, and gifts)
- Activities that manipulate suppliers to benefit the purchasing organization (overstating the seriousness of a problem to obtain concessions from the supplier, threatening the use of a second source, using the organization's economic clout, and permitting information on bids from other suppliers)

FOODSERVICE PURCHASING MANAGER'S

CODE OF ETHICS

I, , recognized as a Foodservice Purchasing Professional, hereby subscribe and agree to abide by the following Principles and Standards of Foodservice Purchasing Practice. To wit my signature of agreement appears below.

I. I will consider the interest of my company in all transactions. Acting as its representative, I will carry out and believe in its established policies.

II. I recognize good business practices can be maintained only on the basis of honest and fair relationships.

III. I agree to make commitments for only that which I can reasonably expect to fulfill.

IV. I will provide a prompt and courteous reception to all who request a legitimate business appointment.

V. I will avoid comments which may discredit or otherwise harm legitimate competition. Likewise, information received in confidence will not be used by me to obtain an unfair advantage in competitive transactions.

VI. I will strive to develop specifications and standards which will enable all qualified sources to compete for the business without prejudice.

VII. I will not accept, nor encourage, the giving of gifts or entertainment where the intent is to sway my decision in favor of the donor versus other qualified competitors.

VIII. I abhor all forms of bribery and will act to expose same whenever encountered.

IX. I will always strive to be up to date on products, materials, supplies, and manufacturing processes which will ensure my company receives the proper quality at the most beneficial cost when it is required by the operator.

X. Recognizing the National Restaurant Association and its Foodservice Purchasing Managers are engaged in activities designed to enhance the development and standing of Foodservice Purchasing, I hereby agree to support and participate in its programs.

I, the undersigned, hereby subscribe.

Witness:
Chairman, NRA Foodservice Purchasing Managers

Developed by the Foodservice Purchasing Managers

Figure 5.33. Code of ethics of Foodservice Purchasing Managers Group.
Source: National Restaurant Association. Used by permission.

Ethics Management

Ethical conduct should be an organizational priority. By implementing the management functions of planning, organizing, staffing, leading, and controlling, management can be sure that ethics are established formally and explicitly into daily organizational life. Hogner (1987) stated that when planning is applied to managing ethics, management is translating societal expectations for business performance and behavior into policies and procedures. Organizing requires structuring of the organization including a network to ensure that information relevant to ethical behavior will be transmitted.

Staffing requires that recruitment and selection procedures be consistent with the ethical code of the organization. Finally, controlling links all the functions together. To control ethical conduct, performance standards must be established and continually monitored, and appropriate appreciative or corrective action taken.

Leadership has been cited as the principal mechanism for increasing ethical performance in business. Nielsen (1989) concluded that leadership can be used effectively in organizational ethics issues. Depending upon the circumstances, including personal courage, one can choose to act and be ethical both as a leader and an individual. Being part of or leading ethical change is generally the more constructive approach, although many times the only short-term effective approach is for an individual to intervene against others or the organization. Dire consequences for the individual may result, however, from standing up to and opposing unethical behavior.

MATERIALS MANAGEMENT

The concept of **materials management** has been well expressed by Dillon (1973) as the unifying force that gives interrelated functional subsystems a sense of common direction. Its goal is to transform materials or physical resources that enter the system into an output that meets definite standards for quantity and quality. An organization using the concept has a single manager responsible for the planning, organizing, motivating, and controlling of all activities principally concerned with the flow of materials into an organization (Leenders & Fearon, 1992).

Zenz and Thompson (1993) gave another definition, derived from the basic concept but expressed in operational terms: Materials management is an organizational concept of centralized responsibility for those activities involved in moving materials into, and in some cases through, the organization. While the activities vary, they usually include purchasing, receiving, storage, inventory control, and traffic, with production control and related activities sometimes included, depending on the product or process involved. Of all activities performed in the materials management cycle, supplier relationships are perhaps the most susceptible link (Handfield, 1993). The dependency of a foodservice operation on outside sources for products often leaves it at the mercy of suppliers' performance. Buyers should promote an environment of mutual trust in which suppliers can help improve the current purchasing process.

The trend toward the inclusion of materials management in organizational structure has been motivated primarily by the need to control the overall cost of materials and to minimize those areas of conflict that have traditionally resulted in excessive inventories. Use of the materials management concept, although initiated in industrial manufacturing, has spread to foodservice-related industries, such as major frozen food processors and commercial bakeries. A materials manager has an important role in Stouffer's frozen food division. Moreover, a number of larger hospitals now have a materials manager in a staff position with advisory responsibilities for all activities involving moving materials through the organization.

Fundamentally, a materials manager is a consultant to all departments or units on the movement of materials through the organization. In smaller organizations that cannot justify additional staff, the materials management activity could be performed by top-level management. Other similar staff positions in the organization are held by personnel, facility, and operational (financial, time, communication) resource managers with the same type of responsibilities as the materials manager.

CHAPTER SUMMARY

This summary is organized by the learning objectives.

1. Purchasing is the acquisition of products for an operation. Receiving is the process of ensuring that products delivered are what were ordered. Storage involves holding products at the proper temperature until they are needed for production. Inventory control is the technique of maintaining assets at desired levels.
2. A specification is a written statement detailing the required quality attributes of a product to be purchased. A specification may be written as a technical specification where quality can be measured by testing. Approved brand specifications indicate product quality by specifying a specific brand. A third type of specification is a performance specification, in which quality is measured by the effective functioning of the item. Examples are found in Appendix A.
3. Quality grades pertain to the palatable qualities of a product, such as tenderness and juiciness. Yield grades are related to the predicted yield of useable product from its source.
4. Informal purchasing involves verbal dialogue between the buyer and seller to quote prices and place an order. Formal purchasing involves a bid process to determine price.
5. Controls are needed in purchasing, receiving, and storage to help safeguard the foodservice operation's assets. Controls include having written documentation of orders made to vendors, assigning different personnel to perform purchasing and receiving functions, moving food products quickly from the loading dock to secure storage, locking storage areas, maintaining proper temperature in storage areas, and practicing first in, first out (FIFO) techniques.

TEST YOUR KNOWLEDGE

1. What are the four areas in the procurement subsystem?
2. How are receiving, storage, and inventory linked with purchasing?
3. Describe the five different components of the marketing channel.
4. What are five U.S. departments that help ensure quality food, and how do they ensure quality in the foodservice market?
5. Explain what a buyer has to know and do before purchasing products for a foodservice business.

CLASS PROJECTS

1. Groups of two to three students schedule meetings with purchasing managers of two different foodservice operations. Write a short essay comparing and contrasting how each does purchasing, what procedures they have in place for receiving and storage, and what each has for controls in its procurement process.

Professional Organization Profile

National Restaurant Association

MISSION

To represent, educate, and promote a rapidly growing industry that is comprised of 900,000 restaurant and foodservice outlets employing 12.2 million people

WHO BELONGS TO THE ORGANIZATION

Members represent more than 315,000 restaurants and foodservice establishments and include restaurant owners and managers, foodservice distributors and suppliers, foodservice consultants, faculty members, and students. Student memberships are available.

ADVANTAGES OF MEMBERSHIP

Members have access to an association website with its online network and industry information service, are represented in Washington on legislative issues, receive discounts on training materials, and can attend association-sponsored seminars and exhibits.

WEBSITE

www.restaurant.org

MEET THE PRESIDENT

Craig Miller, president 2005–2006, is president and CEO of Ruth's Chris Steak House. He began his career as a teenage dishwasher at the fine-dining Neptune Seafood Restaurant and has worked in the restaurant industry for more than 40 years, including executive positions with Uno Corporation, Furr's Restaurant Group, and Darden Restaurants.

2. Visit a local foodservice establishment. Ask to visit its storage areas. Observe and document how the following food items are purchased and stored: ground beef, canned vegetables, frozen vegetables, lettuce, bread, milk, cleaning chemicals, paper products, and china.
3. Visit a local foodservice supplier. Tour its storage facilities. Ask if the supplier will do a can cutting for you of some of the products it sells.

WEB SOURCES

www.restaurant.org	National Restaurant Association Foodservice Purchasing Managers
www.ism.ws	Institute for Supply Management
http://www.foodbuy.com/foodbuy/	Foodbuy Purchasing Services
www.rimag.com	*Restaurants & Institutions* magazine
www.foodservicesearch.com/FoodManagement	*Food Management* magazine
www.foodservicesearch.com/RestaurantHospitality	*Restaurant Hospitality* Magazine
www.fsdmag.com	*Foodservice Director* magazine
www.usda.gov	U.S. Department of Agriculture
www.oznet.kus.edu	Kansas Extension Service
www.ext.nodak.edu	North Dakota Extension Service
www.foodsafety.gov	USDA food safety

BIBLIOGRAPHY

Allen, R. L. (1992). New Senate bill backs seafood inspection plan. *Nation's Restaurant News, 26*(17), 3, 59.

Amos, S. M. (1992). Looking beyond the "specs"—growing together through profitable partnerships. *The Consultant, 25*(2), 69.

Anderson, C. (1998, January 1). Meat inspection system questioned. *Post Register,* 1.

Andrews, K. R. (1989). Ethics in practice. *Harvard Business Review, 5,* 99.

Bales, W. A., & Fearon, H. E. (1993). *CEOs'/presidents' perceptions and expectations of the purchasing function.* Tempe, AZ: Center for Advanced Purchasing Studies.

Broihier, C. (1996). Decoding the new menu—labeling regulations. [Online]. Available: http://www.restaurant.org/RUSA/1996/9610p36.htm

Burros, M. (1997). Trying to get labels on genetically altered food. [Online]. Available: http://www.genetic-id.com/nyt.htm

Carter, J. R., & Narasimhan, R. (1993). *Purchasing and materials management's role in total quality management and customer satisfaction.* Tempe, AZ: Center for Advanced Purchasing Studies.

Catalog of food specifications: A technical assistance manual. (1992). Vol. I. (5th ed.). Dunnellon, FL: Food Industry Services Group in cooperation with the U.S. Department of Agriculture, Food & Nutrition Service.

Cheney, K. (1993). Tempest in a test tube. *Restaurants & Institutions, 103*(5), 76–79, 84, 88, 90, 94.

Contract purchasing: A technical assistance manual. (1992). Vol. II. Dunnellon, FL: Food Industry Services Group in cooperation with the U.S. Department of Agriculture, Food & Nutrition Service.

Cook, R. L. (1992). Expert systems in purchasing: Applications and development. *International Journal of Purchasing and Materials Management, 28*(4), 20–27.

Derr, D. D. (1993). Food irradiation: What is it? Where is it going? *Food & Nutrition News, 65*(1), 5–6.

Dietary Managers Association. (1991). *Professional procurement practices: A guide for dietary managers.* Lombard, IL: Author.

Dillon, T. F. (Ed) (1973). Materials Management: A convert tells why. *Purchasing* 74(5): 43–45, 47.

Dion, P. A., Banting, P. M., Picard, S., & Blenkhorn, D. L. (1992). JIT implementation: A growth opportunity for purchasing. *International Journal of Purchasing and Materials Management, 28*(4), 32–38.

EDF Fact Sheet. (1995). Genetically engineered foods: Who's minding the store? [Online]. Available: http://www.edf.org

Ellram, L. M., & Pearson, J. N. (1993). The role of the purchasing function: Toward team participation. *International Journal of Purchasing and Materials Management, 29*(3), 3–9.

Etherton, T. D. (1993). The new bio-tech foods. *Food & Nutrition News, 65*(3), 1–3.

Food and Drug Administration (FDA). (1995). The new food label [Online]. Available: http://vm.cfsan.fda.gov/~lrd/newlabel.html

The Food and Drug Administration: An overview. (1997). [Online]. Available: http://www.fda.gov/opacom/hpview.html

FDA issues food biotech guidelines. (1992). *Food Insight,* 6–7.

Fearon, H. E. (1988). Organizational relationships in purchasing. *Journal of Purchasing and Materials Management, 24*(4), 2–12.

Food facts for food service supervisors: A technical assistance manual, (1992). Vol. III. (2nd ed.). Dunnellon, FL: Food Industry Services Group in cooperation with the U.S. Department of Agriculture, Food & Nutrition Service.

Food Industry Services Group in cooperation with the U.S. Department of Agriculture, Food and Nutrition Service. (1992). *Guidelines for the storage and care of food products: A technical assistance manual,* 3rd ed. Dunnellon, FL: Author.

Food Safety and Inspection Service (FSIS). (1995). Focus on: Egg products. [Online]. Available: http://www.fsis.usda.gov/OA/pubs/eggprod.htm

Food Safety and Inspection Service (FSIS). (1998). Importing meat and poultry to the United States: A guide for importers and brokers. [Online]. Available: http://www.fsis.usda.gov/OA/programs/imports.htm

Food Safety Project: Iowa State University Extension. (1998). Food irradiation. [Online]. Available: http://www.exnet.iastate.edu/pages/families/fs/rad/irhistory.html

Geller, A. N. (1991). Rule out fraud and theft: Controlling your food-service operation. *Cornell Hotel and Restaurant Administration Quarterly, 32*(4), 55–65.

Germain, R., & Droge, C. (1998). The context, organizational design, and performance of JIT buying versus non-JIT buying firms. *International Journal of Purchasing and Materials Management, 34*(2), 12–18.

Goebeler, G. (1989). Distributor viewpoint: Maintaining relationships. *Restaurant Business, 88*(10), 56.

Gregoire, M. B., & Strohbehn, C. H. (2002). Benefits and obstacles to purchasing food from local growers and producers. *Journal of Child Nutrition and Management.* [Online]. Available: http://docs.schoolnutrition.org/newsroom/jcnm/02spring/gregoire

Gunn, M. (1992). Professionalism in purchasing. *School Food Service Journal, 46*(9), 32–34.

Gunn, M. (1995). *First choice, a purchasing systems manual for school foodservice.* University, MS: National Foodservice Management Institute.

Handfield, R. B. (1993). The role of materials management in developing time-based competition. *International Journal of Purchasing and Materials Management, 29*(1), 2–10.

Heinritz, S., Farrell, P. V., Giunipero, L., & Kolchin, M. (1991). *Purchasing: Principles and applications,* 8th ed. Englewood Cliffs, NJ: Prentice Hall.

Hogner, R. H. (1987). Ethics in the hospitality industry: A management control system perspective. *File Hospitality Review* 5(1): 34–41.

IFIC. (1997). Questions and answers about food irradiation. [Online]. Available: http://ificinfo.health.org/qanda/qairradi.htm

Kapoor, T. (1991). A new look at ethics and its relationship to empowerment. *Hospitality & Tourism Educator, 4*(1), 21–24.

Keegan, P. O. (1993). 'Healthy' menus may fall victim to labeling act. *Nation's Restaurant News, 27*(24), 1, 100.

Keiser, J. R., DeMicco, F., & Grimes, R. N. (2000). *Contemporary management theory: Controlling and analyzing costs in foodservice operations,* 4th ed. Upper Saddle River, NJ: Prentice Hall.

Kim, I. Y., Finley, D. H., Fanslow, A. M., & Hsu, C. H. C. (1992). *Inventory control systems in foodservice organizations: Programmed study guide.* Ames: Iowa State University Press.

Kolchin, M. G., & Giunipero, L. (1993). *Purchasing education and training: Requirements and resources.* Tempe, AZ: Center for Advanced Purchasing Studies.

Kotschevar, L. H., & Donnelly, R. (1998). *Quantity food purchasing,* 5th ed. Upper Saddle River, NJ: Merrill.

Kunkel, M. E. (1993). Position of The American Dietetic Association: Biotechnology and the future of food. *Journal of the American Dietetic Association, 93*(2), 189–192.

Lary, B. K. (1991). Employee theft: Tracking and prosecuting perpetrators of on-the-job theft. *Restaurants USA, 11*(7), 18–22.

Leenders, M. C., & Fearon, H. E. (1992). *Purchasing and materials management,* 9th ed. Homewood, IL: Irwin.

Liddle, A. (1988). Distribution: A new era dawns. Part 3. *Nation's Restaurant News, 22*(29), F1, F6, F10.

Longrée, K., & Armbruster, G. (1996). *Quantity food sanitation,* 5th ed. New York: Wiley.

Lorenzini, B. (1992). The secure restaurant. Part II: Internal security. *Restaurants & Institutions, 102*(25), 84–85, 90, 92, 96, 102.

Martin, R. (1993). Meat inspection proposal spurs price hike fears. *Nation's Restaurant News, 27*(13), 9, 103.

Mayo Clinic. (1994). New on the menu soon? [Online]. Available: http://www.mayohealth.org/mayo/9403/htm/menu_tab.htm

McProud, L. M., & David, B. D. (1976). Applying value analysis to food purchasing. *Hospitals, 50*(18), 109–113.

Montag, G. M., & Hullander, E. L. (1971). Quantitative inventory management. *Journal of the American Dietetic Association.* 59(4): 356–361.

Miles, L. D. (1972). *Techniques of value analysis and engineering.* New York: McGraw-Hill.

Muller, E. W. (1992). *Job analysis identifying the tasks of purchasing.* Tempe, AZ: Center for Advanced Purchasing Studies.

North American Meat Processors Association. (2002). *The meat buyers guide.* Reston, VA: Author.

National Center for Nutrition and Dietetics. (Spring 1993). The new food label: Standardizing the terms to dispel confusion. *From the Center,* 3.

National Food Brokers Association. (Undated). *How to select brokers: A guide to interviewing and selecting retail, foodservice, and industrial ingredients brokers.* Washington, DC: Author.

National Food Service Management Institute. (1992). *Impact of food procurement on the implementation of the Dietary Guidelines for Americans in child nutrition programs: Conference proceedings.* University, MS: Author.

National Livestock and Meat Board. (1991). *The Meat Board's lessons on meat.* Chicago: Author.

National Restaurant Association. (April 1989). *A review of U.S. food grading and inspection programs. Current issues report.* Washington, DC: Author.

National Restaurant Association. (2002). Industry at a glance. [Online]. Available: www.restaurant.org

National Restaurant Association, Educational Foundation. (1992). *Applied foodservice sanitation,* 4th ed. Chicago: Author.

National Restaurant Association, Educational Foundation. (1995). *Serving safe food certification coursebook.* Chicago: Author.

Nielsen, R. P. (1989). Changing unethical organizational behavior. *Academy of Management Executive, 3*(2), 123–130.

NOAA Fisheries: Conserving our nation's living ocean. (1998). [Online]. Available: http://kingfish.ssp.nmfs.gov/

Paparella, M. W. (1998). More on food irradiation. [Online]. Available: http://www.shorejournal.com/9803/mwp0301a.html

Patterson, P. (1993). Roundtable discussions offer buyers insights, helping hand. *Nation's Restaurant News, 27*(24), 80.

Position of the American Dietetic Association: Food irradiation. (2000). *Journal of the American Dietetic Association. 100,* 246–253. [Online]. Available: www.eatright.org

Puckett, R. P., & Miller, B. B. (1988). *Food service manual for health care institutions.* Chicago: American Hospital Publishing.

Reck, R. R., & Long, B. G. (1983). Organizing purchasing as a profit center. *Journal of Purchasing and Materials Management* 19(4): 2–6.

Reid, R. D., & Riegel, C. D. (1989). *Purchasing practices of large foodservice firms.* Tempe, AZ: National Association of Purchasing Management.

Restaurants find bottom-line benefits in pilot voluntary seafood inspection program. (1993). *Washington Weekly, 13*(23), 7.

Rubin, K. W. (1993). The pros & cons: Food irradiation. *FoodService Director, 6*(2), 90.

Ruggles, R. (1993). 'Pharmers' to cultivate more food choices. *Nation's Restaurants News, 27*(24), 62.

Savidge, T. F. (1996). Putting your eggs in one basket. *School Foodservice and Nutrition, 50*(4), 26–28.

Schwartz, W. C. (1989). Employee theft: Guess who's leaving with dinner. *Restaurants USA, 9*(8), 24–29.

Seafood Safety Regulations. (1995). [Online]. Available: http://nic2.hawaii.edu/uhlib2/aqua/inspect-reg.html

Seideman, T. (1993). Bar codes sweep the world. *American Heritage Invention and Technology, 8*(4), 56–63.

Spears, M. C. (1999). *Foodservice procurement: Purchasing for profit.* Upper Saddle River, NJ: Prentice Hall.

Stefanelli, J. M. (1997). *Purchasing: Selection and procurement for the hospitality industry,* 4th ed. New York: Wiley.

Thorner, M. E., & Manning, P. B. (1983). *Quality control in foodservice.* Westport, CT: AVI.

University of Technology Perth Western Australia. (1995). JIT and purchasing. [Online]. Available: http://rolf.ece.curtain.edu.au/~clive/jit/jit.htm

U.S. Department of Agriculture (USDA). (1998). [Online]. Available: http://ams.usda.gov/gac/index.htm

U.S. Department of Agriculture (USDA). (1996). Summary of USDA final rule on pathogen reduction and HACCP. [Online]. Available: http://www.usda.gov/news/releases/1996/07/0362a

U.S. Department of Agriculture, Food and Nutrition Service. (1975). *Food storage guide for schools and institutions.* Washington, DC: U.S. Government Printing Office.

USDA Food Safety and Inspection Service. (1998). [Online]. Available: http://www.fsis.usda.gov/

Wagner, M. (1992). Group purchasing survey: Purchase groups buy goods worth over $15 billion. *Modern Healthcare,* 39, 40, 42, 46, 48–50.

Williams, A. J., Lacy, S., & Smith, W. C. (1992). Purchasing's role in value analysis: Lessons from creative problem solving. *International Journal of Purchasing and Materials Management, 28*(2), 37–42.

Zenz, G. J., & Thompson, G. H. (1993). *Purchasing and the management of materials,* 7th ed. New York: Wiley.

CHAPTER 6

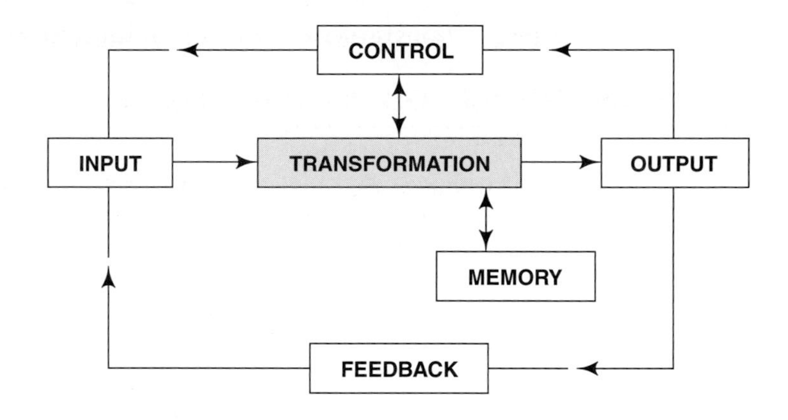

Food Production

Enduring Understanding

- Production of menu items in the needed quantity, with the desired quality, at the appropriate price is a primary goal of foodservice operations.
- Use of standardized recipes will ensure a quality product in the correct quantity.
- Portion control is necessary for control of costs.
- The foodservice industry is energy intensive.

Learning Objectives

After reading and studying this chapter, you should be able to

1. Describe the food production subsystem.
2. Forecast the quantity to prepare using at least two different methods.
3. Prepare a production schedule.
4. Explain the function of an ingredient room.
5. Adjust the quantity of a recipe.
6. Identify pieces of foodservice equipment and describe how they might be used.
7. Differentiate between direct and indirect energy.
8. Define energy terms such as ampere, British thermal unit, horsepower, kilowatt, therm, volt, and watt.
9. Describe procedures for reducing energy consumption.

Food production is a key element in the transformation process in a foodservice operation. Having outputs of quality food and satisfied customers depends on success in the food production process. In this chapter, you will learn about what is involved in the food production process and what it will take for you as a manager to direct successfully the work in this process. You will be introduced to concepts such as forecasting, recipe standardization, and energy control. We also will describe control techniques used in the production functional subsystem.

FUNCTIONAL SUBSYSTEM: FOOD PRODUCTION

Food production: Preparation of menu items in the needed quantity and the desired quality at a cost appropriate to the particular operation.

In the simplest possible terms, the objective of **food production** is the preparation of menu items in the needed quantity and with the desired quality, at a cost appropriate to the particular foodservice operation. *Quantity* is the element that distinguishes production in foodservices from home or family food preparation. *Quality,* an essential concomitant of all food preparation, becomes an extremely vital consideration in mass food production due to the number of employees involved. Quality includes not only the aesthetic aspects of a food product but also nutritional factors and microbiological safety. Cost, of course, determines whether or not a product should be produced for a specific customer.

After procurement, production is the next major subsystem in the transformation element of the foodservice system; it is highlighted in Figure 6.1. Because of the increased use of partially processed foods, such as peeled and sliced apples, any prepreparation will be done in the production unit. **Production** in the generic sense is the process by which products are created. In the context of foodservice, production is the managerial function of converting food purchased in various stages of preparation into menu items that are served to customers.

In foodservice operations today, production is no longer considered merely cooking in the kitchen. It involves planning and controlling ingredients, production methods, food quality, labor productivity, and energy consumption. In essence, foodservice managers responsible for production are resource managers, and they may be designated as such in some organizations.

Planning for production is the establishment of a program of action for transformation of resources into products and services. The manager identifies the necessary resources and determines how the transformation process should be designed to produce the desired products and services. Once this process has been developed, planning must be integrated with the other managerial functions of organizing and controlling.

Planning, organizing, and controlling are overlapping managerial functions, however, and cannot be considered separately. For example, the foodservice manager and the production supervisor might have established a schedule of preparation times to prevent vegetables from being overcooked (controlling), but then suddenly they must revise the production schedule (planning) because an essential employee went home sick and a critical task had to be reassigned. The content of jobs must be analyzed to be sure all tasks are covered (organizing).

PRODUCTION DECISIONS

Decisions are made each day in foodservice operations concerning the necessary quantities to produce and the standards of quality that must be maintained within the limitations of costs. In foodservice operations, as in industry, managers must estimate future events. Thus, forecasting, planning, and production scheduling are important elements for decision making.

All these planning decisions must be made within the constraints of the existing facility. Too often, in a hospital or nursing home, the number of patients or

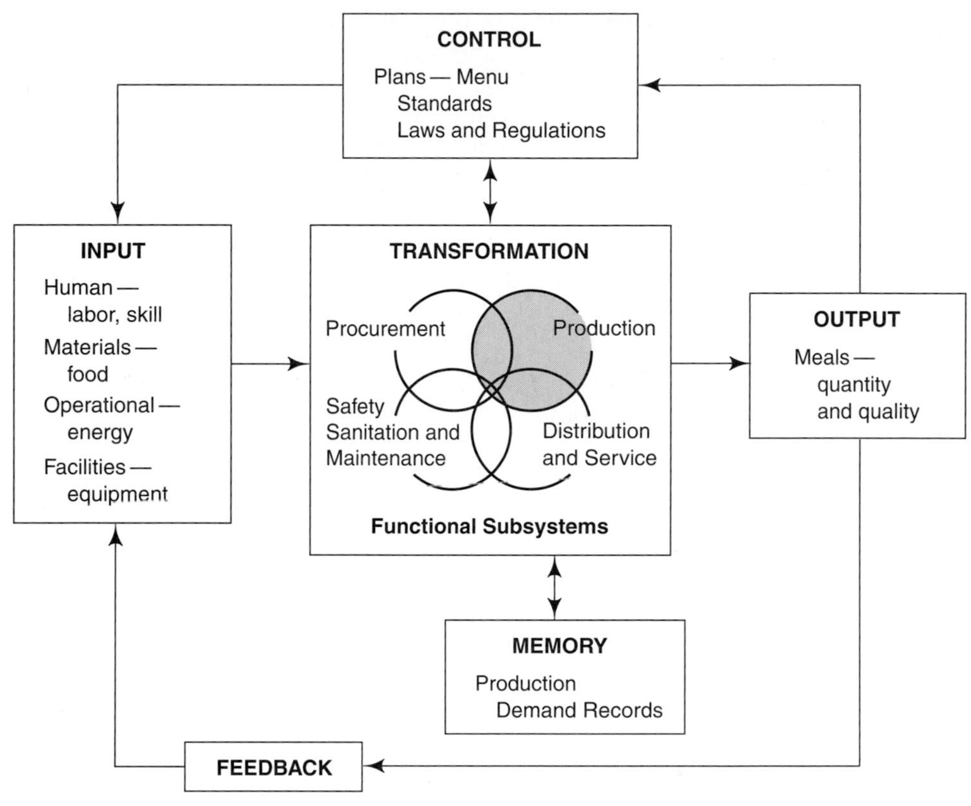

Figure 6.1. Foodservice systems model with the production function highlighted.

residents is increased, but the capacity of the equipment in the kitchen is not. If the anticipated demand exceeds the present capacity, then the facility must be expanded, future production curtailed, or more ready prepared foods purchased to handle the increased demand.

Production planning primarily is the effective synthesis of quantity, quality, and cost objectives. The objective of the production subsystem is to transform human, material, facility, and operational resources into outputs. The secondary objectives focus on the following (Adam & Ebert, 1992):

- Product/service characteristics
- Process characteristics
- Product/service quality
- Efficiency:
  - Effective employee relations and cost control of labor
  - Cost control of material
  - Cost control in facility use
- Customer service:
  - Producing quantities to meet expected demand
  - Meeting the delivery date for products or services

The *characteristics of the product* or menu items depend upon the type of operation, such as short orders in a limited-menu restaurant or hotel coffee shop, individual item selections in a full-service restaurant, or a fixed menu in a school. For example, a ground meat patty might be served as a grilled hamburger on a bun for a limited-menu restaurant, a charbroiled ground steak for a full-service restaurant, or an oven-baked hamburger for a nursing home.

Production process characteristics include the method of food preparation, ranging from grilling to broiling to baking. The process and product characteristics are closely related because both determine the quality of the product and service. *Efficiency of the process* depends upon the control of costs for labor, material, and facility use. All of these secondary objectives lead to customer service.

Obviously, planning cannot be effective without reasonably accurate forecasting of future demand quantities. Finally, standards of quality must be established for all products. Maintenance of quality is a cost factor because of employee training, inventory control of both raw and prepared food items, and sanitation programs.

The cost element in planning is the result of the correlation of food, storage, issue, and production costs with labor, facility, and energy costs. These elements must be considered in all planning. Whenever planning goes beyond a day-to-day basis, forecasting becomes absolutely necessary.

PRODUCTION FORECASTING

Forecasting: Art and science of estimating events in the future, which provides a database for decision making and planning.

Forecasting is the art and science of estimating events in the future and provides the database for decision making and planning. The art of forecasting is the intuition of the forecaster, and the science is the use of past data in a tested model. Both are required to estimate future needs. Forecasting is described as a function of production and constitutes the basis for procurement.

Production Demand

Forecasting is a function of production but also is needed for procurement. Food products must be available for producing menu items for customers. The primary result of forecasting should be customer satisfaction; customers expect to receive what they ordered. In addition, the foodservice manager is concerned with food cost; both overproduction and underproduction affect the bottom line.

Overproduction: Production of more food than is needed for service.

Overproduction, the production of more food than is needed for service, generates extra costs because the salvage of excess food items is not always feasible. Leftover prepared food spoils easily and requires extreme care in handling and storage. Even though some leftover foods might be salvageable by refrigeration, certain foods may break down and lose quality. An example is a custard or cream pie that must be held under refrigeration for microbiological safety but develops a soggy crust quickly and cannot be served. Policies and procedures for the storage of overproduced food items should be well defined and rigorously enforced.

Attempts to reduce overproduction costs by using a leftover high-priced food as an ingredient in a low-cost menu item reduce profits. For example, using leftover rib roast in beef stew, soup stock, or beef hash, all of which could be prepared with less expensive fresh meat, is difficult to justify. In addition to the higher food cost, planning and carrying out these salvage efforts incurs higher labor costs that could have been avoided had overproduction not occurred. Customers often suspect that leftovers are being used, which can be damaging to the image of a foodservice operation.

Underproduction: Production of less food than is needed for service.

Underproduction, the production of less food than is needed for service, can increase costs as much as overproduction. Customers will be disappointed if the menu item is unavailable, and they often have difficulty in making another selection. Furthermore, underproduction may involve both additional labor costs and often the substitution of a higher-priced item.

A wise manager will insist that a similar backup item be available when underproduction occurs. For example, in a university residence hall foodservice, if the grilled meat patties run out, an excellent replacement would be frozen minute steaks, quickly grilled. Such a substitution certainly would increase customer satisfaction even though it hurts the bottom line.

Quantity Demand

The desire for an efficient foodservice operation requires that the production manager know the estimated number of customers or the number of servings of each menu item in time to order from the procurement unit. Good forecasts are essential for managers in planning smooth transitions from current to future output, regardless of the size or type of the foodservice (i.e., schools, hospitals, or restaurants). Forecasts vary in sophistication from those based on historical records and intuition to complex models requiring large amounts of data and computer time. Choosing a forecasting model that is suitable for a particular situation is essential.

Historical Records

Adequate historical records constitute the basis for most forecasting processes. Often, past customer counts, number of menu items prepared, or sales records are used to determine the number of each menu item to prepare. These records must be accurate and complete, or they cannot be extended into the future with any reliability.

Effective production records should include

- Date and day of the week
- Meal or hour of service
- Notation of special event, holiday, and weather conditions, if applicable
- Food items prepared
- Quantity of each item prepared
- Quantity of each item served

An example of a historical record for a catered party at a wedding reception is shown in Figure 6.2. This seasonal menu also could be used for other types of receptions.

Caterers, both social and employed by an organization, must keep accurate records of the amount of food for each event to prevent underproduction or overproduction of menu items on the next similar occasion. Catering is a profit enterprise, and reliable past records are essential because events are not repetitive; elaborate forecasting methods generally are not feasible.

Date: June 7, 2002 Time: 4:30 - 6:30 pm Event: Retirement reception for director of pharmacy
Number to be served: 300

Menu Item	Amount Prepared	Amount Served	Comments
Chilled spiced shrimp	20#	20#	All consumed by 5:30
Cocktail sauce	3 qt	2.5 qt	
Individual wild mushroom strudel	300	200	
Sherry sauce	2 qt	1.5 qt	
Carved beef tenderloin	40#	40#	All consumed but lasted for entire event
Cocktail buns	450	425	
Horseradish sauce	2 qt	2 qt	
Sundried tomato mayonnaise	2 qt	1 qt	
Smoked gouda and roasted vegetable quesadilla	250	250	All consumed by 6:00
Vegetable tray	3 trays	2.5 trays	
Chocolate and raspberry petit fours	300	250	
Associated tea cookies	50 doz	42 doz	

Comments: Weather was sunny, 80°+; actual attendance: approximately 275

Figure 6.2. Example of historical production record.

Although production unit records reveal the vital information on menu items served to customers, production is by no means the only organizational unit that should keep records. Only by cross-referencing records of sales with those of production can a reliable historical basis for forecasting be formalized. Records of sales will yield customer count patterns that can be useful for forecasting. These data can be related to the number of times customers select a given menu item or the daily variations induced by weather or special events.

Historical records in the production unit provide the fundamental base for forecasting quantities when the same meal or menu item is repeated. These records should be correlated with those kept by the purchasing department, which include the name and performance of the supplier and price of the food items.

Forecasting Models

Selecting a forecasting model for a foodservice operation can be a difficult task for a manager. The easiest method is to guess how many people are expected or how much of each menu item is needed. Amazingly, many chefs and cooks can guess quite accurately, especially if the customer count is approximately the same at each meal or the menu is static. But when customers can choose where they want to eat, guessing does not always work. A more scientific method for forecasting is needed. Forecasting models are available and should be researched before deciding which one to use.

Criteria for a Model

Numerous forecasting models have been developed during the past three decades, but, as one might expect, the trend has been toward sophisticated models using computer-based information systems. According to Fitzsimmons and Sullivan (1982), the factors deserving consideration when selecting a forecasting model are

- Cost
- Required accuracy
- Relevancy of past data
- Forecasting lead time
- Underlying pattern of behavior

Cost of Model. The cost of a forecasting model involves the expenses of both development and operation. The developmental costs arise from constructing the model, validating the forecast stability, and, in the case of large operations, writing a computer program. In some cases, educating managers in the use of the model is another cost element. Operational costs, including the cost of making a forecast after the model is developed, are affected by the amount of data and computation time needed. More elaborate models require large amounts of data and thus can be very expensive.

Accuracy of Model. The quality of a forecasting model must be judged primarily by the accuracy of its predictions of future occurrences. An expensive model that yields accurate forecasts might not be as good a choice as a cheaper and less accurate model. This is a decision the foodservice manager must make.

Relevancy of Past Data. In most forecasting models, the general assumption is that past behavioral patterns and relationships will continue in the future. If a clear relationship between the past and the future does not exist, the past data will not be relevant in developing forecasts. In these cases, subjective approaches, such as those that rely heavily on the opinions of knowledgeable persons, may be more appropriate.

Lead Time. Lead time pertains to the length of time into the future that the forecasts are made. Usually, these times are categorized as short-, medium-, or long-term. The choice of a lead time depends on the items being forecast: A short-term lead will be chosen for perishable produce, and a medium- or long-term lead is suitable for canned goods.

Pattern of Behavior. As stated previously, many forecasting models depend on the assumption that behavioral patterns observed in the past will continue into the future and, even more basic, that actual occurrences follow some known pattern. These patterns, however, may be affected by random influences, which are unpredictable factors responsible for forecasting errors. Not all forecasting models work equally well for all patterns of data; therefore, the appropriate model must be selected for a particular situation.

Types of Models

Forecasting models have been classified in numerous ways, but the three most common model categories are

- Time series
- Causal
- Subjective

A model in one classification may include some features of the others. In all methods of forecasting, trends and seasonality in the data must be considered.

Time Series Model. The frequently used time series forecasting model involves the assumption that actual occurrences follow an identifiable pattern over time. Although time series data have a specific relationship to time, deviations in the data make forecasting difficult. To reduce the influence of these deviations, several methods have been developed for smoothing the data curve.

The time series models are the most suitable for short-term forecasts in foodservice operations. Actual data may indicate a trend in a general sense but not give forecast information. To make the past data useful, variations must be reduced to a trend line that can be extended into the future. Two time series models, moving average and exponential smoothing, are used more frequently in forecasting for foodservice than any other type, although causal models may be used as well.

The most common and easiest of the time series models is the **moving average forecasting model.** The process begins by taking the average of the number of portions sold for the last five or more times the menu item was offered as the first point on the trend line. The second point on the line is determined by dropping the first number and adding the most recent number of portions sold to the bottom of the list and then calculating another average. The repetitive process continues for all data.

An example of the moving average model is shown in Table 6.1. Data are for the number of hamburgers sold over the last 10 days. A 5-day moving average is used. The first 5-day moving average is calculated by adding the number sold for each of those days and dividing by 5, giving an average of 176 hamburgers. The next moving average is calculated by adding the number sold for days 2 through 6

Table 6.1 Example of a Moving Average

Day	Number of Hamburgers Sold	5-Day Moving Average
1	150	
2	180	
3	185	176
4	170	186
5	195	187
6	200	182
7	185	183
8	160	180
9	175	
10	180	

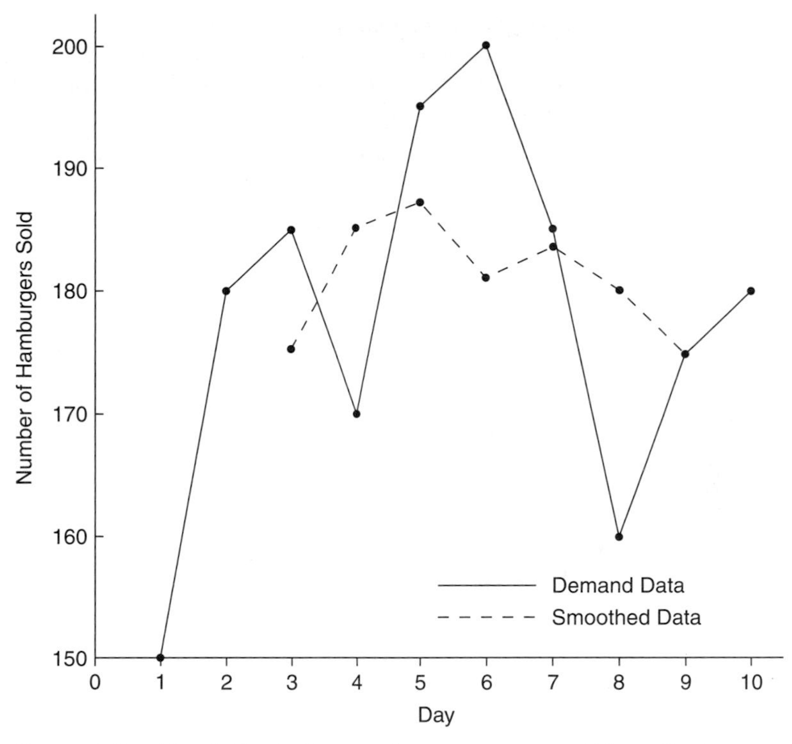

Figure 6.3. Graph illustrating moving average smoothing effect.

and dividing by 5. The procedure is repeated by dropping the earliest day's data and adding the most recent day's data for a total of 5 days.

Demand data and moving average values plotted on the graph (Figure 6.3) illustrate the smoothing effect of the model. Note that the smoothed data curve eliminates the daily variations in demand and thus indicates a trend of the past. Smooth data is an average of what happened in the past, but the person in charge of determining number of servings needed must take into consideration special events, holidays, weather, and other events that may occur in forecasting the actual amount needed for a particular day.

The **exponential smoothing forecasting model** is a popular time series model that can be set up on a computer spreadsheet. It is very similar to the moving average model except that it does not uniformly weigh past observations. Instead, an exponentially decreasing set of weights is used, giving recent values more weight than older ones. Also, the only data required are the weights, the alpha judgment factors, that are applied to the most recent values including the last customer demand and the last forecast, thus eliminating the need to store historical data (Makridakis & Wheelwright, 1989). Alpha (α) is the judgment factor, or smoothing coefficient, and indicates how well the manager believes the most recent data represent current customer count or number of sales.

The judgment factor, α, is a number between 0 and 1 and is used to adjust for any errors in previous forecasts. α is the weight assigned to the most recent customer demand and $1 - \alpha$ is the weight for the most recent forecast. When α has a value close to 1, the new forecast will include a substantial adjustment for any error that occurred in the preceding forecast. Conversely, when α is close to 0, the new forecast will not show much adjustment for the error from the one before. The value of α has been tested in foodservice, and if no major changes occur in the data, customer demand for succeeding weeks is not expected to differ greatly from the past (Messersmith & Miller, 1992). An α of 0.3 is commonly used for demand, leaving 0.7 for the forecast. The most recent forecast values are multiplied by the $1 - \alpha$ quantity, which places a greater weight on recent values. The $1 - \alpha$ quantity acquires exponents in increasing order as the forecast is repeated, thus decreasing weights of older values and having a lesser influence on the trend curve than more recent data. The mathematical expression for this smoothing model is

$$F_t = [\alpha D_{t-2}] + [(1 - \alpha)F_{t-1}]$$

where

α = a constant, usually between 0.1 and 0.3 (judgment factor)

F_t = smoothed value at time t (new forecast)

D_{t-1} = actual observed value at time $t - 1$ (last demand)

F_{t-1} = preceding smoothed value (last forecast)

Stated in words, this forecast equation is

$$\text{New forecast} = \begin{bmatrix} \text{judgment} \\ \text{factor} \end{bmatrix} \times \begin{bmatrix} \text{last} \\ \text{demand} \end{bmatrix} \times \begin{bmatrix} 1 - \text{judgment} \\ \text{factor} \end{bmatrix} \times \begin{bmatrix} \text{last} \\ \text{forecast} \end{bmatrix}$$

Applying the exponential smoothing model to data from Table 6.1 to forecast amount to prepare for day 7 would look as follows:

$$\left(\begin{matrix} \text{judgment} \\ \text{factor} \end{matrix} \times \begin{matrix} \text{last} \\ \text{demand} \end{matrix}\right) + \left(\begin{matrix} 1 - \text{judgment} \\ \text{factor} \end{matrix} \times \begin{matrix} \text{last} \\ \text{forecast} \end{matrix}\right) = \begin{matrix} \text{forecast for} \\ \text{day 7} \end{matrix}$$

$$(.3 \times 200) + (.7 \times 187) = 191$$

Fitzsimmons and Sullivan (1982) summarized exponential smoothing as a popular technique for short-term demand forecasting for the following reasons:

- All past data are considered in the smoothing process.
- More recent data are given more weight than older data.
- The technique requires only a few pieces of data to update a forecast.
- The model can be easily programmed and is inexpensive to use.
- The rate at which the model responds to changes in the underlying pattern of data can be adjusted mathematically.

Causal Model. Causal forecasting models, like time series models, are based on the assumption that an identifiable relationship exists between the item being forecast and other factors. These factors might include selling price, number of customers, market availability, and almost anything else that might influence the item being forecast. Causal models vary in complexity from those relating only one factor, such as selling price, to items being forecast to models using a system of mathematical equations that include numerous variables.

The cost of developing and using causal models is generally high, and consequently they are not used frequently for short-term forecasting, such as for perishable produce. They are, however, popular for medium- and long-term forecasts, such as for canned goods.

The most commonly adopted causal models are called **regression analysis forecasting models.** Following standard statistical terminology, the items being forecast are called dependent variables, and the factors determining the value of the dependent variables are called the independent variables. Regression models require a history of data for the dependent and independent variables to permit plotting over time. Once this is done, the regression process involves finding an equation for a line that minimizes the deviations of the dependent variable from it. Two principal kinds of regression models are linear and multiple.

In linear regression the word **linear** signifies the intent of the analysis to find an equation for a straight line that fits the data points most closely. In conventional statistical terminology, the item being forccast is called the **dependent variable** (Y), and the factors that affect it are called **independent variables** (X).

In the analysis, historic demand data for a single variable will result in a derived equation from a linear regression process in the form of a straight line:

$$Y = a_0 = a_1X$$

in which a_0 and a_1 are numerical constants determined by the regression analysis. As shown in Figure 6.4, a_0 is the intercept of the line on the Y axis and a_1 is the slope of the line. In use, X will be a single independent variable quantity. Data points in the figure are the Y (dependent variable) values for specific values of X (independent variable). Preliminary plotting of the variables on graph paper would be advisable to ascertain if they could be represented reasonably by a straight line. The forecasting value rests on the assumption that the linear relationship between the variables will continue for a reasonable time in the future, or quite simply that the line may be extended. Use of the equation requires only substitution of an anticipated future value for X and then solution for Y, which is the forecasting

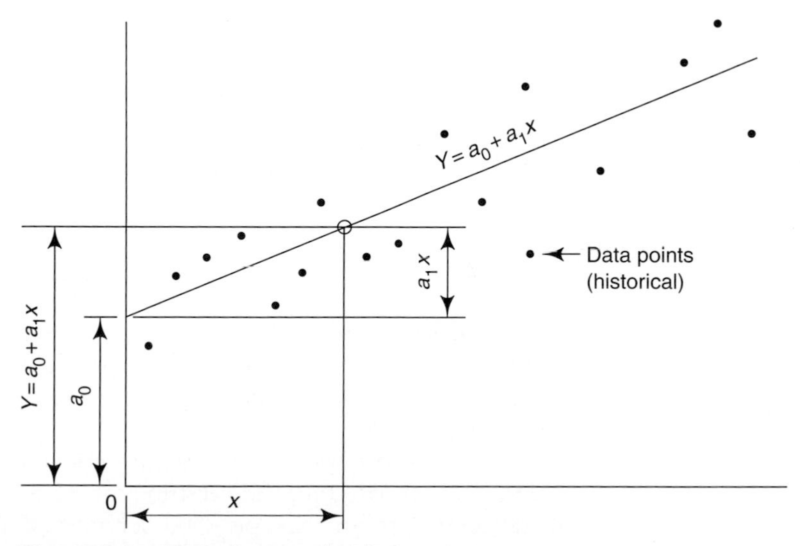

Figure 6.4. Typical regression line.

quantity. Examples of independent variables in hospital foodservice are total number of patient trays served, patient census, cafeteria customer capacity, number of employees, number of patients on regular diets, and number of patients on each modified diet. For example, roast beef might be a popular item for a foodservice, and the relationship between the historic number of patient trays and the pounds of roast beef could yield a regression equation. To forecast beef demand, an anticipated future count of trays would then be inserted into the equation as X to solve for Y, the pounds of roast beef needed.

If determination of the effects of more than one independent variable ($X_1, X_2, \ldots X_n$) on the dependent one is desired, the process is called multiple regression and the derived equation will have the following form:

$$Y = a_0 + a_1X\{1\} + a_2X\{2\} + \ldots + a_nX_n$$

Multiple regression analysis is quite complex, and a good computer program is needed for the solution.

Subjective Model. A subjective forecasting model generally is used when relevant data are scarce or patterns and relationships between data do not tend to persist over time. In these cases, little relationship exists between the past and the long-term future. Forecasters must rely on opinions and other information, generally qualitative, that might relate to the item being forecast.

One of the subjective forecasting models is the **Delphi technique,** which involves a panel of experts who individually complete questionnaires on a chosen topic and return them to the investigator. Results of the first questionnaire are summarized and returned to the panel for revision. Questionnaires are revised successively until a simple majority agreement is reached. The Delphi method can be time-consuming and expensive and is not especially suitable for foodservice forecasting.

Other qualitative forecasting techniques include market research, panel consensus, visionary forecast, and historical analogy. Market research is a systematic and formal procedure for developing and testing hypotheses about actual markets. Panel consensus is based on the assumption that a group of experts can produce a better forecast than one person. This differs from the Delphi method by requiring free communication among the panel members. A visionary forecast is characterized by subjective guesswork and imagination. The historical analogy involves comparative analysis of the introduction and growth of new items with similar new-product history.

PRODUCTION SCHEDULING

Production scheduling: Time sequencing of events required by the production subsystem to produce a meal.

Production scheduling in foodservice operations can be defined as the time sequencing of events required by the production subsystem to produce a meal. Scheduling occurs in two distinct stages—planning and action—and is essential for production control.

In the *planning stage,* forecasts are converted into the quantity of each menu item to be prepared and the distribution of food products to supervisors in each work center. As an example, 500 servings of grilled marinated chicken breasts, pasta with pesto, asparagus spears in a red pepper ring, green salad with artichoke hearts and diced tomatoes, sour dough rolls, and frozen yogurt with fresh strawberries have been forecast for a special dinner. The foodservice director assigns the production of the 500 servings of chicken, pasta, and asparagus to the supervisor of main production, the green salad to the salad unit, and the rolls and dessert to the waitstaff.

Supervisors in each unit assume the responsibility for the *action stage* by preparing a production schedule. Each item is assigned to a specific employee and the time to start its preparation is recorded on the schedule. Careful scheduling assures that the food is prepared for service without lengthy holding and deterioration in quality. Supervisors give feedback to the manager by writing comments on the production schedule.

In small operations in which only one cook and perhaps an assistant are on duty at a time, the foodservice manager might also need to assume the responsibility for the action stage. Every foodservice operation, however small, must have a daily production schedule to control both labor and food costs.

Production Schedule

Production schedule: Written plan for production for a specific date and/or meal.

The **production schedule**, frequently called the production worksheet, is the major control in the production subsystem because it activates the menu and provides a test of forecasting accuracy. The menu should be based on standardized recipes.

The production schedule is highly individualized in various foodservices and may vary from a one-page form for manual completion to a computer program printout. Regardless of the form, certain basic information must be included on each schedule. The unit, production date, and meal should be identified along with other pertinent information such as actual customer count, weather, and special events. In addition, the following information must be included to make it a specific action plan:

- Employee assignments
- Preparation time schedule
- Menu item
- Over- and underproduction
- Quantity to prepare
- Substitutions
- Actual yield
- Additional assignments
- Special instructions and comments
- Prepreparation

The sample production schedule in Figure 6.5 is from a large university residence hall foodservice. Note that the general information previously mentioned is displayed at the top of the worksheet and that the production area is marked as the specific destination for this schedule. The meal count at the conclusion of production also is recorded, which validates this schedule as part of historical data. Specific headings previously listed constitute the column headings.

The production schedule generally is posted on a bulletin board in the unit. The name of the employee in the left-hand column readily enables personnel to find designated duties. The menu item column identifies the recipe by name. Often the unit supervisor will distribute the recipes, either in card form or as a computer printout, to the appropriate employees at the time the schedule is posted.

Unit Main Production

Date 1/22/99 | Meal Count 2153

Weather Fair

Meal ____ Bkf. ____ Lunch X Dinner | Comments Basketball Game

Employee	Menu Item	Quantity to Prepare	Actual Yield	Instructions	Time Schedule	Left over Amount	Run-out Time	Substitution	Cleaning Assignment
Wege Whatley	Country Fried Steak	1200	1220	Use 2 tilting fry pans and oven number 3	Begin frying 2:15 See Frying Time schedule	35 servings	—	—	Whatley—tilting fry pans
Lundin	Giant Rolled Tostados	1000	1020	Serve open face on cafeteria line and customers will roll their own		50 tortillas 10 lbs meat mixture 1 gal cheese sce.	— —	— —	Lundin—slicer and attachments
McCurdy	Whipped Potatoes	1200	1150	If necessary use instant as a back up	Begin steaming potatoes at 3:00	12 lbs	—	—	
McCurdy	Cream Gravy	2000	1600	Make 4 batches 600–600–400–400 (if needed) Serve over both steak and potatoes		2 gal	—	—	
Wege	Mexican Rice	900	900	Use 12 × 10 × 4 in. pans	See Baking Time schedule	18 lbs	—	—	
Mockery	Refried Beans	850	850	Add bean liquid as needed to maintain a moist product		12 lbs	—	—	Mockery—oven number 1, shelves and doors
Mockery	Broccoli Spears	1000	850	Season with melted margarine	Begin 4:00	0	6:00	2½ lbs Asparagus spears	
Mockery	Yogurt Cup	20	20	Serve whole container—blueberry, cherry, rum raisin, plain	Prepare based on demand	8	—	—	

Prepreparation

Employee	Menu item	Quantity	Instructions	Employee	Menu item	Quantity	Instructions
McCurdy	Roast Beef	600 lbs	Pan beef in baking pans, cover and refrigerate	Lundin	(Omelet) pasteurized eggs	30 lbs*	Pour into 60 qt mixer bowl
Mockery	Hard Cooked Eggs	5 doz.	For garnish on spinach	Wege	Ham	10 lbs	Dice for omelets

* 1 case of eggs = 30 lbs pasteurized liquid whole egg.

Figure 6.5. Residence hall foodservice production schedule.
Source: Kansas State University Housing and Dining Services. Used by permission.

The quantity to prepare is the forecast amount for each menu item. The actual yield is the portion count produced by the recipe. Standardized recipes should include portion size and count on the recipe. This information also may be placed on the production schedule. Note that the actual yield indicates overproduction for some items and underproduction for others. The instructions column, completed by the unit supervisor, gives special information and comments on equipment to be used and service instructions. In addition, this column should contain any specific information not included in the recipe, such as for refried beans, "Add bean liquid as needed to maintain a moist product," and for broccoli spears, "Season with melted margarine."

The time schedule, completed by the unit supervisor, is intended to assure that the various menu items will be produced for service at the desired time. It also has references to preparation methods standardized in this main production unit for a particular item, such as country fried steak and Mexican rice.

The leftover amount column indicates over- or underproduction. Underproduction is indicated by a zero and a time entry in the runout time column. The substitution column is used when an item is underproduced, as the broccoli spears were.

In this residence hall foodservice, cleaning assignments are in addition to regularly scheduled cleaning. The name of the person responsible is recorded in the cleaning assignment column. Instructions for prepreparation for the following meal, whether the same day or next, are in the same form as the production schedule. They begin with the name of the person and are followed by the menu item, quantity, and special instructions. The items listed under prepreparation will be on the schedule for the following meal.

This text emphasizes production scheduling because it is an important element of production control that affects the cost of materials, labor, and energy. Regardless of the perfection of the schedule and the assignment of employees to implement it, however, the production employees are the ones who make the schedule work. Realization of this simple fact implies the value of production employee meetings.

Batch cooking: Cooking smaller quantities of menu items as needed for service.

Batch cooking is a variant of production scheduling but is not always done in foodservice operations. In batch cooking, the total estimated quantity of menu items, often vegetables, is divided into smaller quantities, placed in pans ready for final cooking or heating, and then cooked as needed. The example shown in Table 6.2 is a time schedule developed for steaming rice for the dinner meal in a university residence hall foodservice. It illustrates the way in which production can be

Table 6.2 Example of Time Schedule for Batch Cooking of Rice

Buttered Rice (800 servings)
Steam at 5# pressure, 30 min uncovered

Time	Quantity[a]
3:30 pm	1 pan
3:45 pm	1 pan
4:00 pm	2 pans
4:30 pm	4 pans
4:45 pm	2 pans

[a] 4# rice/12 × 20 × 4 in. pan = 80 servings, 10 pans = 800 servings

scheduled to meet the customer demand throughout the meal service time with assurance that a fresh product is being served.

High-speed equipment available today, such as convection steamers and combi-ovens, has made batch cooking feasible for a broad range of menu items. Because many vegetables do not hold up well in a heated service counter, they are frequently prepared using batch cookery. Grilled, deep fried, and broiled items are examples of other products that should be cooked in small quantities to meet service demands.

Production Meetings

Foodservice managers in small operations and unit supervisors in large ones should hold a meeting daily with employees in the production unit. Ordinarily, these production meetings can be rather short, but when a menu is changed, more time is required to discuss new recipes and employee assignments. In foodservice operations that serve breakfast, lunch, and dinner, these meetings are generally scheduled after lunch, when activity in the production unit is minimal.

During these meetings, production unit employees can be encouraged to discuss the effectiveness of the schedule just completed. Problems such as underproduction and suggested corrective measures should be recorded for the next time the menu appears in the cycle.

The meeting should conclude with a discussion of the production schedule for the following three meals. At this time, the employees should review recipes for the various menu items, possible substitutions, and prepreparation for the following day. Free discussion of workloads is appropriate for such meetings and can be a morale builder for the employees who really make the schedule work.

INGREDIENT CONTROL

Ingredient control is a major component of quality and quantity control in the production subsystem and a critical dimension of cost control throughout the foodservice system. The process of ingredient control begins with purchasing, receiving, and storage of foods and continues through forecasting and production.

Ingredient room: Ingredient assembly area designed for measuring ingredients to be transmitted to the various work centers.

Standardized recipe: Recipe that consistently delivers the same quantity and quality of a product when followed precisely.

Two major aspects of ingredient control are ingredient assembly and use of standardized recipes. An **ingredient room**, or ingredient assembly area, is designed for measuring ingredients to be transmitted to the various work centers.

The development and use of **standardized recipes** greatly facilitate purchasing and food production. When adjusted to an accurate forecast quantity, these recipes provide assurance that standards of quality will be consistently maintained. A well-planned program beginning with the standardization of recipes and production procedures needs to be developed individually for each foodservice operation.

INGREDIENT ASSEMBLY

Concepts related to receiving, storage, and inventory control are important components of ingredient control, particularly issuing from storage. Clear policies and procedures control the issue and assembly of all food and supplies, from delivery to service, by requiring proper authorization for removal of products from storage and by issuing only required quantities for production and service.

Advantages of Centralized Ingredient Assembly

Some type of issuing is used in most foodservice operations. The ingredient room contributes to cost reduction and quality improvement by stopping production employees from withdrawing large quantities of products from storage whether or not they are needed. Use of the ingredient room has many advantages, including the redirection of cooks' skills away from the simple tasks of collecting, assembling, and measuring ingredients to production, garnishing, and portion control. By limiting access to ingredients, over- and underproduction of menu items can be eliminated, thus controlling costs.

The concept of an ingredient room dates from the late 1950s. Flack (1959), one of the first to implement a central ingredient room, reported reduced labor cost as a major benefit. In recent years, as competition has increased and the available labor pool has shrunk, managers have shown an increased interest in quality, cost controls, and more efficient use of labor. As a result, many are adopting the ingredient control concept within their foodservice operations (Buchanan, 1993).

Centralized Ingredient Control

Traditionally in foodservice operations, individual cooks obtain recipe ingredients from storerooms, refrigerators, and freezers. Ingredients are issued in cases, boxes, or bags. In production areas, ingredients that are not currently being used generally are stored in storage bins or on shelves. For example, a cook may keep sugar and flour bins under the work counter and spices and condiments above on a shelf.

Keeping track of unused portions of issue units, especially perishable products, provides a challenge for the cook or foodservice manager and leads to decreased control. For example, 5 pounds of frozen mixed vegetables left over from a 30-pound case must be held and used the next time mixed vegetables are on the menu. With the vast number of ingredients used in foodservice operations today, controlling partially used packages of food items can become a major challenge. A cook may decide to add the extra mixed vegetables to the soup kettle rather than return the unneeded amount to storage. Such a practice alters the recipe and adds to cost. This is only one example of problems caused by the traditional method of issuing ingredients.

With centralized ingredient control, the cook is issued only the 25 pounds needed for the forecast production demands on the day of service. The excess of 5 pounds is held in frozen storage in the ingredient room until the next time mixed vegetables are needed for a recipe or as a vegetable on a future menu. Control of unused portions is facilitated because storage is located centrally rather than in various work units throughout the kitchen.

As Buchanan (1993) indicated, food production includes basically two functions: prepreparation and production. Traditionally, cooks have performed both functions. Focusing the cooks' efforts and attention on direct production tasks and away from the simple tasks of prepreparation, which can be assigned to less-skilled employees, can lead to operational efficiencies.

In addition, less-skilled employees can develop skill in performing prepreparation tasks, thereby reducing labor cost. Combining tasks for two or more recipes using similar ingredients is another efficiency. For example, chopped onions may be needed for both meat loaf and a sauce at lunch and for both soup and salad on the dinner menu. By centralizing prepreparation, all the onion that is needed can be chopped at one time and divided into separate batches for each of the four recipes.

Function of the Ingredient Room

The primary function of the ingredient room is to coordinate assembly, prepreparation, measuring, and weighing of the ingredients to meet both the daily production needs and the advance preparation needs of recipes for future meals. Specific activities vary among foodservice operations, but ingredient rooms generally operate 24 hours in advance of production needs (Buchanan, 1993).

An ingredient room may be limited to premeasuring only dry and room-temperature ingredients, or it may be a center in which all ingredients, whether they are at room temperature, refrigerated, or frozen, are assembled, weighed, and measured. The availability of appropriate equipment, such as prepreparation equipment and low-temperature storage, will help determine the activities to be performed in an ingredient room. A freezer in or near the assembly area is required, for example, if ingredient room employees withdraw and thaw frozen products in advance.

Storeroom employees generally assemble full cases and unopened cans for delivery to production. Partial amounts from cans or cartons should be weighed and measured in the central ingredient assembly area. This practice can eliminate waste in a foodservice operation by avoiding partially used cans or cartons of the same product in several locations. For example, canned tomato sauce may be needed by both the main production and salad units; with a central ingredient room, only one partially used can would need to be stored.

Meat products, which have been ordered according to production demands, may or may not be handled in the ingredient room. For example, preportioned meats that have been ordered in a quantity appropriate for the production demand may go directly from receiving to production. Greater control is possible, however, if these products were distributed through the ingredient room.

Prepreparation tasks, such as cleaning and slicing or dicing vegetables or breading and panning meat items, may be done in a prepreparation area or the ingredient room. Today, many of these products, especially produce, are purchased already prepped.

After all ingredients for each recipe have been weighed, measured, chopped, or otherwise prepared, each ingredient is packaged and labeled. Ingredients for each recipe are then transported with a copy of the recipe to the appropriate work unit or held until the scheduled distribution time.

Many computer programs already in existence support the ingredient room (Buchanan, 1993). Recipe amounts that are adjusted and printed by computer are easier to read and are more likely to be accurate than when these adjustments are done manually and are hand printed. This and other types of computer-adjusted recipes increase the speed and productivity of the ingredient room employee and eliminate product errors and costly mistakes. Computers also can produce consolidated ingredient lists by individual ingredients or by production area. In a computer-assisted operation, the recipe can be adjusted and peel-off labels printed for each ingredient. These labels, which usually have the ingredient name and quantity, facilitate marking the ingredient packages for each recipe.

Ingredient Room Organization

In the design of a new foodservice facility, an ingredient room that can be locked should be located between the storage and production areas. In an existing facility, the ingredient room may be located in or near a storeroom, combined with the prepreparation area, or put in a designated part of the production unit.

Necessary large equipment includes refrigeration, which should be in or near the ingredient room area, and a water supply. Trucks or carts are needed for assembly and delivery of recipe ingredients and portable bins for storing sugar, flour, and other dry ingredients. A worktable or a counter is required, with shelving over or near it for such products as spices. An example of an ingredient room worktable is shown in Figure 6.6.

Scales are the most essential pieces of equipment for an ingredient room. A countertop scale and a portion scale are required for weighing the various types of recipe ingredients. Other equipment will vary depending on the specific functions assigned to the ingredient room, such as a slicer, vertical/cutter mixer, food waste disposal, and mixer.

Ingredient Room Staffing

According to Dougherty (1984), in an operation without an ingredient room, production employees spend about one-third of their time determining needs, obtaining supplies, and weighing and measuring ingredients. By centralizing these

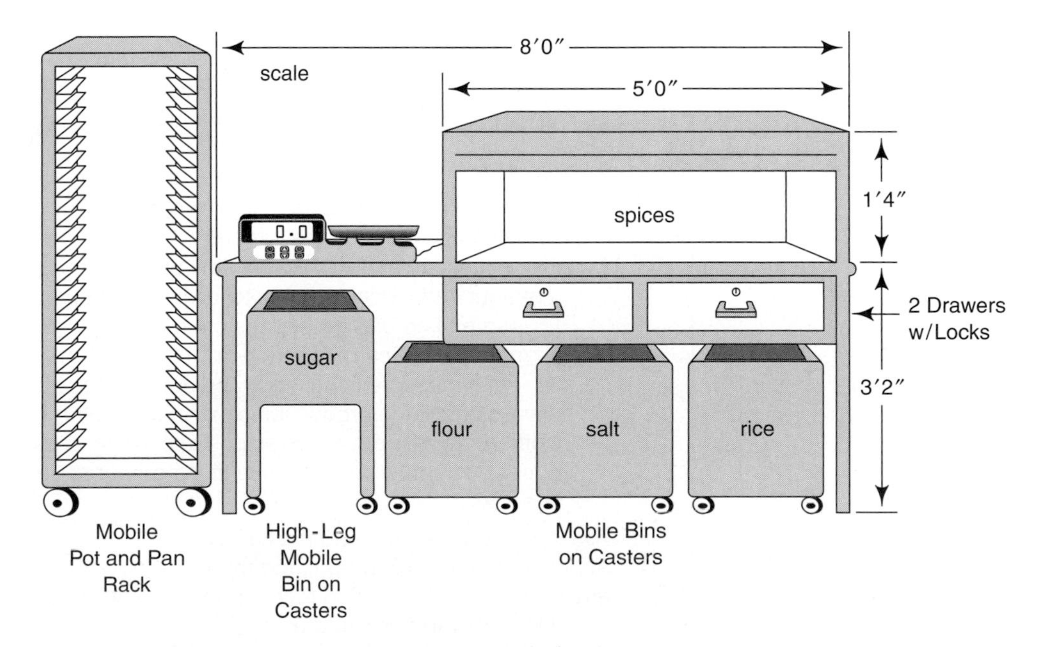

Figure 6.6. Ingredient room worktable arrangement.

activities, production employees are free for higher-level skill tasks, which allows management to reassign less-skilled employees from production to the ingredient room. In smaller operations in which a full-time employee is not needed for ingredient assembly, schedules of production employees can be arranged to permit them to weigh and assemble ingredients. As Dougherty (1984b) indicated, the more activities are centralized, the greater will be the realized benefits.

Employees assigned to the ingredient room must be literate, able to do simple arithmetic, and familiar with storage facilities. They are often responsible for receiving, storage, and ingredient assembly. Qualifications and training, therefore, must be specific to each of these areas of responsibility. Training should include the following areas:

Figure 6.7a.
Job description.
Source: From "Issue Control and Ingredient Assembly System," by D. A. Dougherty. In J. C. Rose (Ed.), *Handbook for Health Care Food Service Management* (pp. 203–204), 1984, Rockville, MD: Aspen Publishers.

INGREDIENT ASSEMBLY CLERK

Name: ______________________ Coffee Break: ________________

Hours: ______________________ Lunch Hour: ________________

Coffee Break: ________________

GENERAL RESPONSIBILITIES AND DUTIES

I. Storeroom Requisition
Pick up storeroom requisition from main kitchen office. This list is a consolidation of all ingredients needed for one day's production of menu items to be assembled in the ingredient room. Check all items against the inventory on hand in the ingredient assembly area and mark out supplies not needed with a felt pen. (For example, sugar will be issued in a 100# bag. The total amount will not be used in any one day. Annotate this requisition, give it to the storeroom clerk for filling and delivery.)

II. Production Schedule and Production Recipes
Pick up production schedule and production recipes from the main kitchen office.
   A. Production recipes. There will be one recipe for each menu item that will be for use by the ingredient assembly clerk. These recipes have been adjusted according to the forecast.
   B. Production schedule. This form will be prepared by the food production manager and will include the following information:
      1. name of employee responsible for product preparation
      2. name of recipe or menu item
      3. time that ingredients are to be delivered to specified production area
   Always check the production schedule to determine how to plan your work. Arrange production recipes according to the times listed on the production schedule so that those to be prepared first are weighed and packaged first.

III. Assembling, Weighing, and Packaging
Accurately assemble, weigh, and measure ingredients according to the amounts on the production recipe printout. Set up ingredients on carts according to the following procedure.
   A. Choose a cart large enough to accommodate the ingredients for each recipe.

- Environmental conditions required to store specific foods
- Ventilation and humidity factors in dry storage
- Safety precautions in handling and storage of nonfood items and toxic materials
- Sanitation standards to prevent contamination or deterioration of foods during storage
- Security measures to ensure against pilferage
- Weighing and measuring procedures

The job description for an ingredient room clerk depends on the activities included in the procedures for the ingredient room. An example of a job description is shown in Figure 6.7.

Figure 6.7b.
Continued

B. Using masking tape and felt-tip pen, label trays and shelves with recipe name. All menu items set up on cart should have approximately the same delivery time.

C. Package ingredients as follows:
   1. Paper cups and lids. Use for small amounts of liquid items, such as Worcestershire sauce, lemon juice, salad oil.
   2. Plastic bags (2 to 3 sizes). Use for all wet ingredients, such as frozen eggs, fresh vegetables. Fasten bags securely and label with felt marker. Write name of ingredient and weight on bag.
   3. Glassine sandwich bags. Use for small amounts of spices or other dry ingredients.
   4. Paper bags (assorted sizes). Use for dry ingredients.
   5. Counter pans. Use for larger quantities of ingredients, such as bread crumbs, flour, celery. Cover the pan with either a lid or plastic wrap.
   6. Original container. Always leave items in the original container whenever possible, such as no. 10 cans or wax paper around margarine.
   7. Jars (½ and 1 gal). For vinegar and salad oil.

   Put small quantities of ingredients for the same recipe on 18-inch × 26-inch sheet pan. After all items are weighed, labeled, and covered, place them on assigned cart shelves. If the cart is not to be delivered immediately, push it into a walk-in refrigerator if any items require refrigeration. Meat items, such as ground beef, should not be put on the cart until just before delivery.

   The production recipe printout should be delivered with the assembled ingredients to the production area. If necessary, inform the cook that the ingredients have been delivered. Ingredients may be transferred from the carts to a table when practical.

IV. General Responsibilities
You are responsible for
A. storing and refrigerating opened cans and using these opened cans first;
B. storing items and rotating stock in the ingredient assembly area;
C. using proper sanitation procedures;
D. washing the equipment that you have used;
E. keeping your work area clean and orderly at all times; and
F. other duties as assigned.

The following factors are among those to consider in scheduling personnel for ingredient assembly:

- Size of operation
- Frequency and time of deliveries
- Size of ingredient room and location of other storage areas
- Type, number, and complexity of menu items to be assembled
- Number of workstations to be supplied
- Schedule for delivery of ingredients to production and serving areas
- Extent of prepreparation performed in ingredient assembly area

The larger the operation, the more complex the menu; and the greater the amount of preparation work, the more employees needed.

RECIPES

Recipe: Formula by which weighed and measured ingredients are combined in a specific procedure to meet predetermined standards.

A **recipe** is a formula by which weighed and measured ingredients are combined in a specific procedure to meet predetermined standards. The recipe is actually a written communication tool that passes information from the foodservice manager to the ingredient room and production employees. In addition, the recipe is an excellent quality and quantity control tool, constituting a standard for each item on the menu that meets customer and management approval. Cost for each recipe can be easily computed because the ingredients and the amount will be the same each time the recipe is used.

Recipes should have a format that is easily understood by the personnel who are responsible for the production and presentation of menu items to customers. Once a recipe has been tested repeatedly and accepted by management and customers, it becomes a standardized recipe and always gives the same results.

Format

Most recipes are written in a definite pattern or style that is identified as a format. For most effective use, all recipes in a particular foodservice operation should be in the same format. This uniformity of style simplifies recipe use by cooks.

Large quantity recipes generally differ in format from home recipes, which have a list of ingredients followed by procedures. The cook in a foodservice operation is more likely to make errors if required to read alternately from the top and bottom of the recipe. A block format, in which the ingredients are listed on the left side of the recipe and the corresponding procedures directly opposite them on the right, generally is used for quantity recipes. In a complete block format, horizontal lines separate each group of ingredients with procedures from those of the next, and vertical lines separate the ingredient, amount, and procedure columns. A modified version, in which only horizontal lines separate the required ingredients for each procedure, often is used. An example of a modified block format is the recipe for lasagna in Figure 6.8. This recipe format is suitable for both recipe cards and computer printouts.

Specific information should be included on each recipe to simplify its use by those preparing and serving the food. Generally, recipes include the following information:

- Name of food item
- Total yield
- Portion size and number of portions
- Cooking time and temperature, if required
- List of ingredients in order of use
- Amount of each ingredient by weight, measure, or count
- Procedures
- Panning or portioning information
- Serving and garnishing suggestions
- Food safety (HACCP) guidelines

The important basic information pertaining to the detailed recipe is shown in the heading of the lasagna recipe (Figure 6.8). It includes the yield, portion size, oven temperature, and baking time. Ingredients are listed in order of use in four procedural groups. Amounts are given to the right of the ingredients. Standard U.S. weights and measures are generally used for all ingredients except small amounts of spices and oil, which are given in the common household units of teaspoons (tsp) and tablespoons (Tbsp).

Nutritive values per portion are provided. Production, service, and storage procedures that will prevent or reduce the hazards of potentially hazardous foods are included. In addition, food safety standards are given for the menu items in the notes on the bottom of the recipe.

Recipes for production greater than the 48 portions in the lasagna example could be written entirely in standard measures without the use of teaspoons, cups, and other household units. Many computer software programs have the capability of calculating quantities in metric, but, so far, these measures are not commonly used in the United States. Some programs will calculate only in weights, some only in measures, but the majority of the large programs calculate in both. In small foodservice operations, measures may be used rather than weights, even though weighing is more accurate than measuring, for determining the proper amount of an ingredient.

Corresponding procedures for each group of ingredients are printed directly opposite them on the right side of the recipe. The layering of the ingredients in lasagna is very important and is detailed in the last procedure. Oven temperature, baking time, and portioning instructions are repeated for the convenience of the cook. Procedures should be checked for clarity and explicitness, permitting a cook to prepare a perfect product without asking managers for further explanation. In recipes that require the use of equipment such as mixers, time and speed should be specified. For example, the procedure for combining the first three ingredients in a cake might be "cream shortening, sugar, and vanilla on medium speed 10 minutes."

Additional information may be added at the bottom of recipes, such as the approximate nutritive values per portion and variations of the recipe. Special serving instructions, such as garnishing and portioning suggestions and storage requirements before and after service, often are included. If recipe cards are used, these additions can be printed at the bottom or on the back. The ingredient and procedure portion of a recipe, however, should never be printed on the back of the card. For long recipes, a second card or page should be used.

Before adopting a particular format, variations might be tried to give the cooks an opportunity to choose the one they like best. Flexibility in recipe formats is not

LASAGNA

Yield: 48 portions or 2 pans 12 × 20 × 2 inches *Portion:* 6 oz
Oven: 350° F *Bake:* 45 minutes-1 hour

Ingredient	***Amount***	***Procedure***
Ground beef	5 lb AP (3 lb EP)	Cook beef, onion, and garlic until meat reaches an internal temperature of 155° F. Drain off fat.
Onions, finely chopped	2 lb 12 oz (EP)	
Garlic, minced	1 oz EP	
Tomatoes, chopped	3 qt	Add tomatoes, seasonings, and herbs to ground meat mixture. Simmer for about 45 minutes.
Tomato puree	3 cups	
Tomato paste	4½ cups	
Salt	2 oz	
Pepper, black	1 tsp	
Basil leaves, dried	3 Tbsp	
Oregano leaves, dried	3 Tbsp	
Parsley, chopped (fresh)	1½ oz	
Frozen Lasagna sheets	3 lb 8 oz	
Mozzarella cheese, shredded	4 lb	Combine cheeses.
Parmesan cheese, grated	10 oz	Layer into two greased 12 × 20 × 2-inch pans according to the following directions for each pan: 1. Meat sauce, 2 lb
Cottage cheese, cream style	2 lb 8 oz	2. Lasagna sheets: 12 oz (3 sheets) 3. Cottage cheese, 1 lb 5 oz 4. Parmesan cheese: 2½ oz 5. Mozzarella cheese: 12 oz 6. Meat sauce: 2 lb 8 oz 7. Repeat steps 2-6 8. Mozzarella cheese: 8 oz
		Cover with foil. Bake at 350° F for 1-1½ hours. Let stand 15–20 minutes before cutting. Do not cover Cut 4 × 6.

Approximate nutritive values per portion ***Calories*** 329

Amount/portion	%DV	Amount/portion	%DV	Amount/portion	%DV		%DV		%DV
Total Fat 18 g	27%	**Cholest.** 70 mg	23%	**Total Carb.** 37 g	12%	**Vitamin A**	22%	**Calcium**	34%
Sat. Fat 9 g	46%	**Sodium** 1100 mg	46%	Fiber 3-4 g	14%	**Vitamin C**	27%	**Iron**	18%
Protein 33 g				Sugars 8 g					

Percent Daily Values (DV) are based on 2000-calorie diet.

Notes ■ Potentially hazardous food. *Food Safety Standards:* Hold food for service at an internal temperature above 140°F. Do not mix old product with new. Cool leftover product quickly (within 4 hours) to below 40°F. See p. 44 for cooling procedures. Reheat leftover product quickly (within 2 hours) to 165°F. Reheat product only once; discard if not used.

Figure 6.8. Example of the format of a typical quantity recipe.
Source: Molt, M. *Food for Fifty.* Copyright © 2002 by Prentice Hall. Reprinted by permission of the publisher.

possible if a computer program is used. The foodservice manager should compare recipe formats in software packages before purchasing one. Once a format is chosen, all recipes should be printed in that style. In converting to a new format, a good method is to adapt the most used recipes first and then gradually extend the conversion to the entire file. The production manager will need to conduct in-service training sessions on the new format for production employees.

Recipes for use at a workstation should be in large print that is easily readable at a distance of 18 to 20 inches; large file cards or 8½-by-11-inch paper must be used. Recipes for the ingredient room or production unit should be in a plastic cover while in use and in some type of rack at the workstations. Because of the larger print, production employees should not have to pick up recipes for a closer look.

The recipe name is generally in bold letters, either in the middle or to the side, at the top of the recipe card. In most operations, a file coding system is established for quick access to each recipe. For example, major categories can be established for menu items, such as beverages, breads, cakes, cheese, cookies, eggs, fish, meat, pies, poultry, salad, sandwiches, soups, and vegetables. Each recipe can be coded with a letter designating the menu item category and a sequential number that identifies the individual recipe. Whatever system is developed should be easy for employees to follow. Today, many foodservices use color coding of major categories to make identification easier.

Methods of maintaining a master file of recipes vary with the foodservice operation. Many organizations have implemented computerized foodservice management systems that provide a database of recipes and allow for recipe adjustments, purchase orders, and costing to be completed quickly. The importance of keeping a backup file of all recipes in a computerized operation cannot be emphasized enough.

Recipe Standardization

The ideal of every manager is to have recipes that consistently deliver the same quantity and quality product when followed precisely. Printed recipes from various sources will not guarantee uniform products in every foodservice. Variations in ingredient characteristics, customer demands, personnel, and equipment may require alterations to the recipe or even preclude its successful use. Production procedures are complicated and difficult to establish because many people are involved, each of whom has definite ideas about how a product should be prepared. **Recipe standardization**, or the process of tailoring a recipe to suit a particular purpose in a specific foodservice operation (Buchanan, 1993), is one of the most important responsibilities of a production manager.

Recipe standardization: Process of tailoring a recipe to suit a particular purpose in a specific foodservice operation.

Standardization requires repeated testing to ensure that the product meets the standards of quality and quantity that have been established by management. Food cost and selling price cannot be correctly calculated unless recipes are standardized to use only specific ingredients in known amounts to yield a definite quantity. A standardized recipe must be retested whenever a small change is made in an ingredient, such as substitution of frozen or dehydrated vegetables for fresh in a recipe for beef stew.

Standardizing recipes is a time-consuming task, and managers must be convinced that the process is worthwhile. Conviction follows from the realization that the most successful foodservice operations use only recipes that have been developed specifically for them. Probably the best example of the use of standardized recipes is in the multiunit limited-menu chains. Each batch of hot biscuits, pizza dough and toppings, fried chicken, and French fries is the same in each unit of the chain every day. This essential uniformity often is assured by recipes available to only a few employees in the main ingredient room of the corporate commissary. The packaged ingredients are sent to each unit throughout the region, the nation, and even the world. Without this stringent control, these operations could not maintain national and international reputations for quality.

Among the advantages for using standardized recipes are that they (Buchanan, 1993)

- Promote uniform quality of menu items. All products should be the same high quality.
- Promote uniform quantity of menu items. Recipes should produce a specific, designated quantity.
- Encourage uniformity of menu items. Servings of the same menu item should have the same size and appearance.
- Increase productivity of cooks. Clarifying procedures on recipes improves the efficiency of cooks.
- Increase managerial productivity. Recipes eliminate guesswork in food ordering and questions asked by cooks, thus freeing managers to concentrate on satisfying the customer.
- Save money by controlling overproduction. Waste is controlled if only the estimated number of portions is produced.
- Save money by controlling inventory levels. More just-in-time purchasing is possible, thus increasing cash flow.
- Simplify menu item costing. Precise calculation of serving costs is vital in establishing selling prices.
- Simplify training of cooks. Recipes with detailed procedures provide an individualized training program for new cooks and chefs.
- Introduce a feeling of job satisfaction. Cooks know that menu items will always be the same quality.
- Reduce anxiety of customers with special dietary needs. A nutrient analysis can be done on each recipe and ingredients can be identified for those with allergies or other health concerns.

Although standardized recipes offer many advantages in a foodservice operation, the key to success is ensuring that recipes are followed carefully and consistently each time an item is produced. Because the human element can be a major variable in product quality and uniformity, employee supervision and training are critical.

The use of standardized recipes has some limitations. Even when they are followed, standardized recipes will not improve a product made from inferior ingredients; good specifications for quality products are essential (Mitani & Dutcher, 1992). Standardization also cannot eliminate the variation found in food. For example, climate, degree of maturity, growing regions, and age of products can affect the menu item. Ingredient substitutions will affect the final product; the recipe has to be standardized again before the menu item is served to customers. Recipes must be standardized for each foodservice operation. Foodservice managers should review and update previously standardized recipes to reflect changes in the organization.

A standardized recipe is one that has been developed for use by a given foodservice operation and has been found to produce consistent results and yield each time it is prepared. The terms *quantity recipe* and *standardized recipe* often are confused. Any recipe that produces 25 servings or more is termed a **quantity recipe.** Quantity recipes are not standardized, however, until they have been adapted to an individual operation.

Gregoire and Henroid (2002) described the recipe standardization process as a cycle of three phases: recipe verification, product evaluation, and quantity adjustment (see Figure 6.9). All recipes start in the recipe verification phase and then move through the product evaluation and quantity adjustment phases. These three phases are repeated until a recipe is standardized (i.e., produces a consistent product).

Recipe Verification

The verification phase includes four major processes that should be completed before the product is evaluated:

- Review components of the recipe.
  - Recipe title
  - Recipe category
  - Ingredients
  - Weight/measure for each ingredient
  - Preparation instructions
  - Cooking temperature and time
  - Portion size
  - Recipe yield
  - Equipment to be used
- Make the recipe.
- Verify the recipe yield.
- Record changes on the recipe.

Typically a manager would work with one recipe at a time. The review of each of the recipe components should be completed before the recipe is made to help ensure that all necessary information is available for the person who will make the recipe.

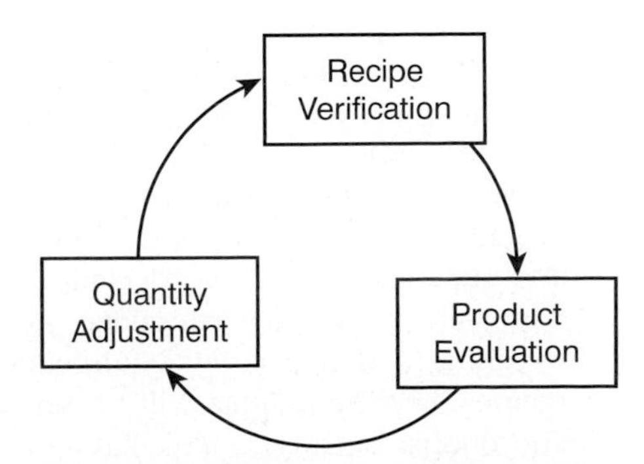

Figure 6.9. Recipe standardization cycle.

Product Evaluation

Product evaluation follows the recipe verification phase and is a critical part of the recipe standardization process. Product evaluation is used to help determine the acceptability of recipe by foodservice managers and staff and customers.

Typically an informal type of evaluation is completed when the product is first prepared. The **informal evaluation** is done to help determine whether to proceed with further standardization of the product. According to Gregoire and Henroid (2002), this evaluation usually focuses on issues such as

- Is the visual appearance of the product acceptable?
- Is the flavor of the product one that customers might enjoy?
- Are the ingredients in the recipe easy to obtain?
- Is the cost per serving within department guidelines?
- Is the labor time needed to make the product within department guidelines?
- Is the equipment needed to prepare the product available?
- Do employees possess the skills needed to prepare the product?

If foodservice staff members believe a recipe has potential for use in the operation, then a formal evaluation is conducted. A **formal evaluation** typically includes the following procedures (Gregoire & Henroid, 2002):

- Selecting a group of staff members and customers as a taste panel
- Choosing or developing an evaluation instrument
- Preparing a sample recipe
- Setting up the sampling area with drinking water, eating and drinking utensils, napkins, and evaluation forms and pencils
- Having participants sample and evaluate products
- Summarizing results
- Determining future plans for the recipe based on evaluation results

Quantity Adjustment

Three procedures have been developed for the quantity adjustment of recipes: the factor method, the percentage method, and direct reading measurement tables. Computer software programs for recipe adjustments are being used in many foodservice operations today to help simplify the quantity adjustment process.

Factor Method. To increase a recipe using the factor recipe adjustment method, the ingredients are generally converted to whole numbers and decimal equivalents; that is, 2 lb 10 oz would be converted to 2.625 lb, but rounded to 2.6 lb for use. Typically, only one decimal place is used in a recipe (e.g., 2.6 lb) unless the original amount is less than 1 pound, in which case two places would be shown (e.g., 0.62 lb). Whenever possible, liquid measures also should be stated in weights; however, liquid measurements may be converted to decimal equivalents of a quart or gallon. Table 6.3 provides data for converting ounces into decimals of a pound and cups and quarts into decimal parts of a gallon. After ingredient quantities are converted to the new amounts, they may be changed into units suitable for the weighing and measuring equipment used in the foodservice operation. In most recipes, dry ingredients will be stated in pounds and ounces and liquids in cups and quarts. Because scales may not be accurate for weighing very small amounts, recipe ingredients of 1 oz or less often are stated in tablespoons or teaspoons.

Table 6.3 Decimal Conversions for Weights (in pounds) and Volume Measures (in gallons)

(Pounds) Weight Measure			(Gallons) Volume Measure		
oz	**lb.**	**Decimal Unit**	**cup**	**qt**	**gal**
½		.031	½		
1		.063	1	¼	
1½		.093	1½		
2	⅛	.125	2	½	
2½		.156	2½		
3		.188	3	¾	
3½		.218	3½		
4	¼	.250	4	1	1/4
4½		.281	4½		
5		.313	5	1¼	
5½		.343	5½		
6	⅜	.375	6	1½	
6½		.406	6½		
7		.438	7	1¾	
7½		.469	7½		
8	½	.500	8	2	½
8½		.531	8½		
9		.563	9	2¼	
9½		.594	9½		
10	⅝	.625	10	2½	
10½		.656	10½		
11		.688	11	2¾	
11½		.719	11½		
12	¾	.750	12	3	3/4
12½		.781	12½		
13		.813	13	3¼	
13½		.844	13½		
14	⅞	.875	14	3½	
14½		.906	14½		
15		.938	15	3¾	
15½		.969	15½		
16	1	1.000	16	4	1

Note: This chart cannot be used to determine decimal parts of cups or quarts, only gallons and pounds.
Source: The American Dietetic Association. *Quantity Food Preparation: Standardizing Recipes and Controlling Ingredients,* 3rd ed. © 1993, Used by permission.

Converting a recipe by using the factor method requires the following:

- Changing ingredient amounts to whole numbers and decimals
- Dividing the desired yield by the recipe yield to determine the conversion factor
- Multiplying all recipe ingredients by the conversion factor
- Reconverting the decimal unit back into pounds and ounces or quarts and cups
- Rounding off amounts to quantities simple to weigh or measure and within an acceptable margin of error
- Checking math for possible errors

To illustrate use of the factor method, assume a college residence hall foodservice has ham loaf on the menu. The recipe in the file is for 50 portions; however, the forecast production demand is 250 portions. Using the procedure outlined previously, the ham loaf recipe would be adjusted to the desired number of servings by first converting the ingredients, as appropriate, to decimal equivalents, as shown in Table 6.4.

The conversion factor to adjust the base recipe from 50 to 250 portions would be determined by dividing the 250 by 50; the resulting factor is 5. The next step is to multiply the ingredient amounts by the factor. To assist the cooks in using the recipe, ingredients for 250 portions of ham loaf would be stated in pounds and ounces or quarts and cups, as shown in the last column in Table 6.5.

Percentage Method. In the percentage method, measurements for ingredients are converted to weights and then the percentage of the total weight for each ingredient is computed. The number of portions is forecast, which provides the basis for determining the ingredient weights from the ingredient percentages. Formulas that have been converted to percentages need not be recalculated. This method allows adjustment to the portion size or forecast and permits a shift of ingredients to be done easily.

Percentages can be readily determined using a desk calculator and with computer assistance they are made even easier. Molt (2001) described the following step-by-step method for recipe adjustment via the percentage method.

Step 1 Convert all ingredients from measure or pounds and ounces to tenths of a pound. Make desired equivalent ingredient substitutions, such as frozen whole eggs for fresh or powdered milk for liquid.

Step 2 Total the weight of ingredients in a recipe after each ingredient has been converted to weight in the edible portion (EP). For example, the weight of carrots or celery should be the weight after cleaning and peeling. The recipe may show both AP (as purchased) and EP weights, but the edible portion is used in determining the total portion weight.

Step 3 Calculate the percentage of each ingredient in the recipe in relation to the total weight. (*Note:* The sum of all percentages must equal 100.)

$$\frac{\text{Individual ingredient weight}}{\text{Total weight}} \times 100 = \text{Percentage of each ingredient}$$

Table 6.4 Conversion to Decimal Units

Ingredients	Amount	Decimal Units
Ground ham	4 lb	
Ground beef	4 lb	
Ground pork	4 lb	
Chopped onion	2 oz	0.125 lb
Black pepper	½ tsp	0.5 tsp
Bread crumbs	1 lb	
Eggs	1 lb 8 oz	1.5 lb
Milk	1 qt	

Table 6.5 Conversion from Decimals

Ingredients	50 Portions	50 × 5 = 250 Portions	Conversion from Decimals
Ground ham	4 lb	20 lb	20 lb
Ground beef	4 lb	20 lb	20 lb
Ground pork	4 lb	20 lb	20 lb
Chopped onion	0.125 lb.	0.625 lb	10 oz
Black pepper	½ tsp	2.5 tsp	2½ tsp
Bread crumbs	1 lb	5 lb	5 lb
Eggs	1.5 lb	7.5 lb	7 lb 8 oz
Milk	1 qt	5 qt	5 qt

Step 4 Check the ratio of ingredients. Standards of ingredient proportions have been established for many items. The ingredients should be in proper balance before going further.

Step 5 Establish the weight needed to provide the desired number of servings, which will be in relation to pan size, portion weight, or equipment capacity.

Examples include the following:

- Total weight must be divisible by the weight per pan.
- A cookie portion may weigh 0.14 lb per serving; therefore, 0.14 times the number of desired servings equals the weight needed.
- Recipe total quantities should be compatible with mixing bowl capacity.

Use the established portions, modular pan charts, or known capacity equipment guides to determine batch sizes. The weight of each individual serving is the constant used in calculating a recipe.

Handling loss: Decrease in the yield of a recipe because of preparation process.

Step 6 **Handling loss** must be added to the weight needed and may vary from 1 to 10%, depending on the product. Similar items produce predictable losses that with some experimentation can be accurately assigned. The formula for adding handling loss to a recipe is as follows:

$$\text{Total weight needed} = \frac{\text{Desired yield}}{100\text{ percent} - \text{Assigned handling loss percent}}$$

Example: Yellow cake has a 1% handling loss. Desired yield is 80 lb of batter for 600 servings.

$$100\% - 1\% = 99\% \text{ or } .99$$

$$\text{Total weight needed} = \frac{80\text{ lb}}{.99}$$

Total weight needed equals 80.80 lb of batter.

Table 6.6 Original Recipe, Brownies

Ingredients	Amount
Eggs	12
Sugar	2 lb
Fat, melted	1 lb
Vanilla	¼ C
Cake flour	12 oz
Cocoa	8 oz
Baking powder	4 tsp
Salt	2 tsp
Nuts, chopped	12 oz

Source: "Recipe Standardization and Percentage Method of Adjustment" by H. McManis and M. Molt, 1978, *NACUFS Journal,* p. 40. Used by permission.

Step 7 Multiply each ingredient percentage number by the total weight to give the exact amount of each ingredient needed. After the percentages of each ingredient have been established, any number of servings can be calculated and the ratio of ingredients to the total will be the same. As in the factor method, one decimal place on a recipe is shown unless the quantity is less than 1 pound, in which case two places are shown.

Step 8 Unless scales are calibrated to read in pounds and tenths of a pound, convert to pounds and ounces or to measure.

Reviewing the process of expanding a recipe will help illustrate recipe adjustment by this method. Tables 6.6, 6.7, and 6.8 demonstrate how to expand a recipe using the percentage method. The result is a brownie recipe for 60 portions of 0.12 lb each.

Table 6.7 Calculate Percent, Brownies

Percent*	Ingredients	Measure	Pounds
20.34	Eggs	12	1.32
30.82	Sugar	2 lb	2.00
15.41	Fat, melted	1 lb	1.00
1.70	Vanilla	¼ C	0.11
11.56	Cake flour	12 oz	0.75
7.71	Cocoa	8 oz	0.50
0.49	Baking powder	4 tsp	0.0316
0.41	Salt	2 tsp	0.127
11.56	Nuts, chopped	12 oz	0.75
100%			6.4886

*Individual ingredients weights = Ingredient percents total weight.

Source: "Recipe Standardization and Percentage Method of Adjustment" by H. McManis and M. Molt, 1978, *NACUFS Journal,* p. 40. Used by permission.

Table 6.8 Recipe for 60 Servings, 0.12 lb Each, 3% Handling Loss

Percent	Ingredients	Pounds
20.34	Eggs	1.51
30.82	Sugar	2.29
15.41	Fat, melted	1.14
1.70	Vanilla	0.13
11.56	Cake flour	0.86
7.71	Cocoa	0.57
0.49	Baking powder	0.04
0.41	Salt	0.03
11.56	Nuts, chopped	0.86
100%		7.42*

*Calculations:
60 × .12 lb each serving = Batter needed (7.2 lb).
3% Handling loss = 7.2/0.97 = 7.42 lb.
7.42 lb × each individual ingredient percent = Amount of each ingredient.

Source: "Recipe Standardization and Percentage Method of Adjustment" by H. McManis and M. Molt, 1978, *NACUFS Journal,* pp. 40–41. Used by permission.

Direct Reading Measurement Tables Method. The third method of recipe adjustment uses direct reading measurement tables. These tables have the advantage of being simple and quick to use and require no mathematical calculations. Tables have been developed for both measured and weighed ingredients.

Buchanan (1993) developed tables for adjusting weight and volume of ingredients in recipes that are divisible by 25. Beginning with weights and measures for 25 portions, incremental values are given in these tables for various magnitudes up to 500. Use of these tables allows adjustment of recipes with a known yield in one of the amounts indicated to desired yields divisible by 25. An excerpt from a direct reading table for adjusting weight ingredients of recipes is shown in Table 6.9. For example, the amount of ground beef needed for 225 portions, using a recipe designed to produce 100 portions, can be determined easily using the table. If the 100-portion recipe requires 21 pounds of beef, then reading to the right for the amount for 200 portions (42 lb) and the left for 25 portions (5 lb 4 oz) the total of 47 lb 4 oz can be determined quickly.

Table 6.9 Excerpt from a Direct Reading Table for Adjusting Weight Ingredients of Recipes Divisible by 25

25	50	75	100	200	300	400	500
5 lb	10 lb	15 lb	20 lb	40 lb	60 lb	80 lb	100 lb
5 lb 4 oz	10 lb 8 oz	15 lb 12 oz	21 lb	42 lb	63 lb	84 lb	105 lb
5 lb 8 oz	11 lb	16 lb 8 oz	22 lb	44 lb	66 lb	88 lb	110 lb
5 lb 12 oz	11 lb 8 oz	17 lb 4 oz	23 lb	46 lb	69 lb	92 lb	115 lb
6 lb	12 lb	18 lb	24 lb	48 lb	72 lb	96 lb	120 lb

Source: The American Dietetic Association. *Quantity Food Preparation: Standardizing Recipes and Controlling Ingredients,* 3rd ed. © 1993, Used by permission.

Converting from Weight to Measure

Most ingredients in quantity recipes are given in weights, but if volume measurements (teaspoon, tablespoon, cups, quarts, or gallons) are to be used, tables such as the following from Molt (2001) are available that assist in converting from weights to measure.

- *Food weights and approximate equivalents in measure,* for example, 1 pound (weight) or 3 cups (measure) cooked cubed chicken.
- *Basic equivalents in weights and measures,* for example, 8 fluid ounces (weight) or 1 cup of milk (measure).
- *Weight (1–16 ounces) and measure equivalents for commonly used foods,* for example, 2 ounces (weight) or 1/4 cup (measure) flour.
- *Guide for rounding off weights and measures,* for example, more than 1/2 cup but less than 1 cup rice (measure) rounded to closest full teaspoon or convert to weight. This table aids in rounding fractions and complex measurements into amounts that are as simple as possible to weigh or measure while maintaining the accuracy needed for quality control.
- *Ounces and decimal equivalents of a pound,* for example, 8 ounces (weight) or 0.5 of a pound of butter (measure). This table is used when increasing or decreasing recipe sizes. The multiplication or division of pounds and ounces is simplified if the ounces are converted to decimal equivalents of a pound.

Adapting Home-Size Recipes

Although good quantity recipes are readily available today, managers often prefer to develop their own formulations, rely on the expertise of cooks who may prepare an item without a written recipe, or adapt a home-size recipe to quantity production. In a residence hall or school foodservice, students may bring recipes from home and request that items be prepared. A nursing home resident might have a favorite item and share the recipe with the cook as a possible selection for the menu.

Special considerations are necessary in adjusting a recipe designed for 6 to 8 servings to an appropriate quantity for 100 servings or more. The following suggestions are given for expanding recipes from home to quantity size:

- Know exactly what ingredients are used and in what quantity.
- Make the recipe in the original home-size quantity following instructions exactly and noting any unclear procedures or ingredient amounts.
- Evaluate the product for acceptability to determine if the recipe has potential for expansion.
- Proceed in incremental stages in expanding the recipe, keeping in mind the quality and appearance of the original product. Evaluate quality at each stage, deciding if modifications are necessary as the recipe is adjusted.
- Determine handling or cooking losses after increasing the recipe to an amount close to 100 servings; usually 5 to 8% loss is typical. The actual yield of the recipe should be reviewed carefully. Mixing, cooking, and preparation times should be noted, especially for producing the item in quantity, because these items increase substantially for quantity production.
- Check ingredient proportion against a standard large quantity recipe for a product of similar type to assess balance of ingredients.

- As with standardized recipes, evaluate products using taste panels and customer acceptance assessments before recipes are added to the permanent file.

The recipe adjustment method varies with the computer software program, which should be checked at the time of purchase. Research, using various computer programs for recipe adjustment, has shown a great variation in the quantity of each ingredient and, therefore, the quality of the product (Lawless & Gregoire, 1987–88; Lawless, Gregoire, Canter, & Setser, 1991). The factor method is most commonly used for recipe adjustment in a computer program.

Careful evaluation is important in adjusting home-size recipes to quantity production. Some recipes that are suitable for service at home are simply not practical to make in quantity because of the time constraints of large-scale operations. If extensive labor time is required, a product may be too costly for most foodservice operations.

QUANTITY FOOD PRODUCTION

Production of quality food in quantity involves a highly complex set of variables. Quantity food production varies widely with the type of operation and foodservice. Many types of foodservice operations and their different objectives were described in Chapter 1. The one- or two-meal-a-day pattern of a school foodservice operation, for example, presents a different production challenge than the 24-hour-a-day, quick-service restaurant. Because the menu is the basic plan for the foodservice system, planning in the food production subsystem depends on menu item selections.

In Chapter 4, four basic foodservices were described: conventional, commissary, ready prepared, and assembly/serve. Differences in processing foods in these various foodservices were indicated. Obviously, those using completely prepared foods requiring only thawing and heating prior to service have different production demands than conventional foodservices in which foods are prepared from scratch.

Quantity food production involves control of ingredients, production methods, quality of food, labor productivity, and energy consumption, all of which are critical to controlling costs. The true test of the production plan is whether the food is acceptable to the customer, produced in the appropriate quantity, microbiologically safe, and prepared within budgetary constraints.

Quantity is the primary element that introduces complexity to food production in the foodservice system. Producing food for 100 people requires much more careful planning and large-scale equipment than does preparing food for a small group at home. In operations serving several thousand, the complexity is drastically increased. Large-scale equipment and mechanized processes are requisites to producing vast quantities of food to serve large numbers of customers.

The foodservice manager must understand the principles of quantity food production. Expertise in food preparation techniques and equipment usage will enable the manager to perform competently in planning and evaluation of foodservice operations. As one expert in foodservice management stated, quantity food production is the nuts and bolts of the foodservice industry.

For many years, cooking methods were classified only as moist heat and dry heat, and equipment was purchased to perform these functions. Steamers still are manufactured to provide moist heat and broilers, fryers, and ovens to provide dry heat. Production equipment is probably the highest cost item in a foodservice operation budget. In our competitive environment, shopping to save a few extra dollars often yields far less savings than does finding the best value. This is often referred to as the **sweet spot** in pricing—the size or capacity of equipment from which the buyer gets the best value and lowest cost per menu item produced. Much of the cost is for the labor required to make the equipment; downsizing standard size models often yields far less savings than expected. A half-size convection oven is priced only 10 to 15% less than a full-size unit. Value should be considered in terms of performance or capacity per dollar invested. Selecting equipment based on that criterion can offer far more savings than just the lowest price.

Sweet spot: Point of best value at lowest cost.

OBJECTIVES OF FOOD PRODUCTION

Food is cooked for three primary reasons:

- Destruction of harmful microorganisms, thus making food safer for human consumption
- Increased digestibility
- Change and enhancement of flavor, form, color, texture, and aroma

First, cooking at proper temperatures can destroy pathogens. The amount of heat required to kill a particular microorganism depends on such factors as time, method, type of food, and type and concentration of the organism. Adequate cooking is a major factor in foodservice sanitation, but proper handling before and after cooking is also critical.

Many foods become more digestible as a result of cooking. For example, protein in cooked meat is more digestible than in raw meat. Raw starch in foods such as potatoes and flour will gelatinize during cooking or baking and become more digestible.

Nutrient value of some foods can be decreased if they are improperly cooked. Vegetables, for example, often are handled improperly during prepreparation or production, thus causing vitamin or mineral loss. The amount of cut surface and length of time between prepreparation and production affect nutrient retention. Nutrients also may be leached out in the amount of water, length of cooking time, and cooking temperature. Proper attention to time and temperature control is important in preserving quality of food, especially the nutrient value. The foodservice manager should be well versed in concepts of food science to understand changes that occur in food during production.

The aesthetic quality of food can be enhanced by cooking; food can be ruined or be made less palatable by improper procedures. The quality of any cooked food depends primarily on the following four variables:

- Type and quality of raw ingredients
- Recipe or formulation for the product
- Expertise of production employees and techniques used in preparation
- Method and duration of holding food items in all stages from procurement through service

Cooking may enhance, conserve, develop, or blend flavors, as in a sauce or soup. Preservation of color is another aesthetic objective in cooking food; for example, the dull color of overcooked vegetables makes them unappealing. Also, overcooked cabbage, cauliflower, and broccoli become mushy and develop a strong flavor, thus losing appeal.

The contrasts between the qualities of meat prepared properly and improperly provide another vivid example of the effect of cooking on the aesthetic quality of food. Consider the juicy, flavorful quality of a properly broiled steak as compared to one that is dry, tough, and overdone. Baked products also are affected greatly by improper preparation and cooking techniques; tenderness is affected by overmixing, and the effects of over- and underbaking are obvious.

These examples all emphasize the importance of adequate controls throughout the pre-preparation, production, and service processes. The holding stage, particularly in cook-chill or cook-freeze foodservices, is another critical component affecting the aesthetic quality and acceptability of food.

METHODS OF PRODUCTION

Many different processes are involved in production of food for service. Preparation may be as simple as washing and displaying the food, such as fresh fruit, or as complex as the preparation of a lemon meringue pie. Production may include cooking, chilling, and freezing processes or a combination.

Cooking is scientifically based on principles of chemistry and physics. Properties of many ingredients used in food production cause reactions of various types. For example, baking powder, when exposed to moisture, gives off carbon dioxide, and the combination of egg, liquid, and oil produces an emulsion identified as mayonnaise.

Heat Transfer

Heat is the factor that causes many reactions to occur, and the type and amount of heat greatly affect the resulting product. Heat is transferred in four ways (Spears, 1999):

- Conduction
- Convection
- Radiation
- Induction

Conduction: Transfer of heat through direct contact from one object or substance to another.

Conduction is the transfer of heat through direct contact of one object or substance with another. Transfer can occur in any of the three states of matter: solid, liquid, or vapor. Metals, as a group of solids, are good conductors; however, different metals conduct heat at different rates. For example, copper, iron, and aluminum are effective conductors for cooking vessels; stainless steel developed from iron is not as effective. In cooking by conduction, the heat is first transferred from a heat source, usually gas or electricity, through a cooking vessel to food. Conduction is the dominant means of heat transfer in grilling, boiling, frying, and, to some degree, baking and roasting. In pan broiling or grilling a steak, for example, the heat is transferred from the source to the pan or grill and then directly to the meat. In pan frying, fat is the transfer agent between the pan and the food.

Convection: Distribution of heat by the movement of liquid or vapor; may be either natural or forced.

Convection is the distribution of heat by the movement of liquid or vapor and may be either natural or forced. Natural convection occurs from density or temperature differences within a liquid or vapor. The temperature differences cause hot air to rise and cool air to fall. Thus, in a kettle of liquid or a deep fat fryer, convection keeps the liquid in motion when heated.

Forced convection is caused by a mechanical device. In convection ovens and convection steamers, for example, fans circulate the heat, which is transferred more quickly to the food, causing faster cooking. Another means of achieving the convection effect is to circulate the food, rather than the heat. A reel oven with shelves that rotate much like a Ferris wheel is an example of moving the food rather than the air. Stirring is another form of forced convection in which heat is redistributed to prevent concentration of heat at the bottom of the container. For example, in cooking a sauce or pudding, stirring is important not only to speed up the cooking but also to prevent scorching and burning.

Radiation: Generation of heat energy by wave action within an object.

Radiation pertains to the generation of heat energy by wave action within an object. The waves do not possess energy but induce heat by molecular action upon entering food. Infrared and microwave are the two types of radiation used in food production. **Infrared** waves have a longer wavelength than visible light does. Broiling is the most familiar example of infrared cooking. In a broiler, an electric or ceramic element heated by a gas flame becomes so hot that it emits infrared radiation, which cooks the food. High-intensity infrared ovens are designed to heat food even more rapidly. Infrared lamps are commonly used in foodservice operations for holding food at a temperature acceptable for service. For example, in restaurants, infrared lamps are frequently placed over the counter where cooks set the plates of food for pickup by servers.

Microwaves have a very short length and are generated by an electromagnetic tube. In use, microwaves penetrate partway into the food and agitate water and/or fat molecules. The friction resulting from this agitation creates heat, which in turn cooks the product. Because microwave radiation affects only water and fat molecules, a dry material will not become hot in a microwave oven. Thus, disposable plastic or paper plates can be used for heating or cooking some foods. Most microwaves penetrate only about 2 inches into food, and heat is transferred to the center of large masses of food by conduction. Microwave cooking is not a predominant method in foodservice operations, because cooking time increases proportionately with the quantity of the food to be cooked. Thus preparing multiple servings of a product in a microwave oven may take as much as or more time than preparing that same item in an oven or steamer. Microwave ovens are widely used, however, for heating prepared foods for service. For example, microwave units are frequently available in hospital galleys, and vending operations often have small microwave ovens for heating sandwiches and soups. Cooking units are now available with microwave and convection functions that can be used singly or together. Microwaves provide thawing and heating energy, and circulating air provides surface color and texture. For example, potatoes may be cooked more rapidly using both microwave and convection, yielding the texture and flavor of a baked potato.

Induction: Use of electrical magnetic fields to excite the molecules of metal cooking surfaces.

Induction is the use of electrical magnetic fields to excite the molecules of metal cooking surfaces (Riell, 1992). Induction-heat burners that cook magnetically have been developed. The burner has no open flame and the burner surface does not get hot; rather molecules in the pan are activated which produce the heat to cook food. According to foodservice operators, induction heating is fast, even, and clean. The units do not require ventilation.

Production Methods and Equipment

Cooking methods are classified either as moist heat or dry heat. Dry heat methods are those in which the heat is conducted by dry air, hot metal, radiation, or a minimum amount of hot fat. Moist heat methods involve the use of water or steam for the cooking process. Different cooking methods are suitable for different kinds of food. For example, tender cuts of meat should be prepared using a dry heat method, and a tougher cut, such as that used for stew, should be cooked using moist heat.

Quantity food production and equipment are closely related and should be discussed together. Both are important components of cooking technology. Essentially, the equipment available dictates the choice of cooking methods. Almost every technical problem relating to cooking involves either method or equipment. The choices of cooking method and equipment are vital because of their effect on many aspects of daily operations, such as labor scheduling, productivity, product quality, speed of service, sanitation and maintenance, energy conservation, menu flexibility, and cost control. The technology of cooking has become more complex over the past few decades, primarily because of innovations resulting in increased efficiency and sophistication of equipment. Combination cooking methods apply both dry and moist heat to the menu item. Equipment today is identified as moist heat, dry heat, and multifunctional.

Moist Heat

Moist heat methods: Heat is conducted to the food product by water or steam.

The most common **moist heat methods** of cooking are boiling, simmering, stewing, poaching, blanching, braising, and steaming. These methods are similar with only slight differences (Spears, 1999). To boil, simmer, stew, or poach means to cook a food in water or a seasoned liquid. The temperature of the liquid determines the method. To *boil* means to cook in a liquid that is boiling rapidly. Boiling generally is reserved for certain vegetables and starches. The high temperature toughens the proteins of meat, fish, and eggs, and the rapid movement breaks delicate foods.

Steam-Jacketed Kettles

Simmering or *stewing* means cooking in a liquid that is boiling gently, with a temperature of 185 to 205°F. Most foods cooked in liquid are simmered even though the word *boiled* may be used as a menu term, such as boiled corned beef and cabbage. These production methods are frequently done in a **steam-jacketed kettle** (Figure 6.10). The steam-jacketed kettle works like a double boiler. One bowl is sealed inside another with a 2-inch space between them. The jacketed area provides space between the two bowls into which steam is introduced and provides the necessary heat to cook the product in the kettle. The amount/pressure of the steam is adjusted with a knob; the more pressure, the hotter the surface and the faster the food product will cook. Most kettles today are *self-contained* with boiler to make their own steam rather than being *direct-steam* models connected to a building steam supply. Steam-jacketed kettles also can be connected to a chilled

Figure 6.10. Steam-jacketed kettle.
Source: Groen, Dover Industries Company. Used by permission.

water supply that will fill the jacketed area with chilled water to help quickly cool ingredients in the kettle. Steam-jacketed kettles often are designed with a pouring lip and mounted on trunnions to allow the contents of the kettle to be easily poured into serving pans.

To *poach* is to cook in a small amount of liquid that is hot but not actually bubbling. The temperature range is about 160 to 180°F. Poaching is used to cook delicate food, such as fish, eggs, or fruit. Because poached foods are bland, the chef has to create maximum flavor without using high-calorie, -fat, and -sodium sauces for healthful cooking.

Blanching is defined as cooking an item partially and briefly, usually in water, although some foods, such as french fries, are blanched in hot fat. Two methods are used for blanching in water. To dissolve blood, salt, or impurities from certain meats and bones, the item is placed in cold water, brought to a boil, simmered briefly, then cooled by plunging in cold water. Blanching is also used to set the color of and destroy enzymes in vegetables or to loosen skins of vegetables and fruits for easier peeling. For this latter purpose, the item is placed in rapidly boiling water, held there until the water returns to a boil, and quickly cooled in cold water.

Braising involves cooking food in a small amount of liquid, usually after browning it. In today's price-conscious economy, chefs are showcasing the more flavorful, less tender, and less costly cuts of meat by preparing braised beef menu items. Braised meats usually are browned in a small amount of fat, which gives a desirable appearance and flavor to the product and sauce. With today's emphasis on healthful cooking, cubed meat can be quickly seared in hot stock, assuming a brown color without taking on added fats, and then finished at a lower temperature. The liquid often is thickened before serving by reducing it over hot heat rather than using a thickening agent. In some recipes for braised items, no liquid is added because the item cooks in its own moisture. Braising may be done on the range or in the oven, although today a covered tilting skillet is frequently used. Other terms describing the braising process are *pot roasting, swissing,* and *fricasseeing*.

Heat transfer rate: The amount of heat from one substance to another in a given amount of time in a given space measured in British thermal units (Btu).

Steamers

To *steam* is to cook foods by exposing them directly to steam. Steam, the gaseous form of water, is often used for cooking in foodservice operations because it is a very efficient heat transfer medium. The **heat transfer rate** of steam is 300 Btu as

psi: pounds per square inch, used to measure steam pressure.

compared to 7 Btu in a convection oven and 38 on a griddle. **Pressure steamers** (2–5 **psi** low pressure or 15 psi high pressure) work by trapping and removing air that causes steam pressure to build (Spears, 1999). The pressureless convection steamer (Figure 6.11) circulates steam in the cooking chamber to shorten the cooking time. Because the cavity is not under pressure, a pressureless steamer's door may be opened at any time during the process to check cooking progress or to remove or add food. In a **pressureless convection steamer,** heat transfer from steam to food is accomplished by forced convection caused by a fan inside which encircles the food, thus cooking it without pressure. Because of a continuous venting system in pressureless steamers, the result is fast, gentle cooking at a low temperature, 212°F. Pressureless steamers are well suited for a wide variety of foods, from fresh to loose pack frozen or frozen block vegetables. Because the steam is continuously vented, unwanted flavor transfer from one food to another is eliminated.

Dry Heat

The major methods of cooking without liquid are roasting, baking, oven frying, broiling, grilling, barbecuing, rotisserie cooking, and frying, including sauteeing, pan frying, and deep fat frying. Some foodservice managers contend that including foods cooked in oil as dry cooking does not make sense because oil is wet. Perhaps three methods of cooking should be recognized: cooking in liquids, cooking in dry heat, and cooking in oil. In this text, however, frying in oil will be included under dry heat (Spears, 1999).

Broilers

A **broiler** has its heat source above the rack that holds the food, usually meat, poultry, and seafood. The food is placed 3 to 6 inches from the heat, depending on the type and intensity of the heat (Spears, 1999). The temperature required depends on the amount of fat, tenderness, or thickness of the food item. Juices and fat from the cooking food drop to a drip pan below the food, reducing the likelihood of fires during cooking. Charbroiling has become popular in many foodservice operations, especially in steak houses and quick-service hamburger

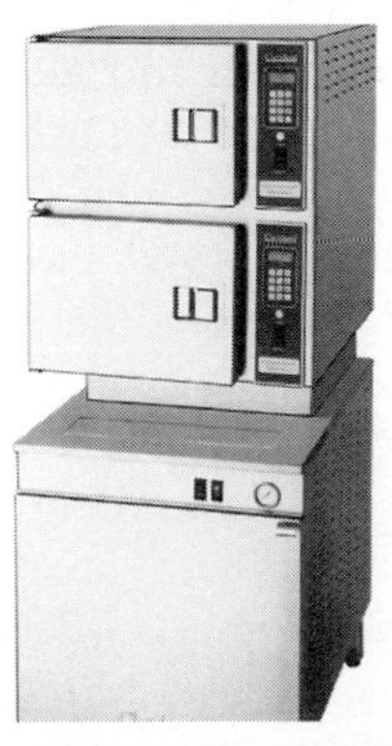

Figure 6.11. Pressureless convection steamer.
Source: Cleveland Range Co., Inc. Used by permission.

Figure 6.12. Charbroiler.
Source: U.S. Range. Used by permission.

establishments. A **charbroiler,** as shown in Figure 6.12, can be heated with gas, electricity, charcoal, or wood. Gas and electric models have a bed of ceramic briquettes above the heat source and below the grid that holds the food. Because the heat source is from below, it technically is a grilling and not a broiling method. Because juices and fat from the cooking food drop onto the heat source, flames and smoke often occur during cooking, which can cause safety concerns. The **salamander** and **cheesemelter** are specialty broilers that are much smaller and often are mounted above a range. The two units differ in that the cheesemelter has a much lower heat output and is used for finishing (such as melting cheese) rather than cooking foods.

Grilling, griddling, and pan broiling are all dry heat cooking methods that use heat from below. Grilling is taking on an international flavor that is exciting customers. Asian teriyakis, Tex-Mex fajitas, Middle Eastern kebabs, and all-American cowboy-grilled steaks are highlights. **Grilling** is done by placing food items on an open grid over a heat source, which may be an electric or gas-heated element, ceramic briquettes, or exotic woods and flavored chips. **Griddling** involves the cooking of food on a flat surface (griddle) that is heated from below by gas or electricity. The griddle cooking surface, or plate, can be cast iron, polished steel, cold rolled steel, or chrome finish plates.

Lang Manufacturing Company introduced double-sided grilling technology with the **Clamshell** and Add-On Clamshell hood that attaches to the foodservice's existing equipment (Figure 6.13). Rapid cooking action from the hood's infrared broiler and the grilling heat from below drive the natural juices in the food to the center and decrease food shrinkage, resulting in juicy meat in half the time. For example, New York steaks that take 7 minutes to broil can be served after only 4 minutes and chicken that takes 6 minutes to broil takes only 3 minutes. Of course, cooking times may vary depending on the thickness and weight of the raw product. A newer piece of grilling equipment is the Mongolian barbecue. It is a round, flat-surface griddle, 4 feet in diameter, that often rotates. At Mongolian barbecue restaurants, guests select foods they want cooked and watch as staff cook the items on the large griddle.

Barbecue-style meats can be cooked in conventional equipment or specialized barbecue equipment as follows (Cooking Equipment, 1986):

Figure 6.13. Clamshell.
Source: Lang Manufacturing Company. Used by permission.

- Basic open pit (steel or concrete wood-burning firebox under a grate, some with rotisserie spits)
- Cylindrical or kettle-shaped smokers with domed lids, some with pans
- Large-capacity, closed-pit, wood-burning rack or reel ovens with fireboxes to the rear
- Gas or electric upright rack ovens that cook at temperatures from 175 to 250°F (smoke is dispersed by wood-charring or a smoke-concentrate device)
- Pressurized barbecue smokers that combine heat (350–425°F) with pressure (12–15 psi) and smoke from wood chips

Deep Fat Fryers

A **deep fat fryer** has a frying kettle or frying bin of oil heated by gas, electric, or infrared burners into which foods are immersed in a fryer basket to be cooked (Figure 6.14). Tremendous improvement has been made in deep fat frying equipment for foodservice operations (Spears, 1999). Three of the most important developments are precise thermostatic control, fast recovery of fat temperature (permitting foodservice operations to produce consistent-quality fried food rapidly), and the automatic filtering of the oil to remove food particles and extend the useful life of the cooking oil. The frying kettle is divided into two zones, the *cooking zone,* which is above the heating elements, and the *cold zone,* which is the space below the cooking element where crumbs and debris fall. Because this space is cooler than the cooking oil, these particles are less likely to burn. **Heat recovery time**, the time it takes for the oil to return to the optimal cooking temperature after food has been added, will greatly impact the total cooking time in a fryer.

Heat recovery time: Amount of time for oil in fryer to return to optimum cooking temperature after food is added.

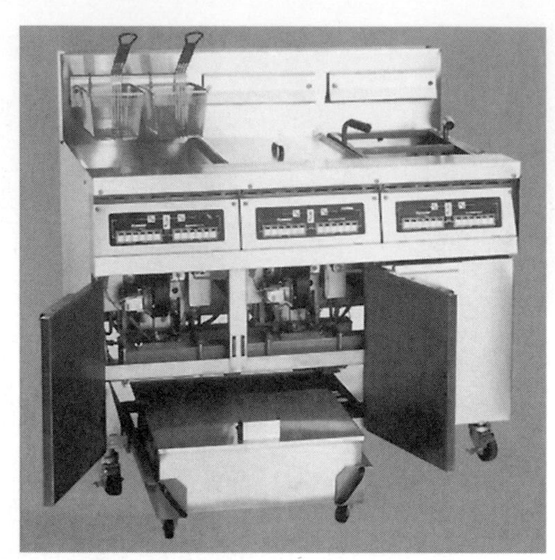

Figure 6.14. Deep fat fryer.
Source: Frymaster. Used by permission.

Ovens

Roasting or baking uses a combination of all three modes of heat transfer: conduction, convection, and radiation. Heat is conducted through the pan to the food, and natural or forced convection circulates the air. Heat also radiates from the hot walls of the cooking chamber. A wide range of oven types is used in foodservice operations. The three main categories are hot-air ovens, infrared broilers, and microwave ovens. The first questions to ask before buying an oven are

- What are we cooking?
- Where are we cooking it?

Many ovens worked fine for a specific type of food 15 or 20 years ago, when most menu items were prepared from scratch. Today in many foodservice operations, more prepared foods and frozen ingredients that require a different type of preparation are being used. Many types of ovens are available, and finding the right one for the operation takes hours of research (Figures 6.15 and 6.16).

- A **range oven** is part of a stove and is located under the cooking surface. Primarily used in small operations.
- A **deck oven** (Figure 6.15a) has traditionally been the standby of hot-air ovens. It is so named because pans of food usually are placed directly on a ceramic or stainless steel deck (bottom of the oven) of an oven that is either 8 or 12 inches high. Typically several deck oven units are stacked on top of each other. Deck ovens typically have separate temperature controls for heating units in the top and bottom of the oven, allowing for more controlled baking of bakery items such as breads, rolls, and desserts.
- A **convection oven** (Figure 6.15b) has a fan on the back or side wall that creates currents of air within the cooking chamber. It has more space and holds two to three times as much food, reduces cooking time by 30%, and cooks at 25 to 35°F lower temperatures, thus conserving energy.

(a) **Stack Deck Oven**
Source: Blodgett. Used by permission.

(b) **Convection Oven**
Source: Blodgett. Used by permission.

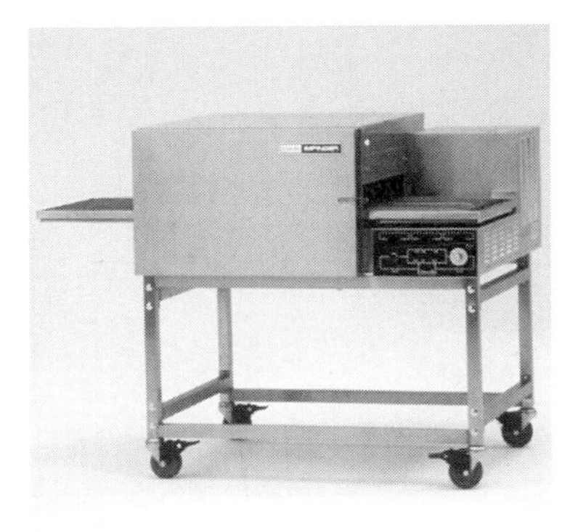

(c) **Impinger Oven**
Source: Lincoln Foodservice Products. Used by permission.

(d) **Microwave Oven**
Source: Panasonic-Commercial Food Service, www.panasonic.com/cmo. Used by permission.

(e) **Smoker Oven**
Source: Cookshack. Used by permission.

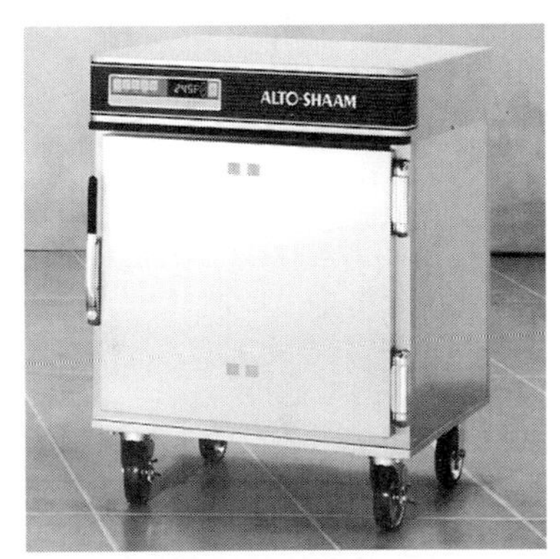

(f) **Low-Temp Cooking and Holding Oven**
Source: Alto-Shaam. Used by permission.

Figure 6.15. Commercial ovens.

- A **conveyor oven** often is called a pizza oven in quick-service operations. Food travels on a moving conveyor belt through the oven, cooking as it moves through the oven. The **impinger** (Figure 6.15c) is a conveyorized oven that blasts high-velocity, heated air into the oven cavity, allowing food to cook much more quickly than in other ovens. Heat sources include infrared, quartz, gas, and electric, which can produce oven temperatures of 600°F.
- A **microwave oven** (Figure 6.15d) is used more for heating prepared foods for service than for cooking foods. Microwaves are generated by an electromagnetic tube that produces microwaves that penetrate partway into the food and agitate water and/or fat molecules in the food. It is this agitation of molecules that actually cooks the food.
- A **smoker oven** (Figure 6.15e) is an electric, compact-size oven with racks to smoke up to 100 pounds of meat at a time. It uses wood chips to produce smoke, and the meat is rubbed with spices and barbecue sauce to give it a piquant flavor.

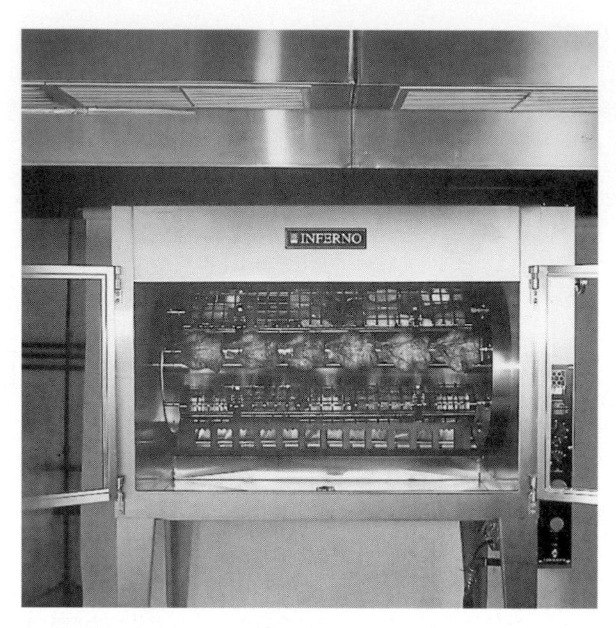

Figure 6.16. Rotisserie multicooking in action.
Source: Hardt Equipment. (1999). Used by permission.

- A **low-temp cooking and holding oven** (Figure 6.15f) has cooking temperatures from 142 to 325°F and holding temperatures from 60 to 200°F. Yield of meats can be increased up to 25% and energy consumption reduced by 30 to 50% when cooked in these ovens.
- A **rotisserie oven** (Figure 6.16) contains rows of metal spits or baskets on which food is placed. The spits or baskets are rotated and the food slow cooked in the oven's warm, usually moist, cavity. The cooking source can be gas, electricity, wood, or charcoal. These ovens were popularized by Boston Market and now often are found in grocery stores as well.
- A **wood-fired oven** burns wood in a well-insulated cavity to heat stone or brick blocks on which the food items are cooked. Wood burned in the ovens usually is chosen based on the smoky flavor it will impart on the food items cooked in the oven, how easy it is to light, and its moisture content (Katsigris & Thomas, 1999).
- A **FlashBake oven** uses a combination of intense light and infrared energy to cook foods quickly. Products such as pizza, chicken breasts, and french fries can be cooked in much less time (2 minutes for a chicken breast, 1 minute for french fries).

Multifunction Equipment

Cutting down on the square footage of a foodservice is a goal in both commercial and onsite foodservice operations. The range, often referred to as a stove, probably was the first piece of multifunction foodservice equipment. Space is expensive and should be used for increasing revenue rather than for production. The amount of labor hours required to prepare food can be decreased if employees do not have to walk miles every day. Three pieces of cooking equipment are manufactured to alleviate many of the space and labor problems in a foodservice operation: the combination convection oven/steamer or combi-oven, the tilting fry or braising pan, and the convection/microwave oven.

- The **combi-oven** (Figure 6.17) directs the flow of both convected air and steam through the oven cavity to produce a super-heated, moist internal atmosphere. Combi-ovens are considered one of the hottest foodservice tools on the market today (Spears, 1999). Industry insiders call them a revolution in cooking. Meat, seafood, poultry, vegetables, and even delicate meringues, pastries, and breads with a crusty surface can be prepared in this oven. Four cooking methods are combined in one unit: convection, steam, convection plus continuous steam, and convection plus cycled steam. The advantage of this combination is its versatility, which permits menu expansion with a single piece of equipment, conservation of valuable floor space, and faster cooking with minimal shrinkage and maximum retention of flavor, color, and nutrients. When steam is added to convection-cooked roasts, the meat stays more moist (Fellin, 1996).
- The **tilting skillet or tilting frypan** (Figure 6.18) is a floor-mounted rectangular pan with a gas- or electric-heated flat bottom, pouring lip, and hinged cover. It is considered the most versatile of all kitchen equipment, because it combines the advantages of a range, griddle, kettle, oven, stock pot, bain marie, and frying pan. These skillets can cook batches of food, such as pasta,

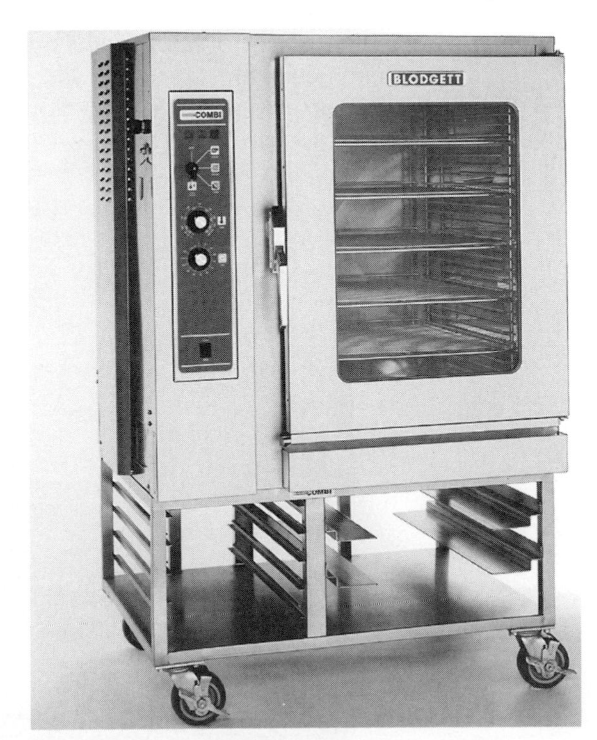

Figure 6.17. Combi-oven.
Source: Blodgett. Used by permission.

Figure 6.18. Tilting skillet.
Source: Groen, Dover Industries Company. Used by permission.

stew, gravies, sauces, and multi-ingredient entrées, and individual orders of bacon and eggs, hamburgers, or steaks.

- The **convection/microwave oven** (Figure 6.19) is a multifunction piece of equipment with convection and microwave capabilities. It can use convection air alone, microwave energy alone, or a combination of the two. It really is a convection oven that browns, bakes, and roasts and a microwave that steams

Figure 6.19. Convection/microwave oven.
Source: Amana Appliances. Used by permission.

and cooks foods. Hot-air jets from a convection oven provide surface color and texture, and microwave energy provides major thawing and heating of the product. By combining these two functions, a turkey or roast can be cooked in a much shorter time than usual, and the product will be juicier on the inside and brown on the outside. A beautiful brown, juicy, broiler chicken can be ready to serve in 18 minutes.

PRODUCTION CONTROLS

Control is the process of ensuring that plans have been followed. Therefore, the essence of control is comparing what is set out to do with what was done and taking any necessary corrective action. Control has not been effective until action has been taken to correct unacceptable deviations from standards.

In essence, then, *quality control* means assuring day-in, day-out consistency in each product offered for consumption. Quantity control, simply stated, means producing the exact amount needed—no more, no less. Each type of control is directly related to control of costs and thus to profit in a commercial operation or to meeting budgetary constraints in a nonprofit establishment. Over- and underproduction create managerial problems and have an impact on cost. Time and temperature, product yield, portion control, and product evaluation all relate directly to quality, quantity, and indirectly to control of costs.

Time and Temperature Control

Time and temperature are critical elements in quantity food production and must be controlled to produce a high-quality product. Excess moisture loss will occur in most products if the cooking time is extended even with a correct temperature. Most food production equipment today has timing devices as an integral part of construction. Equipment is available with computer controls that can be programmed by the manufacturer or operator to control cooking times.

Temperature is recognized as the common denominator for producing the correct degree of doneness. To assure this degree, temperature gradations vary dramatically for different food categories, depending on the physical and chemical changes that occur as food components reach certain temperatures. When foods are subjected to heat, the physical state is altered as moisture is lost. Chemical changes are complex and differ greatly among various food products. An area in which progress is being made is the development of more sophisticated ways of measuring "doneness," such as moisture readings, vapor-content analyzers, and optical sensors to detect color (Riell, 1992).

The proper control of temperature often is dependent on thermostats, which control temperature automatically and precisely. A well-designed thermostat should not be affected by changes in ambient temperatures but instead should respond quickly to changes on the sensing element and be easily set. The functioning of the thermostat should be checked periodically.

Standardized recipes should state temperatures for roasting and baking. Figure 6.20 shows the temperatures commonly used in food production, in both centigrade and Fahrenheit. Tables should be developed for cooking time and

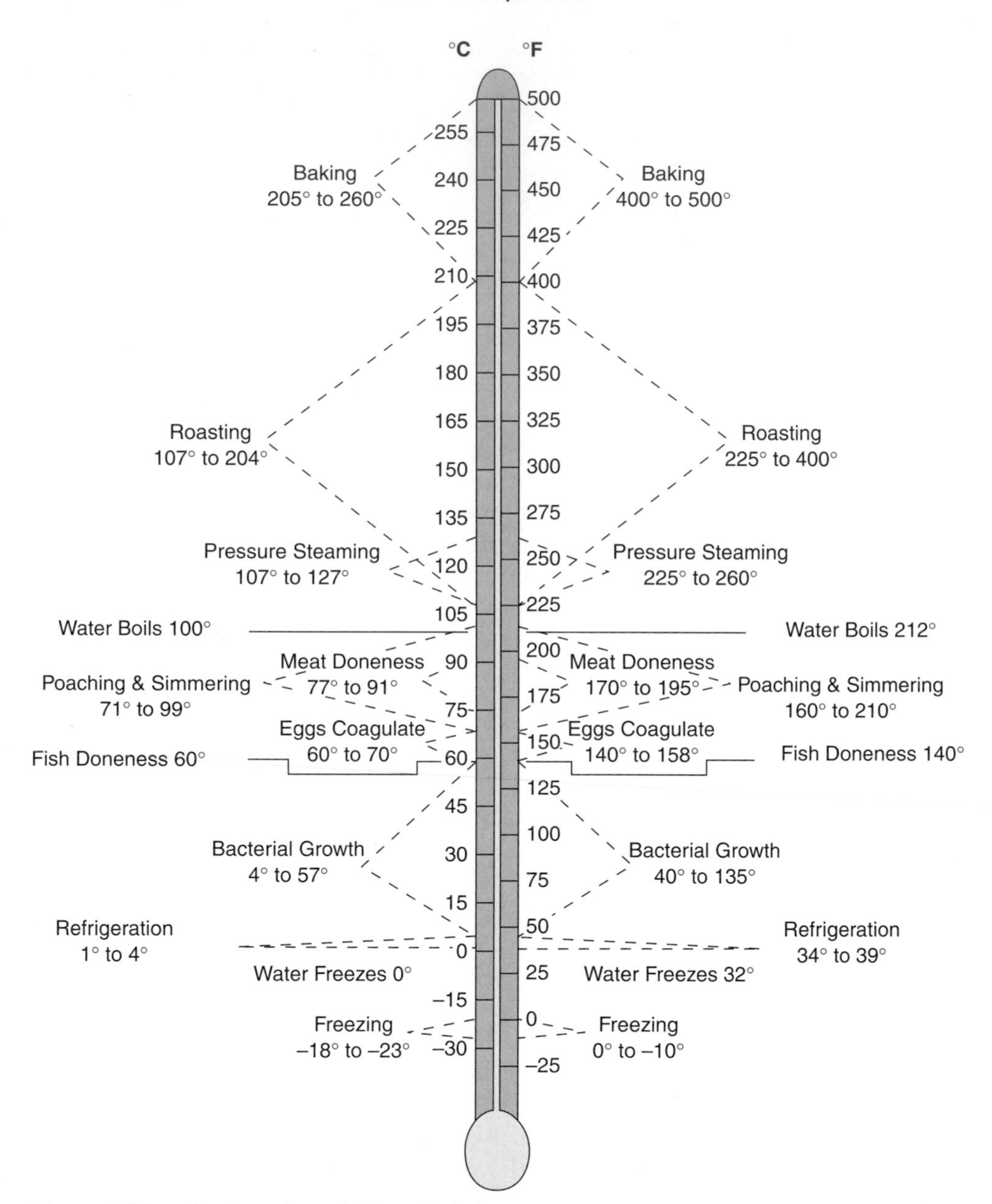

Figure 6.20. Centigrade and Fahrenheit thermometer showing temperatures commonly used in the kitchen.
Source: Adapted from *Food Preparation for the Professional* (p. 108), by D. A. Mizer, M. Porter, and B. Sonnier, 1978, New York: John Wiley & Sons. Reproduced by permission.

temperature for items on the menu, including end-point temperatures for various products. For example, the degree of doneness for meat is best stated in terms of end-point temperature.

Various types of thermometers are needed for determining end-point temperatures or for checking the temperature of foods being held. A meat thermometer, for example, should be used to determine the internal temperature of meat. The chart in Table 6.10 shows the internal temperatures of meats at three levels of doneness:

Table 6.10 Internal Temperatures of Cooked Meats

Raw Products	Internal Temperature
Ground Products	
Hamburger	160°F
Beef, pork, lamb	160°F
Chicken, turkey	160°F
Beef, Lamb, Veal	
Roasts & steaks	
medium-rare	145°F
medium	160°F
well-done	170°F
Pork	
Chops, roasts, ribs	
medium	160°F
well-done	170°F
Hams, fresh	160°F
Sausage	160°F
Poultry	
Chicken, whole & pieces	180°F
Duck	180°F
Turkey—not stuffed	180°F
Whole	180°F
Breast	180°F
Dark meat	180°F
Stuffing, cooked separately or in the bird	165°F
Eggs	
Fried, poached	Yolks & white are firm
Casseroles	160°F
Sauces, custards	160°F

Source: USDA/FDA (1997).

Carry-over cooking: The increase in internal meat temperature after removal from the heat source.

rare, medium, and well done. In meat cookery, however, **carry-over cooking** must be taken into account, especially for large cuts of meat. Carry-over cooking means that the internal temperature of the meat will continue to rise even after the meat is removed from the oven. This phenomenon occurs because the outside of the roasting meat is hotter than the inside, and heat continues to be conducted into the meat until the heat is equalized throughout the piece.

To prevent *E. coli* breakouts, hamburgers should be cooked thoroughly to 160°F. USDA's Food Safety and Inspection Service (FSIS) emphasizes using a meat thermometer, based on research that indicates some ground meat may turn prematurely brown before a safe internal temperature of 160°F has been reached (Meat Industry Insights, 1998). New research shows that premature browning may be more prevalent than originally thought and may occur under normal consumer handling conditions. Other research findings show that some ground meat patties cooked to 160°F or above may remain pink in color for a variety of reasons. FSIS alerts consumers that the color of meat is no longer considered a reliable indicator of ground beef safety. A meat thermometer is the most reliable way to reduce the risk of foodborne illness. An "instant read" thermometer should be used toward the end of the cooking time and register a temperature in about 15 seconds. The meat thermometer should penetrate the thickest part of the hamburger. If a beef patty is not thick enough to check from the top, the thermometer may be inserted sideways.

Time and temperature are closely related elements in cooking. Some foods, like prime rib roast, can be cooked for a longer period of time at lower temperature to achieve the desired degree of doneness; others, like strip steak, should be cooked for a shorter time at higher temperatures.

The integral nature of time-temperature relationships is perhaps best illustrated with baked products in which accurate timing and temperature control are critical to producing high-quality products. For example, a cake baked at too high a temperature will cook too quickly on the outside before the baking powder or soda has produced expansion in the batter. As a result, the cake will crack and become too firm before it is baked internally.

Product Yield

Yield: Amount of product resulting at the end of the procurement/production process.

As purchased (AP): Amount of food before processing.

Edible portion: Amount of food available for eating after preparation and/or cooking.

Yield is the amount of product resulting at the completion of the various phases of the procurement/production/service cycle. It usually is expressed as a definite weight, volume, or serving size. For most foods, losses in volume or weight occur in each phase, although a few foods, such as rice and pasta, increase in volume during production. **As purchased (AP)** is the amount of food bought before processing to give the number of edible portions required to serve a specific number of customers. The AP weight of meat, fish, and poultry decreases before being cooked for many reasons, such as the removal of skin and bones or trimming of fat. What results is the **edible portion (EP)**, which is the weight of a menu item without skin, bones, and fat available to serve the customer after it is cooked. During cooking, shrinkage occurs, up to 35% for a roast. Carving subtracts another 5% before serving to the customer (Kotschevar & Donnelly, 1999). Buyers must factor in these losses to determine the as served (AS) yield, which is important for menu item pricing. Only delicatessen cold meat and poultry products are 100% edible; the cost of bones, skin, and cooking loss is included in the purchase price. The *Food Buying Guide for Child Nutrition Programs* (2002), published by the USDA for use in child nutrition programs, provides data on the edible yield percentages of various foods to assist food buyers in planning amounts to purchase. The cooked yields of selected meat, poultry, and fish products are listed next:

Product	Yield
Ground beef (less than 30% fat)	1 lb AP = .70 lb cooked, drained lean meat
Pork spare ribs	1 lb AP = .3916 lb cooked lean meat
Chicken breast	1 lb AP = .47 lb cooked, boned, chicken meat without skin
Turkey	1 lb AP = .47 lb cooked turkey without skin

As an example, if a 4-ounce grilled boneless and skinless chicken breast is desired, an 8.5-ounce breast with rib bones and skin should be purchased. To calculate the amount of chicken breast with rib bones and skin to purchase, divide the 4-ounce EP by the 47% (0.47) yield.

$$\text{4-ounce EP} \div 0.47 = \text{8.5-ounce AP}$$

The EP price is important to the foodservice manager. Assume the AP price of the chicken breast with bones and skin is $0.106 per ounce. To calculate the EP price of serving a 4-ounce boneless, skinless chicken breast ($0.424 for 4 ounces of chicken), divide the AP price ($0.90) by the EP yield percentage (.47).

AP price ÷ EP yield = EP price

or

$0.424 ÷ 0.47 = $0.90

Fresh fruits and vegetables, which often are peeled, seeded, and cooked, decrease in weight before being consumed, as shown.

Product	**Yield**
Apples	1 lb AP = .78 lb ready-to-serve raw, cored, or peeled apple
Bananas	1 lb AP = .64 lb peeled banana
Cantaloupe	1 lb AP = .47 lb ready-to-serve raw cantaloupe
Watermelon	1 lb AP = .61 lb ready-to-serve raw watermelon
Broccoli	1 lb AP = .81 lb ready-to-cook broccoli
Carrots	1 lb AP = .70 lb ready-to-cook or serve carrots
Iceberg lettuce	1 lb AP = .76 lb ready-to-serve lettuce

The recipe in *Food for Fifty* (Molt, 2001) for seven 9-inch apple pies (56 servings) requires 12 pounds of peeled, cored apple slices. A raw apple has a 78% yield. To calculate the number of pounds of whole apples to purchase for the recipe, divide 12 pounds of apples by 78%, or 0.78.

12 lb apples EP ÷ 0.78 = 15.4 lb apples AP

If the AP price for one pound of apples is $0.80, the buyer then can calculate the EP price.

EP price = $0.80 ÷ 0.78 = $1.03 per pound

In addition to losses during prepreparation, losses may occur during actual production, in portioning, or in panning for baking. **Cooking losses** also account for decrease in yield of many foods, primarily because of moisture loss. As discussed previously in this chapter, the time and temperature used for cooking a menu item will affect the yield.

Handling loss occurs not only during production but also during portioning for service. McManis and Molt (1978) state that handling losses must be considered in determining desired yield. They indicate that cooking and handling loss may vary from 1 to 30%, depending on the product being produced and served. In the brownie recipe included in Table 6.8, a 3% handling loss was reported to illustrate the percentage method of recipe adjustment.

When preparing food in quantity, these losses can have a cumulative impact on the number of portions available from a recipe. That loss must be considered when estimating production demand. Total yield and number of portions should be stated in a standardized recipe, taking into account changes in yield that occur from as purchased (AP) to edible portion (EP) to serving yield.

Table 6.11 Percent Yield by Weight of Potato Baked by Two Methods

Cultivar	Method 1[a] (%)	Method 2[b] (%)
Irish cobbler	83	97
Katahdin	77	95
Russet Burbank	76	94

[a]Baked in 204°C (400°F) oven, skin rubbed with oil.
[b]Baked in 204°C (400°F) oven, skin covered with aluminum foil.

Source: "Guidelines for Determining and Reporting Food Yields," by R. H. Matthews, 1976, *Journal of the American Dietetic Association,* 69, p. 398. Used by permission.

Equipment and cooking procedures also affect food yield. As Table 6.11 shows, the cooking method can have a major impact on resulting yield. A russet Burbank potato baked in foil resulted in 94% yield by weight compared with 76% for one baked without foil.

Less data are available, however, on handling losses during the production and service of food. Studies should be conducted to determine these data and thus have accurate information on production and service yields. In computer-assisted management information systems, data on yields from purchase to service are especially important if the computer is to produce reliable information for ordering, recipe adjustment, production planning, cost, and nutrient composition of food items.

Portion Control

Portion control: Service of same size portion to each customer.

Portion control is one of the essential controls in production of food in quantity. In essence, it is the achievement of uniform serving sizes, which is important not only for control of cost but also for customer satisfaction. Customers of any food service establishment are concerned about value received for price paid, and they may be dissatisfied if portions are not the same for everybody. In a commercial cafeteria, for example, if one person is served a larger portion of spaghetti than another, customers may believe they have been treated unfairly even though the smaller portion may be adequate to satisfy their appetites.

Achieving portion control results from following several important principles:

1. Foods should be purchased according to detailed and accurate specifications to assure that the food purchased will yield the expected number of servings. Often, products are ordered by count or number, with a definite size indicated. As an example, the food buyer may order 150 4-ounce portions of round steak for a Swiss steak recipe. An alternate way of controlling portion size is buying products in individual serving sizes. Examples are individual boxes of cereal, butter and margarine pats, packaged crackers, and condiment packets.

2. Standardized recipes should be used. A standardized recipe will include information on total number and size of portions to be produced from the recipe. Following recipe procedures carefully during prepreparation, production, and service ensures that the correct number of portions will result. For example, cooking a roast at too high a temperature will result in excess moisture loss and decreased yield.

Table 6.12 Disher Equivalents

Disher Number[a]	Approximate Measure[b]	Approximate Weight[b]	Suggested Use
6	10 Tbsp (2/3 cup)	6 oz	Entrée salads
8	8 Tbsp (1/2 cup)	4–5 oz	Entrées
10	6 Tbsp (3/8 cup)	3–4 oz	Desserts, meat patties
12	5 Tbsp (1/3 cup)	2 1/2–3 oz	Croquettes, vegetables, muffins, desserts, salads
16	4 Tbsp (1/4 cup)	2–2 1/4 oz	Muffins, desserts, croquettes
20	3 1/5 Tbsp	1 3/4–2 oz	Muffins, cupcakes, sauces, sandwich fillings
24	2 2/3 Tbsp	1 1/2–1 3/4 oz	Cream puffs
30	2 1/5 Tbsp	1–1 1/2 oz	Large drop cookies
40	1 1/2 Tbsp	3/4 oz	Drop cookies
60	1 Tbsp	1/2 oz	Small drop cookies, garnishes
100	Scant 2 tsp		Tea cookies

[a]Portions per quart.
[b]These measurements are based on food leveled off in the disher. If food is left rounded in the disher, the measure and weight are closer to those of the next-larger disher.
Source: Adapted from *Food for Fifty* (p. 39) by M. Molt, 2001, Upper Saddle River, NJ: Prentice Hall. Used by permission.

3. Managers and employees should know the size and yield of all pans, dishers, and ladles. The pans and serving utensils should be selected in the standardized recipe. Many portion control utensils, such as dishers, "spoodles," and ladles, are on the market to help foodservice employees accurately and consistently portion foods. The approximate yield and typical uses for various sizes of dishers are given in Table 6.12. A number is commonly used to indicate the size of these utensils, which specifies the number of servings per quart when leveled off. For example, a level measure of a number 8 disher yields eight servings per quart, each portion measuring about 1/2 cup. Dishers are used during production to ensure consistent size for such menu items as meatballs, drop cookies, and muffins and during service to ensure correct serving size. Spoodles combine the ease of serving with a spoon and the portion control of a ladle into one handy utensil. Vollrath has developed a color coding system to help employees choose the right size (Figure 6.21). The handles come in 11 different colors; for example, a 1 oz utensil has a black handle, 2 oz a blue handle, 3 oz a pink handle, 6 oz a turquoise handle, and 8 oz an orange handle. If the menu calls for 2 ounces of corn, the manager can tell the employee to use a blue spoodle, or for 1/2 cup of ice cream to use a gray disher. The approximate weight and measure for various sizes of ladles are given in Table 6.13.

Because these utensils are subject to hard usage, they need to be good quality and NSF listed. They should be made of 18-8, Type 304 stainless steel to prevent corrosion, ensure long-lasting service, and be dishwasher safe. The handle of a disher should be one piece and extended to provide a sure grip for comfortable use. An ergonomically approved thumb tab to reduce operator fatigue should be specified for natural hand movement. Spoodles need handles that are designed for hours of comfortable use and should remain cool to the touch up to 230°F and up to 180°F for ladles.

Other types of serving and cooking equipment include spoons, tongs, hamburger and pancake turners, and spatula/scrapers. They all need to be selected carefully because they last for a long time. For example, spoons should be made of 18-8, Type 304 stainless steel and can be solid, perforated, or slotted. Many have

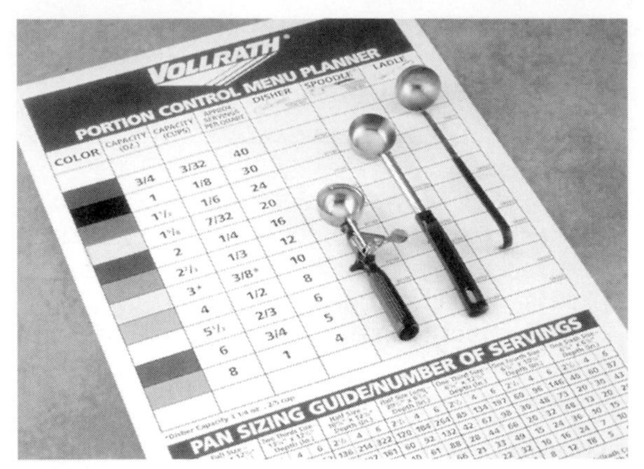

Color coding system

Color coded spoodles

Color coded ladles

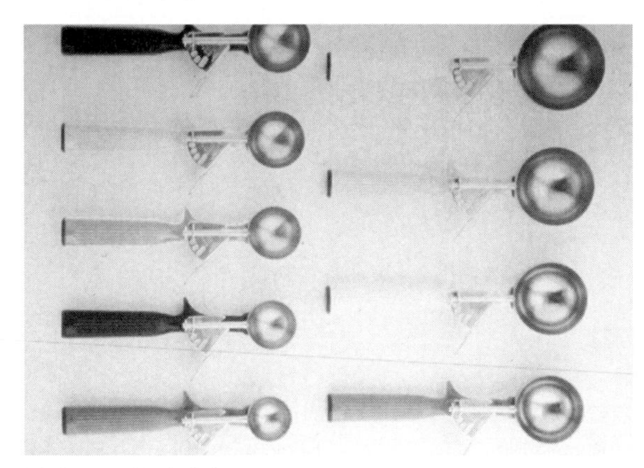

Color coded dishers

Figure 6.21. Color-coded utensils.
Source: The Vollrath Company, L.L.C. Used by permission.

plastic handles that resist heat to 230°F. Many other types of small equipment must be purchased according to menu requirements.

Scales of many kinds are used throughout a foodservice because weight control is essential from the time the food enters the operation from the supplier until it is served. The use of scales during receiving and production has been discussed in

Table 6.13 Ladle Equivalents

Approximate Weight[a]	Approximate Measure[a]	Approximate Portions per Quart	Suggested Use
1 oz	1/8 cup	32	Sauces, salad dressings
2 oz	1/4 cup	16	Gravies, some sauces
4 oz	1/2 cup	8	Stews, creamed dishes
6 oz	3/4 cup	5	Stews, creamed dishes, soup
8 oz	1 cup	4	Soup

[a]These weights and measures are based on food leveled off in the ladle. If food is left rounded in the ladle, the weight and measure are closer to those of the next-larger ladle.

Source: Adapted from *Food for Fifty* (p. 40) by M. Molt, 2001. Upper Saddle River, NJ: Prentice Hall. Used by permission.

previous chapters. Portion scales are an important tool in quantity control in both the production and service units. Cooks and servers should have portion scales available for checking serving sizes.

Slicers are a valuable piece of equipment to assist in ensuring portion control because foods can be sliced more evenly and uniformly than can be done by hand. An example of a slicer is shown in Figure 6.22. A device on a slicer permits adjusting the blade for the desired degree of product thickness. Portion control guides in a foodservice operation should include the correct setting on the slicer for various food items. Automatic slicers are available that do not require the operator to return the food holding device manually, which frees the worker to continue other tasks while food is being sliced. The slicer is used most commonly for meat and meat products; however, it can also be used for vegetables, cheese, bread, and many other items. It is important, however, to clean and sanitize a slicer thoroughly between food products. In portioning sliced foods for service, a portion scale should be readily available for a periodic check on portion size.

To illustrate how lack of control can create food shortages and drive up food cost, assume the server is not careful in serving the chicken and vegetable stir-fry entrée and heaps, rather than levels off, the product in the ladle. The result of yielding only four ladles per quart and not the intended five has several ramifications. The recipe calculated to serve 600 will serve only about 480, and the desired food cost of $1.00 per serving will increase to $1.25. If a product is substituted because of the 120-portion shortage, additional cost may be incurred; the substitution may be a quickly prepared item that is more expensive than the chicken and vegetable stir-fry entrée, resulting in further cost overruns.

Figure 6.22. Food slicer.
Source: Hobart Corporation. Used by permission.

Table 6.14 Portion Control Guide for Selected Menu Items in an Employee Cafeteria

Menu Items	Portion Size
Breakfast Items	
Bacon	2 slices
Cereal, hot	4 oz ladle
Eggs, soft cooked or fried	2
Eggs, scrambled	#16 disher
English muffins	2 halves
Sausage link	2 links
Syrups or sauces	2 oz ladle
Luncheon Entrées	
Baked beans	1 #12 disher
Barbecue meat on bun	1 #16 disher
Meat, egg, or fish salad on sandwich or cold plate	1 #12 disher or 2 #24 dishers
Soups, stews, chili, chowders	1 6 oz ladle
Stuffed green peppers	#8 disher stuffing in half pepper
Dinner Entrées	
Chicken, country fried	1 breast and 1 wing OR 1 thigh and 1 leg
Chicken, curry	4 oz ladle
Fried shrimp	5 if 18–20/lb
Meatballs	2 meatballs, #16 disher
Stronganoff or beef strips in sauce or gravy	4 oz ladle
Potatoes, Vegetables, and Substitutes	
Bread dressing	1 #12 disher
Broccoli, spears	2–3 spears
Okra, whole, french fried	5 pieces
Sweet potatoes, mashed	#12 disher
Tater tots	10–12 pieces
Salads	
Cottage cheese, with fruit or vegetable on plate	#24 disher
Deviled eggs	2 halves
Potato salad	#12 disher
Tomato wedge or slice	3 for salad (6 per tomato)
Desserts	
Cobblers	18″ × 26″ pan, cut 6 × 10
Ice cream and sherbets	#12 disher
Jumbo cookie	#20 disher (before baking)
Pies	9″ tin, 8 per pie

Foodservice operations should develop portion control guides to be used by production and service personnel. Table 6.14 shows typical portion sizes for selected menu items served in an employee cafeteria. Training and supervision are necessary to ensure that proper tools and techniques are followed.

Operating in the black: When revenue minus expenses is a positive value.

Foodservice managers must realize that portion control is essential for **operating in the black.** All food items must be purchased on the basis of weight and measure, and everything served on the basis of specified weight or portion size. Portion control is critical, not only for cost control but also for customer satisfaction. The key steps in ensuring portion control are purchasing by exact specifications, using standardized recipes, and using proper tools and techniques during all aspects of production and service.

ENERGY USE

America's energy crisis in the mid-1970s prompted Congress in 1977 to pass legislation that required the periodic development of energy plans. The first National Energy Plan was developed in 1977 and has been updated several times since then. National energy legislation passed in 2002 created new efficiency standards to reduce electricity use and offered tax incentives for energy efficiency improvements in homes. High energy costs in 2006 renewed interest in energy efficiency.

In many onsite foodservices, utility costs are not considered as important as other costs because they generally are absorbed by the organization's general operating budget and not charged directly to the foodservice operation. Energy used in the foodservice areas of hospitals and schools was between 5 and 10% of the total used for the building (Mason, Shanklin, Hee Wie, & Wolfe, 1999). In restaurants 2.8% of costs are for utilities (Wixson, 2006). The big expenses in restaurant operations are food and labor (Figure 6.23). A school foodservice energy study was conducted in 1994 to determine how much energy was used to produce a school meal. A short summary of the findings from this survey is provided in Figure 6.24.

The foodservice industry is as dependent on steady sources of energy as are many manufacturing industries. The final product, food that is ready to eat, depends greatly on energy-consuming equipment: refrigerators, freezers, ovens, ranges, fryers, holding equipment, dish machines, water heaters, and air conditioning. The space in which personnel work and customers are served must be lighted, heated, cooled, and ventilated. In many foodservice operations, transportation also may be a major consumer of energy. The commissary foodservice, for example, is highly dependent on transportation, and any boosts in gasoline costs will increase the total meal cost.

In the foodservice systems model discussed in Chapter 1, the elements of the foodservice system were outlined, identifying utilities as an operational resource or input into the system. Obviously, heating, cooling, and lighting are important in each of the functional subsystems: procurement, production, distribution and service, and sanitation and maintenance. The quality of the products and services, or the outputs, requires energy-dependent equipment and processes. As with any resource, control measures are needed to ensure effective use of energy in the foodservice system (the control element). Various records (the memory element) are needed to monitor energy use.

To illustrate the dependence of the foodservice system on energy, consider the crisis a manager faces when confronted with a power failure at a critical time during production or service. Loss of power, if prolonged, also has tremendous implications for loss of food in refrigerators or freezers.

An abundance of energy resources has been available in the United States through most of its history. However, an ever-increasing demand for energy has led to increased reliance on an imported supply, particularly for crude oil.

Oil and gas constitute almost three-fourths of U.S. energy consumption; both are fossil fuels and thus nonrenewable resources. One reason for the U.S. energy problem is that the United States relies heavily on oil and gas as sources of fuel. Projections suggest that this reliance on oil and gas will decrease. These fuels are expected to continue to be the major sources of energy in this country, however, still constituting almost 50% in 2025.

Utility services represent a major controllable expense category for restaurant operators, according to the National Restaurant Association's *Restaurant Industry*

	Full-Service Restaurants (Average Check per Person under $10)*	Full Service Restaurants (Average Check per Person $10 and Over)*	Limited-Service Restaurants*
	Percent	Percent	Percent
Where It Came From			
Food Sales	88.6	77.2	98.0
Beverage Sales (alcoholic)	11.4	22.8	2.0
Where It Went			
Cost of Food Sold	29.6	26.8	31.1
Cost of Beverages Sold	3.4	6.8	0.5
Salaries and Wages	30.1	27.9	26.4
Employee Benefits	3.7	4.2	2.5
Direct Operating Expenses	5.9	6.6	5.7
Music and Entertainment	0.45	0.5	0.1
Marketing	2.2	2.5	5.0
Utility Services	3.4	2.6	2.8
Restaurant Occupancy Costs	4.5	5.0	6.4
Repairs and Maintenance	1.7	1.8	1.6
Depreciation	2.1	2.0	2.2
Other Operating Expense/(Income)	0.7	0.1	2.3
General and Administrative	3.0	4.4	2.8
Corporate Overhead	1.9	1.8	1.6
Interest	0.5	0.8	1.2
Other	0.3	0.5	0.2
Income before Income Tax	6.6	5.7	7.6

*All figures are weighted averages based on 1997 data. Amounts are reflected as a percentage of total sales.

Figure 6.23. Restaurant industry operations report.
Source: National Restaurant Association and Deloitte & Touche LLP. Used by permission.

Operations Report, 2004. Full-service operators with lower average per-person checks have the highest utility services costs as a ratio to total sales (4.4%). As the average check increases, the ratio of utility services costs to total sales decreases, to 3.1% of sales for full-service restaurants with an average per-person check of $25 or more. Restaurant operations, particularly quick service, reported electricity as their largest utility expense.

	Average BTUs per Meal
Ready-prepared cook-chill preparation system with a satellite	1,016
Ready-prepared cook-chill preparation system with on-site service	3,683
Conventional preparation system with on-site service	3,622
Conventional preparation system with a satellite	2,037

Figure 6.24. Energy use in school foodservice operations.
Source: Messersmith, A. M., Wheeler, G. M., & Rousso, V. (1994). Energy used to produce meals in school food service. *School Food Service Research Review, 18*(1), 29–37. Used by permission.

A national survey of restaurants reported the following information regarding utilities, including electricity, gas, water, and waste removal (*Restaurant Industry Operations Report,* 2004):

Restaurant Type	Median per Seat
Limited service, average check per person under $6.50	$393
Limited service, average check per person $6.50 and over	$235
Full service, average check per person under $15	$258
Full service, average check per person $15 and over	$250

A study of seven east coast limited-menu to full-service restaurants found the following energy use (Electric Foodservice Council, 1996):

28.0% for climate control
17.8% for hot water
13.5% for lighting
5.8% for refrigeration
34.9% for food preparation

Refrigeration and cooking equipment accounted for 40% of the restaurants' energy usage.

The United States uses a mix of primary energy sources to heat, cool, illuminate, and power its commercial, industrial residential, and transportation sectors. According to U.S. Department of Energy data, in 2005 (www.eia.doe.gov), oil provided approximately 40% of the U.S. energy, coal about 23%, and natural gas more than 20%. Energy resources are sold and billed by different units: electricity in kilowatt-hours (kWh), natural gas in therms or thousands of cubic feet (MCF), and fuel oil by the gallon or barrel. To compare these units for energy content and cost, the British thermal unit (Btu) is used. One Btu contains enough energy to heat 1 pound of water 1 degree Fahrenheit. A common yardstick measures is millions of Btus (MMBtus). The approximate current average cost for industrial users and commercial users is as follows (Marketplace: Statistics, 1999):

Energy Source	1999	1998	Percent Change	Current per MMBtu
Industrial Users				
Natural gas, per MCF	$2.05	$3.20	236%	$1.99
Electricity, per kWh	$0.433	$0.437	<1%	$128.08
Fuel oil (number two)	$0.48	$0.63	229%	$3.46
Commercial Users				
Natural gas, per MCF	$5.54	$6.14	218%	$5.39
Electricity, per kWh	$0.744	$0.761	22%	$218.07
Fuel oil (number two)	$0.51	$0.69	226%	$3.68

Direct energy: Energy expended within the foodservice operation to produce and serve meals.

Table 6.15 provides a list of direct and indirect energy expenditures within a foodservice operation. **Direct energy** refers to energy expended within the foodservice operation to produce and serve menu items at safe temperatures. It is

Table 6.15 Direct and Indirect Energy Expenditures within Foodservice Operations

Direct Energy Expenditures	Indirect Energy Expenditures
Food storage	Food waste disposal
—dry, refrigerated, frozen	Sanitation procedures
Preprocessing	—personnel, equipment
Heat processing	Optimal working environment
Food packaging	—light, heat, air-conditioning
Food storage following heat processing	—physical facilities
—heated, chilled, frozen	
Heat processing of precooked menu items	
Food distribution to units	
Food service	

Source: Unklesbay, N. Energy consumption and school foodservice systems. *School Food Service Research Review, 1*(1):9, 1977. Used by permission.

required for any storage, heating, cooling, packaging, reheating, distributing, or serving functions to be performed for any menu item prepared for service within a facility.

Indirect energy: Energy expended to facilitate functions that use direct energy.

Indirect energy refers to energy expended to facilitate functions that use energy directly. It supports the other necessary functions, like waste disposal, sanitation, and maintenance of optimal work environment, that are involved with the production and service of menu items.

Unklesbay and Unklesbay (1982) named three vital aspects of food quality that should be assured by the effective use of direct energy:

- Sensory quality
- Food safety
- Nutrient retention

Table 6.15 identifies eight categories of activities in which energy is directly used to ensure food quality. As stated previously, however, the activities occurring in a specific operation depend upon the market form of foods and the type of foodservice.

The need for energy conservation should be obvious. Most businesses know the benefits of conserving energy. They also realize the importance of developing a plan for the energy challenges they face, but in today's fast-paced workplace, that is often difficult. Changes occur rapidly and time is scarce. Energy-efficient equipment can be expensive or offer a long payback period. Upper management can be difficult to convince that energy-efficient measures are worthwhile. According to many energy managers, a majority of businesses are indeed addressing the subject—or at least looking into it.

ENERGY CONSERVATION

The program ENERGY STAR® is a voluntary partnership among U.S. organizations, the U.S. Environmental Protection Agency (EPA), and the U.S. Department of Energy to promote energy efficiency in buildings and homes. Organizations that join the partnership follow a proven, cost-effective strategy to save money by reducing the total energy consumption of their buildings. The EPA provides participants in Energy Star buildings unbiased technical information, customized support services, public relations assistance, and access to a broad range of resources and tools. Participants can display an ENERGY STAR® logo in their buildings and can benchmark their operations with others. A Web site, www.energystar.gov, contains information about the program.

The U.S. Department of Energy predicts a savings of $200 billion by 2010 if organizations and consumers take advantage of low-cost energy efficiency opportunities (ENERGY STAR, 2002). Figure 6.25 provides 10 tips from the EPA's Energy Star Small Business division to encourage organizations to evaluate their energy use and take action that can bolster profits and benefit the environment (Traub, 1999).

The energy conservation checklist in Figure 6.26 (Mason, Shanklin, Hee Wie, & Wolfe, 1999) outlines measures that can be used within the operation to reduce energy. These measures are related to food preparation, refrigeration, lighting, HVAC, sanitation and water, and office and administration. Periodic use of the checklist by

#1 TIP: Set back thermostats at night and on weekends. Cost: Nothing. Savings potential: About $1,800 per year for a 33,000-square foot office building with a thermostat set at 75°F, 24 hours per day during the heating months.

#2 TIP: Install light-emitting diode (LED) exit signs. They are 100 percent longer lasting than conventional exit signs and the most energy efficient of their kind. Cost: About $70 per sign to retrofit. Savings potential: About $24 per year, per sign, excluding maintenance costs.

#3 TIP: Lower the thermostat on water heaters. Cost: Nothing. Savings potential: About $24 per year, if the setting is reduced by 10°F and operates at the lowest allowable temperature for normal use.

#4 TIP: Turn off your water heater overnight and on weekends. Cost: Nothing, or $30 if you buy an automatic timer to do it for you. Savings potential: As much as $54 per year depending on how much hot water you use.

#5 TIP: Install an occupancy sensor in hallways, bathrooms, and other areas where lights can be left off most of the time. Cost: Between $25 and $80 annually. Savings potential: About $40 per year depending on where the sensor is installed.

#6 TIP: Turn off computers and other office equipment when they're not being used. Cost: Nothing. Savings potential: As much as $44 per year, per computer depending on what cost per kilowatt-hour.

#7 TIP: Clean all air filters monthly. Cost: Modest to clean, and about $2 to replace one. Savings potential: As much as $60 per filter, depending on the size of the HVAC system.

#8 TIP: Check for drafts coming from doors and windows; caulk and weather-strip as necessary. Cost: $5 or less. Savings potential: About $2 per fixed draft.

#9 TIP: Open drapes or blinds in colder months to let the sunshine warm offices and give the thermostat a break. Do the opposite in the warmer months. Cost: Nothing. Savings potential: About $5 per window depending on location and time of year.

#10 TIP: Adjust thermostats for colder weather. Set them at the lowest temperature allowable that keeps employees and customers comfortable. Cost: Nothing. Savings potential: About $1,000 per year for a small office building open Mon.-Fri., 7 A.M. to 7 P.M.

Figure 6.25. Top 10 tips provided by the EPA's Energy Star Small Business division to evaluate energy use and take action.
Source: Energy User News (1999) and U.S. EPA. "The Conservation Question: Is Your Business Doing Its Part?" Vol. 24(2), 49. Used by permission.

Area	Conservation Practice
Food Preparation	—Cook in largest volume possible. —Cook at lowest temperature possible. —Reduce excess heat loss by monitoring preheat times and cooking temperatures. —Reduce peak loading of electrical equipment. Examples: Use high-energy-demand equipment sequentially rather than simultaneously. Schedule energy-intense cooking, such as baking and roasting, during nonpeak demand time. —Heat only the portion of griddle or equipment to be used. —Use hot tap water for cooking, whenever possible, except in areas where water contains concentrates of heavy metals. —Turn off equipment when not in use. —Keep grills, ovens, and other cooking equipment clean to promote better and more efficient heat transfer. —Write menus to include food items that do not require cooking when possible, especially during warm weather. —Schedule equipment use for nonpeak hours if possible. —Replace inefficient cooking equipment. —Install timers on fans and exhaust systems. —Perform maintenance on all cooking equipment on a regular basis. —Use convection ovens instead of conventional ovens when possible. —Use a timer when baking to prevent unnecessary opening of oven doors. —Load ovens to capacity when possible but do not allow pans to touch oven walls. —Check oven doors for proper closure. —Be sure indicator lights are working on all equipment. —Filter fat in fryers before it congeals, so it does not have to be reheated. —Be sure numbers on control knobs can be seen. —Set equipment to proper cooking temperature. Increasing the temperature uses more energy and may affect food quality.
Refrigeration	—Open doors as seldom as possible. —Turn off lights in equipment when not in use. —Allow hot foods to cool briefly in accordance with safe food-handling practices before refrigerating. —Store frozen and refrigerated foods promptly. —Store items off of floor and away from walls and ceiling to allow for maximum air circulation. —Clean condensers frequently. —Keep thermometers properly calibrated. —Check seals on refrigerators and freezer coors to ensure efficient cooling and freezing. —Install open-door alarms on walk-in doors. —Use vinyl air curtains or air blowers over doors of walk-ins. —Disconnect or remove lights in display cases when not needed. —Develop a regular maintenance schedule for all refrigeration equipment. —Check door gaskets and closures for damage and replace as needed. —Schedule equipment defrost cycles during off-peak demand periods.

Figure 6.26a. Energy conservation checklist.

	—Locate refrigeration and freezing equipment away from heat-producing equipment (e.g., ovens and fryers).
Lighting	—Turn off lights when not in use. —Adjust lighting intensity in areas where close work is not done (e.g., storerooms, aisles). —Limit night lighting for cleaning purposes. —Limit areas of building available for use during holidays and "down periods." —Install timing mechanisms, dimmers, or automatic photocell devices. —Color code light-control panels and switches according to a predetermined schedule of when lights should be turned on or off. —Install energy efficient lightbulbs.
HVAC	—Lower thermostat to 68°F in winter; raise to 78°F in summer. —Adjust duct registers to give the most efficient airflow within a room and balanced airflow between kitchen and service areas, if located adjacent to each other. —Stagger start-up time for individual HVAC units to reduce demand for kilowatt use on your system and to eliminate unnecessary cooling or heating during hours before operation opens. —Open windows to cool when possible. —Shield windows to decrease heat loss or gain. —Clean grease filters on a regular basis. —Operate hood exhaust units only when needed.
Sanitation and Water	—Turn off exhaust fan when not required. —Fill dishwasher to capacity. —Locate hot water booster within 5 feet of dishwasher to avoid heat loss in the pipes. —Install a spring-operated valve on your kitchen and restroom faucets to save water. —Repair leaking faucets, steam pipes, and valves. Implement an effective maintenance and cleaning program for all equipment. —Use hot water only when necessary. —Insulate hot water-holding tanks and hot water pipes to reduce heat loss. —Perform regular service and maintenance on water heaters. —Delime dishmachine on a regular basis.
Office and Administration	—Turn off office equipment including copiers and computers when not needed. They generate heat in addition to using energy. —Screen savers do not save energy; they save only the screen. —Buy copy and fax machines with "power down" or "stand-by" features that operate when machines are idle. —Limit the number of days for delivery of products to decrease the amount of fuel used by suppliers. —Carefully plan department deliveries to conserve fuel and time. —Encourage employees to carpool to and from work. —Use solar-powered calculators instead of battery or electric. —Use manual staplers and pencil sharpeners. —Purchase recycled products. —Be sure all doors to the facility close tightly to improve HVAC efficiency.

Figure 6.26b.
Source: Mason, D. M., Shanklin, C. W., Hee Wie, S., & Wolfe, K. (1999). *Environmental Issues: Impacting Foodservice & Lodging Operations.* Kansas State University. Used by permission.

the individual or committee assigned responsibility for the energy conservation program will provide data for assessing the program's effectiveness.

ENERGY MANAGEMENT

Control of energy costs requires maintaining a recordkeeping system for tracking utility costs and monitoring equipment use. Energy use for lighting, heating, ventilating, and air conditioning of the facilities must also be monitored and controlled.

Preparing and maintaining appropriate records will facilitate identification of the amount and cost of energy use and of any trends that may be developing. Energy use can then be related to the number of customers, and energy cost can be examined in relation to sales volume, which will provide indicators of the success or failure of the energy management program. Monitoring the use and cost of water can also be accomplished, thus providing an analysis of all utilities (Thompson, 1992).

Dollar costs of energy are, of course, important, and comparisons of current costs with those of prior years will provide one basis of comparison. Because of rising utility rates, however, those figures may not be as meaningful as use data. Actual use, therefore, should also be analyzed. An analysis of energy use and cost is an important management function. The glossary in Figure 6.27 lists some of the key terms related to energy consumption.

Glossary

Ampere—a unit of measure of an electric current.

British thermal unit (Btu)—a unit of heat energy. The amount of heat required to raise 1 lb. of water 1 degree Fahrenheit.

Demand charge—based on the highest electric energy use over some fixed period of time during the billing cycle, usually 15 or 30 minutes. As the peak demand determines the size of power generating plants, transmission lines and other equipment needs, it is designed to pay for the utility company's investment.

Demand charge discount—rewards a customer who has a reasonable constant monthly electrical demand. For example, a customer may earn a discount for maintaining a billing demand in excess of 75 percent of his highest maximum demand in the preceding months.

Energy charge—determined by the electric energy used (kilowatt hours). It is designed to cover the base fuel expenses necessary in the generation of electrical energy.

High load factor discount—rewards a customer who uses his demand for many hours of operation. It is a percentage index comparing the actual kilowatt hours used with the maximum billing demand times the total hours in a billing period.

Horsepower—a unit of power equal to 746 watts.

Kilowatt—one thousand watts. Most electrical foodservice equipment is rated in kilowatts.

Kilowatt hour—a measure of work (kilowatt × hours). The unit recorded by electric meters and utilized for billing.

Purchased energy or fuel adjustment—an incremental adjustment to rates reflecting changes in costs of purchased energy or fuel since current rates became effective. It is usually expressed as a fractional amount added to the current rate.

Therm—a unit used for measuring natural gas equal to 100,000 Btu.

Volt—the push that moves the electric current.

Watt—a unit of power. One watt equals the flow of one ampere at a pressure of one volt (watts = volts × amperes).

Figure 6.27. Glossary of energy-related terms.
Source: Energy Management System, 1982. Chicago: National Restaurant Association. Used by permission.

The three forms in Figures 6.28, 6.29, and 6.30, developed by the National Restaurant Association (NRA), are examples of records that should be maintained regularly to track energy use and costs. Typical data for a restaurant have been entered on each form along with explanations of the data, which provide an illustration for a straightforward method of tracking utility consumption and cost. Electricity and gas use have been converted to MBtus to provide a standard base for comparison; one MBtu is equal to 1,000 Btus.

As emphasized in the technical bulletin on energy management prepared by the NRA (1982), the information provided by these records will point out any "leaks" in the utility system of an operation, such as

- Development of poor operating practices
- Malfunctioning of equipment
- Structural damage to the building
- Misreading of meters
- Improper billing by utility companies

In recording data each month, comparable periods should be used to permit comparisons with prior years. For example, if the reading date for the electric meter is on the fifth day of the month, the consumption for a particular month should be for the period from that day to the fifth of the following month.

One factor to be taken into account in analyzing utility use is extremities in weather. Some winters are colder and some summers warmer than others; deviations in use may be explained by the additional demand on heating or air conditioning systems. One way of measuring these variances is the use of degree days, which are deviations of the mean daily temperatures from 65°F. In comparing the prior year's utility use with the current year's, a comparison in the number of degree days might explain deviations.

The form in Figure 6.28 is a worksheet designed to develop information required for the forms shown in Figures 6.29 and 6.30. After utility bills have been received for the current month, the Figure 6.28 worksheet should be completed. The form in Figure 6.29 will assist in monthly tracking of utility consumption and in comparing the current year's consumption of gas, electricity, and water with the prior year's. The form in Figure 6.30 will enable the manager to track total energy use and utility cost and to compute ratios that will assist in analyzing use and cost in relation to volume and sales. These ratios are MBtus per customer and utility cost as a percentage of sales.

Figure 6.31 is a simple flow chart that shows the steps for establishing an energy management program in a facility. Support from the facility's management or maintenance personnel should be obtained before beginning an energy management program. Much of the information needed to make decisions or changes will have to be obtained from other departments or individuals. Figure 6.32 will help the manager estimate the amount of energy required to operate a specific piece of electric or gas equipment.

Employee training is one of the keys to a successful energy conservation program, and the employees are the key to the success of an equipment operation schedule. Ultimately, they are the ones responsible for turning equipment on and off at appropriate times. Before initiating an equipment operation schedule, employee training sessions are needed to present information on the objectives of the

FORM A: MONTHLY UTILITY WORKSHEET

Month January ,19 99

				MBTU[a]	Cost
Gas		cubic feet	×		
1937	therms		× 100	193,700	$900.71[b]
Electricity 20,833	kWhr		× 3,412	71,082	$1470.81[c]
Fuel Oil #2	gals.		× 140		
Propane	gals.		× 91.6		
Steam	lbs.		× 1		
Water 67,200	gals.				$100.80[d]
TOTAL				264,782 MBTU	$ 2472.32

Customer count for month 16,800

MBTUs per customer = Total MBTUs / Customers = 264,782 / 16,800 = 15.8

Sales for month $45,360

Utility costs as percent of sales = Total utility cost / Sales = $2,472.32 / 45,360 = 5.4%

[a] MBTU = 1000 BTU
[b] 1000 cubic feet (MCF) = $4.65
[c] 1 kilowatt hour (kWH) = $0.0706
[d] 1 gallon water = $0.0015

1. Enter the month and year. In this example the information is for January 1999.
2. From your gas bill, obtain the consumption figure. It may be expressed in therms, cubic feet or 100 and 1000 multiples of cubic feet. If gas is metered in 100 cubic feet (CCF) the multiple for conversion to MTBU is 100. If gas is metered in 1000 cubic feet (MCF) the multiple for conversion is 1000. In this example the gas is metered in therms and 1937 therms is equal to 193,700 MBTU. The cost, also obtained from the bill, is $900.71. The conversion values given are average values and satisfactory for these calculations.
3. The total number of kilowatt hours used during the month was 20,833 and the conversion to MBTU provides a value of 71,082. The total cost (including demand charges and any other miscellaneous charges) is $1470.81.
4. Water is metered in gallons or cubic feet. The total consumption during the month was 67,200 gallons at a total cost of $100.80. This includes all other charges as leak insurance and sewage.
5. These are the total figures for MBTU and dollar costs obtained by adding each of the two columns. In this example the totals are 264,782 MBTUs and $2472.32.
6. The customer count for this period is 16,800. It may be more appropriate in your operation to use number of transactions, number of parties or whatever unit is most suitable for your restaurant.
7. MBTUs per customer is obtained by dividing the total MBTUs by the customer count. In the example, dividing 264,782 by 16,800 results in 15.8.
8. Sales for the month were $45,360 obtained from bookkeeping records. The total cost of utilities as determined on this form was $2472.32. Dividing $2472.32 by $45,360 results in a utility cost percentage of 5.4%.

Figure 6.28. Form for recording monthly utilities.
Source: Adapted from Energy Management System, 1982. Chicago: National Restaurant Association. Used by permission.

FORM B: UTILITY CONSUMPTION AND TRACKING

Year 1999

	Gas therms (1)			Electricity (2) Use KWH			Electricity (2) Demand KW			Water gallons (3)			(4)		
Month	1998	1999	Percent Change	1998	1999	Percent Change	1998	1999	Percent Change	1998	1999	Percent Change	1998	1999	Percent Change
JAN	2041	1937	–5.1	23,617	20,833	–118	612	59.5	–2.7	73,600	67,200	–8.7			
FEB															
MAR															
APR															

1. Enter your gas consumption figures in this section. In the example, consumption for January 1998 was 2041 therms while 1937 was the figure for January 1999. This is a 5.1 percent reduction which is indicated in the third column.
2. Electricity use is recorded in this section. Kilowatt hours were 23,617 and 20,833 for January 1998 and January 1999 respectively which represents a 11.8 percent reduction. Demand expressed in kilowatts decreased from 61.2 to 59.5 for a downward change of 2.7 percent. Significant increases in demand may indicate the need to stagger equipment operation or may also point to malfunctioning electrical devices and controls.
3. Water consumption is entered in this portion. Water consumption was 73,600 gallons in January 1998 and 67,200 gallons in January 1999 for a 8.7 percent reduction. Increases in water consumption may indicate system leaks, improperly operating faucets and faulty valves.
4. This section is to be used for other energy sources as propane, fuel oil or steam.

Figure 6.29. Form for analyzing and tracking utility consumption.
Source: Adapted from Energy Management System, 1982. Chicago: National Restaurant Association. Used by permission.

FORM C: UTILITY CONSUMPTION, TRACKING AND COST

Month	Total MBTU ①			Total Utility Cost ②			MBTU per customer ③			Utility Cost as Percent of Sales ④		
	1998	1999	Percent Change	1998	1999	Percent Change	1998	1999	Percent Change	1998	1999	Percent Change
JAN	284,681	264,782	–7.0	2162.57	2472.32	14	16.6	15.8	–4.8	5.1	5.4	–0.3
FEB												
MAR												
APR												

1. The information for this section is obtained from the monthly utilities worksheet (Form A) where all energy used is converted to the common unit of BTUs. The 264,782 MBTU figure for January 1998 of the example is from Form A while the 284,681 MBTU is the like figure for January 1998. There has been a 7.0 percent reduction in the total number of BTUs from one January to the next.
2. Total utility costs, including water for January 1999, was obtained from Form A and is $2472.32. The prior January figure was $2162.57 which indicates a change of 14 percent.
3. This section relates energy consumption to number of customers. The 15.8 MBTU per customer figure was calculated in Form A and from historical data 16.6 was the prior year figure indicating a reduction of 4.8 percent.
4. Utility costs as a percent of sales were 5.4 for January 1999. This figure was obtained from Form A. For January 1998, the figure was 5.1 which indicates a change of –0.3 percent. While total utility costs remained almost constant, utility costs as a percentage of sales decreased because of larger sales volume.

Figure 6.30. Form for computing cost and use indicators for utility consumption.
Source: Adapted from Energy Management System, 1982. Chicago: National Restaurant Association. Used by permission.

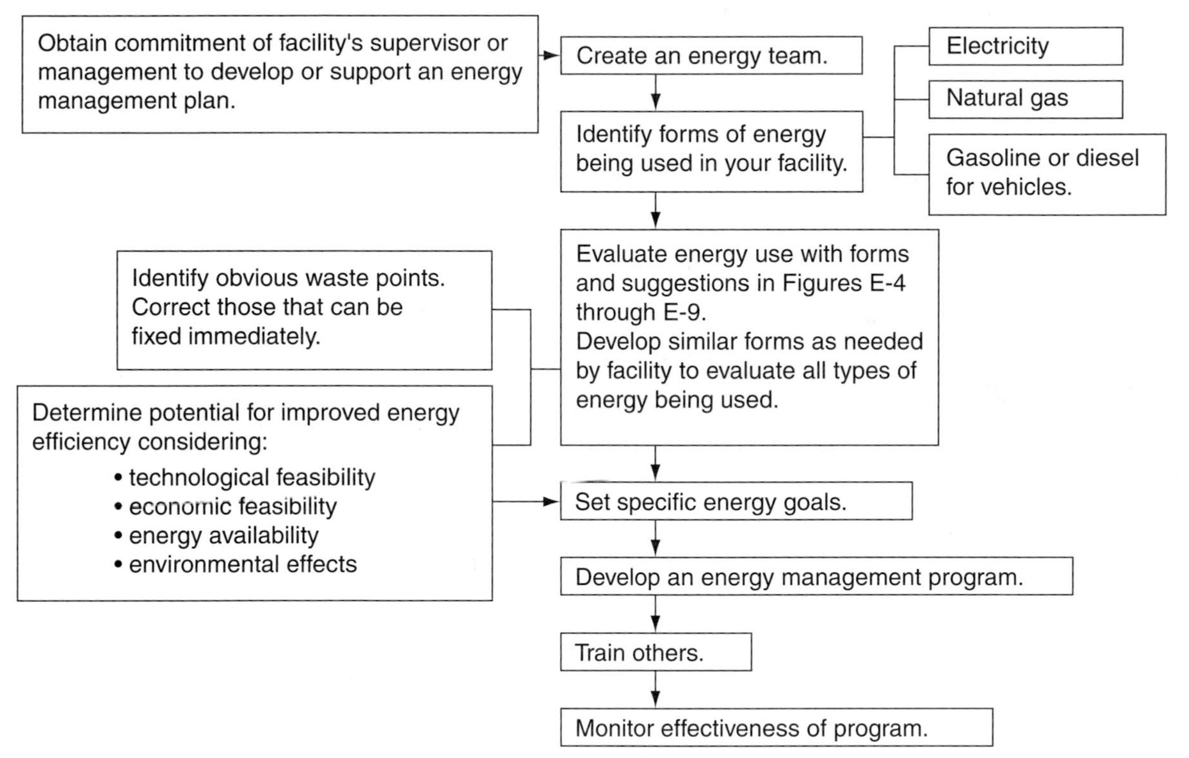

Figure 6.31. Energy management plan flow chart.
Source: Mason, D. M., Shanklin, C. W., Hee Wie, S., & Wolfe, K. (1999). *Environmental Issues: Impacting Foodservice & Lodging Operations.* Kansas State University. Used by permission.

program and on its importance for controlling operational costs and to elicit cooperation and support.

To assist in executing the equipment scheduling aspect of the energy management program, the NRA has developed a series of labels that can be mounted on or close to various pieces of equipment to remind employees of various energy conservation issues (Figure 6.33). Data presented earlier in this chapter indicated that refrigeration and cooking equipment are responsible for approximately 40% of the energy used in restaurants. Therefore, the importance of equipment control in these operational areas should be obvious.

Energy conservation tricks for reducing overhead expenses should be explored by foodservice managers, who have been dealing with increasing energy use over the past several decades. The first big item has been air conditioning dining rooms because customers demand comfort in both summer and winter (Patterson, 1993). The drive for comfort also has extended to the kitchen and is becoming a union bargaining position in most areas. Many electric utility companies are providing analyses of air conditioning demand that involve monitoring temperatures and compressors and cutting off units during slow periods while keeping temperatures within the comfort zone.

Refrigerator compressors and condensing coils generate a lot of heat. The amount of air conditioning can be decreased by 5 to 10% in the kitchen by placing

Equipment	Electricity requirement						
	Amperes[a]	Volts[a]	Kilowatts × (kW)[b]	Hours operated per day	Days operated per week	KWh per week[c]	1,000 BTU per week[d]

[a]Information found on equipment name plate.

[b]Kilowatt—multiply amperage by voltage divided by 1,000 (AMP Volts ÷ 1,000). If horsepower unit appears on name plate, multiply by .746 for kilowatts. If wattage appears on name plate, divide by 1,000 for kilowatts.

[c]Kilowatt-hours per week-measure of energy use per week. Kilowatt × hours per day × number days per week. This information can be used to determine energy cost. Multiply kilowatt-hours per week by cost per kilowatt-hours to determine energy cost.

[d]One thousand BTUs per week. 1000 BTU = (kWH × 3,412) ÷1,000

Equipment	Gas equipment rated input					
	BTU per hour[e]	Cubic feet per hour[f]	Hours operated per day	Days operated per week	Cubic feet per week[g]	1,000 BTU per week[h]

[e]Information found on equipment name plate.

[f]Cubic feet per hour = BTU per hour ÷ 1,000

NOTE: The BTU content of natural gas varies somewhat, but this figure will provide satisfactory results for this analysis.

[g]Cubic feet per week = cubic feet per hour × hours per day × days operated per week

This information can be used to determine energy cost. Cubic feet × cost per unit = $_

If billing is in therms, divide total cubic feet by 100 for number of therms.

[h]One thousand BTUs consumed per week = (cubic feet × 1,000) ÷1,000)

Figure 6.32. Estimating energy use of equipment.
Source: Mason, D. M., Shanklin, C. W., Hee Wie, S., & Wolfe, K. (1999). *Environmental Issues: Impacting Foodservice & Lodging Operations.* Kansas State University. Used by permission.

air conditioners in remote locations outside the building or in the basement. The same procedure can be used for serving and dining areas. A heat pump water heater uses not only heat generated by cooking equipment but also returns cool air to the kitchen. Managers are always looking for ways to reduce costs of energy in cooking equipment; manufacturers are responding by beginning to build insulated equipment. The surprise is that manufacturers have found that keeping heat inside

Figure 6.33. Examples of energy conservation reminders.
Source: National Restaurant Association Catalog of Publications. Used by permission.

the equipment makes cooking more efficient. High-efficiency gas burners in energy-efficient equipment produce more Btus from the same amount of gas; infrared heating in fryers has the same results. Induction heating in electric equipment promises not only more efficiency but also lower kitchen temperatures.

Water conservation is not only good for the environment, but it also saves money (Thompson, 1992). The problem is how to conserve water without cutting back on customer service. Many restaurateurs in California and Nevada have experienced droughtlike conditions in recent years and have had to conserve water by serving it only if requested by customers. Employees need to be trained to conserve water in cooking and cleaning. Fifteen gallons of water are not needed to cook 4 pounds of pasta. Hoses for cleaning floors or sidewalks can be replaced by buckets of water and a deck brush. Check dripping faucets; a little drip could flush $1,000 down the drain in a year. The objective is to use water as efficiently as possible while maintaining high sanitary standards.

Concern about energy costs has stimulated research in the energy use of foodservice operations; a number of studies are listed in the bibliography of this chapter. Much of the work has been conducted at the University of Missouri–Columbia and at the University of Wisconsin–Madison. Research has focused on energy modification of recipes, energy use of various types of equipment, energy demand with differing oven loads, and energy use in the different foodservice operations. This research is yielding valuable data to assist foodservice managers in understanding energy use in foodservice operations and designing energy control programs.

A study on energy consumption and cost for the production of school foodservice meals was funded by the Division of Applied Research of the National Food Service Management Institute (Messersmith, Wheeler, & Rousso, 1994). An energy audit was conducted to determine the cost of energy used by foodservice equipment to produce one meal and the amount of energy used per square foot of production space in four school foodservice operations, each in a different district and with a different foodservice operation. The following types of foodservice operations were used in the study:

- Conventional onsite production and delivery
- Conventional onsite production and one satellite service center
- Ready prepared cook-chill onsite production and delivery

The fourth operation, located in the district administration building, is a ready prepared cook-chill commissary with 90 satellite centers in which approximately 29,000 meals are prepared each day. Energy consumption in the commissary and one satellite operation was used in the study.

Each operation was audited 2 days in the spring and 2 days in the fall. The schools' utility meters were read at half-hour intervals during the same period that energy equipment data were collected in the kitchens. Calculations for electric, gas, and steam energy were converted to Btus for comparison. The overall average per meal was 2,590 Btus; the average energy cost per meal was $0.13. Energy per square foot of production space averaged 432 Btus. These results indicate that an energy conservation program could help the bottom line of a school foodservice operation.

Professional Organization Profile

American Culinary Federation

MISSION

To make a positive difference for culinarians through education, apprenticeship, and certification, while creating a fraternal bond of respect and integrity among culinarians everywhere

WHO BELONGS TO THE ORGANIZATION

Members are professional chefs, cooks, and bakers engaged in the planning, preparation, service, and supervision of foods, pastries, and beverage, in commercial kitchens, hotels, clubs, restaurants, institutions, and schools, as well as serving as culinary research and development consultants, experimental chefs, food directors, and culinary educators. Student memberships are available.

ADVANTAGES OF MEMBERSHIP

Members have access to an association website with its online network, receive the monthly publication, *The National Culinary Review,* and the *Center of the Plate* newsletter, can attend association-sponsored seminars and exhibits, can participate in certification and training programs and competitions, and have access to educational grants.

WEBSITE

www.acfchefs.org

MEET THE PRESIDENT

John Kinsella, president 2005– , is the president and CEO of Kincom, Inc., and senior chef instructor at Midwest Culinary Institute in Cincinnati. He authors the culinary text, *Professional Charcuterie.* He is one of only 58 chefs in the United States holding distinction as a Certified Master Chef.

CHAPTER SUMMARY

This summary is organized by the learning objectives.

1. The food production subsystem involves the transforming of food items from their purchased state to menu items ready to serve to the customer.
2. Several methods for forecasting were described in this chapter, including moving average, exponential smoothing, regression, intuition, and historical records. Use the information in the text to guide your forecasting.

3. Production schedules typically are prepared for a given day or meal. The schedule commonly includes the menu items to be prepared, the quantity to prepare, and who will be responsible for the preparation. The schedules also may include information such as recipe number, time to prepare and/or serve the item, and preparation work to be completed.
4. An ingredient room is designed as a central location where the quantity of food product needed to produce a given recipe in a specified quantity is assembled onto carts. These ingredient quantities then are delivered to the production area for use by the cooks.
5. Several techniques exist for adjusting a recipe; the most common are the factor method, the percentage method, and the direct reading table method. Each is described in detail in this chapter. You can follow these directions while adjusting your recipe.
6. A variety of types of foodservice equipment exists. Many of the more common pieces were included as figures in this chapter. Review these figures to help familiarize yourself with some of the common pieces of foodservice equipment.
7. Direct energy is the energy used to produce and serve food, such as the energy required by the production and service holding equipment. Indirect energy, such as that in sanitation and waste disposal, is that which is used to facilitate functions that use direct energy.
8. The ampere is a unit of measure of electric current; the British thermal unit is a unit of heat energy; one horsepower is a unit of power equal to 746 watts; a kilowatt is 1,000 watts; a therm is a unit for measuring natural gas; voltage is the push that moves electric current; and a watt is a unit of electric power.
9. Energy can be saved in a variety of ways in a foodservice operation. Some examples include the following: Turn on ovens only when needed and shut off when finished; keep refrigerator and freezer doors closed while gathering products from inside the unit; shut off lights in areas not being used; and only run water for needed products—never leave water running.

TEST YOUR KNOWLEDGE

1. Explain how overproduction and underproduction can affect cost.
2. What are two ways to control quantity demand?
3. Why is the setup of a recipe so important to the foodservice industry?
4. Why is the ingredient room so important to the foodservice industry and what goes on in the ingredient room?
5. What are the sources of energy the foodservice industry uses?
6. Why is it important to use energy conservation in the foodservice business?
7. Describe how to create a well-executed energy management plan.

CLASS PROJECTS

1. In groups of two or three, visit a local foodservice operation, observe production, and talk with the foodservice manager. Write a short paper summarizing

how forecasting, production scheduling, and recipe adjustment are done at that operation.

2. Obtain a day's menu from a local foodservice operation. Role-play with class members the role the manager would have in conducting a production meeting with staff about the menu.
3. Create a game that will help other students learn concepts related to production.
4. Work with a local foodservice manager and assist in standardizing a recipe for that operation.
5. Schedule a visit to a local foodservice operation. Conduct an energy audit using the checklist provided in this chapter. Discuss the results with the manager.
6. Invite a representative from the local energy company to come to class to discuss energy use and conservation in foodservice operations.

WEB SOURCES

www.acfchefs.org	American Culinary Federation, Inc.
www.affi.com	American Frozen Food Institute
www.aibonline.org	American Institute of Baking
www.asbe.org	American Society of Baking
www.ciachef.edu	Culinary Institute of America
www.energystar.gov	ENERGY STAR® program
www.fesmag.com	*Foodservice Equipment and Supply* magazine
www.fermag.com	Foodservice Equipment Reports
www.nafem.org	National Association of Foodservice Equipment Manufacturers

BIBLIOGRAPHY

Adam, E., & Ebert, R. (1992). *Production and operations management: Concepts, models, and behaviors,* 5th ed. Englewood Cliffs, NJ: Prentice Hall.

Armentrout, J. S. (Ed.). (2000). *The professional chef's® techniques of healthy cooking,* 2nd ed. New York: Wiley.

American Society for Testing & Materials. (1988). *American Society for Testing & Materials: Manual on sensory testing, methods. ASTM Special Technical Publication No. 434.* Philadelphia: Author.

Association endorses new EPA "green" utility program. (1992). *Washington Weekly, 12*(36), 2–3.

Aulbach, R. E. (1988). *Energy and water resource management,* 2nd ed. East Lansing, MI: Educational Institute of the American Hotel and Motel Association.

Beasley, M. A. (1991). Improving operational efficiencies—Part II. *Food Management, 26*(8), 52, 57.

Bendall, D. (1997). Ranges and griddles. *Food Management, 32*(3), 89, 91, 94.

Bowers, J. (Ed.). (1992). *Food theory and application,* 2nd ed. New York: Macmillan.

Buchanan, P. W. (1993). *Quantity food preparation: Standardizing recipes and controlling ingredients,* 3rd ed. Chicago: The American Dietetic Association.

Cooking equipment: Barbecue and Smoke ovens (1986). *Foodservice Equipment Supplies Specialist Magazine,* 39(5): 35.

Dale, J. C., & Kluga, T. (1992). Energy conservation: More than a good idea. *Cornell Hotel and Restaurant Administration Quarterly, 33*(6), 30–35.

Decareau, R. V. (1992). Microwaves in foodservice. *Journal of Foodservice Systems, 6,* 257–270.
Dougherty, D. A. (1984a). Forecasting production demand. In Rose, J. C. (Ed.), *Handbook for health care food service management* (pp. 193–199). Rockville, MD: Aspen Publishers.
Dougherty, D. A. (1984b). Issue control and ingredient assembly systems. In Rose, J. C. (Ed.), *Handbook for health care food service management* (pp. 199–205). Rockville, MD: Aspen Publishers.
Electric Foodservice Council. (1996). *Training manual.* Fayetteville, GA: Author.
ENERGY STAR (2002). [Online]. Available: http://www.energystar.gov
Federal Energy Administration, Food Industry Advisory Committee. (1977). *Guide to energy conservation for foodservice.* Washington, DC: Government Printing Office.
Fellin, T. A. (1996). Oven overview. *School Foodservice and Nutrition, 50*(9), 63, 66.
Fitzsimmons, J. A., & Sullivan, R. S. (1982). *Service operations management.* New York: McGraw-Hill.
Flack, K. E. (1959). Central ingredient room simplifies food preparation and cuts costs. *Hospitals,* 33(17): 125, 128, 132.
Frable, F. (1996). Finding the 'sweet spot' in equipment pricing. *Nation's Restaurant News, 30*(27), 26, 28.
Garey, J. G., & Simko, M. D. (1987). Adherence to time and temperature standards and food acceptability. *Journal of the American Dietetic Association, 87,* 1513–1518.
Giese, J. (1992). Advances in microwave food processing. *Food Technology, 46*(9), 118–123.
Gilleran, S. (1991). Braising brings rich flavors to beef. *Restaurants & Institutions, 101*(22), 113–114, 121, 124.
Gisslen, W. (2002). *Professional cooking,* 5th ed. New York: Jossey-Bass.
Gregoire, M. B., & Henroid, D. (2002). *Measuring success with standardized recipes.* Washington, DC: United States Department of Agriculture.
Harrington, R. E. (1991). A kilowatt saved. . . . *Restaurants USA, 11*(2), 11.
Jernigan, A. K., & Ross, L. N. (1989). *Food service equipment,* 3rd ed. Ames: Iowa State University Press.
Jernigan, B. S. (1981). Guidelines for energy conservation. *Journal of the American Dietetic Association, 79*(4), 459–462.
Katsigris, C., & Thomas, C. (1999). *Design and equipment for restaurants and foodservice.* New York: Wiley.
Klein, B. P., Matthews, M. E., & Setser, C. S. (June 1984). *Foodservice systems: Time and temperature effects on food quality.* North Central Regional Research Publication No. 293. Urbana-Champagne: University of Illinois.
Kotschevar, L. H. (1990). *Standards, principles, and techniques in quantity food production,* 4th ed. New York: Wiley.
Kotschevar, L. H., & Donnelly, R. (1999). *Quantity food purchasing,* 5th ed. Upper Saddle River, NJ: Prentice Hall.
Kreith, F., & Burmeister, G. (1993). *Energy management and conservation.* Denver: National Conference of State Legislatures, pp. 7–9, 353.
Lampi, R. A., Pickard, D. W., Decareau, R. V., & Smith, D. P. (1990). Perspective and thoughts on foodservice equipment. *Food Technology, 44*(7), 60, 62, 64–66, 68–69, 132.
Lawless, S. T., & Gregoire, M. B. (1987–88). Selection of computer software for recipe adjustment. *NACUFS Journal, 13*(1), 24–27.
Lawless, S. T., Gregoire, M. B., Canter, D. D., & Setser, C. S. (1991). Comparison of cakes produced from computer-generated recipes. *School Food Service Research Review, 15*(1), 23–27.
Lorenzini, B. (1992). Cookware for ethnic fare. *Restaurants & Institutions, 102*(22), 153, 156.
Makridakis, S. G. (1990). *Forecasting, planning, and strategy for the 21st century.* New York: Free Press.
Makridakis, S., & Wheelwright, S. C. (Eds.). (1989). *Forecasting methods for management,* 5th ed. New York: Wiley.

Marketplace: Statistics, trends, and energy data. (1999). *Energy User News, 24*(2), 49.
Mason, D. M., Shanklin, C. W., Hee Wie, S., & Wolfe, K. (1999). *Environmental issues: Impacting foodservice & lodging operations.* Manhattan, KS: Kansas State University.
McManis, H., & Molt, M. (1978). Recipe standardization and percentage method of adjustment. *NACUFS Journal,* 35–41.
McWilliams, M. (1993). *Foods: Experimental perspectives,* 2nd ed. New York: Macmillan.
Meat Industry Insights. (1998). [Online]. Available: http://www.pb.net/spc/mii/970604.HTM
Messersmith, A. M., & Miller, J. L. (1992). *Forecasting in foodservice.* New York: Wiley.
Messersmith, A. M., Moore, A. N., & Hoover, L. W. (1978). A multi-echelon menu item forecasting system for hospitals. *Journal of the American Dietetic Association, 72,* 509–515.
Messersmith, A. M., Rousso, V., & Wheeler, G. (1993). Energy management in three easy steps. *School Food Service Journal, 47*(9), 41, 42, 44.
Messersmith, A. M., Rousso, V., & Wheeler, G. (1994). School food service energy. *School Food Service Research Review, 18*(1), 38–44.
Messersmith, A. M., Wheeler, G., & Rousso, R. (1994). Energy used to produce meals in school food service. *School Food Service Research Review, 18*(1), 29–36.
Miller, J. J., McCahon, C. S., & Miller, J. L. (1993). Foodservice forecasting: Differences in selection of simple mathematical models based on short-term and long-term data sets. *Hospitality Research Journal, 16*(2), 95–102.
Miller, J. L., & Shanklin, C. S. (1988). Forecasting menu item demand in foodservice operations. *Journal of the American Dietetic Association,* 88, 443–446, 449.
Miller, J. L., Thompson, P. A., & Orabella, M. M. (1991). Forecasting in foodservice: Model development, testing, and evaluation. *Journal of the American Dietetic Association, 91,* 569–574.
Mitani, J., & Dutcher, J. (1992). Standardized recipes: Are they worth the hassle? *School Food Service Journal, 46*(8), 70–72.
Mizer, D. A., Sonnier, B., Drummand, K. D., & Porter, M. (1999). *Food preparation for the professional,* 3rd ed. New York: Wiley.
Molt, M. (2001). *Food for fifty,* 11th ed. Englewood Cliffs, NJ: Prentice Hall.
National Restaurant Association. (1979). *Sanitation operations manual.* Chicago: Author.
National Restaurant Association. (1982). *Energy management system.* Chicago: Author.
National Restaurant Association. (1988). *Foodservice and energy to the year 2000.* Washington, DC: Author.
National Restaurant Association. (1992). *Foodservice manager 2000. Current issues report.* Washington, DC: Author.
National Restaurant Association. (2004). *Restaurant Industry Operations Report.* Washington, DC: Author.
Palmer, H. H. (1972). Sensory methods in food-quality assessment. In Paul, P. C., & Palmer, H. H. (Eds.), *Food theory and application* (pp. 727–738). New York: Wiley.
Palmer, J., Kasavana, M. L., & McPherson, R. (1993). Creating a technological circle of service. *Cornell Hotel and Restaurant Administration Quarterly, 34*(1), 81–87.
Patterson, P. (1993). Energy-saving tricks reduce overhead expenses. *Nation's Restaurant News, 25*(17), 56, 65.
Payne-Palacio, J., & Theis, M. (2001). *West & Wood's introduction to foodservice,* 9th ed. Upper Saddle River, NJ: Prentice Hall.
Puckett, R. P., & Miller, B. B. (1988). *Food service manual for health care institutions.* Chicago: American Hospital Publishing.
Riell, H. (1992). Equipment's cutting edge. *Restaurants & Institutions, 12*(4), 22–23.
Ryan, N. R. (1991). Combination cooking. *Restaurants & Institutions, 101*(7), 73–74.
Setser, C. S. (1992). Sensory evaluation. In Kamel, B. S., & Stauffer, C. E. (Eds.), *Advances in baking technology* (pp. 254–291). New York: VCH Publishers.
Shanklin, C. W., & Hoover, L. (1993). Position of The American Dietetic Association: Environmental issues. *Journal of the American Dietetic Association, 93,* 589–591.

Spears, M. C. (1999). *Foodservice procurement: Purchasing for profit.* Upper Saddle River, NJ: Prentice Hall.

Spertzel, J. K. (1992). Combi ovens: A revolution in cooking. *School Food Service Journal, 45*(8), 94, 96.

Stevens, J. W., & Scriven, C. R. (2000). *Manual of equipment and design for the foodservice industry,* 2nd ed. New York: Van Nostrand Reinhold.

Stone, H., & Sidel, J. L. (1992). *Sensory evaluation practices,* 2nd ed. New York: Academic Press.

Terrell, M. E., & Headlund, D. B. (1997). *Large quantity recipes,* 4th ed. New York: Wiley.

Thompson, P. K. (1992). Saving water can save you money. *Restaurants USA, 12*(4), 10–11.

Traub, J. (1999). The conservation question: Is your business doing its part? *Energy User News, 24*(2), 15–16.

Unklesbay, N., & Unklesbay, K. (1982). *Energy Management in Foodservice.* Westport, CT: AVI.

U.S. Department of Agriculture, Food and Nutrition Service. (2002). *Food buying guide for Child Nutrition Programs.* Available online at www.nal.usda.gov:8001/FB6/buying guide.html

Van Warner, R. (1993). Government reaction to food safety should quell fears, but be prudent. *Nation's Restaurant News, 27*(16), 21.

Williams, L. (1998). A new spin for rotisseries. *Restaurants & Institutions, 108*(26), 101–102.

CHAPTER 7

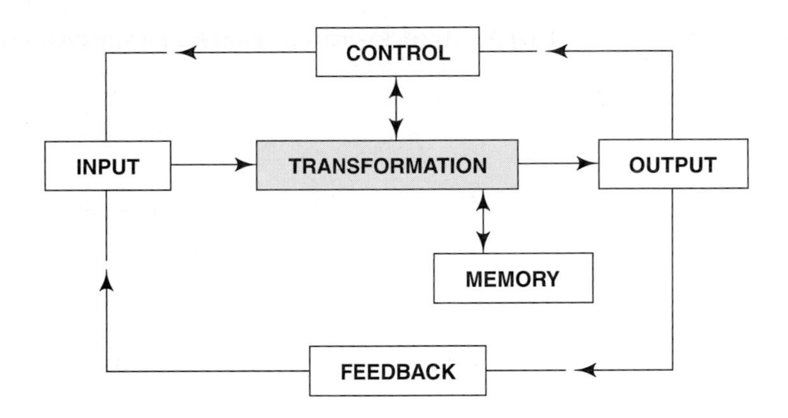

Distribution and Service

Enduring Understanding

- Quality service is a key element for a foodservice operation's success.

Learning Objectives

After reading and studying this chapter, you should be able to:

1. Explain the distribution and service subsystem.
2. Differentiate between centralized and decentralized service.
3. Describe benefits and constraints of various methods of distribution.
4. Compare and contrast counter, table, tray, and self-service.
5. Compare and contrast service and experience economies.

Once food has been prepared in the production subsystem, it must be distributed to service areas and served to the customer. Attention to details in this process, or lack of it, can mean the difference between success and failure in a foodservice operation. In this chapter, we will review the distribution and service subsystem. Suggestions for service success are given.

FUNCTIONAL SUBSYSTEM: DISTRIBUTION AND SERVICE

Distribution: Movement of food from production to service.

Distribution and service is the third subsystem in the transformation element of the foodservice system (Figure 7.1). **Distribution** involves getting food from production to the point of service. Service is the presentation of food to the customer. Depending on the type of foodservice operation, distribution may or may not be a major function. **Service,** however, is a major component of all types of foodservice operations. Vending machines serve customers who want a snack or a quick meal, as does a waitperson in a fine dining restaurant under leisurely conditions.

Distribution is a major concern in hospital foodservices in which patients are served in individual rooms located on many floors and often in separate buildings. Ensuring that the appropriate food is sent to the appropriate place for service to a particular patient is a complex process, which is further complicated by the need to ensure that the food is at the right temperature and is aesthetically appealing. In contrast, in limited-menu restaurant operations, where customers pick up the menu

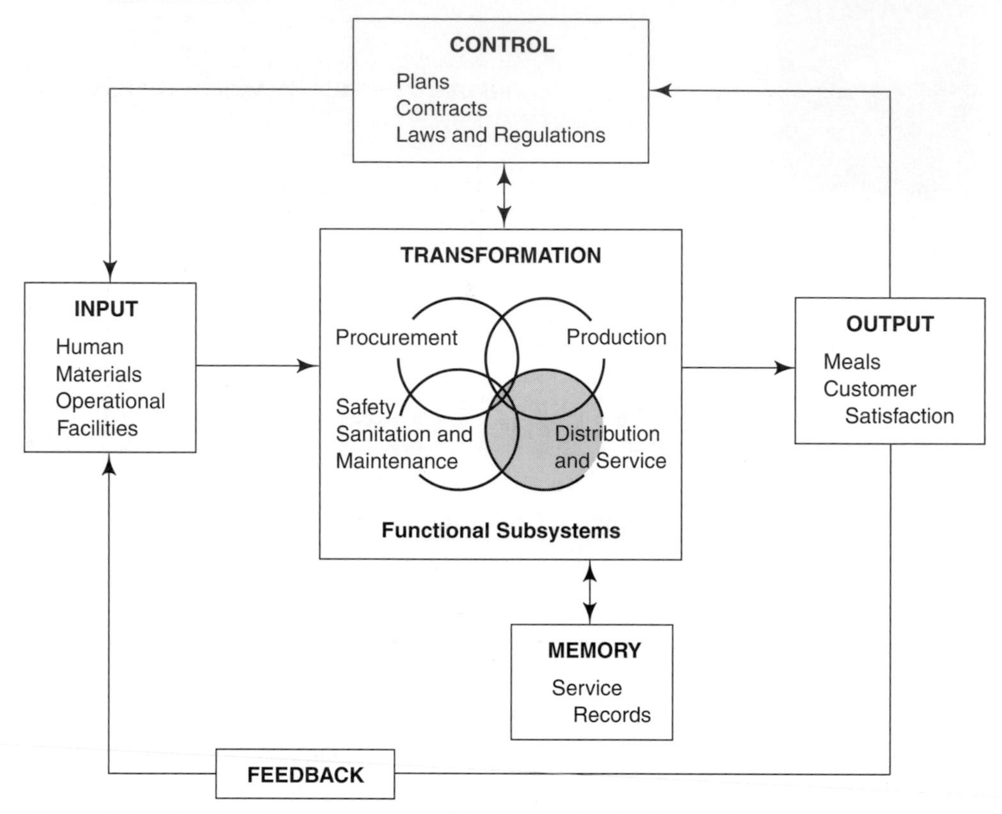

Figure 7.1. Foodservice systems model with the distribution and service subsystem highlighted.

items directly after production and either go off premises for consumption or to a table in the facility, distribution is relatively simple. In fact, distribution and service become the responsibility primarily of the customers and not the employees.

Take-out and home-delivered foods have become an important source of everyday meals. Creative restaurateurs and other food retailers who offer take-out options are, in a sense, becoming American consumers' kitchen-on-the-go. The National Restaurant Association reports that carryout and delivery account for nearly 60% of total restaurant traffic (www.restaurant.org).

Service takes many forms in a foodservice establishment, from that in the upscale fine dining restaurant involving several highly trained employees to that in the many self-service operations—cafeteria, vending, or buffet. The method, speed, and quality of the services provided impact the success of a foodservice establishment. The quality of food may be excellent, the sanitation of the establishment above reproach, and the procurement and storage of food ideal, but if service is lacking, customers will rate the operation as poor.

DISTRIBUTION

The distribution of food from production to the customer depends primarily on four factors:

- type of production system in use
- degree of meal preassembly prior to service
- physical distance between production and service
- amount of time between completion of production until the time of service

Foodservice managers must consider each when evaluating distribution options for their operation. Figure 4.1 (p. 87) illustrates the various process steps that can occur between production and service, creating the need for the distribution function.

Food that is prepared using conventional production and delivered immediately to the customer, as occurs in many restaurant operations, does not require special distribution equipment (Figure 4.3). The temperature and quality of the food are maintained because of the limited time between the completion of production of the food and its service to the customer.

As the time between the completion of production and the time of service increases and/or the distance between the two increases, the options for distribution practices and equipment also increase. Hot- and cold-holding equipment is needed to maintain the proper temperature for various menu items between the time of production and service. Adherence to critical control points for proper serving temperatures is critical during the distribution process. If standards are not met, utensils and equipment must be washed, rinsed, and sanitized and the food product reheated to 165°F. Depending on the service areas, this holding equipment may be stationary or mobile. Some equipment is versatile and can be used for distribution, holding, and service. For example, the mobile modular serving units shown in Figure 7.2 could be used for transporting food for a catered function in a dining room away from the main kitchen and for holding the food until time of service. The units then provide a service counter for self- or waitstaff service.

In some operations, most commonly those using a commissary or base kitchen conventional production system as described in Chapter 4, heated and chilled prepared foods must be transported some distance from production to service. Adding the transportation process necessitates having equipment designed for maintaining temperatures during transportation. In some operations, electrically heated or cooled carts or trucks are used for this transport (Figure 7.2). Other operations use insulated carts for this transportation. Proper HACCP monitoring becomes increasingly important during this process. Temperatures should be recorded before items leave the production area, after they arrive at the satellite unit, prior to the start of service, and periodically throughout service to ensure the safety of foods served.

As shown in Figure 4.1, the process of meal assembly adds another step between production and service and greatly increases the distribution equipment options. Meal assembly may be centralized or decentralized.

In a facility using centralized meal assembly, food trays are assembled for service at a central location close to the main production area. An example of a centralized tray assembly unit is shown in Figure 7.3. The layout uses mobile equipment, which has been widely accepted because of its flexibility and the ease of facility maintenance that it provides. This setup can be readily rearranged or moved for cleaning. A tray slide is an integral component of a centralized tray assembly operation. Meal trays are moved along the tray slide, allowing the placement of food products on the tray at stations positioned along the tray slide. The tray slide could involve manually pushing trays along a tray slide; having skate

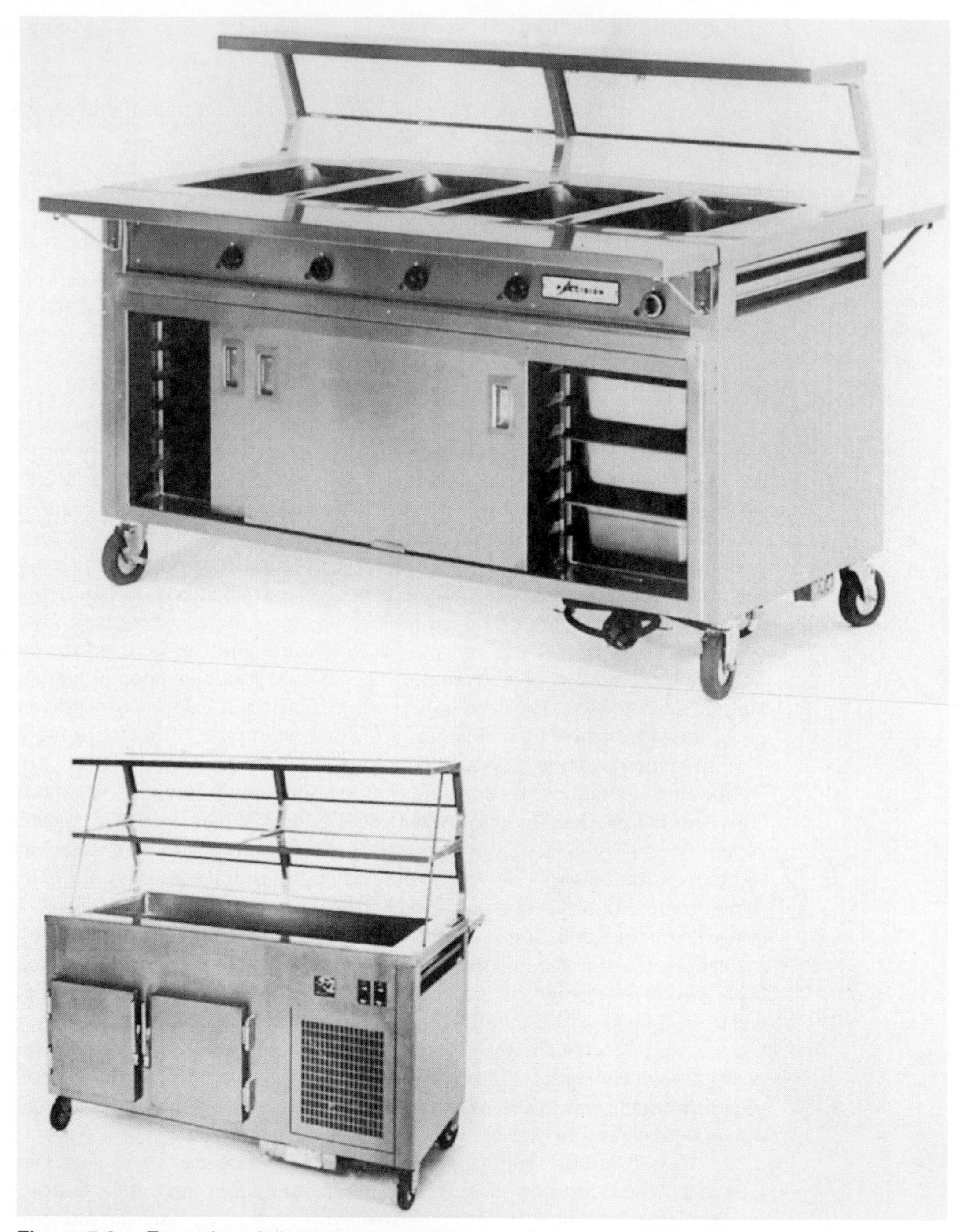

Figure 7.2. Examples of distribution and service equipment: *top,* hot serving unit with heated storage base; *bottom,* cold serving unit with refrigerated storage base.
Source: Courtesy of Precision Metal Products, Inc., Miami, Florida.

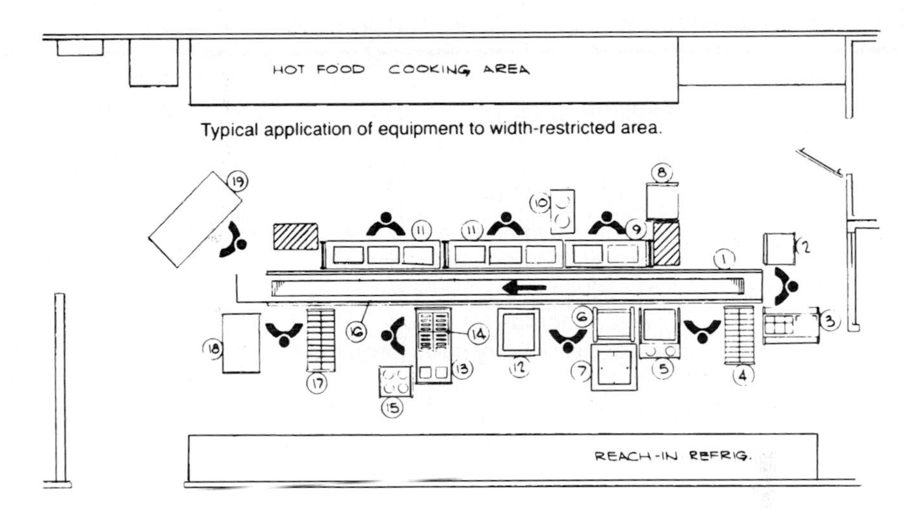

Figure 7.3. Centralized tray assembly unit.
Source: © 1984, Caddy Corporation of America, Pittman, New Jersey. Used by permission.

wheels or rollers on the tray slide to facilitate the movement of trays; or, more commonly, the use of a motorized belt made of fabric, metal slats, or rubber band-vayors. Motorized belts have been designed as straight line or circular.

Once trays have been assembled, they are placed in some form of cart for transportation to the service area. A variety of techniques are used to maintain food temperatures during the transportation process. Transportation carts may be motorized, pushed manually by employees, or moved with special moving equipment.

Meal trays assembled with foods produced in a conventional food production operation will need to have equipment designed for maintenance of both hot and cold temperatures. Methods of heat retention are described in Table 7.1. Heat maintenance can be achieved by placing plates of hot food items on specially designed bases that continue to generate heat during the transportation process or insulated bases designed to maintain current food temperatures. Delivery systems using these bases usually do not include procedures for maintaining the temperature of cold food items. Food trays using heated or insulated bases typically are transported in closed carts, which do not have either heat or refrigeration. Insulated trays are used in many operations. Although these trays do not generate additional heat or refrigeration, they are designed with the insulation ability to maintain hot and cold temperatures for a period of time.

Several types of carts have been developed for maintaining both the hot and cold temperatures of food items. In some, often termed split-tray carts, all of the food items are placed on a single tray with hot foods on one half of the food tray and cold items on the other. This tray is inserted into a special cart that will maintain heat to half of the tray and refrigeration on the other half. Match-a-tray types of carts require that hot food items be placed in a heated compartment separate from the refrigerated portion of the tray. Tray delivery personnel then need to "match" the hot food items with the correct cold food tray prior to serving the meal tray.

In facilities using either a cook-chill or cook-freeze production system, a reheating process may be added between production and service (Figure 4.1). Table 7.2 details benefits and constraints of various reheating methods. Traditionally this

Table 7.1 Benefits and Constraints of Various Meal Distribution Methods

Type of Meal Distribution	Benefits	Constraints
Hot Thermal Retention		
Heated base (pellet, unitized base, induction heat base)	Support equipment and system operation are uncomplicated. No requirement for a special plate: any standard-size china. No special delivery cart is required.	Provisions for maintenance of cold items such as milk, salads, gelatin, ice cream are not made. Hot food cannot be held for a long period of time (more than 45 minutes). Additional service ware pieces need to be inventoried, stored, transported, and washed. Induction heat bases: difficult to determine if heating process initiated.
Insulated components	Only the dinner plate and food are insulated; there are no special bases to heat. Simple in operation. No burn hazard to the attendant or customer. No special delivery cart is required.	Additional service ware pieces need to be inventoried, stored, transported, and washed. Attractive insulated components are often taken home by patients as useful mementos of their hospital experience.
Heat support cart	Foods remain heated until tray is removed for service to the patient.	The potential for maintenance/repair problems is high. Carts can be heavy and difficult to maneuver. No provisions are made for maintenance of cold food items at proper temperatures.
Hot and Cold Thermal Retention		
Split tray	Centralized supervision and control of the meal assembly process. No reassembly of tray components is required in the service areas. Good temperature retention of both hot and cold items. System accommodates late trays within a reasonable period.	Cart is heavy and bulky. A motorized version may be required if any ramps are to be negotiated. Carts are difficult to sanitize. Initial cost of the cart is high and maintenance costs can be high. Due to the relatively heavy weight and limited maneuverability, carts and wall surfaces are subject to damage.
Match-a-tray	Same as described for split tray except that decentralized assembly of meal trays is required prior to service.	Same as described for split tray. Additional labor prior to service is needed to reassemble the complete patient meal.
Insulated tray	Maintains hot and cold zones well without external heat or refrigerant sources. Simplicity of transport is achieved. Does not require a heavy, enclosed delivery cart. Stacked trays protect and insulate food. Less load on the dishwashing facility due to disposables. No complex components to repair, replace, or maintain.	Purchase of special disposable dishes results in higher operational costs. Food holding time is limited to 45 minutes. Long-range cost could be substantially higher than other systems due to disposable and lease costs. Hot foods may take on a "steamed" appearance in the hot compartment due to relatively small volume and lack of venting. Possible adverse patient reaction to eating from a compartmentalized tray.

Type of Meal Distribution	Benefits	Constraints
		Trays can be difficult to sanitize completely due to deep cavity construction. Top and bottom tray compartments do not nest; more storage area required. Rigid presentation and placement of dishes is a limitation of the system.
Insulated components	Only the dinner plate and food are insulated. No bases to heat. Hot and cold foods are placed in insulated containers. There is no burn hazard to the attendant or customer. Cold food items can be held longer than 30 minutes. No special insulated delivery cart is required.	Additional service ware pieces need to be inventoried, stored, transported, and washed. Attractive insulated components are often taken home by patients as useful mementos of their hospital experience. Hot food holding time is limited to 30 minutes.
Cold Thermal Retention/Food Reheating		
Refrigerated carts with conduction heat units	Centralized supervision and control of the meal assembly process. No reassembly of tray components in service area. Good refrigerated temperature retention. Good reheating of hot food items.	Carts can be difficult to sanitize. Initial cost of carts is high and maintenance cost can be high. Hot beverages must be added just prior to service. All hot food items must fit on plate or bowl to be on conduction base.
Split cart—refrigerated and convection heat	Same as described for refrigerated carts with conduction units. All food items, including hot beverages, can be placed on tray in central assembly area.	Cart is heavy and bulky. Initial cost of cart and refrigeration/reheating units very high. Space needed near point of service for refrigerated heating units.
No Thermal Support		
Covered tray	Tray is a simple standard unit. Equipment cost of the system is low.	Requires an immediate and responsive transportation system. High labor component is required for transportation process. No thermal support is available for entrée and other food items.

Source: Adapted from "State-of-the-Art Review of Health Care Patient Feeding System Equipment" in *Hospital Patient Feeding Systems* (pp. 168–172) by P. Hysen and J. Harrison, 1982. Washington, DC: National Academy Press.

Table 7.2 Benefits and Constraints of Various Heat Processing Methods

Method	Benefits	Constraints
Microwave ovens	Food is cooked very rapidly. "On-demand" patient feeding can be achieved.	Food is easily overcooked, and some foods tend to rethermalize unevenly, leaving hot and cold spots. Food does not brown, causing some foods to have an unnatural appearance. Trained operator is required to rethermalize all food products. Employee training is essential to the success of the program. Maintenance of microwave ovens can be a significant cost factor.
Convection ovens	Oven cavities can accommodate 12 to 30 meals at a time; thus higher efficiency can be achieved in the rethermalization and reassembly process as compared to a microwave system.	Speed is increased as compared to a conventional still air oven; however, the process is not as fast as a microwave oven. Some food products experience excessive cooking losses; in others, there is a thickened surface layer on the food from the rethermalization process. Some food products do not rethermalize to a uniform temperature.
Conduction heat units	Equipment cavities can accommodate 12 to 24 meals at a time; thus all meals are ready for service at same time.	Cooking surfaces get very hot; employee burns possible. Cooking surfaces can be hard to clean when food is cooked on to surface. Reheating time can be 45 minutes.

reheating often took place in galley kitchens close to the point of service in microwave or convection ovens. Tray delivery personnel heated the food items for a meal tray before serving the tray. Equipment innovation has resulted in the development of carts that allow this reheating to occur in the carts used to transport assembled meal trays. One cart design has chilled plates or bowls of food being placed on food trays with cut-out openings. These openings allow for dishes to have contact with conduction heat plates when placed in special carts. These heating units heat the plate or bowl and its contents while the remainder of the tray remains chilled in the refrigerated cart. Separate conduction heat units also are available in which all food items to be heated are placed on heated shelves that transfer the heat to the plates or bowls of food. The tray delivery personnel must then assemble the trays prior to service. A newer concept incorporates the split-tray concept with convected heat, allowing cold foods on half of the tray to be held at refrigerated temperatures while the hot food items on the other half of the tray are reheated in a convected heat oven.

In **decentralized meal assembly,** the food products are produced in one location and transported to various locations for assembly at sites near the customer. Equipment to maintain proper temperatures—food warmers, hot food counters, and/or refrigerated equipment—must be provided at each location. Because some foods, such as grilled or fried menu items, do not transport or hold well, some cooking equipment may be available in the service units for these difficult-to-hold foods.

In **centralized meal assembly,** food items are prepared and assembled on trays before being transported to other locations for service. Even in centralized meal assembly, a few items such as coffee and toast may be prepared at point of service. Centralized assembly has the advantages of eliminating double handling of food and facilitating supervision of meal assembly because the activity takes place in one location rather than throughout the facility. In addition, centralized assembly allows for standardization of portions, uniformity of presentation, and decreased waste. Finally, less staff time is needed, and the space occupied by decentralized kitchens can be used for other purposes. Decentralized meal assembly is still used in some institutions, however, because it offers the advantage of less time between meal assembly and service to patients, allowing for potentially higher-quality food. Decentralized facilities also offer greater flexibility in providing for individual customer needs and in making last-minute substitutions and changes.

Depending on the layout and design of the facility, a combination of meal assembly and distribution methods may be used. Some facilities may serve customers in groups onsite, conduct centralized meal assembly of individual meals, and transport foods to satellite units for service offsite.

CATEGORIES OF SERVICE

Service can be categorized in a variety of ways; in fact, a number of combination services exist. Service will be categorized as table service, counter service, self-service, tray service, take-out service, and delivery for discussion in this text. The table service restaurant with a self-service salad bar is an example of a combination of service types.

Service of food and beverages is one of the most diverse activities imaginable, assuming many forms and occurring in a wide variety of places, at all hours of the day and night. Because of today's lifestyles, options can range from fine service with tableside preparation, to coffee and doughnuts in the factory, to hot dogs at the beach.

Table Service

Table service is a very common form of service in the commercial segment of the industry. Table service can be very simple or extremely elaborate; its distinguishing characteristic is service by a waitperson. In most table service operations, a hostess, host, or maître d'hôtel is responsible for seating guests in the dining room.

The most common method of table service in the United States, often referred to as *American-style* service, involves plating the food in the kitchen or service kitchen and then presenting it to the guest. In more elaborate service, often referred to as *French style,* food is prepared at the table—as with bananas Foster or steak Diane. Another type of table service is called *family style,* in which food is brought to the table on platters or in bowls by the waitstaff and then passed around the table by guests. Restaurants featuring country fried chicken or barbecue ribs will frequently feature family-style service, as do some elementary schools and nursing homes.

A well-trained and courteous waitstaff and other service employees are the keys to successful table service operations. In upscale restaurants offering sophisticated service, the job of the waitstaff is highly specialized and truly an art.

Counter Service

Counter service often is found in diners, coffee shops, drug store fountains, and other establishments in which patrons are looking for speedy service. People eating alone can join others at a counter and enjoy the companionship. The most common counter arrangements are shown in Figure 7.4. These arrangements provide not only fast service for a customer but also efficiency for the establishment. The counter attendant is usually responsible for taking the orders, serving the meals, busing dishes, and cleaning the counter and may even serve as cashier except at peak periods.

Self-Service

Today, self-service foodservice operations cover a wide spectrum; cafeteria service is one of the most commonly used forms. Self-service is characteristic of the limited-menu restaurant industry, with counter pick-up service, take-out, and drive-through window service the most common approaches. Other self-service operations include buffets, vending machines, refreshment stands in recreational and sports facil-

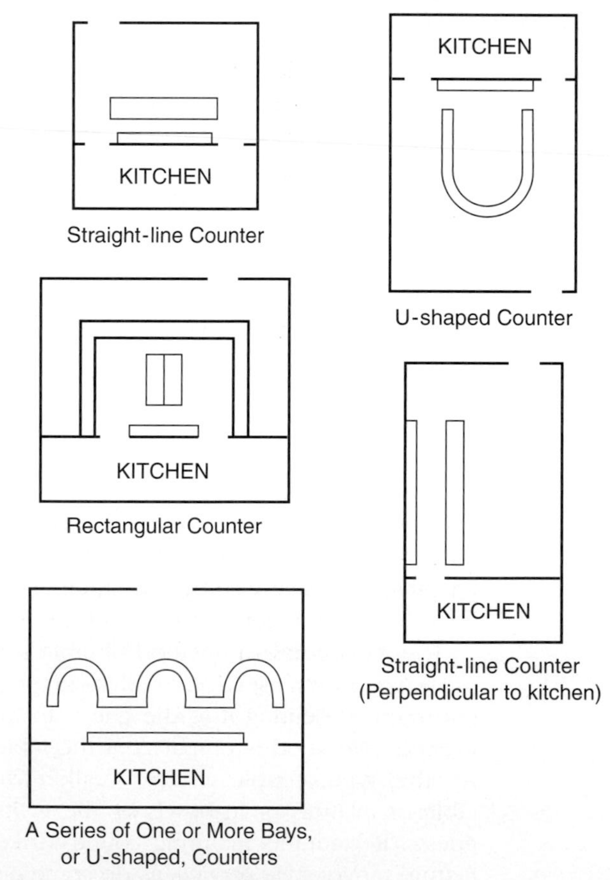

Figure 7.4. Various layouts for counter service.
Source: How to Manage a Restaurant or Institutional Food Service (p. 105) by J. W. Stokes, 1982. Dubuque, IA: William C. Brown. Used by permission.

ities, and mobile foodservice units that range from the small, hot dog cart rolled down the street by the operator to sophisticated operations in motorized vans equipped for preparing a variety of menu items. Today's round-the-clock eating patterns in every imaginable place have created a demand that self-service satisfies.

Cafeteria service is characterized by advance preparation and self-service of most if not all menu items onsite. It is the predominant form of service used in onsite foodservice and employee-feeding operations. Self-busing of trays and dishes is also a common practice as a means of reducing labor costs. Traditionally, commercial cafeterias placed a great deal of emphasis on food display, merchandising, and marketing of menu items. Managers of onsite foodservices are incorporating many of these same strategies to help make their operation more profitable.

The straight-line counter, which may vary greatly in length, is the most common cafeteria counter arrangement. Generally, the length varies with the quantity and variety of menu items instead of being dependent on the number of persons to be served.

An alternative arrangement to the straight-line counter is the hollow square, sometimes called the scramble system. In this layout, the various stations or food counters are positioned to form three or four sides of a square, with space between the counters and perhaps a center island. This layout allows customers to move from one station to another without being held up by the entire line. The hollow square layout not only decreases lines but also permits more people to be served in a smaller space.

Buffet service has enjoyed increasing popularity in recent years in all types of foodservice. A buffet is a type of service where guests obtain all or a portion of their food from a buffet table. Periodic scheduling of buffets in a college residence hall foodservice, an employee cafeteria in a hospital, or an industrial foodservice operation can serve as a monotony breaker and a means of creating goodwill.

Buffet service enables a facility to serve more people in a given time with fewer employees. The usual procedure in commercial operations is for guests to select the entrée, vegetables, and salad from the buffet table before going to the dining table set with flatware, napkins, and water.

Vending machines, dubbed the silent salesperson with a built-in cash register, annually move billions of dollars of products and services to customers around the world. Selling items from machines is nearly as old as recorded history, but the impact of vending machines on the U.S. economy was not recognized before the middle 1940s. Even though the external customer communicates only with the vending machines, employees, the internal customer, work behind the scenes to ensure that customer needs are being met. Temperatures must be recorded daily on all perishable food, and strict adherence to the coding, product handling, and rotation procedures must be maintained. Sanitation procedures and schedules must be developed and checked. Employees servicing the machines also must be trained in customer relations because they represent the vending operation when interacting with the customer.

Many organizations use a contract company rather than setting up their own vending operation. The contract should be reviewed periodically as should the accident, liability, and hazard insurance carried by the vending supplier (Beasley, 1990). Also, the supplier's compliance with city, county, state, and other regulatory agency standards should be checked. Before selecting a supplier, it can be helpful to visit the headquarters to see how and under what conditions the food is prepared.

Competitive bidding has proven beneficial to the purchaser of the service, who can negotiate commission rates and replacement of equipment when necessary.

Foodservice directors who operate their own vending business have the opportunity to tailor a vending program to their customers' needs. These directors should respond to new market trends and technological advances in vending, as they do in their cafeteria and catering programs (Beasley, 1993). They must be innovative and creative in finding ways to make vending a profit center.

As markets shift and change, customer needs and preferences also change (Friedland, 1997). A study on snack vending revealed that customers wanted the following improvements: brand name items, healthier selections, new items, and greater variety. These improvements and attractive machines are enhancing the perception of vending. Multiple choice machines and a beverage machine often are teamed together to provide variety for the customer.

Vending machines have been modified to accept credit and debit cards. Other new payment options include the use of charge cards that permit customers or a department to be billed. The use of a debit card permits the customer's balance on hand to be reduced after each transaction.

Tray Service

Tray service, in which food is carried on a tray to a person by a foodservice employee, is used primarily in healthcare institutions and for in-flight meal service in the airline industry. Room service, in which food is served on a tray or on a cart in a customer's hotel or motel room, is a variation of tray service.

In airline service, food is produced in a commissary by a food contractor that provides meals or snacks according to airline specifications. Specialized tray assembly equipment is tailored to the needs of the operation. In flight, thermal support is needed for heat processing of menu items and cold support for chilled items. Controls are required to ensure that the proper number of meals is provided and to monitor other items, ranging from dishes and flatware to individual tea bags. As an illustration of the complexity of airline service, the food may be loaded onto a plane at its departure location and the empty trays unloaded at the destination.

Many hospitals are shifting their traditional centrally assembled tray service process to an on-demand room service program. This room service program allows hospital patients to call the foodservice department and order their meals when they are ready to eat, similar to what a guest might do when staying in a hotel. The food and nutrition services department at Memorial Sloan Kettering Cancer Center in New York City changed to an on-demand room service delivery process for its patients in 2001 and patient satisfaction and patient meal consumption increased. Cox (2005) indicated that the department's Press Ganey satisfaction scores increased from the 24th percentile to the 99th percentile and patient food consumption increased from 29 to 88% of patients consuming half of their food or more when the hospital implemented the on-demand room service option for patients.

Maintaining appropriate food temperatures and food quality are particular challenges with tray and/or room service. Often food is placed on the tray at a location separate from the point of service, requiring that food be transported to the point of service. Usually the service staff will have limited contact with the customer; training of service staff should focus on ways to use this brief interaction to create a positive experience for the customer.

Take-Out Service

One of the rapidly growing areas of foodservice operations is take-out service. American consumers are cooking less at home in part because of an increase in number of family members who work outside of the home and the pressure to balance work and family life. Take-out service allows consumers to purchase food at one location and then enjoy that food in a location of their choice. According to Sloan (2005), more than half of all Americans purchase food each week through take-out or delivery services; 20% of restaurant orders are made from a vehicle. Quick-service restaurants are the primary provider of take-out food. Many onsite foodservice operations now offer home meal replacement programs, in which food items are packaged and sold as meals for consumption in the home. Often customers will call in advance to place a food order so that the food items can be quickly purchased from the foodservice operation.

Delivery Service

Delivery service is another rapidly growing aspect of foodservice operations. Delivery service involves transporting prepared food items from the foodservice operation to the customer. Typically, delivery service is either to a customer's home or office location. A fee often is charged for this service. One challenge foodservice operators face in the delivery process is ensuring that proper food temperatures are maintained during the delivery process. Transportation equipment, such as a special cart or a motorized vehicle, usually is needed to facilitate the delivery process.

SERVICE MANAGEMENT

Service management is a philosophy, a thought process, a set of values and attitudes, and a set of methods, according to Albrecht and Zemke (1995). Transforming an organization to a customer-driven one takes time, resources, planning, imagination, and tremendous commitment by management.

Service is a word, like *quality,* that means different things to different people at different times. Service can be thought of as an American way of life, and in America each individual holds power.

Total Quality Service

Albrecht (1993, 1995) is concerned that many service organizations are adopting the total quality management (TQM) philosophy, which is based on numbers and work processes with little emphasis on customer value. He suggests that the emphasis be changed from management to service to create a **total quality service (TQS)** model. The emphasis then is to take the long view and focus on the reason the organization exists, which is to serve. The TQS philosophy emphasizes that all quality standards and measures should be customer referenced and should help people guide the organization to deliver outstanding value to its customers. Quality standards should be means to an end but not ends in themselves.

Since World War II, the world has been embracing a quality paradigm. Albrecht (1993) defines a **paradigm** as a mental frame of reference that dominates the way people think and act. The quality movement has focused primarily on zero defects in a product and very little on service quality. In the twenty-first century, the distinction between product and service will become obsolete. Those terms will be replaced by **total customer value,** the combination of the tangible and intangible experienced by customers at the various moments of truth that become their perception of doing business with an organization (Albrecht, 1993). Quality must start with the customer, not with the product or work process that creates it. A **paradigm shift,** which Barker (1992) defined as a new set of rules, from a quality to a total customer value paradigm is occurring in organizations.

Eating a meal in a restaurant often is a one-time, special event for a customer, but serving it is a repetitive, mundane occurrence for an employee. Customers cannot forget a bad experience; even if employees agree to do better the next time, it is too late. Good service may satisfy a customer but not always give total customer value. Exemplary service, however, delights customers by totally exceeding their expectations (Marvin, 1992). Customers keep a mental score and assign a subconscious point value to their experience. The more positive the experience, the higher the score. If the score is higher for one restaurant than for its competitors, the high-score operation becomes the restaurant of choice. If a competitor has a higher score, the original restaurant is in trouble. Any foodservice operation and its service are only as good as its staff.

Mystery shoppers: Persons unknown to customers and employees who eat at a restaurant and evaluate their own experiences and those of other customers.

The U.S. Department of Commerce has found that more than 90% of dissatisfied customers will drift over to the competition, but not always silently (Bode, 1993). This customer will tell as many as nine other people about the bad experience. To prevent this, more and more restaurant owners are hiring **mystery shoppers** who not only record their own experiences but also those of other diners. The mystery shopper must be able to turn a narrative into a snapshot so the management can picture what the dining experience was like. Mystery shopping services are available in many cities. A representative visits the restaurant and then develops a mystery shopper evaluation form. The representative hires shoppers to pose as customers, and they evaluate the restaurant as they dine. "An example of the type of form that might be completed by a mystery shopper is shown in Figure 7.5.

The mission of the mystery shopper can be divided into three phases: monitoring, motivation, and maintenance. Monitoring occurs because employees never know who the shopper is and when she or he is in the restaurant. Motivation of employees is much simpler if the continual flow of feedback from the mystery shopper leads to higher salaries for good performance. Finally, maintenance of superior customer service is possible with fewer visits from the mystery shopper.

The Staff

Delivering consistent, quality service requires having a well-trained staff. Staff employees need to have knowledge of service procedures, a friendly and concerned-for-customer attitude, and an ability to perform the needed service tasks. Many foodservice managers are empowering employees to make decisions that will contribute to a positive customer experience.

Cross-training is a technique being used by foodservice managers to involve employees in the total customer value concept. It usually results in a loyal staff because employees have the opportunity to understand how the foodservice operation works and to find out what is happening in each unit (Weinstein, 1992). Some operations have established a cross-training program in which a front-of-the-house

McBIZ CORPORATION
SHOWBIZ PIZZA/CHUCK E. CHEESE'S
MYSTERY SHOPPER EVALUATION FORM

Shopper Code: ___________
Male:_____ Female:_____
Points Possible: 1,000
Points Earned: ___________

Location: ______________________________

Date of Visit: ______ *Time Entered:* ______ *Time of Exit:* ______ *Day:* ________

Level of Business: Full: ________ 3/4 Full: ________ 1/2 Full: ________ 1/4 Full: ________

Office use only

Please circle "Y" for yes and "N" for no.

GUEST SERVICES (20%)

Server's Name:

Upon delivery of food did the employee:

Office use only			No.	Item
	Y	N	47.	Ask if anything else was needed
	Y	N	48.	Deliver plates with pizza
	Y	N	49.	Employee checked back to inquire about satisfaction or talk with you while you were seated during your meal
	Y	N	50.	Employee offered to take away dirty plates, empty trays, etc. or box left-overs while you were seated
	Y	N	51.	Tables were cleared and wiped quickly after other guests left table **(Do not penalize if guests are in other areas of the restaurant)**
	Y	N	52.	Manager was visible to you during your visit
	Y	N	53.	Manager talked or interacted with guests
	Y	N	54.	Manager was professional and well groomed
	Y	N	55.	Employees appeared to stay focused on their job responsibilities and the needs of guests
	Y	N	56.	Employees were helpful, pleasant, courteous, and friendly to guests
	Y	N	57.	Your child had an opportunity to hug Chuck E. Cheese (Live, walk-around, costumed mascot)
	Y	N	58.	An employee at any time during your visit mentioned the fan club sign-up program to you
	Y	N	59.	All employees were wearing name tags
	Y	N	60.	All employees appeared neat, well groomed and in uniform
	Y	N	61.	Red or blue shirt
	Y	N	62.	Khaki pants or shorts
	Y	N	63.	Red or blue visor
	Y	N	64.	White tennis shoes

Comments ____

Total Possible: 80 Multiply ×2. Points Scored: _______ Guest Points: _______

Figure 7.5. Excerpt from a mystery shopper evaluation form.
Source: Adapted from McBiz Corporation, Topeka, Kansas. Used by permission.

employee starts as a buser at a minimum wage and progresses rapidly to a runner and finally a head runner at a higher salary. Then the employee works in the kitchen at yet a higher salary to see how it operates, learns the computer program, and observes how management handles relationships between front- and back-of-the-house employees. Kitchen employees are given the same opportunity in the dining areas. This cross-training can break down barriers between employees in the front and back of the house, creating a climate that adds to total customer value. For example, cooks who have been cross-trained begin to realize that demands of the waitstaff are not personal demands but are demands of customers.

In some of the fine dining restaurants patterned after those in Europe, service is considered an honorable profession and a career (Ryan, 1993). All new staff, regardless of experience, must go through an apprenticeship program that might last a year to become a fully qualified waitperson. Other operations often have a rigorous and lengthy training program that uses written tests covering a general knowledge of the restaurant and its foods and wines and essays on hypothetical situations that might happen in the restaurant.

Burnout, emotional exhaustion, and loss of enthusiasm for the job are common among many employees in the workforce. Satisfaction of employees with their jobs should be thoroughly examined by management because it can have a great effect on the quality of service. Often managers take advantage of good employees and overload them with tasks. Employees should be praised when they do a good job.

The Special Customer

Customers have been mentioned many times in this text—their demographics, lifestyles, and menu preferences. The importance of pleasing customers has been discussed in every chapter, but what about customers who don't fit the typical pattern but fit into special groups and have to, or choose to, eat away from home? How is the foodservice industry taking care of steady customers, solo diners, customers with small children, and customers with disabilities? Are restaurants, school and college foodservices, and hospitals treating them the same as other customers?

Restaurant customers often are considered transient, especially during vacation times when they stop to eat in a restaurant on the way to their destination. Most restaurants could not survive if customers only visit during vacations or for special events. Restaurateurs find many ways to thank their regular customers for being loyal and steady customers. A chain of restaurants in California rewards its customers who dine regularly at their restaurants with priority reservations by giving them a plastic card with special telephone numbers. When that telephone rings, the front-of-the-house employee knows a steady customer is calling for a reservation, usually for the same day. Top priority is given to that customer.

A restaurant in Virginia Beach has created a T.L.C. (The Local Customer) program. Locals qualify as frequent diners by eating at the restaurant 10 times during off-season, which runs from September through February. They become card-bearing regulars, which qualifies them to reserve a table during peak season, when reservations are not accepted. They also receive bonuses including a complimentary special, such as an appetizer or a glass of wine. The secret of getting customers to return is to make them feel special.

Staff in many operations are being trained to be more sensitive to solo diners. Dining room designers have come up with ideas to make these customers comfortable. Booths are desired by many of these customers, who usually come prepared with a book or magazine and appreciate more space. Another design idea that gives flexibility to single diners is a banquette, which is an upholstered sofa that runs along a wall and forms a long seating space. Small tables for one to four customers are placed in front of the banquette. Tables for one are no longer being placed in corners or outside the kitchen door, but are being distributed throughout the area. Single-dining areas that cater to travelers and local singles are being incorporated into some main dining rooms. Also, place settings often are laid out at the bar, especially at lunchtime, for businesspeople dining alone.

Many restaurants now have children's menus and games to keep children busy while parents eat. Some foodservice operations are even offering child care to customers.

People with disabilities often were not able to eat away from home because restaurants and transportation were not accessible to them, especially if they were in wheelchairs. Hearing and visual impairments also are considered disabilities. With the passage of the Americans with Disabilities Act (ADA) on January 26, 1992, public accommodation rules took effect. Front door, aisle, table, self-service, and restroom access must be available to customers with disabilities. ADA rules apply not only to restaurants but also to all onsite foodservice operations. Hospital and nursing home foodservice managers have had more experience in feeding people with disabilities than have any other managers, but they still must comply with the ADA rules.

Managing Service

Romm (1989) stated that the business of the restaurant takes place in the social space created between the guest and the server. Improving the consistency and quality of the service staff and treatment of customers can be a problem facing the foodservice industry. Managers would like to generate friendly behavior that will be perceived as authentic between the waitstaff and customers. This process requires two steps. First, more attention must be focused on service employees, and their place in the business structure should be reexamined; and second, techniques being used to change employees' behavior need to be examined. Service employees are expected to nurture and entertain customers, but management seldom nurtures employees. The most common response to a need for increasing profits is to reduce payroll expenses by reducing labor hours; yet the employee is expected to deliver good service to customers.

According to Martin (2002), achieving quality customer service requires excellence in both the procedural and personal dimensions of customer service. His model of quality service describes four basic patterns of customer service:

- **Freezer.** A pattern of service in which there is poor procedural and personal service; it conveys the message, "We don't really care about you."
- **Factory.** A pattern of service that is skewed toward procedural efficiency; service may be timely and efficient but employees are cold and impersonal, leaving customers with the impression, "You are a number. We are here to process you as efficiently as we can."

- **Friendly Zoo.** A pattern of service in which employees are very friendly, genuine, and caring but service is slow, inconsistent, and disorganized; it sends a message to the customer that "We are trying hard, but we don't really know what we are doing."
- **Quality Customer Service.** A pattern of service in which both personal and procedural dimensions are handled well; it conveys the message, "We care about you, and we deliver."

The procedural dimension of service quality focuses on the type and timing of service and includes timeliness, incremental flow, anticipation, communication, feedback, accommodation, and organization and supervision. The personal dimension focuses on the style of service and includes three key indicators: attitude, verbal skills, and behavior.

Martin (2002) suggests a six-step approach to quality customer service. The desired outcome is not only the proper service knowledge and attitude but also server behavior patterns necessary for providing quality service. Martin's six-step approach is as follows:

1. Understand customer procedural and personal service expectations.
2. Establish a quality service culture and leadership climate.
3. Institute clear and concise service-delivery standards.
4. Incorporate service standards into organizational systems.
5. Assess progress and reward successes.
6. Continually work on improving quality service.

No matter how well service employees have been trained, they will be challenged at some point to handle service failure situations. According to Ford and Heaton (2000), how the service recovery is handled is much more important to customers than the original failure. A company's ability to recover from a service failure impacts customer likeliness to return to that operation. Martin (2002) recommends a seven-step gracious problem-solving process to help increase the potential for a positive service recovery:

1. The service provider **LISTENS** carefully to the complaint or problem.
2. The service provider **REPEATS** the complaint or problem to get acknowledgment that the customer has been heard correctly.
3. Somewhere along the way, the service provider **APOLOGIZES** to the customer, regardless of who is responsible for the problem or complaint.
4. The service provider **ACKNOWLEDGES** the guest's feelings (anger, frustration, disappointment, etc.). This is an important step that helps establish a nondefensive problem-solving approach.
5. The service provider **MAKES** problem solving a two-way process by asking the customer how he or she would like the problem resolved.
6. The service provider **EXPLAINS** what action can be taken to solve the problem or revert a wrong into a right.
7. The service provider **SAYS** "thank you" to the guest for bringing the problem or complaint to his or her attention.

Tipping

Tipping waitstaff has been a common practice in table and counter service restaurants and for room service delivery in the United States. In Europe, a common practice is to add a service charge to restaurant or hotel bills. Tipping is becoming more common for employees who deliver foods to office or home locations. Tips are voluntary; service charges are not. The service charge is a predetermined amount added to each customer's check and is considered part of the restaurant's gross receipts subject to income tax; it is not a tip. The employer is under no obligation to give the service charge to the waitstaff. In most cases, however, all employees benefit from it by increases in wages. Tips left in addition to service charges belong to the waitstaff and are treated the same as all voluntary tips by wage and tax laws.

The foodservice manager must be aware of federal and state legislation and regulations that have provisions covering tipping and tipped employees. Employee tips are a source of income. This income must be reported to all applicable agencies. Although ultimately the responsibility of tip reporting is an employee's, foodservice operations can be required to pay payroll tax on tip income if unreporting is found.

The Internal Revenue Service (IRS) has established three types of voluntary compliance agreements for reporting tips: Tip Reporting Alternative Commitment (TRAC), Tip Rate Determination Agreement (TRDA), and Employee-Designed Tip Reporting Alternative Commitment (EMTRAC). Each varies according to whether employees sign agreements, what type of employee training is needed, how tips will be reported, and what information is provided to the IRS. Additional information is available at www.irs.gov.

Service Controls

Handling and controlling guest checks is another concern of managers in using a waitstaff. The first element of effective control is to ensure that all menu items are charged to the customer. A duplicate check procedure is the most widely used; the waitperson writes the order on a customer check, simultaneously preparing a carbon duplicate that is submitted to the production area to obtain menu items, or the waitperson enters the order into a computer system, which generates information for the back of the house and ultimately the customer check. The second element of control is to charge proper prices for menu items. Electronic cash registers with preprogrammed prices and computerized systems are widely used for this reason. The assurance that all checks are accounted for is the third element of control. This can be accomplished by having all customer checks numbered sequentially, keeping reserve stocks of checks in locked storage, issuing checks to waitstaff in numerical order, and then sorting the used customer checks periodically into numerical order to determine if any are missing.

Many restaurants accept credit cards as a means of payment. Most restaurant chains use the swipe method of authorizing credit card transactions. The credit card number is used to authorize the transaction and check the card against a computer file of lost or stolen cards within seconds. Customers ordering home or office delivery of food by telephone can give the order taker their card number, and authorization is obtained while they wait. Delivery employees for some restaurants carry a portable addressograph to imprint card information on the check.

EXPERIENCE ECONOMY

Pine and Gilmore (1999) contend that we have moved from a service economy to an experience economy. In their book *The Experience Economy,* the authors describe the movement of the U.S. economy from one based on commodities to one in which value is gained by experiences. As an example, they describe the commodity economy as one in which people grew/raised their own food products such as vegetables and livestock. In the late 1800s, we began to manufacture goods and people began to buy food products at the local market or grocery store. This launched what the authors refer to as a goods economy. By the 1950s, we had moved into the service economy, where people would go out to restaurants to eat their food. The experience economy, which the authors believe is the next phase, involves creating an experience for the customer such as that created in places like Hard Rock Café, Rainforest Café, and Starbucks.

Creating an experience for customers in restaurants and onsite foodservice operations involves engaging them. Pine and Gilmore describe this engagement using two dimensions: participation and connection/environmental relationship. Participation ranges from passive to active and environmental relationship from absorption (mental engagement) to immersion (physical involvement).

Using these two dimensions, Pine and Gilmore propose four realms for creating an experience: entertainment, education, escapism, and estheticism. *Entertainment* (passive participation and absorption connection) provides a way to help people enjoy themselves while eating. The animated stage characters that perform at Chunk E Cheese Pizza are one such example. *Education* (active participation and absorption connection) involves experiences that engage the mind. For example, signage in a hospital cafeteria sharing nutrition information and information about the local growers/producers who supplied the food being served are ways in which education can occur with a dining experience. A guest experiencing *escapism* would actively participate and be immersed in the experience. The Mars 2112 restaurant offers such an experience; guests are "transported" to the planet Mars for their dining experience. The final experience realm, *estheticism,* involves passive participation but immersion in the environment. A meal at Rainforest Café is an esthetic experience.

Foodservice managers should consider ways to incorporate the four E's into their operations. Moving beyond the service economy to the experience economy will require managers to consider ways to incorporate entertainment, education, escapism, and estheticism into their dining operation.

CHAPTER SUMMARY

This summary is organized by the learning objectives.

1. Distribution and service is one of the functional subsystems in the foodservice system model. Distribution involves getting food from production to the point of service. Service is the presentation of food to the customer.
2. Centralized service involves assembling plates of food for service in a location adjacent to the preparation area. Decentralized service involves transporting

Professional Organization Profile

Society of Foodservice Management

MISSION

To provide professional development through research and information, continuing education, and member interaction in a collaborative environment

WHO BELONGS TO THE ORGANIZATION

Members include major corporate liaison personnel and independent operators, as well as professionals with national and regional foodservice contract companies, consultants, foodservice suppliers, and university faculty interested in the onsite foodservice industry. Student memberships are available.

ADVANTAGES OF MEMBERSHIP

Members have access to a members-only section of the association's website with links to the membership directory and industry information; receive a monthly online newsletter, *FastFacts,* highlighting innovations being explored in foodservice operations; can attend association sponsored seminars, conferences, workshops and other networking sessions.

WEBSITE

www.sfm-online.org

MEET THE PRESIDENT

Russ Benson, president 2006–2007, is vice president of guest strategies for Parkhurst Dining Services. He began his foodservice career working in restaurants while in high school. During his 20-plus years in the industry, he has held several food and beverage management positions, including positions with Hyatt Hotels, the Chicago Four Seasons Hotel, and MBNA corporate dining. Russ offers the following advice to students: "Continually learn other disciplines within the industry and get involved in industry associations, community events and always keep up with trade press."

food in bulk to a location separate from the production area and assembling plates of food in that distant location.
3. Table 7.1 provides details regarding the benefits and constraints of various meal distribution systems.
4. Several methods exist for serving food to customers. Counter service involves customers sitting on stools on one side of a counter and being served by staff who work on the opposite side of the counter. Table service involves customers being seated at a table and a waitstaff member taking food orders and serving the meals. Tray service involves an employee carrying food to a customer on a tray, and self-service occurs when customers select the food items they would like from display cases.
5. The service economy focused on the service of food to customers in a dining operation. The experience economy expands the service to include an entertainment, educational, escapism, or estheticism experience.

TEST YOUR KNOWLEDGE

1. Discuss how type of production system, degree of meal preassembly, physical distance from production to service, and amount of time between production and service impact the distribution and service subsystem.
2. Describe the differences among at least three different methods for reheating food items.
3. Explain three categories of service.
4. What is TQS? Why is it important to a foodservice operation?
5. How does a foodservice manager control quality service?

CLASS PROJECTS

1. Divide into teams of two to three students. Each team should visit a different type of foodservice operation (i.e., school, restaurant, hospital, nursing home, etc.). Observe the distribution and service process at that operation, write a summary of the process, and share with the class.
2. Invite the manager of a local foodservice operation to speak to the class about how he or she trains staff members about quality service.
3. Develop a questionnaire to evaluate service at a foodservice operation. Visit a local foodservice operation as a mystery shopper and evaluate the service.
4. Work with a local foodservice operation to design entertainment, educational, escapism, or estheticism experiences for the operation.

WEB SOURCES

www.irs.gov	U.S. Internal Revenue Service
www.isixsigma.com	iSixSigma Service Quality
www.dsgstars.com	DSG Associates Mystery Shoppers
www.nafem.org	North American Foodservice Equipment Manufacturers

BIBLIOGRAPHY

Albrecht, K. (1993). *The only thing that matters.* New York: HarperCollins.

Albrecht, K. (1995). *At America's service.* Homewood, IL: Dow Jones-Irwin.

Albrecht, K., & Zemke, R. (1995). *Service America! Doing business in the new economy.* Homewood, IL: Dow Jones-Irwin.

Anticipating paradigm shifts. (1993). *Food Management, 28*(1), 92.

Barker, J. A. (1992). *Future edge: Discovering the new paradigms of success.* New York: William Morrow.

Beasley, M. A. (1990). Ways vending can work for you. *Food Management, 25*(8), 42, 46.

Beasley, M. A. (1993). Vending: New paths to success. *Food Management, 28*(3), 32.

Bode, D. (1993). The secret service of professional mystery shoppers. *Nation's Restaurant News, 27*(35), 66.

Boorstin, D. J. (1992). Service in America. *Restaurants & Institutions, 102*(11), 70, 74, 82.

Boss, D. (1992). A diversity of individuals. *Food Management, 27*(9), 16.

Brault, D. (1993). Customer time at table unchanged. *Restaurants USA, 13*(3), 37–38.

Buzalka, M. (1998). Canteen brings QSR brands to vending. *Food Management, 33*(7), 20–21.

Cox, S. (2005). On-demand room service. In Putting the WOW! in Dietetics: An Introduction to Customer Satisfaction. [Online]. Available: www.eatright.org

Crosby, M. A. (1992). Bouillabaisse, burgers, and babysitting. *Restaurants USA, 12*(6), 13–16.

Cullen, N. C. (2001). *Life beyond the line.* Upper Saddle River, NJ: Prentice Hall.

Fintel, J. (1991). How to keep them coming back. *Restaurants USA, 11*(5), 14–15.

Follin, M. (1991). Two for the road: Double drive-thru restaurants. *Restaurants USA, 11*(3), 26–28.

Ford, R. C., & Heaton, C. P. (2000). *Managing the guest experience in hospitality.* Albany, NY: Delmar.

Friedland, A. (1997). Machine cuisine: Vending as the technology of convenience. *Food Management, 32*(7), 30–33, 36, 38.

Frinstahl, T. W. (1989). My employees are my service guarantee. *Harvard Business Review, 67*(4), 28–32.

Gotschall, B. (1989). Catering prevails when funding fails. *Restaurants & Institutions, 99*(14), 148–152.

Heskett, J. L. (1986). *Managing in the service economy.* Boston: Harvard Business School Press.

How does service drive the service company? (1991). *Harvard Business Review, 69*(6), 146–150, 154, 156–158.

Hysen, P., Boehrer, J., Greenberg, A., Noseworthy, E., Prentkowski, D., Wilson, T., & Boss, D. (1993). Anticipating paradigm shifts in foodservice. *Food Management, 28*(1), 100–103, 106–108.

Hysen, P., & Harrison, J. (1982). State-of-the art review of health care patient feeding system equipment. In *Hospital patient feeding systems* (pp. 159–192). Washington, DC: National Academy Press.

Iwamuro, R. (1992). Carryout and delivery from tableservice restaurants. *Restaurants USA, 12*(10), 48–51.

Keegan, P. O. (1993). McD's playground concept grows by 'Leaps & Bounds.' *Nation's Restaurant News, 27*(11), 3, 65.

Lorenzini, B. (1992). Fire up burnt-out employees. *Restaurants & Institutions, 102*(7), 27–28.

Martin, W. B. (2002). *Quality service.* Upper Saddle River, NJ: Prentice Hall.

Martin, W. B. (1986a). Defining what quality service is for you. *Cornell Hotel and Restaurant Administration Quarterly, 26*(4), 32–37.

Martin, W. B. (1986b). *Quality service: The restaurant manager's bible.* Ithaca, NY: Cornell University, School of Hotel Administration.

Marvin, B. (1992). Toward exemplary service. *Restaurants & Institutions, 102*(11), 86–87, 93, 95, 97.

National Restaurant Association and National Center for Access Unlimited. (1992). *Americans with Disabilities Act: Answers for foodservice operators*. Washington, DC: Authors.

Pickworth, J. R. (1988). Service delivery systems in the food service industry. *International Journal of Hospitality Management, 7*(1), 43–61.

Pine, B. J., & Gilmore, J. H. (1999) *The experience economy*. Boston, MA: Harvard Business School Press.

Prewitt, M. (1992). Wanted: Singles. *Nation's Restaurant News, 26*(50), 27–28.

Reichheld, F. F. (1993). Loyalty-based management. *Harvard Business Review, 71*(2), 64–73.

Roach, D. (1997). IRS claims to be on trac with tip agreements. *Restaurants USA, 17*(6), 8.

Romm, D. (1989). "Restauration" theater: Giving direction to service. *Cornell Hotel and Restaurant Administration Quarterly, 29*(4), 31–39.

Ryan, N. R. (1991). Food to go. *Restaurants & Institutions, 101*(6), 84–85, 88, 92, 98.

Ryan, N. R. (1993). They give service with a style. *Restaurants & Institutions, 103*(13), 71, 74, 76.

Sanders, E., Paz, P., & Wilkinson, R. (2002). *Service at its best*. Upper Saddle River, NJ: Prentice Hall.

Schuster, K. (1992). Managing cultural diversity. *Food Management, 27*(9), 118–119, 122–123, 128–129.

Shapiro, B. P., Rangan, V. K., & Sviokla, J. J. (1992). Staple yourself to an order. *Harvard Business Review, 70*(4), 113–122.

Sherer, M. (1993). Anticipating paradigm shifts in institutions. *Food Management, 28*(1), 93–96.

Sloan, A. E. (2005). Top 10 global food trends. *Food Technology*. 59(4): 20–32.

Stern, G. M. (1990). Table for one. *Restaurants USA, 10*(3), 14–16.

Stern, G. M. (1991). Service in the 1990s; The way to grow your business. *Restaurants USA, 11*(5), 11–13.

Weinstein, J. (1992). The accessible restaurant. *Restaurants & Institutions, 102*(9), 96–98, 102, 104, 110, 117.

CHAPTER 8

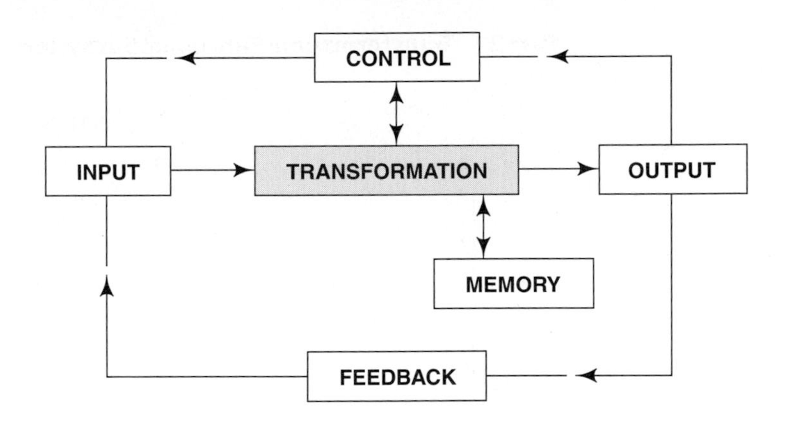

Safety, Sanitation, and Maintenance

Enduring Understanding

- Handwashing and control of time/temperature of food items are the most critical elements of food safety.
- Planning and monitoring are important elements to successful sanitation, maintenance, and risk management programs.

Learning Objectives

After reading and studying this chapter, you should be able to

1. Describe components of the safety, sanitation, and maintenance subsystem.
2. Differentiate the terms clean and sanitary.
3. Compare and contrast methods of solid waste management.
4. Develop policies for helping ensure food, employee, and customer safety.

Safety, sanitation, and maintenance are critical in a foodservice operation. As a future foodservice manager, you will be responsible for ensuring the safety of your employees and customers. An important component of this safety will be the sanitation and maintenance of your equipment and facility. In this chapter, we will discuss the safety, sanitation, and maintenance subsystem of the foodservice system model. Emphasis will be placed on food safety and techniques that you as a manager can implement to help ensure food safety in your operation. Concepts such as risk management and solid waste management also will be discussed.

FUNCTIONAL SUBSYSTEM: SAFETY, SANITATION, AND MAINTENANCE

Safety, sanitation, and maintenance is the last major functional subsystem in the foodservice system, and it permeates all other subsystems (Figure 8.1). The receiving and storage areas need to be checked after each delivery and thoroughly cleaned daily to prevent vermin and rodent infestation. Food safety concerns in each of these areas should be addressed in the facility's HACCP program.

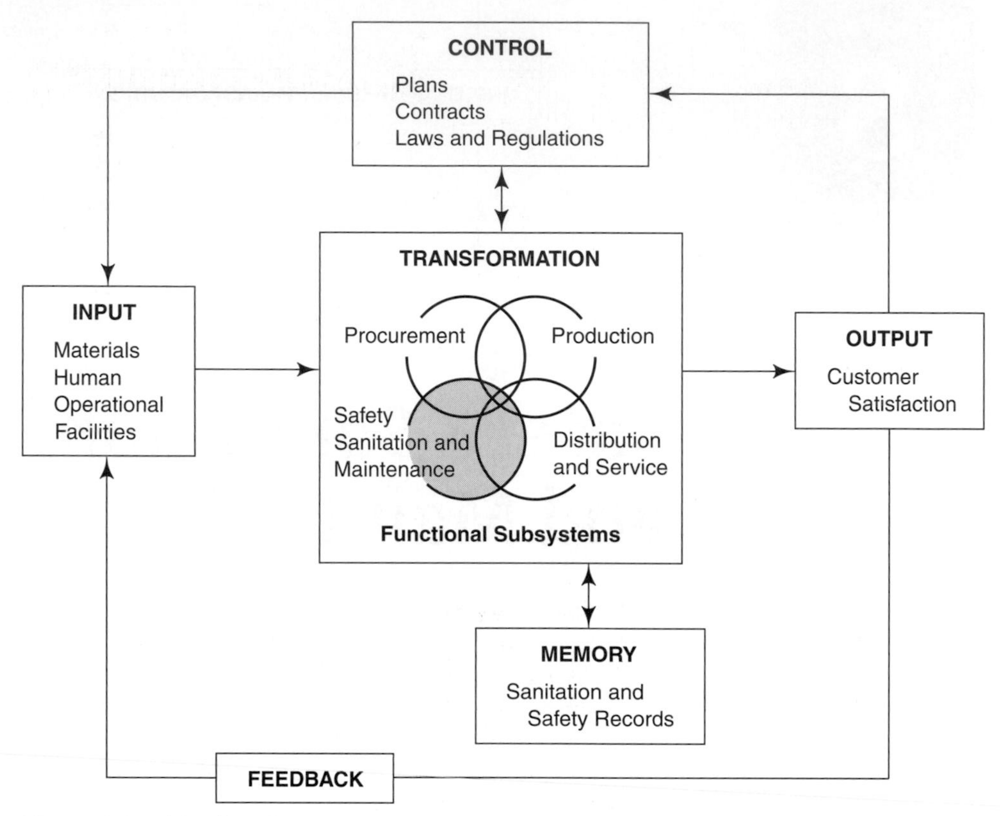

Figure 8.1. Foodservice systems model with the safety, sanitation, and maintenance subsystem highlighted.

Foodservice utensils, dishes, equipment, and facilities require continuous cleaning. Every time a meal is produced and served, food safety practices must be followed and cleaning and sanitizing must be done again. Delivery equipment and service areas also require cleaning and safe food-handling procedures. Cleaning today is complicated. Employees must know what cleaning agent works on what type of soil and whether or not a brush, mop, sponge, scouring pad, or rag should be used (Patterson, 1993).

Ensuring safety in a foodservice operation is a major responsibility of the manager of that operation. This responsibility includes not only the safety of the food served but also the safety of employees and guests of that operation. Risk management activities often focus on safety issues.

In comparison with many industrial jobs, those in foodservice are relatively safe occupations. A National Restaurant Association analysis of Bureau of Labor Statistics data found that the number of nonfatal occupational injuries at eating and drinking places in 2002 was 4.6 per 100 full-time employees, which was a decline from 8.4 in 1990 and was well below the average for all industries and less than half as compared to some other service industries such as nursing and personal care facilities (12.6), air transporatation (12.3), and motor vehicle and equipment (12.1) (Kim, 2004). A foodservice facility, however, has many potential hazards; minor injuries from cuts and burns are common, and more serious injuries occur all too frequently. The quantity of hot foods handled, type of equipment used, and the frequently frenetic pace of a foodservice operation require that safety consciousness is a high priority. Accident prevention must be a priority for foodservice managers because accidents may involve injury or even death of employees or customers.

Maintenance of equipment and facilities is important. The safety of surroundings often is related to cleaning and maintenance practices. Two examples are spills that are not cleaned up properly, which may cause people to fall, and grease buildup in the hoods over the production equipment, which is a major cause of fires in foodservice operations.

FOOD SAFETY

Food safety training and the ongoing management of the food safety system are the most critical functions that occur in the food production facility. Proper food safety is important not only in the United States but around the world. Foodborne illness does not respect borders. A food safety crisis thousands of miles away can affect you and your business just as if it happened a couple of miles away.

The Centers for Disease Control and Prevention (CDC) warns that mishandling of food has caused many foodborne outbreaks in this segment of the industry where food is prepared and served to the public. Food protection began in the early 1900s when the United States Public Health Service (PHS) studied the role of milk in the spread of disease. Model codes were developed to help state and local governments start and maintain effective programs for prevention of foodborne illnesses.

The Food and Drug Administration (FDA), Food Safety and Inspection Service (FSIS), and CDC jointly published *Food Code (2005)*. It is a reference document for regulatory agencies responsible for overseeing food safety in retail outlets such as restaurants, grocery stores, and institutions (for example, nursing homes and child care centers). It may be adopted and used by agencies at all levels of government that are responsible for managing food safety risks. The model *Food Code* provisions are designed to be consistent with federal food laws and regulations and are written for ease of legal adoption at all levels of government. Since publication of the first version of the *Food Code* in 1997, important progress has been made in efforts to monitor and prevent foodborne diseases and ensure that consumers are provided the safest possible foods.

The model *Food Code* is neither federal law nor federal regulation. Rather, it represents the FDA's best advice for a uniform system of regulation to ensure that food is safe and properly protected. HACCP plans as described in the *Food Code* must include flow diagrams, product formulations, training plans, and a corrective action plan. Plans will be reviewed onsite, with records available to judge the adequacy of past corrective actions.

Without question, public awareness has made consumers more cautious about the handling of food at home (Featsent, 1998). The public's once blind trust of the food supply is gone and food phobia is taking its place. Customers are scrutinizing their food and the places where it is purchased, looking for any indication that it might not be safe to eat. Foodservice managers must eliminate any perception of food safety risk in their operations.

The CDC is expanding its prevention efforts to focus on heading off new foodborne pathogens, specific causes of diseases such as bacteria or viruses that can be spread globally by foods tainted with low-level contamination. The CDC has developed the PulseNet system, which will help public health experts determine whether the illnesses are from the same strain or from a common exposure source. PulseNet is a national network of public health laboratories that perform a "finger printing" on bacteria that may be foodborne. Pulse Net provides an early warning system for outbreaks of foodborne disease by having bacteria "finger printing" data on a central CDC computer that is linked to state and local health department (see www.cdc.gov/pulsenet).

Food Spoilage

Spoilage: Denotes unfitness for human consumption due to chemical or biological causes.

Food spoilage is a difficult term to define because people have different concepts about edibility. Generally, however, **spoilage** is understood to denote unfitness for human consumption due to chemical or biological causes. Longrée and Armbruster (1996) identified the following criteria for assuring foods are fit to eat:

- the desired stage of development or maturity of the food
- freedom from pollution at any stage in production and subsequent handling
- freedom from objectionable chemical and physical changes resulting from action of food enzymes; activity of microbes, insects, and rodents; invasion of parasites; and damage from pressure, freezing, heating, or drying
- freedom from microorganisms and parasites causing foodborne illnesses

Many opportunities exist for food contamination in a foodservice operation. Food is often prepared far ahead of service, thus permitting time for bacterial growth (Minor, 1983). The possibility of transmission of disease or harmful substances carried by vectors (flies, cockroaches, weevils, mice, and rats) and by foodservice staff is much greater than in the home. The potential for contamination also is more likely wherever groups of people congregate. Employees may be a hazard if they are not trained in safe food-handling practices.

Today, foodborne illnesses are recognized as a major health problem in the United States. The CDC estimates put the number of cases of foodborne disease in the United States at 76 million each year (information from www.cdc.gov, 2005). Nationally, foodborne illness kills an estimated 5,000 people each year. According to the CDC, reported foodborne outbreaks were caused by mishandling food in commercial and onsite foodservices where ready-to-eat food is prepared and served to the public. Hand washing is the most important way to prevent the spread of infection. According to economists of the Economic Research Service (ERS), the most costly foodborne bacterial pathogens are *Campylobacter, Salmonella, E. coli,* and *Listeria* monocytogenes. The ERS estimated that the economic costs of medical care, productivity losses, and premature death from these pathogens were $6.9 billion (Crutchfield & Roberts, 2000). The CDC tracks the epidemiological character of human diseases and has suggested the following to improve food safety:

- more training for food handlers
- better protocols for investigating foodborne outbreaks
- hazard analyses in food operations
- improved data on how pathogenic organisms are carried and spread, how food preparation contributes to proliferation, and how food handling contributes to an outbreak

Microbial agents are not the only cause of foodborne illness in foodservice operations. Chemicals, herbicides, pesticides, antibiotics, metals, and hormones may contaminate food. In addition, some foods can be poisonous, such as certain varieties of mushrooms. Over the past several years, contamination of the food supply by chemicals or other agents has been a major news topic. Although many incidents have been overemphasized, others represent serious potential health hazards.

All foods deteriorate, some more rapidly than others. Degrees of perishability require food-handling practices that will maintain safety for consumption. The clients of a foodservice establishment have every reason to expect that the food served will be safe and wholesome. Customers can initiate legal action against a foodservice for not protecting their safety. The foodservice organization has a tremendous responsibility for safeguarding its customers. Food-handling practices, beginning with procurement and continuing throughout the production and service of food, must be designed with the safety of the public in mind.

The major types of food spoilage are microbiological, biochemical, physical, and chemical. The extent of contamination of some foods may be difficult to determine from their appearance, odor, and taste, as with inadequately refrigerated potato salad on a salad bar. In other foods, mold, discolored or altered appearance, off-odors, or off-flavors are obvious signs of contamination.

An in-depth discussion of food microbiology is beyond the scope of this text. However, an overview is presented from the standpoint of controlling practices to ensure production of microbiologically safe food.

Microbiological Spoilage

Thousands of species of microorganisms have been identified. The three most common forms are bacteria, molds, and yeast. They are found everywhere that temperature, moisture, and substrate favor life and growth. Some species are valuable and useful in preserving food, producing alcohol, or developing special flavors if they are specially cultured and used under controlled conditions. Other microbial activity, however, can be a primary cause of food spoilage. Food spoilage caused by microorganisms is called **microbiological spoilage.**

Certain microorganisms and parasites are transmitted through food and may cause illnesses in people who ingest contaminated items. According to Longrée and Armbruster (1996), microorganisms causing foodborne illnesses include bacteria, molds, yeasts, viruses, rickettsiae, protozoa, and parasites, such as trichinae. They noted, however, that although most microorganisms producing foodborne illness are bacteria, less than 1% of all bacteria can be considered "enemies of man and many are his friends."

Bacteria

Bacteria are microscopic, unicellular organisms of varying size and shape, including spherical, rod, and spiral. According to the CDC, the most commonly recognized foodborne infections are those caused by the bacteria *Campylobacter, Salmonella,* and *Escherichia coli* O157:H7 (information from www.cdc.gov, 2005). In most instances, the presence of bacteria cannot be seen in food, even if the contaminants are present in sufficient number to produce foodborne illness. Sometimes, however, bacterial contamination may cause a turbid appearance or a slime on a food surface that is visible to the human eye.

FAT TOM is the acronym used to identify the conditions that impact the growth of bacteria (McSwane, Rue, Linton, & Williams, 2004). Those conditions are F (food), A (acid), T (temperature), T (time), O (oxygen), and M (moisture). Although requirements for growth vary among different types of bacteria, all bacterial cells pass through various phases. When the multiplication of bacteria is steady, the number of cells produced over a certain period of time can be plotted. Figure 8.2 shows a typical bacterial growth curve.

Food is the most important condition needed for bacterial growth (McSwane et al., 2004). Foods high in protein or carbohydrate are the most supportive of bacterial growth.

pH value: Degree of a food's acidity or alkalinity.

The degree of a food's acidity or alkalinity, expressed as **pH value**, also affects bacterial growth. The pH value represents the hydrogen ion concentration and is expressed on a scale from 0 to 14, with 7 expressing neutrality. Values below 7 indicate acidity; those above 7 indicate basic or alkaline materials. Bacteria vary widely in their reaction to pH. Although some are quite tolerant to acid, they generally grow best at a pH near neutral, so acid is frequently used in food preservation to suppress bacterial multiplication. Multiplication of the organisms causing food infections and foodborne illnesses are supported in slightly acid, neutral, and slightly alkaline food materials (Longrée & Armbruster, 1996). The pH of some common foods is listed in Figure 8.3.

Microorganisms have specific temperature requirements for growth. At its optimum temperature, a cell multiplies and grows most rapidly, but a cell will also grow within the minimum and maximum temperatures around its optimum. Foods that require time and temperature control because they are capable of supporting growth of pathogenic microorganisms or toxin formation are termed **potentially hazardous foods** (time/temperature control for safety foods) (*Food Code,* 2005).

Potenially hazardous food (time/temperature control for safety food): Food items that require temperature control because they are capable of supporting growth of pathogenic microorganisms or toxin formation.

Various types of bacteria respond differently to temperature. In general, spores of microorganisms are more heat resistant than vegetative mature cells, which are dormant and asexual. Some bacteria form spores inside the wall of their cells when they mature. Spores are more resistant to high heat, low humidity, and other adverse conditions than are vegetative bacteria cells. They may remain dormant for long periods of time and germinate when conditions are favorable into new, sensitive, vegetative cells.

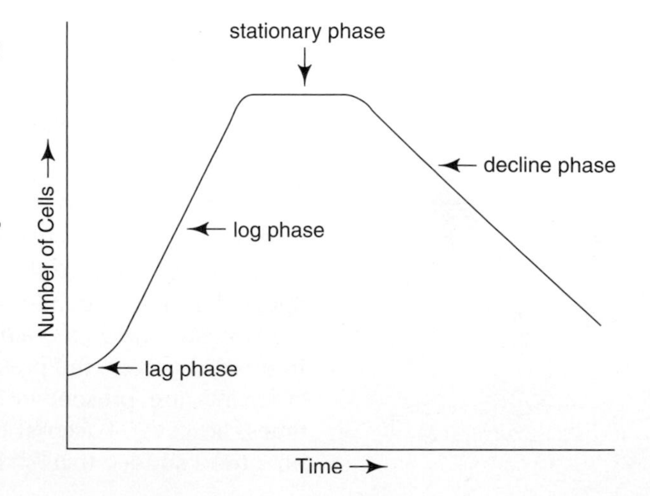

Figure 8.2. Stages in the growth curve of bacteria. (1) Lag phase—no multiplication occurs. (2) Log phase—accelerated growth occurs. (3) Stationary phase—competition for nourishment causes slowdown in multiplication and death of some bacteria. (4) Decline phase—bacterial cells die more quickly due to lack of nutrients or their own waste products. *Source:* From *Applied Foodservice Sanitation* (p. 21) by the Educational Foundation of the National Restaurant Association, 1992, Chicago: Author.

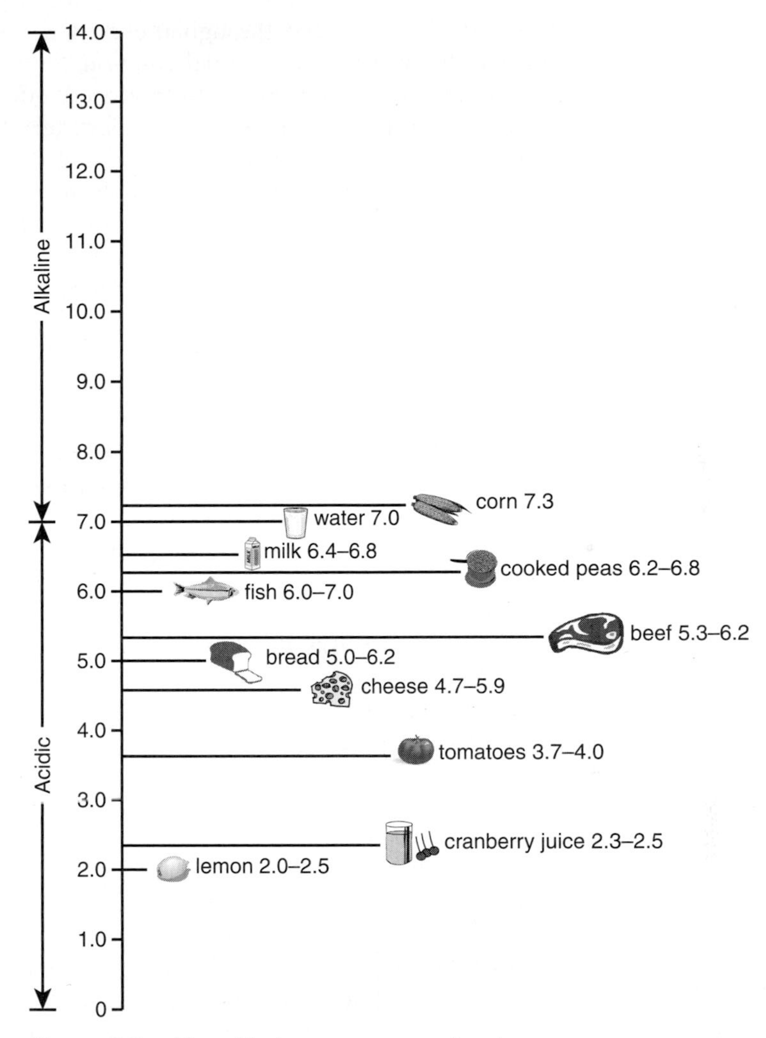

Figure 8.3. The pH of some common foods.

Thermal death time: Time required at a specific temperature to kill a specified number of vegetative cells or spores.

The heat resistance of microorganisms is their **thermal death time**, or the time required at a specified temperature to kill a specified number of vegetative cells or spores under specific conditions. Thermal death depends on the age of the organism, temperature to which it is exposed, length of time for which heat is applied, presence of moisture, and nature of the medium. Thus, time and temperature are important in preserving microbiological quality in foods.

Time also is a critical component in bacterial growth as a single cell can generate more than a million new cells in a few hours time. Bacteria need about four hours to produce enough cells to cause illness.

Aerobic bacteria: Bacteria that need oxygen to grow.

Anaerobic bacteria: Bacteria that reproduce without oxygen.

Bacteria differ in their need for oxygen for growth. **Aerobic bacteria** need oxygen to grow; **anaerobic bacteria** reproduce without oxygen.

Multiplication of bacteria is affected by available moisture in food. The amount of water available to support bacterial growth is termed water activity (A_w). Bacteria need a water activity higher than 0.85 to grow (McSwane et al., 2004). Water becomes less available through the presence of solutes such as salt and sugar, through freezing, or through dehydration.

Various inhibitors have a pronounced effect on bacterial multiplication and death. According to Longrée and Armbruster (1996), inhibitors may be integral in the food, developed during processing as a product of the microorganism's metabolism or added purposely by the processor. The benzoic acid in cranberries and lysozyme in egg whites, for example, are natural inhibitors of these foods. Alcohol produced in the growth and fermentation of yeast, in fruit juices, or in the production of wine is an example of an inhibitory substance that may accumulate and become toxic.

Molds and Yeasts

Molds and yeasts are other common forms of microorganisms important in food sanitation. Molds are larger than bacteria and more complex in structure. In general, they grow on a wide range of substrates—moist or dry, acid or nonacid, high or low in salt or sugar. Molds also grow over an extremely wide range of temperature, although the optimum temperature is between 77°F and 86°F. Because mold growth may appear as highly colored, cottony, powdery, or fuzzy tufts and patches, it is probably the most common type of spoilage that can be identified by the naked eye.

Yeasts are not known to cause foodborne illnesses, but they may cause spoilage of sugar-containing foods. They are unicellular plants that play an important role in the food industry, particularly in the fermentation or leavening of beer, wine, and bread. Yeasts can induce undesirable reactions, however, resulting in sour or vinegary taste.

Viruses

Viruses are small pathogens that multiply in the living cells of the host but not in cooked food. They are capable of causing diseases in plants, animals, and humans. Viruses can be carried in food and water, but they multiply only in the living cell. In many respects, viruses resemble bacteria in that the right temperature, nutrients, moisture, and pH are necessary for effective growth and reproduction. Examples of human diseases caused by viruses are influenza, poliomyelitis, chickenpox, and hepatitis, some of which have been associated with foodborne outbreaks. Many viruses are inactivated by high temperatures (149°–212°F) and by refrigeration.

The Norovirus (formerly known as the *Norwalk virus*) is rapidly increasing as a health threat. Outbreaks have occurred through water contaminated with sewage, raw shellfish harvested from polluted growing areas, and feces of infected food handlers who have not scrubbed their hands after using the toilet. In 1982, an infected employee in a bakery caused some 3,000 illnesses by mixing uncooked butter cream icing in a giant vat with his bare arms up to the elbows in the mixture. The CDC estimates that Norovirus and similar viruses account for more than half of all foodborne outbreaks of gastroenteritis (information from www.cdc.gov, 2005). Approximately 181,000 cases occur annually, with no known associated deaths. Norovirus multiplies in the cells of the human intestinal tract, passes through, and cannot grow again until it reaches another human. Unable to multiply in food, as bacteria do, it must wait for ingestion by a person.

Other Microorganisms

Rickettsiae include such human diseases as typhus fever, Q fever, and Rocky Mountain spotted fever. Cows infected with the organism causing Q fever may transmit the disease to humans through their milk if it has not been properly heat treated. Like viruses, rickettsiae multiply in living tissues only.

Protozoa may be carried by food and cause illness when ingested. Usually microscopic in size, protozoa are unicellular, animal-like forms that are distributed widely in nature, especially in sea water, but also in lakes, streams, and soil. Amoebic dysentery is caused by one of the pathogenic protozoa that can be spread by water and food. Nearly all animals and humans carry protozoa in their intestinal tracts, which has implications for food-handling practices of foodservice employees.

Parasites include *trichinae, tapeworms,* and *roundworms*. Trichinosis is a foodborne disease that affects the muscles of the body. Anyone who eats undercooked meat from infected animals can develop trichinosis; pork, pork products, and wild animal meat are primary sources of trichinella (New York City Department of Health Bureau of Communicable Disease, 1998). This disease is preventable, however, if food is cooked to a proper end-point temperature. The 2002 Pork Producers Council and the USDA Food Safety and Inspection Service recommend that pork should be cooked to an end temperature of 160°F. If the pork is cooked in a microwave oven, it must be heated to an internal temperature of 170°F because postcooking standing times are necessary to allow all parts of the meat to reach the temperature. The U.S. Department of Agriculture (USDA), however, has developed processing methods for pork products that assure a safe product. Problems with trichinae can be prevented by purchasing pork from approved sources and by adequate cooking.

Tapeworms are parasitic intestinal worms of flattened tapelike form that may cause disease in humans when larva-infested meat, either beef or pork, is ingested. Tapeworm contamination can be prevented by appropriate sanitary measures during agricultural production and by the federally regulated inspection of cattle carcasses for evidence of tapeworms. Proper meat processing procedures—especially during boning, which involves a great deal of hand contact—also are important to prevent contamination.

The current popularity of raw seafood dishes, like sushi, sashimi, and ceviche, and undercooked fin fish has introduced a new source of tapeworm and roundworm (*Anasakis*) infestation. *Anasakis* can be particularly devastating because the parasite attaches itself to the wall of the digestive organs and requires surgery to dislodge it. These parasites are destroyed by cooking or freezing. The *Food Code* (2005) indicates that fish that is not to be cooked thoroughly must be frozen to −31°F and stored at −4°F or below for 24 hours, frozen to −31°F or below, and stored for 15 hours or frozen to −4°F or below and stored for 168 hours (7 days). In addition, the foodservice operator must keep a record of the process on file for 90 days.

Biochemical Spoilage

Biochemical spoilage is caused by natural food enzymes, which are complex catalysts that initiate reactions in foods. Off-flavors, odors, or colors may develop in foods if enzymatic reactions are uncontrolled. When fruit is peeled and exposed to air, for example, undesirable browning occurs due to enzyme activation by oxygen. Enzymes may have desirable effects as well, such as the natural tenderizing or aging of meat.

Enzyme formation can be controlled in much the same manner as microorganisms. Heat, cold, drying, addition of inhibiting chemicals, and irradiation are the principal means for controlling and inactivating natural food enzymes (Thorner & Manning, 1983).

Physical Spoilage

Temperature changes, moisture, and dryness can cause physical spoilage of food. Excessive heat, for example, breaks down emulsions, dehydrates food, and destroys certain nutrients. Severe cold also causes various kinds of deterioration; certain starches used in preparing sauces or gravies, for instance, may break down at freezing temperatures.

Excessive moisture may lead to various types of spoilage. For example, excess moisture may support the growth of mold and bacteria. It also may cause physical changes, such as caking, stickiness, or crystallization, which may affect the quality of a food item. A powdered beverage product, for instance, may not be dispensed properly in a vending machine if the moisture level is too high.

A physical hazard is the danger of particles that are not supposed to be in a food product (National Restaurant Association [NRA] Educational Foundation, 2004). Chips of glass or metal from broken glasses or enamelware dishes are obviously dangerous. Metal curls from a worn-out can opener can fall into the food when the can is being opened. Dangers caused by physical contaminants may result from tampering incidents, particularly with soft-packed food items. Food items delivered to the foodservice operation should be rejected if evidence of tampering is seen.

Chemical Spoilage

Chemical spoilage may result from interaction of certain ingredients in a food or beverage with oxygen or light. A chemical hazard is the danger posed by chemical substances contaminating food all along the food supply chain (NRA Educational Foundation, 2004). The reaction of incompatible substances can lead to chemical spoilage, such as the effect of certain metals on foods. As an example, the zinc often used in galvanized containers may render poisonous such acid foods as fruit juices, gelatins, sauerkraut, and tomatoes. The following four kinds of chemical hazards in addition to pesticides can occur in a foodservice operation:

- contamination of food with foodservice chemicals, such as detergents and sanitizers
- use of excessive quantities of additives, preservatives, and spices
- acidic action of foods with metal-lined containers
- contamination of food with toxic metals

Pesticides are chemicals that kill or discourage the growth of pests, which are defined as organisms that cause damage to food, making it inedible, unappealing, or unsafe (Chaisson, Petersen, & Douglass, 1991). Pesticides typically are applied to crops growing in the field but also may be applied after harvest to prevent insect or mold infestation during transport or storage. Much research is being conducted on ways to reduce reliance on applied pesticides.

Integrated pest management (IPM) is an alternate approach being used in agriculture today to control pests. In addition to using pesticides carefully, IPM incorporates the latest agricultural technologies and biological controls, including pest predators and pest diseases, to decrease the amount of pesticide used. Computerized forecasting may be used to predict disease and weather conditions. Plant breeding and genetic engineering may reduce the use of applied pesticides by manipulating genes to "program" plants to make high levels of their own pesticides. The USDA and FDA are responsible for monitoring the food supply to ensure that residue levels are within tolerance limits.

Foodservice chemicals, including detergents, polishes, caustics, and cleaning and drying agents, are poisonous to humans and should never come in contact with food. Labels should be read carefully for directions on how to use and store under safe conditions away from food.

Use of additives and preservatives is still being debated by scientists and legislators. Excessive amounts of certain food additives and preservatives have caused illness, but the reasons have not yet been determined. One foodborne illness that remains in question is that which results from the use of too much monosodium glutamate (MSG), a food additive that serves as a flavor enhancer. Because it is often heavily used in Chinese and Japanese foods, apparent reactions to MSG have been called the "Chinese Restaurant Syndrome." MSG also is very high in sodium. Most of the symptoms after ingestion are subjective. They include a feeling of tightening of the face and neck skin, tingling sensations, dizziness, and headache. MSG apparently affects only persons with sensitivity to MSG (Cody & Kunkel, 2002). Federal law requires that MSG be listed on the label of any product to which it is added. Foodservice managers should avoid adding MSG to recipes.

Food irradiation is classified as a food additive and is regulated by the FDA. It controls microbes responsible for foodborne illness and extends the shelf life of refrigerated foods, such as fresh fruits and vegetables, by delaying ripening. It also extends the shelf life of stored foods like spices and dried herbs.

Preservatives used to preserve the flavor, safety, and consistency of foods have been linked to food contamination. Excessive use of sulfites has affected a number of individuals who are sensitive to this compound, particularly asthmatics. The FDA prohibits the use of sulfites on raw fruits and vegetables that are to be served or sold to customers.

Several food additives or preservatives, when used in excessive amounts, have caused illness. Nitrites, for example, are preservatives used by the meat industry to prevent growth of certain harmful bacteria and as a flavor enhancer (NRA Educational Foundation, 2004). Scientists have established a link between cancer and nitrites when meat containing them is overbrowned or burned. As a result, the meat industry has decreased levels of nitrites in meats. In the 1980s, a number of food-related illnesses, allergic in type, were traced to sulfites used on fresh fruits and vegetables, shrimp, and dried fruit. For packaged foods, proper labeling now is required. Restaurants and supermarkets now are using lemon juice or citric acid for preserving color in these foods, especially in salad bars (Longrée & Armbruster, 1996).

Chemical contamination can occur when high-acid foods are prepared or stored in metal-lined containers. Poisoning may result if brass or copper, galvanized, or gray enamelware containers are used. Fruit juices should never be stored in enamelware coated with lead glaze or tin milk cans. Cases of poisoning have been recorded that are attributed to use of improper metal utensils. Sauerkraut, tomatoes, fruit gelatins, lemonade, and fruit punches have been implicated in metal poisonings.

Toxic metals also have been implicated in food poisoning cases. Copper may become poisonous when it is in prolonged contact with acid foods or carbonated beverages. The vending industry recently has voluntarily discontinued all point-of-sale carbonation systems that do not completely guard against the possibility of backflow into copper water lines (backflow may dissolve the copper). Also, food such as meat placed directly on cadmium-plated refrigerator shelves may be rendered poisonous.

FOODBORNE PATHOGENS

Foodborne pathogen: Virus, microorganism, or other substances that causes disease.

A **foodborne pathogen** is an infecting agent, any virus, microorganism, or other substance that causes disease (AP Dictionary of Science and Technology, 1996). Table 8.1 presents a tabulation of information for various foodborne pathogens.

Table 8.1 Major Foodborne Pathogens

	Bacteria		
	Salmonella spp.	*Shigella spp.*	*Listeria monocytogenes*
Onset Time	6–48 hours	12–50 hours	3–70 days
Duration of Illness	2–3 days	Indefinite, depends on treatment	Indefinite, depends on treatment, but has high fatality in the immuno-compromised
Symptoms	Abdominal pain, headache, nausea, vomiting, fever, diarrhea	Abdominal pain, diarrhea, fever, chills, dehydration, vomiting	Nausea, vomiting, headache, fever, chills, backache, meningitis
Source	Water, soil, domestic and wild animals; also humans, especially as carriers	Human feces, flies	Humans, domestic and wild animals, fowl, soil, water, mud
Associated Foods	Poultry and poultry salads, meat and meat products, milk shell eggs, egg custards and sauces, and other protein foods	Potato, tuna, shrimp, turkey and macaroni salads, lettuce, moist and mixed foods, milk and milk products	Unpasteurized milk and cheese, vegetables, poultry and meats, seafood, and prepared, chilled, ready-to-eat foods
Spore Former	No	No	No
Prevention	Avoid cross-contamination, thoroughly cook poultry to at least 165°F, cool cooked meats and meat products quickly and properly, avoid fecal contamination from food handlers by practicing good personal hygiene	Avoid cross-contamination, avoid fecal contamination from food handlers by practicing good personal hygiene, use sanitary food and water sources, control flies, rapidly cool foods	Use only pasteurized milk and dairy products, cook foods to proper temperatures, avoid cross-contamination, clean and sanitize surfaces

"Food poisoning," as outbreaks of acute gastroenteritis are popularly called, is caused by microbial pathogens that multiply profusely in food. These attacks are either foodborne intoxications or foodborne infections. Their symptoms, which are frequently violent, include nausea, cramping, vomiting, and diarrhea.

Foodborne Intoxication/Infection

Foodborne intoxications: Caused by toxins formed in food prior to consumption.

Foodborne infection: Caused by activity of large numbers of bacterial cells carried by the food into the gastrointestinal tract.

Foodborne intoxications are caused by toxins formed in the food prior to consumption, and **foodborne infections** are caused by the activity of large numbers of bacterial cells carried by the food into the gastrointestinal system of the victim. The symptoms from ingesting toxin-containing food may occur within as short a period of time as 2 hours. The incubation period of an infection, however, is usually longer than that of an intoxication.

Bacteria			
Staphylococcus aureus	*Clostridium perfringens*	*Bacillus cereus*	*Clostridium botulinum*
1–6 hours 1–2 days	8–22 hours 24 hours, lingering symptoms 1–2 weeks	1/2–5 hours; 8–16 hours 6–24 hours	12–36 hours Several days to a year
Nausea, vomiting, diarrhea, dehydration	Abdominal pain, diarrhea	Nausea and vomiting; diarrhea, abdominal cramps	Vertigo, visual disturbances, inability to swallow, respiratory paralysis
Humans (skin, nose, throat, infected sores); also, animals	Humans (intestinal tract), animals, and soil	Soil, cereal crops	Soil, water
Reheated foods, ham and other meats, dairy products, custards, meat, egg, potato salads, cream-filled pastries, and other protein foods	Cooked poultry and meat products that have been improperly cooked, held or cooled before serving	Rice and rice dishes, custards, seasonings, dry food mixes, spices, puddings, cereal products, sauces, vegetable dishes, meatloaf	Improperly processed canned goods of low-acid foods, garlic-in-oil products, grilled onions, stews, meat/poultry loaves
No	Yes	Yes	Yes
Avoid contamination from bare hands, exclude sick food handlers from food preparation and serving, practice good personal hygiene, practice sanitary habits, proper heating, cooling, and refrigeration of food	Use careful time and temperature control in cooling and reheating cooked meat dishes and products	Use careful time and temperature control and quick chilling, proper reheating	Do not use home-canned products, use careful time and temperature control for sous vide items and all large, bulky foods, keep sous vide packages refrigerated, purchase garlic-in-oil in small quantities for immediate use, cook onions only on request, rapidly cool leftovers

Source: Adopted from *Serv Safe Coursebook* by the Educational Foundation of the National Restaurant Association, 1999, Chicago: Author and the *Bad Bag Book* by the Center of Food Safety and Applied Nutrition (see www.cfsan.fda.gov), 2002.

Salmonella

Salmonella spp. frequently has been associated with foodborne illnesses and causes salmonellosis. The bacterium does not release toxins into the food in which it multiplies; rather, the ingested cells continue to multiply in the intestinal tract of the victim, causing illness. The primary source of *Salmonella* is the intestinal tract of carrier animals. A carrier appears to be well and shows no symptoms or signs of illness but harbors causative organisms. Various insects and pets may be reservoirs of *Salmonella*. Food animals are important reservoirs, especially hogs, chickens, turkeys, and ducks. The disease salmonellosis is spread largely by contaminated food and is believed to be one of the major communicable diseases in this country.

A number of raw and processed foods have been found to carry *Salmonella,* especially raw meat, poultry, shellfish, processed meats, egg products, and dried milk. Meat mixtures, dressings, gravies, puddings, and cream-filled pastries are among the menu items frequently indicated in salmonellosis. Food handlers and poor sanitation practices are often associated with outbreaks. Care must be exercised in production, storage, and service to ensure that food is not held for long periods at warm temperatures, cooled slowly, or cut on contaminated surfaces.

Shigella

Shigella spp. is another bacteria that causes foodborne illness, shigellosis, sometimes called bacillary dysentery. It is an infection that occurs 1 to 7 days after the ingestion of the bacteria. Humans are the prime reservoir for *Shigella.* Carriers excrete *Shigella* in their feces and transmit the bacteria to the food if they do not wash their hands properly. Flies also are thought to carry the bacteria. Foods involved are raw produce and moist-prepared foods, such as potato, tuna, turkey, and macaroni salads that have been handled with bare hands during preparation. Shigellosis can be prevented if employees wash their hands after using the toilet, if food is rapidly cooled, and if flies are controlled.

Listeria monocytogenes

Listeria monocytogenes is the bacterium responsible for listeriosis and is widely distributed in nature. It has been isolated from feces of healthy human carriers and sheep, cattle, and poultry. It has been detected in cow's milk and has been isolated from unwashed leafy vegetables and fruit and soil. Also, bacterium has been found in dairy and meat processing factories with some degree of frequency. Listeriosis disease has been linked to consumption of contaminated delicatessen food, milk, soft cheeses (like Mexican-style feta, Brie, Camembert, and blue-veined cheeses), and undercooked chicken.

Staphylococcus aureus

Staphylococcus aureus, a bacterium commonly referred to as *staph* or *S. aureus,* is one of the principal causative agents in foodborne illness. It grows in food.

Staphylococcal intoxication is a fairly frequent cause of foodborne illness, with foods high in protein the usual culprits. Cream pies, custards, meat sauces, gravies, and meat salad are among the products most likely to be involved in foodborne intoxication. The appearance, flavor, or odor of the affected food items are not noticeably altered.

Temperatures must be carefully controlled to prevent multiplication of staphylococci in food. The organism multiplies even under refrigeration if temperatures are not sufficiently low or if the cooling process does not proceed rapidly enough.

Clostridium perfringens

Clostridium perfringens is a common inhabitant of the intestinal tract of healthy animals and human beings and occurs in soil, sewage, water, and dust. The infected food has invariably been held at room temperature or refrigerated in a large mass for several hours. Meats, meat mixtures, and gravies are frequently implicated. Overnight roasting of meat is, therefore, not recommended because of the low temperatures often used, and careful reheating of leftover meat is important. Prevention of *C. perfringens* multiplication can be achieved by refrigerating foods at 40°F or below or holding them at 145°F or higher. In addition, rapid cooling of cooked foods is an important practice.

Bacillus cereus

Foodservice managers are beginning to be concerned about the *B. cereus* toxin, which is found in soil and, therefore, gets into many foods once thought to be safe (NRA Educational Foundation, 2004). *B. cereus* bacteria are found in grains, rice, flour, spices, starch, and in dry mix products such as those used for soups, gravies, and puddings. They also have been found in meat and milk. Time and temperature are very important in preventing rapid increase in the vegetative bacteria and development of spores. Foods should not be held at room temperature for any period of time, but should be chilled to at least 40°F within 4 hours after preparation. Foods stored in the refrigerator need to be in shallow pans with a food depth of less than 2 inches and should be used as quickly as possible after preparation.

Clostridium botulinum

Clostridium botulinum produces a toxin that affects the nervous system and is extremely dangerous. The disease called botulism is the food intoxication caused by this bacteria. Improved food processing techniques have led to greatly reduced incidence of botulism, although inadequately processed home-canned foods are still frequently associated with botulism. Meats, fish, and low-acid vegetables have been found to support toxin formation and growth.

Precautions for avoiding botulism include procuring foods from safe sources, rejecting home-canned products and low-acid products, destroying canned goods with defects such as swells or leaks, storing foods under recommended conditions, and using appropriate methods for thawing frozen foods.

In addition to improperly processed products, other suspicious foods include smoked, vacuum-packed fish; garlic products packed in oil; grilled onions; baked potatoes; turkey loaf; and stew. Sous vide products offer a potential risk because they are vacuum packaged. Soil-grown vegetables, particularly potatoes, can be prime carriers of this toxin.

Other Foodborne Pathogens

Microbiologists continue to find new foodborne pathogens. Some of the pathogens are bacteria; others are viruses or parasites. Several of the emerging foodborne pathogens are described in Table 8.2.

Campylobacter jejuni

Campylobacter jejuni was a well-known pathogen in veterinary medicine before it was considered a human pathogen. It is now recognized as one of the most common causes of gastroenteritis in humans, with a ranking similar in importance to *Salmonella*. A pathogen of cattle, sheep, pigs, and poultry, it is present in the flesh of these food animals and thus may be introduced into the kitchen with the food supply.

This pathogen must be expected to be present in areas where raw meats and poultry are handled even when the techniques of sanitary handling are practiced.

Table 8.2 Emerging Pathogens That Cause Foodborne Illness

	Bacteria		
	Campylobacter jejuni	*Escherichia coli*	*Vibrio parahaemolyticus, Vibrio vulnificus*
Onset Time	3–5 days	12–72 hours	4–96 hours
Duration of Illness	1–4 days	1–3 days	1–8 days
Symptoms	Diarrhea, fever, nausea, abdominal pain, headache	Bloody diarrhea; severe abdominal pain, nausea, vomiting, diarrhea, and occasionally fever	Diarrhea, abdominal cramping, vomiting, headache, fever, chills
Source	Domestic and wild animals	Humans (intestinal tract); animals, particularly cattle	Fish and shellfish (especially from the Gulf of Mexico)
Associated Foods	Raw vegetables, unpasteurized milk and dairy products, poultry, pork, beef, and lamb	Raw and undercooked beef and other red meats, imported cheeses, unpasteurized milk, raw finfish, cream pies, mashed potatoes, and other prepared foods	Raw or improperly cooked oysters or shellfish from contaminated water
Spore Former	No	No	No
Prevention	Avoid cross-contamination, cook foods thoroughly	Cook beef and red meats thoroughly, avoid cross-contamination, use safe food and water supplies, avoid fecal contamination from food handlers by practicing good personal hygiene	Avoid raw or undercooked seafood, purchase seafood from approved sources

In addition to flesh foods, raw milk has been implicated in outbreaks of gastroenteritis caused by *Campylobacter.* The variety of menu items that have been involved in foodborne outbreaks attests to the fact that multiplication of the organism is possible in many different foods and food mixtures, including cake icings, eggs, poultry, and beef.

Escherichia coli

Escherichia coli 0157:H7, or *E. coli,* a bacteria that can be transmitted by eating raw or undercooked ground beef, was identified as the cause of death of four children who had eaten undercooked hamburger patties served in a limited-menu restaurant (McLauchlin, 1993). This bacteria is most likely found on the surface of meat and is killed when cuts such as steaks are grilled (McCarthy, 1993). Grinding meat, however, transfers the bacteria from the surface to the inside of the product, making *E. coli* more difficult to kill.

Bacteria	**Virus**		
Yersinia enterocolitica, Yersinia pseudotuberculosis	Norovirus	Hepatitis A	Rota virus
24–48 hours	24–48 hours	10–50 days	1–3 days
days to weeks	1–3 days	1–2 weeks	4–8 days
Abdominal pain, vomiting, diarrhea, headache	Nausea, vomiting, diarrhea, abdominal pain, headache, and low-grade fever	Sudden onset of fever, fatigue, headache, nausea, abdominal pain, jaundice	Abdominal pain, vomiting, diarrhea, mild fever
Soil, water, pigs, wild rodents	Humans (intestinal tract), contaminated water	Humans (intestinal tract) contaminated food and water	Humans (intestinal tract), contaminated water
Raw or partially cooked meat (beef, pork, lamb), oysters, fish, raw milk	Raw vegetables, prepared salads, raw shellfish, and water contaminated from human feces	Water, shellfish, salads, ice, cold cuts, fruit and fruit juices, vegetables, milk and milk products	Water, ice, foods that do not have further cooking after handling such as salads, fruits, raw vegetables)
No	No	No	No
Thoroughly cook foods minimize cross-contamination, properly clean and sanitize facilities	Use safe food and water supplies, avoid fecal contamination from food handlers by practicing good personal hygiene, thoroughly cook foods	Obtain shellfish from approved sources, prevent cross-contamination by proper hand washing by employees, use sanitary water sources	Use sanitary water sources, prevent cross-contaimination by proper handwashing by employees

Source: Adapted from *ServSafe Coursebook* by the Educational Foundation of the National Restaurant Association, 1999, Chicago: Author and the *Bad Bag Book* by the Center of Food Safety and Applied Nutrition (see www.cfsan.fda.gov), 2002.

Scientists are having difficulty in understanding some of the emerging pathogens, including *E. coli 0157:H7,* that can survive for weeks in acidic foods. Puzo (1998) predicts that consumers and media will be far less forgiving of future mistakes that could lead to spectacular outbreaks and their illnesses, hospitalizations, and in some cases deaths. Almost any amount of *E. coli 0157:H7* can make anyone terribly ill. Until the final decades of the twentieth century, the government's food safety regulations were simple and illnesses were not broadcast to the public. Today, however, the public is concerned about the safety of food and is demanding that the government take care of the problem. Congress has responded with additional funding. Hudson Foods, Inc., was the first corporate victim of foodborne illnesses. Some of its ground beef tested positive for *E. coli 0157:H7,* and it recalled 25 million pounds of the meat. Within 12 weeks, the profitable family-owned company was dissolved.

Norovirus

Norovirus is a viral illness caused by poor personal hygiene among infected food-handlers (NRA Educational Foundation, 2004). Because it is a virus, it does not reproduce in food but remains active until the food is eaten. Norovirus is found in the feces of humans. It is transmitted through contaminated water and human contact, raw vegetables fertilized by manure, salads, raw shellfish, eggs, icing on baked pastries, and manufactured ice cubes and frozen foods. It has been identified as the cause of several incidents of foodborne illness on cruise ships, sometimes causing the early return of the ship to port.

Lesser Known Parasites

Cyclospora is a microscopic parasite composed of one cell. People of all ages are at risk for infection (Healthbeat, 1997). *Cyclospora* infection often is found in people who live or travel in developing countries. Time between becoming infected usually is one week. It *Cyclospora* has all the elements of an emerging infectious disease (Puzo, 1998). Scientists do not know where it lives between outbreaks or much about its disease-causing characteristics. Food is probably its vehicle, and fresh produce and imported food are the culprits. Only recently has *cyclospora* become a foodborne hazard in the United States. The USDA's Economic Research Service (www.ers.usda.gov) reports that imports are very common in the American diet. Nearly 80% of all the fish and shellfish consumed in the United States in 2002 was imported, as were 31% of fruits, juices, and nuts; 9.6% of vegetables; and 9.5% red meat.

Imported foods, particularly produce, have been linked to a growing number of foodborne illness outbreaks. Foodservice operators can avoid potentially contaminated food items by using reputable producers and suppliers, who should comply with growing and transportation standards set by the Produce Marketing Association. Produce should be inspected upon delivery; the delivery truck should be clean and the packaging intact. Employees should be trained in proper procedures, such as the following hand-washing procedures:

- Wash hands thoroughly and vigorously with soap and hot water after using the bathroom and before handling food.

- Wash all produce thoroughly with strong running water, using fingers or a brush to rub the surface clean.
- Pay special attention to stem areas, where bacteria can thrive, and other crevices, such as the skin of a cantaloupe or the folds of lettuce leaves.
- Keep produce, including cut foods, refrigerated as close to service as possible.

Workers need to know that fruits and vegetables can harbor bacteria and other pathogens.

Cryptosporidium is a microscopic parasite in water, often drinking water (Hannahs, 1995). More than one-half million people have been affected by it.

Vibrio cholerae 01 and 0139 remain epidemic in many parts of Central and South America, Asia, and Africa (Food and Drug Administration, 1995). People traveling in cholera-affected areas should not eat food that has not been cooked and is not hot, particularly fish and shellfish, and should drink only beverages that are carbonated or made from boiled or chlorinated water.

Hepatitis is a common disease that affects the liver, causing inflammation (Access Health, 1996). The liver does not function normally. Hepatitis A, B, and C are the three main types of viral hepatitis. Type A, or infectious hepatitis, is one of the most contagious types and often occurs in children and young adults. It is caused by

- Drinking polluted water
- Eating food cooked or washed in polluted water
- Eating contaminated food cooked or washed in polluted water
- Touching a contaminated cup or eating utensils and then putting hands in the mouth or touching a cut or open sore
- Eating shellfish, clams, mussels, and oysters that live in polluted waters

Prions

One of the newest foodborne disease concerns is a group of organisms termed prions, PROteinaceous INfectious particle. According to Cody and Kunkel (2002), prion proteins are small glycosylated protein molecules found in brain cell membranes. Prion diseases, often termed transmissible spongiform encephalopathy (TSE), are infectious diseases of the brain that can occur in both animals and humans. The disease is termed bovine spongiform encephalopathy (BSE) (mad cow disease) in cattle, scrapie in sheep, chronic wasting disease in deer and elk, and Creutzfeld Jakob disease (CJD) or Gerstmann Sträussler syndrome (GSS) in humans. Prions are extremely resistant to heat; heating to 100°C often does not inactivate them. Prion diseases are transmissible between species. The period of time between infection and appearance of clinical symptoms often is more than 10 years in humans (www.prionics.ch, 2005). The disease course in humans includes behavioral changes, ataxia, progressive dementia, and death (Cody & Kunkel, 2002).

CONTROLLING MICROBIOLOGICAL QUALITY OF FOOD

The ultimate goal of the sanitation program in a foodservice operation is to protect customers from foodborne illness. The role of the foodservice manager is to take responsibility both for serving safe food to customers and for training employees on a continual basis.

Control of Food Quality

Control of the microbiological quality of food must focus on the food itself; the people involved in handling food, either as employees or customers; and the facilities and equipment, including both large and small equipment. The legal fees, medical claims, lost wages, and loss of business associated with foodborne illness can be overwhelming (NRA Educational Foundation, 2004). Good sanitation will prevent foodborne illness breakouts, maintain customer goodwill, and keep the bottom line from bottoming out.

Quality of Food

Clearly, the condition of the food brought into the facilities is a critical aspect to consider. Time-temperature control also is critical during storage, production, and service.

Condition of Purchased Food. Possibilities for contamination of food before it is purchased include contaminated equipment, infected pests and animals, untreated sewage, unsafe water, and soil, as shown in Figure 8.4. After purchase, possibilities of contamination exist in storage, preparation, and service of food. Following human consumption, illness occurs. Figure 8.5 illustrates the possible transmission routes from infected persons through respiratory tract discharges, open sores, cuts, and boils, or through hands soiled with feces into food being prepared. The consumed food then completes the transmission to other persons.

Foods traditionally sold in the grocery refrigeration case have had a good safety record until the 1990s, when they were implicated in a number of serious foodborne disease outbreaks, such as salmonellosis from pasteurized milk and listeriosis from a specific brand of Mexican-style cheese. Uncertainty caused by these episodes is intensified by the emergence of newly recognized pathogens that grow at refrigeration temperatures.

Concern is growing about the microbiological safety of the new generation of refrigerated foods, which include such foods as frozen or restaurant dinners; frozen or delicatessen entrées; dry, frozen, or canned pasta; fresh, refrigerated salads; canned, frozen, or dry gravies; canned or dry soups; and frozen, canned, or refrigerated cooked meat. Unlike traditional refrigerated foods, new generation products do not have a readily apparent preservation system. Many of these products have an extended shelf life, but the packaging may enhance growth of pathogens. Other risks are inadequate temperature control for safety and extended shelf life, a poor distribution system, and partial processing, which prevents warning of hazards by destruction of spoilage flora.

Relationship of Time and Temperature. Reducing the effect of contamination is largely a matter of temperature control in the storage, production, and service of foods. The growth of harmful organisms can be slowed or prevented by refrigeration or freezing, and the organisms themselves can be destroyed by sufficient heat. More than half of all foodborne outbreaks are caused by inadequate cooking and improper holding of food (see www.cdc.gov). Minimum, maximum, and optimum temperatures vary for the various pathogenic microorganisms; in general, however, they flourish at temperatures between 41° and 135°F (Figure 8.6). This temperature range is commonly called the food danger zone because bacteria multiply rapidly

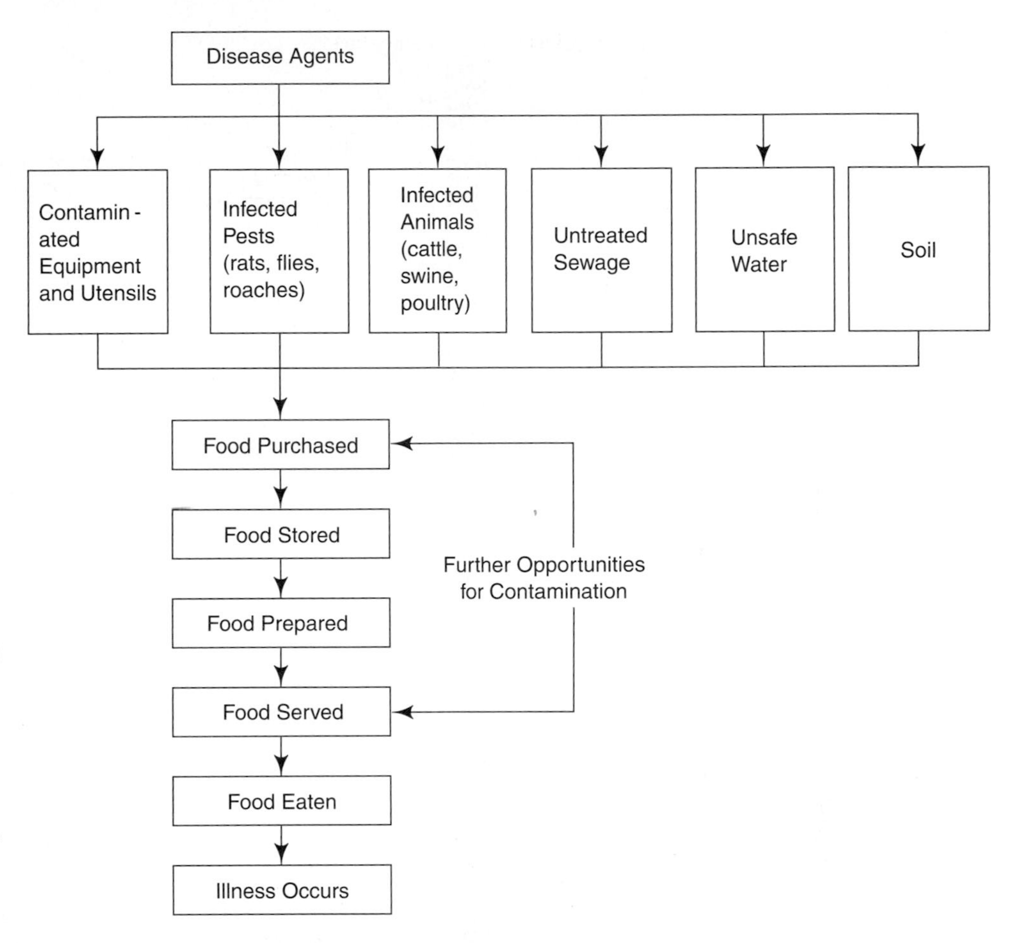

Figure 8.4. Transmission of a foodborne illness from an intermediate source to food and on to humans.
Source: From *Applied Foodservice Sanitation,* 3rd edition (p. 18), 1985, by the Educational Foundation of the National Restaurant Association, Chicago: Author. Used by permission.

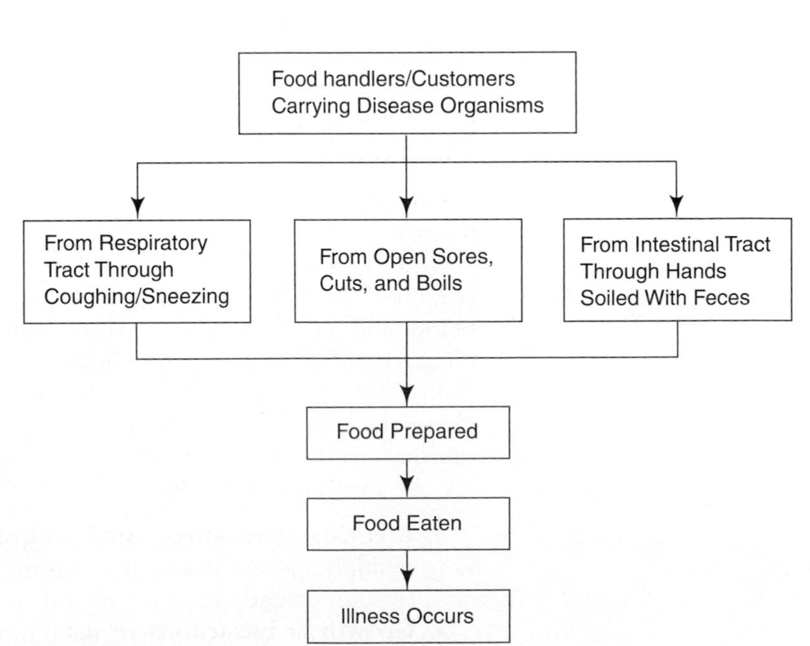

Figure 8.5. Transmission of a foodborne illness from infected human beings to food and back to other human beings.
Source: From *Applied Foodservice Sanitation* (p. 9) by the Educational Foundation of the National Restaurant Association, 1992, Chicago: Author.

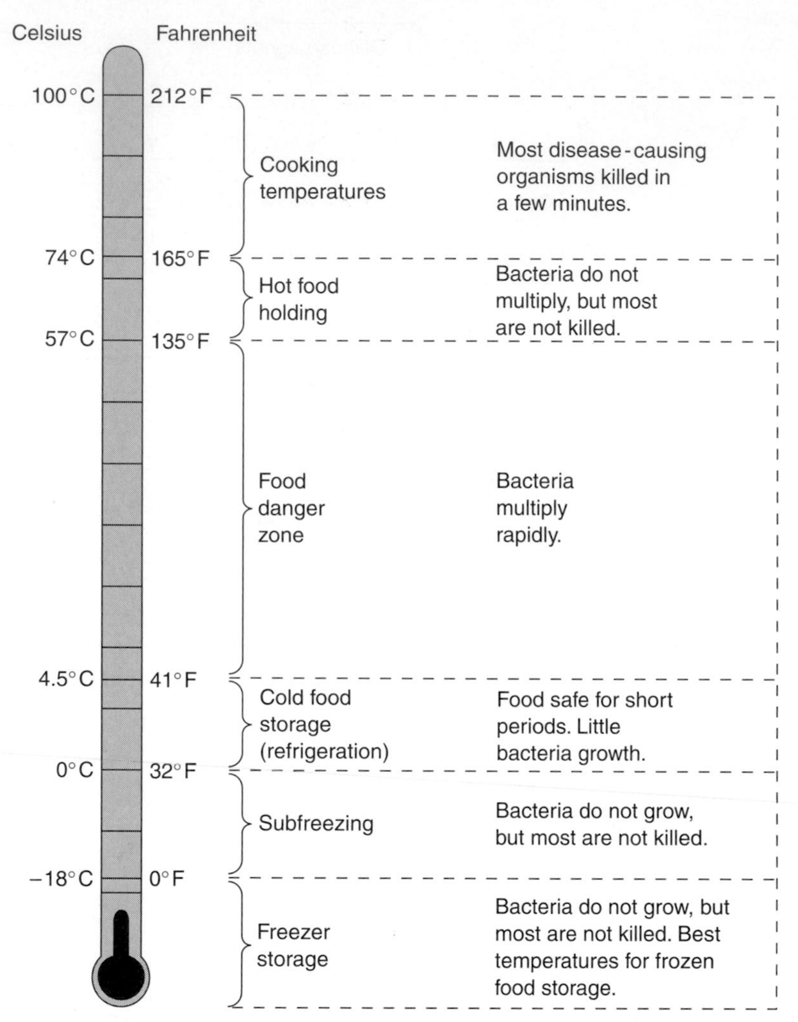

Figure 8.6. Important temperatures in sanitation and food protection.
Source: Adapted from *Keeping Food Safe to Eat,* Home and Garden Bulletin No. 162, 1970, Washington, DC: U.S. Department of Agriculture. Revised based on *Food Code 2005*.

within it. *Food Code* 2005 indicates that potentially hazardous food (time/temperature control for safety food) should be cooled within 2 hours from 135°F to 70°F; and within 6 hours from 135°F to 41°F or less (www.cfsan.fda.gov). Safe temperatures, then, as applied to potentially hazardous food, are those of 41°F and below and 135°F and above. Thus, both time and temperature are critical in handling food to preserve microbiological quality.

In foodservice operations, ingredients and partially and fully prepared menu items are subjected to a wide range of temperatures. In relating food-handling practices to the time and temperature effects on microorganisms in foods, the following are temperature zones at which foods are handled:

- **Freezing, defrosting, and chilling zone.** Temperatures should prevent multiplication of organisms causing foodborne intoxications and infections over an extended storage period.
- **Growth or hazardous zone.** Temperatures allow bacterial multiplication.

- **Hot holding zone.** Temperatures are aimed at preventing multiplication, but usually do not kill the organism.
- **Cooking zone.** Temperatures should be sufficiently high to destroy bacterial cells within a short period of time.

Employee and Customer

Employee personal hygiene and good food-handling practices are basics of a sanitation program in a foodservice facility. One major risk is that unsanitary employees can contaminate, or infect by bacteria, food products that go from receiving to service (Beasley, 1993). When interviewing prospective employees, managers should note their personal grooming habits and appearance. Employees who practice poor hygiene at home and at work are the third most frequently cited cause of outbreaks of foodborne illnesses, according to the NRA Educational Foundation (1995). Foodservice managers, therefore, should emphasize the importance of food safety and sanitation to the employee before hiring and then after hiring, when an educational process should begin. Because the most critical aspect of personal cleanliness is frequent and thorough hand washing, proper methods should be emphasized in training programs. Using gloves can give a false sense of security because the food handler might not change the gloves after handling a contaminated product, resulting in cross-contamination of other products. Managers need to conduct daily inspections of all employees to ensure that proper sanitation practices are being followed.

Control of contamination from customers is more difficult, but various aspects of facility design or policies and procedures can assist in this arena. Sneeze guards on a cafeteria service counter or salad bar can help reduce the spread of bacteria as can isolation procedures for a patient hospitalized with a highly communicable disease.

Researchers at the University of California have placed salad bars under surveillance to see how sanitary they are (ADEC Environmental Sanitation and Food Safety, 1997). In one study, 60% of the customers were observed committing at least one infraction of good sanitation practices while serving themselves at the salad bar. Customers often were seen doing the following:

- Spilling food around hard-to-reach containers
- Dipping their fingers into salad dressing for a sample lick
- Eating from their plates while waiting in the serving line
- Ducking their heads underneath the sneeze guards for better access to the food
- Refilling their soiled plates with second helpings

Salad bars can contain healthy, nutritious food or an unhealthy mixture of foodborne bacteria. Foodservice employees and customers need to make sure that salad bars are not the cause of the bacteria infestation.

Facilities and Equipment

Restaurants need to be the cleanest places on earth. If not, the potential for the transmission of disease is great (Weinstein, 1991a). The design and construction of foodservice facilities and equipment have a major impact on the effectiveness of the sanitation program; maintenance of the facilities to ensure a sanitary environment is also critical. Weinstein (1991a) noted that no computer program for a restaurant's physical plant can replace a broom, a mop, some detergent and water, a scrub brush, a pair of hands, and some old-fashioned elbow grease for cleaning and sanitation.

Proper cleaning of work surfaces and equipment is an important component of any food safety program. The *Food Code* (2005) stipulates that room temperature food contact surfaces used in the preparation of potentially hazardous foods need to be cleaned at least every 4 hours.

Food Safety Programs

A well-designed food safety program monitors all food production activities for errors in handling and eliminates those errors. Critical control points first must be identified before a Hazard Analysis Critical Control Point (HACCP) model can be selected. As everyone working in foodservice knows, serving wholesome, tasty, safe food to customers is one of the foodservice manager's primary goals.

The Los Angeles County Department of Health Services' new food-handler certification program specifies a minimum of 4 hours of training approved by the county health department. Operators are required to file applications for certification and pay a fee. Under the new health inspection rating system, restaurants with scores of 90% or higher would receive an "A" grade, 80 to 90% a "B" grade, and 70 to 79% a "C" grade. Detailed inspection cards and grade must be posted in the facility. The law also requires restaurants to post the location and phone number of the local health department office, permitting customers to file complaints about sanitation.

Hazard Analysis Critical Control Point Model

The Hazard Analysis Critical Control Point (HACCP) concept refers to a model developed initially for quality control in the food processing industry, with special emphasis on microbial control. Critical control points are those steps in production processing in which loss of control would result in an unacceptable safety risk. HACCP is a preventive approach to quality control, identifying potential dangers for corrective action.

The HACCP program was developed in 1971 for the National Aeronautics and Space Administration (NASA) to be sure food fed to astronauts in outer space is absolutely safe. The system had to ensure zero defects in handling food during processing. It had to correct errors before they happened rather than sample finished products to identify those with high levels of contamination (Dulen, 1998).

The original HACCP model was modified for use in foodservice operations by Bobeng and David (1978) to include not only microbiological but also nutritive and sensory quality. They applied the model to quality control of entrée production in conventional, cook-chill, and cook-freeze hospital foodservice operations. HACCP models were developed during three phases: selection of control points, using flow diagrams; identification of critical control points; and establishment of monitors for control.

In 1988, the National Advisory Committee on Microbiological Criteria for Foods (NACMCF) was formed to provide guidance and recommendations to the Secretary of Agriculture and the Secretary of Health and Human Services regarding the microbiological safety of foods. The NACMCF, an advisory committee chartered under the U.S. Department of Agriculture (USDA), includes participants from the USDA (Food Safety and Inspection Service), the Department of Health and Human Services (U.S. Food and Drug Administration and the Centers for Disease Control and Prevention), the Department of Commerce (National Marine Fisheries Service), the Department of Defense (Office of the Army Surgeon General), academia, industry, and state employees. In November 1992, NACMCF defined seven widely accepted principles to consider when developing a HACCP plan.

- Principle 1: Conduct a hazard analysis.
- Principle 2: Determine the critical control points (CCPs).
- Principle 3: Establish critical limits.
- Principle 4: Establish monitoring procedures.
- Principle 5: Establish corrective actions.
- Principle 6: Establish verification procedures.
- Principle 7: Establish recordkeeping and documentation procedures.

Hazard: Unacceptable contamination of food.

Critical control points (CCPs): Locations in the food product flow where mishandling of food is likely to occur.

At each step in the flow of food through operation, risk, which is a chance that a condition in foodservice will lead to a hazard, can occur (Spears, 1999). A **hazard** is considered an unacceptable contamination. As risks are determined, procurement unit heads need to identify in the receiving and storage areas **critical control points (CCPs),** defined as a step or procedure in a foodservice process at which control can be applied and a food safety hazard can be prevented, eliminated, or reduced to acceptable levels. The objective is to identify the points during the production process where food is most likely to be contaminated (Dulen, 1998).

Having a HACCP program in place is required by federal agencies, accrediting bodies, or local health departments for many foodservice operations. The USDA (2005), the National Restaurant Association (2004), the National Food Service Management Institute (2002), and the Iowa State University (ISU) online HACCP Information Center (www.iowahaccp.iastate.edu/sections/foodservice.cfm) have developed manuals and materials to help foodservice managers develop HACCP plans and procedures for their foodservice operations.

According to the ISU HACCP Center, the basis of a strong HACCP program is having the necessary prerequisite programs in place. Those prerequisite programs include

- Sanitation standard operating procedures
- Quality management
- Employee education and training
- Personal hygiene
- Safe food-handling and storage practices
- Temperature monitoring
- Specifications and suppliers
- Food recalls and disaster plans
- Equipment monitoring and calibration
- Preventative maintenance programs
- Integrated pest management

Standard Operating Procedures (SOPs): written, step-by-step instructions for routine tasks.

One of the prerequisite programs is documentation of **Standard Operating Procedures (SOPs)**. SOPs are written, step-by-step instructions for routine tasks. SOP checklists and samples can be found at the online ISU HACCP Information Center. An example of an SOP for handwashing is included in Figure 8.7.

In an HACCP program, temperatures must be monitored and recorded. Placing deep containers of hot foods, especially soups and stocks, in refrigeration units is one of the most often-cited cases of food safety risk. Not only does hot food not reach recommended safe temperatures within the specified time frame, but the heat from the containers can raise the temperatures in refrigerators. Hot foods should be put into shallow pans before refrigeration to speed the cooling. Blast chillers can help reduce very quickly the temperature of these foods. Results of

Safe Food
Iowa State University
School HACCP Project

School District: ____________________
Department: ____________________
Policy No: ____________________

Standard Operating Procedure

Handwashing

Policy: All food production personnel will follow proper handwashing practices to ensure the safety of food served to children.

Procedures: All employees in school foodservice should wash hands using the following steps:

1. Wash hands (including under the fingernails) and forearms vigorously and thoroughly with soap and warm water (a temperature of at least 110°F is required) for a minimum of 20 seconds.
2. Wash with soap – either liquid or powder soap.
3. Use a sanitary nail brush to get under fingernails.
4. Wash between fingers thoroughly.
5. Use only hand sinks designed for that purpose. Do not wash hands in sinks in the production area.
6. Dry hands with single use towels or a mechanical hot dryer. (Retractable cloth towel dispenser systems are not recommended.) Turn off faucets in a sanitary fashion using a paper towel in order to prevent recontamination of clean hands.

The unit supervisor will:

1. Monitor all employees to ensure that they are following proper procedures.
2. Ensure adequate supplies are available for proper handwashing.
3. Follow up as necessary.

Policy last revised on: ______

Figure 8.7. Example of a standing operating procedure for handwashing.
Source: www.iowahaccp.iastate.edu/sections/foodservice.cfm

study on cooling confirm these recommendations. Olds and Sneed (2005) found that a 3-gallon pot of chili placed in a walk-in refrigerator took more than 24 hours to cool from 135°F to 70°F (Olds & Sneed, 2005). When hot chili was placed in a 2-inch pan in the walk-in refrigerator, it cooled from 135°F to 70°F in 7 hours. A 2-inch pan of chili cooled in a blast chiller dropped from 135°F to 70°F in less than 2 hours.

After determining the critical control points, methods have to be established to avoid breakdowns in those problematic areas. Monitoring must be in place to make sure the controls are working successfully. Detailed recordkeeping, such as temperature checks, also is part of the HACCP, as is verification of cleanliness (for example, conducting laboratory tests for bacteria). The HACCP program requires a lot of recordkeeping, which may present challenges for small operations. HACCP implementation does not eliminate the risk of foodborne illness. Contamination problems may be reduced, but the possibility of mishandling food remains real throughout the food chain.

By charting the flow of food through the operation, points can be identified where contamination or growth of microorganisms can occur. Often similar food items (like cold meat sandwiches or cream soups) can be grouped together under one HACCP plan as they will follow the same flow through the operation and have the same CCPs. Implementation of HACCP programs is responsible for making thermometers more sophisticated than they have ever been. The same thermometer often was used for finding out the temperature in a refrigerator or an oven, and results were seldom analyzed. Currently, thermometers are becoming very specialized; for example, some models are designed for ovens, deep fat fryers, and coffee. All potentially hazardous foods should be prepared according to specific HACCP guidelines. The minimum number of thermometers needed in a foodservice operation are the digital pocket test, refrigerator/freezer/dry storage, hot holding, and meat thermometers shown in Figure 8.8. The fifth one, the HACCP Manager, is designed to record temperature, time, and location for any manufacturing or food preparation process that requires accurate recordkeeping; a downloading feature allows the user to graph and chart data to review and analyze for corrective action or required recordkeeping.

Many foodservice operations have supplemented the use of thermometers with a disposable product called T-Sticks. T-Sticks are multipurpose sensor sticks used to monitor food temperatures and the temperature in the dishwasher's final rinse section. They help promote food safety in restaurants and other foodservice operations. They are relatively inexpensive, and employees who might not take the time to track down a thermometer find them easy to use. T-Stick 140 Plus is used for monitoring food temperatures on hot lines or steam tables; food must be held at 140°F or higher to stop growth of harmful bacteria. It turns green at 142°F to 144°F for a margin of safety. T-Stick 160 monitors the cooking temperature of hamburger, ground meat, fish, pork, and eggs and verifies temperatures in the final rinse section of the dishwashing machine. It turns black if the temperature reaches 160°F. An illustration of how to use the T-Stick 160 for cooking hamburgers is shown in Figure 8.9.

With the exception of certain cultured foods, such as blue-veined cheese and yogurt, only pathogen-free ingredients with low levels of microbial count should be purchased. The purchasing manager should check sanitary practices in

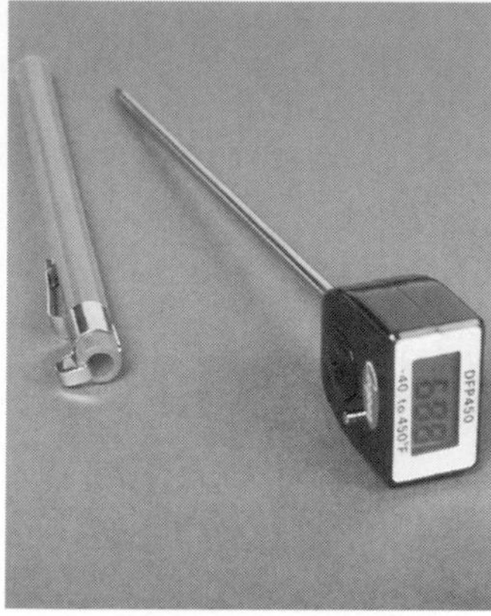

Digital Pocket Test Thermometer.

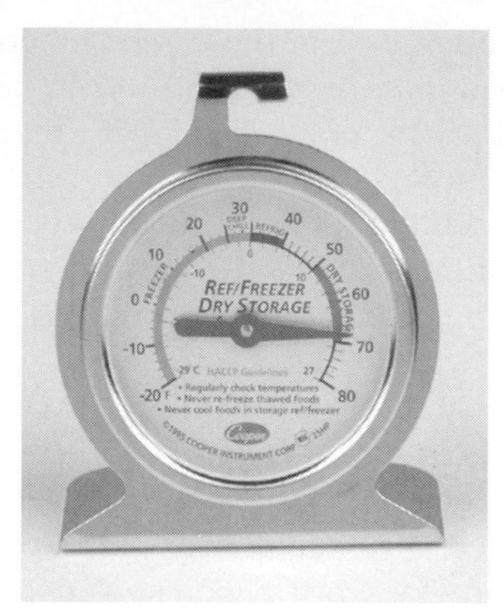

Refrigerator/Freezer/Dry Storage Thermometer.

HACCP Manager.

Hot Holding Thermometer.

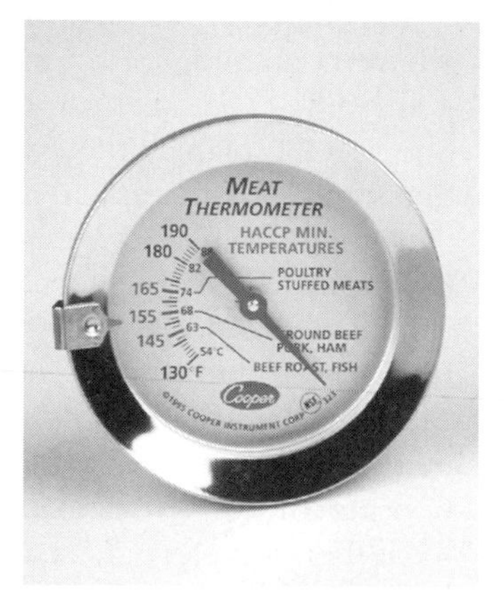

Meat Thermometer.

Figure 8.8. Foodservice operations thermometers.
Source: Cooper Instrument Corporation. Used by permission.

Figure 8.9. Monitoring the temperature of hamburgers with a T-Stick 160 while cooking.
Source: ECOLAB®. Used by permission.

1 Insert white plastic-coated end of T-Stick into the center of the hamburger to be tested. **Do not remove the protective plastic coating. Wait 5 seconds.**

2 Remove T-Stick from hamburger. If plastic-coated end has turned black, temperature of food has reached 160°F (71°C). T-Stick can be discarded.

3 If plastic-coated end still white, food has not yet reached 160°F (71°C). Cook further and repeat steps 1 and 2, using the same T-Stick in the same hamburger.

warehouse and transportation vehicles before selecting a supplier. If food safety is not monitored before purchasing, other areas may use contaminated products that will affect the safety of products served to customers.

The Iowa HACCP Web site (http://www.iowahaccp.iastate.edu/sections/foodservice.cfm) contains a variety of resources to help a foodservice manager develop a HACCP program. Included are a case study detailing the process in a school, SOPs, forms for monitoring temperatures, and employee training materials.

Seafood Safety

On December 18, 1997, the FDA mandated that safety assurance personnel be trained to follow HACCP procedures in the seafood industry (Alaska Sea Grant College Program, 1997). The previous inspection program reacted to problems, but the new plan is more proactive by requiring processors to identify and continuously monitor production steps that have the greatest safety hazards, including toxins, chemicals, pesticides, and decay. Processors will be required to show their HACCP plan to FDA inspectors.

Although fishing vessels, carriers, and retailers are exempt from the regulations, processors are responsible for knowing where the seafood comes from and its condition when it leaves the plant (Allen, 1996). This is an expensive system, but the seafood industry believes it will pay off in the end. Many seafood industries already have an HACCP program in operation and have found it to be cost effective. Long John Silver's, the 1,500-unit, quick-service seafood chain based in Lexington, Kentucky, trains its suppliers in seafood safety, and company officials make frequent trips to trawlers and factories where their produce is caught and processed. In addition, all products served in the chain's restaurants are tested at one of the company's three seafood audit laboratories.

Bioterrorism

The terrorism attacks in the United States on September 11, 2001 prompted legislation and changes in operational practice to better protect the U.S. food supply. Congress passed and President Bush signed into law on June 12, 2002 the Public Health Security and Bioterrorism Preparedness and Response Act of 2002 (the Bioterrorism Act). Title III of the act focuses on protecting the safety and security of the food supply. A U.S. Department of Homeland Security was formed at part of the Homeland Security Act of 2002 in part to help reduce the vulnerability of the United States to terrorist attacks. The nation's food and water were identified as potential targets for terrorist attackes.

In September 2005, the Department of Homeland Security (DHS), U.S. Department of Agriculture (USDA), Food and Drug Administration (FDA), and the Federal Bureau of Investigation (FBI) began collaborations with private industry and the states in a joint initiative termed the Strategic Partnership Program Agroterrorism (SPPA) Initiative, which is designed to protect the nation's food supply.

A number of organizations have developed food biosecurity guidelines for foodservice operations. The FDA Center for Food Safety and Applied Nutrition developed food security preventive measures guidance for food processors and retailers (see www.foodsafety.gov). The guidance documents identify measures that can be taken by foodservice operators to minimize the risk of food being subjected to tampering or criminal or terrorist actions. Guidance is provided for management of food security, physical security, employee hiring and supervision, computer system security, raw materials and packaging security, operations, and finished products. Foodservice operators are encouraged to implement an Operations Risk Management (ORM) process to prioritize the preventive measures that are most likely to have the greatest impact on reducing the risk of food security problems. USDA's Food and Nutrition Service released *Biosecurity Checklist for School Foodservice: Developing a Biosecurity Management Plan* (2004) (see http://schoolmeals.nal.usda.gov/Safety/biosecurity.pdf) to assist school foodservice directors with their food biosecurity planning.

Procedures for Complaints

No foodservice manager is immune to outbreaks of foodborne illness (Cheney, 1993). A cook might fail to heat up the grill to the correct temperature, a refrigerator might go on the blink, employees might forget to wash their hands before cutting meat or produce, and a supplier might deliver a contaminated product. Customers who believe their health has been harmed by food eaten in the foodservice establishment have a right to take the manager to court (NRA Educational Foundation, 1995). The customer might have a legitimate grievance, and managing this crisis correctly could be difficult.

When someone complains of foodborne illness, it is a good idea to have that person complete a complaint report similar to the one shown in Figure 8.10. This will ensure that the right questions are asked even if the business is hectic at the time. Cheney (1993) suggested the following steps for receiving a complaint:

- Obtain all the pertinent information including the names and addresses of all party members, the employee who served the meal, the date and time of the customer's visit, and the suspect meal.
- Remain concerned and polite, but do not admit liability or offer to pay medical bills.
- Never suggest symptoms, but let the complainant tell his or her own story.
- Record the time that the symptoms started, which will help in identifying the disease and determining the foodservice operation's responsibility.
- If possible, try to get a food history of all the meals and snacks eaten before and after the person ate the suspect meal.
- Never offer medical advice; gather information but do not interpret symptoms.

The NRA diagrammed a step-by-step procedure for crisis management of foodborne illness complaints, which is shown in Figure 8.11.

Foodborne Illness/Complaint Report

Complainant name: ______________________ Phone ____________ Work

Address: ______________________ Phone ____________ Home

____

Others in party? ______________________ (Get names and addresses) Use back of form if additional space is needed.

____

____

____

Onset of symptoms: Date ______________ Time ______________

Symptoms: ❑ Nausea ❑ Diarrhea ❑ Fever ❑ Blurred vision

❑ Vomiting ❑ Dizziness ❑ Headache ❑ Abdominal cramps

Other ______________________

Medical treatment: Doctor (Hospital) ______________________

Name Address Phone number

Suspect meal: ______________________

Location: ______________________

Time & date: ______________________

Identification (brand name, lot number): ______________________

Description of meal: ______________________

Leftovers: ______________________ (Refrigerate, *do not freeze*)

Other foods or beverages consumed before or after suspect meal:

Date	Time	Location	Description

Other agencies notified:

Agency	Person to contact	Phone

Remarks:

Report received by: ______________ Date: __________ Time: __________

Referred to: ______________________

Figure 8.10. Foodborne illness/complaint report.
Source: From *Make a S.A.F.E. Choice: A New Approach to Restaurant Self-Inspection* (p. 17) by the National Restaurant Association, 1991, Washington, DC: Author.

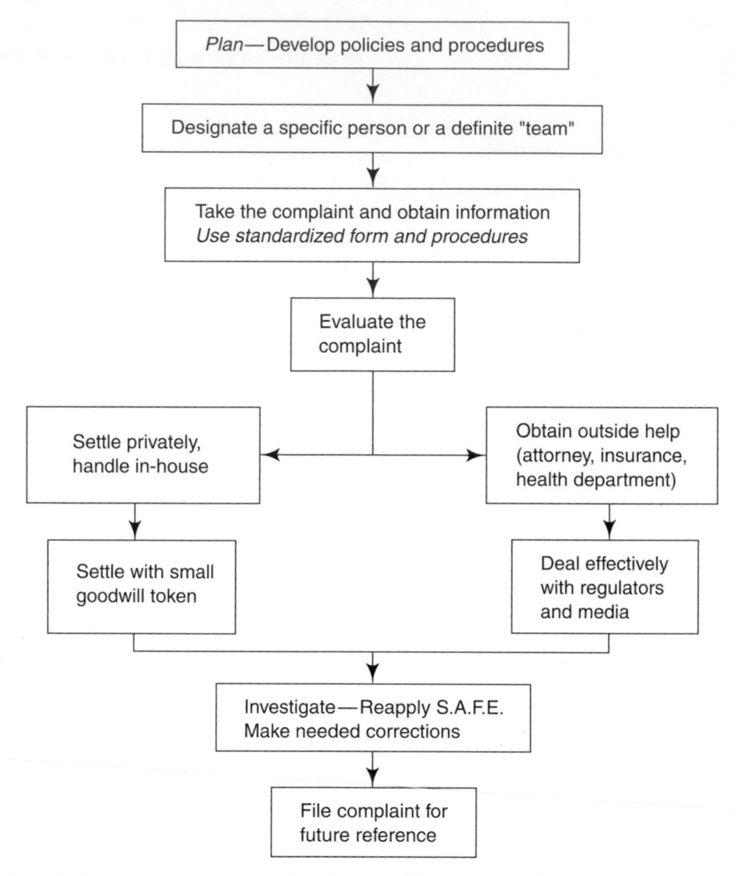

Figure 8.11. Crisis management—foodborne illness complaints.
Source: From *Make a S.A.F.E. Choice: A New Approach to Restaurant Self-Inspection* (p. 19) by the National Restaurant Association, 1991, Washington, DC: Author.

SANITATION REGULATIONS AND STANDARDS

The protection of the food supply available to the consumer is the responsibility of governmental agencies at the federal, state, and local levels. Trade associations and institutes, professional societies, and private associations and foundations are especially concerned about microbiological standards of food products.

Regulations

In Chapter 5, key federal agencies involved in the wholesomeness and quality of food from producer to purchaser were discussed. The sanitation and service of food after it is purchased is controlled largely by state and local agencies and private organizations.

Role of Governmental Agencies

The U.S. Public Health Service (PHS) and its subdivision, the FDA, both of which are agencies within the U.S. Department of Health and Human Services, are charged specifically with protecting the health of all Americans and providing essential human services, especially for those who are least able to help themselves (Department of Health and Human Services, 1998). The PHS identifies and controls health hazards, provides health services, conducts and supports research, and develops training related to health.

Two agencies within the PHS specifically related to sanitation regulations and standards are the CDC and the FDA. The CDC investigates and records reports of foodborne illness and is charged with protecting public health by providing leadership and direction in the control of diseases and other preventable hazards. The Bureau of Training within the CDC develops programs for control of foodborne diseases in the foodservice industry. The CDC is responsible for providing assistance in identifying causes of disease outbreaks, including foodborne illnesses.

The FDA directs its efforts toward protecting the nation's health against unsafe and impure foods, unsafe drugs and cosmetics, and other potential hazards. The FDA has been established as the regulatory agency with responsibility for food safety in the United States, and the Food Safety and Inspection Service (FSIS) has the same responsibility under the USDA. The FDA uses laboratory analysis of foods to detect microbiological and chemical toxicants, adulteration, and incorrect labeling; the USDA does onsite inspections of meat and poultry operations and checks animal health, sanitation, and product labeling (Cody & Kunkel, 2002).

The FDA is responsible for regulating interstate shipment of food, as discussed in Chapter 5, and for inspecting foodservice facilities on interstate carriers such as trains, planes, and ships operated under the U.S. flag. In addition, although local agencies have the primary responsibility for inspecting foodservice establishments, the FDA assists these agencies by developing model codes and ordinances and providing training and technical assistance. The FDA enforces the Federal Food, Drug, and Cosmetic Act and its amendments, Pesticide, Food Additives, Color Additives, and Labeling. The FDA's jurisdiction has been redefined as new concerns have arisen. The FDA also administers the Egg Product Inspections, Bioterrorism, Fair Packaging and Labeling, and Tea Importation Acts. The FSIS administers the Poultry Products Inspection, Egg Products Inspection, and Federal Meat Inspection acts.

Many state and local governments have adopted the U.S. Public Health Service codes in establishing standards of performance in sanitation for foodservice establishments. These codes generally require that employees have medical examinations to determine their qualifications to handle food safely and have a foodhandler's permit, which usually requires a short training program on sanitation practices.

State and local health agencies act to ensure that foodservice establishments (Longrée & Armbruster, 1996)

- are equipped, maintained, and operated to offer minimal opportunities for food hazards to develop
- use food products that are wholesome and safe
- are operated under the supervision of a person knowledgeable in sanitary food-handling practices

State or local agency officials make periodic inspections to foodservice operations to compare their performance with standards of cleanliness and sanitation. Any deficiencies must be corrected before the next inspection. The agency generally has the authority to close an operation with an inordinate number of deficiencies in meeting the sanitation standards.

The Environmental Protection Agency (EPA), another agency within the U.S. Department of Health and Human Services, also has responsibility in certain areas related to sanitation in foodservice establishments. The EPA endeavors to comply with environmental legislation and control pollution systematically by a variety of research, monitoring, standard setting, and enforcement activities. Of particular interest to the foodservice industry are programs on water standards, air quality, pesticides, noise abatement, and solid waste management.

Role of Other Organizations

A number of other organizations are active in upgrading and maintaining the sanitary quality of various food products and establishing standards for foodservice operations. For example, many trade and professional organizations serving various segments of the food industry have established sanitary standards for food processing operations.

The Educational Foundation of the **National Restaurant Association** has exerted aggressive leadership in developing standards and promoting training in foodservice sanitation. It has developed a crisis management program, identified as ServSafe®, that concentrates on three areas of potential risk: food safety, responsible alcohol service, and customer safety (NRA Educational Foundation, 1995). ServSafe® courses focus on the manager's role in assessing risks, establishing policies, and training employees. ServSafe® programs include a textbook and employee training materials for each subject, such as study guides, videos, and an employer's kit with a leader's guide and other teaching aids. Everyone who completes the course and satisfactorily completes a certification examination administered by the Education Foundation is eligible for an Educational Foundation ServSafe® certificate.

In the healthcare industry, the **Joint Commission on Accreditation of Healthcare Organizations** (JCAHO) has encouraged high standards of sanitation by including assessment of sanitary practices in its accreditation standards and visits. The **American Dietetic Association** (ADA) has emphasized food safety for many years and has published *Food Safety for Professionals: A Reference and Study Guide* (Cody & Kunkel, 2002) and *Pesticides in Food: A Guide for Professionals* (Chaisson, Petersen, & Douglass, 1991) and *Disaster and Emergency Preparedness in Foodservice Operations* (Puckett & Norton, 2003).

The **National Association of College and University Food Services** (NACUFS), the trade association for foodservice professionals at institutions of higher education, publishes a standards manual. These standards were designed to be used as a self-monitoring program for improving operations and as part of a voluntary peer review program. Sanitation, safety, and maintenance comprise a major section of the manual.

The **School Nutrition Association** (SNA) formerly the American School Food Service Association (AFSFA), has provided leadership in the school foodservice segment of the industry in promoting good sanitation practices and has been active in providing employee training. The SNA's *Keys to Excellence* (ASFSA, 1995) details sanitation and safety standards. School foodservice directors can do an assessment of their operation using the keys at www.schoolnutrition.org

The **American Public Health Association** (APHA), a professional society representing all disciplines and specialties in public health, has a standing committee on food protection that establishes policies and standards for food sanitation. Several other associations for professionals in the area of food protection and sanitation also promote standards and enforcement procedures in food safety and sanitation.

Members of the **National Council of Chain Restaurants** have formed an ad hoc group, the Food Safety Council, charged with helping formulate government policies and alleviating customer concerns (Allen, 1993). The council wants to project a positive active role in shaping food safety for the future. The group also will participate in developing new federal food safety policies. The food safety committee planned to participate in the public hearings held by FSIS concerning the development of a new method to reduce pathogens in meat and poultry products. Finally, the group will serve as a watchdog for the foodservice industry by collecting and disseminating information on food safety problems at state and local levels.

The **National Sanitation Foundation** (NSF) is one of the most influential semiprivate agencies concerned with sanitation. The NSF, organized by a group of industrial leaders and public health officials, is a nonprofit, noncommercial organization whose mission is to develop and administer programs relating to public health and the environment in the areas of service, research, and education (National Sanitation Foundation, 1997). NSF, along with industry officials, businesses, federal and state regulatory agencies, and the public, develops nationally uniform and voluntary consensus standards for products and services, including

- Drinking water treatment units
- Foodservice equipment
- Vending machines
- Swimming pool equipment
- Refuse compactor systems
- Waste water treatment equipment
- Plastic piping system components
- Water conservation systems

Manufacturers may request that the NSF evaluate their equipment, and they will receive an NSF Testing Laboratory Seal of Approval for equipment meeting NSF standards.

EMPLOYEE SAFETY

Accident: Unexpected event resulting in injury, loss, or damage.

An **accident** is frequently defined as an event that is unexpected or the cause of which was unforeseen, resulting in injury, loss, or damage. An accident is also an unplanned event that interrupts an activity or function. Although they may or may not be the result of negligence, many accidents can be prevented. Safety is every employer's responsibility. Accidents do not just happen—something causes them, and the majority are controllable (Somerville, 1992).

A foodservice operation should have an accident prevention program that seeks to eliminate all accidents, not just those resulting in personal injury. Accidents are expensive and can result in increased insurance premiums, lost productivity, wasted time, overtime expenses, workers' compensation claims, potential lawsuits, and human suffering. Accidents can also result in a fine or legal action if provisions of the Occupational Safety and Health Act (OSHA) are violated.

Many aspects of safety are related to construction and maintenance of the structure and equipment. For example, floors and wiring should be in good repair, and adequate lighting should be provided in work areas, corridors, and outside the facility. Exits should be clearly marked, nonslip flooring materials used, and all equipment supplied with necessary safety devices. Also, fire extinguishers of the appropriate type should be readily available throughout the foodservice facility. The basic traffic flow should be designed to avoid collisions.

Most accidents are the result of human error. Employees may lift heavy loads incorrectly, leave spills on the floor, walk across freshly mopped floors, fail to use safety devices on foodservice equipment, block passageways, or fail to clean greasy filters regularly. Many other unsafe practices can be added to this list. Obviously, then, training is an important part of a safety program. Employees should be taught to prevent accidents by learning to recognize and avoid or correct hazardous conditions. The first day on the job is the best time to start educating a new employee about safety procedures and equipment handling (Spertzel, 1992b).

Congress legislated the Occupational Safety and Health Act in 1970. The purpose of the act is "to assure, so far as possible, every working man and woman in the Nation safe and healthful working conditions, and to preserve our human resources."

OSHA allows a compliance officer to enter a facility to determine adherence to standards and to determine if the workplace is free of recognized hazards. During an OSHA inspection, some of the specific conditions for which the compliance officer will be searching include the following:

- Accessibility of fire extinguishers and their readiness for use
- Guards on floor openings, balcony storage areas, and receiving docks
- Adequate handrails on stairs
- Properly maintained ladders
- Proper guards and electrical grounding for foodservice equipment
- Lighted passageways, clear of obstructions
- Readily available first-aid supplies and instructions
- Proper use of extension cords
- Compliance with OSHA posting and recordkeeping requirements

Citations are issued by an OSHA area director upon review of the compliance officer's inspection report if standards or rules have been violated. Several kinds of violations are possible, which may involve fines or legal action if the violation is sufficiently serious.

More fires start in foodservice than in any other kind of business operation (NRA Educational Foundation, 1992). Managers must check their operations regularly and must establish procedures for handling any hazards that could start fires. Hot oil in fryers can burst into flames at its flammable limit of between 425°F and 500°F and be the source of a fire, or it can increase the severity of a fire that is started another way. Also, oil in ventilation systems and on walls, equipment, and other surfaces is highly flammable. Hoods over ranges and filters that are not cleaned regularly provide an ideal environment for a grease fire (Bendall, 1992). A good solution for high-volume restaurants is the extractor ventilator, which is a series of baffles on the hood to extract grease through a centrifugal action. Some have an automatic wash-down feature to clean the inside of the hood with detergent and hot water at scheduled times. Tests have shown that some of these ventilators can remove more than 90% of the grease from the air.

The National Fire Protection Association has identified ABC classes of fire, which are described in Figure 8.12 along with the types of extinguishers for each. The foodservice manager should know the differences among the extinguishers and purchase the proper kind. The local fire department usually is willing to demonstrate extinguishers.

CLASSES OF FIRES

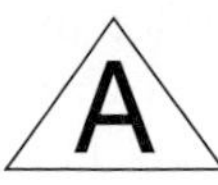

Class A fires include: wood, paper, cloth, cardboard, plastics
Examples: a fire in a trash can; a fire that catches drapes or tablecloths in the dining room

Class B fires include: grease, liquid shortening, oil, flammable liquids
Examples: a fire in a deep fryer; spilled flammable chemicals in kitchen

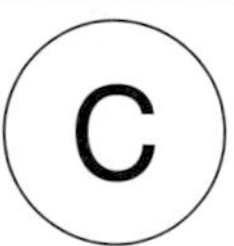

Class C fires include: electrical equipment, motors, switches, and frayed cords
Examples: a fire in a toaster; a fire in the motor of a grinder

CLASSES OF FIRE EXTINGUISHERS

A	AB	BC	ABC
Use on Class A fires only	**Use on Class A and Class B fires only**	**Use on Class B and Class C fires only**	**Use on Class A, Class B, or Class C fires**
Water, stored pressure Water pump tank Multipurpose dry chemical, stored pressure Multipurpose dry chemical, cartridge operated Aqueous film-forming foam (AFFF), stored pressure Halon, stored pressure	Aqueous film-forming foam (AFFF), stored pressure* *Not recommended for deep fryer fires	Dry chemical, stored pressure Dry chemical, cartridge operated Carbon dioxide, self-expelling* Halon, stored pressure* *Not recommended for deep fryer fires	Multipurpose dry chemical, stored pressure* Multipurpose dry chemical, cartridge operated* Halon, stored pressure *Not recommended for deep fryer fires

Figure 8.12. Description of the classes of fires and fire extinguishers.
Source: Adapted in *Applied Foodservice Sanitation* (p. 266) by the Educational Foundation of the National Restaurant Association, 1992, Chicago: Author. From *Safety Operations Manual* by the National Restaurant Association, 1988. Used by permission.

In addition to fire extinguishers, heat and smoke detection devices and some form of fire protection, such as water mist or dry/wet chemicals, should be installed over cooking equipment. Water mist operates from the building's water sprinkler, which has an unlimited supply of water, and is effective in suffocating all types of fires. Dry/wet chemicals in containers are piped to outlet nozzles above each piece of equipment; once the chemicals are discharged, they have to be replaced immediately to provide continuing fire protection. In many states, a state fire marshal has responsibility for approving the design and construction of buildings from a safety and fire protection standpoint. Also, OSHA inspections are designed to assure safe and healthful working conditions. The foodservice manager should augment any external inspections with a periodic internal review, or self-inspection, of safety conditions and practices, followed by any necessary in-service training. Although small foodservice operations cannot afford to hire a safety coordinator, most can appoint a staff member to coordinate safety efforts.

Insurance companies can be an important safety resource in which the service is either included in the premium or is available at a small charge (Somerville, 1992). This service includes establishing a safety program or reinforcing an existing one. Some insurers conduct audits for the operation and help with employee training by providing safety manuals or films and videos. A comprehensive safety audit includes a thorough inspection of the facility from the sidewalks in. Almost every restaurant that has installed a facility self-inspection report has reduced accidents significantly and has had sizable savings on insurance premiums. Godfather's Pizza saved $500,000 in 1 year.

Ergonomics: Study of how workers interact with their work environment.

Although many aspects of safety are concerned with construction and design of facilities, safe practices of employees are also a critical element in a safety program. Ergonomics is another factor of work safety. **Ergonomics** examines how workers interact with their work environment, including equipment, the workstation, and climate; it influences such factors as lighting and footwear, which in turn influence safety. Recommendations such as storing heaviest items on middle shelves to reduce back strain is an example of an ergonomic employee safety recommendation.

Equipment manufacturers have developed equipment with built-in features such as safety valves on pressure steamers and guards on slicing and chopping machines (DOL Officially Declares, 1991). The Department of Labor went one step further and issued regulations to officially prohibit 16- and 17-year-olds from using power-driven food slicers in restaurants, especially quick-service operations. Obviously, safety training must have major emphasis in both initial and in-service employee training. Many resource materials on safety and accident prevention are available, such as those of the National Safety Council and the American Red Cross. Also, personnel from state and local fire prevention agencies are often available as speakers. The National Restaurant Association has published a variety of workplace safety training materials and mini-posters (Figure 8.13) to help foodservice managers continually improve the safety in their operation.

CUSTOMER SAFETY

Many of the factors discussed for employee safety also apply to customer safety. A crack in the sidewalk, an exit door that does not open, grease on the dining room floor, or a cup of hot coffee that is dropped can cause customers to have serious accidents that end in litigation. Customer safety is the responsibility of the foodservice manager and employees. Emergency action procedures should be included in the employee manual and during employee training sessions (NRA Educational Foundation, 1992). A foodservice operation always should have a complete first-aid kit. Some states also have laws specifying the supplies that must be included in this kit. OSHA requires that a restaurant either have a kit equipped according to the advice of a company physician or have physical or telephone access to community

Figure 8.13. Examples of mini-posters on safe practices in foodservice operations.
Source: National Restaurant Association, Chicago. Used by permission.

emergency services. Ideally, a foodservice operation should have present at all times an employee who is trained and certified in first aid, including how to administer the Heimlich maneuver and how to give cardiopulmonary resuscitation (CPR).

Approximately 60% of all choking incidents occur in restaurants (Herlong, 1991). Prior to 1974, when the Heimlich maneuver was introduced, about 20,000 choking fatalities occurred each year; currently about 2,000 to 3,000 occur. Laws on first-aid training requirements vary by state. Some states require only that restaurants post Heimlich maneuver instructional diagrams where all employees can see them, but others require formal training for foodservice employees as well as posting of instructions. The National Restaurant Association, the American Red Cross, and the Heimlich Institute all provide charts and instructional materials on the Heimlich maneuver.

SANITATION

A properly designed foodservice facility is basic to maintain a high standard of sanitation. The first requirement for a sanitary design is cleanability, which means the facility has been arranged so that it can be cleaned easily. Equipment and fixtures should be arranged and designed to comply with sanitation standards, and trash and garbage isolated to avoid contaminating food and attracting pests.

Clean: Free of physical soil and with an outwardly pleasing appearance.

Sanitary: Free of disease causing organisms and other contaminants.

For a facility to be clean is not enough; it must also be sanitary. Although the two words are often used synonymously, **clean** means free of physical soil and with an outwardly pleasing appearance—a glass that sparkles, silver that shines, and a floor that is free from dust and grime. These objects may look clean on the surface but may harbor disease agents or harmful chemicals. **Sanitary** means free of disease-causing organisms and other contaminants. Cleaning and sanitizing are both issues of concern in the maintenance of foodservice facilities and equipment, and together they form the basis for good housekeeping in foodservice operations.

Sanitization

Sanitization is critical for any surface that comes in contact with food, which includes, of course, all dishes, utensils, pots, and pans.

Some sanitizing agents are toxic to humans as well as to bacteria and are, therefore, acceptable for use only on nonfood contact surfaces. Other agents may not be toxic but may have undesirable flavors or odors, which make them unacceptable for use in foodservice operations. The three most common chemicals used in sanitizing are chlorine, iodine, and quaternary ammonia. Table 8.3 summarizes the properties of these sanitizers and outlines procedures for their use.

The receiving area should be designed for ease in cleaning. The floor should be of material that can be easily scrubbed and rinsed and have adequate drains and a water connection nearby to permit hosing down the area. Storage for cleaning supplies should be located conveniently. Since insects tend to congregate near loading

Table 8.3 Chemical Sanitizing Agents

	Chlorine	Iodine	Quaternary Ammonium
Minimum concentration			
—For immersion	50 parts per million (ppm)	12.5 ppm	200 ppm
—For power spray or cleaning in place	50 ppm	12.5 ppm	200 ppm
Temperature of solution	75°–115°F/24–46°C	75°–120°F/ 23.9–48.9°C Iodine will leave solution at 120°F/48.9°C	75°F/23.9°C+
Time for sanitizing			
—For immersion	7 seconds	30 seconds	30 seconds; however, some products require longer contact time; read label
—For power spray or cleaning in place	Follow manufacturer's instructions	Follow manufacturer's instructions	
pH (detergent residue raises pH of solution so rinse thoroughly first)	Must be below 8.0	Must be below 5.0	Most effective around 7.0 but varies with compound
Corrosiveness	Corrosive to some substances	Noncorrosive	Noncorrosive
Response to organic contaminants in water	Quickly inactivated	Made less effective	Not easily affected
Response to hard water	Not affected	Not affected	Some compounds inactivated but varies with formulation; read label. Hardness over 500 ppm is undesirable for some quats
Indication of strength of solution	Test kit required	Amber color indicates presence; use test kits to determine concentration	Test kit required. Follow label instructions closely

Source: Adapted from *Serv Safe Coursebook* (pp. 11–7) by the Educational Foundation of the National Restaurant Association, 1999. Chicago: Author. Used with permission.

docks, adequate screening must be provided. Outdoor bug zappers, which are electrocutor traps used to destroy flying insects, should never be used in a foodservice operation. These traps can create more problems than they solve. They don't control the problem, and they kill beneficial insects such as ladybugs, ground beetles, and mayflies. To reduce the number of insects around outside entrance ways, sodium vapor or yellow lights that cut out the blue and ultraviolet spectrum should be used. Also, outside sticky traps to attract insects are more effective and safer to use than bug zappers.

Floors in the dry storage area must be easy to clean and slip resistant to prevent accidents. External walls and subfloors should be well constructed, insect- and rodent-proof, and insulated. Walls and ceilings should be painted light colors, have a smooth surface that is impervious to moisture, and be easy to wash and repair. Products must never be stored on the floor; they should be stored on shelves or pallets to permit frequent floor cleaning. If steam lines, duct work, and hot water lines must pass through the dry storage area, they should be insulated.

Cleanability that promotes sanitation is a significant need in walk-in refrigerators and freezers. Hard-surface, easy-to-clean floors, walls, and fixtures should be of smooth, nonabsorbent material. Drains to remove scrubbing water and condensate should be located inside walk-ins. Finally, uniform ventilation and adequate lighting should be provided in these units as an aid in maintaining sanitary conditions. Any deficiencies must be corrected before the next inspection. Local health agencies generally have authority to close an operation that has an inordinate number of deficiencies in meeting sanitation standards. Immediate action also is required if the violation is extremely dangerous.

Audits of Sanitation Standards

Evaluation of the maintenance of foodservice sanitation standards is accomplished in two ways: external and internal audits of facilities and practices. An **external audit** may be performed by governmental or nongovernmental agencies; an **internal audit** is the responsibility of the management of a foodservice organization and may be part of a self-inspection program for sanitation or a component of a broader total quality management program.

External Audits

External audits are performed by federal, state, and local governmental agencies, as indicated in the previous discussion of the role of these agencies in monitoring sanitation in foodservice establishments. Many state and local governments have adopted Public Health Service (PHS) codes in establishing standards of performance in sanitation for foodservice establishments. State and local health agencies act to ensure that foodservice establishments

- Are operated under the supervision of a person knowledgeable in sanitary food-handling practices (Longrée & Armbruster, 1996)
- Are equipped, maintained, and operated to offer minimal opportunities for food hazards to develop; and
- Use ingredients and food products that are wholesome and safe.

The government has been interested in securing the sanitary quality of food for many years. Officials with state and local agencies make periodic inspections to compare performance of foodservice operations with standards of cleanliness and sanitation. The FDA has developed the *Food Code* to assist health departments in developing regulations for a foodservice inspection program (FDA, 2005). Some state and local agencies develop their own codes. The FDA recommends inspections at least every 6 months, although the frequency is determined by the local agency.

Sanitarian: Health officials or inspector who is trained in sanitation principles and methods.

A **sanitarian**, often referred to as a health official or inspector, is an individual trained in sanitation principles and methods and public health (NRA Educational Foundation, 1992). The foodservice manager and unit head should accompany the sanitarian during inspection. They should take advantage of the experience and expertise of the sanitarian by asking questions; employees should be encouraged to do the same. They should take notes during the inspection and be willing to correct problems. Problems should be discussed with employees and corrected immediately. Foodservice managers and employees should welcome a visit from the sanitarian and not resent it if they are truly dedicated to serving safe food to customers.

In some areas, HACCP principles have replaced traditional health department inspections that stress the appearance of the facility and spot-checking temperatures. Inspectors trained in these principles examine the procedures related to the flow of food from receiving to service and may verify critical control points for each step. Many state and local ordinances are patterned on the Model FDA Food Establishment Inspection Report, shown in Figure 8.14. Definitions for the compliance items and directions for marking the inspection report can be found in *Food Code* 2005 (see www.cfsan.fda.gov/~dms/fc05-toc.html).

A good sanitation program and well-trained employees result in safe food and are reflected in a good sanitation report (NRA Educational Foundation, 2004). Based on their years of experience, sanitarians generally can offer advice on correcting violations. According to the FDA *Food Code,* corrective actions must be taken on all violations. The larger violations should have a time frame approved by the inspector. The sanitarian usually has the authority to close the operation if violations are excessive and dangerous to public health.

The inspection process in a foodservice facility may begin before a facility is built, as many jurisdictions require a review of plans and specifications for new construction or extensive remodeling. Once a facility is completed, inspection visits are usually conducted prior to issuance of permits to operate. After a foodservice operation has opened, inspections will occur periodically, depending on the workload of the responsible agency and severity of violations at previous inspections. The growth of the foodservice industry has not been matched by expansion of the capacity of health agencies to monitor operations.

Internal Audits

A foodservice organization should have its own program of self-inspection as a means of maintaining standards of sanitation. In organizations with a TQM program, an audit of sanitation practices should be one of its major components. Employees can be an important part of an internal audit when they are empowered to take corrective action if a critical control point is violating the code. A voluntary food safety

FORM 3-A

Food Establishment Inspection Report

Page _____ of _____

As Governed by State Code Section	No. of Risk Factor/Intervention Violations		Date
	No. of Repeat Risk Factor/Intervention Violations		Time In
	Score *(optional)*		Time Out

Establishment	Address	City/State	Zip Code	Telephone
License/Permit #	**Permit Holder**	**Purpose of Inspection**	**Est. Type**	**Risk Category**

FOODBORNE ILLNESS RISK FACTORS AND PUBLIC HEALTH INTERVENTIONS

Circle designated compliance status (IN, OUT, N/O, N/A) for each numbered item — Mark "X" in appropriate box for COS and R

IN=in compliance **OUT**=not in compliance **N/O**=not observed **N/A**=not applicable **COS**=corrected on-site during inspection **R**=repeat violation

	Compliance Status		COS	R
		Supervision		
1	IN OUT	Person in charge present, demonstrates knowledge, and performs duties		
		Employee Health		
2	IN OUT	Management awareness; policy present		
3	IN OUT	Proper use of reporting, restriction & exclusion		
		Good Hygienic Practices		
4	IN OUT N/O	Proper eating, tasting, drinking, or tobacco use		
5	IN OUT N/O	No discharge from eyes, nose, and mouth		
		Preventing Contamination by Hands		
6	IN OUT N/O	Hands clean and properly washed		
7	IN OUT N/A N/O	No bare hand contact with ready-to-eat foods or approved alternate method properly followed		
8	IN OUT	Adequate handwashing facilities supplied & accessible		
		Approved Source		
9	IN OUT	Food obtained from approved source		
10	IN OUT N/A N/O	Food received at proper temperature		
11	IN OUT	Food in good condition, safe, and unadulterated		
12	IN OUT N/A N/O	Required records available: shellstock tags, parasite destruction		
		Protection from Contamination		
13	IN OUT N/A	Food separated and protected		
14	IN OUT N/A	Food-contact surfaces: cleaned & sanitized		
15	IN OUT	Proper disposition of returned, previously served, reconditioned, and unsafe food		

	Compliance Status		COS	R
		Potentially Hazardous Food (TCS food)		
16	IN OUT N/A N/O	Proper cooking time and temperatures		
17	IN OUT N/A N/O	Proper reheating procedures for hot holding		
18	IN OUT N/A N/O	Proper cooling time and temperatures		
19	IN OUT N/A N/O	Proper hot holding temperatures		
20	IN OUT N/A	Proper cold holding temperatures		
21	IN OUT N/A N/O	Proper date marking and disposition		
22	IN OUT N/A N/O	Time as a public health control: procedures & records		
		Consumer Advisory		
23	IN OUT N/A	Consumer advisory provided for raw or undercooked foods		
		Highly Susceptible Populations		
24	IN OUT N/A	Pasteurized foods used; prohibited foods not offered		
		Chemical		
25	IN OUT N/A	Food additives: approved and properly used		
26	IN OUT	Toxic substances properly identified, stored, used		
		Conformance with Approved Procedures		
27	IN OUT N/A	Compliance with variance, specialized process, and HACCP plan		

Risk factors are food preparation practices and employees behaviors most commonly reported to the Centers for Disease Control and Prevention as contributing factors in foodborne illness outbreaks. **Public health interventions** are control measures to prevent foodborne illness or injury.

GOOD RETAIL PRACTICES

Good Retail Practices are preventative measures to control the introduction of pathogens, chemicals, and physical objects into foods.

Mark "X" in box if numbered item is **not** in compliance — Mark "X" in appropriate box for COS and/or R — COS=corrected on-site during inspection — R=repeat violation

			COS	R
		Safe Food and Water		
28		Pasteurized eggs used where required		
29		Water and ice from approved source		
30		Variance obtained for specialized processing methods		
		Food Temperature Control		
31		Proper cooling methods used; adequate equipment for temperature control		
32		Plant food properly cooked for hot holding		
33		Approved thawing methods used		
34		Thermometers provided and accurate		
		Food Identification		
35		Food properly labeled; original container		
		Prevention of Food Contamination		
36		Insects, rodents, and animals not present		
37		Contamination prevented during food preparation, storage & display		
38		Personal cleanliness		
39		Wiping cloths: properly used and stored		
40		Washing fruits and vegetables		

			COS	R
		Proper Use of Utensils		
41		In-use utensils: properly stored		
42		Utensils, equipment and linens: properly stored, dried, handled		
43		Single-use/single-service articles: properly stored, used		
44		Gloves used properly		
		Utensils, Equipment and Vending		
45		Food and nonfood-contact surfaces cleanable, properly designed, constructed, and used		
46		Warewashing facilities: installed, maintained, used; test strips		
47		Nonfood-contact surfaces clean		
		Physical Facilities		
48		Hot and cold water available; adequate pressure		
49		Plumbing installed; proper backflow devices		
50		Sewage and waste water properly disposed		
51		Toilet facilities: properly constructed, supplied, cleaned		
52		Garbage/refuse properly disposed; facilities maintained		
53		Physical facilities installed, maintained, and clean		
54		Adequate ventilation and lighting; designated areas use		

Person in Charge (Signature)	Date:
Inspector (Signature)	Follow-up: YES NO (Circle one) Follow-up Date:

Figure 8.14. Food establishment inspection report.
Source: Food Code, 2005.

program of self-inspection will assure the government and the public that the foodservice operation is protecting the safety of food in each step of production.

Employee Training in Food Sanitation

Trained personnel who have good personal hygiene habits and follow recommended food-handling practices are critical to an effective sanitation program. Equally important are strong leadership by management, provision of appropriate tools and equipment, and continual follow-up. The time, money, and effort that go into a sanitation program are wasted if the foodservice staff is not knowledgeable about appropriate sanitation practices (Longrée & Armbruster, 1996).

The education of foodservice managers should include microbiology and sanitation to equip them for leadership in design and implementation of inservice sanitation programs. Continuing education is needed to maintain competency.

Professional associations have an important role in upgrading sanitation practices in the foodservice industry. As mentioned earlier in this chapter, several professional associations are concerned with training standards and programs for foodservice employees. The certification programs of several organizations, like the American School Food Service Association and the Dietary Managers Association, include knowledge and competency in sanitation practices as a major component. A major contributor to the upgrading of sanitation practices is the Educational Foundation of the NRA through its national uniform sanitation, training, and certification plan for foodservice managers termed ServSafe®.

Training of nonmanagerial employees is the responsibility of the foodservice manager. Initial training in safe food-handling practices and personal hygiene is needed for new employees.

Many teaching aids are available to assist the manager in conducting training programs on safe food handling. Federal and state health agencies and commercial and private organizations have videotapes, films, slides, posters, and manuals available free or at low cost. Many foodservice organizations also have developed comprehensive manuals for training and as a reference for their own personnel.

Training materials, mini-posters, and other training aids are available from many companies and organizations. The mini-posters shown in Figure 8.15 are examples of materials available from the Integrated Food Safety Information Delivery System (www.profoodsafety.org). The posters are available in 14 different languages to help train foodservice employees whose first language is not English. Online food safety lessons are available for training employees at the Iowa State University Extension's Food safety Web site (www.extension.iastate.edu/foodsafety). Online information is available at the U.S. government food safety Web site (www.foodsafety.gov) and the online food safety answers line (www.foodsafetyanswers.org).

Ware Washing

Ware washing: Process of washing and sanitizing dishes, glassware, flatware, and pots and pans either manually or mechanically.

Ware washing is the process of washing and sanitizing dishes, glassware, flatware, and pots and pans either manually or mechanically. Sinks, dishmachines, and pot and pan washing machines are the most common equipment for this process. Specialized equipment, such as flatware washers and glassware washers, is available.

Employee Handwashing

1. Wet hands with hot, running water
2. Apply soap
3. Rub hands for at least 20 seconds
4. Clean under fingernails and between fingers
5. Rinse hands thoroughly under running water
6. Dry hands

Section 2-301.12, 1999 Food Code

For Additional Information Contact
Your Local Health Department

NO Smoking
NO Drinking
NO Eating

In Food Preparation or
Food Handling Areas

Section 2-401.11, 1999 Food Code

For Additional Information Contact
Your Local Health Department

Always Remember

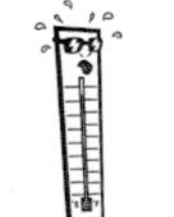

Keep Hot Foods Hot!
Maintain hot foods at a temperature of
140°F (60°C) or hotter

Keep Cold Foods Cold!
Maintain cold foods at a temperature of
41°F (5°C) or colder

Section 3-501.16, 1999 Food Code

For Additional Information Contact
Your Local Health Department

Figure 8.15. Sanitation posters.
Source: Integrated Food Safety Information Delivery System (www.profoodsafety.org) signs made possible from grant # FD-R-001731-01 from the Food and Drug Administration.

Dishmachines

Many different brands of dishmachines are on the market today. Most manufacturers have a series of machines starting with simple models to very sophisticated equipment. The dishwashing process, whether manual or machine driven, consists of scrapping, prewashing, washing, sanitizing, and air drying. Although dishmachines are the most reliable way to clean and sanitize dishes and utensils, many problems can occur if machines are not installed or operated correctly (Table 8.4).

Table 8.4 Dishwashing Problems and Cures

Symptom	Possible Cause	Suggested Cure
Soiled Dishes	Insufficient detergent	Use enough detergent in wash water to ensure complete soil removal and suspension.
	Wash water temperature too low	Keep water temperature within recommended ranges to dissolve food residues and to facilitate heat accumulation (for sanitation).
	Inadequate wash and rinse times	Allow sufficient time for wash and rinse operations to be effective. (Time should be automatically controlled by timer or by conveyor speed.)
	Improperly cleaned equipment	Unclog rinse and wash nozzles to maintain proper pressure-spray pattern and flow conditions. Overflow must be open. Keep wash water as clean as possible by prescraping dishes, etc. Change water in tanks at proper intervals.
	Improper racking	Check to make sure racking or placement is done according to size and type. Silverware should always be presoaked, placed in silver holders without sorting. Avoid masking or shielding.
Films	Water hardness	Use an external softening process. Use proper detergent to provide internal conditioning. Check temperature of wash and rinse water. Water maintained above recommended temperature ranges may precipitate film.
	Detergent carryover	Maintain adequate pressure and volume of rinse water, or worn wash jets or improper angle of wash spray might cause wash solution to splash over into final rinse spray.
	Improperly cleaned or rinsed equipment	Prevent scale buildup in equipment by adopting frequent and adequate cleaning practices. Maintain adequate pressure and volume of water.
Greasy Films	Low pH. Insufficient detergent. Low water temperature. Improperly cleaned equipment	Maintain adequate alkalinity to saponify greases; check detergent, water temperature. Unclog all wash and rinse nozzles to provide proper spray action. Clogged rinse nozzles may also interfere with wash tank overflow. Change water in tanks at proper intervals.
Streaking	Alkalinity in the water. Highly dissolved solids in water	Use an external treatment method to reduce alkalinity. Within reason (up to 300–400 ppm), selection of proper rinse additive will eliminate streaking. Above this range external treatment is required to reduce solids.
	Improperly cleaned or rinsed equipment	Maintain adequate pressure and volume of rinse water. Alkaline cleaners used for washing must be thoroughly rinsed from dishes.
Spotting	Rinse water hardness	Provide external or internal softening. Use additional rinse additive.
	Rinse water temperature too high or too low	Check rinse water temperature. Dishes may be flash drying, or water may be drying on dishes rather than draining off.
	Inadequate time between rinsing and storage	Allow sufficient time for air drying.
Foaming	Detergent: dissolved or suspended solids in water	Change to a low sudsing product. Use an appropriate treatment method to reduce the solid content of the water.
	Food soil	Adequately remove gross soil before washing. The decomposition of carbohydrates, protein, or fats may cause foaming during the wash cycle. Change water in tanks at proper intervals.
Coffee, tea, metal staining	Improper detergent	Food dye or metal stains, particularly where plastic dishware is used, normally requires a chlorinated machine washing detergent for proper destaining.
	Improperly cleaned equipment	Keep all wash sprays and rinse nozzles open. Keep equipment free from deposits of films or materials which could cause foam buildup in future wash cycles.

Source: Recommended Field Evaluation Procedures for Spray-Type Dishwashing Machines, 1982, Ann Arbor, MI: National Sanitation Foundation. Used by permission.

In choosing the size of a dishmachine to purchase, check the manufacturer's data chart that gives the maximum mechanical capacity of the machine. A factor of 70% should be used to determine what actually happens in the dishroom. Seldom is the maximum attainable. Production of clean dishes will vary depending on the type and efficiency of the dishroom layout, traffic flow, type and length of time the food soil has remained on the dishes, relative hardness of water, skill of the dishmachine operator, and fluctuations in flow of soiled dishes. Suppliers might contend that dishmachine sizing and layout is an art, not pure science.

To illustrate some of the differences in dishmachines, the Hobart brand will be used as an example (Figure 8.16a–d). Dishmachines generally are classified by the number of tanks they have. Door rack conveyor and flight-type continuous-racking automatic conveyor dishmachine are the major categories (Spears, 1999).

- **Single tank.** The single tank model shown in Figure 8.16a is designed for a corner installation and has two doors that can be manually or automatically opened and one combined wash-and-rinse tank. It holds a rack of dishes that does not move. Dishes are washed by a detergent and water from below. Recent models also have a rotating wash arm in the top of the machine. A tall version of this machine is available (Figure 8.16b). One of its features is that it is 27 in. tall and 20 in. wide and the interior can hold 18 in. × 26 in. sheet pans or up to a 60-quart mixing bowl.
- **Rack conveyor.** Dishes are still racked in the dishmachine models shown in Figure 8.16c. After dishes are scrapped and sorted, they are placed in racks designed for plates, cups, or glasses. **Scrapping** is a dishwasher term used for disposing of fragments of discarded or leftover food. The racks with soiled dishes are put on a conveyor and come out at the other end clean and dry. The machine has one, two, or three tanks. The two-tank machine has prewash and power-wash tanks, and the three-tank machine has a heavy-duty power prewash, power-wash, and power-rinse. The prewash tank has powerful jets that use overflow detergent water from the power-wash tank to quickly strip soil from the dishes.
- **Flight-type continuous conveyor.** This dishmachine is especially popular in high-volume operations. As shown in Figure 8.16d, plates and trays are placed between rows of plastic pegs on a conveyor; smaller items such as glasses, cups, and flatware are racked before sending them through the machine.

Scrapping: Disposing of fragments of discarded or leftover food in the dishwashing process.

Pot and Pan Washers

Many smaller pots and pans can be washed in a dishmachine. The scraping with a knife or spatula and soaking required for burned-on food particles are usually done at the pot and pan sink close to the production areas. A common procedure is to transport pots and pans that have been prerinsed to the dishmachine for washing and sanitizing after the bulk of the dishwashing has been completed.

In large-volume operations, special pot- and pan-washing machines are used for this labor-intensive task. The machines are heavy duty and capable of cleaning cooked-on foods off pots and pans. Pot washing is quite different from dishwashing because pressurized hot water is sprayed directly on the soiled surface.

A piece of equipment called **Power Soak** (Figure 8.17) is considered the easiest way to clean pots and pans. Power Soak capitalizes on the natural scouring

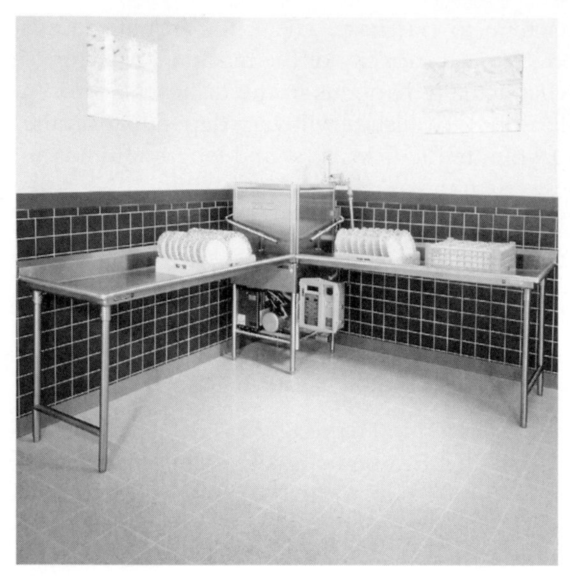

Figure 8.16a. Corner, single-tank dishmachine.
Source: Hobart Corporation. Used by permission.

Figure 8.16b. Single-tank dishmachine.
Source: Hobart Corporation. Used by permission.

Figure 8.16c. Rack conveyor one-tank and rack conveyor two-tanks dishmachines.
Source: Hobart Corporation. Used by permission.

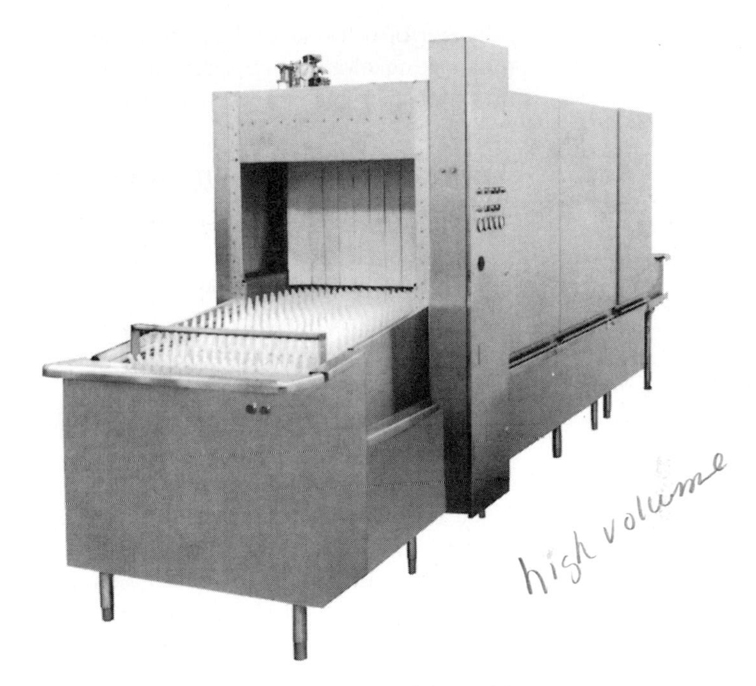

Figure 8.16d. Flight-type continuous conveyor dishmachine.
Source: Hobart Corporation. Used by permission.

abilities of high-turbulence, heated water. Maintaining an optimum cleaning temperature of 115°F loosens soil while powerful jets blast clinging particles away. Dirty pots and pans are literally water-blasted clean, thus eliminating scrubbing by hand. Power Soak is so powerful that it can clean dirty hood filters and oven parts. The "power" behind Power Soak is in its recirculating wash pump, which dispatches more than 300 gallons of water every minute. At the beginning of the first shift of employees, the tank is filled with warm water, and detergent is added. Because the pans are prerinsed and scraped, the water and detergent should success-

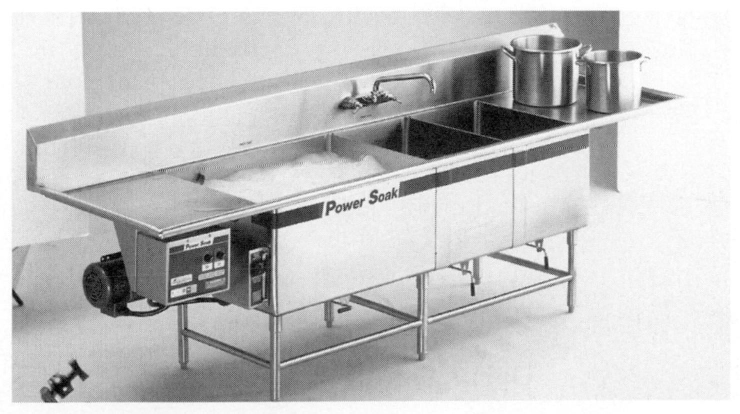

Figure 8.17. Power Soak.
Source: Courtesy of Metcraft, Inc. Used by permission.

fully perform for several hours. A heating element keeps the temperature at 115°F; the heating element automatically turns on with the presence of water and turns off when the wash tank is emptied.

Sanitation of Kitchen and Dining Areas

Sanitation and maintenance of all foodservice facilities and equipment require constant diligence on the part of the foodservice staff and management. The management of a foodservice operation must exercise constant supervision because cleaning and maintenance tasks generally are considered unpleasant, but they are extremely important in maintaining standards of sanitation. Regularly scheduled training programs on proper cleaning procedures should be established.

Design for sanitation must begin when the facility is being planned. Floors, walls, and ceilings must be constructed for easy maintenance, and the arrangement and design of the equipment and fixtures should facilitate cleaning. Having equipment on wheels and using quick disconnects on gas equipment will facilitate movement of equipment for cleaning. Facilities for proper disposal of trash and garbage are necessary to avoid contaminating food and attracting pests.

In general, the following procedures should be followed regularly in maintenance of floors:

- Spills should be wiped up promptly to avoid tracking and to eliminate a safety hazard.
- Regular schedules for cleaning floors should be established. Floors subjected to heavy traffic and food spills, such as in the production areas, must be scrubbed at least daily and hosed, stripped, and steamed periodically for more thorough cleaning.
- Floor care equipment, including brooms, mops, and vacuums, should be cleaned regularly.

Dish Storage

Handling and storage of clean dishes are important aspects of a sanitation program in a foodservice operation. All dishes and utensils must be stored dry and in clean, dust-free areas above the floor and protected from dust, mop splashes, and other forms of contamination. Mobile equipment designed for storage of various types of dishes and glassware is ideal. Metal carts are being replaced with high-density plastic with a rigid polyurethane foam core (Figure 8.18). Both are seamless and equipped with hand grips on all four sides and have four swivel neoprene casters (two with brakes). The top one is a 4-column unit that holds dishes from 4 1/2 in. to 12 5/8 in. in diameter. The bottom cart is unique because it holds 9 columns instead of the traditional 8 columns. Each column holds up to 60 dishes for a total capacity of 540 dishes. The high-density polymer shell makes dish dollies resistant to cracking, peeling, or clipping. They also have integrally molded bumpers. In addition, they are available in attractive colors.

Garbage and Trash Disposal

Garbage and trash must be handled carefully in a foodservice operation because of the potential for contaminating food, equipment, and utensils and for attracting insects and other pests. The manager needs to establish procedures for handling

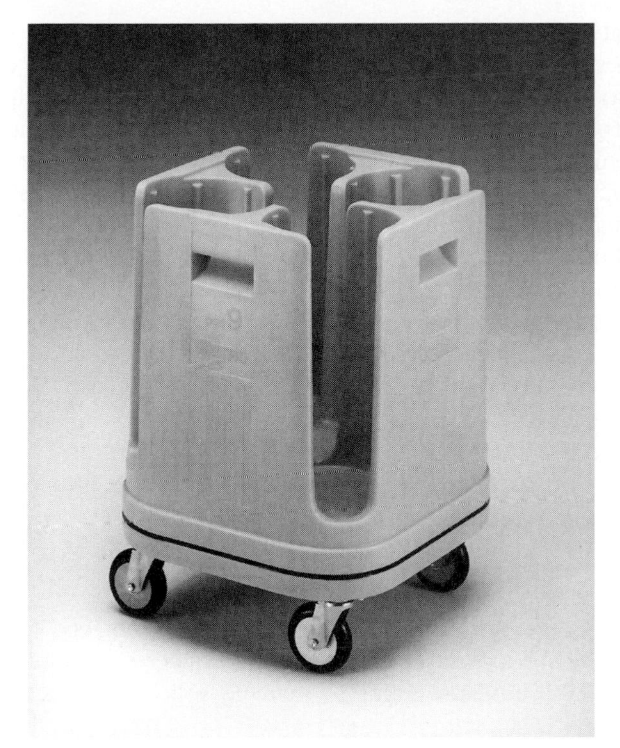

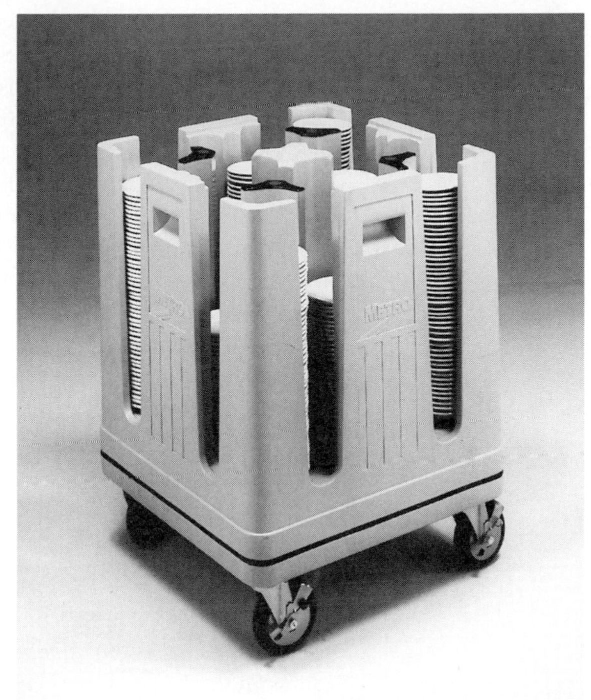

Figure 8.18. Polymer "poker chip" dish dolly and adjustable polymer "poker chip" dish dolly.
Source: Inter Metro Industries Corporation (1999). Used by permission.

garbage and trash within the operation and then disposing of the solid waste into the environment.

Procedures for Handling

The following general rules apply to trash and garbage handling in the foodservice operation:

- Garbage and trash containers must be leakproof, easily cleanable, pestproof, and durable with tight-fitting lids. Today, plastic bags frequently are used for lining containers to facilitate disposal.
- Garbage and trash should not be allowed to accumulate anywhere but in containers.
- Garbage and trash should be removed from production areas on a frequent basis for appropriate disposal.
- Garbage storage areas should be easily cleanable and pestproof. If long holding times for garbage are required, these areas should be refrigerated to prevent decomposition, odor, and infestation by vermin.
- A garbage can washing area with hot water and a floor drain should be located away from food production and storage areas.

Mechanical devices are used in most foodservice facilities to assist in garbage and trash disposal. At a minimum, garbage disposal units should be available in prepreparation, dishwashing, and pot and pan washing areas. **Pulpers** are replacing garbage disposal units if water consumption and sewage use are concerns in the community. A pulper works somewhat like a garbage disposal except that it is designed especially for the disposal of cardboard, paper, and food waste. The pulper hydrates products into a slurry in a shredding device and then presses water out of it. The waste becomes a semidry, degradable pulp ready for disposal; the excess water is recycled in the pulping tank for reuse. Solid waste can be reduced by 85%, which means less space is used in a landfill. Pulpers are a good alternative when disposals cannot be used. Pulpers can handle paper trays, foam, foil, corrugated boxes, bones, food scraps, and some plastics. Pulpers require a lot of maintenance and cleaning. At the end of the night, the pulper must be broken down and cleaned; otherwise the wad of garbage would become harder than a rock. When considering disposals or pulpers, buyers should check the latest water/sewer department regulations.

Mechanical trash compactors (Figure 8.19) are used for dry bulky trash, such as cans and cartons. Compacting reduces the volume of trash to one fifth of its original bulk.

Solid Waste

A broad area of ethics involves the responsibility of a company to society, often termed *social responsiveness.* Most foodservice managers are concerned about environmental issues and are establishing programs to manage solid waste in their

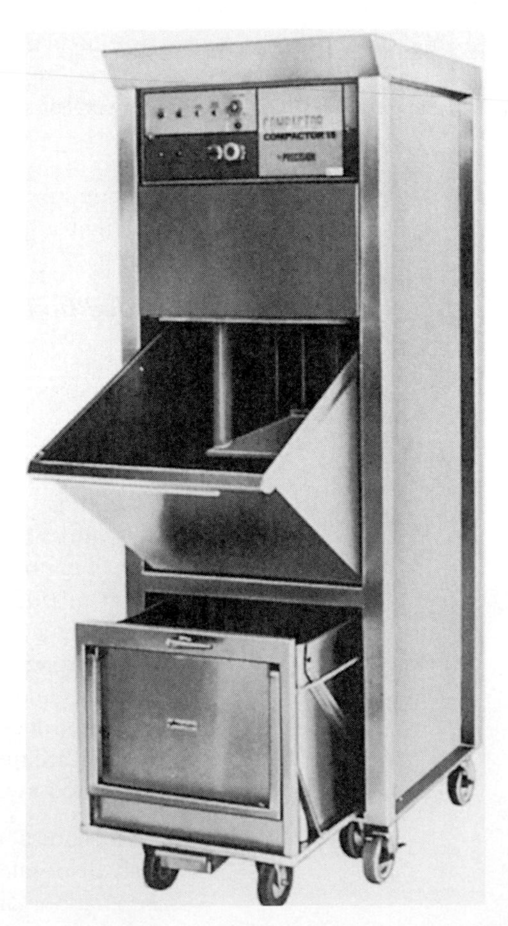

Figure 8.19. Trash compactor.
Source: Precision Metal Products, Inc., Miami, FL. Used by permission.

operations. Pollution prevention or reduction at the source was declared America's top priority in the Pollution Prevention Act passed by Congress in 1990. Pollution that cannot be prevented should be recycled in an environmentally safe manner. A goal to recycle at least 25% of the solid waste in America was established as part of this act. The Environmental Protection Agency developed a hierarchy of priorities for addressing solid waste management. The goal is to prevent pollution before it occurs (Mason, Shanklin, Hee Wie, & Wolfe, 1999). The hierarchy includes

- Source reduction
- Recycling, including composting
- Landfilling
- Incineration

Managing solid waste efficiently and effectively requires planning. An integrated solid waste management program includes

- Menu design and planning
- Purchase specifications
- Food production practices
- Service methods
- Portion control
- Waste-product disposal methods
- Consumer education
- Employee training

Generally about 60% to 70% of the solid waste discarded from a foodservice operation is service related (Hollingsworth, Shanklin, & Cross, 1995). Service waste includes food, napkins, straws, and condiment packaging. The remaining 30% to 40% is from the food production and preparation areas.

The American Dietetic Association position on conserving the environment (Shanklin & Hackes 2001) encourages environmentally responsible practices that minimize the quantity of waste that is generated. The position paper indicates that an average of 0.32 to 1.47 lb. of food and package waste are discarded per meal served in foodservice operations. Foodservice managers are encouraged to develop integrated waste management systems that include source reduction, recycling, and waste combustion to reduce the amount of waste going to landfills.

Every foodservice facility should have a solid waste management program. Many operations place all waste into a common dumpster. Others use a combination of garbage disposals and dumpsters. Some operations recycle part of their waste, whereas others compost some of their organic waste, such as food and paper.

Hospitals, nursing homes, schools, prisons, colleges and universities, and other onsite foodservice operations across the country are successfully recycling and composting their solid waste. *Food Management* conducted an informal survey of onsite foodservice operators about their waste management practices (King, 1995). The results showed that most (82%) of the onsite foodservice operators were

- Recycling
- Practicing source reduction
- Incinerating
- Composting

Some operators indicated that they would recycle but had no centers in their areas. The most common items being recycled were cardboard (84%), aluminum

(79%), and paper (78%). The most common source reduction method reported was a reusable mug or cup program (74%). In addition, operators were decreasing the use of disposables and purchasing in bulk. Challenges noted by operators included lack of recyclers in the area (39%) and lack of customer interest (32%). A survey (Su, Mason, & Shanklin, 1994) of Kansas school foodservice operators revealed the following:

- The majority of operators use garbage disposals to process food waste from the kitchen area.
- Most operations place service waste in the dumpster.
- Almost half of those surveyed collapse cardboard boxes before discarding.
- About 70% of the operators surveyed do not recycle. Lack of recycling centers in the area was the most common reason.

Recycling: Act of removing materials from solid waste stream for reprocessing into valuable new materials and useful products.

Composting: Controlled application of the natural process of organic degradation.

Combustion: Form of solid waste recycling in which the energy value of combustible waste materials is recovered.

Recycling is the second method for decreasing waste in foodservice operations. The Foodservice and Packaging Institute (1991) defines **recycling** as the act of removing materials from the solid waste stream for reprocessing into valuable new materials and useful products. Many of the traditional packaging materials, including paper, metals, plastic and glass, can be recycled. **Composting** is the controlled application of the natural process of organic degradation, according to the Composting Council (Crosby, 1993). Because of its contact with food, foodservice paper is ideal for composting. A commercial composting plant accelerates natural biodegradation, converting mixed organic waste to a nutrient-rich soil conditioner in great demand in agriculture and horticulture.

Combustion, or incineration, is a form of solid waste recycling in which the energy value of combustible waste materials is recovered (Council on Plastics and Packaging in the Environment, 1991). Modern waste-to-energy plants reduce the volume of waste going to landfills by 80 to 90% while generating electricity and revenue for users.

Some organizations and communities are being forced to explore alternative methods for solid waste management. Regionalization and transfer stations are becoming viable operations for many counties and municipalities. Combining resources may result in cost savings for the communities involved in regionalization plans.

Regionalization occurs when two or more counties or governments combine resources to site a landfill or develop a system for waste disposal for their communities. Some communities use transfer stations to manage their solid waste. After being collected and compacted at a transfer station, the solid waste is taken to a landfill in another area of the state or in a nearby state. Some communities are now transporting garbage more than 100 miles for legal disposal (Mason, Shanklin, Hee Wie, & Wolfe, 1999).

Bioremediation companies take advantage of a technological breakthrough to offer a biological solution to an old and nagging restaurant problem, clogged drains and grease traps (Yaffar & Dibner, 1992). Restaurant managers can use a drain cleaning company or a plumber whenever an emergency occurs or can hire a bioremediation company that can establish a chemical or biological preventive maintenance program. The **biological solution** is to use bacteria to break down animal fats and food products that clog drains. Naturally occurring organisms that grow on grease, flour, and other foods have been isolated. Once the food source is depleted, the organisms die and the environment returns to its original natural state. Engineering bacteria in the future is a possibility, but first the EPA must decide if releasing a new organism into a foodservice operation is acceptable. Using chemical or biological solutions on a regular basis provides preventive maintenance for a serious problem and is a known fixed cost in the budget.

Employee and Guest Facilities

Locker rooms should be provided for employees to change clothes. Individual lockers with locks are needed for storing street clothes and personal effects when employees are working. Adequate space and good lighting are necessary for changing clothes and for employee safety. Floors should be tile laid in cement or other nonabsorbent materials, especially in the toilet and hand-washing sink areas. Employee facilities should be located near the work area.

Guest rest rooms should be easily accessible from the dining room. In most commercial foodservice operations, the rest room foyer is decorated in harmony with the dining room. Walls might be papered and the floor might be carpeted in this area, but the toilet and handwashing sink areas should have tile floors and walls.

Keeping both employee and guest rest rooms clean can be a major management problem. Frequent management inspections are required, especially during changing of employee shifts or guest meal times, when traffic volume is the highest. These inspections are too important to delegate to an employee. Many customers will not return to a foodservice operation if they are dissatisfied with the cleanliness of a rest room. Employee rest rooms should be maintained at the same quality as those for guests.

MAINTENANCE

Facilities and equipment are important factors in any HACCP-based program. Poorly designed facilities and equipment make cleaning and sanitization difficult (NRA Educational Foundation, 2004). Public health, building, and zoning departments have the power to regulate the building of a facility and to approve plans before new construction or remodeling can be done. The U.S. Department of Justice and the Equal Opportunity Commission, in accordance with the Americans with Disabilities Act, have guidelines for facility design that include access to facilities by disabled customers and employees.

Foodservice managers should look for the **National Sanitation Foundation** (NSF) International mark or the **Underwriters Laboratories'** (UL) sanitation classification listings of commercial foodservice equipment that comply with those of NSF International (Figure 8.20). One of the standards of NSF International is that equipment is easy to clean. The familiar UL symbol is well known to building officials, electrical inspectors, public health officials, and consumers. UL has over 70 years' experience evaluating commercial gas and electric cooking appliances and refrigeration, food preparation, and processing equipment for manufacturers seeking safety and sanitation certification and energy-efficiency evaluations. UL staff members are experts at carrying out a variety of product investigations of foodservice industry equipment. In addition, UL has developed nearly 20 safety standards for the commercial foodservice industry (Underwriters Laboratories Inc., 1993).

Figure 8.20. Underwriters Laboratory® certification mark and National Sanitation Foundation® international certification mark on foodservice equipment.

Equipment meeting these standards can display the blue UL sanitation certification mark (Figure 8.20).

Local authorities provide checklists of desirable or necessary features for good sanitation. Well-planned preventive maintenance and pest control, therefore, are essential to keep the facilities and equipment free of contamination.

Preventive Maintenance

Preventive maintenance: Keeping equipment and facilities in a good state of repair.

Preventive maintenance is the process of keeping equipment and facilities in good repair. It has two aspects: regular cleaning schedules and standard procedures, and the preventive and corrective maintenance of foodservice equipment and facilities.

Some cleaning should be performed every day and included in the daily tasks of specific employees. Other cleaning tasks may be scheduled on a weekly, monthly, or less frequent basis, as appropriate to the operation, but must be done regularly for proper maintenance of the facilities. In some instances, specific scheduling of additional employees or perhaps specialized cleaning crews may be required. Whatever the schedules, however, proper tools, equipment, and cleaning materials are basic to an effective facility maintenance program and must be on hand as required.

All cleaning tasks should be combined in a master schedule that includes a list of what is to be cleaned, when each task should be done, how the task should be performed, and who has the assigned responsibility. Table 8.5 illustrates a partial master cleaning schedule for a food production area.

Specific cleaning procedures need to be developed to supplement the master schedule. Employees should be instructed in the procedures for cleaning foodservice equipment and in what the proper cleaning devices and materials are. Equipment cleaning procedures must be sufficiently detailed and presented in a step-by-step procedure to ensure that the correct process is followed and that any special precautions are heeded. Table 8.6 presents the procedure for cleaning a

Table 8.5 Sample Cleaning Schedule (Partial) for Food Preparation Area

Item	What	When	Use	Who
Floors	Wipe up spills	As soon as possible	Cloth, mop and bucket, broom and dustpan	________
	Damp mop	Once per shift, between rushes	Mop, bucket	________
	Scrub	Daily, closing	Brushes, squeegee bucket, detergent (brand)	________
	Strip, reseal	January, June	See procedure	________
Walls and ceilings	Wipe up splashes	As soon as possible	Clean cloth, detergent (brand)	________
	Wash walls	February, August		________
Work tables	Clean and sanitize tops	Between uses and at end of day	See cleaning procedure for each table	________
	Empty, clean and sanitize drawers, clean frame, shelf	Weekly, Sat. closing	See cleaning procedure for each table	________
Hoods and filters	Empty grease traps	When necessary	Container for grease	________
	Clean inside and out	Daily, closing	See cleaning procedure	________
	Clean filters	Weekly, Wed. closing	Dishwashing machine	________
Broiler	Empty drip pan; wipe down	When necessary	Container for grease; clean cloth	________
	Clean grid tray, inside, outside, top	After each use	See cleaning procedure for each broiler	________

Source: Applied Foodservice Sanitation, 4th edition (p. 217), by the Educational Foundation of the National Restaurant Association, 1992, Chicago: Author. Used by permission.

food slicer along with important safety precautions. Similar procedures need to be developed for all pieces of equipment. Manufacturers' instructions can be useful in developing these procedures.

Bacteriological counts on dishes, utensils, and equipment need to be performed on a regular basis as a check on sanitization. In some localities, health inspectors may perform these tests. In most onsite foodservices, these tests are performed by personnel in the facility. In a healthcare facility, for example, laboratory personnel usually have responsibility for bacteriological examinations in the dietetic services department on a regular basis without advance notice.

Keeping records on equipment maintenance often is a manager's headache; rising costs of these services add to the headaches (Frable, 1996b). Most foodservice managers have few records of the age, condition, service history, or maintenance requirements for equipment in their operations. Records for the number of repairs and cost of each provide crucial information when considering whether to repair or replace equipment. Another benefit of keeping inventory and service records is tracking warranties. The manager who has no records might be paying for parts and labor that are covered by extended warranties purchased with the equipment.

Some type of electronic database can help track this warranty information. An equipment maintenance program is a good way of documenting age, condition, and reliability of all equipment and mechanical systems. Some database software

Table 8.6 Sample Cleaning Procedures for a Food Slicer

When	How	Use
After each use	1. Turn off machine.	
	2. Remove electric cord from socket.	
	3. Set blade control to zero.	
	4. Remove meat carriage. Turn knob at bottom of carriage.	
	5. Remove the back blade guard. Loosen knob on the guard.	
	6. Remove the top blade guard Loosen knob at center of blade.	
	7. Take parts to pot-and-pan sink, scrub.	Manual detergent solution, gong brush.
	8. Rinse.	Clean hot water, 170°F (76.7°C) for 1 minute. Use double S hook to remove parts from hot water.
	9. Allow parts to air dry on clean surface.	
	10. Wash blade and machine shell by swabbing. CAUTION: PROCEED WITH CARE WHILE BLADE IS EXPOSED.	Use brush dipped in detergent solution or use a bunched cloth, folded to several thicknesses. Wear steel-reinforced gloves.
	11. Rinse by swabbing.	Clean hot water, clean bunched cloth.
	12. Sanitize blade, allow to air dry.	Clean water, chemical sanitizer, clean bunched cloth.
	13. Replace front blade guard immediately after cleaning shell, Tighten knob.	
	14. Replace back blade guard. Tighten knob.	
	15. Replace meat carriage. Tighten knob.	
	16. Leave blade control at zero.	
	17. Replace electric cord into socket.	

Source: Adapted in *Applied Foodservice Sanitation* (p. 219) by the Educational Foundation of the National Restaurant Association, 1992, Chicago: Author. From *Sanitary Techniques in Foodservice* (pp. 102–103) by K. Longrée and G. G. Blaker, 1982, New York: John Wiley. Used by permission.

permits an unlimited amount of equipment to be entered, which allows for scheduling of periodic maintenance and repairs while keeping notes on each.

Pest Control

Controlling pests in a foodservice operation is of critical importance. The presence of rodents and insects can be a serious problem because both are sources of contamination of food, equipment, and utensils. Many pests carry disease-causing organisms and cause considerable spoilage and waste. Also, nothing will turn off customers more than the sight of a mouse or cockroach running across the floor in the dining room. Pests can be controlled by following these three basic rules (NRA Educational Foundation, 2004):

- Deny pests food, water, and a hiding or nesting place.
- Deny pests access to the facility.

- Work with a licensed pest control operator (PCO) to eliminate pests that do enter.

A HACCP-based program will help ensure that potential contamination from pests does not threaten food safety. The foodservice manager should develop an integrated pest management (IPM) program that combines preventive tactics and control methods to reduce pest infestations (NRA Educational Foundation, 1999). Although roaches, flies, rats, and mice are the most common pests in a foodservice facility, beetles, moths, and ants may be a problem as well. Cockroaches head the list of the most frequent and bothersome insect pests and can infest any part of a foodservice operation. They carry disease-causing organisms such as *Salmonella,* fungi, and viruses. A PCO should be called immediately when cockroaches are seen because only a trained person can handle the situation. Cockroaches hide and lay eggs in places that are dark, warm, moist, and hard to clean, such as behind stoves and refrigerators and in cracks between ceiling tiles. All incoming shipments of foodstuffs should be inspected and any roach-infested goods refused.

Flies are a greater menace to human health than cockroaches. They transmit foodborne illnesses because they feed on human and animal wastes and garbage. Flies enter the facility primarily through outside doors or other external openings. Control can be facilitated by having tight-fitting and self-closing doors, closed windows, and good screening. Screened or closed storage for garbage is also important. Control, however, should be handled by a licensed PCO.

All openings to the outside must be protected against the entrance of rats and mice. Again, constant surveillance for signs of the presence of rodents is needed, and control of rats and mice should be left to a licensed professional PCO. These are formidable pests, and their control must be entrusted to professionals (NRA Educational Foundation, 2004).

Pesticides can be toxic, and some may cause fire or explosions if not handled properly. They must be used with extreme care and labeled and stored properly, away from food storage areas. Because pesticides can be hazardous to humans, the foodservice manager should rely on a PCO's knowledge as part of an IPM solution to a consistent pest problem. In the long run, good sanitation practices are the best form of pest control.

RISK MANAGEMENT

Safety, sanitation, and maintenance are important components in preventing accidents and illness in foodservice operations. They also are key components of what has become known as risk management.

Risk: Possibility of loss or injury.

Risk management: Discipline dealing with possibility that some future event will cause harm to an organization.

Risk is defined as the possibility of loss or injury. **Risk management** is a discipline for dealing with the possibility that some future event will cause harm to an organization. According to Tom Cippolone, director of risk management for Darden Restaurants, the role of the risk manager is to create the safest environment possible for employees and customers (Carlino, 2002). A risk manager often is responsible for overseeing operational security, ensuring workplace safety for workers and customers, managing litigation, and helping improve the bottom-line performance of the organization.

The Nonprofit Risk Management Center (2002) suggests that risk management basically involves answering three questions:

- What can go wrong?
- What will we do (both to prevent harm from occurring in the first place and to deal with the aftermath of an "incident")?
- If something happens, how will we pay for it?

Risk managers usually are charged with identifying potential areas of risk for an organization and then planning strategies to help reduce the likelihood of that risk. John Pinkerton, the person in charge of the risk management for Hard Rock Café, indicates that his unit's role is protection of people, property, and profits (Carlino, 2002).

Employee training is critical to the reduction of risks in an organization. Lack of proper employee training is the primary cause of injuries and the resulting damages, whether fiscal, physical, or mental (King, 2002). Employee training should include not only how to perform required aspects of the assigned job safely, but also how to be observant for situations that could cause harm to others, such as spills on floors, unprotected hot or sharp edges, and so on.

The role of risk management does not belong only to the person in an organization identified as the risk manager. All managers should be observant of their operation and of the potentials for loss or injury that might be in their operation.

CHAPTER SUMMARY

This summary is organized by the learning objectives.

1. The safety, sanitation, and maintenance subsystem includes activities that relate to food, employee, and customer safety and the sanitation and safety of foodservice facilities and equipment.
2. The terms clean and sanitary have different meanings. Clean is to be free from soil. Sanitary is to be free of disease causing organisms.
3. Several methods exist for solid waste management from a foodservice establishment. Items such as cardboard, paper, plastic, glass, and aluminum can be recycled; food products can be composted; and many items can be incinerated and the energy recovered for other uses.
4. The text includes many examples of policies for helping ensure food, employee, and customer safety. One example of a food safety policy is, "Food items should not be in the temperature danger zone (41°F–140°F) for more than 4 hours." An example of an employee safety policy is, "All spills shall be wiped up immediately." An example of a customer safety policy is, "Exits from the dining room shall be clearly identified."

TEST YOUR KNOWLEDGE

1. Explain the four types of food spoilage, and give examples of each.
2. What is a foodborne pathogen? Give three examples and explain each.
3. What is sanitation's role in the management of food quality?
4. Describe what regulations are in place to help with food safety.

Professional Organization Profile

International Association of Food Protection

MISSION

To provide food safety professionals worldwide with a forum to exchange information on protecting the food supply

WHO BELONGS TO THE ORGANIZATION

Members include more than 3,000 food safety professionals from more than 50 countries. Members work in industry, government, and academic institutions. Student memberships are available.

ADVANTAGES OF MEMBERSHIP

Members have access to an association website with a section for members only; receive the monthly publication, *Journal of Food Protection,* and monthly issues of the Food Protections Trends newsletter; can purchase food safety materials at a reduced price; and can attend association sponsored conferences and seminars.

WEBSITE

www.foodprotection.org

MEET THE PRESIDENT

Frank Yiannas, president 2006–2007, is Director of Safety and Health Programs at Walt Disney World. His pioneering work in developing food safety training and HACCP implementation at Disney lead to the corporation receiving the Black Pearl Award for corporate excellence in food safety from the International Association for Food Protection.

5. What is role of maintenance in the foodservice industry?
6. What act was passed by Congress in 1970 to ensure the safety of employees? How does this act ensure customer safety?
7. What is risk management? Why is it important to a foodservice operation?

CLASS PROJECTS

1. In groups of two or three people, research a foodborne pathogen and develop a handout for your classmates about what they, as future foodservice managers, should know about this pathogen.

2. Invite a local sanitarian to visit class and discuss how a sanitation inspection is done.
3. Work with a local school or assisted living foodservice service director and identify standard operating procedures that the operation does yet have documented (see www.iowahaccp.iastate.edu/sections/foodservice.cfm). Write at least two Standard Operating Procedures for that operation.
4. Using the Food Establishment Inspection Report form found in the *Food Code* 2005, work with a local foodservice director to complete an inspection of that operation.

WEB SOURCES

www.fsis.usda.gov	USDA Food Safety and Inspection Service
www.dhs.gov	U.S. Department of Homeland Security
www.foodsafety.gov/%7Edms/foodcode.html	*Food Code* 1997, 1999, 2001, 2005
www.cfsan.fda.gov	FDA Center for Food Safety and Nutrition
www.foodsafety.gov	U.S. Government Food Safety website
http://www.iowahaccp.iastate.edu/sections/foodservice.cfm	HACCP Information Center, Iowa State University
www.extension.iastate.edu/foodsafety	Iowa State University Extension Food Safety website
www.foodsafetyanswers.org	Online food safety answers
www.fightbac.org	Partnership for Food Safety Education
www.nal.usda.gov/fsrio/	Food Safety Research Information Office at the National Agriculture Library
www.foodsafety.cas.psu.edu	Pennsylvania State University Food Safety website
www.profoodsafety.org	Integrated Food Safety Information Delivery System
www.cdc.gov/foodsafety	Centers for Disease Control Food Safety website
www.nsf.org	National Sanitation Foundation
www.ul.com	Underwriters Laboratory
www.otherwhitemeat.com	National Pork Council
http://www.fsis.usda.gov/About%5FFSIS/NACMCF/	National Advisory Committee on Microbiological Criteria for Food
www.prionics.ch	Prionics AG Corporation

BIBLIOGRAPHY

Access Health, Inc. (1996). Hepatitis. [Online]. Available: http://www.yourhealth.com/ahl/1952.html

ADEC Environmental Sanitation and Food Safety. (1997). Safety at the salad bar. [Online]. Available: http://www.state.ak.us/dec/deh/sanitat/sansalad.htm

Alaska Sea Grant College Program. (1997). HACCP news releases. [Online]. Available: http://www.uaf.edu/map/haccp_releases.html

Allen, R. L. (1993). NCCR creates group to battle food fears. *Nation's Restaurant News, 27*(24), 7.

Allen, R. L. (1996). Foodservice operators cheer FDA's new HACCP seafood-safety program. *Nation's Restaurant News, 30*(2), 3, 61.

American School Food Service Association. (1995). Keys to excellence. Alexandria, VA: Author. [Online]. Available: www.schoolnutrition.org

American School Food Service Association. (1998). A history of the American School Food Service Association. [Online]. Available: http://www.asfsa.org/who/history.htm

AP Dictionary of Science and Technology. (1996). Pathogen. [Online]. Available: http://www.europe.apnet.com/inscight/05021997/pathoge7.htm

Beasley, M. A. (1993). Food safety through hygiene. *Food Management, 28*(5), 36.

Bendall, D. (1992). Keeping your kitchen safe and secure. *Restaurant Hospitality, 76*(11), 124, 126.

Boss, D., & King, P. (1993). SFM's waste management survey. *Food Management, 28*(3), 42, 44.

Bobeng, B. J., & David, B. D. (1977). HACCP models for quality control of entrée production in foodservice systems. *Journal of Food Protection, 40*(9), 632–638.

Bobeng, B. J., & David, B. D. (1978). HACCP models for quality control of entrée production in hospital foodservice systems. I: Development of Hazard Analysis Critical Control Point models II: Quality assessment of beef loaves utilizing HACCP models. *Journal of the American Dietetic Association, 73*(5), 524–535.

Bryan, F. L. (2001). Conducting effective foodborne disease investigations. *Food Safety Magazine, 7*(1), 23–26, 45–47.

Byers, B., Shanklin, C. W., Hoover, L. C., & Puckett, R. (1994). *Food service manual for health care institutions.* Chicago: American Hospital Publishing.

Carlino, B. (2002). Risk managers: Defusing explosive issues ranks as No. 1 priority. *Nation's Restaurant News.* Available online at www.nrn.com

Chaisson, C. F., Petersen, B., & Douglass, J. S. (1991). *Pesticides in food: A guide for professionals.* Chicago: The American Dietetic Association.

Cheney, K. (1993). Managing a crisis. *Restaurants & Institutions, 103*(13), 51, 56, 58, 62, 66.

Cody, M. M., & Kunkel, M. E. (2002). *Food safety for professionals: A reference and study guide* (2nd ed.). Chicago: The American Dietetic Association.

Council on Plastics and Packaging in the Environment. (1991). *Questions and answers on plastics, packaging and the environment.* Washington, DC: Author.

Crosby, M. A. (1993). Composting: Recycling restaurant waste back to its roots. *Restaurants USA, 13*(1), 10–11.

Crutchfield, S. R. & Roberts, T. (2000). Food safety efforts accelerate in the 1990s. *Food Review, 23*(3), 44–49. Available online at www.ers.usda.gov

Department of Health and Human Services. (1998). HHS: What we do. [Online]. Available: http://www.hhhs.gov/about/profile.html

Dkystra, J. J., & Schwarz, A. R. (1991). The race against germs. *School Food Service Journal, 45*(6), 92, 94, 98.

DOL officially declares meat slicers off-limits to 16–17-year-olds. *Washington Weekly, 11*(47), 4.

Dulen, J. (1998). HACCP becomes reality. *Restaurants & Institutions, 108*(3), 90.

Dykstra, J. J., & Schwarz, A. R. (1991) The race against germs. *School Food Service Journal, 45*(6), 92, 94, 98.

Environmental Protection Agency. (1988). *EPA report to Congress: Solid waste disposal in the United States.* Vols. 1 and 2. Washington, DC: U.S. Government Printing Office.

Environmental Protection Agency, Office of Solid Waste. (1989). *The solid waste dilemma: An agenda for action. Final report of the Municipal Solid Waste Task Force.* Washington, DC: U.S. Government Printing Office. EPA/530-SW-89-019.

Farquharson, J. (1998). Education, training key. *Food Management, 33*(9), 70.

FAS Online (1998). U.S. fruit and vegetable import: calendar year 1997. [Online]. Available: http://www.fas.usda.gov

Featsent, A. W. (1998). Food Fright! Consumers' Perceptions of Food Safety versus Reality. *Restaurants USA, 18*(6), 30–34.

Fontana, A. J. (2001). Water activity's role in food safety and quality. *Food Safety Magazine, 7*(1), 19–21, 57.

Food and Drug Administration. (1995). Center for Food Safety and Applied Nutrition. [Online]. Available: http://vm.cfsan.fda.gov/~MOW/cholstuf.html

Food and Drug Administration. (2005). U.S. Department of Health and Human Services/2005 Food Code. [Online]. Available: www.cfsan.fda.gov/~dms/fc05-toc.html

Foodservice & Packaging Institute, Inc. (1991). *Foodservice disposables: Should I feel guilty?* Washington, DC: Author.

Foodservice injury rates hover near average. (1992). *Washington Weekly, 12*(3), 4.

Frable, F. (1996a). New, low-cost pulpers merit a rethinking of waste disposal. *Nation's Restaurant News, 30*(5), 39, 51.

Frable, F. (1996b). Manage equipment maintenance, replacement with software. *Nation's Restaurant News, 30*(5), 39, 51.

Frumkin, D. (2002). Foodservice takes aim on violence in the workplace. *Nation's Restaurant News.* Available online at www.nrn.com

Government regulation of food safety: Interaction of scientific and societal forces. A scientific status summary by the Institute of Food Technologists' expert panel on food safety and nutrition. (1992). *Food Technology, 46*(1), 73–80.

GRAS status: What's in a name? (1992). *Food Insight* (March/April), 6–7.

Hannahs, G. (1995). Cryptosporidium parvum: An emerging pathogen. [Online]. Available: http://www.kenyon.edu/depts/biology/bio38/hannahs/crypto.htm

Hazard analysis and critical control point principles and application guidelines. (1998). *Journal of Food Production, 61*(9), 1246–1259. [Online]. Available: www.fsis.usda.gov/OPHS/nacmc

Herlong, J. E. (1991). The Heimlich maneuver: Part of a new standard of service? *Restaurants USA, 11*(2), 17–19.

Hollingsworth, M. D., Shanklin, C. W., & Cross, N. (1995). Waste stream analysis in seven selected school foodservice operations. *School Food Service Research Review, 19*(2), 81–87.

Illinois Department of Public Health. (1997). Healthbeat. [Online]. Available: http://www.idph.state.il.us/public/hb/hbcyclo.htm

Industry shows signs of improvement on a slippery issue. (1998). *Nation's Restaurants News.* Available online at www.findarticles.com

Kim, K. (2004). Restaurant industry's commitment to professionalism results in decrease in workplace injuries. Available online at www.restaurant.org

King, P. (1992). Implementing a HAACP program. *Food Management, 27*(12), 54, 56, 58.

King, P. (1993). Recycling & source reduction. *Food Management, 28*(1), 54–55, 58, 60.

King, P. (1995). Garbage wars '95: A mid-decade report. *Food Management, 30*(6), 82–86.

King, P. (2002). On-site feeders say customer safety is first and foremost. *Nation's Restaurant News.* Available online at www.nrn.com

Kleiner, A. (1991). What does it mean to be green? *Harvard Business Review, 69*(4), 38–42, 44, 46–47.

Liston, J. (1989). Current issues in food safety—especially seafoods. *Journal of the American Dietetic Association, 89*(7), 911–913.

Liston, J. (1990). Microbial hazards of seafood consumption. *Food Technology, 44*(12), 56, 58–62.

Longrée, K., & Blaker, G. G. (1982). *Sanitary techniques in foodservice* (2nd ed.). New York: Wiley.

Longrée, K., & Armbruster, G. (1996). *Quantity food sanitation* (5th ed). New York: Wiley.

Lorenzini, B. (1992). Institutional giants face the waste issue. *Restaurants & Institutions, 102*(7), 80–81, 88–89, 92, 96.

Mason, D. M., Shanklin, C. W., Hee Wie, S., & Wolfe, K. (1999). *Environmental issues: Impacting foodservice & lodging operations.* Manhattan, KS: Kansas State University.

McCarthy, B. (1993). All about burgers. *Restaurants and Institutions* 103(7):89.

McLauchin, A. (1993). Restaurants urged to cook ground beef thoroughly. *Restaurants USA* 13(3):7.

McSwane, D., Rue, N. R., Linton, R., & Williams, A. G. (2004). *Food safety fundamentals.* Upper Saddle River, NJ: Prentice Hall.

Minor, L. J. (1983). *Sanitation, safety & environmental standards.* Westport, CT: AVI.

National Advisory Committee for Microbiological Criteria for Foods. (1989). *HACCP principles for food production.* Washington, DC: U.S. Department of Agriculture, Food Safety and Inspection Service.

National Association of College and University Food Services. (1998). About NACUFS. [Online]. Available: http://www.nacufs.org/nacufs/about/

National Association of College and University Food Services. (1991). *National Association of College and University Food Services professional standards manual* (2nd ed.). East Lansing: Michigan State University.

National Food Service Management Institute. (2002). *HACCP for Child Nutrition Programs.* Oxford, MS: Author.

National Pork Producers Council. (2002). Cooking today's pork. [Online]. Available: http://www.otherwhitemeat.com

National Restaurant Association. (1979). *Sanitation operations manual.* Chicago: Author.

National Restaurant Association. (1981). *Safety operations manual.* Chicago: Author.

National Restaurant Association. (1984). *Sanitation operations manual.* Chicago: Author.

National Restaurant Association. (1986). *A restaurateur's guide to consumer food safety concerns. Current issues report.* Washington, DC: Author.

National Restaurant Association. (1989). *The solid waste problem. Current issues report.* Washington, DC: Author.

National Restaurant Association. (1991). *Make a S.A.F.E. choice: Sanitary assessment of food environment.* Washington, DC: Author.

National Restaurant Association. (2002). *A practical approach to HACCP.* Chicago: Author.

National Restaurant Association Educational Foundation (1992). *Applied Foodservice Sanitation* (3rd ed.) Chicago, IL: Author.

National Restaurant Association, Educational Foundation. (1995). SERVSAFE® Serving Safe Food: Certification Coursebook. Chicago: Author.

National Restaurant Association, Educational Foundation. (2004). Serv Safe® Coursebook 3rd Ed. Chicago: Author.

National Sanitation Foundation. (1982). *Recommended field evaluation procedures for spray-type dishwashing machines.* Ann Arbor, MI: Author.

National Sanitation Foundation. (1997). The National Sanitation Foundation: The standard for performance. [Online]. Available: http://www.culigansystems.com/nsf.html

New York City Department of Health Bureau of Communicable Disease. (1998). Trichinosis. [Online]. Available: http://www.ci.nyc.ny.us/nyclink/html

Nonprofit Risk Management Center. (2002). Risk Management Basics. [Online]. Available: www.nonprofitrisk.org

Olds, D. A., & Sneed, J. (2005). Cooling rates of chili using refrigerator, blast chiller, and chill stick cooling methods. [Online]. Available: http://docs.schoolnutrition.org/newsroom/jcnm/05spring/olds/index.asp

Opitz, A. (1992). Recycling: Making a world of difference. *School Food Service Journal, 46*(2), 24–25, 28, 30.

Patterson, P. (1993). Clean it or close it: Business demands strong sanitary focus. *Nation's Restaurant News, 27*(9), 44.

Pence, J. (1992). The greening of food service in North America. In *Food service and the environment: Conference proceedings* (pp. 1–12). Wellington, New Zealand: Victoria University of Wellington.

Puckett, R., & Norton, C. (2003). *Disaster and emergency preparedness in food service operations.* Chicago: American Dietetic Association.

Puzo, D. (1998). Food safety: The thin blue. *Restaurants & Institutions, 108*(11), 94–95, 100, 102, 104, 106.

Quinton, B., & Weinstein, J. (1991). Operations: Who's leading the green revolution? *Restaurants & Institutions, 101*(30), 32–35, 38, 44, 45, 54.

Riell, H. (1993). Pulpers get a new life at age 40. *Restaurants USA,* 13(3), 9.

Saving the planet. What school food service needs to know about going green. (1992). *School Food Service Journal, 46*(2), 20–21.

Shanklin, C. (1991). Solid waste management: How will you respond to the challenge? *Journal of the American Dietetic Association, 91,* 663–664.

Shanklin, C., & Hackes, B. (2001). Position of The American Dietetic Association: Dietetics professionals can implement practices to conserve natural resources and protect the environment. *Journal of the American Dietetic Association, 101,* 1221–1227.

Shanklin, C. W., & Hoover, L. (1993). Position of The American Dietetic Association: Environmental Issues. *Journal of the American Dietetic Association, 93,* 589–591.

Solid waste management resource guide. (1992). *School Food Service Journal, 46*(2), 42.

Somerville, S. R. (1992). Safety is no accident. *Restaurants USA,* 12(7), 14–18.

Spears, M. C. (1999). *Foodservice procurement: Purchasing for profit.* Upper Saddle River, NJ: Prentice Hall.

Spertzel, J. K. (1992a). The great dishroom debate: Permanentware vs. disposables. *School Food Service Journal, 46*(2), 36–38.

Spertzel, J. K. (1992b). Safety is no accident. *School Food Service Journal, 46*(4), 50–53.

Su, A., Mason, D. M., & Shanklin, C. W. (1994). Kansas school foodservice operators solid waste management practice. Unpublished raw data.

Testin, R. F., & Vergano, P. J. (1990). Packaging in America in the 1990s: *Packaging's role in contemporary American society: The benefits and challenges.* Herndon, VA: Institute of Packaging Professionals.

The complete HACCP manual for institutional food service operations (2nd ed.). (1997). Dunkirk, NY: Food Service Associates.

Thorner, M. E., & Manning, P. B. (1983). *Quality control in foodservice* (rev. ed.). Westport, CT: AVI Publishing.

Underwriters Laboratories, Inc. (1993). *Foodservice equipment.* Northbrook, IL: Author.

Unklesbay, N. (1977). Monitoring for quality control in alternate foodservice systems. *Journal of the American Dietetic Association, 71*(4), 423–428.

Unklesbay, N., Maxcy, R. B., Knickrehm, M., Stevenson, K., Cremer, M., & Matthews, M. E. (1977). Foodservice systems: Product flow and microbial quality and safety of foods. North Central Regional Research Bulletin No. 245. Columbia: University of Missouri-Columbia Agricultural Experiment Station.

U.S. Department of Agriculture. (1970). *Keeping food safe to eat, Home and Garden Bulletin No. 162.* Washington, DC: Author.

U.S. Department of Agriculture. (2004). Biosecurity Checklist for School Foodservice: Developing a Biosecurity Management Plan. [Online]. Available: http://schoolmeals.nal.usda .gov/Safety/biosecurity.pdf

U.S. Department of Agriculture. (2005). Guidance for school food authorities: Developing a school food safety program based on the process approach to HACCP principles. [Online]. Available: http://www.fns.usda.gov/cnd/Lunch/Downloadable/ HACCPGuidance.pdf

U.S. Department of Health, Education, and Welfare, Public Health Service, Food and Drug Administration. (1978). *Food service sanitation manual: 1976 recommendations of the Food and Drug Administration.* DHEW Publication No. (FDA) 78-2081. Washington, DC: Government Printing Office.

U.S. Food and Drug Administration. (2001). *Food Code 2001.* Washington, DC: Author. Available online at www.cfsan.fda.gov/~dms/fcol-toc.html

U.S. Food and Drug Administration Center for Food Safety and Applied Nutrition. (2002). The Bad Bug Book. [Online]. Available: www.cfsan.fda.gov

Way, L. (1992). OSHA test promotes foodservice safety. *Restaurants USA, 12*(4), 7.

Weinstein, J. (1991a). The clean restaurant. I: Physical plant. *Restaurants & Institutions,* 101(12), 90–91, 94, 96, 98, 100, 106–107.

Weinstein, J. (1991b). The clean restaurant. II: Employee hygiene. *Restaurants & Institutions, 101*(13), 138–139, 142, 144, 148.

Weinstein, J. (1992a). Greening of the giants. *Restaurants & Institutions, 102*(19), 146–150, 152, 156.

Weinstein, J. (1992b). The accessible restaurant. *Restaurants & Institutions, 102*(9), 96–98, 102, 104, 110, 117.

Wolf, I. D. (1992). Critical issues in food safety, 1991–2000. *Food Technology, 46*(1), 64–70.

Yaffar, E. A., & Dibner, M. D. (1992). Grease-eating microbes: A high-tech solution to a low-tech problem. *Cornell Hotel and Restaurant Administration Quarterly, 33*(6), 84–90.

CHAPTER
9

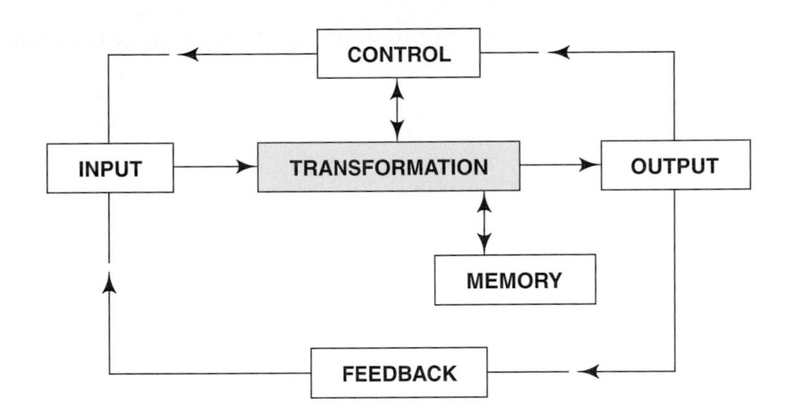

Management Principles

Enduring Understanding

- Management requires human, technical, and conceptual skills.
- Managers perform a variety of roles and functions.
- Successful managers are effective and efficient.
- Managers help set organizational culture.

Learning Objectives

After reading and studying this chapter, you should be able to

1. Define management-related terms such as effectiveness, efficiency, authority, responsibility, and span of control.
2. Describe the roles and functions of managers.
3. Differentiate among technical, human, and conceptual skills.
4. Draw an organizational chart for an organization.

Management is a critical element in helping transform inputs of the foodservice system into outputs. In this chapter, management concepts are reviewed and applied to the foodservice operation. The importance of management functions in the transformation element of the foodservice system is emphasized. Ways to structure the organization also will be discussed.

THE MANAGEMENT PROCESS

Management: Process of integrating resources for accomplishment of objectives.

Management has been defined as a process whereby unrelated resources are integrated into a total system for accomplishment of objectives. Management, involving the basic functions of planning, organizing, staffing, leading, and controlling, also was described as the primary force that coordinates the activities of subsystems within organizations. Management was explained by Robbins and Coulter (2005) as the process of coordinating work activities so that they are completed efficiently and effectively by working with and through other people. All these definitions underscore how important it is that managerial activity be directed toward achieving

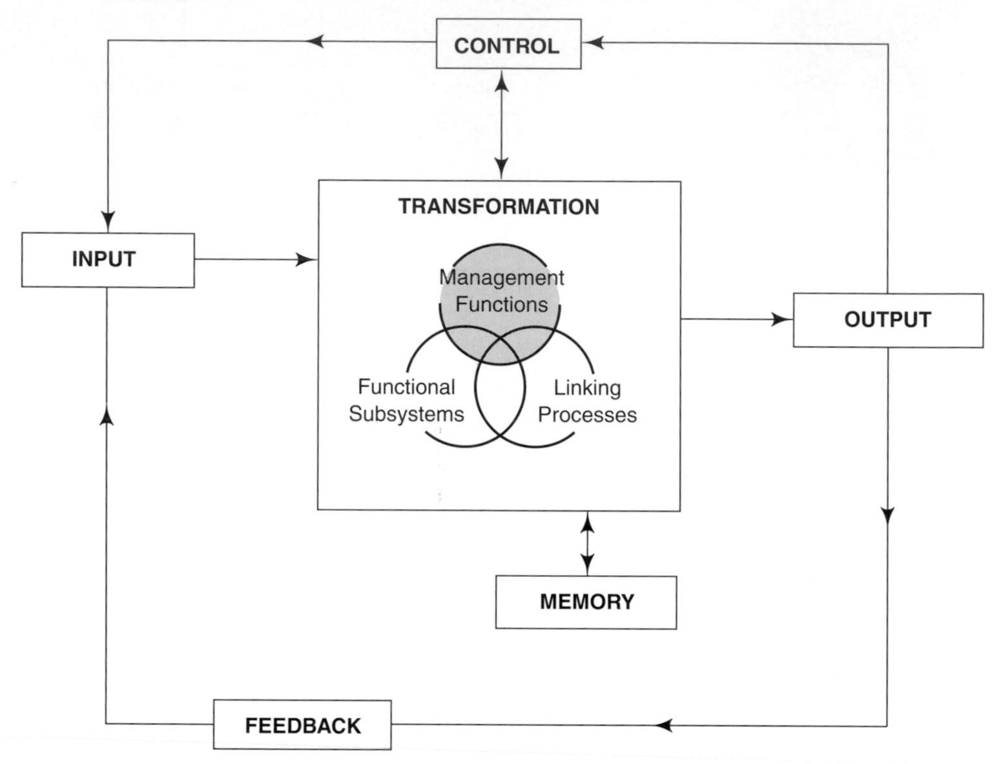

Figure 9.1. Management as part of the foodservice systems model.

the goals and objectives of the organization. As shown in Figure 9.1, management is part of the transformation process, turning inputs into outputs.

Managing Organizations

Although this book focuses on managing foodservice organizations, management concepts have broad applications because much of an individual's activity takes place within an organizational context. We are all members of a family, which is the basic unit of our society and the first organization in which most of us interact as members. We have spent a great deal of time in educational institutions and in informal groups. Such groups develop spontaneously when several people with mutual interests pursue a common objective, which could be a shopping expedition, a fishing trip, or a picnic. Work accounts for a large part of our time; professional organizations, social clubs, and churches provide activities that encompass much of our leisure time.

Organization: Group of people working together in a structured and coordinated way to achieve goals.

The tendency to develop cooperative and interdependent relationships is a basic human characteristic. All organizations, ranging on a continuum from informal ad hoc groups to formal, highly structured organizations, require managing. An **organization** is defined as a group of people working together in a structured and coordinated way to achieve goals (Griffin, 2001). Resources come together in an organization; the manager is responsible for coordinating them in a sensible way by acquiring, organizing, and combining resources to accomplish goals. Management is a set of activities (planning, organizing, staffing, directing, and controlling) directed at an organization's resources (human, materials, facilities, and operational) for achieving goals effectively and efficiently (Griffin, 2001).

Managerial Efficiency and Effectiveness

Management requires coordination of human and material resources while maintaining concern for morals, ethics, and ideals. Goals are determined by values and preferences, but the method for reaching them must be socially and morally acceptable. A manager's job is unpredictable and full of challenges, but it is also filled with opportunities to make a difference.

Many of the situations that contribute to managers' uncertainty about how to act stem from the environment in which organizations function. Of particular interest are those specific groups that are likely to affect the organization, such as owners, competitors, customers, suppliers, and regulators (Griffin, 2001).

Authority: Delegation from top to lower levels of management and the right of managers to direct others and take action because of their position.

Responsibility: Obligation to perform an assigned activity or see that someone else performs it.

Accountability: A state of being responsible to one's self, to some organization, or even to the public.

Efficiency: "Doing things right."

Effectiveness: "Doing the right things."

Authority, responsibility, and accountability are concepts important to the process of management. **Authority** is delegated from the top level to lower levels of management and is the right of a manager to direct others and take actions because of his or her position in the organization. **Responsibility** is the obligation to perform an assigned activity or see that someone else performs it. Because responsibility is an obligation a person accepts, it cannot be delegated or passed to another; essentially, the obligation remains with the person who accepted the responsibility. **Accountability** is the state of being responsible to one's self, to some organization, or even to the public. In the systems context, management was described as a process for accomplishment of objectives, implying, therefore, that accountability is an integral aspect of the managerial role.

Managers must show results in an era when scarce resources are an increasing concern. Efficient and effective use of these resources to produce desired results is a requisite for a viable organization. In contemporary jargon, **efficiency** is described as "doing things right," and **effectiveness** as "doing the right things."

According to Robbins and Coulter (2005), managerial efficiency, the ability to get things done correctly, is getting the most output from the least amount of input. Foodservice managers who can reduce the cost of food products to attain goals are acting efficiently. Effectiveness, in contrast, is the ability to choose appropriate objectives; an effective foodservice manager selects the right things to accomplish certain ends, such as interviewing customers to determine if quality expectations have been met.

The foodservice manager who plans a menu featuring grilled orange roughy when the customer would prefer fish and chips may be performing efficiently but not effectively. No amount of efficiency can compensate for lack of effectiveness. Drucker (1964), one of the first management authorities to discuss efficiency and effectiveness in relation to managerial performance, stated that the question is not how to do things right but how to find the right things to do. Thus, effectiveness is at the heart of accountability.

TYPES OF MANAGERS

The term *manager* has been used up to this point to refer to anyone who is responsible for people and other organizational resources. An organization, however, has different types of managers with diverse tasks and responsibilities. Managers can be classified by the level of their jobs in the organization and by the nature of their organizational responsibilities.

Managerial Levels

Most organizations have first-line, middle, and top managerial levels (Figure 9.2). First-line, or first-level, managers generally are responsible for supervising employees. In the foodservice organization, these managers usually are referred to as foodservice supervisors. Functional responsibilities may be indicated as part of their title. For example, in a college residence hall foodservice, first-line supervisors may be assigned to production, service, or sanitation.

In Chapter 1, a model (see Figure 1.3) was presented that showed three levels of an organization: technical, organizational, and policy making. Figure 9.2 shows how the management levels work within the organizational levels. First-line managers function at the technical core and are responsible for day-to-day operational activities.

Middle management may refer to more than one level in an organization, depending on the complexity of the organization. The primary responsibility of middle managers is to coordinate activities that implement policies of the organization and to facilitate activities at the technical level. Middle managers direct the activities of other managers and sometimes those of functional employees. This level of management also is responsible for facilitating communication between the lower and upper levels of the organization, and it functions at the organizational level.

In a college residence hall foodservice, one or more middle managers may be responsible for coordinating each of the various units that make up the foodservice operation. Multiple foodservice centers, each with a unit manager and often an assistant manager, are essential because of the size of the campus and the convenience of the customer. Within each of the units, the first-line managers report to the unit managers, who in turn report to the campus foodservice director.

Top managers make up the relatively small group of executives that control the organization. They develop the vision for the organization's future, are responsible for its overall management, establish operating policies, and guide organizational

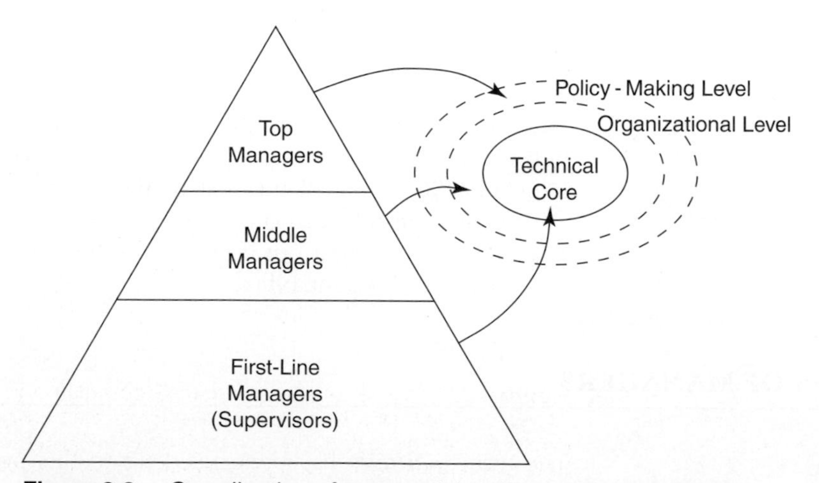

Figure 9.2. Coordination of managerial and organizational levels.

interaction with the environment. These managers operate at the policy-making level of the organization. Multiunit managers in restaurant chains are responsible for policy implementation, sales promotions, facility appearance and maintenance, financial control, and personnel management (Umbreit, 1989).

Educational Preparation of Managers

The lower-level managers in foodservice organizations are often employees who work up through the ranks of the organization and may not have formal management training. In the healthcare segment of the industry, however, many first-line supervisors have completed a 1- to 2-year training program for dietary managers at vocational technical schools, junior or community colleges, or by correspondence. Increasingly, in healthcare organizations, dietetic technicians are being employed as first-line supervisors because of the responsibility of patient nutritional care. Generally, a dietetic technician has completed a 2-year associate degree program.

Historically, middle- and upper-level managers in onsite foodservices have been the most likely to be professionally educated. The current trend, however, is to recruit managers with college degrees, especially in corporate offices of multiunit restaurant operations. Top managers of these operations or university foodservices generally are identified as chief executive officers (CEOs). Although the term *administrator* is frequently used, the titles of president or CEO are becoming more popular.

TQM Managerial Levels

Changes in the foodservice organization will be imperative if managers are planning to adopt a TQM philosophy. The traditional organization model is a pyramid with first-line managers as a base (see Figure 9.2). Inverting the pyramid provides a model for an organization committed to TQM implementation, as shown in Figure 9.3

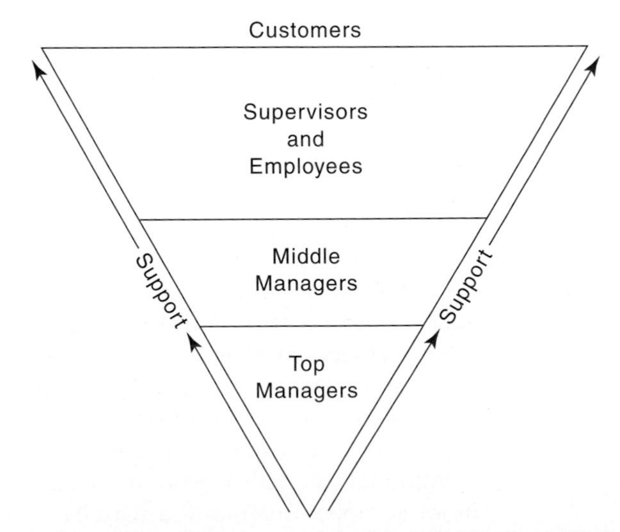

Figure 9.3. Model for an organization committed to TQM implementation.
Source: Adapted from "Employee Involvement in the Quality Process" by J. J. Gufreda, L. A. Maynard, and L. N. Lytle, 1990. In *Total Quality: An Executive's Guide for the 1990s* (p. 168), edited by E. C. Huge for Ernst & Young. Homewood, IL: Dow Jones-Irwin. Reprinted by permission of the publisher.

(Gufreda, Maynard, & Lytle, 1990). Note that employees have been added to this new management model. The first-line managers and employees become the most important workers in the organization because they are producing menu items and serving customers. Top and middle managers in their new roles as planners, coaches, and facilitators should support and guide the supervisors and employees.

Top managers must focus on creating a vision for the future of the foodservice organization by developing a change strategy (Huge & Vasily, 1990). The goal of the organization will be to satisfy customers, exciting them about the food and service. Dessler (2002) indicated that companies are reacting to competitive and technical change by

- Creating smaller organizational units
- Forming cross-functioning teams
- Empowering employees to make decisions
- Reducing the number of organization levels
- Placing an emphasis on vision and values
- Finding ways to take advantage of the Internet

Robbins and Coulter (2005) suggested that organizations of the future will be flexible, customer oriented, skills focused, dynamic, and involvement oriented. In many cases, traditional top-management activities will need to be shared and performed by others in the organization.

Managers are now asked to create an environment that encourages all employees to solve problems and make improvements and that empowers them to implement solutions. Managers should share business or competitive information with all employees because they need to understand where their organization stands in terms of profit and loss and market share. Only then can employees make good decisions that fit into the mission of the organization (Gufreda, Maynard, & Lytle, 1990).

General and Functional Managers

Earlier in this chapter, managers were classified by their level in the organization; managers also can be classified according to the range of organizational activities for which they are responsible. In this second classification, managers can be considered either general or functional managers.

A **general manager** is responsible for all the activities of a unit. In a restaurant, everything that happens on a specific shift is the responsibility of the general manager on duty. A **functional manager** is responsible for only one area of organizational activity, such as the bar. If the bartender is absent, the bar manager must make arrangements for coverage. If, however, the bar manager is absent, the general manager would be responsible for covering the position or appointing someone to take the manager's place.

Although a small organization may have only one general manager, a larger, more complex organization may have several. A college foodservice director and unit managers and assistant unit managers at university foodservice centers typically are all considered general managers. Depending on the size of the units, two or more functional managers may be responsible for various areas of activity within each of the units.

ROLES OF MANAGERS

Mintzberg (1980) described the manager's job in terms of various roles, which he referred to as organized sets of behaviors identified with a position. He depicted the manager's position as being composed of 10 different but closely related roles, shown in Figure 9.4. The formal authority of a manager gives rise to interpersonal, informational, and decisional roles.

Interpersonal Roles

Interpersonal roles of figurehead, leader, and liaison focus on relationships. The *figurehead* role has been described by some management experts as the representational responsibility of management. By virtue of a manager's role as head of an organization or unit, ceremonial duties must be performed and may involve a written proclamation or an appearance at an important function. For example, a manager's ceremonial tasks may include greeting a group of touring dignitaries or signing certificates for a group of employees who have completed a training program.

The manager in charge of an organization or unit also is responsible for the work of the staff; this constitutes the *leader* role. Functions of this role range from hiring and training employees to creating an environment that will motivate the staff. Mintzberg (1975) contended that the influence of the manager is seen most clearly in the role of leader. Although formal authority vests the manager with great potential power, leadership determines, in large measure, how much is realized. A manager must encourage employees and assist them in reconciling personal needs with organizational goals.

The manager also must assume the interpersonal role of *liaison* by dealing with people both inside and outside the organization. Managers must relate effectively to peers in other departments of the organization and to suppliers and clients. Depending on a manager's level in the organization, responsibility for liaison relationships will vary. In Mintzberg's (1975) research, 44% of the time that company chief executives spent with people was spent with people outside their organizations. The liaison role is important in building a manager's information system.

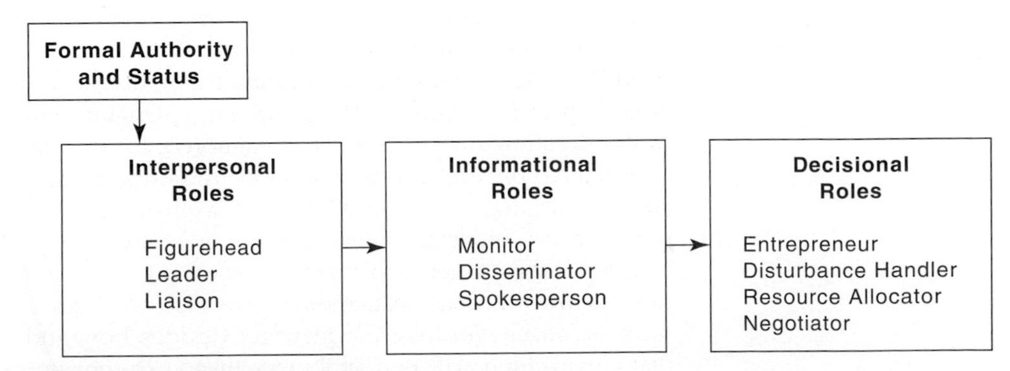

Figure 9.4. Managerial roles.
Source: Adapted from "The Manager's Job: Folklore and Fact" by H. Mintzberg, 1975, *Harvard Business Review, 53*(4), p. 55. Used by permission.

Informational Roles

Mintzberg (1975) suggested that communication may be the most important aspect of a manager's job. A manager needs information to make sound decisions, and others in a manager's unit or organization depend on information they receive from and transmit through the manager. According to Mintzberg (1975), the **informational roles** of a manager are those of monitor, disseminator, and spokesperson.

As *monitor,* the manager constantly searches for information to use to become more effective. The manager queries liaison contacts and subordinates and must be alert to unsolicited information that may result from the network of contacts previously developed. The manager collects this information in many forms and must discern implications of its use for the organization.

In the *disseminator* role, the manager transmits information to subordinates who otherwise would probably have no access to this information. An important aspect of this role is to make decisions concerning the information needs of staff members. The manager must assume responsibility to disseminate information that helps staff members become well informed and more effective.

The *spokesperson* role of the manager is closely akin to the figurehead role. In the spokesperson role, the manager transmits information to people inside and outside the organization or unit. For example, the director of dietetics in a hospital should keep the administrator up to date about problems in the department, and the food and beverage manager in a hotel should relay information to the general manager. The spokesperson role may also include providing information to legislators, suppliers, and community groups.

Decisional Roles

The manager occupies the major role in decision making within the organization. Because of vested formal authority, a manager may commit the unit to new courses of action and determine unit strategy. As Mintzberg (1975) indicated, informational roles provide a manager with basic inputs for decision making. The **decisional roles** include those of entrepreneur, disturbance handler, resource allocator, and negotiator.

As *entrepreneur,* the manager is the voluntary initiator of change. The entrepreneur role may involve, for example, a decision to change the menu after networking with other restaurateurs or customers.

In the role of *disturbance handler,* the manager responds to situations that are beyond his or her control. In this role, the manager must act because the pressures of the situation are too severe to be ignored; for example, a strike looms, or a supplier fails to provide goods or services. Although a good manager attempts to avoid crisis situations, no organization is so well run or systematized that every contingency in the uncertain environment can be avoided. Disturbances may arise because poor managers ignore situations until a crisis arises; good managers also must deal with occasional crises.

As *resource allocator,* the manager decides how and to whom the resources of the organization will be distributed. In authorizing important decisions, the manager must be mindful of the needs of the unit while considering priorities of the overall operation. Such decisions often will require compromise.

In the *negotiator* role, the manager participates in a process of give and take until a satisfactory compromise is reached. Managers have this responsibility because only they have the requisite information and authority to develop complex contracts with suppliers or less formal negotiations within the organization. For example, the unit manager of a limited-menu restaurant chain might negotiate with the parent company about local advertising.

MANAGEMENT SKILLS

Katz (1974) identified three basic types of skills—technical, human, and conceptual—which he said are needed by all managers. The relative importance of these three skills varies, however, with the level of managerial responsibility. Katz defined a **skill** as an ability that can be developed and that is manifested in performance. He described the manager as one who directs the activities of others and undertakes the responsibility for achieving certain objectives through these efforts. Technical, human, and conceptual skills are interrelated, but they are examined separately in the following paragraphs.

Technical Skill

A **technical skill** involves an understanding of, and proficiency in, a specific kind of activity, particularly one involving methods or techniques. Such skill requires specialized knowledge, analytical ability, and expertise in the use of tools and procedures. Managers need sufficient technical skill to understand and supervise activities in their areas of responsibility. For example, the foodservice manager must understand quantity food production and operation of equipment. Managers must have technical expertise to develop the right questions to ask subordinates and the abilities to evaluate operations, train employees, and respond in crisis situations.

Human Skill

Human, or **interpersonal, skill** concerns working with people and understanding their behavior. Human skill, which requires effective communication, is vital to all the manager's activities and must be consistently demonstrated in actions. As Katz (1974) indicated, human skill cannot be a "sometime thing." Such skillfulness must be a natural, continuous activity that involves being sensitive to the needs and motivations of others in the organization.

Katz (1974) described two aspects of human skill: leadership within the manager's own unit and skill in intergroup relationships. This description of human skill is similar to Mintzberg's (1975) interpersonal roles of leader and liaison. Both authors emphasized the importance of a manager working effectively with staff within the organizational unit and with people outside the unit. The campus foodservice director described previously must work effectively with unit managers within the department and the housing director, head of maintenance, and campus purchasing director.

Conceptual Skill

Conceptual skill is the ability to view the organization as a whole, recognizing how various parts depend on one another and how changes in one part affect other parts. Conceptual skill also involves the ability to understand the organization within the environmental context; a good example is the relationship of the organization to other similar organizations and to suppliers within the community. It also includes understanding the impact of political, social, and economic forces on the organization. From this description, conceptual skill is obviously a systems approach to management. A manager needs conceptual skill to recognize how the various forces in a given situation are interrelated to ensure that decisions are made in the best interest of the overall organization.

In summary, Katz (1974) stated that effective management depends on three basic skills: technical, human, and conceptual. Adequate technical skill is needed to accomplish the mechanics of the job, sufficient human skill is necessary in working with others to enable development of a cooperative effort, and conceptual skill is required to recognize interrelationships of factors involved in the job.

Managerial Levels and Skills

Although all three skills are important at every managerial level, the technical, human, and conceptual skills used by managers vary at different levels of responsibility (see Figure 9.5). Technical skill is most important at the lower levels of management, identified as nonsupervisory and supervisory by Hersey, Blanchard, and Johnson (2000), and it becomes less important in the higher levels. The nonsupervisory level includes employees who participate in on-the-job training of other employees. The foodservice production supervisor at the supervisory level, for example, is called upon to use technical skills frequently in supervising employees in daily operations. These technical skills are important in evaluating products, in training employees, and in problem solving.

The middle manager at the managerial level uses technical skills in performing the tasks of evaluating operations and selecting employees who have appropriate skills to perform various jobs. Also, in crisis situations, the middle manager's technical skills may be called into action. Top-level managers at the executive level, al-

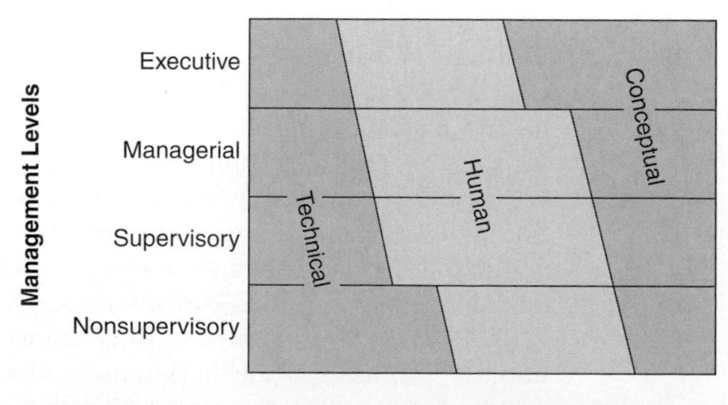

Figure 9.5. Management skills necessary at various organizational levels.
Source: Adapted from *Management of Organizational Behavior* (p. 8) by P. Hersey and K. Blanchard, 1993, Englewood Cliffs, NJ: Prentice Hall. Copyright 1993. Used by permission.

though generally not involved in daily operations, need understanding of technical operations to enable effective planning.

Human skill, the ability to work effectively with others, is essential at every level of management, as reflected in Figure 9.5. The first-line manager, who is responsible for daily supervision of operating employees, must be effective in guiding and leading these individuals to accomplish the activities for which they are responsible. These employees must be motivated to produce quality products, to serve customers cheerfully, and to wash dishes properly. Morale and satisfaction are important to each employee's effective performance.

Middle managers, because of their pivotal role in the organization, must be especially accomplished in human skills. These managers must effectively lead their own groups and appropriately relate to other parts of the organization. At the top level, the manager must be equally effective in dealing with people outside the organization.

The importance of conceptual skill increases with movement up the ranks of the organization. The higher a manager is in the hierarchy, the greater the manager's involvement in broad, long-range decisions affecting large parts of the organization. At this level, conceptual skill becomes the most important one for successful performance.

MANAGEMENT FUNCTIONS

The work performed by managers has been described in many ways. One of the more common is to organize managerial work into what has been called "functions." The five management functions are planning, organizing, staffing, leading, and controlling. Managers perform these functions in the process of coordinating activities of the subsystems of the organization.

Planning

Plans, which are the result of the managerial process of planning, establish organizational objectives and set up procedures for reaching them. Plans provide for acquiring and committing resources to attain objectives and for assigning members their activities. Plans also provide standards for monitoring performance of the organization and taking corrective action when necessary.

Definition of Planning

Planning: Management functions of determining in advance what should happen.

Planning is defined as determining in advance what should happen. Planning is essential as a manager organizes, staffs, leads, and controls. For example, a supervisor or manager prepares the menu, which is a basic plan that indicates the organization of the food preparation unit (organizing), the number (staffing) and assignments (leading) of employees, and the quality and cost of the product (controlling).

A hierarchy of plans is shown in Figure 9.6. The initial plans are the goals and objectives of the organization, thus providing the basis for objectives of the various subsystems. **Goals** represent the desired future conditions that individuals, groups, or organizations strive to achieve (Kast & Rosenzweig, 1985). **Objectives** are

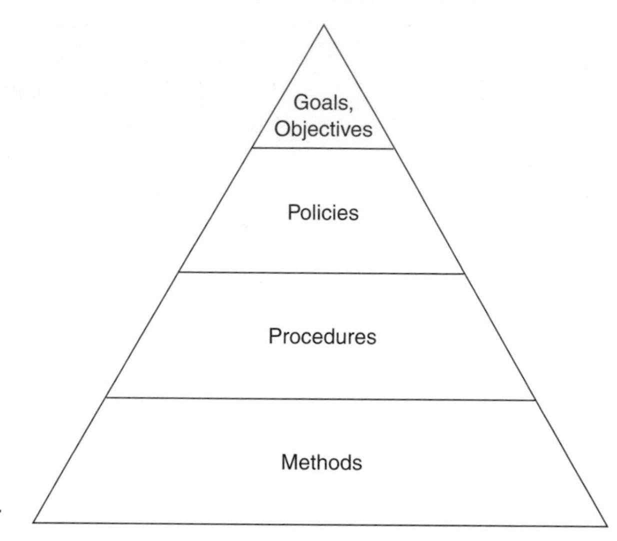

Figure 9.6. Hierarchy of plans.

merely goals, or end points, and set the direction for all managerial planning (Fulmer, 1988). Once objectives are determined, specific plans such as policies, procedures, and methods can be established for achieving them in a more systematic manner. **Policies** are the guidelines for action in an organization, and **procedures** and **methods** define steps for implementation.

Policy: General guide to organized behavior developed by top-level management.

Procedure: Chronological sequence of activities.

Method: Details for one step in a process.

An organization will have only a few broad plans but many specific plans, as depicted in Figure 9.7. District school foodservices, for example, may have only two broad goals, one concerned with provision of nutritious meals within federal and state guidelines and budgetary constraints, and another with nutrition education. Many policies would be needed, however, to achieve these goals and assure uniformity of operations throughout the various schools in the district. An even greater number of procedures would be needed to give school foodservice employees specific instructions on implementation of policies.

Dimensions of Planning

Kast and Rosenzweig (1985) identified four dimensions of planning: repetitiveness, time span, level of management, and flexibility. All are important components to the planning process.

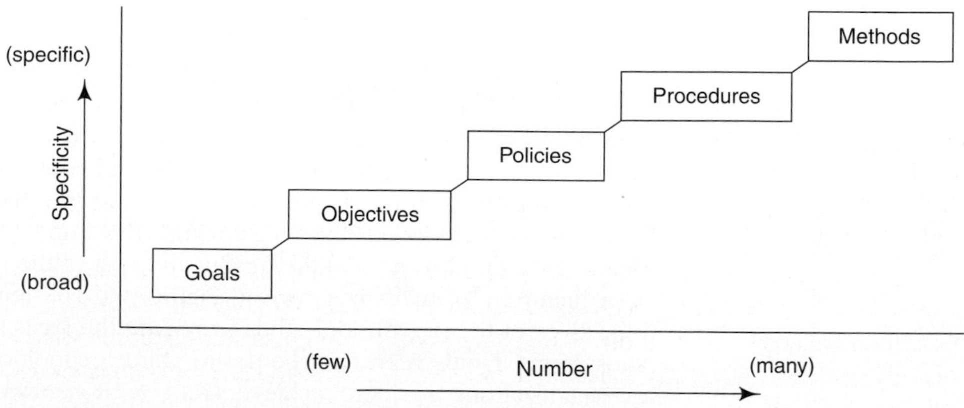

Figure 9.7. Relationship of number and specificity of plans.

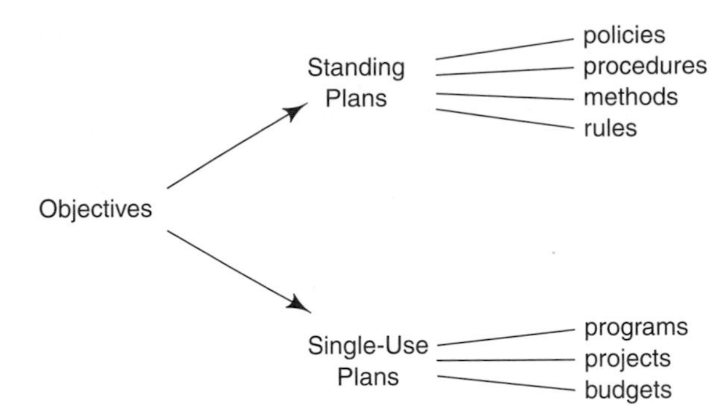

Figure 9.8. Relationship of standing and single-use plans to objectives.

Repetitiveness. Figure 9.8 shows the relationship of standing and single-use plans to organizational objectives. **Standing plans,** or plans for repetitive action, are used over and over again; **single-use plans,** also called single purpose, are not repeated but remain as part of historical records of the organization. Standing plans result in policies, operating procedures, methods, and rules, all of which are a primary cohesive force connecting the various subsystems of an organization.

A **policy,** the broadest of the standing plans, is a general guide to organizational behavior developed by the governing body or top-level management. Organizations should have a wide variety of policies covering the most important functions. Frequently these policies are formalized and available in policy and procedure manuals. Characteristics of policies and procedures are listed in Table 9.1.

Procedures and methods establish more definite steps for the performance of certain activities and are developed especially for use at the technical level of the system. A **procedure** shows a chronological sequence of activities; a **method** is even more detailed, relating to only one step of a procedure.

Rules: Specification of action, stating what must or must not be done.

Rules specify action by stating what must or must not be done, whenever or wherever they are in effect. Some examples of rules are prohibitions against smoking and the requirements to wear a specific uniform and hair restraint in the production area of a foodservice operation. A policy and its procedures for a hospital department of dietetics are shown in Figure 9.9.

An advantage of the standing plan is that it ensures uniformity of operations throughout the system. Once established, understood, and accepted, the standing plan provides similarity of action in meeting certain situations; on the negative side, however, standing plans may create resistance to change.

Management by exception is an important concept in relation to standing plans. Although standing plans serve as guidelines for decision making, upper levels of management must become involved whenever the application of policy is questioned.

Table 9.1 Characteristics of Policies and Procedures

Policies	Procedures
Guide decision making throughout the organization	Specify guides to action
Delimit an area within which a decision can be made	Delineate steps in descending order for completion of a task
Activate goals and objectives of the organization	Order sequential actions for performance of workers
Give direction for action	

Single-use, or single-purpose, plans are designed to attain specific objectives, usually within a relatively short period of time. A single-use plan in a foodservice organization might be a major program for the design, development, and construction of a central food-processing facility for a restaurant chain; a plan for a "monotony breaker" in a college residence hall foodservice; or a plan for a New Year's Eve celebration at a country club.

Time Span. The time span for planning refers to short-range versus long-range planning. Short-range, or operational, planning typically covers a period of 1 year or less. The operating budget for a year is one example of a short-range plan. Long-range planning in most organizations encompasses a 5-year cycle; however, a longer time span may be essential for some aspects of planning, such as a major building program. Long-range planning begins with an assessment of the current conditions and projections about changes. Managers must be able to see the connections between actions in one place and consequences in another (Kanter, 1992). Effective long-range planning requires a mission statement of the long-range vision of the organization.

The model for long-range planning shown in Figure 9.10 indicates the progression from premising to planning to implementing and reviewing the resulting plan. In the premising phase, the basis for the plan is considered in terms of the mission and opportunities of the organization. The planning phase consists of developing

POLICY: SCHEDULING TIME OFF

Policy Statement: Prior supervisory approval is required when scheduling time off in order for an employee to obtain a satisfactory attendance record. Work is scheduled to satisfy department workload and to accommodate employee's need for time off. This policy supplements the hospital-wide absenteeism policy.

Procedure:

1. Time off should be requested from the supervisor at least two weeks prior to the posting of the work schedule in which the absence will occur.
    - Time off will be granted whenever possible, depending on the needs of the department and the availability of adequately trained substitute workers.
    - In general, no more than one employee from any area will be granted time off at any one time. Requests usually will be considered on a first come, first serve basis.
2. Occasionally, time off will be requested after the work schedule has been posted. In such cases, supervisory approval will be granted for the requested time off only if ALL the following conditions are met:

    The substitute employee

    - is asked to work by the person initiating the request;
    - has worked the position previously;
    - does not accrue overtime except with supervisor's permission; and
    - is not allowed to work more than six (6) days in a row, or three (3) weekends in a row, because of the change.

Figure 9.9. An example of a policy and its procedures for a small hospital.

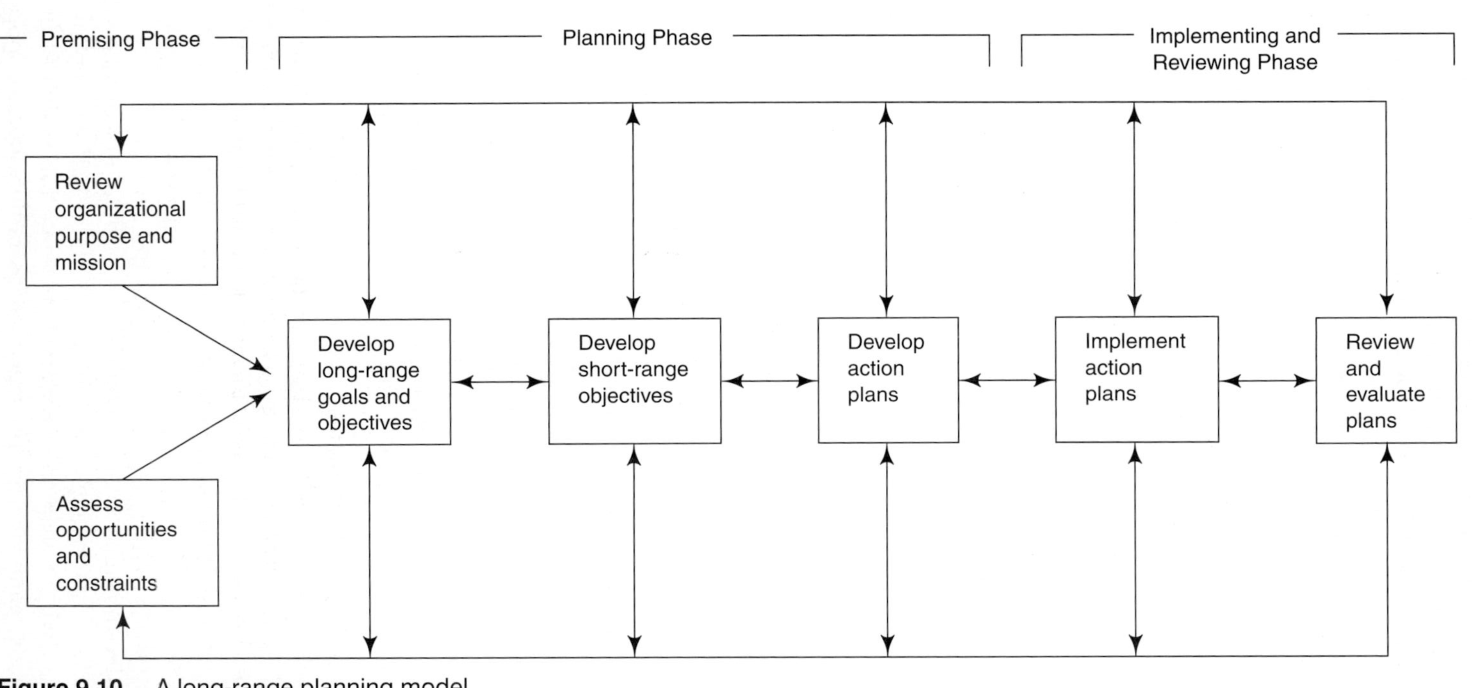

Figure 9.10. A long-range planning model.

long-range goals and objectives, short-range objectives, and action plans logically leading to implementation of the long-range plan. At this time, a final review and evaluation are necessary and may result in revision.

Strategic planning is a continuous and systematic process in which people make decisions about intended future outcomes, how outcomes are to be accomplished, and how success is measured and evaluated. In the early 1920s, Harvard Business School developed the Harvard Policy Model, one of the first strategic planning methodologies for private businesses. The model defines strategy as a pattern of purposes and policies defining the company and its business. A **strategy** is the thread or underlying logic that holds a business together. The firm weaves purposes and policies in a pattern that unites company resources, senior management, market information, and social obligations. Strategies determine organizational structure and appropriate strategies that will lead to improved economic performance.

As used today, strategic planning has a strong connotation of overcoming obstacles, as can be seen in the derivation of the word strategy from the Greek *stratego,* meaning to plan the defeat of an enemy through effective use of resources. In modern business terminology, an organization must develop a competitive edge over its rivals by planning the effective use of personnel, materials, facilities, and operational resources. The outcome of the strategic planning process is a brief working document that unifies action of participants toward achievement.

Strategic planning and long-range planning for organizations are often used synonymously (Bryson, 1995). While there may be little difference in outcome, in practice they usually differ in four fundamental ways.

- While both focus on an organization and what it should do to improve its performance, strategic planning relies more on identifying and resolving issues, while long-range planning focuses more on specifying goals and objectives and translating them into work programs.
- Strategic planning emphasizes assessment of the environment outside and inside the organization far more than long-range planning.
- Strategic planners are more likely than long-range planners to summon forth an idealized version of an organization and ask how it might be achieved.
- Strategic planning is more action oriented than long-range planning. Strategic planners usually consider a range of possible futures and focus on the implications of present decisions and actions in relation to that range.

An example of strategic planning is the introduction of prepackaged salads by McDonald's (Alva, 1988). Long-range planning was predicated upon the desirability of serving salads; strategic planning was involved with overcoming the flaws of salad bars, namely, the difficulty of quality control and the impossibility of service from a drive-through window. During the strategic planning stage, McDonald's tested various ways to present salads for 10 years before launching its prepackaged products. The goal was quality control and satisfactory service of salads for its drive-throughs, which account for more than 50% of total sales. Originally, McDonald's prepackaging consisted of a single- or half-size serving in a transparent plastic dish with a snap-on cover and a separate foil pouch of dressing. After a massive media campaign, prepackaged salads rapidly became popular. The effect of McDonald's advertising campaign was a public response that induced other limited-menu restaurant operators to follow suit. Burger King, Hardee's, Arby's, KFC, and

Wendy's International currently offer prepackaged salads in response to competitive pressures. In 2004, the salad concept was expanded with the introduction of the premium salads.

Level of Management. A relationship exists between the hierarchy of plans and the level of management involved in the planning effort (Figure 9.11). Generally, top managers, who function at the policy-making level of the organization, as shown in Figure 9.11, are responsible for broad, comprehensive planning involving goals and objectives. Middle managers, at the organizational or coordinative level, are responsible for developing policies; first-line managers at the technical or operational level are responsible for developing procedures and methods.

The differences in planning responsibility associated with managerial levels help explain the required skills distribution needed by managers. In the discussion of the three managerial skills—technical, human, and conceptual—the importance of conceptual skills at top levels of the organization and, conversely, of technical skills at the lower level were emphasized. The responsibility of these managers for broad versus specific operational plans should make these concepts clear. Because long-range and strategic plans are broad and time oriented, upper levels of managers must have conceptual ability, enabling them to view the overall organization in relation to its environment. Similarly, the first-line manager must be well versed on technical operations because of the responsibility for planning daily production and service activities.

To apply these concepts to a foodservice organization, the top management of a large national limited-menu restaurant company is concerned with issues such as identifying sites for new locations, assessing the impact of adding new menu items on costs, revenues, and profits, and projecting capital required for expansion. The manager of one of the units, however, is concerned with scheduling employees, predicting the impact of bad weather on customer traffic, and ordering an adequate amount of frozen yogurt for the next day. As managers move up in the organization, they must develop skills in long-range planning.

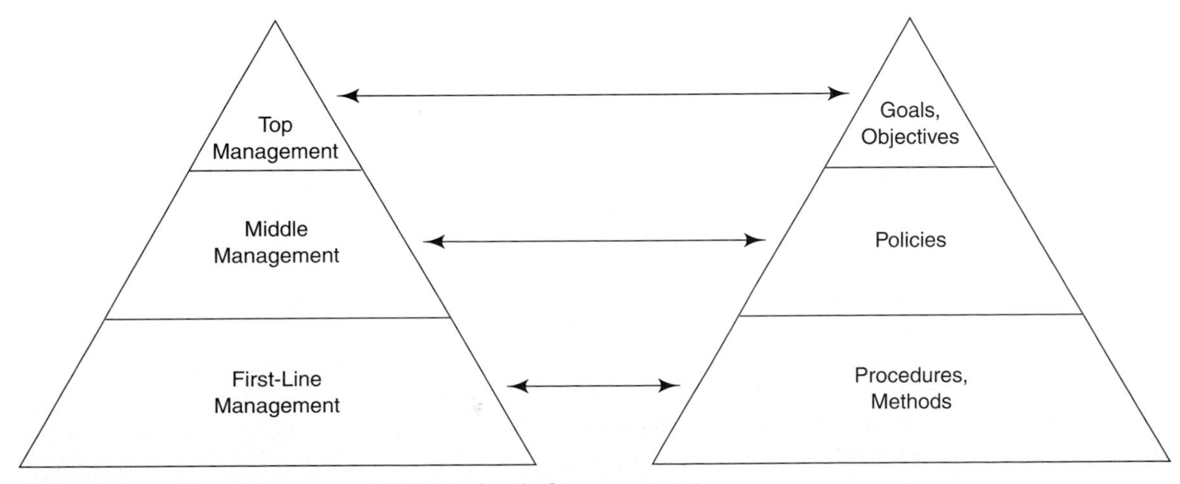

Figure 9.11. Planning responsibilities by level of management.

Flexibility. One of the major considerations in planning is the permissible degree of flexibility. Long-range planning involves decision making that commits resources over an extended period of time. Rapidly changing technology, competitive and market situations, and political pressures make forecasting extremely difficult. Rigid planning at early stages involves the risk of inability to cope with changes. Organizations may have to compromise on rigidity versus flexibility by developing relatively fixed short-range operation plans and more flexible long-range strategic plans.

Organizing

Organizing: Management function of grouping activities, delegating authority, and coordinating relationships, horizontally and vertically.

After developing objectives and plans to achieve them, managers must arrange the work to achieve these objectives. **Organizing** is the process of grouping activities, delegating authority to accomplish activities, providing for coordination of relationships, and facilitating decision making (Robbins & Coulter, 2005). A more complete discussion of the organizing function is included later in this chapter.

Formal Organization

The outcome of organizing is the development of the formal organization, which is usually depicted in the form of a chart. An example of an organizational chart for a university residence hall foodservice department is shown in Figure 9.12. Once managers have established objectives and developed plans to reach them, they must design an organization to activate these plans.

Different objectives will require different kinds of organizations. For example, an organization for a limited-menu restaurant operation will be far different from

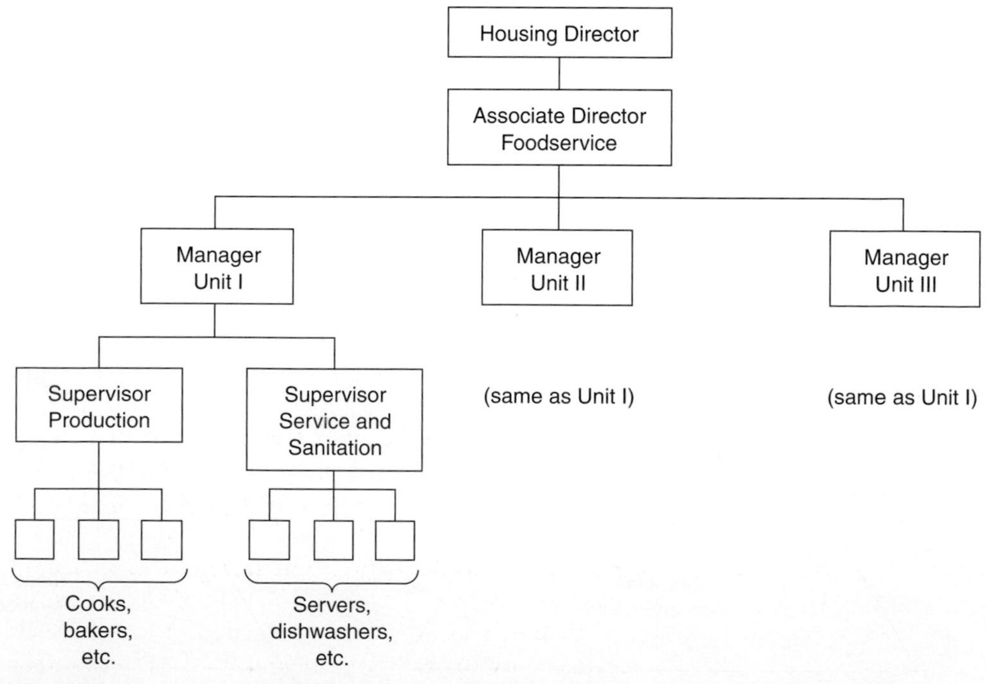

Figure 9.12. Organization chart for a university residence hall foodservice department.

one for an upscale gourmet restaurant. Similarly, the organization of a 50-bed nursing home foodservice department will differ markedly from that of a 500-bed teaching hospital.

The TQM philosophy includes review of organizational structure as an important component. According to Hamilton (1993), a more flexible organizational structure that removes layers of management, empowers employees, tears down communication barriers, and fosters workforce creativity is needed. Managers must have the ability to accomplish a given set of objectives. The process of organizing is determining the way in which work is to be arranged and allocated among organization members for effective attainment of goals.

Concepts of Organizing

Span of management (span of control): Number of employees that can be effectively supervised by one manager.

Organizing is a dynamic process based on two concepts, span of management and authority. **Span of management**, often referred to as **span of control**, is concerned with the number of people any one person can supervise effectively. Because that number is limited, organizations must be departmentalized by areas of activity, with someone in charge of each area. Authority is the basis through which work is directed.

Span of Management. Managers cannot supervise effectively beyond a limited number of persons, and it is difficult to determine the appropriate number. Several factors are involved in determining the proper number, which is referred to as the span of management. The factors are the following:

- **Organizational policies.** Clearly defined policies can reduce the time managers spend making decisions; the more comprehensive the policies, the greater the span of management.
- **Availability of staff experts.** Managers can have increased span if staff experts are available to provide advice and services.
- **Competence of staff.** Well-trained workers can perform their jobs without close supervision, thus freeing competent managers to expand their span of management.
- **Objective standards.** In organizations with objective standards and standardized procedures, workers have a basis by which to gauge their own progress, thus allowing managers to concentrate on exceptions. As a result, larger spans are possible.
- **Nature of the work.** Less complicated work tends to require less supervision than more complicated work. Generally, the simpler and more uniform the work, the greater the possible span.
- **Distribution of workforce.** The number of areas where supervised workers are on duty may inhibit severely a manager's ability to visit all work sites. The greater the dispersion of workers, the shorter is the span.

Authority, Responsibility, and Delegation. Authority is defined as the right of a manager to direct others and to take action because of the position held in the organization. This authority is delegated down the hierarchy of the organization as designated by upper management. A sound organizing effort, therefore, includes defining job activities and scopes of authority for each position in the organization.

As indicated previously, **responsibility** is a concept closely related to authority and refers to an obligation for performing an assigned activity. In accepting a job, a person agrees to discharge the duties or be responsible for their accomplishment by others. Because responsibility is an obligation a person accepts, it cannot be delegated or passed to a subordinate; that is, the obligation remains with the person who accepted the job. Authority, however, must be delegated to enable individuals to carry out their responsibilities or obligations. Without proper authority, first-line and middle managers may find completion of delegated job activities difficult.

Delegation: Process of assigning job activities and authority to a specific employee within the organization.

Delegation is the process of assigning job activities to specific individuals within the organization. Through this process, authority and responsibility are transferred to lower-level personnel within an organization. In a sense, delegation is the essence of management because management has been defined as getting work done with and through other people.

Failure to delegate is a weakness common to many managers. They often do not delegate enough because of the time and effort required to communicate to others. All too frequently, managers believe the saying, "If you want a job done right, do it yourself."

Effective delegation, however, has many advantages. It is one of the most important means managers have of developing the potential of their subordinates; also, as subordinates accept additional responsibility, managers are freed for planning and other tasks that require conceptual skill. For effective delegation to occur, the following three elements are important:

- Specific tasks must be assigned clearly. A manager must communicate the nature of a task to ensure that a subordinate clearly understands it.
- Sufficient authority must be granted. The subordinate must be given the power to accomplish the assigned duties. This authority must be understood by the subordinate and also by others whose cooperation is required to complete the task.
- Responsibility must be created. A sense of accountability must be engendered in the subordinate to ensure responsible completion of the assigned task. The subordinate must be empowered to make decisions on quality issues to meet customer expectations.

Organizing is the division of labor. Within organizations, labor can be divided both horizontally through departmentalization and vertically through the delegation of authority. In designing organizations, line and staff authority relationships are created. Generally, line personnel are in a linear responsibility relationship; a superior has supervision over a subordinate. Staff personnel serve in an advisory capacity to line managers.

With the growing complexity and size of organizations, the staff role has become especially important. Staff personnel may function to extend the effectiveness of line personnel, as the administrative assistant does for an organization president. Other staff personnel provide advice and service throughout an organization. As an example, the personnel manager relieves the line manager of such functions as recruiting and screening new employees. Staff may also provide advice or services to only a segment of an organization. A dietetics consultant in a nursing home provides advice and counsel primarily to the dietetics services department.

The distinction between line and staff is not as clear as it might appear in the foregoing paragraphs. In many organizations, managers have both line and staff responsibilities. In a small foodservice operation, separate staff probably would not be available to perform such functions as personnel and quality control; thus the manager would perform these functions in addition to being responsible for production of goods and services. In larger organizations, middle managers may be advisers to various other units or departments in the organization. For example, a food and beverage manager may serve on a TQM team for an entire hotel.

Mintzberg (1989) agreed that the distinction between line and staff is becoming blurred. He stated that an innovative organization has an administrative component and an operating core. The administrative component consists of line managers and staff experts and the operating core of employees. In the past few years, administrative positions have been eliminated and the responsibility for decision making has been shifted to the operating core. CEOs who support a TQM program have given administrative responsibility in the form of empowerment to operation employees. Their ultimate goal is to satisfy customers while reducing costs.

Staffing

Staffing: Management function of determining the appropriate number of employees needed by the organization for the work that must be accomplished.

The most valuable resources of an organization are its human resources—the people who provide the organization with their work, talent, drive, and commitment. Among the most critical tasks of a manager is **staffing**, the recruitment, selection, training, and development of people who will be most effective in helping the organization meet its goals. Competent people at all levels are required to ensure that appropriate goals are pursued and that activities proceed in such a way that these goals are achieved.

In the organizing process, various jobs in the organization are defined. The staffing process then involves a series of steps designed to supply the right people to the right positions at the right time. This process is performed on a continuing basis because organizational personnel change over time due to resignation, retirement, and other reasons.

In many organizations, staffing is carried out primarily by a personnel department. The responsibility for staffing, however, lies with line managers. Every line manager, even one not involved in recruiting and selecting personnel, is responsible for training, development, and other aspects of staffing.

Staffing consists of several steps (Figure 9.13): human resources planning; recruitment and selection; orientation, training, and development; performance appraisal; and compensation. Closely linked to these steps are a variety of staffing functions concerned with maintenance of the workforce. These functions include promotions, demotions, transfers, layoffs, and dismissals. **Human resources planning** is designed to ensure that the organization's labor requirements are met continuously. It is a process involving both forecasting of staffing needs and analysis of labor market conditions.

Recruitment and **selection** are concerned with developing a pool of job applicants and evaluating and choosing among them. These processes have become increasingly complex because of legislative employment mandates. They are discussed in Chapter 12.

Orientation, training, and **development** are processes designed first to acquaint newcomers with the organization and its goals and policies and to inform

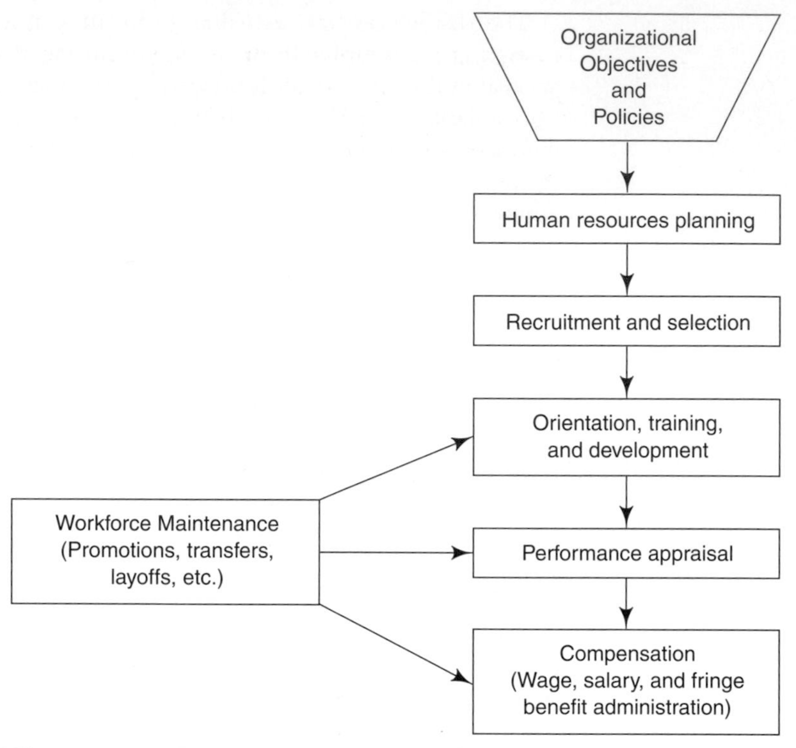

Figure 9.13. Steps in the staffing process.

them of their responsibilities. Later, training is designed to improve job skills, and development programs are used to prepare employees for increased responsibilities.

Performance appraisal is concerned with comparison of an individual's performance with established standards for the job. It also involves determination of rewards for high performance and corrective action to bring low performance in line with standards. Rewards may include bonuses, pay increments, or more challenging work assignments. Additional training is often necessary for low performers.

Compensation encompasses all activities concerned with administration of the wage and salary program. Fringe benefits are an important part of the program and include insurance programs, leave time, and retirement programs.

Directing

Directing: Management function of directing human resources for the accomplishment of objectives.

The directing function of management involves directing and channeling human effort for the accomplishment of objectives. **Directing** is the human resources function particularly concerned with individual and group behavior. All managing involves interaction with people and thus demands an understanding of how we affect and are affected by others. When managers direct others, they use that understanding to accomplish tasks through the work of other members of the organization.

Management and leadership often are thought of as synonymous terms. Directing the work of others is certainly one of the most critical functions of management, and success in management is closely related to success in directing, which also has been described as the interpersonal aspect of managing.

Directing is primarily concerned with creating an environment in which members of the organization are motivated to contribute to achieving goals. It has many dimensions, including morale, employee satisfaction, productivity, and communication. As stated in Chapter 1 in the discussion of personnel satisfaction as a system output, managers must be concerned with assisting employees in achieving personal objectives at work and with coordinating personal and organizational objectives.

The traditional view of the organization is centered on chain of command, negative sanctions, and economic incentives to motivate workers. In the late 1920s, the famous Hawthorne studies conducted by Western Electric and Harvard University researchers revealed that the factors that influenced worker performance included such things as social and psychological conditions, informal group pressure, participation in decision making, and recognition (Roethlisberger & Dickson, 1956). Since that time, the behavioral sciences have added new dimensions to the understanding of motivation and behavior in the workplace. Today, directing is viewed as being concerned with interpersonal and intergroup relationships. The role of the manager includes influencing these relationships to create cooperation and enlist commitment to organizational goals. The evolution of a better-educated workforce today has significantly increased the use of participative management in organizations. In Chapter 10, these and other aspects of leadership, including factors affecting leadership style, are described in more detail.

Controlling

Controlling: Management function of ensuring that plans are being followed.

Controlling is the process of ensuring that plans are being followed. It involves comparing what should be done with what was done and then taking corrective action, if necessary. Controlling must be a continuous process that affects and is affected by each of the other managerial functions. For example, the goals, objectives, and policies established in the planning process become control standards. When comparing performance to these standards, the need for new goals, objectives, and policies may become obvious. Effective organizing, staffing, and leading result in more effective control. Likewise, more effective control also leads to better organizing, staffing, and leading. Within this interrelatedness and interdependency of managerial functions, controlling relates most closely to planning.

Standards created in the planning process define the dimensions of what is expected to happen. These expected performance standards are the criteria that managers use to control performance; in turn, feedback from the controlling process is the information managers use to evaluate and adjust plans (Figure 9.14). The

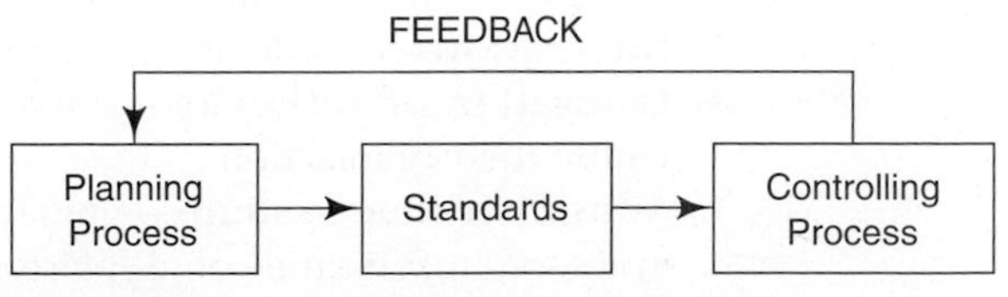

Figure 9.14. The planning-controlling cycle.

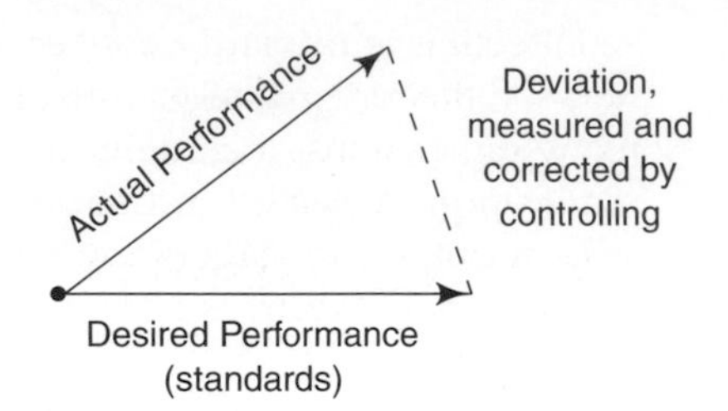

Figure 9.15. The controlling function of management.

controlling function of management involves the following three steps, as depicted in Figure 9.15:

- Measuring actual performance and comparing it with desired performance or standards
- Analyzing deviations between actual and desired performance and determining whether or not deviations are within acceptable limits
- Taking action to correct unacceptable deviations

Taking corrective action is a process that cuts across both the leading and controlling functions because many deviations from expected standards are related to performance of personnel. For example, fewer portions than expected from a particular recipe might be caused by a foodservice worker using an inappropriate portioning tool.

ORGANIZATIONAL STRUCTURE

An organization is defined as a group of people working together in some form of coordinated effort to attain objectives. An ideal organization results in the most efficient use of resources.

The organization structure is based on the objectives that management has established and on plans and programs to achieve these objectives. Different types of structures will be required for traditional and new organizations, each with different objectives.

The Traditional Organization

Kast and Rosenzweig (1985) stated that the **traditional organization** frequently is defined in terms of the following:

- **Organization chart and job descriptions or position guides**—pattern of formal relationships and duties
- **Differentiation or departmentalization**—assignment of various activities or tasks to different units or people of the organization
- **Integration**—coordination of separate activities or tasks
- **Delegation of authority**—power, status, and hierarchical relationships within the organization
- **Administrative systems**—guidance of activities and relationships of people in the organization through planned and formalized policies, procedures, and controls

One of the primary reasons for organizing in the traditional organization is to establish lines of authority, which create order. Without delineation of authority, there may be chaos, in which everyone is telling everyone else what to do.

The New Organization

In the **new organization,** employers are challenged to improve the quality of work life and to develop a corporate, or organizational, culture (Sherman, Snell, & Bohlander, 1997). Employees today are concerned about having a full-time job, but they are also concerned about their personal lives. Day care, pregnancy leave, and family leave have become important to employees.

Quality of Work Life

To improve the **quality of work life (QWL)** in the organization, managers need to look at the way work is organized and the way jobs are designed. Each organization has special problems and designs jobs to solve those problems. Some general guidelines, adapted from the Ontario, Canada, Quality of Life Centre (cited in Sherman, Snell, & Bohlander, 1997), include the following:

- Decisions are made at the lowest possible level.
- Teams of employees are responsible for a complete job.
- Technical and social potential of employees and the organization is developed.
- Quality and quality control are components of production.
- Safety and health of employees are emphasized.
- Immediate feedback of information required to perform a job is available.
- Problems are solved by teams, but responsibility for decisions is shared by all levels of the organization.

Participative management: Involving employees in the decision-making process.

Participative management, leader-member relations, self-managed teams, and team-based leadership are becoming more prominent in all sizes of operations. **Participative management** involves empowering employees to participate in decisions about their work and employment conditions (Sherman, Snell, & Bohlander, 1997). The concept is not new, but it had not been acted upon until recently, primarily because managers feared they would lose their authority. Both managers and employees must realize that if the organization is to survive, they must work together to reduce costs and avoid becoming victims of foreign and domestic competition. Too often problem-solving groups focus too much on participation and neglect other potentially better solutions. Only when they carefully consider all options can groups reach their full potential of goal achievement and quality of life. In addition, leadership skill can be an important impediment to achieving group effectiveness. Some managers who can handle one-on-one situations cannot lead a group. Group leadership requires more skill than leading individuals.

Leader-member relations refer to the nature of the relationship between the leader and work group. If the leader and the group have a high degree of mutual trust, respect, and confidence, and if they like one another, relations are assumed to be good. If there is little trust, respect, or confidence, and if they do not like each other, relations are poor. Naturally, good relations are more favorable (Griffin, 2001).

Most management writers agree that **leadership** is the process of influencing the activities of an individual or a group in efforts toward goal achievements in a given situation (Hersey, Blanchard, & Johnson, 1996). In an organization, a team is a group of workers that serves as a unit, often with little or no supervision (Griffin, 1999). Work accomplishment is from committed people; interdependence in organization purpose leads to relationships of trust and respect (Hersey, Blanchard, & Johnson, 2000). **Team-based leadership,** therefore, is two or more people who interact regularly to accomplish a common purpose or goal to be considered a group. Today, teams are sometimes called **self-managed teams** because they do the daily work (Hersey, Blanchard, & Johnson, 1996). Organizations create teams for many reasons. They give more responsibility to the workers who perform the task. Also, workers are given more authority and decision-making freedom. Ideally, teams will become very cohesive groups and high performers.

Corporate Culture

Corporate culture (organizational culture): Shared philosophies, values, assumptions, beliefs, expectations, attitudes, and norms that knit an organization together.

Culture has become a buzzword in business, but its contribution to the success or failure of a foodservice operation is so important that every owner or manager should learn what culture is and how to use it (Fintel, 1989). **Corporate culture**, or **organizational culture**, is defined as the shared philosophies, values, assumptions, beliefs, expectations, attitudes, and norms that knit an organization together (Sherman, Snell, & Bohlander, 1997).

Every company has a culture, but it is not always well defined. The most successful companies have adopted a positive culture, one that values its employees and treats them as part of a team. Positive cultures have the following qualities in common (Fintel, 1989):

- **Integrity:** Building trust between people in the organization
- **Bottom-up style of management:** Involving employees as part of the team
- **Having fun:** Finding ways both at work and outside of work for fun
- **Community involvement:** Participating in community service programs
- **Emphasis on physical health and fitness:** Practicing a belief that a sound mind goes along with a sound body

Although culture starts at the top, to mean anything it must be passed on to employees at all levels. Managers are encouraged to talk about the organization's culture constantly. Some organizations post signs containing the mission statement or their culture principles for all employees to see. Pizza Hut's culture is called "ownership," the feeling that comes from knowing an employee can affect the company's direction through expertise, innovative ideas, and hard work.

Today the most significant challenge facing restaurant operators is recruiting and retaining quality employees (Eating Place Trends, 1998). Some operators are concerned with the shrinking labor pool but must be proactive to make sure their operation will thrive in today's marketplace (Cain, 1998). Developing a good relationship between management and staff by giving rewards for good job performance, being sure the workers receive training they need, and providing consistent feedback are all important components of a caring culture that enables employees to do their jobs better. The caring culture checklist suggests that foodservice operations

- Have a written mission statement.
- Remind employees of the mission statement.
- Have a hands-on style of management.

- Foster open relationships between management and employees.
- Empower hourly employees.
- Give incentives for superior performance.
- Recognize superior performance.
- Give employees an ownership stake in the business.

Adhering to a caring culture checklist increases the odds that employees will work for a longer tenure and former employees will have a positive view of the industry.

The best place to start in developing a caring culture is to formalize a mission statement and to be sure that each employee understands and believes in it, thus showing employees how important their performance is to the operation's overall success. When employees understand that the overall success of the operation depends on them, management must then support and guide them to help them reach their workplace goals. This requires a hands-on style of management, a style that facilitates open communication between management and employees. Developing a reward system between employees and management also is part of creating a caring culture.

Evidence of how a company's corporate culture is articulated in mission and value statements can be found on many corporate websites. Olive Garden (www.olivegarden.com), for example, includes the following statement about their company and their employee principles on their website:

> Olive Garden is a family of local restaurants focused on delighting every guest with a genuine Italian dining experience. We are proud to serve fresh, high-quality Italian food, complemented by a great glass of wine. We offer a comfortable, home-like setting where guests are welcomed like family and receive warm, friendly service.
>
> Employee principles:
>
> 1. We are committed to open and honest communication, mutual respect and strong teamwork.
> 2. We are clear on each individual's role, accountabilities and key performance measures.
> 3. We do not compromise standards in selection, training and job performance.
>   - We only hire people with the skills and potential to succeed.
>   - People advance from training only when they have demonstrated the required competencies. Peer interviewing helps us select the right people for the team.
> 4. In the process of making a change we seek the opinions of those closest to the action, listen and value their ideas.
> 5. Everyone should expect regular, ongoing training opportunities to sharpen and advance their skills.
> 6. When accountabilities are not being met, we act quickly.
>   - Feedback
>   - Re-training/Redirection
>   - Assess results with appropriate consequences
> 7. We will achieve the results and share our successes.

Another large international corporation, McDonald's (www.mcdonalds.com), offers the following vision and strategies on its website:

> McDonald's vision is to be the world's best quick-service restaurant experience. Being the best means providing outstanding quality, service, cleanliness and value, so that we make every customer in every restaurant smile. To achieve our vision, we are focused on three worldwide strategies:

- Be the best employer for our people in each community around the world.
- Deliver operational excellence to our customers in each of our restaurants.
- Achieve enduring profitable growth by expanding the brand and leveraging the strengths of the McDonald's system through innovation and technology.

Both companies show a strong commitment to their employees and customers in their statements, yet each has unique characteristic phrases that help differentiate it from other companies.

Division of Labor

Organizing is basically a process of division of labor, which can be divided either vertically or horizontally. Organizing also improves the efficiency and quality of work, as the coordinated efforts of people working together begin to produce a synergistic effect. As defined in Chapter 1, synergy means that the units or parts of an organization acting in concert can produce more impact than by operating separately. Synergism can result from division of labor and from increased coordination, both of which are products of organization. Improved communication also can be a product of organization and its structurally defined channels. In the traditional organization, employee positions can be discussed as either vertical or horizontal. In new management organizations, the division of labor lines are not as clear-cut because employees work as teams and are empowered to make decisions.

Vertical

Vertical division of labor is based on the establishment of lines of authority. In addition to establishing authority at various levels of the organization, vertical division of labor facilitates communication flow.

Chain of command: Clear and distinct lines of authority within an organization. "Who reports to whom."

The **chain of command** has clear and distinct lines of authority that need to be established among all positions in the organization (Griffin, 2001). The chain of command has two components: the unity of command and the scalar principle. **Unity of command** means that the employee reports to only one manager. The **scalar principle** indicates that a clear and unbroken line of authority extends from the bottom to the top position in the organization.

The 1991 organization chart (Figure 9.16) for the Department of Food and Nutrition Services in the Rush-Presbyterian-St. Luke's Medical Center in Chicago, which is licensed for 1,100 patients, illustrates the vertical division of labor. At that time, the department servesd between 6,500 and 7,500 meals per day to patients, in the employee cafeteria, in the Atrium Court Café for visitors, and at a private club, Room 500, which was often used for special events. The main kitchen was responsible for patient foodservice, and the employee cafeteria kitchen also prepared menu items for the Atrium Court Café. Room 500 had its own kitchen. Note the levels of management, especially middle management from the associate director of foodservice administration through the four operation managers, required for this conventional foodservice in a traditional organization.

Authority, the right of a manager to direct others and to take action, is delegated down the hierarchy of the organization. The tapered concept of authority (Figure 9.17) holds that the breadth and scope of authority become more limited at the lower levels of an organization. To apply this concept to the Rush-Presbyterian-St. Luke's Medical Center depicted in Figure 9.16, the manager with the broadest

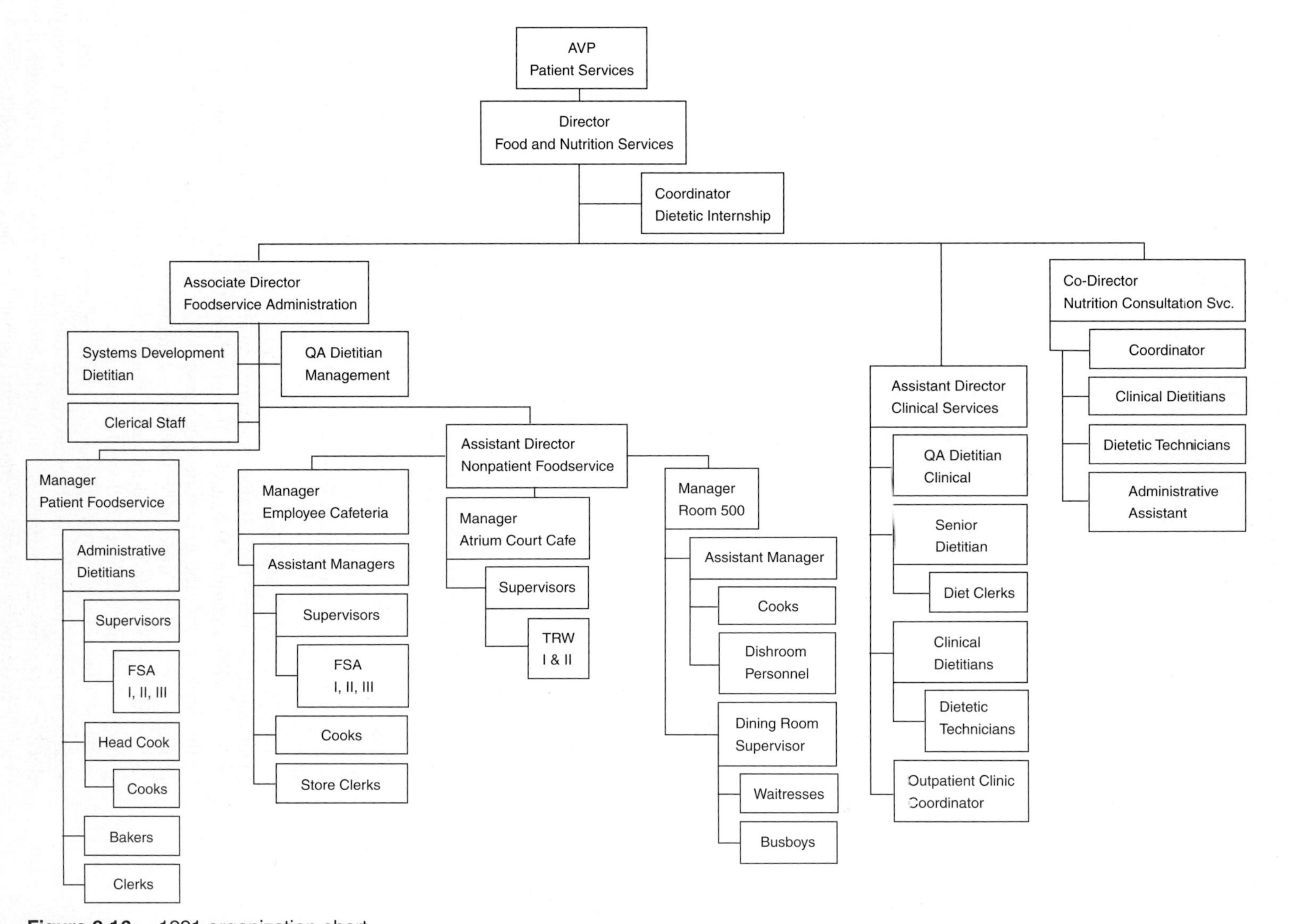

Figure 9.16. 1991 organization chart.
Source: Department of Food and Nutrition Services, Rush-Presbyterian-St. Luke's Medical Center, Chicago, IL. Used by permission.

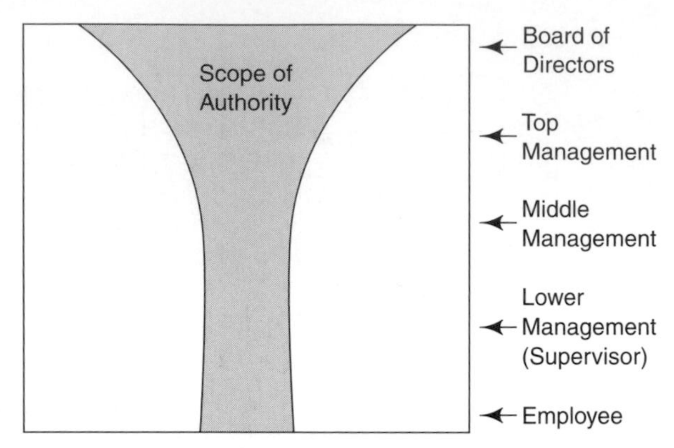

Figure 9.17. Tapered concept of authority.

scope of authority is the associate director for foodservice administration; each succeeding level has a narrower scope.

Through the process of delegation, the authority and responsibility of organization members are established. Delegation is the process of assigning job activities and authority to a specific individual within the organization.

Responsibility is the obligation to perform an assigned activity. Because responsibility is an obligation a person accepts, it cannot be delegated or passed to a subordinate. Managers can delegate responsibilities to subordinates in the sense of making subordinates responsible to them; however, this delegation does not make managers any less responsible to their superiors. When a manager delegates responsibility, the manager does not abdicate any responsibility. For example, Joe, the head cook, cannot say to the production supervisor, "It's all his fault," regarding a product failure of one of the assistant cooks. He must bear responsibility himself.

Authority once delegated, however, is given up by the person who delegated it. According to a principle of organization called the **parity principle,** authority and responsibility must coincide; that is, management must delegate sufficient authority so subordinates can do their jobs. At the same time, subordinates can be expected to accept responsibility only for those areas within their authority.

One of the major considerations affecting delegation of authority is decentralization. The key question is, How much of what authority should be granted to whom and for what purpose? The degree to which an organization is centralized or decentralized is basic to this question. These characteristics are at opposite ends of a continuum.

In a centralized organization, most decisions are made at the top, and lower-level managers have limited discretion in decision making. The degree of centralization/decentralization is related to the number of decisions made at lower levels of the organization, the importance of those decisions, and the amount of checking required for decision making by lower-level managers.

The degree of decentralization varies widely in large organizations. In some organizations, a high degree of decentralization may exist in major functions, but the auxiliary functions of purchasing, accounting, or personnel may be centralized. In a large hospital, for example, the director of dietetics may have authority over production and service functions but limited authority for purchasing because a purchasing department has procurement responsibility for the entire hospital.

Horizontal

The new organizational philosophy rejects the vertical hierarchy that has been popular in American companies for years. Instead, companies with the new philosophy are using a **horizontal division of labor** with an emphasis on encouraging employees to share ideas across all levels and departments (Van Warner, 1993).

American executives are reengineering their organizations in a way that could revolutionize how foodservice managers do business (Van Warner, 1993). Employee teams are revising departments, policies, procedures, and job functions to focus on the customer. Managers and employees are being **cross-trained** to handle multiple jobs, with the entire organization flattened to develop a coaching environment. Cross-training decreases boredom, which could affect the quality of work (Griffin, 2001). The ultimate goal is to create a flexible, more quickly reacting organization that is less distracted by internal problems and bureaucracy. The reengineered organization focuses on the customer, and every person needs to have the goal of providing the customer with quality, value, and service.

The major changes required for becoming a new organization are evolutionary and occur over years, not weeks or months. An excellent example of this evolution is the Rush-Presbyterian-St. Luke's Medical Center, which was introduced earlier to illustrate the vertical division of labor. In order to stay within the budget, a consultant company was hired to eliminate levels of management and, at the same time, increase customer satisfaction in both products and services. The decision was to reduce the number of levels in each department to three. This required changing the medical center from a traditional to a new organization.

In the Department of Food & Nutrition Services, for example, the first change, in April 1993, was to organize the department by functions, each with its own manager: production, service, patient tray assembly, and materials management/computer systems (Figure 9.18). Because the foodservice operation is decentralized and has many locations, this change was not practical and was rescinded. The second change occurred in September 1993, when the decision was made to go back to the natural divisions of the department (Figure 9.19). In this change, functions, such as production and service, are performed within each division—patient, nonpatient, and Room 500 foodservice operations. Employees in each division were encouraged to cooperate with those in others and to share their resources with divisions having similar functions. The goal is to consolidate menus and establish pools of employees—cooks, for example—who have been cross-trained to work in all divisions. Because this plan did not really use the team approach, the hospital initiated another change in November 1993 (Figure 9.20).

Patient foodservices, the largest kitchen in the hospital (Figure 9.19), was chosen to pilot the team approach to decision making. The manager of the patient foodservice kitchen, with coaching from the director of Food and Nutrition Services and with input from the employees, redrew the organization chart to reflect realistically the ways in which teams of employees solve problems. The kitchen was divided into four interrelated functional units—production, trayline, sanitation, and materials management (Figure 9.20). Of the three assistant managers in production, two in trayline, two in sanitation, and one in materials management, the manager designated one in each unit to serve as team coordinator. In this revision, the manager served as a facilitator for the four teams to make sure everyone understood the goals of the organization and the need for effective communication among team

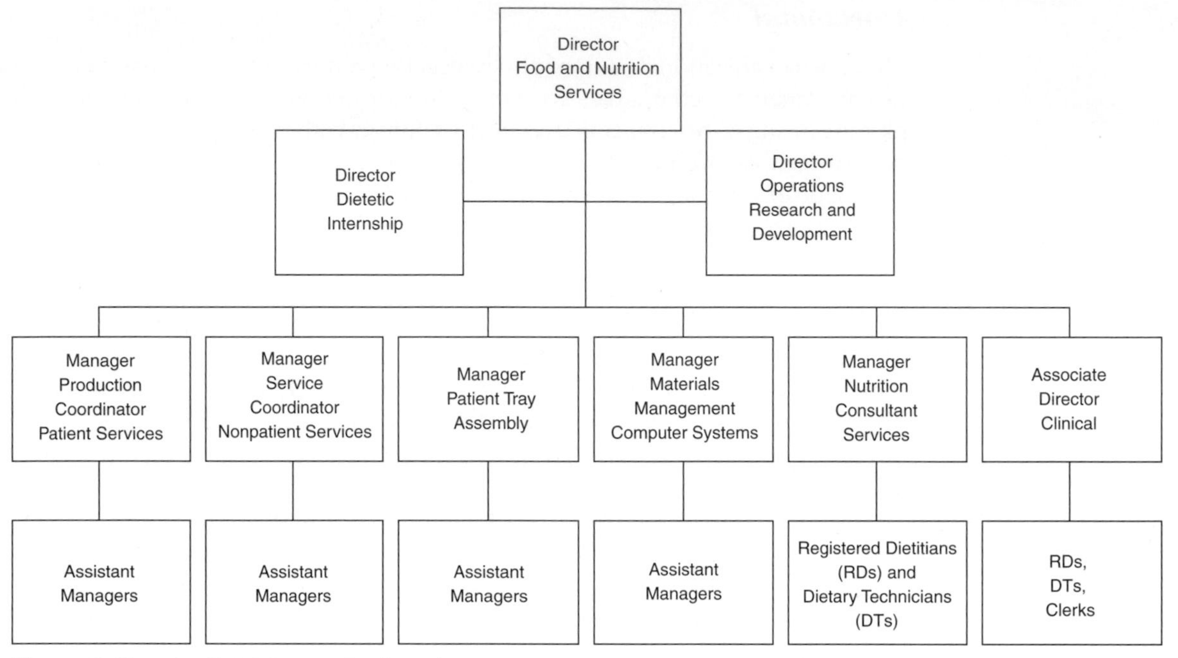

Figure 9.18. April 1993 organization chart.
Source: Department of Food and Nutrition Services, Rush-Presbyterian-St. Luke's Medical Center, Chicago, IL. Used by permission.

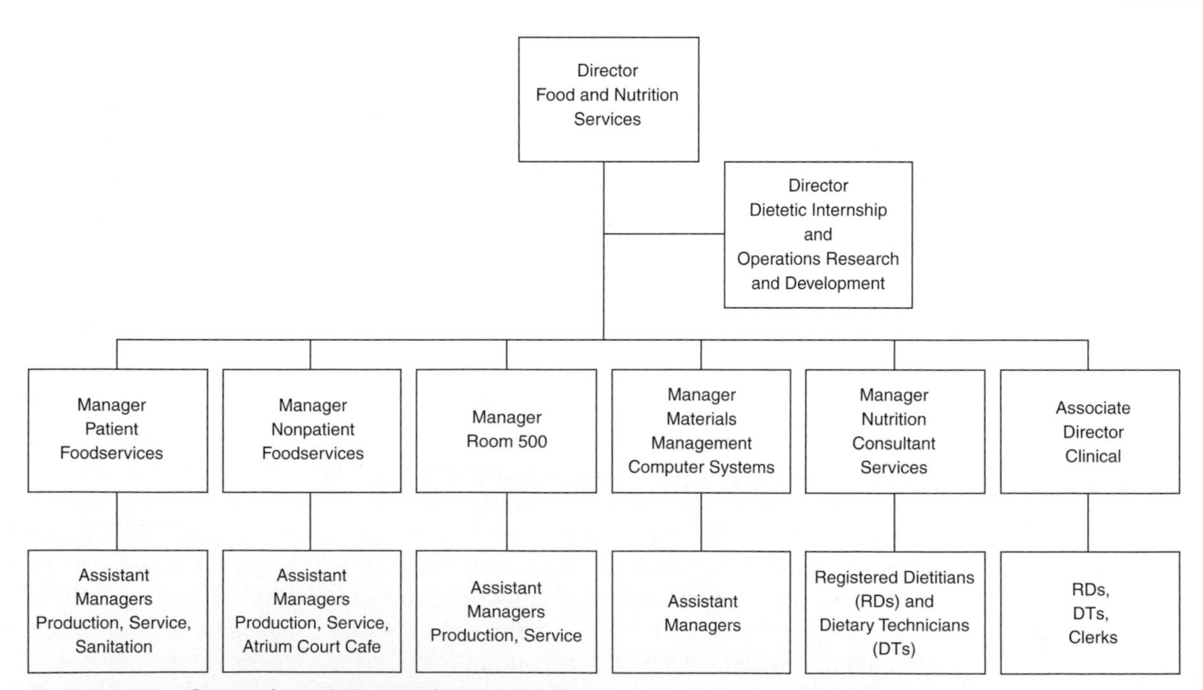

Figure 9.19. September 1993 organization chart.
Source: Department of Food and Nutrition Services, Rush-Presbyterian-St. Luke's Medical Center, Chicago, IL. Used by permission.

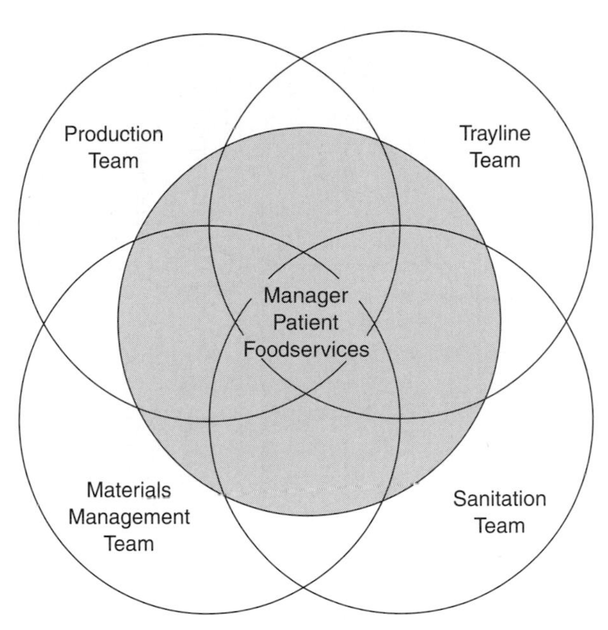

Figure 9.20. November 1993 organization chart.
Source: Department of Food and Nutrition Services, Rush-Presbyterian-St. Luke's Medical Center, Chicago, IL. Used by permission.

members in the patient foodservice division and other divisions. As the teams became self-sufficient, the manager was given other responsibilities, thus requiring the teams to manage the patient foodservice kitchen.

Another change became effective in March 1999 and is shown in Figure 9.21. An associate director of foodservice was added. Although shown in the more traditional hierarchy model, the team approach to operating the units continues.

Labor trends indicate that efforts to make changes will be worthwhile (Van Warner, 1993). Qualified employees are decreasing in numbers, probably because of declining numbers in educational programs. Also, people who do not speak English are becoming a bigger part of the workforce. For foodservice operations, these labor trends mean that more training is needed to spread decision making around instead of depending on a few highly skilled managers. The new organizational structure will need to minimize language, cultural, and educational barriers. Reshaping organizations with empowerment and technology as a base is in order.

Underlying Concepts of Organization

Because managers cannot supervise an unlimited number of subordinates, different areas of organizational activity must be defined, with someone placed in charge of each area. Span of management in the traditional organization refers to the number of subordinates who can be supervised effectively by one manager. This concept, also called span of control, is the basis for the departmentalization process.

Authority, the second concept underlying organization, provides the basis through which managers command work. The source of authority and how it is used are fundamental to the effectiveness of the organization in accomplishing its goals.

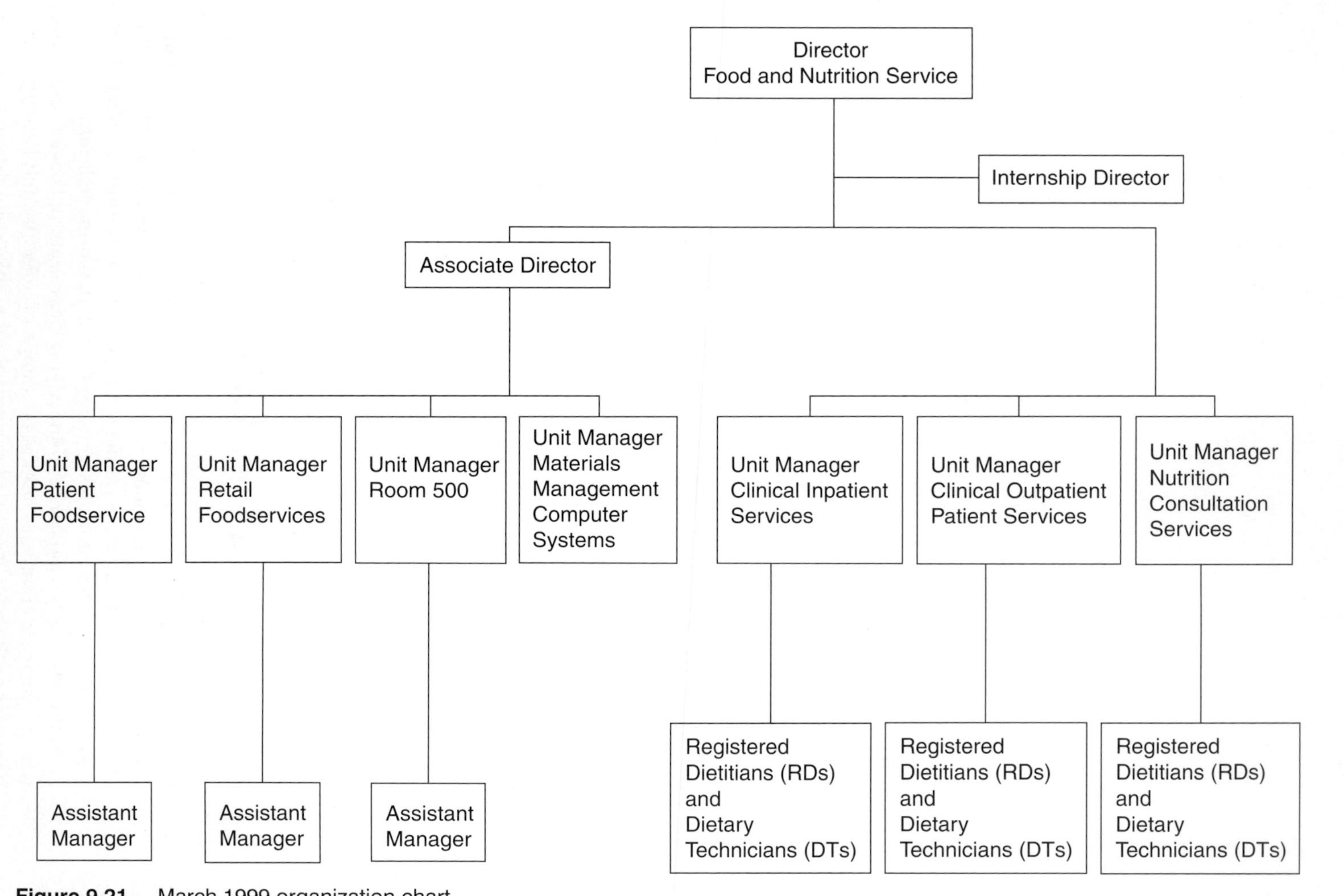

Figure 9.21. March 1999 organization chart.
Source: Department of Food and Nutrition Services, Rush-Presbyterian-St. Luke's Medical Center, Chicago, IL. Used by permission.

Span of Management

The problem of managers being able to manage only a limited number of workers is not unique to any type of organization or any industry. What is the appropriate span? One response could be that it depends on the situation. This situational approach, also called the contingency approach, is applicable in answering many questions about organizations. Organizational policies, availability of staff experts, competency of workers, existence of objective work standards, technology, and nature of the work are among the factors affecting the span of management in specific situations.

The narrower the span of management, the more levels are needed in the organization. Because each level must be supervised by managers, the more levels that are created, the more managers are needed. Conversely, with a wider span, fewer levels and fewer managers are required. Thus, the resulting organizational shape is a tall, narrow pyramid or a shallow, flat, broad pyramid, as illustrated in Figure 9.22. The leadership style and personality of the manager are other influences on span of management.

Early management theorists attempted to define the appropriate number of superior-subordinate relationships in terms of a mathematical formula. In 1933, Graicunas published a classic paper presenting a formula for analyzing the potential number of these relationships (1973, originally published 1933). Based on the work of Graicunas and others, Urwick (1938) stated the concept of span of management as follows: "No superior can supervise directly the work of more than five or, at most, six subordinates whose work interlocks."

During the years since those early publications, researchers have found that a variety of factors influence the appropriate span of management and that this span is not strictly a function of the number of relationships. In addition to those listed previously, the complexity, variety, and proximity of jobs and the abilities of the manager are other factors related to span of management. Today, span of management is a crucial factor in structuring organizations, but no universal formula has been developed for an ideal span (Griffin, 2001).

In a foodservice organization in which many of the workers have low educational levels and limited training, a narrower span of management may be appropriate. If workers are well trained, a wider span is possible and fewer supervisors are needed.

Formal versus Acceptance Authority

Up to this point in this text, authority has been defined as the right of the manager to direct others and to take action. Actually, this definition deals only with one view of authority, that is, formal authority. This view of authority has its roots in the writings of Weber (1947), a German sociologist who in the late 1800s and early 1900s influenced the development of management thought. He viewed authority as "legitimate power," which involves the willing and unconditional compliance of subordinates based on their belief that it is legitimate for the manager to give commands and illegitimate for them to refuse to obey.

Formal authority: Authority that exists because of position in the organization.

Today, two views of authority are generally recognized in the management literature, formal authority and acceptance authority. **Formal authority** is considered a top-down theory because it traces the flow of authority from the top to the bottom of the organization (Figure 9.23). As Haimann, Scott, and Conner (1985)

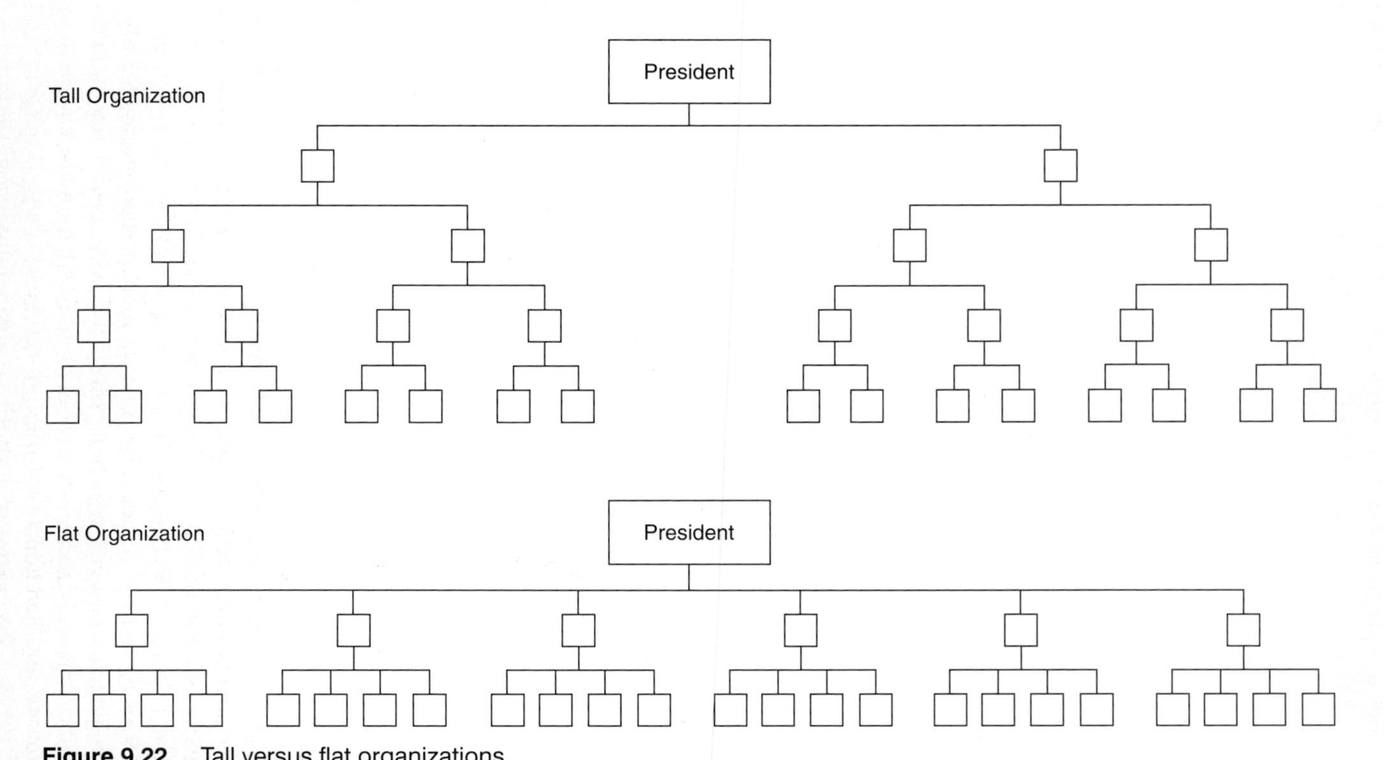

Figure 9.22. Tall versus flat organizations.
Source: Griffin, Ricky W., *Management,* 6th edition. Copyright © 1999 by Houghton Mifflin Company. Used by permission.

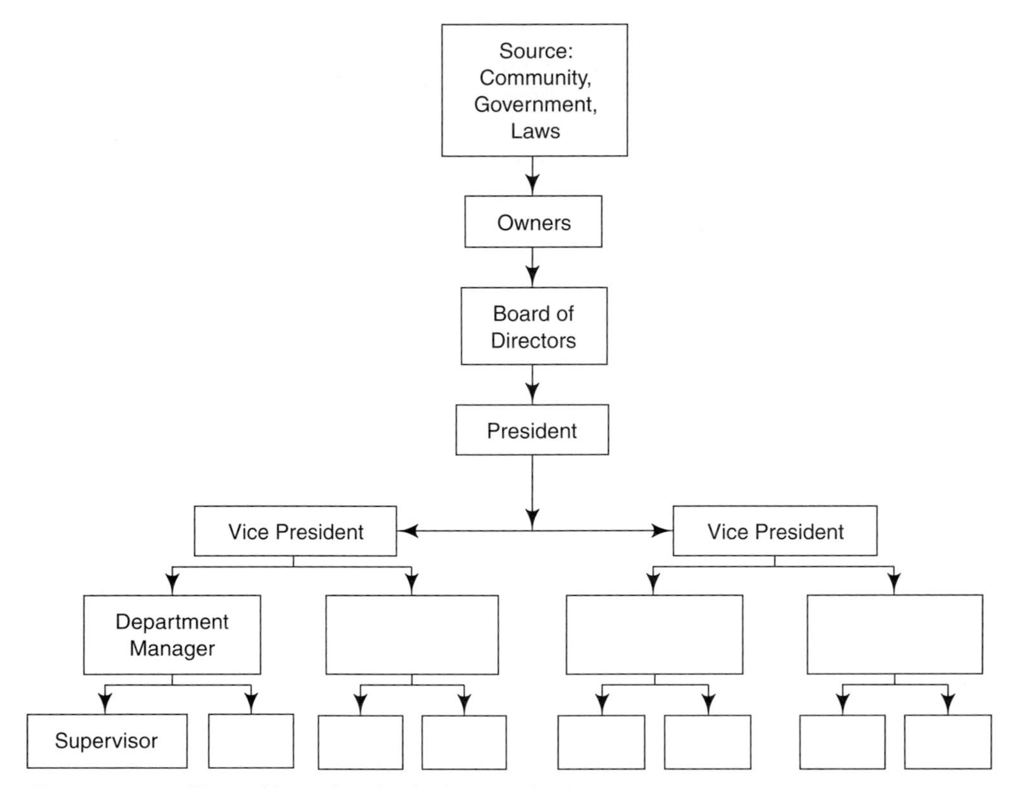

Figure 9.23. Flow of formal authority in organizations.

indicated, society is the ultimate source of formal authority in the United States because of the constitutional guarantee of private property. Obviously, this statement is directly applicable to public sector institutions, such as public schools, state universities, and hospitals operated by the city, county, state, or federal government, but it applies to organizations in the private sector as well.

Formal authority is also referred to as *positional authority,* meaning that authority is derived from the position or office. For example, cooks and other production workers recognize the production manager as having certain authority because of the position he or she holds in the organization.

Acceptance authority: Authority based on employee's acceptance of that authority.

Acceptance authority is based on the concept that managers have no effective authority unless and until subordinates are acquiescent. Although they may have formal authority, this authority is effective only if subordinates accept it. The accept-comply process is related to a subordinate's zone of acceptance. In other words, a manager's order will be accepted without conscious question if it falls within the range of job duties anticipated when the subordinate accepted employment. The employee, however, may refuse to perform a task he or she considers to be outside this range. For example, a cook would willingly prepare any item on the menu, even items added at the last minute, but may be unwilling to help out in the dishroom, feeling that dishwashing is not part of the job.

Authority of competence: Authority based on a manager's competence or expertise.

Acceptance authority is also related to a manager's expertise and personal attributes. **Authority of competence** or expertise is based on technical knowledge and experience. A command may be accepted, not because of organizational title, but because the employee believes the person giving the command is knowledgeable.

This concept underscores the importance of foodservice managers possessing technical skills. If the foodservice manager understands technical operations, such as production methods and equipment operation, foodservice employees view the manager as knowledgeable and generally will be more willing to accept direction.

Subordinates may also accept the authority of the manager because they want to please or help the person giving the command. The charismatic leadership of many well-known historical figures is the epitome of this personal authority.

Departmentalization

One of the first things that happens when people create an organization is that they divide up their work to allow specialization. As the organization grows and tasks become more numerous and varied, this division of labor is formalized into jobs and departments.

Departmentalization: Process of grouping jobs according to some logical arrangement.

Departmentalization, which is the process of grouping jobs according to some logical arrangement, is the most frequently used method for implementing division of labor (Griffin, 2001). Although work units can be structured in a number of ways, all units divide the work and thus establish a pattern of task and authority relationships. This pattern becomes the organizational structure.

In a small delicatessen, for example, a husband and wife may informally share the tasks of preparing sandwiches, salads, and drinks; serving customers; collecting money; wiping tables; washing dishes and utensils; and performing other maintenance duties. They will probably find that each of them will take on the principal responsibility for certain tasks; however, as the business grows, they may need to hire part-time workers to assist at peak periods. These workers will probably be assigned specific duties rather than be responsible for the wide range of duties performed by the husband-wife team. Thus, jobs are created around specialized tasks.

This small business could eventually be the basis for development of a large multiunit national chain of delicatessens throughout the United States. Additional levels of management would be needed, highly specialized jobs created, and formalized relationships required. At the corporate level, departments focusing on specific functions, such as marketing, procurement, and finance, would be created. This illustration is an example of the development of an organization into jobs, levels of management, divisions, and departments.

Departments are commonly organized by function, product, geography, customer, process, equipment, or time. As indicated earlier, the type and size of the organization are key factors influencing the form an organization structure will take.

Functional

Functional departmentalization occurs when organization units are defined by the nature of the work. (The word function is used here to mean organizational functions, such as finance and production, rather than basic managerial functions, such as planning and controlling.) All organizations create some product or service, market these products or services, and finance their ventures. Therefore, most organizations have three basic functions: production, sales, and finance. In the nonprofit foodservice operation, the sales function may be one of clientele service and creation of goodwill, and the finance function may be considered business affairs.

Even in these organizations, however, the need to apply marketing concepts is widely recognized.

The primary advantage of such departmentalization is that it allows specialization within function and provides for efficient use of equipment and other resources. It provides a logical way of arranging activities because functions are grouped that naturally seem to belong together. Each department and its manager are concerned with one type of work.

Product and Service

Under departmentalization by a product or a service, all activities required in producing and marketing them are usually under the direction of a single manager. *Product departmentalization* allows workers to identify with the particular product and encourages expansion, improvement, and diversification. Duplication of functions may be a problem, however, because each division or department may be involved in marketing, production, and so forth. This pattern of departmentalization is not common in the foodservice industry, except perhaps in large conglomerate corporations.

Geography

Departmentalization by territory is most likely to occur in organizations that maintain physically dispersed and autonomous operations or offices. *Geographic departmentalization* permits the use of local personnel and may help create customer goodwill and a responsiveness to local customs.

National restaurant chains often are divided into regional areas, with a regional manager and staff responsible for all the operations in a particular area. For example, several of the large contract foodservice companies are divided into several geographic regions, such as East, Midwest, Northeast, Northwest, and South.

Customer

Another type of departmentalization is based on division by type of customers served. A contract foodservice company, for example, that has divisions for schools, colleges, and healthcare is departmentalized by type of customer. A wholesale firm that distributes products to grocery stores and to hotels, restaurants, and institutions may be subdivided into two corresponding divisions. This approach to departmentalization permits the wholesaler to serve the specialized needs of both the grocer and the foodservice operator.

Other Types

Process, equipment, and time are other bases for departmentalization. Process and equipment are closely related to functional departmentalization. In large foodservice operations, a deep-fat frying section within the production unit would be an example of *process/equipment departmentalization*. A food factory, such as that in a commissary foodservice operation, might be divided into units based on process or equipment because of the specialization needed for the large volume produced in the operation.

Time or *shift* is also a common way of departmentalizing in some organizations. Organizations such as hospitals that function around the clock often organize activities on this basis. Usually, activities grouped this way are first departmentalized on some other basis, perhaps by product or function. Then, within that category, they are organized into shifts. For example, a hospital is departmentalized by functions, such as dietetics services and nursing services; the various departments may then have shifts with a supervisor in charge of each.

Departmentalization is practiced as a means not only of implementing division of labor but also of improving control and communications. Typically, as the organization grows, it adds levels and departments. Coordination is another key objective in departmentalization. The type of departmentalization that is best for an organization depends on its specific needs.

Line and Staff

Line position: A position in the direct chain of command.

Staff position: Position intended to provide expertise, advice, and support for line positions.

In addition to vertical division of labor through delegation and horizontal division of labor through departmentalization, labor may be divided into line and staff. A **line position** is a position in the direct chain of command that is responsible for the achievement of an organization's goals (Griffin, 2001). Line positions include a procurement, production, or service manager. A **staff position** is intended to provide expertise, advice, and support for line positions. A materials or human resources manager is an example of a staff position.

Line employees typically are responsible for production of products and services in the organization, and staff personnel may function in assisting or advising roles. Staff work revolves around the performance of staff activities, the utilization of technical knowledge, and the creation and distribution of technical information to line managers.

Authority

One important difference between line and staff is authority (Griffin, 2001). Line authority is referred to as formal authority created by the organizational hierarchy. Staff authority, however, is based on expertise in specialized activities. Generally, staff personnel provide expert advice and counsel to line managers but lack the right to command them, with two exceptions. First, staff managers exercise line authority over employees in their own departments; second, staff may have functional authority over the line in restricted areas of activity. This functional authority is delegated to an activity and gives managers performing the activity the right to command. Authority granted in this manner, however, is confined to the specialized area to which it was delegated.

The quality control manager, for example, may have functional authority over the work of supervisors in other departments. If inspectors find a product quality problem, they may require the supervisor to suspend production until the problem is corrected.

This example applies directly to a commissary foodservice operation. The microbiologist on the quality control staff may identify a problem with microbial count in a product being produced in a food factory and require that production be curtailed until the source of contamination is identified.

Administrative Intensity

Administrative intensity is the degree to which managerial positions are concentrated in staff positions (Griffin, 2001). A high administrative intensity organization is one with many staff positions relative to the number of line positions; low administrative intensity emphasizes more line positions. Many organizations have reduced their administrative intensity in recent years by eliminating many staff positions.

Organization Chart

The organization chart graphically portrays the organization structure. It depicts the basic relationships of positions and functions while specifying the formal authority and communication network of the organization. The title of a position on the chart broadly identifies its activities; distance from the top indicates the position's relative status. Lines between positions are used to indicate the prescribed formal interaction.

The organization chart is a simplified model of the structure. It is not an exact representation of reality and therefore has limitations. For example, the degree of authority a superior has over a subordinate is not indicated. The chart, however, assists employees of the organization in understanding and visualizing the structure. Charts should be revised periodically because organizations are dynamic and undergo many changes over time.

Responsibility and authority for the preparation, review, and final approval of the organization chart generally lie with top management, although approval may be the responsibility of the board of directors. At the departmental level, the chart may be the responsibility of the department head, although approval may be required from the next level up in the organization.

Vertical organization charts have been the most conventional type, but currently the horizontal chart is being used in the new organizations. In the vertical chart, the levels of the organization are depicted in a pyramid form, with lines showing the chain of command. Special relationships may be indicated by the positioning of functions and lines on the chart. Dotted lines are often used to indicate communication links in an organization. Staff functions may be depicted by horizontal placement between top and department managers. Referring again to the Rush-Presbyterian-St. Luke's Medical Center traditional organization chart (Figure 9.16), the associate director of foodservice administration and the codirector of nutrition consultation services are on the same level, which indicates equal responsibility and authority in the department.

Coordination

Coordination: Process of linking activities of various departments in the organization.

Coordination is a major component of organization structure (Griffin, 2001). **Coordination** is the process of linking activities of various departments in the organization. Job specialization and departmentalization involve breaking jobs down into small units and then combining the units to form a department. The activities in the department then need to be linked and focused on organizational goals. Some of the most useful means for maintaining coordination among interdependent units are horizontal interaction; policies, procedures, and rules; standards; communication; and committees and task forces.

In a large medical center, for example, *horizontal interaction* is required among departments. Nursing service and dietetics service staff often communicate directly rather than through the vertical organization. Such lateral relationships facilitate communication in an organization.

Policies, procedures, and rules ensure consistency in operations and are an important method of coordination. Managers may also establish schedules and other plans to coordinate action. Events are often unpredictable, however, and must be coordinated by managers using their judgment. Overreliance on rules and regulations can create problems in organizations.

Standards need to be developed before good coordination occurs in a foodservice operation. For example, large limited-menu restaurants usually have specific standards regarding production and service of products. One doughnut chain, for instance, requires that all products not sold within 4 hours after frying must be discarded. Specific formulations and frying procedures must also be followed.

Another way in which managers act to coordinate activities in an organization is in a *linking role*. Communication is the responsibility of managers for linking with managers at higher levels in the organization and with others at their own level. This concept was derived from Likert's work (1961) in which he described managers as "linking pins."

Appointment of committees and task forces is a mechanism used in organizations for coordination. These groups serve an important role when problem solving must involve several departments. Problems involving half a dozen departments, for example, can be dealt with efficiently by such groups; otherwise, problems are referred upward through the chain of command.

Committees and task forces are common in all foodservice operations. Many foodservice organizations in metropolitan centers have formed task forces to recommend ways to feed the needy and homeless, and one solution many find is for the organization to contribute to municipal soup kitchens.

MANAGEMENT PRACTICES

There are many management practices that have become commonplace in foodservice operations. Some of the more common are behavior modeling, open door policy, managing by walking around, and making work "fun."

Behavior modeling is sometimes referred to as "do as I do." It means modeling the behavior you expect from your employees. How you treat customers, what food safety practices you follow, how your talk to employees, and so on all can demonstrate to employees the level of quality you expect in your operation.

Managers with an *open door policy* encourage employees to come to their office with ideas, concerns, and questions, The idea behind the concept is that a manager is approachable and has time to listen to employees.

Managing by walking around is a practice where manager walk through their operation on a regular basis talking with employees and supervisors in the process. It provides a way to visually see what is going on in your operation and to visit informally with employees in their work areas.

The importance of having "fun" in the workplace was first popularized by Lundin, Paul, and Christensen (2000) in their book *FISH! A Remarkable Way to Boost Morale and Improve Results,* which describes the daily fun that occurs at

Pike's Place Fish, a fish market in Seattle that has been extremely successful with its joyful atmosphere and great customer service. Collins's (2001) research on characteristics of "great" companies identified "having fun" as one of the important characteristics of such successful companies.

CHAPTER SUMMARY

This summary is organized by the learning objectives.

1. There are several terms used in management. Effectiveness is defined as doing the right things; efficiency as doing things right. Authority is the right of managers to direct the work of others; responsibility is the obligation to perform responsibility and assigned responsibility. Span of control is the number of employees who report to a manager.
2. According to Mintzberg (1975), a manager's position could be organized into one or more of ten roles: figurehead, leader, liaison, monitor, disseminator, spokesperson, entrepreneur, disturbance handler, resource allocator, and negotiator. Managers also are believed to perform several functions: planning, organizing, staffing, directing, and controlling.
3. Katz (1974) identified three basic skills that he believed all managers needed to possess: technical skill, which involves a proficiency in a specific activity; human skill, which includes working with people and understanding their behavior; and conceptual skill, which is the ability to view the organization as a whole.
4. An organization chart depicts graphically the positions in an organization and the hierarchy of these positions. This chapter provides examples of various models of organization charts.

TEST YOUR KNOWLEDGE

1. What is required of a foodservice manager?
2. How can foodservice managers improve the quality of work life for their employees?
3. Give examples of activities performed by foodservice managers that are examples of each of Mintzberg's managerial roles.
4. Describe different ways work is divided among foodservice employees.
5. How are management skills linked with the functions a foodservice manager has to perform?

CLASS PROJECTS

1. Invite a local foodservice manager to class. Ask the manager to discuss his or her role as a manager and to describe examples of performing managerial functions in his or her job.
2. Divide the class into small groups. Ask group members to share with each other examples of how they have used technical, human, and conceptual skills in positions they have held.

Professional Organization Profile

American Society for Healthcare Foodservice Administrators

MISSION

Advancing healthcare foodservice leadership through education, networking, and resources that empower members to accomplish exceptional outcomes

WHO BELONGS TO THE ORGANIZATION

Members include foodservice management professionals, business representatives, consultants, and educators. The organization is an affiliate of the American Hospital Association. Student memberships are available.

ADVANTAGES OF MEMBERSHIP

Members have access to a members-only section of the ASHFSA and AHA websites; receive a quarterly newsletter, *Healthcare Foodservice TRENDS,* highlighting innovations being explored in foodservice operations; have access to the American Hospital Association Resource Center; can attend association sponsored conferences at reduced rates; and are eligible for educational scholarships and professional association awards.

WEBSITE

www.ashfsa.org

MEET THE PRESIDENT

Regina Toomey Beuno, President 2005–2006, is director of food, nutrition, and transport at The Valley Hospital in Ridgewood, NJ.

3. Have the class read the book *FISH! A Remarkable Way to Boost Morale and Improve Results,* and discuss ways the books principles could be incorporated into a foodservice operation.

WEB SOURCES

www.onetcenter.org	U.S. Department of Labor O*NET (Occupational Information Network)
www.aom.pace.edu	Academy of Management online

BIBLIOGRAPHY

Alva, M. (1988). The death of the salad bar. *Nation's Restaurant News* 22(17) F7–F8.

American Institute of Management. (1959). *What is management?* New York: American Management Association.

Boettinger, H. M. (1975). Is management really an art? *Harvard Business Review, 53*(1), 54–55, 57–59, 61, 63–64.

Bryson, J. M. (1995). *Strategic planning for public and nonprofit organizations: A guide to strengthening and sustaining organizational achievement.* San Francisco: Jossey-Bass.

Bryson, J. M. (1999). *Strategic planning for public and nonprofit organizations: A guide to strengthening and sustaining organizational achievement.* San Francisco. Jossey Bass.

Byers, B., Shanklin, C. W., Hoover, L. C., & Puckett, R. (1994). *Food service manual for health care institutions.* Chicago: American Hospital Publishing.

Cain, H. (1998). Here's some good news—now what are you going to do with it? *Restaurants USA, 18*(11), 50.

Chase, R. B., & Aquilano, N. J. (1992). *Production and operations management: A life-cycle approach* (4th ed.). Homewood, IL: Richard D. Irwin.

Collins, J. (2001). *Good to great.* New York: HarperCollins.

Dessler, G. (2002). *A Framework for Management.* Upper Saddle River, NJ: Prentice Hall.

Donnelly, J. H., Gibson, J. L., & Ivancevich, J. M. (1997). *Fundamentals of management* (10th ed.). Homewood, IL: BPI/Irwin.

Drucker, P. F. (1964). *Managing for results.* New York: Harper & Row.

Eating Place Trends: Management Trends. (1998). *Restaurants USA, 18*(11), F5–F7.

Ferguson, D. H., & Berger, F. (1984). Restaurant managers: What do they really do? *Cornell Hotel and Restaurant Administration Quarterly, 25*(1), 27–38.

Fintel, J. (1989). Restaurant cultures: Positive cultures can keep companies healthy. *Restaurants USA,* 9(10), 12–16.

Fulmer, R. M. (1988). *The new management* (4th ed.). New York: MacMillan.

Ghorpade, J., & Atchison, T. J. (1980). The concept of job analysis: A review and some suggestions. *Public Personnel Management, 9,* 134–143.

Graicunas, V. A. (1973). Relationship in organization. In Gulick, L., & Urwick, F. L. (eds.), *Papers on the science of administration* (pp. 181–187). New York: Institute of Public Administration. Original work published 1933.

Griffin, R. W. (2001). *Management* (7th ed.). Boston: Houghton Mifflin.

Gufreda, J. J., Maynard, L. A., & Lytle, L. N. (1990). Employee involvement in the quality process. In Huge, E. C. (ed.), *Total quality: An executive's guide for the 1990s* (pp. 162–176). Homewood, IL: Dow Jones-Irwin.

Hackman, J. R. (1977). Work design. In Hackman, J. R., & Suttle, J. L. (eds.), *Improving life at work: Behavioral science approaches to organizational change* (pp. 96–162). Glenview, IL: Scott, Foresman.

Haimann, T., Scott, W. G., & Conner, P. E. (1985). *Managing the modern organization* (4th ed.). Boston: Houghton Mifflin.

Hamilton, J. (1993). Toppling the power of the pyramid. *Hospitals, 67*(1), 38, 40–41.

Hart, C. W. L., Spizizen, G. S., & Muller, C. C. (1988). Management development in the foodservice industry. *Hospitality Education and Research Journal, 12*(1), 1–20.

Hersey, P., Blanchard, K., & Johnson, D. (2000). *Management of organizational behavior: Utilizing human resources* (8th ed.). Upper Saddle River, NJ: Prentice Hall.

Hirschhorn, L., & Gilmore, T. (1992). The new boundaries of the "boundaryless" company. *Harvard Business Review, 70*(3), 104–115.

Huber, G. P., & McDaniel, R. R. (1986). The decision-making paradigm of organizational design. *Management Science, 32*(5), 572–589.

Huge, E. C., & Vasily, G. (1990). Leading cultural change: Developing vision and change strategy. In Huge, E. C. (ed.), *Total quality: An executive's guide for the 1990s* (pp. 54–69). Homewood, IL: Dow Jones-Irwin.

Hunt, V. D. (1992). *Quality in America: How to implement a competitive quality program.* Homewood, IL: Business One Irwin.

Jones, P. (1993). Foodservice operations management. In Khan, M., Olsen, M., & Var, T. (eds.), *VNR's encyclopedia of hospitality and tourism* (pp. 27–36). New York: Van Nostrand Reinhold.

Kanter, R. M. (1992). The long view. *Harvard Business Review, 70*(5), 9–11.

Kast, F. E., & Rosenzweig, J. E. (1985). *Organization and management: A systems and contingency approach* (4th ed.). New York: McGraw-Hill.

Katz, R. L. (1974). Skills of an effective administrator. *Harvard Business Review, 52*(5), 90–102.

Katzenbach, J. R., & Smith, D. K. (1993). The discipline of teams. *Harvard Business Review, 71*(2), 111–120.

Likert, R. (1961). *New patterns of management*. New York: McGraw-Hill.

Lundin, S. C., Paul, H., & Christensen, J. (2000). *FISH! A remarkable way to boost morale and improve results*. New York: Hyperion.

Mintzberg, H. (1975). The manager's job: Folklore and fact. *Harvard Business Review, 53*(4), 49–61.

Mintzberg, H. (1980). *The nature of managerial work*. Englewood Cliffs, NJ: Prentice Hall.

Mintzberg, H. (1981). Organization design: Fashion or fit? *Harvard Business Review, 59*(1), 103–116.

Mintzberg, H. (1987). Crafting strategy. *Harvard Business Review, 65*(4), 66–75.

Mintzberg, H. (1989). *Mintzberg on management: Inside our strange world of organizations*. New York: Free Press.

Pesci, P. H., Spears, M. C., & Vaden, A. G. (1982). A method for developing major responsibilities and performance standards for foodservice personnel. *NACUFS Journal, 4,* 17–21.

Peters, T. (1992). *Liberation management: Necessary disorganization for the nanosecond nineties*. New York: Knopf.

Primozic, K. I., Primozic, E. A., & Leben, J. (1991). *Strategic choices: Supremacy, survival, or sayonara*. New York: McGraw-Hill.

Robbins, S. P., & Coulter, M. (2005). *Management* (8th ed.). Upper Saddle River, NJ: Prentice Hall.

Roethlisberger, F. J., & Dickson, W. J. (1956). *Management and the worker*. Cambridge, MA: Harvard University Press.

Sherman, A. W., Snell, S., & Bohlander, G. W. (1997). *Managing human resources* (11th ed.). Cincinnati: South-Western Publishing.

Sultemeier, P. M., Gregoire, M. B., & Spears, M. C. (1989). Managerial functions of college and university foodservice managers. *Journal of the American Dietetic Association, 89*(7), 924–928.

Tse, E. C., & Olsen, M. D. (1988). The impact of strategy and structure on the organizational performance of restaurant firms. *Hospitality Education and Research Journal, 12*(2), 265–276.

Ulrich, D., & Wiersema, M. E. (1989). Gaining strategic and organizational capability in a turbulent business environment. *Academy of Management Executive, 3*(2), 115–122.

Umbreit, W. T. (1989). Multi-unit management: Managing at a distance. *Cornell Hotel and Restaurant Administration Quarterly, 30*(1), 53–59.

Urwick, L. F. (1938). *Scientific principles and organizations*. Institute of Management Series No. 19. New York: American Management Association.

U.S. Department of Labor, Employment, and Training Administration. (1991). *Dictionary of occupational titles* (4th ed.). Washington, DC: U.S. Government Printing Office.

Uyterhoeven, H. E. R. (1992). General managers in the middle. *Harvard Business Review, 70*(2), 75–85.

Van Warner, R. (1993). Reinventing the organization. *Nation's Restaurant News, 27*(26), 29.

Weber, M. (1947). *The theory of social and economic organizations*. New York: Free Press.

Weller, N. (1992). Preparing written job descriptions. *Restaurants USA, 12*(2), 20–23.

West, J. J. (1990). Strategy, environmental scanning and firm performance: An integration of content and process in the foodservice industry. *Hospitality Research Journal,* 14(1), 87–100.

CHAPTER 10

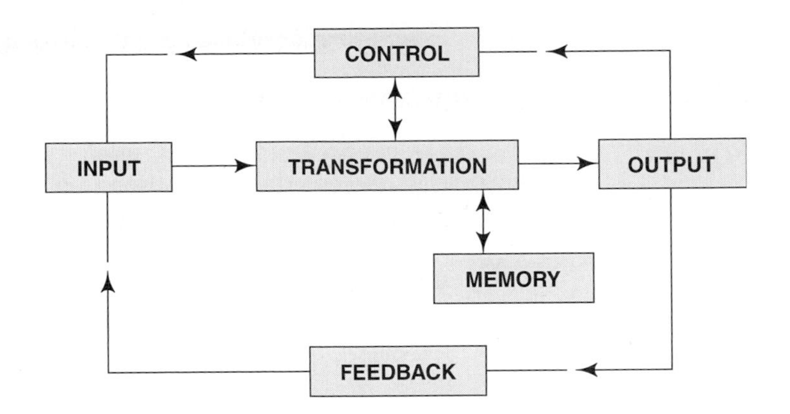

Leadership and Organizational Change

Enduring Understanding

- Leadership and management are different.
- Leadership involves influencing others toward goals.
- Leaders must be change agents.

Learning Objectives

After reading and studying this chapter, you should be able to

1. Describe theories of motivation and their application to foodservice management.
2. Discuss the concepts of job satisfaction, organizational citizenship behaviors, and continuance and affective commitment.
3. Explain the various bases of power that a leader might have.
4. Compare and contrast multiple theories of leadership.
5. Implement change in an organization.

What makes a leader effective? This question has been asked numerous times over the years in attempts to understand the concept of leadership. In spite of intensive research, knowledge of what it takes to be an effective leader is still limited. In this chapter you will learn about some of the theories of leadership and motivation. Organizational change also will be discussed.

Leading is one of the functions that managers perform in the process of coordinating activities of the foodservice system. The leading function involves directing and channeling human effort to accomplish objectives. Therefore, leading is concerned with creating an environment in which members of the organization are motivated to contribute to organizational goals and changes. As shown in Figure 10.1, leadership surrounds the foodservice system and provides a sense of direction for the system.

MOTIVATION AND WORK PERFORMANCE

Motivation is concerned with the causes of human behavior. An understanding of human behavior is important to managers as they attempt to influence this behavior in the work environment. The study of motivation and behavior is a search for answers to perplexing questions about human nature.

Because of the importance of human resources in organizations, managers must have an understanding of behavior, not only to understand the past but also to predict or change the future. Highly motivated employees in a foodservice or any other organization can elicit substantial increases in performance and decreases in such problems as absenteeism, turnover, grievances, low morale, and tardiness.

Meaning of Motivation

Motivation: Inner force that activates or moves a person toward achievement of a goal.

Definitions of motivation usually include such words as aim, desire, intention, objective, goal, and purpose. A definition commonly quoted is that of Berelson and Steiner (1964): **Motivation** is all those inner striving conditions described as wishes, desires, and drives, and is an inner force that activates or moves a person.

The process of motivation can be viewed as a causative sequence:

Needs → Drives or motives → Achievement of goals

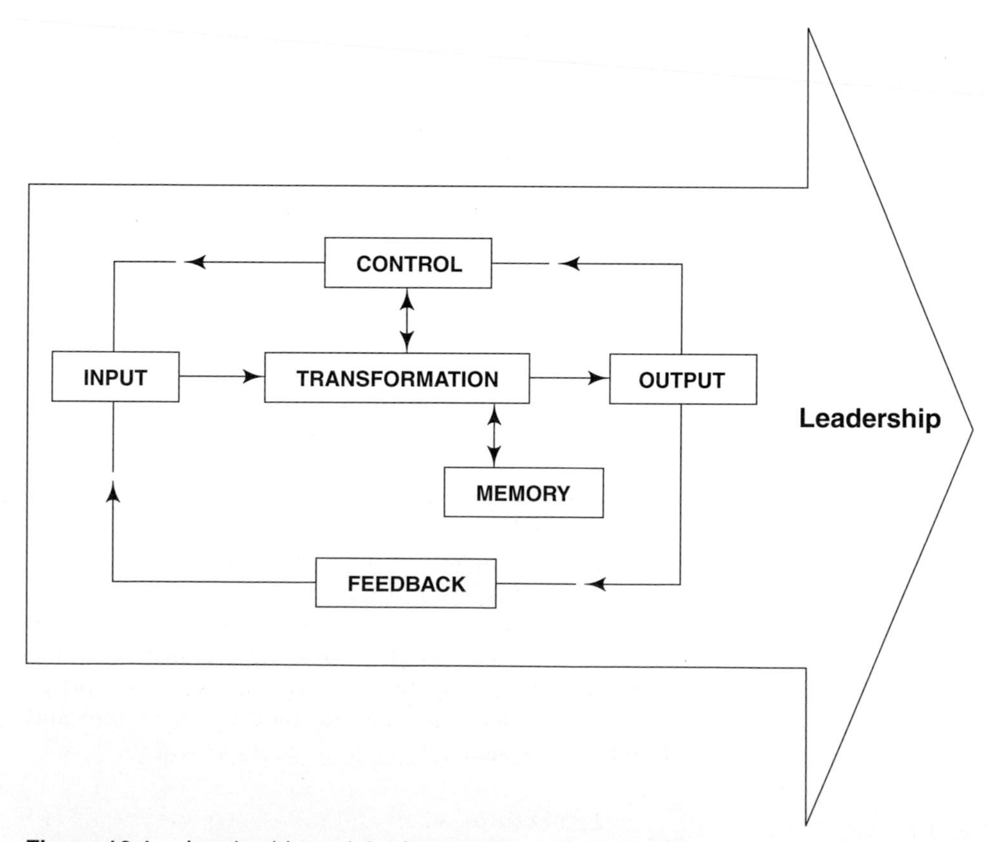

Figure 10.1. Leadership and the foodservice systems model.

In the motivation process, needs produce motives that lead to the accomplishment of goals or objectives. Needs are caused by deficiencies that may be physical or psychological. *Physical needs,* also called innate or primary needs, include food, water, and shelter. *Psychological needs,* also referred to as acquired needs, are those we learn in response to our culture or environment. They include esteem, affection, and power.

Drives or motives are the "whys" of behavior. They arouse and maintain activity and determine the general direction of an individual's behavior. Hersey, Blanchard, and Johnson (1996) explained a **motive** as something within an individual that prompts a person to action.

Motive: Something that prompts a person to action.

Achievement of the goal in the motivation process satisfies the need and reduces the motive. When the goal is reached, balance is restored; however, other needs arise, which are then satisfied by the same sequence of events.

A distinction should be made between positive and negative motivation. Motivation can be either a driving force toward some object or condition (*positive motivation*) or a driving force away from some object or condition (*negative motivation*).

Goals can also be positive or negative. A *positive goal* is desirable and the object of directed behavior, as in an employee's desire to do the best job possible. A *negative goal* is undesirable and behavior is directed away from it; an employee who wishes to avoid censure is an example. Goals depend on an individual's subjective experiences, physical capacity, prevailing norms and values, and the potential accessibility of the goal. Furthermore, an individual's self-perception also serves to influence goals.

Both needs and goals are interdependent, and individuals are not always aware of their needs. In addition, needs and goals are constantly changing. As individuals attain goals, they develop new ones; if certain goals are not attained, they develop substitutes. Individuals who are blocked in attempts to satisfy their needs or achieve goals may become frustrated and exhibit dysfunctional or defensive behavior such as

- **Withdrawal:** When an individual becomes less involved in their work; may be exhibited by apathy, excessive absences, lateness, or turnover. It is one mechanism used to avoid frustrating situations.
- **Aggression:** When an individual directly attacks the source of frustration or another object or party. For example, a foodservice employee who is upset with his or her supervisor may slam and bang the pots and pans as a way of venting frustration.
- **Substitution:** When an individual puts something in the place of the original object. For example, a foodservice employee bypassed for promotion may seek leadership positions in organizations outside the workplace.
- **Compensation:** When a person goes overboard in one area or activity to make up for deficiencies in another.
- **Revert or regress:** When an individual exhibits childlike behavior as a way of dealing with an unpleasant situation. For example, horseplay in the dishroom is an example of regression.
- **Repression:** When an individual loses awareness of or forgets incidents that cause anxiety or frustration.
- **Projection:** When an individual attributes his or her own feelings to someone else. For example, a foodservice employee who is displeased about a

rule or policy may tell the supervisor how upset another employee is rather than admit personal dissatisfaction.

- **Rationalization:** When an individual presents a reason that is less ego deflating or more socially acceptable than the true reason. A baker who blames the oven for poor bakery products is using this defense mechanism.

To some extent, everyone relies on defense mechanisms. Excessive defensive behaviors by employees may be minimized, however, if supervisors encourage constructive behavior. Also, managers who understand defensive behavior should have greater empathy and realize that such behaviors are methods of coping with frustration.

Theories of Motivation

A number of theories of motivation have been developed: need hierarchy, achievement-power-affiliation, two-factor, expectancy, and reinforcement. These theories, described briefly in the following pages, are all different constructs that may prove useful in understanding behavior.

Need Hierarchy

One of the most popular theories of motivation was proposed by Maslow (1943). This theory, frequently referred to as the need hierarchy theory, states that people are motivated by their desire to satisfy specific needs, which are arranged in the following ascending hierarchical order (Figure 10.2).

- **Physiological**—needs of the human body that must be satisfied to sustain life
- **Safety**—needs concerned with the protection of individuals from physical or psychological harm
- **Social**—needs for love, affection, belonging
- **Esteem**—needs relating to feelings of self-respect and self-worth, along with respect and esteem from one's peers
- **Self-actualization**—needs related to one's potential or to the desire to fulfill one's potential

Prepotent need: Need that is dominant over all others.

According to this theory, each need is prepotent or dominant over all higher-level needs until it has been partially or completely satisfied. A **prepotent need** is one that has greater influence over other needs. Also, according to this theory, a satisfied need is no longer a motivator. A prepotent lower-order need, however, might not need to be satisfied completely before the next higher one becomes potent or dominant. For example, the safety need may not have to be satisfied completely before social needs become motivators.

In our society, the physiological and safety needs are more easily and frequently satisfied than other needs. Many of the tangible rewards, such as pay and fringe benefits, that organizations offer are primarily directed to physiological and safety needs.

The strengths of an individual's needs may shift in different situations. During bad economic times, for example, physiological and safety needs may dominate an individual's behavior, whereas higher-level needs may dominate in good economic times. Also, different methods may be used by different individuals to satisfy particular needs.

MASLOW'S HIERARCHY	EXAMPLES OF METHODS FOR SATISFYING NEEDS
Self-Actualization Needs (realizing one's potential growth using creative talents)	Challenging work allowing creativity, opportunities for personal growth, and advancement
Esteem Needs (achievement recognition and status)	Title and responsibility of job, praise and rewards as recognition for accomplishments, promotions, competent management, prestigious facilities
Social Needs (love, belonging, affiliation, acceptance)	Friendly associates, organized employee activities such as bowling or softball leagues, picnics, parties
Safety Needs (protection against danger, freedom from fear, security)	Benefit programs such as insurance and retirement plans, job security, safe and healthy working conditions, competent, consistent, and fair leadership
Physiological Needs (survival needs, air, water, food, clothing, shelter, and sex)	Pay, benefits, working conditions

Figure 10.2. Maslow's need hierarchy and methods of satisfying needs in organizations.
Source: Maslow's hierarchy of needs data from *Motivation and Personality,* 3rd ed., by Abraham H. Maslow. Revised by Robert Frager, James Fadiman, Cynthia McReynolds, and Ruth Cox. Copyright 1954, © 1987 by Harper & Row Publishers, Inc. Copyright © 1970 by Abraham H. Maslow. Reprinted by permission of HarperCollins Publishers, Inc.

Interesting work and opportunities for advancement are means that organizations use to appeal to higher-order needs. Obviously, determining the need level of each individual foodservice employee can be a difficult process, but managers may find this theory useful in attempting to understand their employees' motivations.

Achievement-Power-Affiliation

In his writing on motivation, McClelland (1985) emphasized needs that are learned and socially acquired as individuals interact with the environment. The **achievement-power-affiliation theory** holds that all people have three needs:

- A need to achieve
- A need for power
- A need for affiliation

Achievement Motive. The need for achievement is a desire to do something better or more efficiently than it has been done before. An individual with a high need for achievement tends to have the following traits:

- Responds to goals
- Seeks a challenge but establishes attainable goals with only a moderate degree of risk

- Exhibits greater concern for personal achievement than rewards of success
- Desires concrete feedback on performance
- Takes personal responsibility for finding solutions to problems
- Maintains a high energy level and willingness to work hard

Persons high in the need for achievement tend to gravitate toward managerial and sales positions. In these occupations, individuals are often able to manage themselves and thus satisfy the basic drive for achievement. Individuals with high achievement needs tend to get ahead in organizations because they are producers—they get things done. These individuals are task oriented and work to their capacity, and they expect others to do the same. As a result, the foodservice manager who has a high need for achievement may sometimes lack the human skills and patience necessary to manage employees with lower achievement motivation.

Power Motive. The need for power is basically a concern for influencing people. An individual with a high-power need tends to exhibit the following behavior:

- Enjoys competition with others in situations allowing domination
- Desires acquiring and exercising power or influence over others
- Seeks confrontation with others

McClelland and Burnham (1976) identified two aspects of power: positive and negative. Positive use of power is essential for a manager to accomplish results through the efforts of others in an organization. The negative aspect of power is when an individual seeks power for personal benefits, which may be detrimental to the organization.

Affiliation Motive. The need for affiliation is characterized by the desire to be liked by others and to establish or maintain friendly relationships. A person with a high need for affiliation tends to be one who

- Wants to be liked by others
- Seeks to establish and maintain friendships
- Enjoys social activities
- Joins organizations

McClelland (1985) maintained that most people have a degree of each of these needs, but the level of intensity varies. For example, an individual may be high in the need for achievement, moderate in the need for power, and low in the need for affiliation. Managers should recognize these differing needs in their employees when dealing with them. A foodservice employee with a high need for affiliation, for instance, would probably respond positively to warmth and support, whereas an employee with a high need for achievement would tend to respond to increased responsibility or feedback.

Two-Factor

Herzberg (1966) developed the **two-factor theory** of work motivation, which focuses on the rewards or outcomes of performance that satisfy needs. Two sets of rewards or outcomes are identified: those related to job satisfaction and those related to job dissatisfaction. Those factors related to satisfaction, called *motivators* are related to the environment or content of the job. They include

- Achievement
- Recognition
- Responsibility
- Advancement
- The work itself
- Potential for growth

Those factors related to dissatisfaction, called *maintenance or hygiene factors,* are related to the environment or context of the job. They include

- Pay
- Supervision
- Job security
- Working conditions
- Organizational policies
- Interpersonal relationships on the job.

Based on his research, Herzberg concluded that although employees are dissatisfied by the absence of maintenance factors, the presence of those conditions does not cause motivation. The maintenance factors, he says, are necessary to maintain a minimum level of need satisfaction.

In addition, the presence of some job factors cause high levels of motivation and job satisfaction, but the absence of these factors may not be highly dissatisfying. This second group of factors, which are internal to the job, are a major source of motivation.

Herzberg's research often has been criticized on the basis that the results of his initial interviews can be interpreted in several different ways, that his sample of 200 accountants and engineers is not representative of the general population, and that subsequent research often has not supported his theory. Although Herzberg's theory is not held in high esteem by researchers in the field, it has had a major impact on manager's awareness of the need to increase motivation in the workplace.

An examination of Herzberg's and Maslow's theories of motivation reveals some similarities. Maslow's theory is helpful in identifying the needs or motives; Herzberg's theory provides insights into the goals and incentives that tend to satisfy these needs. The two theories are compared in Figure 10.3. Maslow's physiological, safety, and social needs are motivation factors. The esteem needs, however, involve both status and recognition. Because status tends to be a function of the position a person occupies and recognition is gained through competence and achievement, esteem needs are related to both maintenance and motivation factors. Self-actualization needs in Maslow's conceptualization are related to Herzberg's motivational factors.

Expectancy

Managers should develop an understanding of human needs and the variety of organizational means available to satisfy employees' needs. The needs approach to motivation, however, does not account adequately for differences among individual employees or explain why people behave in many different ways when accomplishing the same or similar goals.

To explain these differences, several expectancy approaches to motivation have been advanced in the past several years. Two of the most prominent ones were

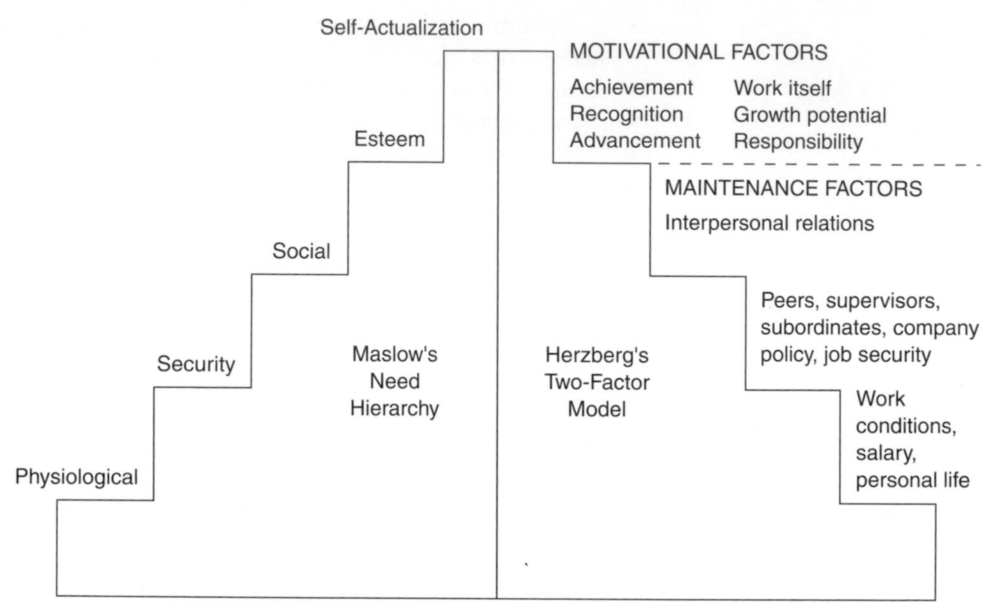

Figure 10.3. A comparison of Maslow's and Herzberg's models of motivation.

developed by Vroom (1994) and Porter and Lawler (1968). **Expectancy theory** attempts to explain behavior in terms of an individual's goals, choices, and expectations of achieving these goals. This theory assumes people can determine the outcomes they prefer and make realistic estimates of their chances of attaining them.

Expectancy theory is based on the belief that people act in such a manner as to increase pleasure and decrease displeasure. According to this theory, people are motivated to work if they believe their efforts will be rewarded and if they value the rewards that are being offered. The first requirement can be broken down into the following two components:

- the expectancy that increased effort will lead to increased performance
- the expectancy that increased performance will lead to increased rewards

These expectancies are developed largely from an individual's experiences.

Valence: Value an employee places on rewards offered by the organization.

The second part of expectancy theory is concerned with the value the employee places on the rewards offered by the organization, also referred to as **valence**. Organizations have tended to assume that all rewards will be valued by employees. Obviously, however, some rewards are more valued than others, and certain rewards may be viewed negatively by some employees.

According to this theory, the factors of expectancy and valence determine motivation, and both must be present for a high level of motivation. In other words, high expectancy or high valence alone will not ensure motivation. If a foodservice employee places a high value on money (high valence) but believes there is little chance of receiving a pay increase (low expectancy), the employee will probably not be highly motivated to work hard because of the low probability of receiving the pay increase.

All employees in an organization do not share the same goals or values regarding pay, promotions, benefits, or working conditions. Managers must consider an employee's goals and values in attempting to create a motivational climate. A key factor in expectancy theory is what the employee perceives as important or of value—"what's in it for me"—not what the manager believes the employee should value.

A major contribution of expectancy theory is that it explains how the goals of employees influence behavior at work. Employee behavior is influenced by assessments of the probability that certain behavior will lead to goal attainment.

Reinforcement

Reinforcement theory, which is associated with Skinner (1971), is often called *operant conditioning* or *behavior modification*. Rather than emphasize the concept of a motive or a process of motivation, these theories deal with how the consequences of a past action influence future actions in a cyclical learning process.

According to Skinner, people behave in a certain way because they have learned at some previous time that certain behaviors are associated with positive outcomes and that others are associated with negative outcomes. Because people prefer pleasant outcomes, they are likely to avoid behaviors that have unpleasant consequences. For example, foodservice employees may be likely to follow the rules and policies of the organization because they have learned during previous experiences—at home, at school, or elsewhere—that disobedience leads to punishment.

The general concept behind reinforcement theory is that reinforced behavior will be repeated, and behavior that is not reinforced is less likely to be repeated. *Reinforcers* are not always rewards and do not necessarily have to be positive in nature. The example cited previously of the foodservice employee wishing to avoid disciplinary action is an avoidance reinforcer. The current emphasis in management practices is on the use of positive reinforcers, including both tangible and intangible rewards, such as pay increases, promotions, or recognition for good performance.

Job Satisfaction

Closely related to motivation is the concept of job satisfaction. While motivation reflects a person's drive to perform, satisfaction reflects that person's attitude in a situation. Satisfaction is largely determined by conditions in the environment and in the situation; motivation is determined by needs and goals.

According to George and Jones (2002), there are many theories and models of job satisfaction, all of which focus on one or more of four components: personality, values, the work situation, and social influence. **Job satisfaction** refers to the individual's mind-set about the job, which may be positive or negative. It is not synonymous with organizational morale. *Morale* is related to group attitudes and job satisfaction is concerned with individual attitudes.

Job satisfaction: An individual's feelings and beliefs about his or her job.

Many managers believe that job satisfaction is positively associated with job performance; that is, the greater the job satisfaction, the greater the performance. Years of research, however, have failed to support such a belief; rather research suggests there is little if any relationship between job satisfaction and job performance.

Research does suggest however, a negative relationship exists between job satisfaction and turnover; that is, as job satisfaction decreases, the likeliness to leave an organization increases. According to George and Jones (2002), workers who are very satisfied with their jobs may never think about quitting; the point of becoming dissatisfied with a job often is the point at which individuals begin searching for a new job.

Organizational Citizenship Behaviors (OCBs): Positive, voluntary behaviors that enhance organizational efficiency.

Job satisfaction also appears to be related to what have been termed **Organizational Citizenship Behaviors (OCBs).** These are behaviors that are voluntary and above and beyond the call of duty. They are not required of organization members; however, they are necessary for organizational efficiencies and include such things as protecting the organization from theft, helping co-workers, making constructive suggestions, and developing one's skills and abilities. Research suggests that satisfied employees are more likely to engage in various forms of OCBs.

Affective commitment: Commitment to an organization because one is happy to be working for the organization, believes in the organization, and wants to do what is best for the organization.

Continuous commitment: Commitment to an organization only because the cost of leaving is too great.

The feelings and beliefs an individual has about his/her job is termed job satisfaction. When these feelings and beliefs are focused more broadly on the organization, the concept organizational commitment is used. Allen and Meyer (1996) describe two types of organizational commitment, affective and continuous. **Affective commitment** is the commitment to the organization that employees display because they are happy to be members of the organization, believe in it and what it stands for, and intend to do what is best for the organization. An employee with high affective commitment stays with an organization because they want to stay. **Continuous commitment** is the commitment that exists when an employee feels the costs of leaving the organization are too great (i.e., loss of seniority, pension, medical benefits, etc). An employee with high continuous commitment stays with an organization because they believe they must stay, not because they want to stay. George and Jones (2002) suggest that when affective commitment is high, employees are more likely to want to do what is good for the organization, more likely to perform OCBs, and less likely to quit. They indicate that when continuance commitment is high, employees usually will not go beyond the call of duty because their commitment is based more on necessity than on a belief in what the organization stands for.

LEADERSHIP

Leadership: Process of influencing activities of an individual or group toward achieving organizational goals.

Leadership is the process of influencing the activities of an individual or group in efforts toward goal achievement. As Hersey, Blanchard, and Johnson (1996) stated, the leadership process (L) is identified as a function of the leader (l), the follower (f), and other situational (s) variables. This relationship can be expressed as follows:

$$L = f(l, f, s)$$

This definition makes no mention of any particular type of organization because leadership occurs in any situation in which someone is trying to influence the behavior of another individual or group. Thus, everyone attempts leadership at one time or another in his or her activities at work, in social settings, at home, or elsewhere.

Dynamic and effective leadership is one major attribute that distinguishes successful from unsuccessful organizations. Leaders are those who are willing to assume significant leadership roles and who have the ability to get the job done effectively. The effective organizational leader is one who can influence people to strive willingly for group objectives.

Leadership and Power

Leaders possess power, and power has various dimensions. The concept of power is closely related to the concepts of leadership and authority because power is one of the means by which a leader influences the behavior of followers.

Traditional Power Structure

While leadership can be viewed as any attempt to influence, power can be described as a leader's influence potential (Hersey, Blanchard, & Johnson, 1996). Power is the resource that enables a leader to induce compliance or influence others. Some authors distinguish between position power and personal power. *Position power* is derived from a person's official position in an organization, whereas *personal power* comes from personal attributes and expertise.

Several other authors have developed more specific power-base classifications. The one devised by French and Raven (1960) is probably the most widely accepted. They proposed five bases of power: legitimate, reward, coercive, expert, and referent power. Information power and connection power are two additional bases identified by other authorities (Hersey, Blanchard, & Johnson, 1996; Raven & Kluganski, 1970). These seven bases of power, or potential means of influencing the behavior of others, are defined as follows:

- **Legitimate power.** Comes from the formal position held by an individual in an organization; generally, the higher the position, the higher legitimate power tends to be. A leader high in legitimate power induces compliance from others because the followers believe this person has the right to give directions by virtue of his or her position.
- **Reward power.** Comes from a leader's ability to reward others. Examples of formal rewards are increases in pay, promotions, or favorable job assignments.
- **Coercive power.** Comes from the authority of the leader to punish those who do not comply. A leader with coercive power can fire, demote, threaten, or give undesirable work assignments to induce compliance from others.
- **Expert power.** Held by those leaders who are viewed as being competent in their job. Knowledge gained through education or experience and a demonstration of ability to perform are sources of expert power. A leader high in expert power can influence others because of their respect for his or her abilities.
- **Referent power** (sometimes called charisma). Based on identification of followers with a leader. A leader high in referent power is generally well liked and admired by others; thus, the leader can influence others because of this identification and admiration.
- **Information power.** Based on the leader's possession of or access to information that others perceive as valuable. This power influences others either because they need the information or want to be a part of things.

- **Connection power.** Based on the leader's connections with influential or important persons inside or outside the organization. A leader high in connection power induces compliance from others who aim at gaining the favor or avoiding the disfavor of the influential connection.

These concepts have application to understanding the power of leaders in any organization, including those in the foodservice industry. An example of this power is a hypothetical but typical situation involving a chef in a country club operation. In our hypothetical example, David Lott is a highly skilled chef who has responsibility for directing the foodservice operations at Stoneyville Country Club. Mr. Lott is considered fair and consistent but an all-business type of supervisor. He is responsible for hiring and firing all foodservice staff and for recommending pay increments and promotions to the club manager.

Mr. Lott's potential means for influencing the foodservice staff—or, in other words, his power—is derived from the position he holds in the organization. Because he is in charge of the foodservice and has authority for punishment and rewards, he has legitimate, coercive, and reward power. Also, from the description of him as a "highly skilled chef," the inference is that he has expert power; from the "all-business" classification, the inference could be made that his referent power is limited.

Although all these dimensions of power are important to leaders, expert and referent powers tend to be related to subordinates' satisfaction and performance, and coercive is the most negative. Expert and legitimate powers appear to be the most important for compliance. In general, depending on the situation, any one of these dimensions is of value to the leader.

Evolving Power Structure

Managerial authority is diminishing and new tools of leadership are emerging (Kanter, 1989). Managers whose power came from hierarchy and who were accustomed to limited personal control are learning to shift their perspectives and widen their scopes. The new managerial work consists of looking outside an area of responsibility to find opportunities and of forming teams from relevant disciplines to address them. This work involves communication and collaboration across functions, units, and other operations with overlapping activities and resources. Title and rank will be less important factors in success in the new managerial work than having the knowledge, skills, and sensitivity to mobilize people and motivate them to do their best. A prime example of this shift may be seen in a hospital in which the patient tray delivery supervisor discusses the means of improving patient service with the nursing supervisor on the unit. Under the older, more autocratic system, the tray delivery supervisor would have discussed improvements with the foodservice manager, who would have followed through with a decision. Direct communication between the two supervisors involves shifting power and produces more immediate effective action.

Philosophies of Human Nature

Assumptions that managers have regarding other people are major factors determining the climate for motivation. To learn how to create an environment conducive to a high level of employee motivation, managers should develop an understanding and philosophy of human nature and its influence on leadership style. A review of the work of McGregor and Bennis (1985) and Argyris (1957) is useful in gaining this understanding.

McGregor's Theory X and Theory Y

McGregor stressed the importance of understanding the relationship between motivation and philosophies of human nature. In observing the practices and approaches of managers, McGregor concluded that two concepts of human nature were predominant. He referred to these as **Theory X** and **Theory Y.** These two distinct concepts—the negative Theory X and the positive Theory Y—relate to basic philosophies or assumptions that managers hold regarding the way employees view work and how they can be motivated. These philosophies are summarized in Figure 10.4.

Theory X suggests that motivation will be primarily through fear and that the supervisor will be required to maintain close surveillance of subordinates if the organizational objectives are to be attained. Furthermore, the manager must protect the employees from their own shortcomings and weaknesses and, if necessary, goad them into action. Although Theory X is by no means without its supporters, it is not in keeping with more current concepts of behavioral science. Theory Y, in contrast, emphasizes managerial leadership by permitting subordinates to experience personal satisfaction and to be self-directed. These contrasting sets of assumptions lead to different leadership styles among managers and different behaviors among employees. Managers who hold to Theory X tend to be autocratic, and those with a Theory Y philosophy tend to be more participative.

Mondy, Sharplin, and Flippo (1988) suggested that if a manager accepts the Theory Y philosophy of human nature, the following practices might be considered seriously: flexible working hours, job enrichment, participative decision making, and abandonment of time clocks. One should not conclude, however, that McGregor

Theory X	Theory Y
1. Work is inherently distasteful to most people.	1. Work is as natural as play, if the conditions are favorable.
2. Most people are not ambitious, have little desire for responsibility, and prefer to be directed.	2. Self-control is often indispensable in achieving organizational goals.
3. Most people have little capacity for creativity in solving organizational problems.	3. The capacity for creativity in solving organizational problems is widely distributed in the population.
4. Motivation occurs only at the physiological and security levels.	4. Motivation occurs at the social, esteem, and self-actualization levels, as well as at the physiological and security levels.
5. Most people must be closely controlled and often coerced to achieve organizational objectives.	5. People can be self-directed and creative at work if properly motivated.

Figure 10.4. McGregor's Theory X and Theory Y.
Source: Hersey, P., Blanchard, K. H., & Johnson, D. E. (1996). *Management of Organizational Behavior: Utilizing Human Resources,* 7th edition, 1996. Reprinted with permission of Center for Leadership Studies, Escondido, CA.

advocated Theory Y as the panacea for all management problems. Although Theory Y is no utopia, McGregor argued that it provides a basis for improved management and organizational performance.

Argyris's Immaturity-Maturity Theory

Review of the work of Argyris (1957) also will assist managers in developing an understanding of human behavior. According to Argyris, a number of changes take place in the personality of individuals as they develop into mature adults over the years (Table 10.1). Further, these changes reside on a continuum, and the "healthy" personality develops along the continuum from immaturity to maturity.

Argyris questioned the assumption that widespread problems of worker apathy and lack of effort in organizations are simply the result of individual laziness. When people join the workforce, he contended, many jobs and management practices are not designed to support their mature personality. Employees who have minimal control over their environment tend to act in passive, dependent, and subordinate ways and, as a result, to behave immaturely.

According to Argyris, treating people immaturely is built into traditional organizational principles such as task specialization, chain of command, unity of direction, and span of control. He stated that these concepts of formal organization lead to assumptions about human nature that are incompatible with the proper development of maturity in human personality. In colloquial terms, the Argyris theory can be summarized as follows: Management creates "Mickey Mouse" jobs and then is surprised with "Mickey Mouse" behavior.

Argyris challenged management to provide a work climate in which individuals have a chance to grow and mature as individuals while working for the success of the organization. He contended that giving people the opportunity to grow and mature on the job allows employees to use more of their potential. Furthermore, although all workers do not want to accept more responsibility or deal with the problems that responsibility brings, the number of employees whose motivation can be improved is much larger than many managers suspect.

The Argyris theory has application to understanding jobs and work behavior in the foodservice industry. Although essential to operations, the industry has many jobs that are routine, repetitive, unchallenging, and boring. The foodservice manager should not be surprised to find that employees are not turned on to such jobs as pot-washer, dishwasher, or general kitchen worker. The challenge to the food-

Table 10.1 Argyris's Immaturity-Maturity Theory

Immaturity	vs.	Maturity
Passive		Increased activity
Dependence		Independence
Behave in a few ways		Capable of behaving in many ways
Erratic, shallow interests		Deeper, stronger interests
Short time perspective		Long time perspective (past and future)
Subordinate position		Equal or superordinate position
Lack of awareness of self		Awareness and control over self

service manager is to enrich jobs to the maximum extent possible. One must acknowledge, however, that the potential is limited for making some jobs in the foodservice operation highly motivating and exciting.

Leadership Effectiveness

Although you may be able to identify a person you believe is a leader, being able to define leadership is much more difficult. A variety of definitions exist for leadership, we will use the following: Leadership is the ability to create an environment in which members of the organization are motivated to help the organization achieve its goals.

An effective leader influences others. Some leaders have formal authority to exert influence on others and are called **formal leaders.** Other leaders may have no formal job authority, yet may exert considerable influence because of special skills or talents. These leaders are termed **informal leaders.** In general, both exist in every organization and both, because of their influence, can help an organization meet its goals.

Trait Concepts in Leadership

Many studies have been conducted and articles written attempting to identify traits of leaders. Traits are personal characteristics that describe how a person thinks, feels, and/or acts.

Maxwell (1999) identified 21 personal characteristics that he believed were needed for someone to be a truly effective leader. Those traits are

- Character
- Charisma
- Commitment
- Communication
- Competence
- Courage
- Discernment
- Focus
- Generosity
- Initiative
- Listening
- Passion
- Positive attitude
- Problem solving
- Relationships
- Responsibility
- Security
- Self-discipline
- Servanthood
- Teachability
- Vision

George and Jones (2002) summarized several research projects and proposed the following as traits that showed the strongest relationship to leadership:

- Intelligence
- Task-relevant knowledge
- Dominance
- Self-confidence
- Energy/activity level
- Tolerance for stress
- Integrity and honesty
- Emotional maturity

Basic Leadership Styles

Early studies on leadership identified three basic styles: autocratic, laissez-faire, and democratic (Rue & Byars, 1989). Responsibility for decision making is the key factor differentiating these leadership styles. Generally, the *autocratic leader* makes most decisions, the *laissez-faire leader* allows the group to make the decisions, and the *democratic leader* guides and encourages the group to make decisions.

In the early work on leadership styles, democratic leadership was considered the most desirable and productive. Current research does not necessarily support this conclusion. Instead, various styles of leadership have been found to be effective in different situations, which are discussed in more detail later in this chapter. The primary contribution of this early research was the identification of the three basic styles of leadership.

Behavioral Concepts of Leadership

When research shifted from an emphasis on personality and physical traits to an examination of leadership behavior, the focus was on determining the most effective leadership style. Many of the studies were conducted at the University of Michigan and Ohio State University. Building on the work at Michigan and Ohio State, and on results from their own research, Blake and Mouton (1964) proposed another leadership behavior model, the Managerial Grid®.

University of Michigan Leadership Studies. Leadership studies conducted at the Institute for Social Research at the University of Michigan were designed to characterize leadership effectiveness. These studies isolated two major concepts of leadership: employee orientation and production orientation (Kahn & Katz, 1960).

Employee-centered leaders were identified by their special emphasis on the human relations part of their job, and production-oriented leaders emphasized performance and the more technical characteristics of work. Results of the Michigan studies showed that supervisors of high-producing sections were more likely to have the following traits:

- Receive general rather than close supervision from their superiors
- Spend more time in supervision
- Give general rather than close supervision to their employees
- Be employee oriented rather than production oriented

Likert (1967), a former director of the Institute of Social Research at the University of Michigan, developed a continuum of leadership styles, from autocratic to participative, based on a summary he prepared of the research on leadership. He proposed the following four basic management styles:

- Exploitive autocratic
- Benevolent autocratic
- Consultative
- Participative

In the *exploitive autocratic* management style, employees are motivated by fear, threats, and punishment and seldom by reward. Almost all decisions are made by top management, and only occasionally does communication move up from employees to the manager. The *benevolent autocratic* style indicates that only certain minor decisions are made by employees, and communication moving upward is generally ignored. Small rewards are given, but threats and punishment are the norm. In *consultative style,* employees gain some confidence. Information flows up and down, but all major decisions come from the top. The *participative* management style operates on a basis of trust and responsibility. Employees discuss the job with their superiors, and communication flows up, down, and laterally; decision making is spread evenly through the organization.

Likert examined the characteristics of communication flow, decision-making processes, goal setting, control mechanisms, and other operational characteristics of organizations and assessed managerial and leadership styles. Results of these studies indicated that participative was the most effective style of management. Its emphasis is on a group participative role with full involvement of the employees in the process of establishing goals and making job-related decisions.

Ohio State Leadership Studies. Beginning in the 1940s, researchers at Ohio State University started a series of in-depth studies on the behavior of leaders in a wide variety of organizations. These studies were conducted about the same time as those at the University of Michigan and used similar concepts. Two dimensions of leadership behavior emerged from those studies (Stogdill, 1974): consideration and initiating structure.

Consideration indicates behavior that expresses friendship, develops mutual trust and respect, and develops strong interpersonal relationships with subordinates. Leaders who exhibit consideration are supportive of their employees, use their employees' ideas, and allow frequent participation in decisions.

Initiating structure indicates behavior that defines work and establishes well-defined communication patterns and clear relationships between the leader and subordinate. Leaders who initiate structure emphasize goals and deadlines, give employees detailed task assignments, and define performance expectations in specific terms.

In studying leader behavior, Ohio State researchers found that consideration and initiating structure are separate and distinct dimensions; a high score on one dimension does not necessitate a low score on the other. In other words, the behavior of a leader can be described as a mix of both dimensions. In depicting various patterns of leader behavior, the two dimensions were plotted on two axes, which resulted in four quadrants describing four leadership styles (Figure 10.5).

In general, researchers found that leaders who showed high consideration for people and initiating structure tended to have higher-performing and more satisfied subordinates than did others. They also concluded that the relationship between these dimensions and leadership effectiveness depends on the group.

The two-dimensional view of leadership is the basis for the Managerial Grid developed by Blake and Mouton (1964, 1978). The two dimensions of the grid are

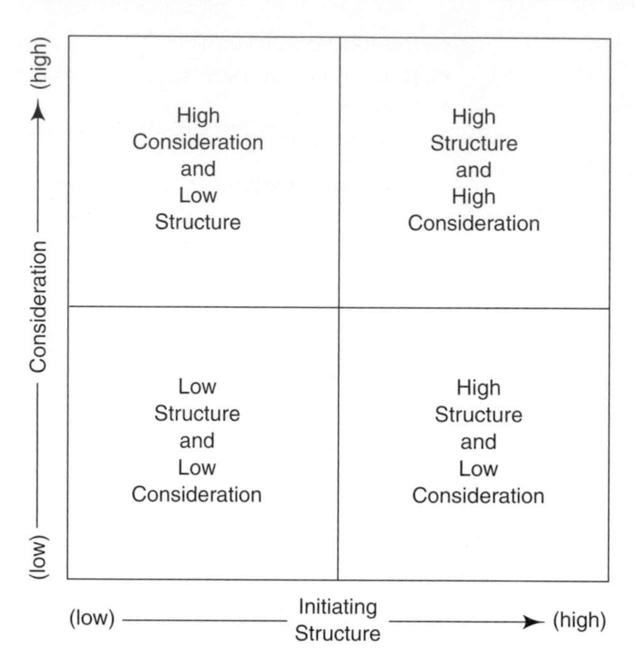

Figure 10.5. The Ohio State leadership quadrants.

concern for people and concern for production, which are similar to the dimensions of the Ohio State model.

Leadership Grid. The Blake and Mouton (1964, 1978) Managerial Grid then became the basis for the Leadership Grid® developed by Blake and McCanse (1991), shown in Figure 10.6. It has 9 possible positions along the vertical and horizontal axes for a total of 81 possible leadership styles, although 5 basic styles similar to those of Blake and Mouton are generally discussed. The Leadership Grid was expanded following years of research to include two additional styles: 9 + 9 paternalistic management, in which reward is promised for compliance and punishment threatened for noncompliance, and opportunistic management, which depends on the style the leader believes will return him or her the greatest personal benefit. A third dimension that was revealed in the research explores the positive and negative motivations underlying the styles.

The Managerial Grid and now the Leadership Grid have been used widely in organization development and have become so popular among some managers that they refer to the styles by number. For example, 9,9, rather than team management, is often used to refer to the leader who has high concern for both people and work.

Leadership Practices Inventory. Kouzes and Posner (2002) developed the Leadership Practices Inventory (LPI) as a tool for measuring practices they found common in leaders. Their work proposes that successful leaders exhibit five best practices:

- **Challenging the Process**—leaders take risks, innovate, and experiment.
- **Inspiring a Shared Vision**—leaders invent the future and enable others to become excited about this vision.
- **Enabling Others to Act**—leaders encourage team work and collaboration and empower others; believed to be the most significant of the five practices.

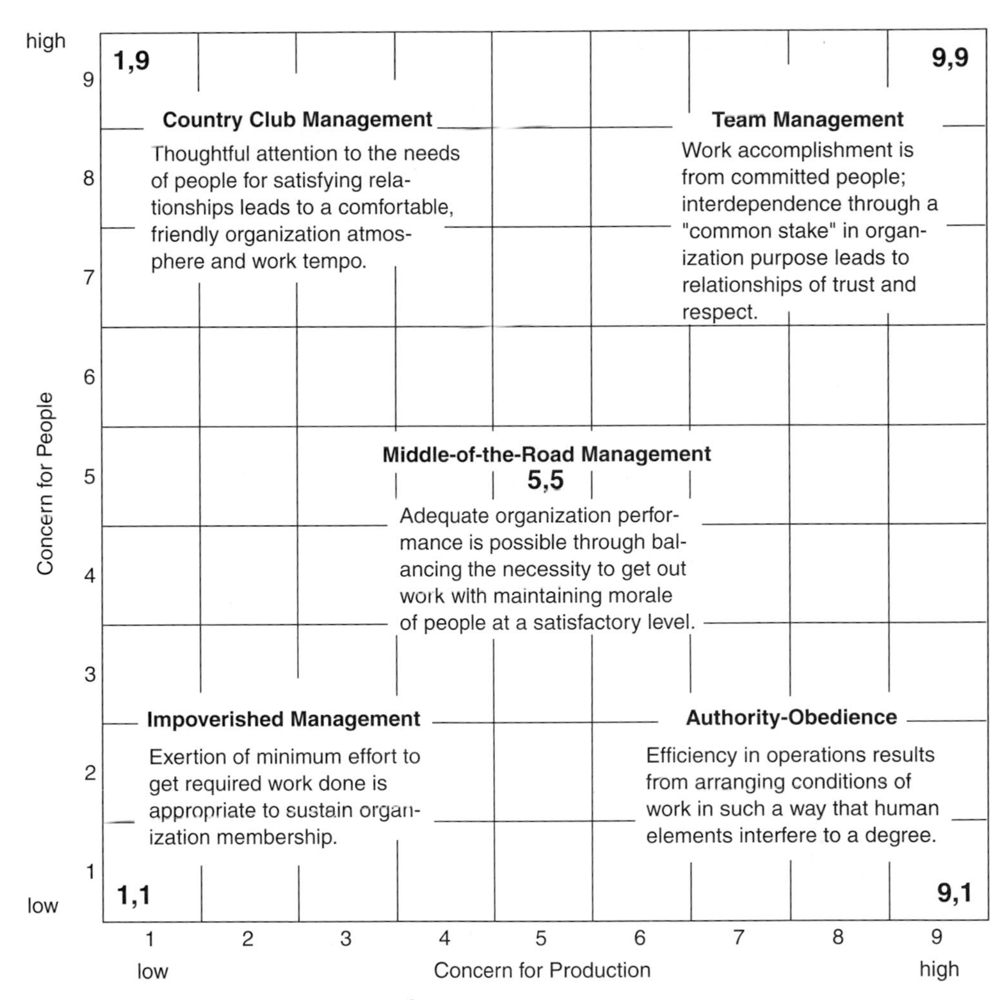

Figure 10.6. The Leadership Grid®.
Source: From *Leadership Dilemmas—Grid Solutions* (p. 29). By R. R. Blake and A. A. McCanse (formerly the Managerial Grid Figure by R. R. Blake and J. S. Mouton), 1991. Houston: Gulf Publishing Company. Copyright © 1991, by Scientific Methods, Inc. Reproduced by permission of the owners.

- **Modeling the Way**—leaders lead by example.
- **Encouraging the Heart**—leaders recognize the work of others and encourage others when they become exhausted, frustrated, or disenchanted.

Arendt and Gregoire (2005a, b) explored leadership behaviors in undergraduate students majoring in dietetics and hospitality management. They found that students perceived themselves as leaders in a variety of settings. Enabling Others to Act was the leadership practice most often reported by students.

Situational and Contingency Approaches

Situational and contingency approaches emphasize leadership skills, behavior, and roles thought to be dependent on the situation. These approaches are based on the hypothesis that behavior of effective leaders in one setting may be substantially different from that in another. The current emphasis in leadership research, which is largely focused on the leadership situation, has shifted because previous attempts to determine ineffective characteristics and behaviors were inconclusive.

One of the first situational approaches to leadership was developed by Tannenbaum and Schmidt (1958). Fiedler (1967) has also made significant contributions to understanding contingency or situational approaches to leadership. Reddin (1967a) and, more recently, Hersey, Blanchard, and Johnson (1996) have developed situational models of leadership. The path-goal leadership model is another contingency approach.

Leadership Continuum. Tannenbaum and Schmidt (1958) developed a continuum, or range, of possible leadership behaviors. Each type of behavior is related to the degree of authority used by the manager and the amount of freedom available to subordinates in reaching decisions. The actions range from those in which a high degree of control is exercised to those in which a manager releases a high degree of control.

This continuum was revised and published again in 1973. The revised model, shown in Figure 10.7, reflects two major changes. The interdependencies between

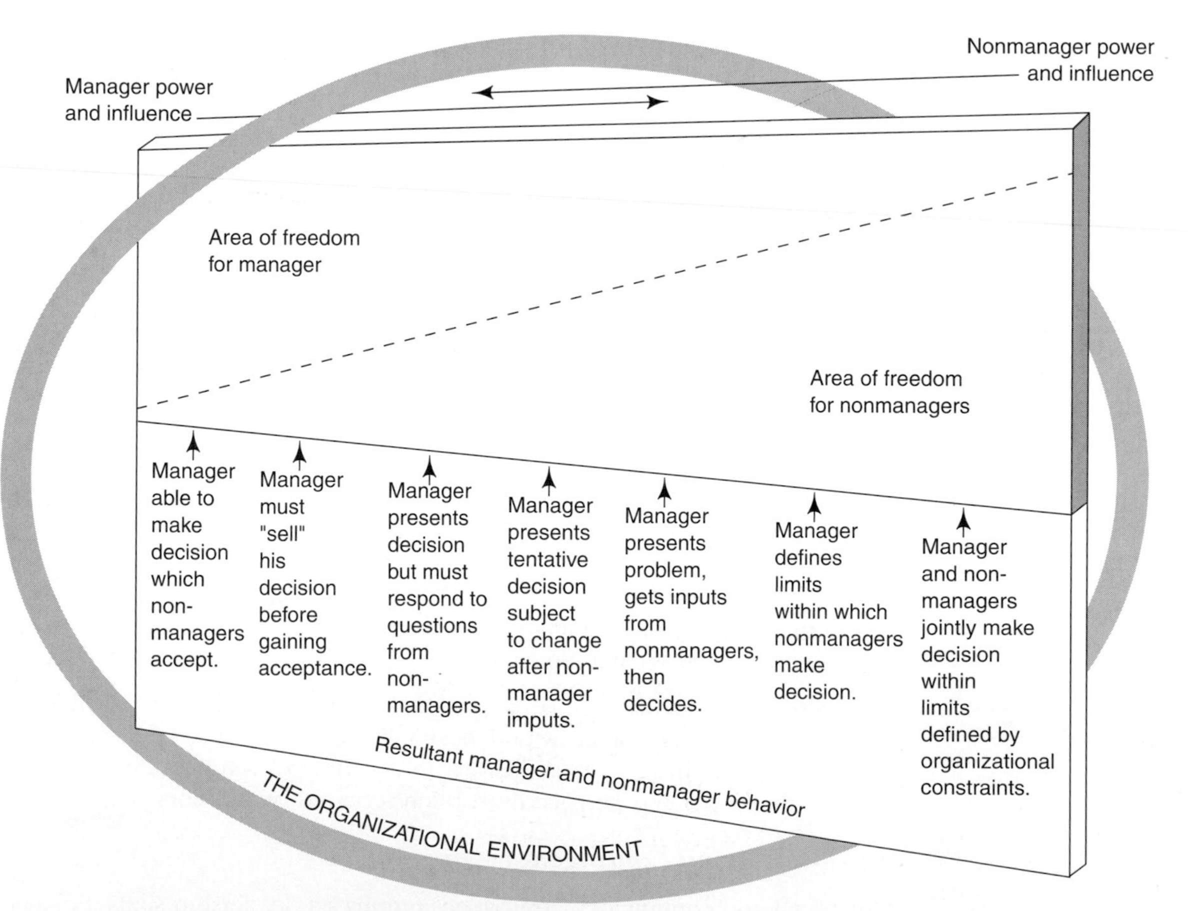

Figure 10.7. Continuum of manager-nonmanager behavior.
Source: Reprinted by permission of *Harvard Business Review.* An exhibit from "How to Choose a Leadership Pattern" by R. Tannenbaum and W. H. Schmidt, 1973, *51*(3), p. 162. Copyright © 1973 by the President and Fellows of Harvard College; all rights reserved.

the organization and its environment were acknowledged, as shown by the circular addition to the diagram depicting the organizational and societal environment. Also, originally, the terms *boss centered* and *subordinate centered* were used. In revising the model, the terms *manager* and *nonmanager* were substituted to denote functional rather than hierarchical differences.

Three forces were identified as affecting the leadership appropriate in a given situation:

- Forces in the manager
- Forces in subordinates or nonmanagers
- Forces in the situation

In Table 10.2, the factors within each of these forces are listed. These forces differ in strength and interaction in different situations; therefore, one style of leadership is not effective in all situations. In fact, the underlying concept of the continuum is that the manager may employ a variety of approaches, which are dependent on the forces operating in a particular situation.

Table 10.2 Forces in the Leadership Situation

Forces in the Manager	Forces in Nonmanagers	Forces in the Situation
Value system: How the manager personally feels about delegating Degree of confidence in staff	Need for independence: Some people need and want direction while others do not	Type of organization: Centralized versus decentralized
Personal leadership inclinations: Authoritarian versus participative	Readiness to assume responsibility: Different people need different degrees of responsibility	Work group effectiveness: How effectively the group works together
Feelings of security in uncertain situations	Tolerance for ambiguity: Specific versus general directions	The problem itself: The work group's knowledge and experience relevant to the problem
	Interest and perceived importance of the problem: People generally have more interest in and work harder on important problems	Time pressure: It is difficult to delegate to subordinates in crisis situations
	Degree of understanding and identification with organizational goals: A manager is more likely to delegate authority to an individual who seems to have a positive attitude about the organization	Demands from upper levels of management
	Degree of expectation in sharing in decision making: People who have worked under subordinate-centered leadership tend to resent boss-centered leadership	Demands from government, unions, and society in general

Source: Reprinted by permission of *Harvard Business Review.* An exhibit from "How to Choose a Leadership Pattern" by R. Tannenbaum and W. H. Schmidt, 1973, *51*(3), p. 162.

For example, a foodservice manager who is generally democratic and involves employees in decision making has to take charge in a crisis situation, such as a kitchen fire, that requires directive and authoritarian leadership.

Successful leaders are keenly aware of the forces most relevant to their behavior at a given time. They can act appropriately in relation to other individuals involved in a situation and to the organizational and social environmental forces. In general, however, Tannenbaum and Schmidt encouraged managers to shift toward more participative approaches to decision making. Some of the benefits of more participative styles are that they

- Raise employees' motivational level
- Increase willingness to change
- Improve quality of decisions
- Develop teamwork and morale
- Further the individual development of employees

A criticism of the continuum is that managers will be viewed as inconsistent or wishy-washy if they react differently in every situation. The key to effective leadership using the continuum is that the leader will behave consistently in situations in which similar forces are operating.

Contingency Approach. Fiedler (1967) developed a leadership contingency model in which he defined three major situational variables. In his theory, Fiedler proposed that these three variables seem to determine if a given situation is favorable to leaders:

- **Leader-member relations**—personal relations with members of the group
- **Task structure**—degree of structure in the task assigned to the group
- **Position power**—power and authority a leader's position provides

Fiedler defined the favorableness of a situation as the degree to which the situation enables the leader to exert influence over the group. The most favorable situation for leaders is one in which they are well liked by the members (good leader-member relations), have a powerful position (high position power), and are directing a well-defined job (high task structure). Fiedler examined task-oriented and relationship-oriented leadership to determine the most effective style.

In both highly favorable and highly unfavorable situations, a task-oriented leader seems to be more effective. In highly favorable situations, the group is ready to be directed and is willing to be told what to do. In highly unfavorable situations, the group welcomes the opportunity of having the leader take the responsibility for making decisions and giving directions. In moderately favorable situations, however, a relationship-oriented leader tends to be more effective because cooperation is more successful than task-oriented leadership for this particular group. Fiedler's work has been important in demonstrating particular styles of leadership that are most effective in given situations.

Leader Effectiveness Model. Hersey, Blanchard, and Johnson (1996) developed a leadership model that has gained considerable acceptance. In their model, task behavior and relationship behavior are used to describe concepts similar to those of consideration and initiating structure in the Ohio State studies. They define the terms in the following ways:

- **Task behavior**—the extent to which the leader engages in spelling out the duties and responsibilities of an individual or group. It includes telling people what to do, how to do it, when to do it, where to do it, and who is to do it.
- **Relationship behavior**—the extent to which the leader engages in two-way or multi-way communication. It includes listening, facilitating, and supportive behaviors (Hersey, Blanchard, & Johnson, 1996).

Recognizing that the effectiveness of leaders depends on how their styles interrelate with the situations in which they operate, an effectiveness dimension was added to the two-dimensional model based on the Ohio State studies (Figure 10.8). The effectiveness dimension was drawn from the work of Reddin (1967a), who contended that a variety of leadership styles might be either effective or ineffective.

By adding the effectiveness dimension to the task and relationship behavior dimensions, Hersey, Blanchard, and Johnson attempted to integrate the concepts of leadership style and situational demands of a specific environment. When the leader's style is appropriate to a given situation, it is termed *effective;* when the style is inappropriate, it is termed *ineffective*. The difference between effective and ineffective styles often is not the behavior of the leader but the appropriateness of the behavior to the environment in which it is used.

As illustrated in their model (Figure 10.9), leadership style in a particular situation can fall somewhere between extremely effective and extremely ineffective; effectiveness, therefore, is a matter of degree. As summarized in Table 10.3, the appropriateness of a leader's style in a given situation is related to the reactions of followers, superiors, and associates. The issue of consistency, which was pointed out with the Tannenbaum and Schmidt leadership continuum, might also be raised

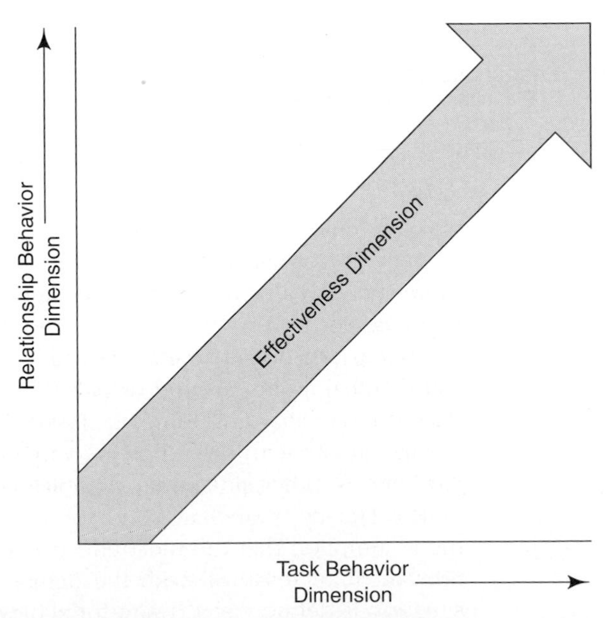

Figure 10.8. The effectiveness dimension of leadership.
Source: Hersey, P., Blanchard, K. H., & Johnson, D. E. (1996). *Management of Organizational Behavior: Utilizing Human Resources,* 7th edition, 1996. Reprinted with permission of Center for Leadership Studies, Escondido, CA.

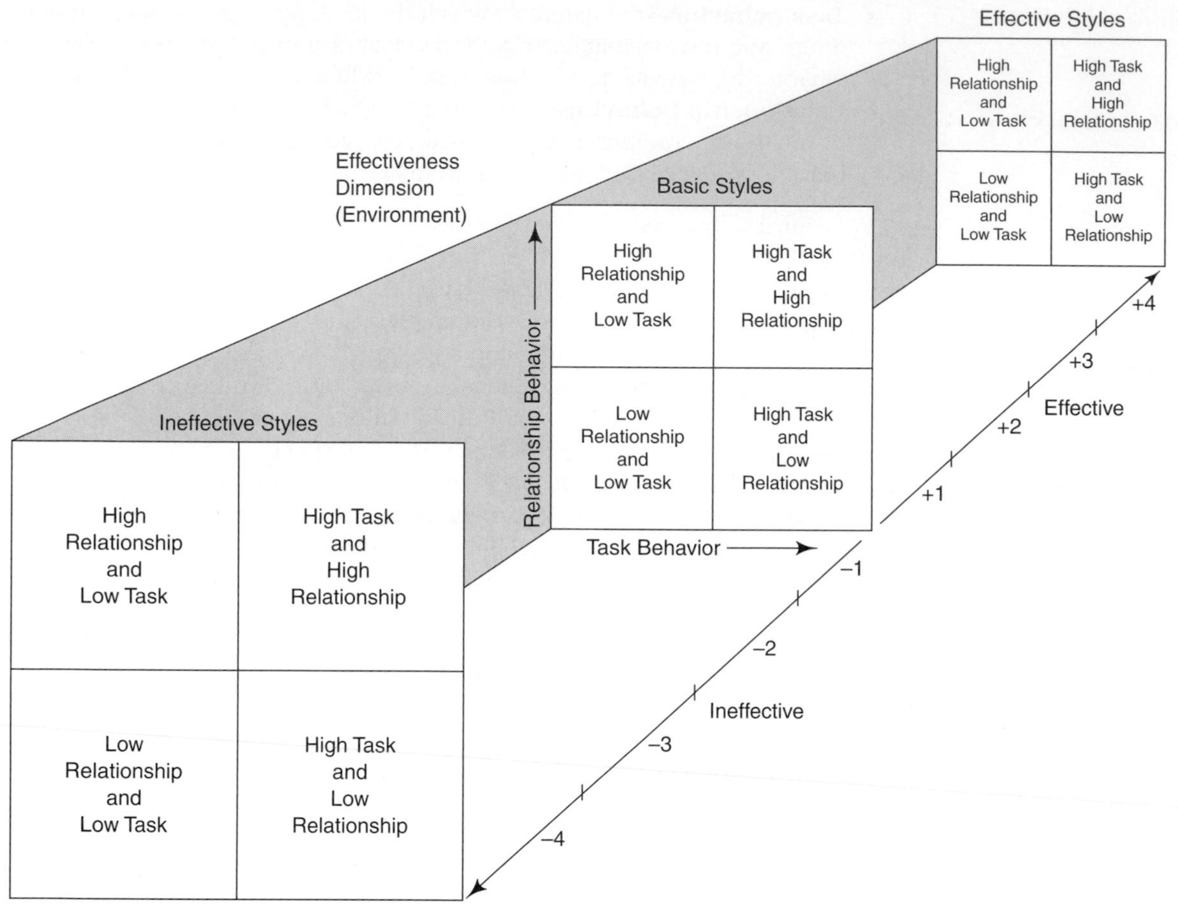

Figure 10.9. Leadership effectiveness model.
Source: Hersey, P., Blanchard, K. H., & Johnson, D. E. (1996). *Management of Organizational Behavior: Utilizing Human Resources,* 7th edition, 1996. Reprinted with permission of Center for Leadership Studies, Escondido, CA.

with the effectiveness model. Hersey, Blanchard, and Johnson concluded that consistency is not using the same style all the time; instead, *consistency* is using the same style for all similar situations and varying the style appropriately as the situation changes.

The task of diagnosing a leadership environment is complex because the leader is the focal point around which all the other environmental variables interact (Hersey, Blanchard, & Johnson, 1996). The leader has a style and expectations and has to interact with the following variables: followers, superiors, associates, and organizations; other situational variables; and job demands.

The Hersey, Blanchard, and Johnson (1996) leadership model also is based on the assumption that the most effective style varies with the followers' level of task-relevant readiness and with the demands of the situation. *Readiness* in the work situation is defined as a desire for achievement based on challenging but attainable goals, willingness and ability to accept responsibility, and education or experience and skills relevant to a particular task. Hersey, Blanchard, and Johnson defined four leadership styles appropriate to various readiness levels of followers in each

Table 10.3 Effective and Ineffective Leadership Styles

Basic Styles	Effective	Ineffective
High Task and Low Relationship Behavior	Seen as having well-defined methods for accomplishing goals that are helpful to the followers.	Seen as imposing methods on others; sometimes seen as unpleasant and interested only in short-run output.
High Task and High Relationship Behavior	Seen as satisfying the needs of the group for setting goals and organizing work, but also providing high levels of socio-emotional support.	Seen as initiating more structure than is needed by the group and often appears not to be genuine in interpersonal relationships.
High Relationship and Low Task Behavior	Seen as having implicit trust in people and as being primarily concerned with facilitating their goal accomplishment.	Seen as primarily interested in harmony; sometimes seen as unwilling to accomplish a task if it risks disrupting a relationship or losing "good person" image.
Low Relationship and Low Task Behavior	Seen as appropriately delegating to subordinates decisions about how the work should be done and providing little socioemotional support where little is needed by the group.	Seen as providing little structure or socio-emotional support when needed by members of the group.

Source: Adapted from *The 3-D Management Style Theory, Theory Paper #2—Managerial Styles,* by W. J. Reddin, 1967. Canada: Social Sciences Systems.

of their various areas of responsibility (Figure 10.10). A person's readiness level gives the manager a good clue about where to begin further development of that individual. A manager desiring to influence an employee in an area in which the employee is unable and unwilling, indicating a low readiness level (R1), must develop the person by closely supervising the staff member's behavior using the "telling" leadership style (S1), as shown in Figure 10.10.

Once this manager has diagnosed the readiness level of the follower as "low," the appropriate style can be determined (Figure 10.11). The manager would construct a right angle from a point on the readiness continuum to the point where it

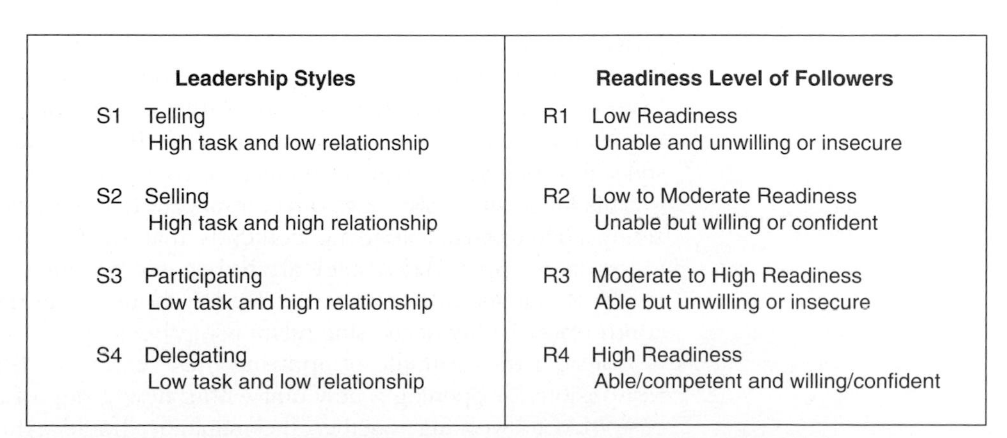

Figure 10.10. Relationship of leadership style and readiness of followers.
Source: Hersey, P., Blanchard, K. H., & Johnson, D. E. (1996). *Management of Organizational Behavior: Utilizing Human Resources,* 7th edition, 1996. Reprinted with permission of Center for Leadership Studies, Escondido, CA.

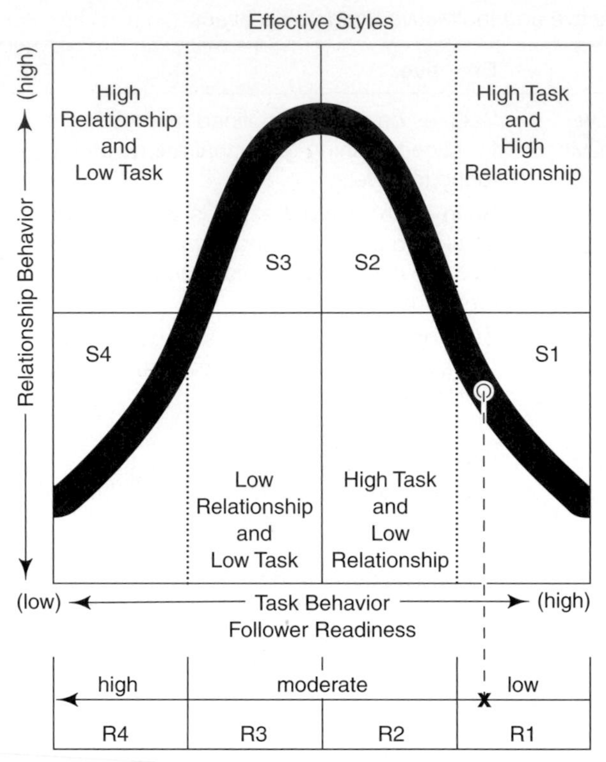

Figure 10.11. Leadership styles appropriate to various readiness levels of followers.
Source: Hersey, P., Blanchard, K. H., & Johnson, D. E. (1996). *Management of Organizational Behavior: Utilizing Human Resources,* 7th edition, 1996. Reprinted with permission of Center for Leadership Studies, Escondido, CA.

meets the curved line in the effective styles portion of the model. The leader should use a telling style (S1) in which the staff member must be told what to do and then shown how to do it.

In summary, Hersey, Blanchard, and Johnson's leadership effectiveness model provides a useful framework for leadership (Figure 10.11). Their model suggests that no one best leadership style is appropriate to meet the needs of all situations. Instead, leadership style must be adaptable and flexible enough to meet the changing needs of employees and situations. The effective leader is one who can modify styles as employees develop and change or as required by the situation.

This leadership model has direct application to leadership in foodservice organizations. Situational leadership contends that strong direction with low-readiness followers is appropriate if they are to become productive. As followers reach high levels of readiness, however, the leader should respond not only by decreasing control but also by decreasing relationship behavior.

Consider this example of applying these concepts. If a foodservice manager is responsible for opening a new unit with a new group of employees who are unaccustomed to working together, the manager should exhibit a directing, or telling, style of leadership. After the unit has been in operation several years and the work group has become accustomed to working as a team, assuming turnover is low and each employee knows his or her job well, a change in leadership style is indicated.

A participative, or delegating, style would probably be more appropriate. The foodservice manager, as with any other person in a leadership role, must be flexible in adapting his or her leadership style to changing situations and conditions.

Path-Goal Leadership Model. Another important contingency leadership concept, which focuses on the leader's effect on the subordinate's motivation to perform, was developed by Evans (1970) and House (1971). The model is based on the expectancy concept of motivation, which emphasizes expectancies and valences, previously discussed in this chapter. The path-goal concept focuses on the leader's impact on the subordinate's goals and the paths to achieve those goals.

As you will recall, expectancies are beliefs that efforts will be rewarded, and valences are the value or attractiveness of those rewards. Basically, the path-goal theory assumes that individuals react rationally in pursuing certain goals because those goals ultimately result in highly valued payoffs to the individual.

Leaders may affect employees' expectancies and valences in several ways:

- Assigning individuals to tasks for which they have high valences
- Supporting employee efforts to achieve goals
- Tying extrinsic rewards (pay increases, recognition, promotion) to accomplishment of goals
- Providing specific extrinsic rewards that employees value

These actions on the part of the leader can increase effectiveness because employees reach higher levels of performance through increased motivation on the job. Additionally, the path-goal theory implies that the degree to which the leader can be effective in eliciting work-goal directed behavior depends on the situation.

The path-goal leadership concept focuses on four types of leader behavior and two situational factors (House & Mitchell, 1974):

- **Directive**—leadership behavior characterized by providing guidelines, letting subordinates know what is expected, setting definite performance standards, and controlling behavior to ensure adherence to rules
- **Supportive**—leadership behavior characterized by being friendly and showing concern for subordinates' well-being and needs
- **Achievement oriented**—leadership behavior characterized by setting challenging goals and seeking to improve performance
- **Participative**—leadership behavior characterized by sharing information, consulting with employees, and emphasizing group decision making

The *situational factors* include subordinates' characteristics, such as locus of control, and characteristics of the work environment, such as structure and complexity of the task. *Locus of control* refers to the tendency of people to rely on internal or external sources. *Internal locus of control* is operational with people who attribute task success or failure to their own strengths and weaknesses, whereas *external locus of control* is characteristic of people with a tendency to attribute success or failure to the nature of the situation around them.

In Table 10.4, effective leader behaviors in relation to various situational factors are summarized. As an application of the path-goal concepts, foodservice employees with a high need for affiliation will be more satisfied with a supportive leader; those with a high need for security will be more satisfied with a directive leader. As

Table 10.4 Interaction of Leader Behavior and Situational Factors

Situational Factors		
Subordinate Characteristics	**Characteristics of the Work Environment**	**Effective Leader Behaviors**
High need for affiliation		Supportive
High need for security		Directive
Internal locus of control		Participative
External locus of control		Directive
	Structured task	Directive
	Unstructured task	Supportive
High growth need strength	Complex task	Participative and achievement oriented
Low growth need strength	Complex task	Directive
High growth need strength	Simple task	Supportive
Low growth need strength	Simple task	Supportive and directive

Source: Organizational Behavior: Applied Concepts (p. 362) by R. D. Middlemist and M. A. Hitt, 1981. Chicago: Science Research Associates. Used by permission.

indicated in the chart, the nature of the task also influences the leadership behavior appropriate to a given situation.

Reciprocal Approaches to Leadership

Reciprocal approaches to leadership focus more on the interactions among leaders and their followers than on characteristics of the leaders themselves. Emotional components of this interaction are recognized as important as well.

Transformational leadership: Leadership that inspires followers to become motivated to work toward organizational rather than personal gain.

Transformational Leadership. The concept of **transformational leadership** was introduced initially by Bernard Bass (Bass, 1985, 1986). According to Bass, transformational leadership occurs when leaders transform or change their followers in ways that lead the followers to

- Trust the leader
- Perform behaviors that contribute to the achievement of organizational goals
- Perform at a high level

George and Jones (2002) designed a model to show this transformational process. As shown in Figure 10.12, a transformational leader has charisma, intellectually stimulates followers, and engages in developmental consideration. As a result, followers have increased awareness of the importance of their tasks and of performing them well; are aware of their needs for personal growth, development, and accomplishment; and are motivated to work for the good of the organization rather than for their own personal gain or benefit.

Transformational leaders are agents of change. They are able to inspire members of the organization to achieve more than they thought possible. According to Robbins and Coulter (2002), transformational leaders pay attention to the concerns and developmental needs of individual followers; they change followers' awareness of issues by helping those followers look at old problems in new ways; and they are able to excite, arouse, and inspire followers to perform extra effort to achieve group goals.

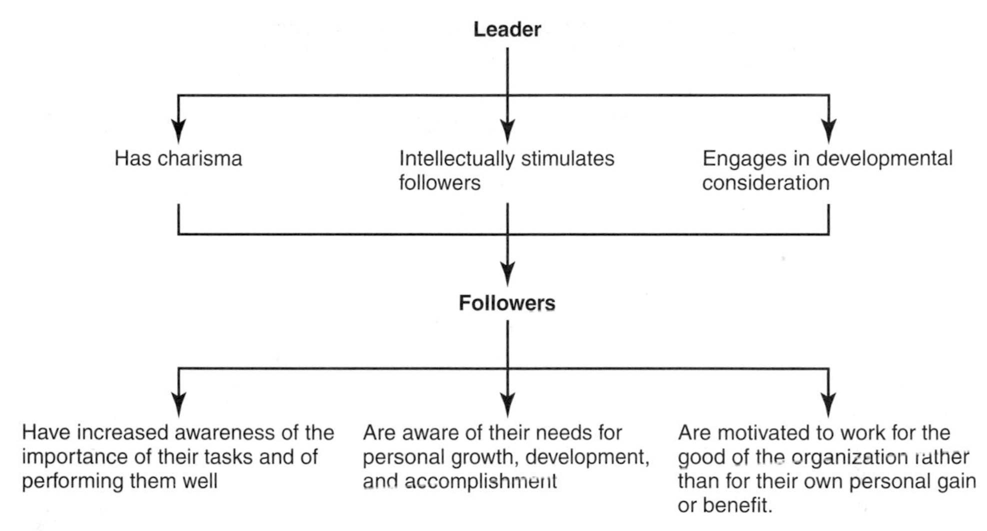

Figure 10.12. Transformational leadership.

Transactional leadership: Leadership that focuses on clarifying roles and responsibilities and use of rewards and punishment to achieve goals.

Transformational leadership often is contrasted with transactional leadership. **Transactional leaders** guide and motivate their followers by clarifying role and task requirements and by use of rewards and punishment. Transformational leadership build on and extends transactional leadership.

Servant Leadership. Greenleaf (1977) introduced the term *servant leader* to describe individuals who were servants first not leaders first; those who worked to be sure that other's needs were met and helped others to grow both physically and emotionally. A servant leader, according to Greenleaf, encourages collaboration, trust, foresight, listening, and the ethical use of power and empowerment.

Emerging Leadership Competencies. Yukl (2002) suggests that several additional leadership competencies have been identified: emotional intelligence, social intelligence, and metacognition. Each involves a cluster of related skills.

Emotional intelligence: Extent to which a person is in tune with his/her own feelings and the feelings of others.

Emotional intelligence has been defined as the extent to which a person is in tune with his/her own feelings and the feelings of others and can manage his/her emotions well in relationships with others (Goleman, 1998). It is the ability to integrate emotions and reasons in such a way that emotions are used to facilitate reasoning and emotions are managed with reasoning. Emotional intelligence involves skills in self-awareness of one's moods and emotions, communication of one's feelings to others, empathy in understanding moods and emotions in others, and self-regulation of emotions into appropriate behavior.

Social intelligence: Ability to determine the requirements for leadership in a particular situation and select the appropriate response.

Social intelligence is the ability to use social perceptiveness and behavioral flexibility to determine the requirements for leadership in a particular situation and select the appropriate response (Yukl, 2002). Social perceptiveness is defined as the ability to understand the functional needs, problems, and opportunities of a person or group and the characteristics, social relationships, and collective processes that will enhance or limit the ability to influence that person or group. Behavioral flexibility is the ability and willingness to modify one's behavior to fit the situation.

Metacognition: Ability to learn and adapt to change.

Metacognition is the ability to learn and adapt to change (Yukl, 2002). It involves the ability to introspectively analyze one's ability to define and solve problems and find ways to improve. It also includes self-awareness of one's strengths and limitations related to skills and emotions.

Primal Leadership. Goleman, Boyatzis, and McKee's (2002) concept of primal leadership stresses that the emotional task of the leader is primal (first and most important). Primal leadership draws heavily on the concept of emotional intelligence and emphasizes the importance of empathic listening and *resonance,* a leader's ability to perceive and influence the emotions of others. The authors reported that results of their research with nearly 4,000 executives showed that leaders who used styles with a positive emotional impact had firms with decidedly better financial returns than those who did not. The authors described six leadership styles within the primal leadership model:

- **Visionary**—leader articulates where the organization is going and attunes the vision to the values of the employees to help them believe in the importance of their work to this vision; viewed as the most effective style
- **Coaching**—leader works one-on-one with employees to help identify their strengths and weaknesses and career goals and build self confidence and autonomy; work is delegated to employees based on their abilities
- **Affiliative**—leader focuses on the emotional needs of employees and promotes harmony, fosters friendly interactions, and nurtures personal relationships
- **Democratic**—leader gains input from others in decision making process; listening is a key strength of this style
- **Pacesetter**—leader has a drive to succeed not based on external rewards but on own high standards of excellence; works in situations where employees are self-motivated, highly competent, and need little direction
- **Commanding**—leader issues orders without asking input; should be used only in crisis situations

Implications of Leadership Theories

From this discussion of leadership theories, one might believe that being a truly effective leader is almost impossible. Effective leadership is difficult to achieve, but managers have succeeded in the effort.

A common thread emerges from observations of effective leaders in all types of organizations. Successful leaders have either analyzed situational factors and adapted their leadership style to them or have altered the factors to match their style. They have vision and are able to inspire offers toward organizational goals.

This overview of the leadership literature should make one major conclusion obvious: No one best style of leadership exists. Leadership should be seen as a function of forces in the leader, in the followers, and in the situation. A range of leader behaviors can be effective or ineffective, depending on styles and expectations of superiors, followers, and associates and on job demands and organizational characteristics.

The successful leader must have a concern for both tasks and people in the work situation, because leadership is defined as a process of influencing activities of individuals in efforts toward goal achievement. Thus, flexibility and adaptability to changing conditions and situations are underlying concepts of effective leadership in all organizations.

Research on chief executive officers and presidents in the onsite foodservice industry discovered four foundations of effective leadership (Cichy, Sciarini, Cook, & Patton, 1991). According to them, effective leaders do the following:

- Develop and provide a complete vision
- Earn and return trust
- Listen and communicate effectively
- Persevere when others give up

Cichy, Sciarini, and Patton (1992) surveyed the top 100 foodservice leaders in the United States for their opinions about those personal qualities associated with effective leaders. The leaders agreed or strongly agreed that the following six keys to effective leadership were important to their leadership style:

- Develop a vision
- Trust your subordinates
- Encourage risk
- Simplify
- Keep your cool
- Invite dissent

The respondents did not attach a high level of importance to a seventh key, "being perceived as an expert." They then ranked the four foundations of leadership from the previous study as follows: (a) trust, (b) vision, (c) communication, and (d) perseverance.

Leadership Development

The process of helping individuals identify and improve their abilities to function in leadership roles is known as leadership development. These programs are based on the assumptions that leadership is essential to organizational success and leadership can be developed in individuals.

Greenberg (2005) identified several techniques that are being used in leadership development programs:

- **360-degree feedback**—getting evaluative input from subordinates, peers, and superiors helps identify aspects of one's leadership that might need to be changed
- **Networking**—interacting with individuals both within and outside the organization can provide important information, build peer relationships, and promote cooperation
- **Executive coaching**—interacting one-on-one with a coach to assess leader strengths and weakness and develop a comprehensive plan for leadership development
- **Mentoring**—receiving support from more experienced colleagues to help in career development
- **Job assignments**—holding positions that provide leadership experiences
- **Action learning**—participating in a continuous process of learning and reflection

Although not a leadership development process per se, the identification of one's personality traits is commonly used in leadership development programs to assist participants learn more about their personality and how they respond to others. The most common personality type testing being used is the Myers-Briggs Type Indicator® (MBTI®) offered through the Center for Applications of Psychological Type (www.capt.org). The MBTI® categorizes personality on four dimensions:

- **Extroversion-Introversion (*EI*)**—Extroverts are outer world oriented and base their perception on people and objects; introverts are inner world oriented and base their perceptions on concepts and ideas.
- **Sensing-Intuition (*SN*)**—Sensing denotes a preference for clear, tangible data; intuition shows a preference for abstract and conceptual.
- **Thinking-Feeling (*TF*)**—Thinking uses logical, analytical, objective decision making; feeling relies of feeling, and personal and social values in decision making.
- **Judgment-Perception (*JP*)**—Judgment uses a judgment process when viewing others in the world; perception denotes a use of a perceptive process about others.

When combined these dimensions allow the possibility of 16 different personality types, which are denoted by the letter for each dimension. An ISTJ, for example, would denote a personality type that is *I*ntroverted, *S*ensing, *T*hinking, and *J*udgment. The MBTI® must be administered by a trained administrator who can then provide interpretation of the categorization results.

COMPARISON OF MANAGEMENT AND LEADERSHIP

The terms *management* and *leadership* often have been used interchangeably to describe the role of someone who has responsibility for overseeing the work of others and directing that work toward achievement of organizational goals. Management author John Kotter (1999) believes that leadership differs from, yet compliments, management. He suggests that

- Management is about coping with complexities, whereas leadership is about coping with change.
- Management organizes and staffs people to achieve goals; leadership focuses on aligning people toward goals and involves communicating and empowering to help achieve goals.
- Management controls people by pushing them in the right direction; leadership motivates them by satisfying basic human needs.

Robbins and Coulter (2002) agree that management and leadership are different. They suggest that managers are appointed to their position and their ability to influence is based on the formal authority inherent in their positions. Leaders differ in that they may either be appointed to their position or may emerge from the group. Leaders are able to influence without having formal authority.

PERSONAL AND ORGANIZATIONAL CHANGE

Change: Movement from one state to another.

Change, the movement from one state to another, is an expected occurrence in both personal life and organizational activity. Change often involves a paradigm shift, a break in an old way of thinking about something to allow one to look at a situation very differently. Effective managers and leaders accept that change should and will occur and work with themselves and others in order to facilitate the rate and outcomes of change.

Personal Change

Personal change focuses on the examination of one's personal characteristics and the development and execution of plans to change one of more of those characteristics. A strong advocate for the importance of personal change is the author Steven Covey. His book *The 7 Habits of Highly Effective People* (Covey, 1989) details ways individuals can change themselves and as a result become more effective. Covey stresses that minor changes in life are made by making changes in attitudes and behaviors but he believes to make a significant change, a change in a person's basic paradigms is needed. Covey suggests these seven habits, if internalized, will help individuals better solve problems, maximize opportunities, and continually learn and integrate other opportunities. The seven habits are

1. Be proactive.
2. Begin with the end in mind.
3. Put first things first.
4. Think win/win.
5. Seek first to understand . . . then to be understood.
6. Synergize.
7. Sharpen the saw (renewal of one's physical, mental, social/emotional, and spiritual dimensions).

Organizational Change

Organizational change: Any substantive modification to some part of the organization.

Organizational change involves moving the organization from one point to another. Managing change is an integral part of a foodservice manager's job.

A variety of external and internal forces for change exist in each organization. Some of these forces include

- Competition
- Governmental laws and regulations
- Economic and political pressures
- Technology
- Employee attitudes
- Workforce demographics
- Introduction of new equipment

Change agent: Person who initiates change.

For change to occur, a catalyst is needed. Managers often serve in this role as catalyst for change in an organization, and as a result they maybe termed **change agents**.

Lewin (1951) was the first to propose a model for successful change in an organization. His three-stage model of change suggested that successful change required *unfreezing* the status quo, *changing* to a new state, and then *refreezing* the organization to make the change permanent. Lewin believed that there were two sets of forces in organizations that were always in opposition to each other: forces for changes and resistance to change. When the forces were in equal balance, the organization would maintain a state of status quo. Creating change in an organization would involve the manager finding ways either to reduce the resistance to change or to increase the forces for change or both.

Robbins and Coulter (2002) contend that Lewin's model, which suggests stability and predictability in the business environment, may not work as well in the uncertain and dynamic business environment of today. They view managing change as similar to having to continually maneuver in uninterrupted white water rapids. Disruptions in the status quo are constant and border on chaos. They suggest that to be effective in such an environment, today's managers will need to find ways to initiate change within their organizations and manage employee and organizational resistance to change.

Individuals resist change for many reasons. The most common reasons are their uncertainty and insecurity about the outcome after the change, concern for personal loss as a result of the change, personal habits, and beliefs that the change may not be in the organization's best interest. Organizations also may have characteristics that impede the change process, such as power bases, organizational structure, and organizational culture. Managers can reduce resistance to change through a variety of techniques including communication, education, participative decision making, empowerment, bargaining, negotiation, manipulation, and coercion.

Spensor Johnson's best-selling book *Who Moved My Cheese?* (Johnson, 1998) provides practical insights for dealing with change and is required reading for employees in many organizations faced with change. Johnson offers several guidelines to help individuals better accept and adjust to change:

- Change happens.
- Anticipate change.
- Monitor change.
- Adapt to change quickly.
- Change.
- Enjoy change.
- Be ready to change quickly and enjoy it again.

CHAPTER SUMMARY

This summary is organized by the learning objectives.

1. A number of theories of motivation have been developed. Most common are Maslow's needs theory, McClelland's achievement-power-affiliation theory, Herzberg's two-factor theory, Vroom's expectancy theory, and Skinner's reinforcement theory. Maslow's theory suggests that people are motivated by their desire to satisfy a hierarchy of basic needs: physiological, safety, social, esteem, and self-actualization. McClelland's theory suggests that people have three

Professional Organization Profile

The National Society of Healthcare Foodservice Management

MISSION

To be the industry's leading advocate of "on-staff" healthcare food and nutrition professionals

WHO BELONGS TO THE ORGANIZATION

Membership includes more than 4,000 Foodservice Operators and Associates. Operator members are "on-staff" food and nutrition experts who are leaders of foodservice departments throughout the healthcare industry; Associate members include food/equipment manufacturers, foodservice distributors, brokerage companies, growers, and consultants. Student memberships are available.

ADVANTAGES OF MEMBERSHIP

HFM provides a network for interaction among practitioners and offers various training tools and skill building efforts. HFM offers a National training conference, networking meetings, a member benchmarking service, a successful operations guide, advocacy and public relations, brand development assistance, membership directory, JCAHO updates, the *Innovator* quarterly newsletter and *Bytes* online newsletter highlighting best practices in the healthcare foodservice market.

WEBSITE

www.hfm.org

MEET THE PRESIDENT

Sharon Cox, president 2005–2006, is the Director of Food and Nutrition Services at Memorial Sloan-Kettering Cancer Center (MSKCC). Her mission is to provide excellent food, enhanced nutrition care, and exceptional service. This dream was realized with the implementation of a hospital-wide room service program at MSKCC in 2002. Since the implementation of this program, the department has received national recognition for achieving a "Press Ganey" ranking in the 99th percentile for meals. Sharon offers the following advice to students: "A career in food and nutrition has endless possibilities, tap into the network and access the resources that are available. Join a professional organization, attend professional meetings, network with leaders in the industry, find a mentor, volunteer, and follow your passion."

needs: a need to achieve, a need for power, and a need for affiliation. Herzberg's theory suggested that maintenance factors such as work conditions, pay, job security, and interpersonal relations will not motivate individuals but will dissatisfy them if they are not met. His motivational factors included factors internal to the job, such as achievement, recognition, advancement, and responsibility. Vroom's expectancy theory suggests that employees are motivated to work if they believe their efforts will be rewarded and if they value the rewards that are being given. Skinner's reinforcement theory suggests that reinforced behavior will be repeated. Foodservice managers can use these motivation theories to help better understand employees and to find ways to structure the work environment to be motivating to employees.

2. Job satisfaction refers to an individual's mind-set about a job, either positive or negative. Organizational citizenship behaviors are positive, voluntary behaviors by employees that are necessary for organizational efficiency. Continuance commitment is the commitment that an employee has to the organization simply because the costs of leaving are too great. Affective commitment is the commitment employees have to an organization because they believe in the organization and want to do what is best for the organization.
3. Many different bases of power have been identified. Personal power comes from a person's attributes and expertise. Positional power is derived from a person's official position in the organization. Legitimate power comes from the formal position held by the individual in an organization. Reward power comes from an individual's ability to reward the work of others. Coercive power comes from the authority of a leader to punish those who do not comply. Expert power is held by leaders who are perceived to be knowledgeable in their job. Referent power (charisma) is the ability to influence others because of having their admiration and respect. Information power is based on the leader's possession of or access to information. Connection power is based on the leader's connections to influential or important persons.
4. Many theories of leadership have been discussed. They can be broadly categorized as trait theories of leadership, which focus on characteristics of an individual that suggest leadership ability; behavioral theories, which focus on behaviors that would be important for a leader; situational and contingency theories, which suggest that the leader behaviors should be contingent upon the situation; and transformational theories, which suggests that leaders transform others to achieve an organization's goals.
5. Change involves moving from one state of being to another. To be effective change agents, managers must reduce employee and organizational resistance to change by using techniques such as communication, education, negotiation, participative decision making, and empowerment.

TEST YOUR KNOWLEDGE

1. Describe how leadership and management differ.
2. How is job satisfaction believed to be related to organizational citizenship behaviors?
3. Describe differences in leadership behaviors under McGregor's Theory X and Theory Y.

4. What have been identified as traits of effective leaders?
5. How are the University of Michigan, Ohio State, and Blake and Mouton leadership grids similar? How do they differ?
6. What is a situational style of leadership?
7. Describe Lewin's model for change. How can this model help manage change in an organization?

CLASS PROJECTS

1. Interview a company leader to learn how he or she motivates employees. What is his or her philosophy of leadership? How has his or her leadership style changed?
2. Read the book *Who Moved My Cheese?*. Discuss how managers can apply concepts in the book when implementing change.
3. Have each student identify one person he or she believes is an effective leader. Using concepts described in the chapter, have each student describe how the person they identified demonstrates leadership characteristics.
4. Arrange to have a trained Myers-Briggs Type Indicator® (MBTI®) test administrator come to class to administer the personality test and help students interpret the results.

WEB SOURCES

www.ccl.org — Center for Creative Leadership
www.cio.com/research/leadership — Leadership and Management Research Center
http://leadership.wharton.upenn.edu/welcome/index.shtml — Wharton Center for Leadership and Change Management
www.capt.org — Center for Applications of Psychological Type (Myers Briggs)

BIBLIOGRAPHY

Allen, N. J., & Meyer, P. (1996). Affective, continuous, and normative committment in organizations: An examination of insticnt validity. *Journal of Vocational Behavior, 49,* 252–276.

Arendt, S. W., & Gregoire, M. B. (2005a). Leadership behaviors of undergraduate dietetics students—practices, contexts, and self perceptions. *Journal of the American Dietetic Association. 105,* 1289–1294.

Arendt, S. W., & Gregoire, M. B. (2005b). Leadership behaviors in hospitality management students. *Journal of Hospitality and Tourism Education.* 17(4): 20–27.

Argyris, C. (1957). *Personality and organization.* New York: Harper & Row.

Badaracco, J. L., & Ellsworth, R. R. (1993). *Leadership and the quest for integrity.* Boston: Harvard Business School Press.

Barrett, E. B., Nagy, M. C., & Maize, R. S. (1992). Salary discrepancies between male and female foodservice directors in JCAHO-accredited hospitals. *Journal of the American Dietetic Association, 92,* 1078–1082.

Batty, J. (1992). Breaking through the glass ceiling. *Restaurants USA, 12*(10), 37–41.
Batty, J. (1993). Finding the balance between work and family life. *Restaurants USA, 13*(3), 26–31.
Bass, B. M. (1985). *Leadership and performance beyond expectations.* New York: The Free Press.
Bass, B. M. (1986). *Transformational leadership: Charistina and beyond* (Technical Report No. 85–90). Binghamton, NY; State University of New York, School of Management.
Bennis, W. (1997). *Why leaders can't lead: The unconscious conspiracy continues.* San Francisco: Jossey-Bass.
Berelson, B., & Steiner, G. A. (1964). *Human behavior: An inventory of scientific findings.* New York: Harcourt, Brace, & World.
Blake, R. R., and McCanse, A. A. (1991). *Leadership Dilemmas—Grid Solutions.* Houston, TX: Gulf Publishing.
Blake, R. R., & Mouton, J. S. (1964). *The managerial grid.* Houston: Gulf Publishing.
Blake, R. R., & Mouton, J. S. (1978). *The new managerial grid.* Houston: Gulf Publishing.
Blake, R. R., & Mouton, J. S. (1984). *The managerial grid III* (3rd ed.). Houston: Gulf Publishing.
Burack, E. H., & Mathys, N. J. (1983). *Introduction to management: A career perspective.* New York: Wiley.
Cichy, R. F., Sciarini, M. P., Cook, C. L., & Patton, M. E. (1991). Leadership in the lodging and non-commercial food service industries. *FIU Hospitality Review, 9*(1), 1–10.
Cichy, R. F., Sciarini, M. P., & Patton, M. E. (1992). Food-service leadership: Could Attila run a restaurant? *Cornell Hotel and Restaurant Administration Quarterly, 33*(1), 47–55.
Covey, S. R. (1989). *The 7 habits of highly effective people.* New York: Simon & Schuster.
Drucker, P. F. (1993). *Managing for the future: The 1990s and beyond.* New York: Truman Tally Books/Dutton.
Evans, M. G. (1970). The effects of supervisory behavior on the path-goal relationship. *Organizational Behavior and Human Performance, 5*(3), 277–298.
Fiedler, F. E. (1967). *A theory of leadership effectiveness.* New York: McGraw-Hill.
French, J. R. P., & Raven, B. (1960). The bases of social power. In Cartwright, D., & Zander, A. (eds.), *Group dynamics: Research and theory* (2nd ed.) (pp. 607–623). Ann Arbor: University of Michigan, Institute for Social Research.
George, J. M., & Jones, G. R. (2002). *Organizational behavior* (3rd ed.). Upper Saddle River. NJ: Prentice Hall.
Goleman, D. (1998) *Working with emotional intelligence.* New York: Bantam Books.
Goleman, D., Boyatzis, R., & McKee, A. (2002). *Primal leadership: Learning to lead with emotional intelligence.* Boston: Harvard Business School Press.
Greenberg, J. (2005). *Managing behavior in organizations.* Upper Saddle River, NJ: Prentice Hall.
Greenleaf, R. (1977). *Servant leadership: A journey in the nature of legitimate power and greatness.* New York: Paulist Press.
Griffin, R. W. (1999). *Management* (6th ed.). Boston: Houghton Mifflin.
Hersey, P., Blanchard, K. H., & Johnson, D. E. (1996). *Management of organizational behavior: Utilizing human resources* (7th ed.). Upper Saddle River, NJ: Prentice Hall.
Herzberg, F. (1966). *Work and the nature of man.* Cleveland: World.
Herzberg, F. (1974). The wise old Turk. *Harvard Business Review, 52*(5), 70–80.
Herzberg, F. (1987). One more time: How do you motivate employees? *Harvard Business Review, 65*(5), 109–120.
Hopkins, D. E., Vaden, A. G., & Vaden, R. E. (1980). Some aspects of organization identification among school food service employees. *School Food Service Research Review, 4*(1), 34–42.
House, R. J. (1971). A path-goal theory of leader effectiveness. *Administrative Science Quarterly, 16*(3), 321–338.

House, R. J., & Mitchell, T. R. (1974). Path-goal theory of leadership. *Journal of Contemporary Business,* Autumn, 81–97.
Johnson, S. (1998). *Who moved my cheese?* New York: Putnan.
Kahn, R., & Katz, D. (1960). Leadership practices in relation to productivity and morale. In Cartwright, D., & Zander, A. (eds.), *Group dynamics: Research and theory* (2nd ed.) (pp. 554–570). Elmsford, NY: Row Peterson.
Kanter, R. M. (1989). The new managerial work. *Harvard Business Review, 67*(6), 85–92.
Kotter, J. P. (1990). What leaders really do. *Harvard Business Review, 68*(3), 103–111.
Kotter, J. P. (1999). *John Kotter on what leaders really do.* Boston: Harvard Business School Press.
Kouzes, J. M. and Posner, B. Z. (2002). *The Leadership Challenge* (3rd ed). San Francisco, CA: Jossey-Bass.
Lewin, K. (1951). *Field theory in social science.* New York: Harper & Row.
Likert, R. (1967). *Human organization: Its management and value.* New York: McGraw-Hill.
Maccoby, M. (1981). *The leader: A new face for American management.* New York: Simon & Schuster.
Maccoby, M. (1995). *Why work: Motivating and leading the new generation* (2nd ed.). New York: Simon & Schuster.
Maslow, A. H. (1943). A theory of human motivation. *Psychology Review, 50,* 370–396.
Maslow, A. H. (1987). *Motivation and personality* (3rd ed.). New York: Harper & Row.
McClelland, D. C. (1985). *The achieving society.* New York: Free Press.
McClelland, D. C., & Burnham, D. H. (1976). Power is a great motivator. *Harvard Business Review, 54*(2), 100–110.
McGregor, D., & Bennis, W. C. (1985). *The human side of enterprise.* New York: McGraw-Hill.
Maxwell, J. C. (1999). *The 21 indispensable qualities of a leader.* Nashville, TN: Thomas Nelson.
Middlemist, R. D., & Hitt, M. A. (1981). *Organizational behavior: Applied concepts.* Chicago: Science Research Association.
Mondy, R. W., Sharplin, A., & Flippo, E. B. (1988). *Management: Concepts and practices* (4th ed.). Boston: Allyn & Bacon.
Naisbitt, J., & Aburdene, P. (1990). *Megatrends 2000: Ten new directions for the 1990s.* New York: William Morrow.
Peters, T. J., & Austin, N. A. (1989). *Passion for excellence: The leadership difference.* New York: Random House.
Porter, L. W., & Lawler, E. E. (1968). *Managerial attitudes and performance.* Homewood, IL: Irwin-Dorsey.
Raven, B. H., & Kluganski, W. (1970). Conflict and power. In Swingle, P. G., (ed.), *The structure of conflict* (pp. 69–109). New York: Academic Press.
Reddin, W. J. (1967a). The 3-D management style theory. *Training and Development Journal, 21*(4), 8–17.
Reddin, W. J. (1967b). *The 3-D management style theory. Theory paper No. 2—Managerial styles.* Fredericton, NB, Canada: Social Sciences Systems.
Retrospective commentary. (1973). *Harvard Business Review, 51*(3), 162–164, 166–168, 170, 173, 175, 178–183.
Robbins, S. P., & Coulter, M. (2002). *Management* (7th ed.). Upper Saddle River, NJ: Prentice Hall.
Rue, L. W., and Byars, L. L. (1989). *Management Theory and Application* (5th ed). Homewood, IL: Richard D. Irwin.
Schwartz, F. N. (1989). Management women and the new facts of life. *Harvard Business Review, 67*(1), 65–76.
Schwartz, F. N. (1992). Women as a business imperative. *Harvard Business Review, 70*(2), 105–113.
Skinner, B. F. (1971). *Beyond freedom and dignity.* New York: Knopf.
Slater, D. (1989). Role models: Inspiring by example. *Restaurants USA, 9*(6), 15–17.

Smith, P. C., Kendall, L. M., & Hulin, C. L. (1969). *The measurement of satisfaction in work and retirement: A strategy for the study of attitudes*. Chicago: Rand McNally.

Stogdill, R. M. (1974). *Handbook of leadership*. New York: Free Press.

Tannenbaum, R., & Schmidt, W. H. (1958). How to choose a leadership pattern. *Harvard Business Review, 36*(2), 95–101.

Vroom, V. H. (1994). *Work and motivation*. San Francisco: Jossey-Bass.

Yukl, G. (2002). *Leadership in organizations* (5th ed.). Upper Saddle River, NJ: Prentice Hall.

Zaleznik, A. (1977). Managers and leaders: Are they different? *Harvard Business Review, 55*(3), 67–78.

Zaleznik, A. (1993). *Learning leadership: Cases and commentaries on abuses of power in organizations*. Chicago: Bonus Books.

CHAPTER

11

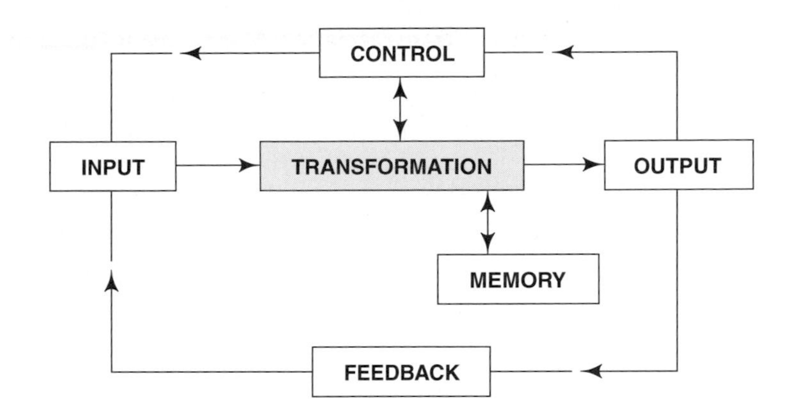

Decision Making, Communication, and Balance

Enduring Understanding

- Effective managers must be good decision makers and communicators.

Learning Objectives

After reading and studying this chapter, you should be able to

1. Discuss the decision-making process and the different types of decisions managers make.
2. Compare and contrast different decision-making techniques.
3. Describe the communication process and identify strategies for improving communication.
4. Define the concept of balance and its role in the foodservice system model.

Decision making, communication, and balance are the linking processes used to help coordinate the work in a foodservice operation toward its goals. In this chapter we will discuss each of these concepts and how each is instrumental to the success of the organization. As a future manager, you will make decisions and communicate to others throughout each work day. Your actions and decisions will help provide the balance needed in an organization.

LINKING PROCESSES

Linking processes are needed to coordinate the activities of the system so they can accomplish the goals and objectives. These linking processes are decision making, communication, and balance (Figure 11.1).

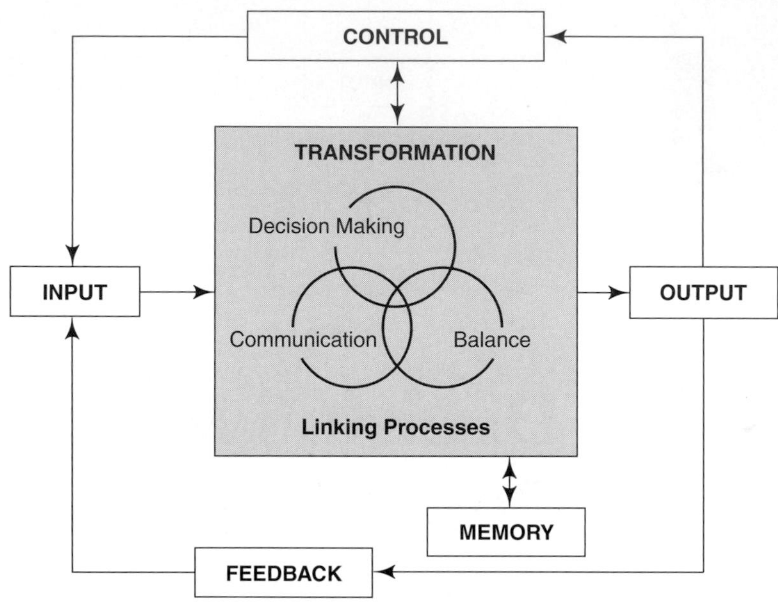

Figure 11.1. Foodservice systems model with the linking processes highlighted.

Decision making is the selection of a course of action from a variety of alternatives, and *communication* is the vehicle whereby decisions and other information are transmitted. *Balance* concerns management's ability to maintain organizational stability, which is related to effective decision making and communication.

Linking processes are integral to management's effectiveness in transforming inputs to outputs. Decision makers need to know more than ever before. With the advent of computers, the capacity to collect and process information has grown, but information is not the same as knowledge. Decision making requires seeking the facts from all the information and relating them to current situations so they become usable knowledge.

The linking processes govern the flow of resources or system inputs. Through the use of decision making and communication, people, money, and equipment are moved through the system.

DECISION MAKING

Managers make decisions for the purpose of achieving individual and organizational objectives. Effective managers must be good decision makers. Decision making involves three primary stages—first, *definition of the problem;* second, *identification and analysis of possible courses of action;* and third, actual *selection of a particular course of action.*

Analyzing the decision processes by these stages illustrates the difference between management and nonmanagement decisions. Managerial decisions encompass all three stages; nonmanagerial decisions are concentrated in the last, or choice, stage. For example, when a customer complains that the steak is too well done, the waitperson may follow the traditional course of action and exchange it for a satisfactory one. An employee who is empowered to make decisions under the TQM philosophy might decide that that remedy doesn't go far enough, however, and remove the steak from the check as well. Management, however, must have thought through alternatives and consequences and then trained employees on making choices.

Types of Decisions

Foodservice managers must make many different types of decisions. Most decisions, however, fall into one of two categories: programmed and nonprogrammed.

Programmed versus Nonprogrammed

Programmed decisions: Decision reached by following established policies and procedures.

Nonprogrammed decisions: Relatively unstructured decision that takes a higher degree of judgment.

Programmed decisions are reached by following established policies and procedures. Normally, the decision maker is familiar with the situation surrounding a programmed decision. These decisions also are referred to as routine or repetitive decisions. Limited judgment is called for in making programmed decisions. These decisions are made primarily by lower-level managers and employees in an organization.

Nonprogrammed decisions are unique and have little or no precedent. These decisions are relatively unstructured and generally require a more creative approach on the part of the decision maker than programmed decisions. Often when dealing with nonprogrammed decisions, the decision maker must develop the procedure to be used. Naturally, these decisions tend to be more difficult to make than programmed decisions. Deciding on a location for a new foodservice operation and selecting a new piece of equipment are examples of nonprogrammed decisions.

Because programmed decisions are concerned mostly with concrete problems that require immediate solutions and are frequently quantitative in nature, they tend to be reached in a short time. Nonprogrammed decisions usually involve a longer time horizon because they tend to focus on qualitative problems and require a much greater degree of judgment. When making such judgmental decisions, managers must frequently rely on wisdom, experience, and philosophic insight rather than on established policies and procedures.

Kinds of decisions are also related to level of management in the organization. Tasks of higher-level managers often have a longer time frame and involve more judgment than those of lower-level managers. Conversely, the time perspective of the lower-level manager is shorter, and the tasks need less judgment. Therefore, higher-level managers tend to make mostly nonprogrammed and lower-level managers mostly programmed decisions (Figure 11.2). The foodservice supervisor, for example, is concerned with day-to-day operational decisions, whereas the director of foodservice is responsible for decisions concerned with the future of the overall organization.

Organizational versus Personal

Nonprogrammed decisions are of two general kinds—organizational and personal. *Organizational decisions* relate to the purposes, objectives, and activities of the organization, and *personal decisions* are concerned with the manager's

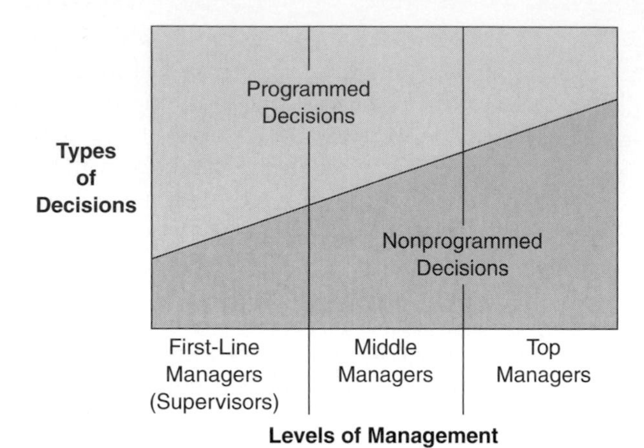

Figure 11.2. Types of decisions according to levels of management.

individual goals. These personal decisions, however, may affect the organization, and vice versa.

A manager may decide to resign from a position in a university foodservice, for example, to take a job in a large commercial foodservice organization. After the move from the university, the commercial company may decide to transfer the manager from one regional office to another. The first decision is a personal one affecting two organizations and the second an organizational decision affecting the individual.

The Decision-Making Process

Foodservice managers make decisions constantly and seldom have time to do a scientific analysis of every problem. Figure 11.3 outlines the steps a manager should follow to make rational and logical decisions. These steps in rational decision making will keep the decision maker focused on facts and guard against inappropriate assumptions and pitfalls.

Recognizing and Defining the Situation

A stimulus, which can be either positive or negative, is necessary for someone to recognize that a problem exists and a decision is necessary. The problem must be clearly defined before action is taken. The foodservice manager should understand the problem and its relationship to other factors in the system.

Identifying Alternatives

Once the decision is recognized and defined, alternatives should be identified. Obvious alternatives should be examined along with those that are creative. The more important the decision, the more attention should be placed on selecting alternatives. When selecting creative alternatives, constraints (legal restrictions, moral and ethical norms, or the power and authority of the manager, for example) might restrict which ones are chosen.

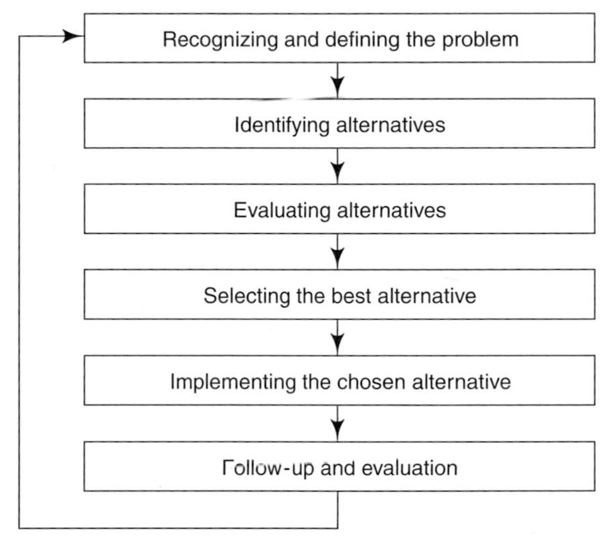

Figure 11.3. Steps in the rational decision-making process.
Source: Griffin, Ricky W., *Management,* 6th edition. Copyright © 1999 by Houghton Mifflin Company. Used with permission.

Evaluating Alternatives

Foodservice managers should evaluate all alternatives to determine if they meet the needs of the operation and the feasibility and consequences of using them. For example, in selecting a supplier, a foodservice manager might consider criteria such as dependability, price, delivery time, and variety of products offered. After identifying the possible choices and collecting information, each supplier should be evaluated on the various criteria and then ranked. This evaluation must be conducted carefully to be sure that the selected alternative is successful. How the decision will affect the operation and what the cost will be are major considerations in selecting an alternative. For big decisions, managers should estimate the prices of the alternatives.

Selecting the Best Alternative

After evaluating the alternatives for meeting the needs of the operation and the feasibility and consequences of using them, probably most alternatives will be rejected. The crux of decision making is choosing the best out of those remaining. The decision maker often has to make a subjective choice among the final choices.

Even under the best of conditions, all feasible alternatives cannot be developed because of human or situational limitations. Rarely can managers make decisions that maximize objectives because complete information is never available about all the possible alternatives and their potential results. Complete knowledge of alternatives and consequences would allow the best possible alternative to be chosen; managers often have to make what March and Simon (1958) call a "satisfying," or satisfactory but not optimal decision, because information is lacking. Also, managers may tend to choose the first satisfactory alternative and discontinue their search for additional alternatives. Another pitfall is to continue the search for alternatives when, in fact, the manager may be avoiding making a decision.

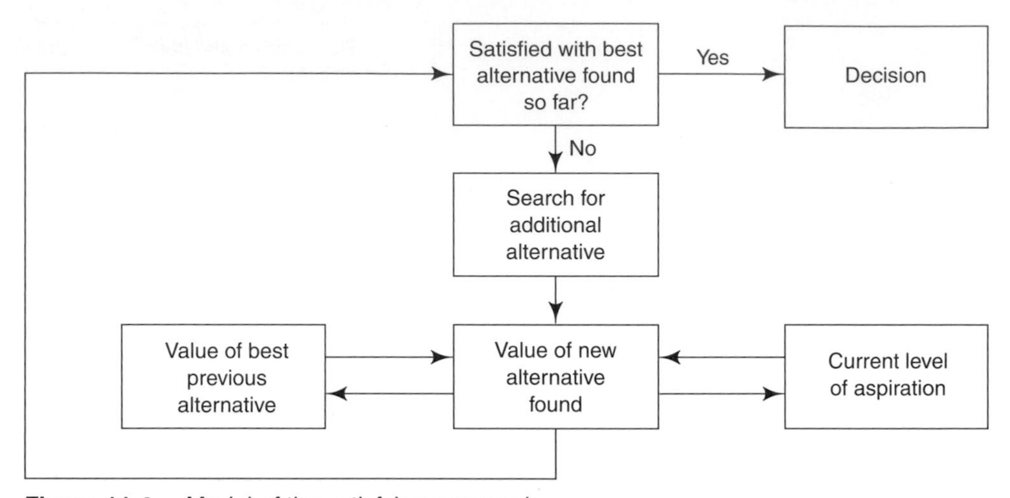

Figure 11.4. Model of the satisfying approach.
Source: Adapted from *Organizations* (p. 49) by J. G. March and A. Simon. Copyright © 1958 by John Wiley & Sons, Inc. Reprinted by permission of John Wiley & Sons, Inc.

Figure 11.4 represents the satisfying approach to decision making. As shown in the model, if the decision maker is satisfied that an acceptable alternative has been found, that alternative is selected. Otherwise, the decision maker searches for an additional alternative. If the value of the new alternative is greater than that of the previous one consistent with the current level of aspiration, it will be chosen. The double arrows indicate a two-way relationship between value and aspirations. The net result determines whether or not the decision maker is satisfied with the alternative. Thus, the manager selects the first alternative that meets the minimum satisfaction criteria and makes no real attempt to optimize; that is, the manager makes a "satisficing" decision rather than the best possible decision.

Implementing the Chosen Alternative

After an alternative is chosen, it should be implemented into the organization, but that can be difficult to do. When implementing a decision, managers must consider people's resistance to change, which might include insecurity, inconvenience, and fear of the unknown. The decision-making process does not end when the decision is made. The decision must be implemented, and the manager must monitor results to ascertain that the selected alternative solves the problem.

Follow-Up and Evaluation

Managers finally have to evaluate the effectiveness of their decision. They need to decide if their chosen alternative was the correct one. If it is not, maybe the second or third alternative would be better or maybe the problem was not correctly defined. Perhaps the chosen alternative is the best, but more time may be needed or perhaps the implementation needs to be revised.

Monitoring outcomes results in feedback that should be considered in future determination of objectives and in decision making. Many good decisions fail because of poor implementation. The decision must be communicated properly, and support among employees must be organized. For example, a computer program for inventory control in a foodservice will have no value until personnel implement it. In addition, the necessary resources, especially personnel, must be assigned for implementation of the program. Managers often assume that once they make a decision, their role is over. This is far from true. Proper implementation of the decision is a critical component in decision making.

Conditions for Making Decisions

Regardless of the approach used, decisions frequently are made at one time for events that will occur at another, and the conditions are seldom identical. Take, for instance, the decision to produce sufficient entrées for the usual 2,000 students who come to lunch on Wednesday in a university residence hall foodservice. This decision may result in excessive overproduction if many students choose not to dash across campus for lunch because of an unexpected downpour occurring between 11:30 A.M. and 12:30 P.M. Although nature obviously is not under the control of the decision maker, it affects the outcome of the decision.

The environment within which the decision maker operates, therefore, affects the decision-making process. Conditions in the environment change and predictions are difficult; yet managers must make decisions based on the information available, even though it may be incomplete or involve factors outside their control. These conditions under which decisions are made are referred to as *certainty, risk,* and *uncertainty* (Figure 11.5). They tend to vary with the time frame that encompasses the decision. The longer the future time period involved in the decision, the less certain are the environmental conditions. The various degrees of certainty in relation to time frame are illustrated in Figure 11.6. As shown, the tendency is to move on the certainty/uncertainty continuum into conditions of risk and uncertainty as the time frame becomes longer.

Conditions of Certainty

Conditions of certainty: Adequate information is available to assure results of decision.

Under **conditions of certainty,** a decision maker has adequate information to assure results. A decision under conditions of certainty involves choosing the alternative that will maximize the objective. For example, in school foodservice, the number of children scheduled to eat lunch in an elementary school is usually determined by teachers soon after school convenes each morning. These data are transmitted to the school lunch manager, who then can determine production needs based on a known number of lunch participants. Guesswork is not involved in estimating the number of students who will eat. Under conditions of certainty, management science techniques, such as linear programming, break-even analysis, and inventory control models, have been used effectively.

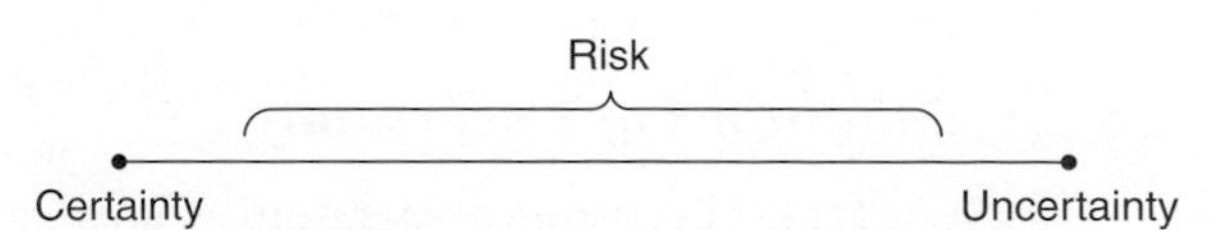

Figure 11.5. Certainty-uncertainty continuum in decision making.

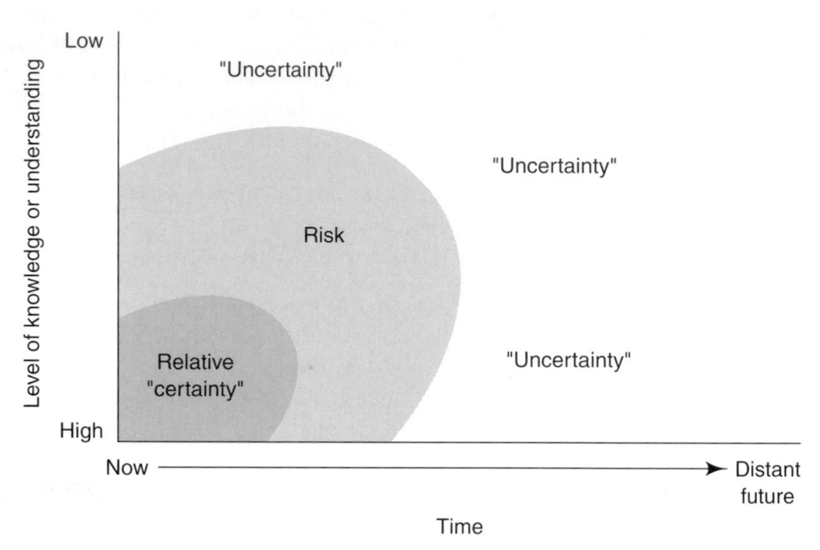

Figure 11.6. Effect of time and level of knowledge on decision-making situations.
Source: Introduction to Management (p. 170) by E. H. Burack and N. J. Mathys. Copyright © 1983 by John Wiley & Sons, Inc. Reprinted by permission of John Wiley & Sons, Inc.

Conditions of Risk

Conditions of risk: Results of decision are uncertain; probability techniques are necessary for estimating the likelihood of events occurring in the future.

Because conditions of certainty are becoming less common in today's complex and rapidly changing world, estimating the likelihood or probability of various events occurring in the future is often the only possibility for planning. This condition is called *risk.* Under **conditions of risk,** various probability techniques are helpful in making decisions. In decision making under risk, managers are faced with the possibility that any one of several things may occur. The assumption is that based on the manager's research, experience, and other available information, the probabilities are known about each of these various states.

In the previous example of the university foodservice and the effect of rain on lunch participation, at the time of production forecasting, the manager might have information about the potential for rain from a weather forecast. The manager also can review past records to determine the predicted effect on weekday meal attendance when rain occurs over the lunch hour. Therefore, while the outcome cannot be known with certainty and the manager is taking a risk in terms of possible over- or underproduction, the weather forecast and prior records may assist in making more accurate projections about participation should an unexpected downpour occur.

Probabilities enable the calculation of an expected value of given alternatives. The expected value of a particular decision is the average of returns that would be obtained if the same decision were made in the same situation over and over again.

Conditions of Uncertainty

When the occurrence of future events cannot be predicted, a state of uncertainty exists. Many changes or unknown facts can emerge when decision time frames are long. To predict what is likely to occur with any degree of certainty, therefore, is quite difficult. In these situations, foodservice managers frequently apply their experience, judgment, and intuition to narrow the range of choices. Input from others may help reduce some of the uncertainty. Involvement of knowledgeable people in the decision process, therefore, may be beneficial. Under **conditions of uncertainty,** some managers will delay decisions until conditions stabilize or will take a path of least risk.

Conditions of uncertainty: Occurrence of future events cannot be predicted.

If a decision must be made, however, even though the decision maker has little or no knowledge about the occurrence of various states of nature, one of three basic approaches may be taken. The first is to choose the alternative that is the best of all possible outcomes for all alternatives. This is an *optimistic approach* and is sometimes called the *maximum approach*. A second approach in dealing with uncertainty is to compare the worst possible outcome of each of the alternatives and select the alternative with the least possible negative outcomes. This is the *pessimistic approach,* sometimes called the *maximin approach*. The final approach is to choose the alternative that has the least variation among its possible outcomes. It is a *risk-averting approach* and may make for effective planning; however, the payoff potential is less than with the optimistic approach.

Decision-Making Techniques

Various techniques have been developed to assist managers in making decisions. Some of these techniques are highly complex and quantitative in nature. In this section, selected techniques are described briefly.

Decision Trees

Decision trees allow management to assess the consequences of a sequence of decisions with reference to a particular problem. The approach involves linking a number of event "branches" graphically, which results in a schematic resembling a tree. The process starts with a primary decision that has at least two alternatives to be evaluated. The probability of each outcome is ascertained, along with its monetary value.

A simplified example of a decision tree is shown in Figure 11.7. The decision in this example concerns the expansion of a restaurant's services to include take-out food. The two alternatives are to expand or not to expand. The restaurateur is faced with the probability of the competitor on the next block also introducing take-out service. To assess the decision in question, the probabilities for all these occurrences should be determined, along with the effect on net income if each were to occur. This determination will permit the restaurant owner to make the decision with the best potential payoff.

Cost-Benefit

Cost-benefit analysis is a technique for comparing the costs and financial benefits of a project or decision. Sometimes a financial value is placed on intangible benefits so they can be considered in the analysis. Before adding any new personnel or programs, most managers want to compare the cost of implementing and maintaining the programs with the increase in performance attributed to them (RGI Cost Benefit Analysis, 1998). The increase in performance produces a payoff measured

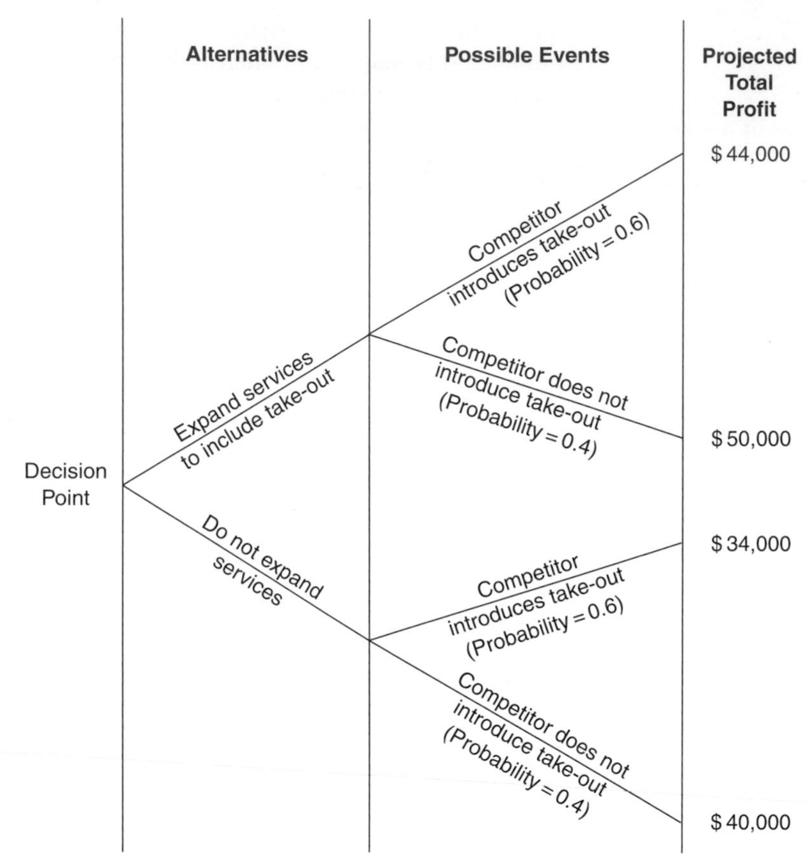

Figure 11.7. Example of a decision tree for making a decision on expansion of restaurant services.

either in dollars or in the achievement of organizational objectives. Several approaches exist for conducting cost-benefit analysis. The use of computers can facilitate calculations and makes information available. Management, however, has to decide what kind of information it needs to avoid information overload.

Cost-Effectiveness

Cost-effectiveness is a technique that provides a comparison of alternative courses of action in terms of their cost and effectiveness in attaining a specific objective. It customarily is used in an attempt to minimize dollar cost, subject to some goal requirement that may not be measurable in dollars, or, in a converse attempt, to maximize some measurement of output subject to a budgetary constraint.

Cost-effectiveness has been given a great deal of emphasis in recent years in public programs and in public institutions. Concern over increasing costs in tax-supported services has led to the use of this type of analysis.

A structure-of-choice model can be used to illustrate the cost-effectiveness concept. Once the desired goal is determined, alternatives for meeting it are listed. The alternatives then are evaluated for both cost and effectiveness and ranked in order of desirability by comparison with defined criteria. The alternative selected

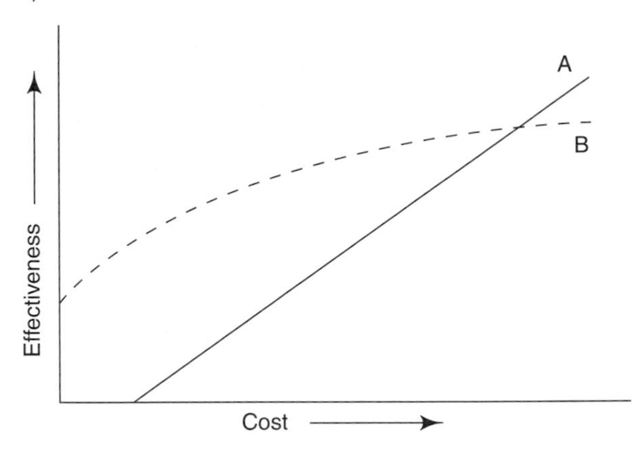

Figure 11.8. Comparison of alternatives in decision making. *Source:* M. C. Spears: "Concepts of Cost Effectiveness: Accountability for Nutrition, Productivity." Copyright The American Dietetic Association. Reprinted by permission from *Journal of the American Dietetic Association,* 68, p. 343, 1976.

may not always be the least costly because potential effectiveness is a major consideration.

Quite often, the final decision is based on selecting the best of two alternatives, either of which is considered acceptable. A simplified comparison of alternatives is illustrated in Figure 11.8. The cost-effectiveness curves for alternatives A and B intersect at a point where, for the same cost, the effectiveness is identified. At a cost less than that at the point of intersection, alternative B is given greater effectiveness; if costs higher than that at the point of intersection can be tolerated, alternative A is preferable. The judgment of the decision maker comes into play in answering the question of whether or not the gain in effectiveness of A over B at a higher cost is worthwhile.

The cost-effectiveness concept can be illustrated by using a product cost analysis for 48 blueberry muffins conducted at Pillsbury's Foodservice Technical Center (Table 11.1). Four methods for preparing blueberry muffins are analyzed for labor and ingredient cost and for meeting the needs of a specific foodservice operation. The goal, which is shown in the structure-of-choice model (Figure 11.9), is to serve a high-quality product that meets customers' expectations at the lowest possible cost. The four alternatives are

- Scratch recipe (Product 1)
- Pillsbury Basic Muffin Mix (Product 2)
- Pillsbury TubeSet® Frozen Batter (Product 3)
- Pillsbury Frozen Baked Muffins (Product 4)

The criteria for the selection are that the ingredients or muffins are available on the market, the final product looks good, the customer is pleased, and the food and labor costs are minimum. The effectiveness of each of the four types of muffins will be determined by a taste panel that includes chefs, managers, and customers; a questionnaire for customers to complete; and customer comments gathered by waitpersons. The cost analysis done by Pillsbury's Foodservice Technical Center is used for the cost component of the cost-effectiveness model. The kitchen for the coffee shop is small but has a convection combi (combination convection steamer-convection oven) but does not have a trained baker. After conducting a cost-effectiveness analysis, the alternatives are ranked and the final decision is to use the Pillsbury Muffin Mix. Muffins can be baked according to demand during the breakfast hours, giving a home-made touch to the product, and the cost of

Table 11.1 Product Cost Analysis for 48 2-oz. Blueberry Muffins

	Prep Time			
Production Step	**Scratch Recipe**	**Pillsbury Basic Muffin Mix[a] (6/5 lb. cartons)**	**Pillsbury TubeSet® Frozen Batter (6/3 lb. tubes)**	**Pillsbury Frozen Baked Muffins (96 muffins)**
Pan prep (paper liners)	2 min.	2 min.	2 min.	0 min.
Ingredient prep/weighing	6¼ min.	1½ min.	0 min.	0 min.
Mixing and scraping bowl	5½ min.	1¾ min.	0 min.	0 min.
Portioning batter into pans	3 min.	3 min.	4 min.	0 min.
Loading oven	¾ min.	¾ min.	¾ min.	0 min.
Total preparation time	17½ min.	9 min.	6¾ min.	0 min.
Baking time	20 min.	20 min.	20 min.	0 min.
Total prep and bake time	37½ min.	29 min.	26¾ min.	0 min.
Labor cost per batch[b]	$2.14 ($0.044 each)	$1.10 ($0.022 each)	$0.83 ($0.017 each)	$0.00
Clean-up time (and cost)[b]	3 min. ($0.0044 each)	1½ min. ($0.0022 each)	0 min.	0 min.
Ingredient cost per muffin[c]	$0.06 each	$0.11 each	$0.147 each	$0.245 each
Total cost per muffin	$0.1084 each	$0.134 each	$0.164 each	$0.245 each

[a]Prepared with 12 oz. of frozen blueberries.
[b]Preparation labor costs based on wages of $7.35/hr.
Labor costs calculated by rounding total preparation time to the nearest minute.
Baking times not included. Clean-up costs based on wages of $4.25/hr.
[c]Based on anticipated distributor price and case yield information.

Source: Grand Metropolitan Foodservice USA, Minneapolis, MN. Used by permission.

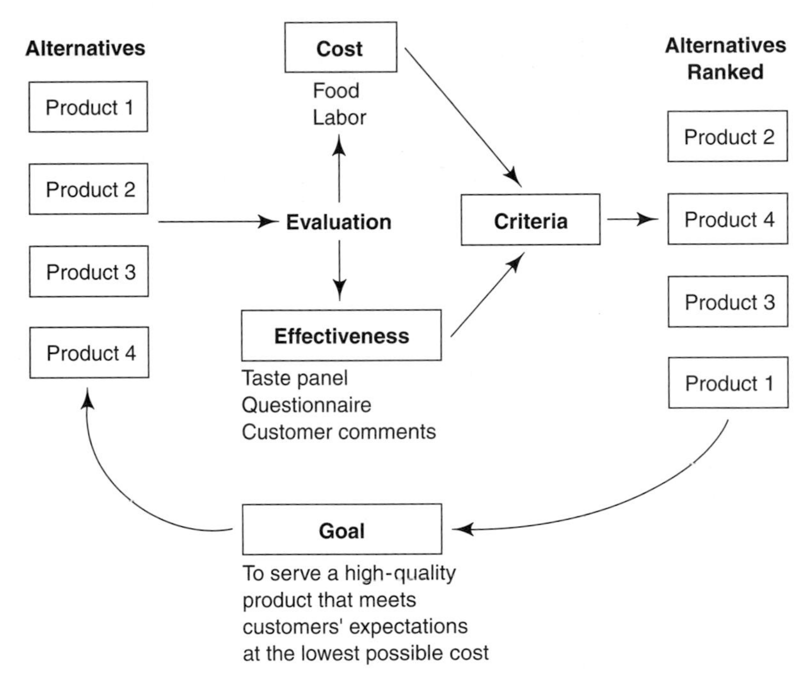

Figure 11.9. Structure-of-choice model for selecting one type of blueberry muffin.

each muffin is low enough to make a good profit for the operation. A small supply of the second alternative, Pillsbury Frozen Baked Muffins, however, would be kept in the freezer for emergencies.

Networks

Network: Graphic representation of a project, depicting the flow and sequence of defined activities and events.

The Program Evaluation and Review Technique (PERT) and Critical Path Method (CPM) are networks for decision making. A **network** is a graphic representation of a project, depicting the flow and sequence of defined activities and events. An *activity* defines the work to be performed; an *event* marks the beginning or end of an activity.

PERT and CPM are two widely known and used management science techniques for planning, scheduling, and controlling large projects. For example, a catering manager could use these techniques to improve the planning and control of a complex function such as a wedding. These two techniques have been combined into a PERT-type system that uses a network to depict the sequence of activities in a particular project. A *project* is defined as a group of tasks that are performed in a certain sequence to reach an objective. In a network diagram, an event is represented by a node (circle), and each activity by a line that extends between two events. Directional arrows are drawn on each line to indicate the sequence in which events must be accomplished.

The first step in managing a project is to specify all the activities required to complete it. The example used by Leitch (1989) is the activity sequence in setting up a banquet and serving guests. In Table 11.2, the sequential activities, each identified by a letter, are listed and described. The immediate predecessor for each

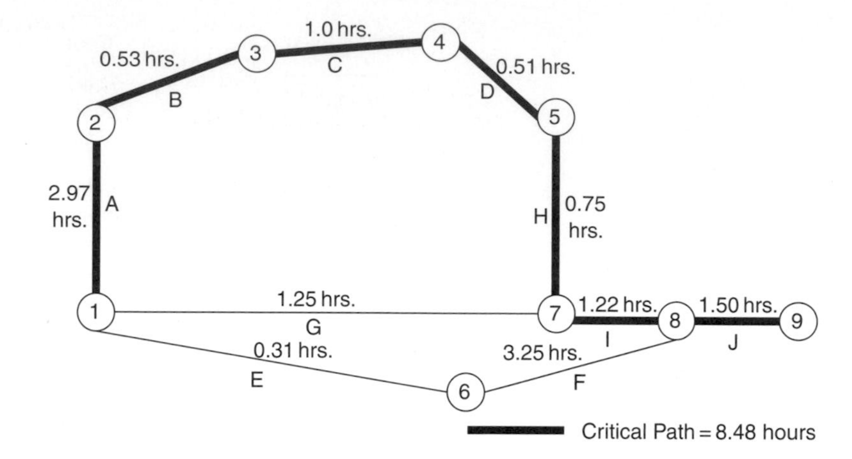

Figure 11.10. Example of a PERT network chart.
Source: Adapted from "Applications of a PERT-Type System and 'Crashing' in a Food Service Operation" by G. A. Leitch, 1989, *FIU Hospitality Review,* 7(2), pp. 66–67. Used by permission.

activity also is listed. For example, activity I, plate salad and serve, cannot begin until activity G, prep salad and store, is completed. Activity A, however, can begin anytime because it does not depend on the completion of other activities.

The next step is to construct the network diagram that connects all these activities. Figure 11.10 is the network representation of the project described in Table 11.2. Because the PERT-type system typically is used for projects in which the time when activities are finished is uncertain, the following three time estimates need to be identified for each activity:

- most optimistic time, or shortest time, assuming most favorable conditions
- most likely time, which implies the most realistic time
- most pessimistic time, or the longest time, assuming the most unfavorable conditions

In the example, a time is estimated for each of the three possible outcomes—optimistic, likely, and pessimistic times—and the mean time in hours is shown in

Table 11.2 PERT Activities

Activity	Description of Activity	Immediate Predecessor
A	Bring tables and chairs up from basement and arrange in hall	—
B	Pick up tablecloths from laundry and place on tables	A
C	Arrange place setting and decorations	B
D	Fill water glasses	C
E	Turn ovens on and perform equipment check	—
F	Prepare and cook main course	E
G	Prepare salad and store	—
H	Seat guests	D
I	Plate salad and serve	G, H
J	Plate dinner and serve	F, I

Source: "Applications of a PERT-Type System and 'Crashing' in a Food Service Operation" by G. A. Leitch, 1989, *FIU Hospitality Review,* 7(2), p. 67. Used by permission.

Figure 11.10. Once the times have been plotted for each activity, the minimum time to set up the banquet and serve the guests can be determined. This is done by adding the amount of time, identified as *time duration,* for each path through the network. The path with the longest duration also is known as the *critical path* because if any activities are delayed, the completion of the project also will be delayed. In the example, the manager should be concerned with activities A, B, C, D, H, I, and J. These activities lie on the critical path and could delay dinner if they are not completed on time. The expected completion time for this event is 8.48 hours (Leitch, 1989). Computer software is commercially available to assist with the drawing of a PERT chart and the identification of the critical path.

Linear Programming

Linear programming is a technique useful in determining an optimal combination of resources to obtain a desired objective. Loomba (1978) contended that linear programming is one of the most versatile, powerful, and useful techniques for making managerial decisions. This concept has been used in solving a broad range of problems in industry, government, healthcare, and education. A few examples of the kinds of decisions that can be determined by linear programming are the optimal product mix, transportation schedule, plant location, and assignment of personnel and machines.

The objective may be to find the lowest cost or highest profit possible from given resources. Linear programming must be considered with recognition of the limitations on its use, however. A general prerequisite is that a linear or straight line relationship exists among the factors involved. In any linear programming problem, the manager must identify a measurable objective or criterion of effectiveness. The constraints must be also specified.

Balintfy, Rumpf, and Sinha (1980) used linear programming techniques to plan the best menus at the lowest cost for school lunches. Balintfy is the pioneer in computer menu planning based on linear programming. In his work, nutrient constraints are specified and aesthetic factors quantified in a model designed to plan least-cost menus. His later work includes food preferences in the menu planning model. There has been limited use of this technique, however, because of the time involved in identifying and quantifying objectives and constraints.

Other Techniques

A variety of other quantitative decision-making techniques have been developed, including game theory, queuing, and simulation models. There has been limited application of these techniques because they often require complex computer models in the foodservice field. As the field develops, however, managers probably will be applying these techniques, as have managers in other industries, to improve decision-making effectiveness.

Game theory introduces a competitive note in decision making by bringing into a simulated decision situation the actions of an opponent. Competition for market share is an example of a problem in which game theory might be used. The assumptions are that all competitors, or players, have the objective of winning the game and that they are capable of making independent and rational decisions. Competitors are presumed to be interested in maximizing gains and minimizing losses. Game theory will show the highest gain with the smallest amount of losses, regardless of what the competitor does.

Queuing theory develops relationships involved in waiting in line. Customers awaiting service or work awaiting inspection in a production line are typical of the problems that may be approached by the methods of queuing theory. The theory balances the cost of waiting lines against the cost of preventing them by expanding facilities. The problem is figuring out the cost of total waiting—that is, the cost of tolerating the queue—and weighing it against the expense of constructing enough facilities to decrease the need for the queue. Sometimes eliminating all delay is more costly than keeping some. The basic framework of queuing is shown in Figure 11.11. Queuing problems can be solved by analytical procedures or simulation.

The concept of **simulation** is to use some device for imitating a real-life occurrence and studying its properties, behavior, and operating characteristics. The device can be physical, mathematical, or some other model for describing the behavior of an occurrence that a manager wishes to design, improve, or operate. For example, operating characteristics of a new piece of equipment can be simulated in a laboratory. Similarly, behavior of an occurrence also can be simulated by experimenting on a mathematical model that represents it. Nettles and Gregoire (1997) used computer simulation to determine staffing patterns for school lunch services, and they showed how varying class arrival times could reduce the time students spent in line. The advantages to using computer simulation are as follows:

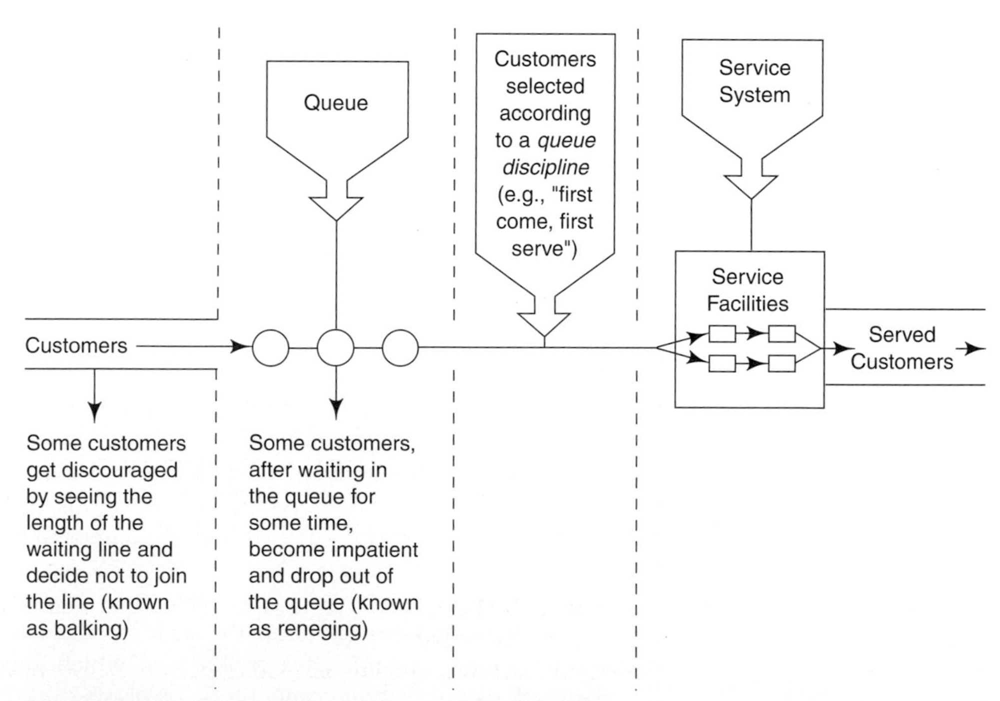

Figure 11.11. Basic framework of a queuing system.
Source: Management: A Quantitative Perspective (p. 429) by N. P. Loomba, 1978, New York: Macmillan. Used by permission.

- Changes in procedures can be examined without disrupting operations.
- The time to complete tasks can be modified and the effect on the operation explored.
- Bottlenecks in the flow of operations can be identified.

Artificial intelligence is a computer program that attempts to duplicate the thought processes of experienced decision makers (Griffin, 1999). **Expert systems** are artificially intelligent computer software programs (Cook, 1992). These systems solve problems by emulating the problem-solving behavior of human experts. Decision making starts with exploring "if . . . then" situations that could occur in solving a problem, and these situations become the knowledge base for the system. The following examples are "if . . . then" statements for a hypothetical steak house's pricing policy. The boldface statement is the decision on pricing policy for the restaurant based on the "if . . . then" situation.

- If the profit on a steak is high and demand is weak, then price policy to the customer is **decrease price.**
- If the profit on a steak is normal and demand is steady, then price policy to the customer is **maintain price.**
- If the profit on a steak is low and demand is strong, then price policy to the customer is **increase price.**

By using many "if . . . then" rules, an expert system can be created that can mimic what is actually occurring in an operation.

In any simulation model, the goal is to understand the behavior of a system by testing the model under a variety of operating conditions. The need to experiment on a real-life system thus is eliminated. Sophisticated computer programs have been developed for mathematical simulations. These models require careful delineation of the components of the system and translation into mathematical terms.

Group Decision Making

Groups make a great number of decisions in most organizations. These groups may be standing or specially designated committees, teams, task forces, or project groups. Often, informal groups may be called together to assist with a particular decision.

Individual versus Group

When should a decision be made by a group rather than an individual? This issue has been greatly debated. According to Burack and Mathys (1983), individual versus group decision making largely depends on factors such as complexity and importance of the problem, time available, degree of acceptance required, amount of information needed to make a decision, and the usual manner in which decisions are made in an organization. They describe three possibilities for managerial decision making:

- **Individual decision.** Managers can make decisions themselves using information available to them.
- **Combination decision.** Managers can make decisions after consulting with others.
- **Group decision.** Managers can allow decisions to be made by the group, of which the manager is usually a member.

Group decision making, then, is used because managers frequently confront situations in which they must seek information and elicit judgments of other people. This is especially true for nonprogrammed decisions. In most organizations, rarely does an individual consistently make this type of decision alone. Problems are usually complex, and solutions require specialized knowledge from several fields. Usually, no one manager possesses all the kinds of knowledge needed.

Group decision making also may be used when two or more organizational units will be affected by the decisions. Because most decisions must eventually be accepted and implemented by several units, involving the groups affected in the decision making may be helpful.

Methods of Group Decision Making

Important decisions are being made in organizations by groups rather than by individuals. Group decision making most often is accomplished within interacting, Delphi, nominal, or focus groups.

Interacting Groups. An interacting group is a decision-making group in which members discuss, argue, and agree upon the best alternative (Griffin, 1999). Existing groups may be departments, work groups, or standing committees. New groups can be ad hoc committees, work teams, or task forces. An advantage of this method is that interacting promotes new ideas and understanding.

Delphi Groups. A Delphi group is used for developing a consensus of expert opinion (Griffin, 1999). A panel of experts, who contribute individually, makes predictions about a specific problem. Their opinions are combined and averaged and then returned to the panel for a second prediction. Members who made unusual predictions may be asked to justify them before sending them to the other members of the panel. When the predictions stabilize, the average prediction represents the decision of the group of experts. The Delphi method is good for forecasting technological breakthroughs but takes too much time and is too expensive for everyday decision making.

Nominal Groups. The nominal group method is a structured technique for generating creative and innovative alternatives or ideas (Griffin, 1999). Members of the group meet together but do not talk freely among themselves like members of interacting groups. The manager presents the problem to group members and asks them to write down as many alternatives for solutions as possible. They then take turns presenting their ideas, which are recorded on a flip chart. Members then vote by rank-ordering the various alternatives. The top-ranking alternative represents the decision of the group, which can be accepted or rejected by the manager.

Focus Groups. A focus group is a qualitative marketing method. It has been used successfully for many years by large, multiunit chains and independents looking for customer feedback (Dee, 1990). The focus group consists of 10 to 20 people brought together for a one-time meeting of about 2 hours to discuss some predetermined aspect of a particular establishment. Men and women are selected to participate if they meet certain criteria, such as being a frequent customer in the restaurant or in a competitor's establishment. Focus groups examine the motivation behind human behavior and, therefore, examine why people act the way they do, not what they do. Bob Evans Farms, Inc., used focus groups to correct sluggish sales in some of its Florida operations, and Burger King used focus groups before launching its chicken sandwich options.

Advantages and Disadvantages

One advantage of group decision making is that more information and knowledge are available. A group can generate more alternatives than can an individual and can communicate the decision to employees in their work group or department.

One of the disadvantages is that group processes sometimes prevent full discussion of facts and alternatives. Group norms, member roles, established communication patterns, and cohesiveness may deter the group and lead to ineffective decisions. Also, group decision making takes time and is expensive. One person may dominate the group. Two phenomena often occurring in decision-making groups that may interfere with good decisions are groupthink and risky shift.

Groupthink: Situation where reaching an agreement becomes more important to group members than arriving at a sound decision.

Risky shift: Tendency of individuals to take more risk in groups than as individuals.

Groupthink occurs when reaching an agreement becomes more important to group members than arriving at a sound decision. In cohesive groups, members often want to avoid being too harsh in judging other members. They dislike bickering and conflict, perceiving them as threats to "team spirit." Janis (1983), who popularized the concept of groupthink, suggested that faulty decision making may occur because group members do not want to rock the boat.

Risky shift is the tendency of individuals to accept or take more risk in groups than they would individually. Riskier decisions may result from group decision making because group members share the risk with others rather than having to bear responsibility individually.

COMMUNICATION

Communication has a major role in determining how effectively people work together and coordinate their efforts to achieve an organization's objectives. All organizations have recognized the importance of communication and have made major expenditures to improve it. Managers must communicate with superiors, other managers, and employees to convey their vision and goals for the organization (Griffin, 1999). These others have to communicate with managers to let them know what is happening in their work life and how they can be more effective.

Early in this chapter, communication was identified as one of the linking processes in organizations that is critical to managerial effectiveness and to the effective functioning of the foodservice system. Mintzberg (1975) found that managers spend a majority of their time communicating, much of which involves verbal communication.

Because of the importance of communication to organizations and to the personal effectiveness of managers, persons in leadership positions must be well versed on the basics of communication and apply good communication techniques in all their activities. Breakdowns in the communication process may lead to employee dissatisfaction, customer dissatisfaction, misunderstanding, misinterpretation, and a whole range of other problems.

Poor communication, although often cited as the reason for an organizational problem, may be merely a symptom of a more serious situation. Good communication is not a panacea for all organizational problems. It will not compensate for poor planning or poor decisions, although plans and decisions must be communicated to a variety of individuals in an organization for implementation. Thus, communication is an extremely important skill for managers; its significance cannot be overstated.

Communication Defined

Communication is the transfer of information that is meaningful to those involved. It also is defined as the transmittal of understanding. Effective communication is the process of sending a message in such a way that the message received is as close in meaning as possible to the message intended (Griffin, 1999). It occurs in many forms, ranging from face-to-face conversation to written memoranda, and involves verbal, nonverbal, and implied messages. Communication in organizations often is viewed from two perspectives: communication between individuals (interpersonal) and communication within the formal organizational structure (organizational). These two basic forms of communication are obviously interdependent and interrelated.

The simplest model of communication is as follows:

Sender → Message → Receiver

Regardless of the type of communication, it includes these three elements.

Communication Process

The simple model does not show the complexity of the communication process. A more sophisticated model is shown in Figure 11.12. The process of communication starts when the sender wants to transmit information to the receiver. The sender has a message, an idea, a fact, or some other information to transmit to someone or some group.

For example, the manager of a privately owned restaurant decides to convert the traditional management style currently being used to the new style and has developed a plan to eliminate some middle-management positions. The manager is eager to tell the restaurant owner about her plan. This idea may have simple or complex meaning to the sender. Meaning is an abstract concept that is highly personal. No direct relationship exists between the symbols and gestures used in communication and meaning, as shown in Figure 11.13. A common problem in communication is the misinterpretation that may result from the receiver not understanding the message in the way the sender intends.

The sender must encode the information to be transmitted into a series of symbols or gestures. For example, the manager might have said to the owner, "I will fire middle managers," or "We will save money," or "I will retrain middle managers for other positions." The manager choose to say, "We will save money by eliminating middle management positions." The encoding process is influenced by the content of the message and the familiarity of the sender and receiver.

After the message is encoded, it is transmitted through the appropriate channel. Channels in an organization include meetings, memos, letters, reports, and telephone calls. Because the owner is located many miles away from the restaurant,

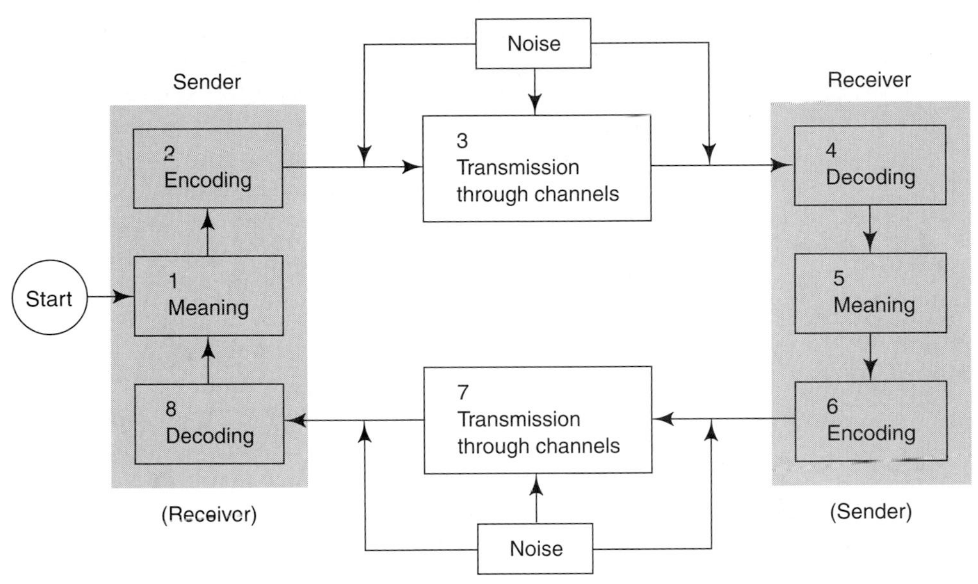

Figure 11.12. The communication process.
Source: Griffin, Ricky W., *Management,* 6th edition. Copyright © 1999 by Houghton Mifflin Company. Used with permission.

the manager talked to him over the telephone. After the message is received, it is decoded into a form that has meaning to the receiver. The owner might have thought, "Great, now I will be able to invest in another restaurant with the money saved," or "This is great news for the company," or "She is trying to get her salary increased." This meaning can vary from concrete to abstract. The owner's actual feeling was that this is great news for the company. Often the meaning requires another response and the cycle from sender to receiver starts all over again when a

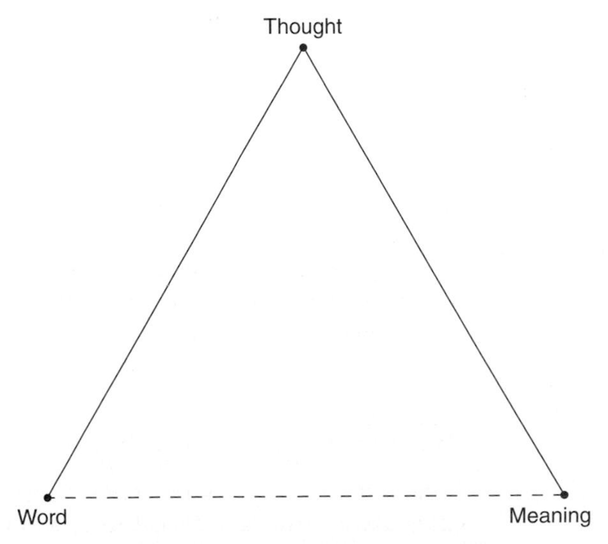

Figure 11.13. Triangle of meaning.

new message is sent back to the original sender. The owner told her immediately that this is great news for the company and made an appointment to meet with the manager in the restaurant within a week to discuss the plan.

Noise refers to all the types of interference that may distort or compete with the message during its transmission; it is shown at the top and bottom of Figure 11.12. Examples of noise are the inability to hear the sender, who is speaking too softly, distortion of the message by extraneous sounds, and inattention of the receiver. If the owner had been thinking about something else and told the manager that they would talk about the plan on his annual visit to the restaurant, she might wonder why she even bothered to try to save money for him. Instead, his decision to make a special trip to discuss the plan with her positively reinforced her idea and her effort to keep him informed.

Communication may be one-way or two-way. In *one-way communication,* the sender communicates without expecting or getting feedback from the receiver. *Two-way communication* exists when feedback is provided by the receiver. Feedback enhances the effectiveness of the communication process and helps to ensure that the intended message is received by allowing the receiver to clarify the message and permitting the sender to refine the communication. One-way communication takes considerably less time than two-way communication, but it is less accurate. If communication must be fast and accuracy is easy to achieve, one-way communication may be both economical and efficient. However, if accuracy is important and the message is complex, two-way communication is almost essential.

The communication skills, attitudes, knowledge, and the social system or culture of both the sender and receiver affect the communication. Differences in these elements between the sender and receiver may lead to communication problems. An obvious example is when the language is different for the sender and receiver. The foodservice manager responsible for supervising employees who speak little or no English would certainly understand this concept, which also has applicability to other types of communication. For example, when electronic communication systems are incompatible, communication breakdown also occurs.

Interpersonal Communication

A common and often incorrect assumption made in communicating with other persons is that the message was transmitted and received accurately. The assumption frequently is made that the message was transmitted effectively and the receiver understood it. This assumption is often incorrect and can be the source of many communication problems between individuals.

Interpersonal communication flows from individual to individual. The objective in interpersonal communication should be to increase the area of understanding (Figure 11.14). Ideally, the maximum overlap of "what was meant" and "what was perceived" is desired. In the following pages, some of the common barriers to effective interpersonal communications are discussed.

Barriers to Communication

"I didn't understand" is a common reply supervisors hear from employees when discussing why results were different from those expected. The dining room may be set up incorrectly, the wrong number of portions may be prepared, or the dietary

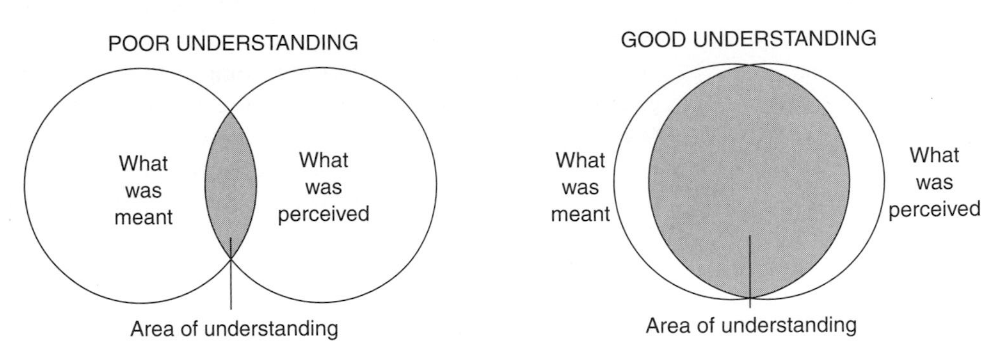

Figure 11.14. Areas of understanding.
Source: Introduction to Management: A Career Perspective (p. 379) by E. H. Burack and N. J. Mathys, Copyright © 1983 by John Wiley & Sons, Inc. Reprinted by permission of John Wiley & Sons, Inc.

aide may come to work at the wrong time. All these problems may be the result of communication breakdown between the supervisor and the foodservice employee.

Sayles and Strauss (1966) and Chaney and Martin (2004) identified the following common barriers in interpersonal communication:

- **Hearing an expected message.** Past experience leads one to expect to hear certain messages that may not be correct in some situations.
- **Ignoring conflicting information.** A message that disagrees with one's preconceptions is likely to be ignored. For example, the foodservice manager who believes the operation has good food and service may ignore customer complaints.
- **Differing perceptions.** Words, actions, and situations are perceived in accordance with the receiver's values and experiences. Different people react differently to the same message.
- **Evaluating the source.** The meaning applied to any message is influenced by evaluation of the source. For example, if the sender is viewed as knowledgeable, the message is interpreted differently than if the sender's knowledge on a particular topic is questioned.
- **Interpreting words differently.** Because of the complexity of language, words have many different meanings. For example, the word dinner may mean a meal at noon to some and an evening meal to others.
- **Ignoring nonverbal cues.** Tone of voice, facial expressions, and gestures may affect communication. In an attempt to convey a message, the sender may ignore the fact that the receiver seems preoccupied, and the message will not be heard.
- **Becoming emotional.** Emotion will affect transmission and interpretation of messages. For example, if the receiver perceives the sender as hostile, the response may be defensive or aggressive, creating a negative effect on the communication.
- **Cultural differences.** Differences in ethnic, religious, and/or social status may impact the understanding of a message.
- **Linguistics.** Different languages, dialects, and accents spoken by the sender and receiver or use of a vocabulary by the sender that is beyond the comprehension of the listener may alter understanding of the message being sent.

Inference may be a barrier or facilitator in communication. As Haney (1986) discussed, inferences are constantly being drawn in communication with other people; that is, conclusions may be drawn based on incomplete information and action taken as a result. If these inferences are incorrect, problems may occur; if correct, the inference may lead to efficiency in communication because needless information is avoided. A risk is involved, however, because of the potential of an inference leading to an incorrect conclusion.

Haney identified another barrier in communication that he referred to as *allness.* It occurs when one unconsciously assumes it is possible to know or say everything about something. Arrogance, intolerance of other viewpoints, and closed-mindedness are consequences of allness. Everyone can identify an individual whom we might call a "know-it-all." This person is the epitome of the allness concept.

Ability to interact effectively with other individuals is affected negatively by allness. Professionals, in particular, need to be sensitive to developing allness in dealing with nonprofessionals. A common fault among persons with highly specialized knowledge is the feeling that they know what's best for the other person.

All these barriers impede interpersonal communication and may lead to more serious problems in relationships. Sensitivity to and awareness of these barriers may assist in improving communication; a number of techniques can also be used to enhance interpersonal communication.

Techniques for Improved Communication

Techniques for improved communication are summarized in Figure 11.15. Using feedback can result in more effective communication because it allows the sender to search for verbal and nonverbal cues from the receiver. Questions can be encouraged and areas of confusion clarified as a result of two-way communication in which effective feedback is involved. Also, face-to-face communication may encourage feedback. People generally express themselves more freely in face-to-face situations or in other two-way communication, such as telephone or e-mail conversations.

Using several channels will improve the chances that a proper message is communicated. For example, following up a verbal message with a written note will serve as a reinforcement.

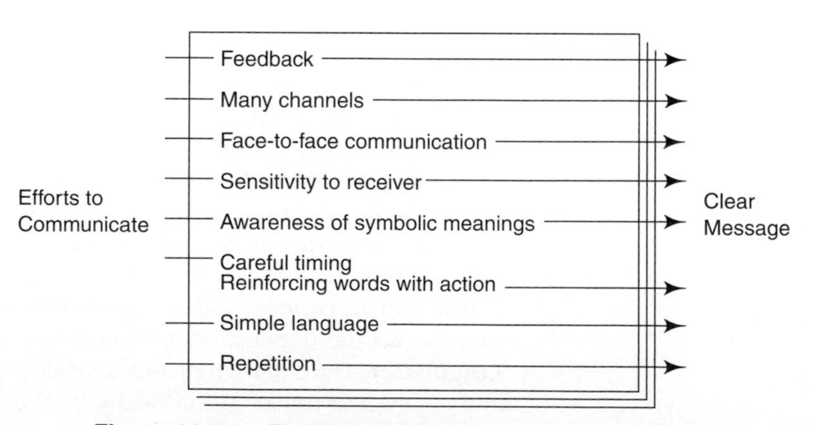

Figure 11.15. Techniques for improved communication.

Sensitivity to the receiver will enable the communicator to adapt the message to the situation. Individuals differ in their values, needs, attitudes, and expectations. Empathy with these differences will facilitate interpersonal communication.

Awareness of symbolic meanings can be particularly important in communication. For example, sensitivity to negative connotations and thus avoiding red flag words, which excite people to anger, is a way to improve interpersonal communication.

"It's not what you say, but what you do" is another tenet for improving communication. This concept, of course, is concerned with timing of the message and reinforcing words with action.

Using direct, simple language and avoiding jargon is another rule for improving communications. Especially important is the need to limit the message to the knowledge level of the receiver. This concept becomes vividly apparent in communication with children, but it may also be important in communicating with adults who may be less willing than children to say "I don't understand." Using the correct amount of repetition can serve as a reinforcement to the receiver's understanding of the message. Unnecessary redundancy or the overuse of cliches, however, may dull the receiver's attention.

Listening is basic to effective communication because receiving messages is as much a part of the process as sending them. Daydreaming and preoccupation with other matters or mentally arguing with points made by the speaker before the talk is finished may preclude individuals from listening. Impatience and lack of interest in the message are other impediments to listening.

Organizational Communication

Factors discussed in relation to interpersonal communication apply to communication within organizations as well. Effective organizational communication involves getting an accurate message from one person to another. The effectiveness of organizational communication, however, is also influenced by several characteristics unique to organizations.

Lesikar (1977) identified four factors that influence the effectiveness of organizational communication: the formal channels of communication, the authority structure, job specialization, and a factor he calls "information ownership." *Formal channels* influence communication effectiveness in space. First, as organizations grow, the channels cover an increasingly larger distance. Second, the formal channels of communication may inhibit the flow of information between levels in the organization. The cook, for example, almost always will communicate problems to the foodservice supervisor. The foodservice supervisor may or may not communicate this information to the unit manager. Although this accepted channel of communication has advantages, such as keeping higher-level managers from being overloaded with too much information, it also has disadvantages. Higher-level managers may not always receive all the information they need for staying on top of operations. With additional levels in an organization, the accuracy of messages up and down the organization also will be affected.

The *authority structure* affects communication because of status and power differences among organizational members. The content and accuracy of communication will be affected by authority differences. For example, conversation between the director of foodservice and a foodservice employee may tend to be a strained politeness or somewhat formal. Neither one is likely to say much of importance.

Job specialization can be both a help and hindrance in communication. It facilitates communication within a work group because members speak the same jargon and frequently develop a group cohesiveness. Communication between work groups, however, may be inhibited. Cafeteria employees may have free-flowing communication, as may also be true of the production employees. Communication between these two groups may not be as effective as that within each group.

The term *information ownership* means that individuals possess unique information and knowledge about their jobs. Such information is a form of power for those who possess it, making them unwilling to share the information with others.

Overcrowded work areas, too many individuals competing for the attention of a manager, the improper choice of a communication channel, and time pressures are other factors that impede organizational communication. For example, a typed letter rather than face-to-face conversation for handling a touchy matter may lead to hurt feelings.

Directions of Internal Communication

Within the organization, managers must provide for communication in four distinct directions: downward, upward, horizontal, and diagonal. Although discussed only briefly in this chapter, managers also must communicate effectively with individuals and groups outside the organization.

Downward. Individuals at higher levels of the organization communicate downward to those at lower levels. The most common forms of communication are job instructions, policy statements, procedure manuals, and official publications of the organization. In addition, middle- and lower-level managers are usually contacted by written memos or with some other official directive. A summary of various channels available to carry information downward is listed in Figure 11.16.

External means, such as radio, television, and newspaper press releases, may be used to communicate not only with employees but also with the general public. For example, a school district closing because of snow is usually announced over the radio and television as a way of communicating to the school district staff, parents, and students.

Upward. An effective organization needs open channels of upward communication as much as it needs downward communication. In large organizations, good upward communication is especially difficult. Suggestion boxes, special meetings, and grievance procedures are devices used for upward communication. Some channels for upward flow of information are summarized in Figure 11.17.

Effective upward communication is important because it provides employees with an opportunity to have a say in what happens in the organization. Equally important, however, is the need for management to receive vital information from lower levels in the organization.

Horizontal. Provision for horizontal flow of communication will enhance organizational effectiveness. Within the foodservice operation, effective horizontal communication between production and service is critical to ensure that quality food is available at the right time in the right place.

Chain of Command
Orders and information can be given personally or in written fashion and transmitted from one level to another. This is the most frequently used channel and is appropriate on either an individual or a group basis.

Posters and Bulletin Boards
Some workers may not read them. This is especially true when the posters or bulletins are not kept current. Thus, this channel may be useful only as a supplementary device.

Company Newsletters or Newspapers
A great deal of information about the company, its products, and policies can be communicated in this way. Readership is increased if some space is allocated to personal items of interest to employees.

Letters and Pay Inserts
Direct mail may be used when top management wants to communicate matters of importance. Since the letter is sent directly to the employee from the company, a reasonable chance exists that it will be read. Inserting a letter with the paycheck ensures that each worker receives a copy.

E-mail
Companies, whose employees all have access to e-mail on their jobs, may send letters or messages in a "paperless" way using e-mail.

Employee Handbooks and Pamphlets
Handbooks are frequently used during the hiring and orientation process as an introduction to the organization. Too often, however, they are unread even when the firm demands a signed statement that the employee is acquainted with their contents. When special systems are being introduced, such as a pension plan or a job evaluation system, concise, well-illustrated pamphlets are often prepared to facilitate understanding and stimulate acceptance.

Annual Reports
Annual reports are increasingly written not only for the stockholders but also for the employee. A worker may be able to obtain information about the firm in this way.

Loudspeaker Systems
The loudspeaker system is used not only for paging purposes but also to make announcements while they are "hot."

Figure 11.16. Formal downward channels of communication.
Source: Adapted from *Management: Concepts and Practices* (pp. 430–432) by R. W. Mondy, A. Sharplin, and E. B. Flippo, 1988, Boston: Allyn & Bacon. Used by permission.

Department head meetings are one way to facilitate interdepartmental communication. The more interdependent the work of organizational units or departments, the greater the need for horizontal communication.

Diagonal. The use of diagonal channels of communication is a way of minimizing time and effort expended in organizations. Having reports and other information flow directly between departments or units that have a diagonal placement in the organization may result in more effective flow of information.

Open-Door Policy
An open-door policy is an established guideline that allows workers to bypass immediate supervisors concerning substantive matters without fear or reprisal. Managers are encouraged to create an environment in which subordinates will feel free to come to them with problems and recommendations.

Suggestion Systems
Many companies have formal suggestion systems. Some have suggestion boxes. When a suggestion system is used, every suggestion should receive careful consideration. Workers should be promptly informed of the results of the decision on each suggestion.

Questionnaires
Anonymous questionnaires sometimes are given to workers in an attempt to identify problem areas within the organization.

The Grievance Procedure
The grievance procedure is a mechanism that gives subordinates the opportunity of carrying appeals beyond their immediate supervisor.

Ombudsperson
An ombudsperson is a complaint officer with access to top management who hears employee complaints, investigates, and sometimes recommends appropriate action.

Special Meetings
Special employee meetings to discuss particular company policies or procedures are sometimes scheduled by management to obtain employee feedback.

Figure 11.17. Formal upward channels of communication.
Source: Adapted from *Management: Concepts and Practices* (pp. 434–436) by R. W. Mondy, A. Sharplin, and E. B. Flippo, 1988, Boston: Allyn & Bacon. Used by permission.

For example, the ordering clerk in a foodservice operation may send requests directly to the purchasing department rather than through the channels of the foodservice department. As long as procedures for this direct communication exist, the function of ordering and receiving goods is facilitated. Communication can be either informal or formal.

Informal Communication

Informal networks develop in organizations to supplement formal communication channels. Oral communication usually occurs in face-to-face conversations, telephone calls, and group discussions (Griffin, 1999). The primary advantage of oral communication is that it promotes quick feedback to verbal questions or agreements and facial expressions. It is easy to use, and paper and pencil or computers are not needed. The major drawback of using oral communications is that distortion of the message can result.

Grapevine: Informal communication channel in which information is passed from one employee to another.

The **grapevine** is another type of informal communication and may facilitate organizational communication and meet the social needs of individuals within the organization. Although the grapevine may filter or distort messages and occasionally transmit rumors and gossip, its speed and accuracy in getting messages through an organization often are very useful.

The effective manager is one who has learned to use the grapevine to advantage and recognizes its importance in fulfilling the communication needs of members of the organization. The grapevine can and should work for the manager. Informal communication systems can help to speed the work-related flow of information by making use of the natural interaction among people in organizations. Ignoring the grapevine and rumors is a dangerous alternative because a rumor usually has some element of truth and may be a symptom of a larger problem.

Formal Communication

Written communication usually gives the appearance of being more formal and authoritative than oral communication. Because the same words are communicated to all who receive these messages, written communication tends to be interpreted more accurately and is often used when consistent action is required and a record of the communication is necessary. It also provides a permanent record of the exchange, and the receiver can refer to it as often as needed. A common problem with written communication, however, is proliferation of paperwork that is of questionable importance. Excessive red tape in organizations is often the result of excessive requirements for written communication.

Today most managers prefer not to use written communications because it inhibits feedback and interchange (Griffin, 1999). When a manager sends a letter, it must be written or dictated, typed, mailed, received, opened, and read. If the message is misunderstood, several days may pass before it is rectified. A telephone call or an e-mail message often can achieve the same outcome in much less time.

Recent breakthroughs in electronic communication have changed organization communication drastically. A site on the Internet is fast becoming an essential part of communications for thousands of foodservice operations (Dinehart, 1997). Each week, additional operations climb aboard the Net, propelling this medium forward on its path to becoming an indispensable channel for information and marketing. Today, teleconferences, in which managers stay at their own locations, are taking the place of face-to-face meetings. Cellular telephones and facsimile (fax) machines and e-mail make communications easier. Many people use cellular telephones to make calls while commuting back and forth to work. Fax machines permit people to use written communications and receive quick feedback. Psychologists are beginning to associate some problems with these communication breakthroughs. People are no longer part of the organizational grapevine and miss out on the informal communications that take place at work. Also, the use of electronic communications makes it difficult to build a strong culture, develop working relationships, and create a mutually supportive atmosphere of trust and cooperatives.

Inter- and Cross-Cultural Communication

As businesses expand to locations around the globe, the importance increases of **intercultural communication** (sometimes termed cross-cultural), the communication between persons of different cultures. According to Chaney and Martin (2004) *culture* deals with the way people live and includes language, physical, and psychological dimensions. Norms, rules, and roles within a culture will impact the encoding and decoding of verbal and nonverbal messages. Having an understanding of not only the language but the customs of those you are communicating with can increase the quality of the communication process.

One does not need to leave the United States to see the impact of culture on the communication process. Each year immigrants from many nations enter the United States and are faced with issues of acculturation. **Acculturation** occurs as one adapts or adjusts to a new and different culture. The acculturation process can occur in one of four ways (Alkhazraji, 1997):

- **Assimilation:** Individuals are absorbed in new culture and withdraws from their former culture.
- **Integration:** Individuals becomes integral part of new culture but maintains the integrity of their previous culture.
- **Separation:** Individuals keep their culture and stay independent of the new culture.
- **Deculturation:** Individuals lose their original culture but do not accept new culture.

Effective intercultural communication requires an understanding of many issues in addition to the nuances of the language of country or region. Chaney and Martin (2004) identified several issues that differ among cultures that can affect the communication process:

- **Paralanguage:** Paralanguage, the rate, pitch, and volume of the voice giving the message, can affect the meaning interpreted from that message. Paralanguage can convey positive or negative emotions and attitudes. For example, in some cultures talking rapidly and loudly is perceived as normal; in others it is thought to show anger.
- **Chronemics:** Chronemics, attitudes about time, vary among cultures. In some, *monochromic time,* focusing on one major activity at a time, is most common; in others, *polychromic time,* working on several major items at the same time, is more common. Attitudes toward time also include perceptions related to punctuality which can vary from a great emphasis being placed on "being on time" in one culture to a "mañana" attitude of putting off to tomorrow what didn't get done today in another culture.
- **Proxemics:** Proxemics, the physical space between individuals when they are communicating, also differs among cultures and may alter interpretation of the message being given. People in the United States tend to need more space, for example, than do persons from Greek, Latin American, or Arab cultures. Hall and Hall (1990) described four zones of interaction: *intimate zone* (less than 18 inches, reserved for very close friends or business handshakes), *personal zone* (18 inches to 4 feet, used when giving instructions or working closely with another), *social zone* (4 to 12 feet, used for most business interaction), and *public distance* (greater than 12 feet, very formal interaction).
- **Oculesics:** Oculesics is the use of eye contact during communication. In some cultures use of direct eye contact indicates listening; in others it is a sign of disrespect and bad manners. The use of direct eye contact and the amount of time direct eye contact is made may send unintended nonverbal messages in the communication process.
- **Olfactics:** Olfactics, or smell, can have a positive or negative nonverbal impact on the communication process. Body, breath, and clothes odors are viewed negatively in some cultures and as natural and inoffensive in others.

- **Haptics:** Communicating through touch or body contact is referred to as haptics. Touch, when used appropriately, can convey support and trust; however, when used inappropriately can be uncomfortable and create a sense of distrust and annoyance.
- **Kinesics:** The use of body movements such as facial expressions, gestures, and posture in the communication process is often termed kinesics. Body movements can express emotion, add emphasis, and provide clarity, but they can also create disrespect, anger, and shame when misinterpreted. The "thumbs-up" gesture, for example is used in the United States as a signal that everything is going OK; that same gesture is considered rude in West Africa.
- **Chromatics:** Colors have different meanings in different cultures and thus use of color can convey an unintended nonverbal message. White, for example, is viewed as peaceful and pure in some cultures and is associated with mourning in others.
- **Silence:** The duration and appropriateness of silence can be interpreted in different ways depending on the culture. Long periods of silence are often not comfortable for U.S. businesspeople but are common in negotiations by Japanese businesspeople.

Robbins (2005) offers four suggestions for reducing misconceptions and misperceptions when interacting with someone from a different culture:

- **Assume differences until similarities proven**—persons for other countries are often very different in their customs, values, and beliefs; fewer errs in communication occur if one assumes many differences exist rather than assuming those from other cultures are very similar to one's own.
- **Emphasize description rather than interpretation or evaluation**—avoid making judgments on the meaning of communication based on how something was said; focus on getting a complete description of what is wanted.
- **Practice empathy**—attempt to see yourself as the listener to better understand his or her values and frame of reference; this will help ensure that the communication sent is more likely to be received accurately.
- **Treat your interpretation as a working hypothesis**—initial assessment of the meanings behind intercultural communications may not be accurate; continually check that your interpretation of the communication is what was intended by the sender.

Written communication also can differ among cultures. Chaney and Martin (2004) offer several suggestions for "internationalizing" written communication in English.

- Use the most common English words, use the most common meaning of those words, and use them in their most common way.
- Select words with few similar or alternate meanings.
- Avoid slang, acronyms, *emoticons* (:-o), and "shorthand" (4 representing for, U for you).
- Conform to rules of grammar, avoid misplaced modifiers, dangling participles, and incomplete sentences.
- Use short, simple sentences.

Negotiation

Negotiation: Process in which two or more parties make offers, counteroffers, and concessions to reach an agreement.

Negotiation, as a form of communication, is a process in which two or more parties make offers, counteroffers, and concessions to reach an agreement. Often the views differ between the two parties in the negotiation process and the result of negotiation involves compromise between them.

Robbins (2005) described two general approaches to negotiation: distributive bargaining and integrative bargaining. **Distributive bargaining** occurs in situations where resources available are fixed and the negotiation focuses on what portion of the resources each will get. In such negotiation each party has a *target point,* the amount he or she would like to get, and a *resistance point,* the lowest acceptable amount that will influence the negotiation discussions. The area of overlap between the target and resistance points of each negotiator is termed the settlement range. The perception is that the party who gets closest to his or her target wins and the other loses.

Integrative bargaining operates in situations where there is a variable amount of resources available and there can be many possible settlement options which can result in perceived "wins" for both parties in the negotiation.

Negotiation may occur in many aspects of a manager's job. Some examples of issues that might be negotiated include

- between a manager and an employee regarding salary and benefits
- between a manager and a supplier over price and delivery issues
- between management and union representatives to finalize a collective bargaining agreement
- between managers of two work units to work out differences in opinions related to job responsibilities/completion

Robbins and Coulter (2002) offered several recommendations for managers to use to become more effective in negotiating:

- **Research the individual with whom you will be negotiating**—understanding the other's position will help you better understand the other's behavior and suggest solutions that might be agreeable.
- **Begin with a positive overture**—negotiating with a positive attitude is more likely to be reciprocated positively and can lead more quickly to agreement.
- **Address problems, not personalities**—concentrate on the negotiations issues, not the personality characteristics of the person(s) with whom you are negotiating.
- **Pay little attention to initial offers**—initial offers tend to be extreme and idealistic.
- **Emphasize win-win situations**—look for integrative solutions; frame options in terms of the other's interests.
- **Create an open and trusting environment**—strategies to create an open and trusting environment include: listen more, talk less; ask questions; focus on position of others; avoid becoming defensive.

BALANCE

Balance: Managerial adaptation to changing environmental, political, social, and technological conditions.

Balance, a linking process mentioned in Chapter 1, refers to managerial adaptations to changing economical, political, social, and technological conditions. Foodservice organizations are the products of the organizing function of management, which provides the mechanism for coordinating and integrating all activity toward accomplishment of objectives. According to Stoner and Wankel (1986), the organizing process involves balancing a company's needs for both stability and change. An organization's structure gives stability and reliability to the actions of its members; both are required for an organization to move coherently toward its goals. Altering the structure of an organization can be a means of adapting to and bringing about change, or it can be a source of resistance to change.

Conditions both outside and inside organizations are changing rapidly and radically today. Economic conditions, availability, and cost of materials and money, technological and product innovations, and government regulations all can shift rapidly. A great variety of external forces, including competitive actions, can pressure organizations to modify their structure, goals, and methods of operation. Pressures for change also may arise from a number of sources within the organization, especially from new strategies, technologies, and employee attitudes and behavior. Internally, employees are changing by placing greater emphasis on human values and questioning authority. They assert their rights under affirmative action, safety, compensation, and other legislation, and they question the fairness of management decisions and actions. Employees desire to improve the quality of their working lives. External and internal forces for change often are linked, especially when changes in values and attitudes are involved.

Organizational design is affected by three types of external environments: stable, changing, and turbulent (Stoner & Wankel, 1986). A *stable environment* has experienced little or no change; product changes seldom occur and modifications are planned in advance. Market demand is predictable, and laws affecting the organization have remained the same for many years. Stable environments are rare due to the advent of many technological changes. Environmental changes, however, can occur in the product, market, law, or technology. These changes do not surprise managers because trends are apparent. Many service industries function in a *changing environment.* The organization is in a *turbulent environment* when competitors market new, unexpected products, when laws are passed that affect businesses in radical ways, and when technology suddenly changes product design or production methods. Few organizations function in a continuously turbulent environment, however; if a radical change does occur, only a temporary period of turbulence is endured before they make an adjustment. Even though many forces for change can be found in organizations, other forces serve to maintain a state of balance. Forces opposing change also are supporting stability or status quo. According to the *force-field theory* of Lewin (de Rivera, 1976), any behavior is the result of a balance between driving and restraining forces. Driving forces push in one direction while restraining forces push in the other, and the result is a reconciliation of the two sets of forces. An increase in the driving forces might increase performance, but they also might increase the restraining forces. For example, the manager of a foodservice in a very busy theme park decided that employees would have to work every weekend to increase profits. This change might result in a decrease of profits, however, if employees' hostility, distrust, and greater resistance cause additional declines in productivity. Employees who want change push, and those who do not push back. Driving forces generate restraining forces. Decreasing restraining forces, therefore, is usually a more effective method of encouraging change than increasing driving forces. The balance concept suggests that organizations have forces that keep performance from falling too low and forces that keep it from rising too high.

Every organization needs to be committed to its product and service (Peters & Waterman, 1988). In addition, it needs to have a strong set of values and a firm vision of the future. Peters and Waterman described the properties of a balanced structure that responds to three needs: to maintain efficiency around the basics, to engage in innovation on a regular basis, and to avoid calcification by ensuring responsiveness to major threats. The structure of the organization is based on three pillars, each of which responds to these needs.

- **Stability pillar**—responds to the need for efficiency and is found in a departmentalized structure
- **Entrepreneurial pillar**—keeps the structure small, which is a requisite for continual adaptiveness and innovation
- **Habit-breaking pillar**—includes a willingness to reorganize frequently, adjusting to various forces

CHAPTER SUMMARY

This summary is organized by the learning objectives.

1. Decision making is the process of selecting a course of action from among alternatives. Managers are involved in a variety of types of decisions. Programmed decisions are those reached by following established policies and procedures; nonprogrammed decisions are relatively unstructured and require judgment because they are unique and cannot be based on established policies and procedures. Managers will make decisions under varying degrees of certainty and risk.
2. A variety of techniques are used to assist the manager with decision making. Some of the techniques involve analysis of information, such as decision trees, which use a patterned sequence of decisions to help arrive at the decision, and cost-benefit and cost-effectiveness analyses, which use quantitative data to help demonstrate the best decision. Other techniques involve mathematical modeling of the process, such as PERT, CPM, and linear programming to provide data for the decision-making process. There also are techniques such as focus groups, the Delphi technique, and nominal groups that involve discussions with individuals in helping decide the best course of action.
3. The communication process involves a sender who encodes a message and sends it to the receiver. The receiver decodes the message. This communication process can be enhanced by using feedback, having many channels of communication, using face-to-face communication, having sensitivity to the receiver,

Professional Organization Profile

Dietary Management Association

MISSION

DMA is the premier resource for food service managers, directors, and those aspiring to careers in Food Service Management.

WHO BELONGS TO THE ORGANIZATION

Members include more than 14,000 food service management professionals who work in hospitals, long-term care facilities, schools, correctional facilities, and other non-commercial foodservice settings. Student memberships are available.

ADVANTAGES OF MEMBERSHIP

Members have access to a members-only section of the association website that provides a link to an online network for interacting with other food service managers across the country; receive a monthly magazine, *Dietary Manager,* highlighting innovations being explored in food service operations; attend association sponsored conferences and workshops to obtain continuing education credit; and are eligible to receive association sponsored scholarships.

WEBSITE

www.dmaonline.org

MEET THE PRESIDENT

Phillis E. Fletcher, CDM, CFPP, President 2005–2006, has worked in the healthcare food service field for the past 28 years. She began her food service career as a dishwasher, trayline worker, and cook in a small rural hospital. She was promoted to Assistant Food Service Director and then Director of the department. Phillis offers the following advice to students: "The food service industry has so many exciting career opportunities. Being able to use your innovation and creativity will open so many doors for you. Remember to always take pride in your profession and sell yourself as the most important person any business could have on their team."

being aware of the symbolic meanings of words, using careful timing, reinforcing words with action, using simple language, and using repetition.
4. Balance is the adaptation by managers to the economical, political, social, and technological conditions surrounding an operation. Balance is a key linking process in helping change inputs to outputs in the transformation process.

TEST YOUR KNOWLEDGE

1. Define the six stages of the decision-making process.
2. How does good communication help with decision making in an organization?
3. What are the advantages and disadvantages of group decision making and individual decision making?
4. How is communication channeled through the downward, upward, horizontal, and diagonal directions of communication?
5. How can a manager help reduce the barriers to communication in an organization?
6. What impact can culture have on the communication process?

CLASS PROJECTS

1. Divide into groups of five people. Have one person make a story and then whisper it to the person next to him or her. This person in turn will whisper it to the person next to him or her. Continue this process until the story has been told to all. Have the last person who heard the story tell it out load. Then have the first person, who made up the story, tell the story as it was originally described. Discuss what caused the breakdown in this communication process. Discuss how communication errors occur in a foodservice operation.
2. Talk to a foodservice manager; ask about how he or she makes decisions. Ask for examples of decisions that have been made during that day. Share the results of the interview with classmates.
3. Invite students from other cultures to speak to the class about customs and practices in their culture related to communication.

WEB SOURCES

www.iabc.com	International Association of Business Communication
www.qpronline.com	QPRonline—an online quality management resource
www.toastmasters.org	Toastmater's International

BIBLIOGRAPHY

Alkhazraji, K. M. (1997). *Immigrants and cultural adaptation in the American workplace.* New York: Garland.

Balintfy, J. L., Rumpf, D., & Sinha, P. (1980). The effect of preference-maximized menu on consumption of school lunches. *School Food Service Research Review, 4*(1), 48–53.

Batty, J. (1993). Communication between the sexes. *Restaurants USA, 13*(6), 31–34.

Berger, E., Ferguson, D. H., & Woods, R. (1987). How restaurateurs make decisions. *Cornell Hotel and Restaurant Administration Quarterly, 27*(4), 49–57.

Brownell, J. (1987). Listening: The toughest management skill. *Cornell Hotel and Restaurant Administration Quarterly, 27*(4), 65–71.

Burack, E. H., & Mathys, N. J. (1983). *Introduction to management: A career perspective.* New York: Wiley.

Chaney, L. H., & Martin, J. S. (2004). *Intercultural business communication.* Upper Saddle River, NJ: Prentice Hall.

Cook, R. L. (1992). Expert systems in purchasing: Application and development. *International Journal of Purchasing and Materials Management, 28*(4), 20–27.

Dee, D. (1990). Focus groups: Finding out why customers act the way they do. *Restaurants USA, 10*(7), 30–34.

de Rivera, J. (Ed.). (1976). *Field theory as human-science: Contributions of Levin's Berlin group.* New York: Gardner Press.

Dinehart, J. W. (1997). The Web: Toy or tool for restaurants? *Restaurants & Institutions, 107*(16), 74, 76.

Etzioni, A. (1989). Humble decision-making. *Harvard Business Review, 67*(4), 122–126.

Griffin, R. W. (1999). *Management* (6th ed.). Boston: Houghton Mifflin.

Hall, E. T., & Hall, M. R. (1990). *Understanding cultural differences: Germans, French, and Americans.* Yarmouth, ME: Intercultural Press.

Hampton, D. R. (1986). *Management* (3rd ed.). New York: McGraw-Hill.

Haney, W. V. (1986). *Communication and interpersonal relations: Texts and cases* (5th ed.). Homewood, IL: Richard D. Irwin.

Hersey, P., Blanchard, K., & Johnson, D. E. (1996). *Management of organizational behavior: Utilizing human resources* (7th ed.). Englewood Cliffs, NJ: Prentice Hall.

Janis, L. L. (1983). Groupthink. In Staw, B. M., (ed.), *Psychological foundations of organizational behavior* (2nd ed.) (pp. 514–522). Glenview, IL: Scott, Foresman.

Leitch, G. A. (1989). Application of a PERT-type system and "crashing" in a food service operation. *FIU Hospitality Review, 7*(2), 66–76.

Lesikar, R. V. (1977). A general semantics approach to communication barriers in organization. In Davis, K. (ed.), *Organizational behavior: A book of readings* (5th ed.) (pp. 336–340). New York: McGraw-Hill.

Lesikar, R. V., & Pettit, J. D. (1994). *Communication theory and application.* Homewood, IL: Richard D. Irwin.

Loomba, N. P. (1978). *Management: A quantitative perspective.* New York: Macmillan.

March, J. G., & Simon, H. A. (1958). *Organizations.* New York: Wiley.

Mintzberg, H. (1975). The manager's job: Folklore and fact. *Harvard Business Review, 53*(4): 49–61.

Mondy, R. W., & Premeaux, S. R. (1992). *Management: Concepts, practices, & skills* (6th ed.). Boston: Allyn & Bacon.

Nettles, M. F., & Gregoire, M. B. (1997). Use of computer simulation in school foodservice. *Journal of Foodservice Systems 4*(3), 143–156.

Peters, T. J., & Waterman, R. H. (1988). *In search of excellence: Lessons from America's best-run companies.* New York: Harper & Row.

RGI Cost Benefit Analysis. (1998). [Online]. Available: http://www.cts.com/browse/rgi/cba.htm

Robbins, S. P., & Coulter, M. (2002). *Management* (7th ed.). Upper Saddle River, NJ: Prentice Hall.

Robbins, S. P. (2005). *Essential of organizational behavior* (8th ed.). Upper Saddle River, NJ: Prentice Hall.

Rogers, C. R., & Roethlisberger, F. J. (1991). Barriers and gateways to communication. *Harvard Business Review, 69*(6), 105–111.

Rue, L. W., & Byars, L. L. (1989). *Management: Theory and application* (5th ed.). Homewood, IL: Richard D. Irwin.

Sayles, L. R. (1989). *Leadership: Managing in real organizations* (2nd ed.). New York: McGraw-Hill.

Sayles, L. R., & Strauss, G. (1966). *Human behavior in organizations.* Englewood Cliffs, NJ: Prentice Hall.

Spears, M. C. (1976). Concepts of cost effectiveness: Accountability for nutrition, productivity. *Journal of the American Dietetic Association, 68*(4), 341–346.

Stoner, J. A. F., & Wankel, C. (1986). *Management* (3rd ed.). Englewood Cliffs, NJ: Prentice Hall.

CHAPTER

12

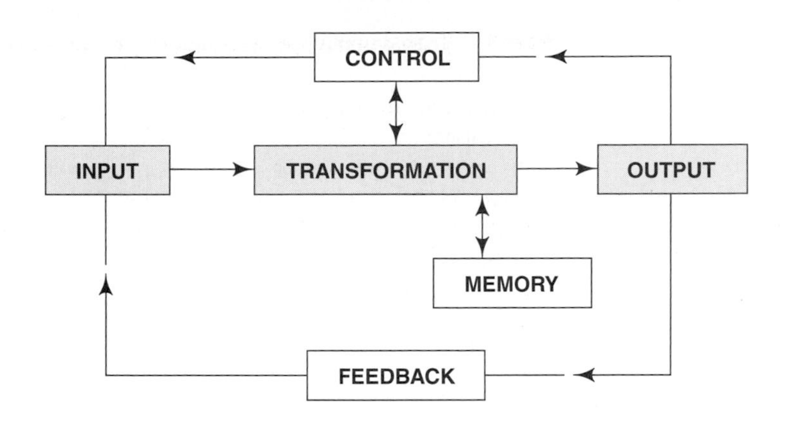

Management of Human Resources

Enduring Understanding

- Human resources are one of the most important inputs of a foodservice operation.
- Careful selection, training, and scheduling of employees can reduce labor costs.

Learning Objectives

After reading and studying this chapter, you should be able to

1. Differentiate among the recruitment, selection, orientation, training, and performance appraisal components of human resource management.
2. Describe the impact of at least five different laws on human resource management.
3. Define terms related to labor relations, such as collective bargaining, union steward, mediation, and arbitration.
4. Analyze the productivity of a work unit and make suggestions for improving the productivity.

The most important resources in a foodservice organization are its human resources. In this chapter, you will learn about the many and varied responsibilities related to managing the human resources of an organization. You will be introduced to the processes of recruitment, selection, orientation, and evaluation of employees. We also will discuss ways to control labor costs through proper staffing, scheduling, and work design.

EVOLUTION OF HUMAN RESOURCES MANAGEMENT

Selection, training, and compensation of employees were considered basic functions of personnel management for many decades. These functions were considered independent, with no regard for their interrelationships. **Human resources management (HRM)** emerged from this traditional viewpoint and today involves an integrative process of recruiting selecting, training, and maintaining the workforce needed to achieve an organization's goals.

Personnel management went from the factory, in which power-driven equipment and improved techniques made manufacturing products cheaper, to mass production, to HRM. Factory jobs were monotonous and often unhealthy and dangerous. In the late 1880s, state legislation and collective bargaining appeared, and working conditions began to improve. Mass production followed in the early 1890s with the use of labor-saving equipment that further improved production techniques. **Scientific management,** defined as the systematic approach to improving worker efficiency based on the collection and analysis of data, appeared in manufacturing (Sherman, Snell, & Bohlander, 1997). Frederick W. Taylor, considered the father of scientific management, Frank and Lillian Gilbreth, and Henry L. Gantt all contributed to this era in which performance standards and time studies were emphasized.

HRM began in the 1920s when the human element was added to management functions. The Hawthorne studies, the human relations movement, and behavioral sciences led to HRM as it is today. Along with HRM came the increase in both federal and state legislation, starting with the National Labor Relations Act in 1935. Human resources managers are responsible for compliance with all state and local laws and regulations that govern work organizations, including foodservice operations. These managers also will need to assume a broader role in organizational strategy and change and at the same time consider the bottom line of the operation.

Moving from a traditional organization to a total quality management culture demands much from the human resources function. Blackburn and Rosen (1993) interviewed human resources professionals at organizations that had received the Malcolm Baldrige Award to determine what kind of contribution they made to the company's effort to receive the award. The evolution from traditional personnel practices to a total quality paradigm in HRM policies is shown in Table 12.1 (Blackburn & Rosen, 1993).

Outcomes of building a TQM culture in the Baldrige award organizations indicate that measurable results in the bottom line have been generally impressive. Increased on-time deliveries, reduced customer complaints, and reduction in new product development time were some of the benefits of a total quality program. As a function of management, human resource management is part of the transformation process in the foodservice system (Figure 12.1). Human resources management encompasses

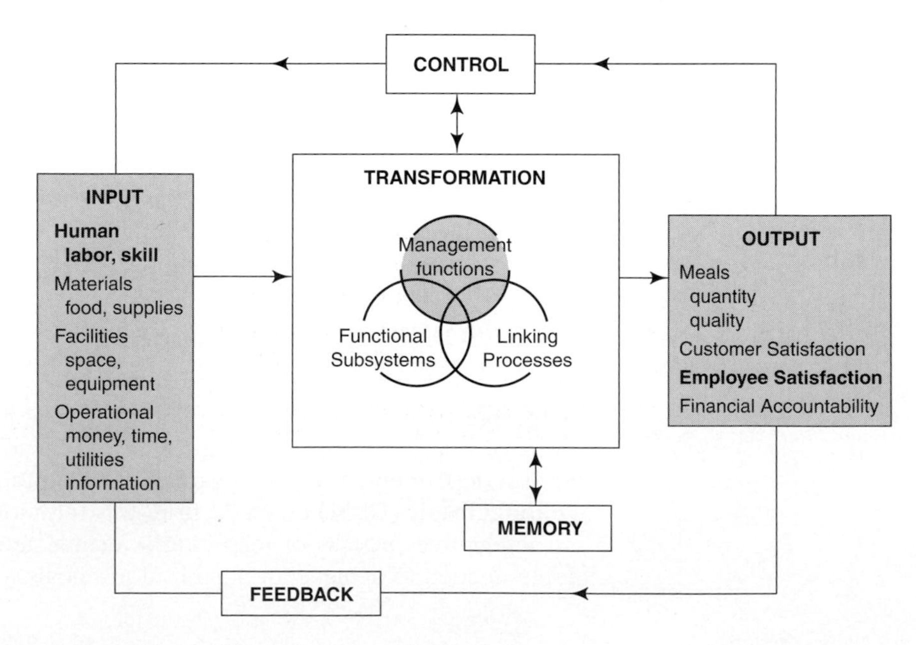

Figure 12.1. Human resources management: A management function in the transformation process.

Table 12.1 Evolution of a Total Quality Human Resources Paradigm

Corporate Context Dimension	**Traditional Paradigm**	**Total Quality Paradigm**
Corporate culture	Individualism Differentiation Autocratic leadership Profits Productivity	Collective efforts Cross-functional work Coaching/enabling Customer satisfaction Quality
Human Resources Characteristics	**Traditional Paradigm**	**Total Quality Paradigm**
Communications	Top-down	Top-down Horizontal, lateral Multidirectional
Voice and involvement	Employment-at-will Suggestion systems	Due process Quality circles Attitude surveys
Job design	Efficiency Productivity Standard procedures Narrow span of control Specific job descriptions	Quality Customization Innovation Wide span of control Autonomous work teams Empowerment
Training	Job-related skills Functional, technical Productivity	Broad range of skills Cross-functional Diagnostic, problem solving Productivity and quality
Performance measurement and evaluation	Individual goals Supervisory review Emphasize financial performance	Team goals Customer, peer, and supervisory review Emphasize quality and service
Rewards	Competition for individual merit increases and benefits	Team/group-based rewards Financial rewards, financial and nonfinancial recognition
Health and safety	Treat problems	Prevent problems Safety programs Wellness programs Employee assistance
Selection/promotion career development	Selected by manager Narrow job skills Promotion based on individual accomplishment Linear career path	Selected by peers Problem-solving skills Promotion based on group facilitation Horizontal career path

Source: "Total Quality and Human Resources Management: Lessons Learned from Baldrige Award-Winning Companies" by R. Blackburn and B. Rosen, 1993, *Academy of Management Executive, 7*(3), p. 51. Used by permission.

the recruitment, selection, orientation, training, supervision, and evaluation of one of the key inputs to the foodservice system—its human resources. Effective human resource management is necessary to ensure that all of the outputs to the system are achieved, including employee satisfaction. Human resources executives, once viewed by some executives as having perfunctory roles in the organization, are now welcomed to positions on strategic policy committees (Blackburn & Rosen, 1993).

HUMAN RESOURCES PLANNING

Human resources planning is the process of anticipating and making provision for the movement of people into, within, and out of an organization (Sherman, Snell, & Bohlander, 1997). The objective is to use these people as effectively as possible and to have available the required number of people with qualifications for positions when openings occur. A strategic plan must be developed, followed by forecasting future employee needs. Finally, a supply and demand analysis is required. Of course, all human resources planning must be done within the legal environment of the United States.

Strategic Plan

The human resources plan must fit into the strategic plan of the organization. When the goals of the organization are established, availability of both the internal and external human resources needs to be determined. The human resources manager should be on the strategic planning team. Members of the team should be aware of current employees' skills, abilities, and knowledge and their and future employees' training needs. They also should know what wages and benefits are required to attract qualified employees.

The human resources manager can be proactive by scanning environmental trends such as economic factors, technological changes, political and legislative issues; social concerns such as child care and educational priorities; and demographic trends of the workforce (Sherman, Snell, & Bohlander, 1997). In a booming economy, fewer job candidates may be available because unemployment may be low; in a depressed economy, organizations may have to cut back on operations and consequently on the number of employees. Technological changes will affect the type of specialized personnel an organization will require. Thus, the organizational environment will define the limits within which a human resources plan must operate. After managers fully understand the jobs that are needed in the organization, they can start planning by assessing trends in employee usage and future organizational plans.

Forecasting Supply and Demand

Human resources forecasting involves determination of the number, type, and qualifications of individuals who will be needed to perform specific duties at a certain time. Historical data can be used to predict demand for employees, such as chefs and waitstaff. Large organizations will require a special planner who uses mathematical models for forecasting employment needs. The internal supply of labor, consisting of the number and type of employees who will be in the operation at some future date, needs to be determined first. Then the external supply, the number and type of people who will be available for hiring in the labor market, must be forecast. The simplest method is to adjust the present staffing levels for anticipated turnovers and retirements.

Supply Analysis

Staffing tables: Pictorial representations of all jobs with the number of employees in those jobs and future employment requirements.

Once future requirements for employees have been forecast, the manager then must determine if numbers and types of employees are adequate to staff anticipated openings. Supply analysis uses both internal and external sources. An internal supply analysis begins with the preparation of **staffing tables,** which are pictorial representations of all organizational jobs with the numbers of employees in those jobs and future employment requirements (Sherman, Snell, & Bohlander, 1997). In conjunction with staffing tables are **skills inventories,** which contain information, usually computerized, on each employee's education, skills, experience, and career aspirations (Griffin, 2001). Forecasting the external supply of labor can be a problem. How does a foodservice manager, for example, predict the number of cooks who will be looking for work in Montana 3 years from now? This information can be calculated from sources such as state employment commissions, government reports, and figures supplied by educational institutions on the number of students in the field.

Job Analysis

Job analysis: Process of obtaining information about jobs by determining the duties and tasks or activities of those jobs.

Job analysis often is referred to as the base of human resource management because the information collected serves so many functions. **Job analysis** is the process of obtaining information about jobs by determining the duties and tasks or activities of those jobs (Sherman, Snell, & Bohlander, 1997). In larger organizations, the job analysis often is performed by a specialist retained for that particular purpose. These professionals, identified as job analysts, most often have an industrial engineering background and have access to many tools for the analysis. The job analyst can observe the employee performing the job, interview the employee, have the employee complete a questionnaire, or keep a log. In smaller organizations, the analysis can be conducted by the supervisor of the job or the employee who holds the job.

The procedure involves a systematic investigation of jobs by following a number of predetermined steps, such as collecting data on tasks, performance standards, and skills required. When completed, job analysis results in a written report summarizing the information from analyzing 20 or 30 individual job tasks or activities. Job analysis is concerned with objective and verifiable information about the actual requirement of a job. Job descriptions and job specifications developed through job analysis should be as accurate as possible to make them valuable to managers who make HRM decisions. The end products of a job analysis are job descriptions and specifications and performance standards.

The Occupational Information Network, O*NET (www.onetcenter.org) is an online source of valuable job information for both managers and potential employees. The site is administered by the U.S. Department of Labor and contains standardized and comprehensive information about more than 1,200 occupational titles, which are based on the Standard Occupational Classification (SOC) system. Included for each occupational title is a SOC number; job definition; list of tasks performed; knowledge, skills and abilities needed; work activities and job content; interests; work values; related occupations; and links to national and state wage and employment data. The site also includes career exploration tools, testing and assessment guides, and occupational outlook information. An example of job title, definition, and tasks performed for the job title of foodservice manager is included in Figure 12.2.

Job Description

Job description: Tasks, duties, and responsibilities of a job, the job's working conditions, and the tools, materials, and equipment used to perform it.

The **job description** lists the tasks, duties, and responsibilities of a job, the job's working conditions, and the tools, materials, and equipment used to perform it (Griffin, 2001). Most job descriptions include at least three sections: job title, identification, and duties. If job specifications are not separate, they are usually placed at the end of the job description. A sample job description for a bartender is shown in Figure 12.3.

11-9051.0 Food Service Managers

Definition:

Plan, direct, or coordinate activities of an organization or department that serves food and beverages.

Tasks:

Plan, direct, or coordinate activities of an organization or department that serves food and beverages.

Monitors compliance with health and fire regulations regarding food preparation and serving and building maintenance in lodging and dining facility.

Plans menus and food utilization based on anticipated number of guests, nutritional value, palatability, popularity, and costs.

Organizes and directs worker training programs, resolves personnel problems, hires new staff, and evaluates employee performance in dining and lodging facilities.

Coordinates assignments of cooking personnel to ensure economical use of food and timely preparation.

Estimates food, liquor, wine, and other beverage consumption to anticipate amount to be purchased or requisitioned.

Monitors food preparation and methods, size of portions, and garnishing and presentation of food to ensure food is prepared and presented in accepted manner.

Monitors budget, payroll records, and reviews financial transactions to ensure expenditures are authorized and budgeted.

Investigates and resolves complaints regarding food quality, service, or accommodations.

Reviews menus and analyzes recipes to determine labor and overhead costs, and assigns prices to menu items.

Establishes and enforces nutrition standards for dining establishment based on accepted industry standards.

Figure 12.2. Definition and tasks performed for job title: foodservice manager.
Source: O*NET, U.S. Department of Labor online occupational information network at www.onetcenter.org.

SAMPLE JOB DESCRIPTION

Job Title: Bartender **Date:** 1/15/99

Department: Bar **Reports To:** Lead Bartender

Prepared By: Hunt (Lead Bartender) **Approved By:** Toms (Mgr)

JOB SUMMARY:
Describe the general purpose of the job or why the job exists.

Mixes and serves alcoholic and nonalcoholic drinks to patrons of bar and service bar following standard recipes.

ESSENTIAL FUNCTIONS:
List clear and precise statements of the responsibilities, job duties and major tasks performed—and percentage of time spent on each.

- Mixes ingredients such as liquor, soda, water and sugar to prepare cocktails and other drinks 30%
- Serves wine, draft beer or bottled beer 30%
- Collects money for drinks served 20%
- Orders liquors and supplies 5%
- Arranges glasses and bottles to make an attractive display 10%
- Slices and peels fruits for garnishing drinks 5%

ACCOUNTABILITIES:
Summarize the major results and key outcomes.

Bar and restaurant patron drink orders are filled quickly (within two minutes) and according to recipe. Daily bar inventories correspond with bar cash-register totals. Bar area is always neat and clean.

QUALIFICATION STANDARDS:
List the personal and professional qualifications required, including skill, experience, education, physical, mental safety, and other requirements.

- Able to operate a cash register and make change
- Stands during entire shift
- Reaches, bends, stoops, shakes, stirs and wipes
- Lifts 30-pound cases about 10 times per shift and carries cases up one flight of stairs
- Frequent immersion of hands in water (every 5 minutes)
- Hazards may include but are not limited to cuts from broken glass, burns, slipping, tripping
- Requires one year experience as a bartender

Figure 12.3. Sample job description for a bartender.
Source: "Preparing Written Job Descriptions" by N. Weller, 1992, *Restaurants USA, 12*(2), p. 21. Used with the permission of the National Restaurant Association.

Job Title

Selection of a **job title** is important to employees because it gives them status and indicates their level in the organization. The title may be used to indicate, to a limited extent, the degree of authority the job possesses. The title "sanitation supervisor" indicates that the job involves more authority than "sanitation worker," and the title "head cook" indicates that the job is higher in the organization than "cook" or "assistant cook." Until recently, the titles of some jobs indicated that the job was for a "man" or a "woman," which implies that the job can be performed only by members of one gender. Thus, a waiter or waitress is now a waitperson.

Job Identification

The **job identification** section of a job description usually follows the job title (Sherman, Snell, & Bohlander, 1997). It includes such information as the departmental location of the job, the person to whom the jobholder reports, and often the number of employees in the department and the SOC code number. This information is found at the top of the sample job description for a bartender (Figure 12.3).

Job Duties

Statements covering **job duties** are usually arranged in their order of importance. These statements should indicate the weight or value of each duty generally measured by the percentage of time devoted to it, as shown in Essential Functions in the bartender job description (Figure 12.3). The description should also state responsibilities, identified as Qualification Standards in the same job description, an explanation of duties, and the equipment used by the employee.

Job Specification

The **job specification** lists the abilities, skills, and other credentials needed to do the job (Griffin, 2001). The job specification, popularly referred to as the job spec, contains a statement of job conditions relating to the health, safety, and comfort of the employee, including any equipment and any job hazard. Knowledge of job content and job requirements is necessary to develop selection methods and job-related performance appraisals and to establish equitable salary rates. In the bartender job description (Figure 12.3), the Qualification Standards are really a job specification that was included in the job description rather than making it a separate document.

One of the problems with job descriptions and specifications is that they are often very poorly written (Sherman, Snell, & Bohlander, 1997). Job duties often are written in vague rather than specific terms. In today's legal environment, federal guidelines and court decisions require that specific performance requirements are based on valid job-related criteria.

Performance Standards

Performance standards: Desired results at a definite level of quality for a specified job.

Performance is the attainment of a desired result and standard at a definite level of quality for a specified purpose; **performance standards,** therefore, define desired results at a definite level of quality for a specified job. Specific details in the standards that can be measured objectively are quality, quantity, and time factors, and they can be grouped under productivity. In many organizations, productivity standards are based on past performance.

In Figure 12.4, an excerpt from a job description for a lead cook is shown. The responsibility and four relevant performance standards have been identified by management with input from the cooks. Standards should be based on work-related behaviors and clearly stated in writing. Activities to be observed and measured should be obvious in the standard. Measuring performance of a standard becomes an unavoidable challenge to the manager in evaluating employees.

According to Byers, Shanklin, Hoover, and Puckett (1994), in most organizations, the performance evaluation system is developed by the human resources department. The mechanics of developing an evaluation instrument and determining how often employees will be evaluated, however, is the responsibility of the manager. The rating scale for each performance standard should be the same as the one used elsewhere in the organization. Generally, descriptive terms such as always, almost always, sometimes, rarely, and never are assigned numbers to make the scores as objective as possible. The employee receives a score for each performance standard and a total score for each responsibility.

The percentage of time spent by the employee in each responsibility gives the weight of relative importance and is determined by the manager or the employees in the unit. A performance evaluation score is determined by multiplying the total score for each responsibility by its weight. Pesci, Spears, and Vaden (1982) developed a methodology for devising performance standards for employees in a university residence hall foodservice. Their work involved defining major responsibilities and related standards for the position of Cook II. Table 12.2 shows the resulting weights for the various areas of responsibility, as based on the priority ratings provided by the cooks themselves.

Job Design

Job design is an outgrowth of job analysis and is concerned with structuring jobs to improve organization efficiency and employee job satisfaction (Sherman, Snell, & Bohlander, 1997). Work needs to be divided into manageable units and eventually into jobs that can be performed by employees. A **job** is a set of all tasks that must be performed by a given employee (Chase & Aquilano, 1992). Each job should be clear and distinct to prevent employees from misunderstanding the job and to help them recognize what is expected of them (Sherman, Snell, & Bohlander, 1997).

Job: Set of tasks to be performed by a given employee.

Job design is a complex function because of the variety of factors that enter into arriving at the ultimate job structure, as shown in Figure 12.5 (Chase & Aquilano, 1992). Decisions have to be made about who will perform the job and where and how it is to be performed. One of the considerations in job design is the quality of work life (QWL) of employees. Job enlargement, job enrichment, job characteristics, and employee work teams are all programs that increase the QWL in organizations.

Job Enlargement

Job enlargement: Increase in the total number of tasks employees perform.

Job enlargement was developed to increase the total number of tasks that employees perform. Tasks are the individual activities that make up a job. Peeling vegetables and measuring or weighing ingredients, for example, are two tasks that

COMMUNITY HOSPITAL

DEPARTMENT OF FOOD AND NUTRITION SERVICES

JOB DESCRIPTION

JOB TITLE: Lead Cook DEPARTMENT: Dietary DATE: September 1999

JOB TITLE OF PERSON TO WHOM REPORTING: Production Mgr. JOB CODE #: 116

JOB PAY: REVISED: FEBRUARY 1997

NO. OF PERSONS SUPERVISED: N/A EDUCATION REQUIREMENTS: High School or Equivalent. Education in Food Service or Cooking Desired.

PRIOR EXPERIENCE REQUIREMENTS: One year experience or equivalent in institutional cooking, preferably in a healthcare setting.

OTHER COMMENTS: Exposed to heat, humidity, steam, cooking odors, refrigerator temperatures and wet floors. Possible job related injuries include serious cuts from knives or power equipment, burns from cooking equipment and strains, sprains, or falls. Work is performed while standing or walking. Occasionally exerts considerable physical effort in moving or lifting of supplies and/or hot food items.

JOB SUMMARY: Works as a team member with one or more cooks in the daily production requirements for patient and/or employee foodservices. Prepares meats, fish, fowl, vegetables, gravies, sauces, soups, salad ingredients and baked goods according to standardized recipes. Assures freshness, proper serving temperatures, and the minimization of food waste.

Responsibilities	Performance Standards
3 Maintains standards of quality as specified by the Department of Dietetics and all basic food handling guidelines as specified by local, state and federal health agencies.	**A** All foods are to be stored at proper temperatures. Cold foods at or below 41°F, hot foods at 140°F or above. Holding and processing temperatures between these ranges should not exceed 4 hours. **B** All foods are to be covered, labeled, and dated when stored. **C** All foods are to be rotated on a first in first out basis in accordance with the department's standards for holding and storing foods. **D** All foods are to be served at the appropriate serving temperature as specified on the steam-table layout diagrams.

Figure 12.4. Excerpt from a job description with performance standards for a lead cook.

Table 12.2 Weights for Each Major Responsibility for Cook II

	Foodservice Unit				
Area of Responsibility	**1**	**2**	**3**	**Overall**	**Final Weights**
	←—% priority[a]—→				%
1. Food production	48	47	43	47	45[b]
2. Equipment care	5	5	6	5	5
3. Storage and handling of food	3	3	4	3	5
4. Employee training	15	16	17	16	15
5. Personal hygiene	3	2	3	3	5
6. Work habits	26	27	27	26	25

[a]Based on priority ratings of tasks within each area of responsibility provided by cooks in each of the food centers.
[b]Weights proposed after review of reliability data and % priority from each food center. Overall % priority data were adjusted to reflect 5% increments for each major responsibility as suggested by Kansas State University, Personnel Services.

Source: "A Method for Developing Major Responsibilities and Performance Standards for Foodservice Personnel," by P. H. Pesci, M. C. Spears, and G. Vaden, 1982, *NACUFS Journal,* pp. 17–21. Used by permission.

are part of an ingredient room worker's job. The rationale for enlarging jobs is that employees become bored if they have to do the same job day after day.

Job Enrichment

Job enrichment: Increase in the variety and number of tasks and control the employee has over the job.

Job enrichment assumes that increasing the variety of tasks is not enough to improve employee motivation. Thus, job enrichment attempts to increase the number of tasks along with the control the employee has over the job. Frederick Herzberg was a proponent of job enrichment, which he believed motivated employees while

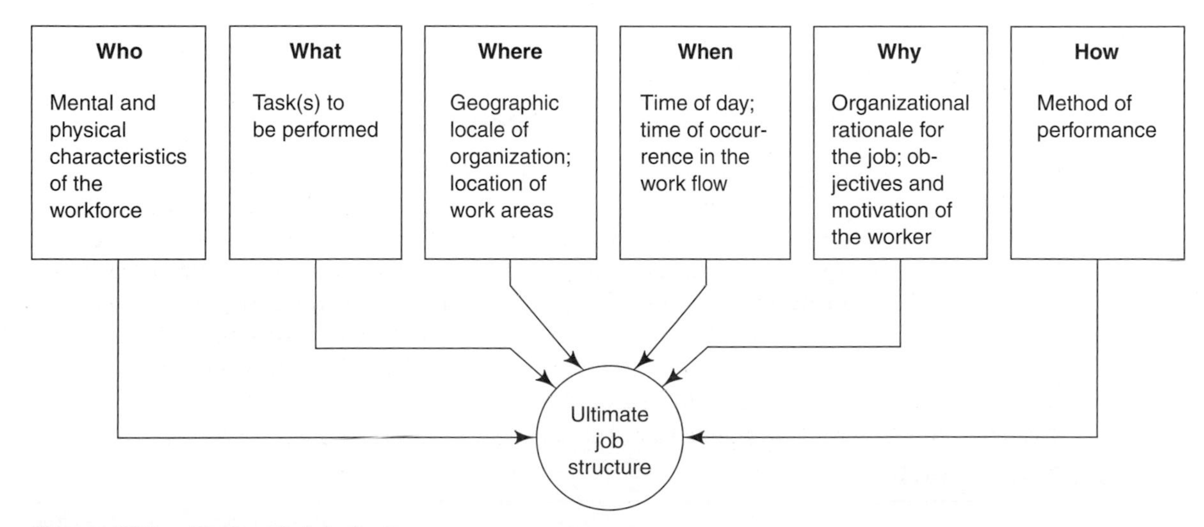

Figure 12.5. Factors in job design.
Source: Production and Operations Management: A Life-Cycle Approach (p. 495) by R. B. Chase and N. J. Aquilano, 1992, Homewood, IL: Richard D. Irwin. Used by permission.

achieving job satisfaction and performance goals. He concluded that the five factors for enriching jobs and motivating employees are

- Achievement
- Recognition
- Growth
- Responsibility
- Performance

These factors apply to the complete job rather than to parts of it. These factors allow employees to assume more decision-making power and become more involved in their own work. Job enrichment could be a component of TQM because employees, who are internal customers, are empowered to make decisions, which leads to job satisfaction.

Job Characteristics

A new area for job design studies was researched when behavioral scientists focused on identifying various job dimensions that would improve the efficiency of organizations and the job satisfaction of employees (Sherman, Snell, & Bohlander, 1997). The classic studies on job characteristics by Hackman (1977) resulted in the definition of five job dimensions or task characteristics:

- Skill variety
- Task identity
- Task significance
- Autonomy and
- Feedback

If these are present to a high degree, Hackman observed that meaningfulness of work, responsibility, and knowledge of actual results of work activities could contribute to work performance and job satisfaction. Hackman concluded that the greater the extent of all five task characteristics in a job, the more likely it is that the job holder will be highly motivated and experience job satisfaction. The job characteristics model works when employees have the desire for the autonomy, variety, responsibility, and challenge of enriched jobs. Otherwise, redesign of jobs almost always fails and employees are frustrated.

Balancing Supply and Demand

After comparing the internal demand for employees with the external supply, managers can make plans to deal with internal shortages or with overstaffing. Demand is based on the forecast, and supply is based on finding employees who have the required qualifications to fill vacancies. If shortages are predicted, new employees can be hired, present employees can be retrained, those retiring can be asked to stay on, or labor-saving methods, such as using more ready prepared food, can be introduced into the organization. If employees need to be hired, the external labor forecast can help managers plan a recruitment program.

LEGAL ENVIRONMENT

Union contracts influence human resources planning if clauses on regulating transfers, promotions, and discharges are included. Government legislation also has a vital role. Significant pieces of legislation that affect human resources management are explained in Table 12.3. Most of these acts have criteria for implementation, such as the number of employees in the organization. Human resources managers in large organizations keep up-to-date on this legislation, but in small operations, which have only one or two managers, managers need to be aware of the legislation that has an impact on their operation.

Equal Employment Opportunities

Equal employment opportunity: Policy of equal employment (nondiscrimination) for all.

Many laws have been passed that address issues of **equal employment opportunities** (nondiscrimination) for all individuals. As shown in Table 12.3, numerous laws have been passed making it illegal to discriminate in hiring based on a person's race, color, sex, religion, national origin, age, disability, or pregnancy.

Title VII of the Civil Rights Act of 1964 created **the Equal Employment Opportunity Commission** (EEOC). The commission consists of five commissioners and

Table 12.3 Major Federal Laws Affecting Human Resource Management

Date	Act	Explanation
1963	Equal Pay Act	Equal pay for equal work regardless of gender
1964	Title VII of the Civil Rights Act (amended in 1974)	Forbids discrimination on the basis of race, color, religion, gender, or national origin; established Equal Employment Opportunity Commission (EEOC)
1967	Age Discrimination in Employment (amended in 1978, 1986)	Forbids discrimination against employees between 40 and 70 years of age
1978	Pregnancy Discrimination (amendment to Title VII of Civil Rights Act)	Broadens gender discrimination to include pregnancy, childbirth, or related medical conditions
1986	Immigration Reform and Control Act	Prohibits discrimination based on citizenship or nationality; requires employers to verify employees are authorized to work in the United States
1988	Worker Adjustment and Retraining Act (WARN)	Requires employers to notify employees of impending layoffs
1990	Americans with Disabilities Act	Prohibits discrimination against persons with physical or mental disabilities or the chronically ill
1990	Immigration Act	Increased levels of immigrations, especially for highly skilled professionals and executives
1991	Civil Rights Act of 1991 (amended previous Civil Rights and Americans with Disabilities Acts)	Provides remedies for intentional discrimination and harassment, expands protection for victims of discrimination, established Glass Ceiling Commission, broadens coverage to include government employees
1993	Family and Medical Leave Act	Gives employees up to 12 weeks unpaid and job-protected leave per year for themselves for the birth or adoption of a child or to care for a spouse, parent, or child with a serious health condition
1994	Uniformed Services Employment and Reemployment Rights Act	Forbids discrimination against individuals who take leave from work to fulfill their military service obligation
1996	Health Insurance Portability and Accountability Act	Allows employees to transfer their coverage of existing illnesses to new employer's insurance plan
1996	Illegal Immigration Reform and Immigrant Responsibility Act	Places severe limitations on persons who remain in United States longer than permitted by their visa and/or who violate their nonimmigrant status

a general counsel all of whom are appointed by the U.S. president. The EEOC makes equal employment policy, enforces employment provisions of civil rights legislation, and rules on discrimination charges that are filed with the EEOC (see www.eeoc.gov). The EEOC works through 51 field offices, more than 90 state and local fair employment practices agencies, and more than 60 tribal employment rights offices.

Harassment

The EEOC has declared sexual harassment to be a form of gender discrimination and thus in violation of Title VII of the Civil Rights Act of 1964 (Batty, 1993). The U.S. Supreme Court ruled unanimously in 1986 that sexual harassment violates Title VII, and the court eliminated the "I didn't know" defense for employers sued for sexual harassment. The Civil Rights Act of 1991 provided appropriate remedies for intentional harassment. On June 26, 1998, the Supreme Court, in *Faragher vs. City of Boca Raton* and *Burlington Industries, Inc., vs. Ellerth,* ruled that employers are vicariously liable for sexual harassment by supervisory employees even if the employer was unaware of the harassment. The Supreme Court has ruled that ignorance is no defense and made employers strictly liable for their supervisors' sexual harassment (Davis, 1998).

The EEOC and similar agencies at the state and local level received 15,889 charges of sexual harassment in 1997. Approximately 1,479 of those complaints involved eating-and-drinking establishments. Preliminary data for 1998 recorded 1,277 such complaints (Ursin, 1999). The EEOC has identified two forms of harassment. *Quid pro quo,* which means "something for something" in Latin, indicates that a manager demands sexual favors or insists that an employee put up with sexually harassing behavior for a job benefit (Batty, 1993). The second form involves a *hostile work environment,* which imposes Title VII liability on the employer if the workplace is rendered offensive by acts of sexual harassment that the employer knew about or should have known about (Hamilton & Veglahn, 1992). Discrimination suits have been brought against restaurant owners who require waitstaff to wear sexually revealing uniforms; uniforms help define a restaurant's concept and create a certain atmosphere (Batty, 1993). Even though women are usually the victims of sexual harassment, some men are beginning to report sexual harassment by their female bosses.

Many restaurant operators have long recognized that having a company policy on sexual harassment is wise. The two aforementioned U.S. Supreme Court decisions have made instituting such a policy "mandatory now," according to Peter Kilgore, legal counsel for the National Restaurant Association (Ursin, 1999). The policy must be communicated to and enforced among all employees. Having a sexual harassment policy buried somewhere that no one can read is not enough. Employers in both of the aforementioned cases had policies developed but they were not discussed with employees.

Immigration

In 1986, Congress passed the Immigration Reform and Control Act (IRCA) to control unauthorized immigration by making it unlawful for a person or organization to recruit or hire persons not legally eligible for employment in the United States.

Employers must comply with the law by verifying and maintaining records on the legal rights of applicants to work in the United States.

IRCA requires employers to complete an I-9 form for each employee to verify the worker's identity and eligibility to work in the United States. The law also prohibits employers of four or more persons from discriminating against employees or job applicants based on their national origin or citizenship status. Many employers feel strongly that immigration paperwork is overwhelming, burdensome, and hard to cope with; they have little knowledge about a federal ban on job bias against workers who may look or sound foreign (Employer Focus Groups, 1993). IRCA made headlines in the early 1990s when journalists reported that Zoë Baird, a nominee for Attorney General, had hired an illegal immigrant as a nanny but neglected to comply with the Immigration Reform and Control Act and notify the Immigration and Naturalization Service about the nanny's status (Cheney, 1993). Subsequently, she withdrew her nomination.

The Immigration Act of 1990 revised U.S. policy on legal immigration and increased levels of immigration especially of highly skilled professionals and executives. In 1996, the Illegal Immigration Reform and Immigrant Responsibility Act was signed into law. This Act placed stricter penalties on those who remain in the United States longer than permitted by their visa or who violate their nonimmigrant status. It also placed added restrictions on benefits for aliens.

Disabilities

The Americans with Disabilities Act (ADA) is comprehensive legislation that creates new rights and extends existing rights for the 43 million Americans with disabilities; it also protects disabled Americans who are not covered under existing laws (Cross, 1993). The purpose of the act is to

- Provide a national mandate to eliminate discrimination against individuals with disabilities
- Provide consistent enforceable standards for those with disabilities
- Ensure that the federal government plays a central role in enforcing the standards
- Invoke congressional authority to address the major areas of discrimination faced by the disabled

The ADA's definition of disability is broad and provides civil rights protection for people in the following three categories (ADA Employment, 1993):

- A physical or mental impairment that substantially limits one or more major life activities, such as walking, seeing, hearing, speaking, learning, or working
- A record of impairments (such as mental, emotional, or physical illness and alcohol or drug addiction) from which people have recovered or are recovering
- Reactions by others implying that people have an impairment, such as severe burns or being rumored to have AIDS

The EEOC has provided examples of conditions that qualify as disabilities, among which are orthopedic, visual, or speech impairments, HIV infection, acquired immune deficiency syndrome (AIDS), epilepsy, mental retardation, and former drug use (Cross, 1993). The ADA does not require hiring disabled persons who are not qualified for a job in terms of skill, education, and experience, but it does require that disabled persons who meet the job qualifications should be given equal consideration for a job and equal treatment on the job (McLauchlin, 1992). The U.S. Justice Department reported that complaints filed against restaurants and bars accounted for just over 11% of the nearly 700 complaints filed in the first 10 months after the ADA's rules took effect (Foodservice Accounts, 1993). The ADA is not to be feared, however, because it is an excellent tool to cement a long-term working relationship with a potential employee (Doyle, 1992). In fact, restaurants are the leading industry in the employment of people with disabilities. The restaurant industry has been making reasonable accommodations for years.

Affirmative Action

Affirmative action: Conscious and proactive effort by organizations with government contracts to ensure employment of minorities and women in proportion to their representation in that organization's relevant labor market.

Affirmative action is a conscious and proactive effort by organizations with government contracts to ensure employment of minorities and women in proportion to their representation in that organization's relevant labor market (Mondy, Noe, & Premeaux, 2002). Affirmative action policies are administered by the Department of Labor's Employment Standards Administration's Office of Federal Contract Compliance Programs (OFCCP) (see www.dol.gov).

While the equal employment opportunity law is largely a policy of nondiscrimination, affirmative action requires employers to analyze their workforce and develop a plan of action to correct areas of past discrimination (Sherman, Snell, & Bohlander, 1997). The Affirmative Action Program (AAP) reaffirms the commitment to nondiscrimination and equal employment opportunity through affirmative action to ensure equal treatment of applicants and employees without regard to race, color, national origin, ancestry, sex, age, religion, or disability (Kansas Department of Administration, 1999). The goal of the AAP is to achieve a workforce that includes a representation of qualified minorities, women, and persons with disabilities that approximates their availability in the state resident workforce. Employers must make an affirmative effort to ensure all persons have an equal opportunity to be informed of and to compete for employment opportunities; and to ensure that all employees have an equal opportunity to compete for promotional opportunities, receive training and enjoy the benefits and privileges of employment. The AAP should include a policy statement that reflects the company's attitudes regarding equal employment opportunity, assign an individual in the organization with responsibility for preparing and implementing the company's AAP, and provide a process for monitoring and reporting on progress (Mondy, Noe, & Premeaux, 2002). An acceptable AAP will include an analysis of whether there are deficiencies in utilization of women and minorities and a plan with a time table for correction of deficiencies.

EMPLOYMENT PROCESS

Once an organization has an idea of its future human resources needs, the next step usually is recruiting new employees. The major phases in the employment process are recruitment, selection, and orientation.

Recruitment

Recruitment: Process of locating and encouraging potential applicants to apply for a job opening.

Recruitment is the process of attempting to locate and encourage qualified, potential applicants to apply for existing or anticipated job openings (Mondy, Noe, & Premeaux, 2002). Efforts are made to inform applicants about qualifications required for the job and career opportunities the organization can give them. Managers must decide whether the job vacancy will be filled by someone inside or outside the organization. The decision depends on availability of employees, the organization's human resources policies, and requirements of the vacant job.

Inside the Organization

Most organizations try to fill vacancies above entry-level positions through promotions and transfers (Sherman, Snell, & Bohlander, 1997). Promotion from within has several advantages:

- Employee morale and motivation are positively affected, assuming such promotions are perceived as equitable.
- Employees can be protected from layoff.
- Employees' work experiences can be broadened.
- Abilities of present employees are used to their fullest extent, which improves the return on investment in employees.
- Promotion from within is usually less expensive than hiring from outside the organization.
- Individuals recruited from within also will be familiar with the organization and therefore may need less training and orientation.

Methods of Finding Candidates. A number of methods are used for recruiting from within. Qualified candidates can be identified by computerized records, job posting and bidding, and recall of those who have been laid off (Sherman, Snell, & Bohlander, 1997). Computer data banks contain complete records and qualifications of each employee. This information can be retrieved in minutes.

Organizations may inform employees about job openings by using job posting and bidding. Vacancy notices can be posted on bulletin boards, in employee publications, in special handouts, and by direct mail and public-address systems. Employees who have been laid off may be recalled to their jobs if economic conditions improve and they left in good standing. Employee layoffs usually are based on seniority or ability. Unions generally recommend seniority because they contend members should be entitled to certain rights proportionate to the number of years on the job. Using seniority may have an impact on women and minorities, who often have less seniority than other employees.

Limitations of Recruiting from Within. Sometimes requirements for a higher-level job may be very different from those of the current job. For example, a foodservice production supervisor may be promoted to unit manager because of excellent job performance in supervising production employees and operations. This same individual may fail, however, as unit manager of the foodservice operation because his or her scope of skills and knowledge may be inadequate for this higher-level position.

Another danger in promoting from within is that new ideas may not be heard. Even if an organizational policy is to promote from within, candidates from the outside occasionally should be considered to prevent inbreeding of ideas and attitudes.

Outside the Organization

Eventually, after promotions and transfers, a person outside the organization is sometimes recruited to fill a vacancy. When a promotion occurs, a chain reaction of promotions follows. The decision a manager faces is at which level a person should be recruited. The labor market varies with the type of job to be filled and the salary to be paid for the job.

External sources for obtaining employees include help wanted advertisements in newspapers, trade journals, and online job posting services. Radio, television, billboards, and posters also have been used. The statement *equal opportunity employer* generally is included in the advertisement.

Each state has an employment agency responsible for the unemployment insurance program; each agency is subject to regulations by the U.S. Employment Services (USES). Unemployed people must register at a local office and be available for employment to receive a weekly unemployment compensation check (Sherman, Snell, & Bohlander, 1997). The local office has a computerized job bank for the area, and the USES maintains a nationwide data bank. These agencies also help employers with employment testing, job analysis, evaluation programs, and community wage surveys. Private employment agencies, which help job seekers find the right job, charge a fee that may be paid by either the employer or the job seeker or both. In contrast, executive search firms, often called headhunters, help employers find the right person for the right job. Fees, paid for by the employer, are usually a certain percentage of the annual salary for the position to be filled.

Educational institutions are good sources for candidates who have little work experience. High school students can fill part-time, entry-level positions. Vocational schools and community colleges provide candidates for technical jobs, and colleges and universities for technical and professional areas. Recruiters for various organizations often are sent to college campuses to seek candidates.

Other recruitment efforts can be helped by recommendations made by current employees or from unsolicited applications. Many professional organizations offer a placement center to members. Labor unions can be an important source of applicants, especially for blue-collar jobs.

Temporary employment agencies are one of the fastest growing recruitment sources (Sherman, Snell, & Bohlander, 1997). Workers do not receive benefits from their temporary employers, and they can be dismissed without filing unemployment insurance claims. Employee leasing is a process whereby an employer terminates a number of employees who then are hired by a company that "leases" employees back to the original organization. The employee leasing company takes care of all the human resources duties and in return is paid a fee, usually a percentage of payroll cost.

In all recruitment activities, organizations must be cognizant of legal requirements. Organizational practices or policies that adversely affect equal employment opportunities for any group are prohibited unless the restriction is a justifiable job requirement.

Selection

Selection: Process of comparing applicant knowledge, skills, and abilities to those required of a position and choosing applicant most qualified.

The **selection** process begins after recruiting applicants for a job. The process includes a comparison of applicant skills, knowledge, and education with the requirements of the job; it involves decision making by the organization and the applicant. Qualifications for each applicant must be compared to the job requirements identified in the job specification. After thorough screening, employers choose an applicant who meets the criteria indicating that she or he will learn the job easily and adjust to it with a minimum of difficulty. Union agreements and civil service regulations that protect employees with job tenure provide management with an additional incentive for making a good selection. Their regulations on discharging protected employees can be very difficult.

The selection process for all workers is protected by equal employment legislation, court decisions, and the *Uniform Guidelines on Selection Procedures* (see www.dol.gov). These guidelines recommend that an employer be able to demonstrate that selection procedures are valid in predicting or measuring performance in a particular job (Sherman, Snell, & Bohlander, 1997). Factors that affect the process include a small pool of applicants, geographic immobility of career couples, and changing staffing needs because of promotion or turnover.

The steps in the selection process are depicted in Figure 12.6. In practice, the actual selection process varies among organizations and among levels in the same

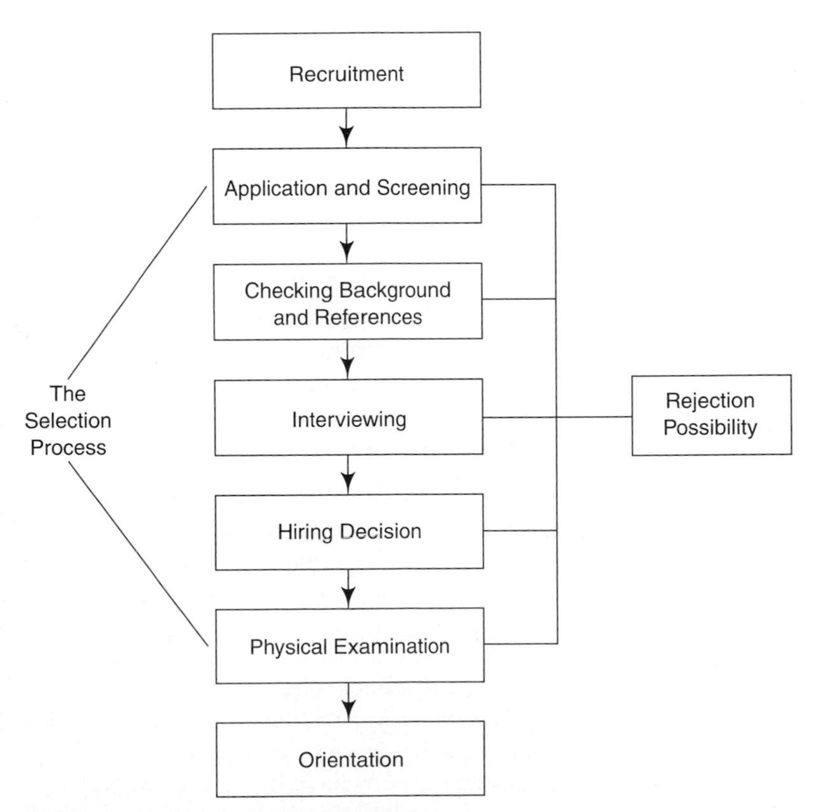

Figure 12.6. Steps in the selection process.

organization. For example, the selection interview for production and service employees may be perfunctory, whereas an extensive process may be necessary for choosing the unit manager for a foodservice operation, starting with submission of a résumé, then a formal application and interview. The selection process should conform to accepted ethical standards, including privacy and confidentiality, and all legal requirements. The intent of the process is to gather from applicants reliable and valid information that will predict their job success and then to hire the candidates likely to be most successful (Griffin, 2001).

Application and Screening

The application form serves three purposes:

- Indicates that the applicant is interested in a position
- Provides the interviewer with basic information to conduct an interview
- Becomes part of the file if the applicant is hired

In organizations with a human resources department, the application and initial screening steps will be conducted by that department.

The application form or blank is used to obtain information that will be helpful in reaching an employment decision. The application collects objective biographical information about an applicant, such as education, work experience, special skills, and references. Questions referring to an applicant's age, gender, race, religion, national origin, or family status may violate provisions of the Civil Rights Act of 1964. Information requested on an employment application should not lead to discrimination against applicants.

In some organizations, preliminary interviews may be used to screen out unqualified applicants. This is a quick way to identify those candidates worthy of more attention (Marvin, 1992). An inadequate educational level or experience record or obvious disinterest may be reasons for eliminating the individual from the applicant pool at this point in the process. The applicant may be asked questions on previous experiences, salary expectations, and other issues related to the job. In foodservice organizations, an important issue to discuss initially is the applicant's willingness to work weekends and holidays or early morning and late evening hours.

Background and Reference Checks

The practice in many organizations is to conduct background and reference checks after the initial screening interview. In others, however, this checking may be done after the employment interview. When the interviewer is satisfied that the applicant meets the qualifications for the job, information about previous employment and other information given by the applicant is investigated.

Most organizations use some combination of mail, telephone, and e-mail correspondence to check references. Telephone checks are used most often because they save time and responses are more candid. Supervisors who know the applicant's work habits and performance usually give the best information. Recently, some organizations have been charged with negligently hiring or retaining employees who later commit crimes (Sherman, Snell, & Bohlander, 1997). These organizations are charged with failure to adequately check references, criminal records, or general background that would have shown the employee's behavior. Inadequate reference checking is one of the major causes of high turnover, employee theft, and white-collar crime.

Interviews

The employment interview has been the most widely used and probably the most important step in the selection process. The interview may be conducted by one person or a group of persons in the organization, as the guide to interviewing (Figure 12.7) shows. Some serious doubts have been raised about the validity of the interview as a selection method because of differences between interviewers.

For managerial and professional personnel, a series of interviews may be conducted. For example, a potential director of a department of food and nutrition services in a large medical center hospital would probably be interviewed by several different administrators, by the professional staff in the department, and perhaps by several of the other department heads. An applicant for a cook's position in this same hospital, however, might be interviewed by only the foodservice supervisor and the production dietitian.

Interviews can be either structured or unstructured. In the **structured interview,** the interviewer asks specific questions of all interviewees. The interviewer knows in advance the questions that are to be asked and merely proceeds down the list of questions while recording the responses. This interview technique gives a common body of data on all interviewees, allows for systematic coverage of all information deemed necessary for all applicants, and provides a means for minimizing the personal biases and prejudices of the interviewer. Structured interviews are frequently used in interviews for lower-level jobs. The **behavioral interview,** a structured-type of interview, includes specific questions designed to provide information about a candidate's typical behavior in a described situation.

The **unstructured interview** allows the interviewer the freedom to ask questions he or she believes are important. Broad questions, such as "Tell me about your previous job," are asked. The unstructured interview may be useful in assessing such characteristics of an individual as ability to communicate and interpersonal skills. Comparison of answers across interviewees is difficult with unstructured interviews, however, because questions may be quite different or asked in a different context. Unstructured interviews are generally used with higher-level personnel in the organization because of the broad nature of these jobs.

More attention is being given to the structured interview because of EEOC requirements (Sherman, Snell, & Bohlander, 1997). This type of interview is more likely to provide the type of information needed for making sound decisions. It helps to reduce the possibility of legal charges of discrimination. Employers need to realize that the interview is highly vulnerable to legal attack, and more litigation in this area can be expected in the future.

Regardless of the method used, the interview must be planned. First, interviewers must acquaint themselves with the job description and specification for the vacant job. Second, they must review the application file carefully, including the information from background and reference checks. Third, for structured interviews, the questions should be prepared in advance; for the unstructured interview, a general outline should be developed.

INTRODUCTION

These guidelines have been prepared to help interviewers conduct fair and objective interviews. An interview should provide as much information as possible about an applicant's potential to perform the duties of a particular position. The most valuable interview is objective and permits the interviewer(s) to determine the knowledge, skills, and abilities of a prospective employee.

INTERVIEW DEVELOPMENT

Form the Interview Team

If feasible, use a team approach. The team approach is preferable because it saves time and allows for comparison of the applicant by the team members. The size of the interview team may vary, but generally two to three members are recommended.

Familiarize the Interviewer(s) With the Position

The interviewer(s) must be familiar with the major duties and responsibilities, and the essential knowledge, skills, and abilities of the position at entry level. Be sure that each interviewer reviews the position description carefully.

Establish Criteria for Selection

The selection criteria must be consistent with the complexity and level of the job. Focus on performance factors that can be demonstrated in the selection procedure. Understand the departmental and organizational goals as they relate to this position. Such criteria must be job-related and might include performance during the interview, relevant training, education and experience, affirmative action goals, etc. Example: To what extent is job success dependent upon effective oral communication skills, on-the-spot reasoning skills, and the ability to effectively present oneself to strangers?

Develop Job-Related Questions

"Nice to know" questions are not permitted! Lawsuits may result from applicants who are rejected on the basis of irrelevant questions asked by interviewers.

Develop Interviewing Strategies

There are may different interviewing strategies. Develop strategies that are appropriate for the position level and skill requirements.

Establish a System to Evaluate the Responses

It might be beneficial to set up a formula for rating or ranking the applicant's responses to the questions based on the selection criteria. Evaluating the responses in this manner will help make the selection process easier and more objective.

INTERVIEW SUGGESTIONS

Preparing Questions

When developing questions, always keep in mind that they must be job-related and appropriate for the complexity and level of the position. It is helpful to weigh the questions based on the importance

Figure 12.7a. A guide for interviewing.
Source: State of Kansas, Department of Administration, Division of Personnel Services. Used by permission.

of each selection criterion. Below are six main categories of questions that are commonly used by interviewers. Different types of questions may be combined to obtain a certain response.

1. **Close-ended questions.** These questions may sometimes be helpful when an interviewer(s) wants to know certain information at the outset or needs to determine specific kinds of knowledge. Example: "Could you name the five specific applications involved in . . .?"
2. **Probing questions.** These questions allow the interviewer(s) to delve deeper for needed information. Example: "Why?," "What caused that to happen?", or "Under what circumstances did that occur?"
3. **Hypothetical questions.** Hypothetical situations based on specific job-related facts are presented to the applicant for solutions. Example: "What would you do if . . ."; "How would you handle . . ."
4. **Loaded questions.** These questions force an applicant to choose between two undesirable alternatives. The most effective way to employ a loaded question is to recall a real-life situation where two divergent approaches were both carefully considered, then frame the situation as a question starting with, "What would be your approach to a situation where . . ."
5. **Leading questions.** The interviewer(s) sets up the question so that the applicant provides the desired response. When leading questions are asked, the interviewer cannot hope to learn anything about the applicant.
6. **Open-ended questions.** These are the most effective questions, yield the greatest amount of information, and allow the applicant latitude in responding. Example: "What did you like about your last job?"

Determining Strategies

Although there are many different interviewing strategies, the following are examples of three different perceptive strategies.

1. **Situational interviewing.** This strategy is based on the assumption that the closer you can get to a real work situation, the better the evaluation will be. The situational interview could involve taking a tour of the workplace and asking the interviewee to actually perform some aspect of the job, or a closely related aspect of the job.
2. **Stress interviewing.** This strategy calls for the use of tough or negatively phrased questions. The interviewer(s) is trying to keep the candidate off balance while evaluating poise and quick thinking under pressure. This style would *not* be suitable if the employee will not face undue stress on the job.
3. **Behavioral interviewing.** The interviewer(s) is looking for a behavioral pattern. All questions are based on the past. The assumption is that a "leopard never changes its spots." The interviewer(s) may get an idea of what action the interviewee might take in the future based on what happened in the past.

Each of these strategies has its strengths and weaknesses. One strategy should not be used exclusively for all interviews. Different position levels might require different interview approaches. The sensible approach is to take the best aspects of each style and combine them to produce a comprehensive strategy.

Note: The interview process should not include the use of a testing device without prior approval from the Division of Personnel Services (K.A.R. 1–6–10).

Figure 12.7b. Continued.

Evaluating Responses

As part of evaluating the responses, the interviewer(s) should review the job description to ensure thorough familiarity with the requirements, duties, and responsibilities of the position. Furthermore, the interviewer(s) should review the work history and relevant educational credentials of each candidate and consider the intangible requirements of the job. Finally, the interviewer(s) should review the selection criteria, evaluate and rate the responses, and rank the applicants based on that criteria.

INTERVIEW PROCESS

Pre-Interview

1. Schedule interviews to allow sufficient time for post-interview discussion, completion of notes, etc.
2. Secure an interview setting that is free from interruptions or distractions.
3. Review applications and resumes provided by the applicants.
4. Provide an accurate position description to each applicant and allow adequate time for reading before the interview begins.

Opening the Interview

1. Review the functions of the agency or unit in which the position is located.
2. Allow the applicant an opportunity to pose questions or seek clarification concerning the position.
3. Explain the interview process to the applicant.

Questioning

1. Question the applicant following the method established in the developing stage.
2. Be consistent with all applicants.
3. Allow the applicant sufficient time to respond to each question.
4. Record any relevant information elicited from the questions.

Closing the Interview

1. Inform the applicant when the decision will be made and how notification will occur.
2. Confirm the date of the applicant's availability to begin work.
3. Confirm the applicant's correct address and telephone number.
4. Give the applicant a final opportunity to raise any questions.
5. Obtain all necessary information from the applicant about references.

Post-Interview

1. Review the selection criteria.
2. Review and complete notes.
3. Avoid prejudgment and discussion of applicants between interviews.
4. Use the selection criteria established in the developing stages.
5. Rank the applicants based on the selection criteria.
6. When possible, decide upon a second and third choice in the event the first choice should decline the offer.
7. Document the basis for the final recommendation.
8. Notify all applicants interviewed of the results prior to announcing the selection.

Figure 12.7c. Continued.

As with other aspects of the employment process, interviews are subject to numerous legal questions. The EEOC does not approve of questions related to race, color, age, religion, gender, or national origin. An interviewer can ask about physical handicaps, however, if the job involves manual labor, but not otherwise. Guidelines for employment inquiries for the state of Kansas are shown in Table 12.4. Tips for conducting an effective interview include the following:

- Establish rapport with the applicant.
- Allow sufficient time for an uninterrupted interview.
- Hold interview in a place where privacy is possible.
- Avoid questions that may be discriminatory.
- Avoid asking questions that can be "yes" or "no."
- Avoid asking leading questions to which the expected response is obvious.
- Ask questions that allow candidates to express themselves.
- Avoid **personal biases.**
- Avoid the **halo effect.**

Personal bias: When personal preferences alter objective decision making.

Halo effect: When a single trait dominates the assessment of another individual.

Medical Examination

A medical examination is the last step before hiring. The Americans with Disabilities Act stipulates that medical examinations must be directly relevant to requirements of a position and cannot be performed until an offer of employment has

Table 12.4 Guidelines for Employment Inquiries in the State of Kansas

	Permissible Inquiries	Inquiries That Must Be Avoided[a]
Name	Questions that will enable work and education records to be checked.	Inquiry about the name which would indicate lineage, ancestry, national origin, descent, or marital status.
Age	If age is a legal requirement, whether applicant meets the minimum or maximum age requirements; upon hire, proof of age can be required.	If age is not a legal requirement, any inquiry or requirement that proof of age be submitted must be avoided. Note: The Age Discrimination in Employment Act, as amended in 1986 prohibits discrimination against persons over age 40. The Kansas Act Against Discrimination prohibits discrimination against persons age 18 and over.
Race or Color	Race may be requested for affirmative action statistical recording purposes. Applicants must be informed that the provision of such information is voluntary.	Any inquiries that would indicate race or color.
Gender	Inquiry or restriction of employment is permissible only where a bona fide occupational qualification exists. (This BFOQ exception is interpreted very narrowly by the courts and EEOC.) The employer must prove that the BFOQ exists and that all members of the affected class are incapable of performing the job.	Any inquiry that would indicate gender.

(*continued*)

Table 12.4 Guidelines for Employment Inquiries in the State of Kansas (*continued*)

	Permissible Inquiries	Inquiries That Must Be Avoided[a]
Marital and Family Status	Whether applicant can meet specified work schedules and/or will be able to travel.	Any inquiry which would reveal marital status; information on applicant's children, child care arrangements, or pregnancy.
Disabilities	Under the provisions of the Kansas Act Against Discrimination, as amended, and the Americans with Disabilities Act of 1990, applicants may be asked if they are able to perform the essential duties of the position with or without reasonable accommodation.	Whether an applicant is disabled or inquiry about the nature or severity of the disability. Inquiries about any association with or relationship to a person with a disability. Note: Except in cases where undue hardship can be proven, employers must make reasonable accommodations for an employee's disability. Reasonable accommodation may include making facilities accessible, job restructuring, modified work schedules, modifying examinations, training materials or policies, acquiring or modifying equipment or devices, or providing qualified readers or interpreters.
Religion	Employers may inform applicants of normal hours and days of work required by the job. Note: Except in cases where undue hardship can be proven, employers must make reasonable accommodations for an employee's religious practices. Reasonable accommodation may include voluntary substitutions, flexible scheduling, lateral transfer, change of job duties, or use of annual or vacation leave.	Any inquiry which would indicate applicant's religious practices and customs.
Address	Address may be requested so that the applicant can be contacted. Names of persons with whom applicant resides may be requested for compliance with the nepotism policy.	Any inquiry which may indicate ethnicity or national origin.
Ancestry or National Origin	Languages applicant reads, speaks, or writes and the degree of fluency if a specific language is necessary to perform the job.	Inquiries into applicant's lineage, ancestry, national origin, descent, birthplace, or native language; how applicant learned a foreign language.
Arrest, Conviction and Court Records	Inquiry into arrest records and actual convictions which relate reasonably to fitness to perform a particular job. ARREST—The employer must consider whether the alleged conduct is job-related, the likelihood that the alleged conduct was actually committed, and the time that has passed since the arrest. CONVICTION—The employer must consider the nature and gravity of the offense(s), the time that has passed since the conviction and/or completion of the sentence, and whether the conduct for which the applicant was convicted is job related.	Ask or check into a person's arrest, court, or conviction record if not substantially related to functions and responsibilities of the particular job in question.

Table 12.4 Guidelines for Employment Inquiries in the State of Kansas (*continued*)

	Permissible Inquiries	Inquiries That Must Be Avoided[a]
Birthplace and Citizenship	If United States citizenship is a legal requirement, inquiry about the citizenship of an applicant is permissible. The Employment Eligibility Verification (Form I-9) must be submitted by those who are hired to provide evidence of identity and employment eligibility.	Any inquiry which would indicate the birthplace of the applicant or any of the applicant's relatives.
Military Service	Type of education and experience gained as it relates to a particular job.	Type of discharge.
Photograph	Statement that a photo may be required after hire for purposes of identification.	Any requirement or suggestion that a photo be supplied before hiring.
Education	Applicant's academic, vocational, or professional education; schools attended.	Any inquiry which would indicate the nationality, racial, or religious affiliation of a school; years of attendance and dates of graduation.
Experience	Applicant's work experience, including names and addresses of previous employers, dates of employment, reasons for leaving, and salary history.	Any inquiry regarding non-job-related work experience.
Financial Status	If required for business necessity, questions concerning financial stability. Examples of agencies that make inquiries into applicants' financial status are the Kansas Highway Patrol, Kansas Bureau of Investigation, and the Kansas Lottery.	If not required for business necessity, questions concerning financial stability.
Notice in Case of Emergency	Name and address of person(s) to be notified in case of accident or emergency may be requested after selection is made.	Name and address of relative(s) to be notified in case of accident or emergency.
Organizations	Inquiry into the organizations to which an applicant belongs and offices held relative to the applicant's ability to perform the job sought. NOTE—An applicant should not be required to provide the name of an organization which will reveal the religious, racial, or ethnic affiliation of the organization.	A list of all organizations to which the applicant belongs.
References	Names and addresses of persons who will provide professional and/or character references for applicant.	Requirement that a reference be supplied by a particular individual.
Relatives	Names of applicant's relatives already employed by the state agency in which employment is sought for compliance with the nepotism policy.	Name or address of applicant's relatives who are not employed by the state agency in which employment is sought.

[a]Any inquiry should be avoided that, although not specifically listed among the above, is designed to elicit information that is not needed to consider an applicant for employment.

Source: State of Kansas, Department of Administration, Division of Personnel Services. Used by permission.

been given (see www.usdoj.gov). A medical examination is not required by all employers, but when used, it can only be used to determine if the applicant is physically able to perform the work.

Health regulations in some cities or counties specify that foodservice workers have food-handler permits or cards prior to employment. Requirements for securing such a permit differ according to local regulations and may include successful completion of a sanitation training program, a test on food-handling practices, and a limited physical exam. Because of the particular concern in foodservice organizations about preventing contamination of food and spreading communicable diseases, blood tests, stool cultures, and TB testing, sometimes are required prior to employment and at periodic intervals thereafter. This practice is especially common in healthcare organizations. Drug testing also is becoming more common.

Hiring Decision

The most critical step in the selection process is the decision to accept or reject applicants for employment. The final decision must be made carefully because of the cost of placing new employees on the payroll, the relatively short probationary period in most organizations, and affirmative action considerations.

When all the information about an applicant has been assembled, a method must be developed for summarizing it. Checklists and summary forms commonly are used as a means of assuring that all the pertinent information is included in the evaluation of an applicant. Applicants are rank ordered according to total score, which provides an objective basis for evaluating them in the decision-making process. Although the decision is not based solely on scores, this technique will force the decision maker to analyze the relevant factors and the candidates in a systematic fashion.

The manager or supervisor to whom the new employee would report generally makes the hiring decision. In large organizations, approvals by higher-level managers may be required. Before a final offer is made, the selection recommendation also may need to be reviewed by affirmative action personnel.

The offer generally will confirm the details of the job, working arrangements, and salary or wages, and specify a time limit in which the applicant must reach a decision. Individuals who are rejected should be notified immediately and given the reason for rejection.

The salary offered should be competitive with those of similar jobs in other organizations in the area and should be compatible with the existing salary structure in the organization. Too low an offer may cause the new employee to feel disgruntled; too high an offer may cause morale problems with current employees.

Orientation

Orientation: Formal process of familiarizing new employees to the organization, job, and work unit.

The recruitment and selection of employees are important steps in the employment process, but the careful planning and decision making will be negated if orientation to the organization and the job is not carried out properly. **Orientation** is the formal process of familiarizing new employees with the organization, job, and work unit (Sherman, Snell, & Bohlander, 1997). Orientation is designed to provide new employees with the information they need to function comfortably and effectively in the organization. Three types of information typically are included in orientation:

- Review of the organization and how the employee's job contributes to the organization's objectives
- Specific information on policies, work rules, and benefits
- General information about the daily work routine

In large organizations, the orientation tends to be more formal than in small organizations, but it is an important process for getting employees off to a good start regardless of the formality of the orientation or size of the organization. If an organization has a human resources department, staff in that department generally provides information about the overall organization policies and benefits. Orientation in the department then focuses on job-related issues.

If new employees are properly oriented, several objectives can be achieved. First, start-up costs can be minimized because new employees may make costly mistakes if they are not properly oriented. Second, anxieties can be reduced. Third, orientation can help create realistic job expectations.

Unfortunately, the importance of an orientation program is underestimated in many organizations. New employees often are given a policy and procedure manual and told to study it until given another assignment. Policies and procedures may have little meaning to new employees who are not familiar with the operation.

The first day of employment is crucial to the success or failure of a new employee. An actual incident that occurred in a hospital foodservice provides an example of how weak the orientation process may be in some situations. On the first day at work, after completing the formal 2-hour group orientation session conducted by staff in the human resources department, Bill, a new dishwasher, walked into the department of food and nutrition services, and the director told him, "There is the dishroom. Go find Joe; he'll show you the ropes." Needless to say, the anxiety of a new employee was probably not reduced by that experience!

Zemke (1989a) stated that creating a first-class orientation program and process is a job worth doing, but doing it well requires time, money, patience, some creativity, and an understanding of what has worked well for others. Puckett (1982) developed the checklist in Figure 12.8 for the initial orientation of new foodservice employees. She cautioned that new employees should not be overloaded with information on the first day and stressed the need for follow-up sessions during the first week, with a review session during the first month on the job. Periodic monitoring and reinforcement during the early weeks on the job also are advisable. The checklist in Figure 12.9 outlines the follow-up orientation that Puckett suggested be held at the end of the second or third week on the job.

Proper orientation pays off in terms of decreased turnover and increased job performance. There is no magic formula for reducing employee turnover, but Hogan (1992) suggested that establishing an orientation program that is consistently followed and does much more than merely explain the mechanics of being an employee is a step in the right direction. For example, the organization's goals, philosophy, and culture should be explained to all new employees.

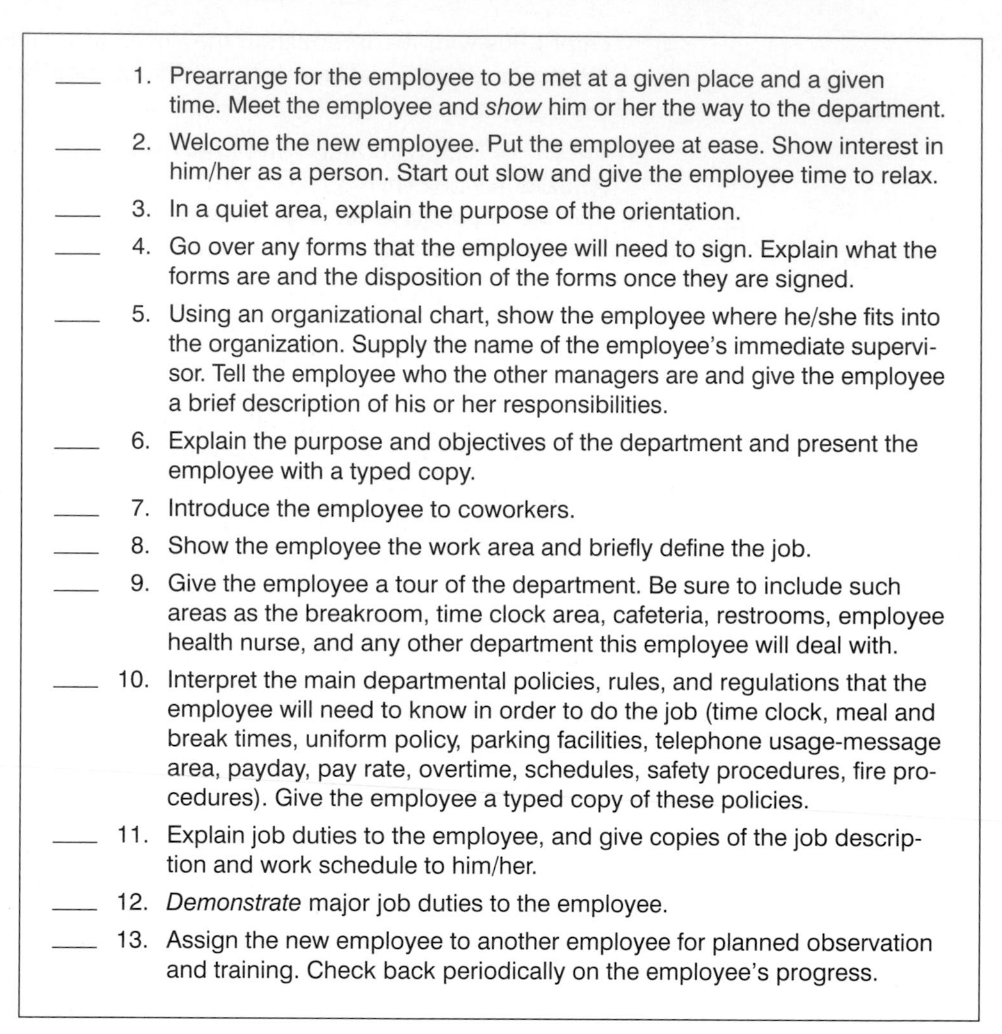

____ 1. Prearrange for the employee to be met at a given place and a given time. Meet the employee and *show* him or her the way to the department.

____ 2. Welcome the new employee. Put the employee at ease. Show interest in him/her as a person. Start out slow and give the employee time to relax.

____ 3. In a quiet area, explain the purpose of the orientation.

____ 4. Go over any forms that the employee will need to sign. Explain what the forms are and the disposition of the forms once they are signed.

____ 5. Using an organizational chart, show the employee where he/she fits into the organization. Supply the name of the employee's immediate supervisor. Tell the employee who the other managers are and give the employee a brief description of his or her responsibilities.

____ 6. Explain the purpose and objectives of the department and present the employee with a typed copy.

____ 7. Introduce the employee to coworkers.

____ 8. Show the employee the work area and briefly define the job.

____ 9. Give the employee a tour of the department. Be sure to include such areas as the breakroom, time clock area, cafeteria, restrooms, employee health nurse, and any other department this employee will deal with.

____ 10. Interpret the main departmental policies, rules, and regulations that the employee will need to know in order to do the job (time clock, meal and break times, uniform policy, parking facilities, telephone usage-message area, payday, pay rate, overtime, schedules, safety procedures, fire procedures). Give the employee a typed copy of these policies.

____ 11. Explain job duties to the employee, and give copies of the job description and work schedule to him/her.

____ 12. *Demonstrate* major job duties to the employee.

____ 13. Assign the new employee to another employee for planned observation and training. Check back periodically on the employee's progress.

Figure 12.8. Orientation checklist.
Source: "Making or Breaking the New Employee" by R. P. Puckett, 1982, *Contemporary Administrator, 5*(10), p. 14. Used by permission.

DEVELOPING AND MAINTAINING THE WORKFORCE

The employment process includes only the initial stages in building an efficient and stable workforce. Employees need continual development if their potential is to be used effectively in an organization. This development process begins with orientation but should be continued throughout employment in the organization.

Employee development has become increasingly vital to the success of organizations. Changes in technology require that employees possess the knowledge and skills to cope with new processes and techniques.

Performance evaluation is another step in the development and maintenance of the workforce. Other staffing functions include promotions, transfers, demotions, separations, and compensation management (the administration of salaries, wages, and fringe benefits).

____ 1. Review of dress code.

____ 2. Review of job description; check for problems.

____ 3. Explanation of performance evaluation procedures.

____ 4. Review of all fringe benefits, sick leave, holidays, annual leave, education policy, seniority, leaves of absence, promotions, grievance procedure, appeals procedure, lay off, merit increases. (If possible, be sure that the employee has a copy of these policies and procedures.)

____ 5. Explanation of accident prevention procedures in case there is an injury.

____ 6. Review of fire plan. Supply copy of procedures.

____ 7. Review of disaster plan. Supply copy of procedures.

____ 8. Review of any other policies or procedures discussed during the first week. Does the employee know the rules and regulations?

____ 9. Review of any other policies or procedures specific to the department that will assist the employee in his/her duties.

____ 10. Affirm that the employee has received and read personnel and departmental policies and procedures.

Figure 12.9. Checklist for follow-up orientation session with a new employee.
Source: "Making or Breaking the New Employee" by R. P. Puckett, 1982, *Contemporary Administrator, 5*(10), p. 15. Used by permission.

Training and Development

Training: Ongoing process of updating skills of an employee.

Training is and must be the responsibility of all managers. The term **training** is frequently used to refer to the teaching of technical skills to employees, and **management development** refers to programs designed to improve the technical, human, and conceptual skills of managers. Many new employees have the knowledge, skills, and abilities for their job, but others might need extensive training before they can perform effectively. The primary purpose of a training program is to help the organization meet its goals. Equally important, however, is helping trainees meet their personal goals.

The need for training new employees and employees promoted to new jobs is obvious; however, training is needed for all employees to maintain standards in a foodservice operation. In healthcare facilities, training for all employees is mandated in federal regulations and accreditation standards. In other segments of the foodservice industry, the positive benefits of training in terms of increased productivity and morale gradually are being recognized.

In addition to training for separate jobs, many employers offer training programs that go beyond the immediate job requirements. Educational assistance for employees is given in some foodservice operations. One fast-food operation in a major city spent $10,000 in 1 year for tuition and other costs for employees attending a local community college; the operation saved much more in reduced turnover.

Another area of educational concern is illiteracy. Business is going to have to find its own solutions to workplace illiteracy (Sherer, 1990), just as it must respond to other social ills. Foodservice, because of its low-skill, low-paying jobs, attracts large numbers of functionally and marginally illiterate employees. Many foodservice managers have started literacy programs for employees and family members. Help is available from local school districts, adult and continuing education centers, state education departments, local colleges and universities, and the federal government's Job Training Partnership Act funds. Regardless of cost, basic skills and English as a second language training programs might become a way of life with restaurant operators. Most operators say the cost is worth it; typical literacy programs foster increased morale, better communications, lower turnover, employee loyalty, and higher productivity.

In large organizations, a training staff may be available to assist with various aspects of a training program. Some training is conducted on a group basis, but individual training also is necessary. A needs assessment should determine what kind of training is needed and where it is needed.

After needs have been identified, managers should then initiate appropriate training efforts. **On-the-job training** methods involve employees learning how to do a particular job while actually doing the job and include job rotation, internships, and apprenticeships. In job rotation, employees are assigned to work on a series of jobs over a period of time, thereby permitting them to learn a variety of skills. In an internship, job training is combined with classroom instruction. In an apprenticeship, employees are assigned to highly skilled coworkers responsible for their training.

Off-the-job training takes place outside the workplace. Methods used in off-the-job training range from laboratory experiences that simulate actual working conditions to other types of participative experiences, such as case studies or role playing, and to classroom activities such as seminars, lectures, and films. The programmed instruction method, in which subject matter is broken down into organized and logical sequences, requires responses by the trainee. Computer-assisted instruction also is used to deliver training material directly through a computer terminal in an interactive format. For training managers, management games can be used to make decisions affecting a hypothetical organization.

Employee work schedules and workloads need to be considered in planning inservice sessions because varied work shifts and alternating days off are common practices in foodservice operations. Training sessions may need to be held at the end of the work day or at the end of one shift and the beginning of another if the daily work routine does not allow time. Employees generally are given extra compensation for attending training sessions if they are scheduled outside the regular work day. Inservice training sessions conducted during the work day usually are fairly short, varying in length from 15 minutes to no more than 1 hour. Whenever possible, discussion should be supplemented with visual aids, demonstrations, or printed materials. Resource persons from other departments or from outside the organization will add interest to the program. Sessions should provide opportunity for employee involvement and should start and end promptly.

Regardless of the training method used, carefully formulated objectives and a well-planned outline are critical to the success of the training effort. An example of a simple form for developing a lesson plan is included in Figure 12.10.

This and Lippitt (1966) summarized several conditions for effective learning that also should be taken into consideration in employee training (Figure 12.11). Managers should be aware of several common pitfalls to avoid. Lack of reinforcement is a common training error; frequently, managers point out mistakes but do not give praise when an employee does the job correctly.

Training Session Title: Work Simplification Date: August 08, 1999

Objective: After completing the training session, the employee will be able to apply work simplification principles to his or her job.

Content Outline	Learning Experience, Activities, Evaluation	References, Resource Material
1. What is work simplification? • Define work simplification. • Identify benefits of work simplification. 2. How do you apply work simplification? • Discuss questions for analyzing present methods—what, why, how, when, where, who. • Identify changes that can be made in present methods based on the analysis, e.g., materials used, finished appearance, steps in assembly. • Discuss types of changes that can be implemented and evaluation of effectiveness of change.	• Class discussion • Demonstration of several work simplification techniques in food production. • Written examination on principles of motion economy. • Observation of work performance.	Neil, C. A.: Working smarter not harder. *School Food Service Journal 37*(10):51, 1983.

Figure 12.10. Form for planning in-service training session.
Source: Neil, C. A., Working smarter not harder. *School Food Service Journal, 37*(10):51, 1983.

The phrase "practice makes perfect" applies to the training process. All too often, managers expect employees to perform a job perfectly the first time, forgetting that practice and repetition are important aspects of learning. Managers also forget that people learn at different rates—some learn rapidly and some more slowly. If a person is a slow learner, he or she is not necessarily a poor performer. Training must take into consideration the differences among individuals.

Development of managers and other professional staff may include continuing education opportunities provided by colleges and universities, professional and trade associations, or other organizations. These may include workshops, short courses, seminars, trade shows, or conventions. Some organizations have programs to assist managers in obtaining degrees either while working or during an educational leave.

Even though some of these continuing education experiences may not be directly related to a manager's job, they may serve to renew enthusiasm, introduce the person to new ideas, or provide new personal contacts that may later be a resource for information. The manager's participation may be fully or partially funded by the organization.

1. Acceptance that all people can learn.
2. The individual must be motivated to learn.
3. Learning is an active, not passive, process.
4. Normally, the learner must have guidance.
5. Approximate materials for sequential learning must be provided: hands-on experiences, cases, problems, discussion, reading.
6. Time must be provided to practice the learning, to internalize, to give confidence.
7. Learning methods, if possible, should be varied to avoid boredom.
8. The learner must secure satisfaction from the learning.
9. The learner must get reinforcement of the correct behavior.
10. Standards of performance should be set for the learner.
11. A recognition that there are different levels of learning and that these take different times and methods.

Figure 12.11. Conditions for effective learning.
Source: "Learning Theories and Training" by L. This and G. Lippitt, 1966, *Training and Development Journal, 20*(4), p. 2. Copyright 1966, Training and Development, American Society for Training and Development. Reprinted with permission. All rights reserved.

No training and development program is complete until evaluation is done. Employees may be asked to evaluate individual and group training in terms of satisfaction, benefit derived, and suggestions for future training. The true test of the effectiveness of any training effort, however, is improvement in employee performance.

Performance Appraisal

Performance refers to the degree of accomplishment of the tasks that make up an individual's job. Often confused with effort, performance is measured in terms of results. **Performance appraisals,** the assessment of an employee's performance during a specified period of time, take place in every organization, although they are not always formal. The success or failure of a performance appraisal program depends upon its philosophy and the attitudes and skills of those who manage it (Sherman, Snell, & Bohlander, 1997). Gathering information is the first step in the process. The information about specific employees must be evaluated based on whether the employees are meeting organizational needs, and the information must be communicated to employees in hopes that it will result in their reaching high levels of performance.

Mondy, Noe, and Premeaux (2002) identified several characteristics of an effective performance appraisal system:

- **Job-related criteria**—the performance review should include only criteria directly related to the job being performed.
- **Performance expectations**—performance expectation should be clearly defined and managers and employees should discuss them in advance of the appraisal period.

- **Standardization**—employees in the same job category under the same supervisor should be appraised using the same instrument and with the same frequency of review.
- **Trained appraisers**—those evaluating the work of others should receive training on issues such as accuracy, consistency, objectivity, and process.
- **Continuous open communication**—feedback on performance should be provided on a continuous basis.
- **Performance reviews**—a specific time line for discussion of employee performance should be established.
- **Due process**—a formal grievance procedure should be established.

Objectives

The primary objectives of a performance appraisal program are

- To provide employees with the opportunity to discuss their performance with the supervisor or manager
- To identify strengths and weaknesses of the employee's performance
- To suggest ways the employee can meet performance standards, if they have not been met
- To assist the employee in setting objectives and personal development plans
- To provide a basis for future job assignments and salary recommendations

Research has shown that performance appraisals are used primarily for compensation decisions (Sherman, Snell, & Bohlander, 1997). Employee placement decisions, such as transfers, promotions, or demotions, usually are based on performance appraisals. Many organizations also are using these appraisals to document employee performance as a protection against possible charges of wrongful termination or unfair employment practices.

Individuals want feedback about their performance, and performance appraisals can provide an opportunity to obtain such information. If performance compares favorably with that of other employees, an individual will obviously react favorably to performance review; however, feedback on poor performance is more difficult to accept.

In actual practice, appraisal programs have often yielded disappointing results. The focus too frequently is on past performance with little attention given to future performance.

Managers or supervisors usually are responsible for appraising employees in their units. They probably are in the best position to do this, but they often do not have time to observe the employees' performance adequately, which results in a less-than-objective appraisal. Often managers have to rely on others to complete an appraisal. Peer appraisals sometimes are used, but confidentiality can become a problem with this method. Managers often ask employees to rate themselves on the appraisal form before the performance interview, at which time the supervisor and the employee discuss job performance and agree on a final appraisal. At this meeting, future performance goals can be established for the employee, and these goals can be used as the basis for the next performance appraisal. The major difficulty with most performance appraisals is that the judgments frequently are subjective, relating primarily to personality traits or to observations that cannot be verified.

Subordinates often are asked to complete a performance appraisal form for their superiors. They can rate performance dimensions, such as leadership, oral communication, delegation of authority, coordination of team efforts, and interest in subordinates, but they should not rate performance of job tasks, such as budgeting and analytical ability (Sherman, Snell, & Bohlander, 1997).

Criteria for measuring performance should be relevant to the job, understandable, and measurable. Job descriptions provide the basis for performance appraisal.

Methods

Commonly used performance appraisal methods include the following: checklist, rating scale, critical incident, management by objectives, and 360-degree feedback. In the *checklist method,* the rater does not evaluate performance but merely records it. A list of statements or questions that are answered by yes/no responses is the basis for the evaluation. For example, a checklist for a chef might include the following items that can be answered with a yes or no: The chef

_____ is concerned about pleasing the customer
_____ checks with waitperson if order is not clear
_____ is concerned about presentation of the food
_____ adheres to food safety principles

A wide variety of **rating scales** have been developed. Rating scales are a very common method of performance appraisal. In the rating scale method, each trait or characteristic to be rated is represented by a scale on which a rater indicates the degree to which an employee possesses that trait or characteristic (Sherman, Snell, & Bohlander, 1997). There are many variations of the rating scale. The differences are to be found in (1) the characteristics or dimensions on which individuals are rated, (2) the degree to which the performance dimension is defined for the rater, and (3) how clearly the points on the scale are defined.

Mixed standard scales are a modification of the basic rating scale. Rather than evaluate a trait according to a scale, the rater is given three specific behavioral descriptions relevant to each trait. The descriptions for the trait should reflect three types of performance; superior, average, or inferior (Sherman, Snell, & Bohlander, 1997). The *behaviorally anchored rating scale* (BARS) consists of a series of 5 to 10 vertical scales, one of each important dimension of job performance anchored by the incidents judged to be critical.

The *critical incident technique* involves identifying incidents of employee behavior (Ingalsbe & Spears, 1979). An incident is considered critical when it illustrates that the employee has or has not done something that results in unusual success or failure on some part of the job. Therefore, critical incidents may include both positive and negative examples of employee performance. Although this method has the advantage of providing a sampling of behavior throughout the evaluation period, it is also a time-consuming technique.

Management by objectives (MBO) is a method for performance evaluation that is used primarily with managerial and professional personnel. Managers and their superiors agree on the objectives to be achieved, usually for a 1-year period. Periodic assessment of progress on the objectives is conducted at several intervals during the year, and objectives are revised if deemed appropriate. The degree to which these objectives are achieved, then, provides the basis for evaluation at the end of the period. MBO provides a great deal of objective feedback but requires an excessive amount of time and paperwork. If they are to succeed, MBO programs should meet several requirements (Sherman, Snell, & Bohlander, 1997):

1. Objectives set at each level of the organization should be quantifiable and measurable for both the long and short term.
2. Expected results must be under the employee's control, and goals must be consistent for each level.
3. Managers and employees must establish specific times when goals are to be reviewed and evaluated.
4. Each employee goal statement must be accompanied by a description of how that goal will be accomplished.

The MBO system is not without its critics. One researcher contends that MBO is a lengthy and costly appraisal system with only a moderate impact on organizational success. Another criticism is that performance data are designed to measure results on a short-term rather than a long-term basis. A recent addition to performance appraisal is the use of 360-degree review. This review involves obtaining evaluations about an individual from superiors, subordinates, peers, and potentially customers. Such feedback provides a much more comprehensive view of performance.

The *360-degree feedback method* involves obtaining evaluation input from individuals in an organization's hierarchy that are both above and below the individual being evaluated. Typically this will include co-workers, subordinates, and superiors who will provide input for a performance review. Yum Brands, Inc., uses the 360-degree feedback approach with all of its employees in its more than 1,000 Pizza Hut, Taco Bell, KFC, A&W, and Long John Silver's franchises (Shuit, 2005).

Appraisal Interviews

The appraisal interview provides the manager with the opportunity to discuss an employee's performance and explore areas of improvement. Also, employees should have the opportunity to discuss their concerns and problems in the job situation. Because the purpose of the interview is to make plans for improvement, the manager should focus on the future rather than the past. In some instances, employees may be asked to complete a performance appraisal form prior to the interview, and this form is used during the interview session for comparison with the supervisor's assessment. Interviews should be scheduled several days in advance and conducted in a private setting because of the sensitivity involved.

The manager should observe the following points during the interview:

- Emphasize strengths on which the individual can build rather than stress weaknesses.
- Avoid recommendations about changing personal traits; instead, suggest more acceptable ways of performing.
- Discuss behaviors that are appropriate or need improvement.
- Concentrate on opportunities for growth within the employee's present job.
- Limit plans for change or growth to a few objectives that can be accomplished within a reasonable period of time.

The appraisal interview is probably the most important part of the performance appraisal. Trying to discuss the employee's past performance and plans for improvement in one session can be difficult. Dividing the interview into two sessions might be the answer.

Personnel Actions

The performance appraisal process provides the basis for various types of personnel actions. If deficiencies in employee performance are noted, plans for improvement should be developed or appropriate disciplinary actions should be used. Changes in job placement may be suggested during evaluation and may include promotions, demotions, transfers, or even separations such as those by dismissal, resignation, or retirement. Leaves with or without pay are other personnel actions for which a manager must plan to maintain an adequate workforce.

Promotion

Promotion: Change in job to one at a higher level in the organization.

A **promotion** is a change of assignment to a job at a higher level in the organization. The new job generally provides higher pay and status and requires more skill and responsibility. Advancement can serve as an incentive for improved performance for employees at all levels of the organization.

Promotions are tied closely to the performance appraisal and should be a way for recognizing good performance; however, individuals should not be promoted to positions in which their abilities are not appropriate. For example, an action-oriented unit or regional manager in a large multiunit foodservice organization who is a good decision maker might be misplaced in a staff position in the central office. Instead, this individual should probably be considered for a higher-level line position.

Two basic criteria for promotions are merit and seniority. Promotions should be fair, based on merit, and untainted by favoritism. Even when promotions are fair and appropriate, problems may occur, such as resentment by employees bypassed for promotion. If an organization is unionized, the union contract often requires that seniority be considered.

Succession planning involves planning for qualified individuals to be available to assume management positions as they become available. Mondy, Noe, and Premeaux (2002) encourage creation of a management database for succession planning. The database would include information such as a manager's current position, professional development programs completed and yet needed, possible positions for promotion to, and possible replacements if this manager were promoted.

Demotion

Demotion: Change in job to one at a lower level in the organization.

A **demotion** consists of a change in job assignment to a lower organizational level in a job involving less skill, responsibility, status, and pay. Employees may be demoted because of reduction in positions of the type they are holding or because of reorganization. Demotion may also be used as a disciplinary action for unsatisfactory performance or for failure to comply with policies, rules, or standards in the organization. Acceptance and adjustment to loss of pay and status are difficult for most employees. Demoted employees may need special supervision or counseling as a result.

When the abilities or skills of a long-service employee decline, the solution may be to restructure a job rather than demote a loyal employee. In some instances the problem is solved by moving the person to a job with an impressive title but little authority or responsibility.

Transfer

Transfer: Change in job to one at approximately the same level elsewhere in the organization.

Transfer involves moving an employee to another job at approximately the same level in the organization with basically the same pay, performance requirements, and status. A transfer may require an employee to change the work group, workplace, work shift, or organizational unit, or even move to another geographic area. A transfer can result from an organizational decision or an employee request. Transfers permit the placement of employees in jobs in which the need for their services is greatest. Also, transfers permit employees to be placed in jobs they prefer or to join a more compatible work group.

Separation

Separation: Voluntary or involuntary termination of a job.

A **separation** involves either voluntary or involuntary termination of an employee. In instances of voluntary termination or employee resignations, many organizations attempt to determine why employees are leaving by asking them to complete a questionnaire or by conducting an exit interview. This interview can be a powerful tool for identifying a variety of personnel-related problems. According to human resources researchers, employees leaving a company can provide insight into problems encountered while in the organization's service. The exit interview provides employers an opportunity to learn why their employees are leaving and to devise plans for correcting turnover.

Involuntary separation, or firing an employee, should be considered an action of last resort because of the investment the organization has in the employee. Training and counseling should be tried before termination is considered. Also, because of legal implications, thorough and complete documentation of poor performance is necessary when firing an employee.

Another type of separation may be the result of a decision to reduce the workforce. In this instance, the layoff may be the result of elimination of jobs, new technology, or depressed economic conditions. Decisions on layoffs should be made carefully, and assistance in finding another job should be given to the laid-off employee, if at all possible.

Employee Discipline

Discipline: Action against an employee who fails to conform to the policies or rules of an organization.

Discipline is the action against an employee who fails to conform to the policies or rules established by the organization. It is used to aid in obtaining effective performance and to ensure adherence to work rules. It also serves to establish minimum standards of performance and behavior. Some of the more common problems with employees that necessitate disciplinary action are listed in Figure 12.12.

Foodservice operators agree that the key to effective discipline is keeping in touch with employees (Boyle, 1987). Good communications begin with clearly defined expectations, which should be discussed during the job interview. Disciplining an employee is not easy, but it can be constructive. Once alerted to a problem,

Attendance Problems

Unexcused absence
Chronic absenteeism
Unexcused/excessive tardiness
Leaving without permission

Dishonesty and Related Problems

Theft
Falsifying employment application
Willful damage to company property
Punching another employee's time card
Falsifying work records

On-the-Job Behavior Problems

Intoxication at work
Insubordination
Horseplay
Smoking in unauthorized places
Fighting
Gambling
Failure to use safe devices
Failure to report injuries
Carelessness
Sleeping on the job
Abusive or threatening language to supervisors
Possession of narcotics or alcohol
Possession of firearms or other weapons

Figure 12.12. Common disciplinary problems.
Source: Personnel Management. The Utilization of Human Resources, 6th edition, by H. J. Chruden and A. W. Sherman, Copyright 1980, South-Western Publishing Co. Used by permission.

the employee may be able to solve it together with the manager, leading to substantial improvement in the employee's overall job performance.

Disciplinary Procedures

Usually, the following several steps are involved in a disciplinary procedure:

- Unrecorded oral warning
- Oral warning with notation in an employee's personnel file
- Written reprimand
- Suspension
- Discharge

Exceptions do occur in the preceding steps, depending on the nature of the employee's offense. For example, a foodservice employee caught in the act of theft would probably be fired immediately, rather than be warned, orally or in writing, or suspended.

A key element in discipline is consistency. An employee should feel that any other employee would receive the same discipline under essentially the same circumstances. Disciplinary action is the consequence of an employee's behavior, and not of a personality conflict with the supervisor. Discipline must be administered in a straightforward, calm way without anger or apology. The manager should avoid argument and administer discipline in private to avoid embarrassing the employee. In a few instances, public reprimand may be necessary to enable a manager to regain control of a situation.

In most organizations, a grievance procedure has been established to ensure that employees have due process in disciplinary situations. In other words, grievance procedures allow an employee to have a fair hearing when he or she believes a personnel action has been administered unfairly. These procedures generally involve a multistep process whereby the employee first requests a review by the next highest level supervisor. If a satisfactory solution is not found at this level, then a higher-level review can be requested. Grievance committees often are appointed to review actions not settled through the chain of command. In unionized organizations, grievance procedures are specified in the union contract and include provisions for hearings by an outside arbitrator when problems cannot be solved within the organization.

Grievances and disciplinary action can be reduced in a number of ways. These methods include the following:

- Preparation of accurate job descriptions and specifications
- Selection of individuals with appropriate qualifications for job requirements
- Development of effective orientation, training, and performance evaluation systems
- Use of good human skills by supervisors

Identifying Causes

An employee's immediate supervisor has the primary responsibility for preventing or correcting disciplinary problems. In attempting to uncover reasons for unsatisfactory behavior, the supervisor must first consider if the employee is aware of certain policies and work rules before initiating disciplinary action. If the employee is aware of these policies, the supervisor then must realize that understanding the causes underlying the problem is as important as dealing with the problematic behavior; any attempt to prevent recurrence requires an understanding of these causes.

Health problems, personal crises, emotional problems, or chemical dependency may be the source of unsatisfactory performance. Marital, family, financial, or legal problems are the predominant personal crisis situations that may affect employee performance.

Stress also can impact job performance. Although most employees can cope with stress, some employees cannot and become troubled employees. Impaired mental health, alcoholism, and drug abuse may be symptomatic of the troubled employee. *Burnout,* a stress-related condition usually caused by working long hours in high-pressure situations, often leads to turnover in the foodservice industry.

Supervisors need to be alert to all these possible causes of problematic behavior. Because supervisors may not be prepared to assist or counsel an employee, many organizations provide them with some combination of training, policy guidance, or counseling. The counseling may be for supervisors or employees, and it may be with a specialized counselor or medical personnel both inside and outside the organization.

Employee assistance program (EAP): Program that provides diagnoses, counseling, and referral for advice or treatment for problems related to alcohol or drug abuse, emotional difficulties, and marital or family difficulties.

Many organizations are beginning to take a more positive approach in dealing with troubled employees. An **employee assistance program (EAP)** typically provides diagnosis, counseling, and referral for advice or treatment, when necessary, for the problems related to alcohol or drug abuse, emotional difficulties, and marital, family, or financial difficulties.

These programs typically include training of supervisors, internal staff counseling, referral services to community agencies, and involvement of the employee's dependents. Organizations are recognizing that discharging a troubled employee may be not only inhumane but also an ineffective solution because similar problems may occur with a new employee. Employee assistance programs provide a cost-effective method for resolving these problems.

Compensation Management

Employee compensation represents a substantial part of the operating costs of an organization. Good working conditions, sound employment practices, and compensation appropriate for an individual's qualifications and the responsibilities of a job are essential for recruitment and retention of capable employees.

Compensation

Compensation: Financial remuneration by employers to employees in exchange for their work.

Compensation is the financial remuneration given by the organization to its employees in exchange for their work. It includes salaries or wages and benefits. *Salary* is the term used to refer to earnings of managerial and professional personnel; *wages* refer to hourly earnings of employees covered by the Fair Labor Standards Act, sometimes called the Minimum Wage or Wage-Hour Law. In establishing the rate of pay for employees, the employer must consider many factors, including wage mix, before making a decision. Benefits are rewards such as insurance that provide security to employees and their family members. Wages and salaries are often referred to as direct compensation because the employee receives them in cash; benefits are termed indirect compensation because they are given to the employee in the form of a plan rather than in cash.

Wage Mix

A *wage mix,* shown in Figure 12.13, is a combination of both external and internal factors that can influence rates at which employees are paid (Sherman, Snell, & Bohlander, 1997). Government legislation also influences wage mix.

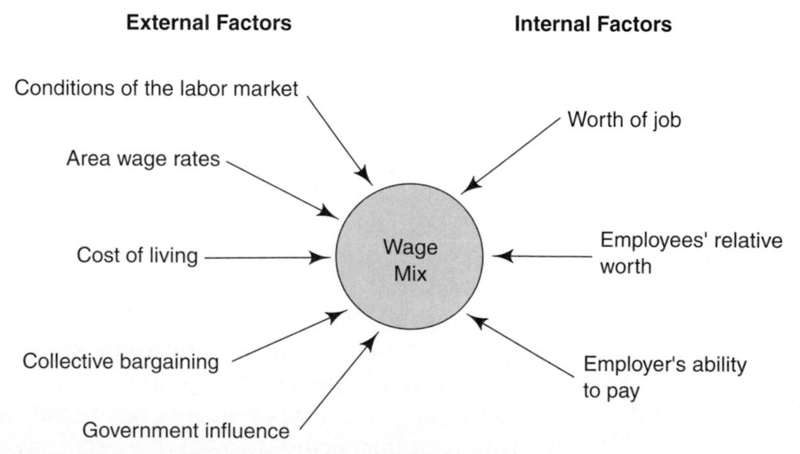

Figure 12.13. External and internal factors affecting the wage mix.
Source: Reproduced from *Managing Human Resources,* 11th edition, by A. W. Sherman, G. W. Bohlander, and S. Snell. Copyright 1998. Reprinted with permission of South-Western Publishing, a division of Thomson Learning. Fax 800-730-2215.

External Factors. The external factors that influence the wage mix include labor market conditions, geographic area, cost of living, collective bargaining, and government influence. Labor market conditions reflect the supply and demand of employees in the geographic area. A wage structure that is comparable with wages in the geographic area should be developed. If this information is not available in the area, a national survey could be used. For example, the U.S. Bureau of Labor Statistics collects and has available online (www.bls.gov) data on employment hours and wages.

Compensation rates during inflation have been increased periodically to permit employees to maintain their purchasing power (Sherman, Snell, & Bohlander, 1997). Wage setters generally use the consumer price index, which is a measure of the average change in prices over time in a fixed so-called market basket of products and services. Labor unions use collective bargaining to increase compensation. The union's goal is to increase real wages by a percentage greater than the consumer price index to improve the purchasing power and standard of living of its members.

Internal Factors. Often, if the organization does not have a compensation program, the worth of jobs is determined by people familiar with them. Pay rates can be influenced by the labor market or collective bargaining. Organizations with compensation programs use job evaluations to aid in setting pay rates. Job evaluation is the process of determining the relative worth of jobs to establish which jobs should be paid more than others. The first step of compensation is a job evaluation to identify the work content and requirements of each job: required responsibilities, education, experience, and the skill necessary for performing duties under the existing conditions.

Two methods used for evaluating jobs are job ranking and job classification. *Job ranking,* in which jobs are arranged on the basis of their relative worth from most to least complex, is the simplest method but is subject to the amount of skill and information that raters have. *Job classifications* group jobs according to a series of predetermined wage classes or grades. This method is designed to place jobs of the same general value in the same pay grade, and it provides a smooth progression for advancement.

Employees' relative worth often is based on the premise that employees who possess the same qualifications should receive the same pay. Differences in performance are rewarded through promotion of various incentive systems. Merit raises should be based on an effective performance appraisal, but often they are given automatically. An employer's ability to pay is based on the willingness of the taxpayer to provide funds in the public sector or by the profits from products and services in the private sector.

Benefits

Benefits are noncash rewards given to employees by their employer as part of their employment. The cost of employee benefits has increased dramatically. According to Mondy, Noe, and Premeaux (2002), a typical worker who earns $30,000 per year will receive approximately $11,700 (almost 40%) indirectly in benefits. Some of these benefits, such as Social Security and workers' compensation, are mandated by law; others, such as health insurance and paid leave, are at the discretion of the company.

Gómez-Mejía, Balkin, and Cardy (2001) categorized benefits into six groups as follows:

- **Legally required benefits**—include Social Security, workers' compensation, unemployment insurance, and family and medical leave.
- **Health insurance**—includes insurance for health, dental, and/or eye care; service may be provided through traditional fee-for-service health insurance plans, health maintenance organizations (HMOs), or preferred provider organizations (PPOs); some plans will offer out-of-network or point-of-service (POS) options.
- **Retirement**—includes plans that will provide income to employees after they retire through either a defined benefit plan, in which the employer agrees to provide a specific level of retirement income, or through a defined contribution plan, such as 401(k), individual retirement account (IRA), or employee stock option plans, which require specific contributions by the employee.
- **Insurance**—includes insurance plans for life, disability, and/or supplemental unemployment.
- **Paid time off (PTO)**—includes payment for time not worked and may include illness, vacation, holidays, personal time, and severance pay.
- **Employee services**—includes benefits such as child care, health club memberships, subsidized company cafeterias, parking privileges, relocation benefits, educational assistance, financial services, discounts on company products, transportation to and from work, clothing reimbursement allowances, employee assistance programs, onsite health services, and concierge services.

Because employees have differing benefit needs, there has been an increase in *flexible benefit packages.* The designs of the flexible plans differ among organizations. Some are designed as *modular plans,* which allow the employee to select from among a list of benefits or different levels of benefits. *Core-plus plans* usually consist of a core of essential benefits that all employees receive and an array of benefit options that employees can select. *Flexible spending accounts* are accounts set up by the employer for individual employees into which they can deposit pretax dollars to use to pay for items such as child care and medical care.

Government Influence

Since the 1930s, a broad spectrum of federal laws has been enacted with significant impact on the compensation practices in organizations. Several laws address compensation relating to nondiscrimination. Others relate to either compensation or benefits; for example, the Fair Labor Standards Act deals with compensation; Social Security, unemployment, and workers compensation are insurance benefits. Private pension plans also are subject to federal regulations under the Employment Retirement Income Security Act (ERISA). In addition, in foodservice and other segments of the hospitality industry, legislation and regulations have been enacted that apply specifically to employees who receive tips. Managers must stay up to date on current provisions of legislation related to compensation and employment practices.

Compensation Regulations. The Fair Labor Standards Act (FLSA) was passed in 1938 and has been amended many times since then. The major provisions of the FLSA are concerned with minimum wage rates, overtime payments, child labor, and equal rights. Through the years, amendments to the act have enlarged the number of work groups covered by the law and have steadily increased the minimum wage. Employees in the foodservice industry were not covered by the minimum wage legislation until the mid-1960s. Employers are required to pay 1 1/2 times the regular rate for all hours worked in excess of 40 during a given week. If employees are given time off for overtime work, it must be 1 1/2 times the number of overtime hours. The FLSA forbids the employment of minors under age 18 in hazardous occupations. Employment of minors under age 16 also is controlled under the act. The Equal Pay Act of 1963 and the federal Age Discrimination Act of 1967 prohibit discrimination of women and employees over the age of 40.

The act also defines specific occupations that are exempt from minimum wage and overtime requirements. Executive, administrative, professional, and outside salespersons are included in the exempt category. Employee exemption depends on the responsibilities and duties of a job and the salary paid. Definitions and requirements for job responsibilities and salary provisions are defined by the U.S. Department of Labor for these exempt categories.

Benefits Regulations. The Social Security Act of 1935 established an insurance program to protect covered employees against loss of earnings resulting from retirement, unemployment, disability, or, in the case of dependents, from the death or disability of the person supporting them. The Social Security program is supported by means of a tax levied against an employee's earnings that must be matched by the employer.

In recent years, the tax rate and the maximum amount of earnings subject to the tax have periodically been adjusted upward to cover liberalization of benefits. In the mid-1960s, the Medicaid and Medicare programs, which provide health insurance benefits, were established by amendments to the Social Security Act.

Employees who have been working in employment covered by the Social Security Act and who are laid off may be eligible for unemployment insurance benefits for a period of up to 26 weeks. To receive these benefits, employees who become unemployed through no fault of their own must submit an application for unemployment compensation with the state employment agency, register for available work, and be willing to accept any suitable employment that may be offered to them. Each state has its own unemployment insurance law that defines the terms and benefits of its program. Funds for unemployment compensation come from a federal payroll tax based on wages paid to each employee. Most of the tax is refunded to the states that operate their unemployment compensation programs in accordance with minimum standards set by the federal government.

The first U.S. legislation providing compensation to workers disabled in industrial accidents was enacted in 1911. Today, workers' compensation insurance covers most American workers. The theory is that the cost of work-related accidents and illnesses should be considered one of the costs of doing business and ultimately should be passed on to the customer. Disabilities may result from injuries or accidents or from occupational diseases such as black lung, radiation illness, and asbestosis. Each state has its own laws and agencies to administer disabilities laws. In general, these laws provide income and medical benefits to victims of work-related accidents, reduce court delays arising from personal injury litigation, encourage employer interest in safety and rehabilitation, eliminate payment of fees to lawyers and witnesses, and promote study of accident causes.

Most of the nonfarm workforce now receives some form of retirement protection through either private pension plans, deferred profit-sharing plans, or savings plans. In recent years, these programs increasingly have been placed under federal regulation. To protect pension plans from failure, the Employment Retirement Income Security Act of 1974, commonly known as ERISA or the Pension Reform Law, was enacted. Although the act does not require employers to have a pension plan, it provides standards and controls for plans. This law establishes employer responsibilities, reporting and disclosure, employee participation and coverage, vested rights in accrued benefit, and other requirements.

Compensatory benefits include paid time off, such as vacation, sick leave, holidays, military and jury duty, and absences due to a death in the family or other personal leave. The Family and Medical Leave Act of 1993 grants family and temporary medical leave under certain circumstances (Department of Labor, 1999). The purpose of this act is

1. To balance the demands of the workplace with the needs of families, to promote the stability and economic security of families, and to promote national interests in preserving family integrity;
2. To entitle employees to take reasonable leave for medical reasons, for the birth or adoption of a child, and for the care of child, spouse, or parent who has a serious health condition;
3. To accomplish the purpose described in 1 and 2 in a manner that accommodates the legitimate interests of employers;
4. To accomplish the purpose described in 1 and 2 in a manner that minimizes the potential for employment discrimination on the basis of sex by ensuring generally that leave is available for eligible medical reasons (including maternity-related disability) and for compelling family reasons, on a gender-neutral basis; and
5. To promote the goal of equal employment opportunity for women and men.

Two pieces of legislation impact the health insurance benefits offered by companies. The Consolidated Omnibus Budget Reconciliation Act (COBRA) of 1985 gives employees the right to continue their health insurance coverage after their employment with a company has ended. Employees and their dependents are entitled to 18–36 months additional coverage from the group health insurance plan offered by the employee's former employer. The employee must pay the full rate for the coverage. In 1996, the Health Insurance Portability and Accountability Act (HIPAA) was passed. This act allows employees the ability to transfer between health insurance plans when changing jobs without a gap in coverage due to preexisting medical conditions.

STAFFING AND SCHEDULING

Staffing: Management function that determines the appropriate number of employees needed by the organization for the work that must be accomplished.

The terms *staffing* and *scheduling* are sometimes used interchangeably; in fact, they refer to separate but interrelated functions. **Staffing** concerns the determination of the appropriate number of employees needed by the operation for the work that must be accomplished. Job analyses and work production standards provide the basis for determining staffing needs.

Scheduling: Assigning employees to specific work hours and days.

Scheduling means having the correct number of workers on duty, as determined by staffing needs. Scheduling involves assignments of employees to specific working hours and work days. The challenge of scheduling is having sufficient staff for busy meal periods without having excess staff during slack periods between meals.

Variables

Staffing and scheduling depend on many factors. Operational differences, such as which meals are served or where the foodservice operation is located, have a great effect on the number of employees and the time they will work. Also, the type of foodservice—conventional, commissary, ready prepared, or assembly/serve—affects staffing and scheduling.

Operational Differences

In foodservice operations, staffing and scheduling can become extremely complex because of the highly variable nature of the business. In a commercial foodservice, for example, the weekend dinner meal is often a peak time, and Tuesday or Wednesday night might attract only a small number of diners. An operation serving primarily a lunch crowd in a business area, however, may have very low volume in the evening. Hospital censuses generally tend to vary seasonally; patient loads are frequently at a low point during vacation months and holiday seasons.

In a university residence hall foodservice, a school lunchroom, and some other foodservice operations, customer participation is much more predictable. In these operations, however, a number of other factors affect participation, such as scheduled campus events or menu offerings. These examples illustrate the staffing and scheduling demands on foodservice managers to ensure adequate personnel without overstaffing.

Scheduling is further complicated by absenteeism, labor turnover, vacations and holidays, days off, and differing skills of employees. In addition to all these considerations, special events and catering functions usually have an impact on the schedule. Determination of minimal staffing requirements is based primarily on providing coverage for all events in the operation.

Type of Foodservice

The type of foodservice is a major determinant of staffing needs in the operation. A distinctive example is provided by contrasting the conventional and the assembly/serve foodservices. Obviously, the staffing needs are much different in a 200-bed hospital with a conventional foodservice operation, in which all or most foods are prepared from scratch, from those in an assembly/serve foodservice, in which foods require little or no processing and fewer less highly skilled employees.

One of the advantages of the ready prepared foodservice operations using cook-chill or cook-freeze methods is that the majority of the employees, especially those in production, can be scheduled on a regular 40-hour, five-day, Monday-through-Friday basis. This schedule is possible, of course, because production is for inventory rather than for immediate service. As a result, only a skeleton staff is needed at mealtime, during weekends, and on holidays, primarily for service, dishwashing, and cleanup. Fewer employees have to work the undesirable early morning, late evening, weekend, or holiday schedules.

Relief Employees

Scheduling only full-time employees to do all the work in a foodservice operation could create some problems. During rush hours, customers would complain that the operation is understaffed, but during slack times the foodservice manager would be concerned that employees are sitting around with nothing to do while increasing labor costs. A solution for many service industries is to hire part-time or temporary employees to help out in busy meal service hours.

Part-Time Employees

In some foodservice operations, most of the staff are part-time employees, a practice particularly prevalent in quick-service restaurants. Part-time employees quite often are not eligible for many benefit programs, such as vacation and sick leave time, holidays, or insurance. In some organizations, part-time employees receive these benefits when hours of employment reach a specified level. Benefits such as vacation and sick leave time may be prorated according to the number of hours worked.

Teenagers or college students traditionally have made up the bulk of the part-time workforce in quick-service operations; changes in the national birthrate, however, have created a short supply of this age group. Employers need to be aware of labor laws affecting teenagers and understand the following federal restrictions when hiring (Employers Should, 1993):

- Workers must be at least 14 years old to work in foodservice jobs.
- Minors aged 14 and 15 are limited to working 8 hours a day and 40 hours a week. During weeks when school is in session, workers under 16 may not work before 7:00 A.M. or after 9:00 P.M.
- Fourteen- and fifteen-year-olds may not hold jobs in which they do repair or maintenance work on machines or equipment; they may not cook (other than at soda fountains, lunch counters, snack bars, or cafeteria serving counters) or bake; and they are not permitted to load or unload goods to and from trucks.
- Finally, no worker under 18 is permitted to operate power-driven meat and food slicers.

All states, many of which have higher standards than federal law, have teenage employment laws as well. Any employer who is covered by both laws must observe the higher standard.

In foodservice operations in which the patronage varies greatly within the week or in establishments involved in catered events, adequate staffing is provided by scheduling persons to work only on busy weekends or to assist with special events. **Split-shift scheduling,** in which employees are scheduled to work during peak hours only, is another way in which foodservice managers attempt to have adequate staffing when they need it and minimal staffing during between-meal, low-volume times. Dining room hostesses, waiters, waitresses, and other service personnel are frequently scheduled to work during the noon meal, but they are off during the afternoon when the dining room may be closed, and then return for the evening meal. Such scheduling must be done carefully, however, because many states have work span laws that require the hours worked to fall within a given span of time.

Temporary and Leased Employees

The growth of the foodservice industry and the shortage of employees for foodservice jobs has prompted some managers to use a more expensive alternative: temporary employees, referred to as *temps*. Many businesses today are using temps from professional temporary employment agencies to fill short-term staffing needs. Agency fees can be more than a third above the standard wage for a position. Some managers justify the extra cost by citing savings in recruitment and employee benefits.

Depending on the agency, employees are available with skills for jobs ranging from dishwasher to executive chef, from hotel administrator to controller or marketing director. Some managers question the quality of temporary employees and the cost of using them on a regular basis. A foodservice operation that needs temporary help calls an agency, explains its requirements, and agrees to an hourly fee. The agency reviews its database of available employees and makes the closest match possible. The employee reports for work, completes the assignment, and is paid by the agency, which then bills the company. Hourly rates range from 50 to 150% above the wage a company would pay a permanent employee. Most of the agencies provide employee benefits, workers' compensation, and liability insurance.

Another idea that was started in Chicago is leasing staff. More than 1,500 leasing companies nationwide employ more than 1 million leased employees in all kinds of businesses (Resnick, 1993). Skyrocketing insurance costs and the cost of complying with government regulations make taking care of employees a difficult task. Restaurant owners can cut costs and headaches and concentrate on operating a restaurant by leasing employees from companies that provide everything from payroll processing to discounted healthcare coverage and workers compensation insurance. The leasing company hires all the employees away from the restaurant and leases them back for a monthly percentage. Under this arrangement, the leasing company handles employee-related tasks, and the client maintains supervisory control of the employees, sets the pay rate, and does scheduling. Some leasing companies even will screen and hire employees. The cost to the client often is lower than it would be if the foodservice operation were responsible. Foodservice managers should choose a leasing company carefully and do a thorough check of its financial health. Should the leasing company fail, foodservice operations are not free of liability for paying healthcare and workers' compensation claims.

Indices for Staffing

Several indices have been developed to assist managers in various segments of the foodservice industry in determining their staffing needs. In Tables 12.5 and 12.6, staffing is shown for a restaurant kitchen and dining room at various patronage levels.

Industry-wide standards for staffing have not been developed to the level of sophistication of those in manufacturing industries, such as the automobile industry.

Table 12.5 Staffing for Kitchen

Jobs to Be Filled	For 0–49 Customers	For 50–99 Customers	For 100–175 Customers	For 175+ Customers
Chef	1	1	1	1
Cook	1	2	3	4
Salads—pantry	1	2	2	3
Dishwasher	1	2	3	3
Potwasher	1	1	1	1
Cleaner	0	1	1	1
Storeroom person	0	1	1	1
Baker	0	1	1	1

Source: Donald Lundberg, *The Management of People in Hotels and Restaurants,* 5th edition.

Sneed and Kresse (1989) summarized **productivity levels,** the meals per labor hour, for various foodservice operations as follows:

Quick-service restaurant	9.5
Fine dining restaurant	1.4
Family restaurant	4.8
Cafeteria	5.5
Acute care facility	3.5
Extended care facility	5.0
School foodservice	13–15.0

These productivity levels reflect industry averages and should serve only as a guide for determining staffing needs and evaluating employee performance. Although productivity levels can be compared to industry averages, comparison of an operation with itself over time could be more valuable (Sneed & Kresse, 1989). Deviations in productivity can be determined and improvement identified by self-comparison.

Approximately 1.55 employees are necessary for everyday coverage of full-time positions (Keiser, DeMicco, & Grimes, 2000). The number of positions to be filled should be multiplied by 1.55 employees per position. Although meals must be served 365 days per year in many foodservice operations, full-time employees generally are available an average of only 236 days a year because of days off and

Table 12.6 Staffing for Dining Room

	Number of Customers								
Jobs to Be Filled	0–37	38–58	59–75	76–95	96–112	113–129	130–145	146–166	167+
Hostess	1	1	1	1	1	1	1	1	1
Waiter/Waitress	2	3	4	5	6	7	8	9	10
Bus person	1	2	2	3	3	3	3	4	5
Bar waiter/waitress	1	1½	1½	2	2	2½	2½	2½	2½

Source: Donald Lundberg, *The Management of People in Hotels and Restaurants,* 5th edition.

benefit days. To determine the actual number of employees needed, the number of full-time positions is multiplied by 0.55 to find the number of relief employees necessary in addition to full-time staff.

Full-time equivalents (FTEs): Number of total hours worked in a week divided by 40 (the number of hours worked by one full-time employee) to determine the number of full-time equivalent employees.

Staffing needs often are stated in **full-time equivalents (FTEs).** Absolute FTEs indicate the minimum number of employees needed to staff the facility; adjusted FTEs take into account the benefit days and days off. A minimum number of FTEs is computed by dividing the total number of hours for operating the foodservice for a period of time by the normal workload hours of one employee. For example, if a restaurant requires 200 hours of labor a day and the normal working period is 8 hours, 200 divided by 8 would equal 25 FTEs. Because this calculation does not include any time off, the actual number will be more than 25 employees.

Meal equivalent: Calculation of numbers of snacks, nourishment, paid meals, into a common number of meals.

A **meal equivalent** is the number of snacks or nourishments or beverages it takes to equal one meal. Meal equivalents often are calculated in healthcare operations where between-meal snacks and nourishments are served to patients, and beverages and between-meal foods are offered to cafeteria customers. Because it takes less labor time to prepare these snacks a formula is needed to equate these "smaller" meals to the main meals. In onsite foodservice operations, the number of meals or snacks is divided by a specific number such as 3, or 4, or 5. In a commercial operation, the meal equivalent may be determined by dividing the total dollar sales for a specific time period by the cost of a prototype meal. For example, a prototype meal might consist of a salad, entrée, bread, dessert, and beverage.

Issues in Employee Scheduling

Foodservice operations have unique challenges for employee scheduling. The hours between breakfast and dinner in a three-meal-a-day operation do not lend themselves to two full shifts. Typically more employees are needed for service at mealtimes and fewer employees are needed between meals. The manager needs to determine the type of work schedule that would be best for the operation.

Types of Schedules

Three types of work schedules—master, shift, and production—must be made by the foodservice manager. The master schedule shows days on and off duty and vacations. The shift schedule will indicate the position and hours worked and may indicate the number of days worked per week; it also lists relief assignments for positions when regular workers are off. The production schedule identifies tasks to be completed for the production of a meal.

Master Schedule. In most foodservice facilities, a master schedule, which includes days off, serves as an overall plan for employee scheduling. Generally, some type of rotation is used for scheduling days off, especially in 6- or 7-day-a-week operations, permitting employees to have some weekend time off on a periodic basis. A policy of every other weekend or every third weekend off is not uncommon. The master schedule provides the basis for developing the weekly, biweekly, or monthly schedule.

Figure 12.14 depicts a master schedule for the cooks' unit in a university residence hall foodservice serving approximately 2,200 meals daily. The schedule is based on a 2-week rotating schedule. The specific hours of employment for each

		WEEK I							WEEK II						
HOURS	NAME	SUN	MON	TUE	WED	THU	FRI	SAT	SUN	MON	TUE	WED	THU	FRI	SAT
5:30–2:00	Johnson, Cindy	X				X					X				X
5:30–2:00	Smith, Bill			X				X	X				X		
6:00–2:30	Erickson, Sue	X				X						X			X
6:00–2:30	Henry, Amanda	X					X				X				X
6:00–2:00	Porter, Jill				X			X	X					X	
5:30–2:10	Vallen, Eric	X	X						X	X					
10:30–7:00	Simkin, Jan			X				X	X				X		
10:30–7:00	Oliver, Hazel	X				X					X				X
10:30–7:00	Hill, Bertha		Lunch	X				X	X				X	Lunch	
10:30–7:00	Ring, Darlene				X		Lunch	X	X					X	
10:30–7:00	Olsen, Maggie		X					X	X					X	
10:30–7:00	Wilson, Art	X				Lunch	X			X					X
10:30–7:00	Miller, Carol	X					X			Lunch	Lunch	X			X

Figure 12.14. Example of a master schedule for a cook's unit.

position are in the first column with the name of the assigned employee in the second column. The assignment for these duties is indicated for each day of week I and week II; days off are indicated by an X. Note that each full-time employee has one weekend off during the 2-week cycle, either Saturday of the first week and Sunday of the second or Saturday of the second week and Sunday of the first. The word lunch on an employee's line means that the cook must help with lunch preparation because the cook assigned to that duty has the day off. A similar master schedule could be made for each work unit in the foodservice operation.

A paid vacation of 2 weeks is common in most organizations and may be extended to three weeks or longer for employees after a defined number of years of service. Holidays generally vary between 8 and 10 days per year. Because many foodservice operations continue to conduct business during holidays and vacation time, scheduling of vacations and holidays can present a special challenge for managers. In fact, in some commercial operations, holidays may be among the highest volume days of the year and require extra staff rather than only a skeleton staff. Employees typically are compensated by being paid at a higher rate when they work a holiday.

Careful vacation planning is needed to avoid higher labor costs in the form of overtime or the excessive use of replacement workers. Managers should attempt to schedule vacations at low-volume periods throughout the year and to stagger them so that only minimal relief staffing will be required. Effective scheduling of vacations can be accomplished by requesting each employee to submit preferences with alternates early in the year. A master vacation schedule then can be prepared that will permit the most effective labor use.

Shift Schedule. The **shift schedule** shows the staffing pattern of the operation. In the schedule shown in Figure 12.15, the staffing for a dishroom operation for three meal periods covering 16 hours has been divided into two basic shifts. For the most part, this rigid shift scheduling is not the most effective approach to scheduling in foodservice operations. For instance, in the example in Figure 12.15, all six dishwashers come on duty at 7:00 A.M., and soiled dishes in any quantity

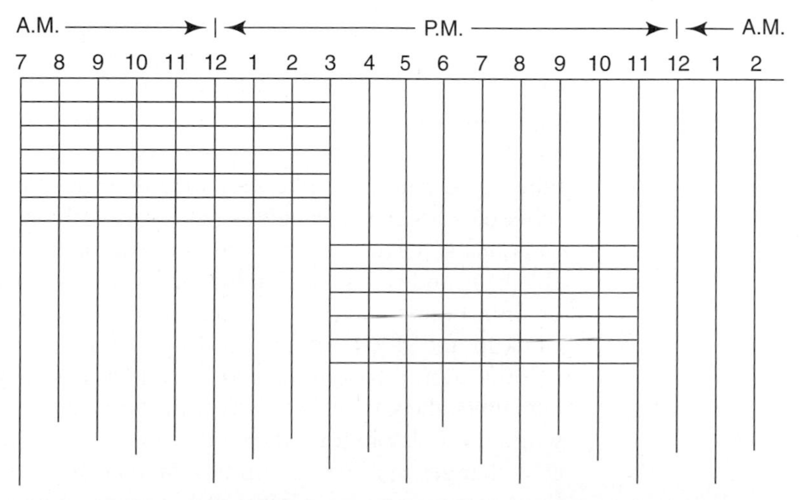

Figure 12.15. Example of a shift schedule for a dishroom operation.

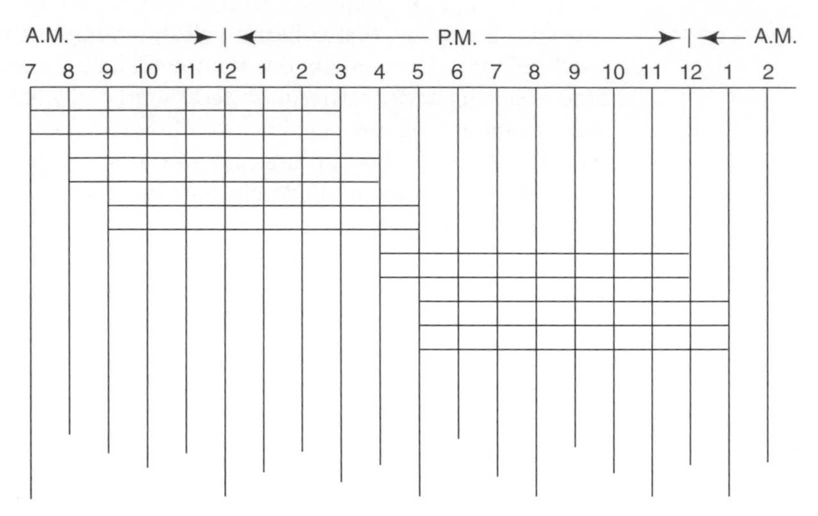

Figure 12.16. Example of a staggered schedule for a dishroom operation.

may not come into the dishroom until 7:30 or 7:45 A.M. One or two of the workers may be required to fill the dishwasher and prepare it for use for the breakfast dishes; the other workers would probably have time that would be difficult to use efficiently in some other way.

A variation of a shift schedule is a **staggered schedule** (Figure 12.16), which provides for employees to begin work at varying times, generally resulting in better use of the labor force. Staggered scheduling usually will lead to reduction in idle time and is more adaptable to the fluctuating pattern of activity in a foodservice operation.

Computer software is available for automated scheduling. The complex nature of the process and variations in the program may require close monitoring by management and an occasional override.

Control of Overtime

Uncontrolled overtime is a key factor in driving up labor cost. In some instances, employees may need to work beyond their normal hours; in other cases, however, supervisors may use overtime as a substitute for proper scheduling and planning. Moreover, employees may try to create opportunities for overtime because of the time-and-a-half wage rate they may receive. With proper staffing and realistic work schedules, overtime becomes necessary only in emergencies.

Policies regarding time cards must be formulated to ensure that employees check in and out at scheduled times, unless authorized for a different time. Most organizations today have a computerized timekeeping system instead of the older clock-in and clock-out system. In these newer systems, employees pass a personal identification card through the recognition slot of a time machine or cash register. Managers then may get daily printouts of the number of hours each employee worked the day before. If overtime was recorded, managers are alerted to the labor cost changes in adequate time to take corrective measures. Controls are needed for legal reasons and for control of labor cost. Labor laws define when overtime pay is due to an employee, and the recorded time from the clock usually provides the official record of time worked for employees covered by overtime provisions.

Overtime that can be anticipated may be controlled by requiring overtime authorization, as shown in Figure 12.17. Approval of overtime is indicated by the signature of designated managerial staff members. Unpredictable overtime required by a breakdown of equipment or unexpected business should require an overtime report filed within 24 hours. The amount of overtime should be shown and the reason stated. The supervisor should sign the completed form.

Alternate Work Schedules

Alternatives to the standard workweek have been a topic of interest and experimentation in recent years. Trends have included the introduction of several discretionary time work schedules and new forms of part-time employment. These new approaches can be grouped into three categories:

- Compressed workweek
- Discretionary working time/flex time
- Part-time employment

Three work patterns characterize the *compressed workweek* trend. A change in days holds the total hours constant but reduces the number of days worked so that employees may work, for example, a 4-day week, 10 hours per day. Some foodservice operations have experimented with this approach, but it has not gained widespread popularity in the industry. The second pattern, a change in hours worked, shortens the number of hours in the week, but the number of days worked remains the same. A 7-hour day, for instance, is characteristic of some organizational scheduling patterns. The third pattern is a change in days and hours, in which both dimensions are changed.

Discretionary working time modifications to the standard workweek include the staggered start system and flexible working hours. In using the *staggered start,* the

Overtime Authorization

Date ______________________

Name ______________________ Unit ______________________

Reason for Overtime ______________________

____

____

Amount of Overtime: ________ Hrs.

____
Signature of Unit Manager

Figure 12.17. Example of an overtime authorization form.

organization or the employees choose, from a number of management-defined options, when the employees wish to start their fixed-hour working day.

Flexible working hours, or *flex time* as the system is sometimes called, is the second major variation of discretionary working time schedules. Generally, the organization defines a range of hours within which employees may select their starting time. A number of variations to flex time systems have been developed. Flex time is impractical for many foodservice operations, however, because of the demands of production and service and scheduled mealtimes.

Task contracting is another alternative to flexible work schedules in which the employee contracts to fulfill a defined task or piece of work. This approach has been used by some foodservice caterers who may contract an individual for preparation of particular food items or for serving a particular function.

Two new variations in part-time employment, job sharing and job splitting, have emerged. In *job sharing,* a single job is divided and shared by two or more employees, each of whom must be capable of performing the entire range of tasks in the job description. In *job splitting,* the tasks that constitute a single job are divided, with subsets of differentiated tasks assigned to two or more employees.

Several of these alternative work schedules have been tried or are now being used in foodservice industries. Others are less applicable because of the unique characteristics of the industry. When experimenting with new scheduling methods, foodservice managers should use a careful process that involves gaining employee acceptance, pilot testing, and carefully evaluating outcomes before making a change.

LABOR RELATIONS

Since the 1800s, labor unions have been an important force shaping organizational practices, legislation, and political thought (Sherman, Snell, & Bohlander, 1997). Today, unions continue to be of interest because of their influence on organizational productivity, global competitiveness of the United States, the development of labor law, and human resources policies and practices. Like most organizations, unions are undergoing changes in both operation and philosophy. Labor has a new role in the United States that includes labor management programs, union company buyouts, and representation by the union on boards of directors.

Labor relations, therefore, are an important function of HRM in many types of organizations. **Labor relations** is a term referring to the interaction between management and a labor union. A closely related term is **collective bargaining,** which focuses on the negotiation for the settlement of the terms of a collective agreement between an employer and a union. Unions are challenging the ability of managers to direct and control the various functions of HRM. For example, union seniority provisions may influence who is selected for job promotions or training programs. Pay rates may be determined by union negotiations, or employee performance appraisal methods may have to be changed to meet union guidelines. Labor relations, therefore, is a specialized function of HRM to which employers must give consideration.

Labor relations: Interaction between management and labor union.

Collective bargaining: Negotiation between management and the union on terms of the collective agreement between them.

Historically, unions have had their greatest success in the manufacturing, mining, transportation, and construction segments of the economy. With the decline in these sectors and the relative growth in the service industries, unions have begun to focus organizational efforts on service employees. Retailing establishments and restaurants, once believed to be immune to unionization because of their largely unskilled labor force, have yielded large numbers of new union members. Legislation in the 1970s removed the previous exemption of nonprofit healthcare institutions from unionization of employees. These developments indicate that managers of foodservice operations must be knowledgeable about labor legislation, collective bargaining, and other aspects of labor-management relations.

Reasons for Joining Unions

Employees often consider union membership because of job, social, and/or political reasons. Dissatisfaction with management because of compensation, job security, or management attitude are the most common reasons employees band together and join a union (Mondy, Noe, & Premeaux, 2002). The power of employees to bargain with their employer on an individual basis and to protect themselves from arbitrary and unfair treatment is limited. Research has indicated that employees join unions because of economic needs and general dissatisfaction with management policies and as a way to fulfill social needs. As a result, many employees find that bargaining collectively with their employer through a union is advantageous. Moreover, this method of bargaining is protected by legislation. When employees are unionized, HRM is subject to negotiation with the union and to the terms of the agreement. Unions often provide a social outlet for employees and can offer opportunities for leadership. Some employees join unions because they work in *union shops,* which require employees to join a union as a condition of employment in states where they are permitted.

Structure and Functions of Unions

Modern unionization has its roots in the guilds of artisans of the Middle Ages. For centuries, people with similar work skills have formed organizations to govern entry to an occupation, define standards of occupational conduct, and regulate employment. In the United States, the first union activity dates from the late 1700s.

In 1955, the American Federation of Labor (AFL), composed primarily of craft unions of, for example, carpenters and plumbers, and the Congress of Industrial Organizations (CIO), made up of industrial unions, merged to form the **AFL-CIO.** Fifty-two national and international unions belong to this federation. The AFL-CIO serves its members by (Sherman, Snell, & Bohlander, 1997)

- Lobbying before legislative bodies on subjects of interest to labor
- Coordinating organizing efforts among its affiliated unions
- Publicizing concerns and benefits of unionization to the public
- Resolving disputes between different unions as they occur

The major difference between national and international unions is that the international union organizes employees and charters local unions in foreign countries. The national union provides technical assistance in negotiating and administering labor contracts, financial assistance during strikes, administration of union-sponsored pension plans and other fringe benefits, training programs for local union officers, and publications. In addition, national unions are generally active in lobbying efforts in Congress and at the state level. Finally, the national union also establishes the rules and conditions under which the local unions may be chartered.

Union steward (shop steward): Union employee who is elected to represent other union members in their relations with an immediate supervisor or other managers.

Officers of local unions are responsible for negotiating local labor agreements and investigating member grievances. The **union steward,** or **shop steward,** represents union members in their relations with an immediate supervisor or other managers. Stewards usually are elected by union members in their department and serve without pay. When stewards represent members at grievance meetings on organizational time, their lost earnings are paid by the union. The **business representative** is hired by the local union to manage the union and also to settle a member's grievance if the steward was not successful.

Several unions exist for foodservice employee. Among the largest national unions are

- Hotel Employee and Restaurant Employees (H.E.R.E) Union
- United Food and Commercial Workers Union
- Service Employees International Union

Union Formation and Growth

Formation of an employee union begins with a determination of whether there is a sufficient number of employees interested in forming a bargaining unit. This interest is determined by the number of employees who sign an *authorization card*. By law, at least 30% of employees in a work group would need to sign an authorization card for an election to be held. Often, union organizers would like at least 50% of employees to sign authorization cards to indicate sufficient interest in the union.

When sufficient interest has been shown through signing of authorization cards, the union can petition the National Labor Relations Board (NLRB) for an election to be held. Typically once an election has been ordered, intense campaigning occurs by both union supporters and company management detailing perceived strengths and concerns about union membership. The NLRB sets the date for the election, monitors the secret ballot process, reports the election results, and certifies the union, if a majority of employees vote for such representation.

The long-term strength of unions depends in part on having sufficient numbers of members to give a large and united voice. Unions use a variety of strategies to help increase their membership numbers, including political involvement, union salting, and use of union organizers (Mondy, Noe, & Premeaux, 2002). Political involvement efforts have focused on support of candidates, education of candidates on labor issues, and formation of strategic alliances with other groups with similar beliefs. **Union salting** is a process where union organizers apply for jobs in a nonunionized operation. Once hired, the organizers work to unionize the employees of that company (Iverson, 2001).

Union salting: Union organizers who apply for positions in non-unionized companies and when hired, work to form a union of the employees.

Government Regulation

In the private sector, four major federal laws regulate employer and union conduct in labor-management relations: the Norris-LaGuardia, Wagner, Taft-Hartley, and Landrum-Griffin acts. In the public sector, state and local government employees are covered by state laws, and federal civil service workers are covered by the Civil Service Reform Act of 1978.

Norris-LaGuardia Act

The Norris-LaGuardia Act of 1932, also known as the Anti-Injunction Act, severely restricted the ability of employers to obtain a federal injunction forbidding a union from engaging in picketing or strike activities. The Norris-LaGuardia Act also nullified yellow-dog contracts, which were agreements that required workers to state they were not union members and promise not to join one.

Wagner Act

The 1935 Wagner Act, formally called the National Labor Relations Act, has had the most significant impact on labor-management relations of any legislation. The act placed the protective power of the federal government behind employee efforts to organize and bargain collectively through representatives of their choice.

The Wagner Act established the right of a union to be the exclusive bargaining agent for all workers in a bargaining unit. A **bargaining unit** is a group of jobs in a firm, plant, or industry with sufficient commonality to constitute an entity that can be represented in union negotiations by a particular agent.

Bargaining unit: Group of jobs represented in the union contract negotiations.

The law declared the following to be unfair labor practices:

- management support of a company union
- discharge or discipline of workers for union activities
- discrimination against workers making complaints to the **National Labor Relations Board (NLRB)**
- refusal to bargain with employee representatives
- interference with the rights of employees to act together for mutual aid or protection

The primary duties identified for the NLRB were to hold secret ballot elections to determine representation and to interpret and apply the law concerning unfair labor practices. The courts may review decisions on unfair labor practices, but the NLRB's decisions on representation elections are final.

Taft-Hartley Act

The Taft-Hartley Act was passed in 1947 over President Truman's veto. The law amends the Wagner Act, although most of the major provisions of the earlier law were retained. The major thrust of the legislation was to balance the powers of labor and management. Before passage of the Wagner Act, employees had little power to organize and bargain, and therefore the early labor legislation restricted only employer activity. Because the bargaining power of unions increased significantly following the passage of the Wagner Act, restraints on union practices were considered necessary.

The following activities were defined as unfair union practices in the Taft-Hartley Act:

- Restraining or coercing employers in the selection of parties to bargain on their behalf
- Persuading employers to discriminate against any employees
- Refusing to bargain collectively
- Participating in secondary boycotts and jurisdictional disputes

- Attempting to force recognition from an employer when another union is already the representative
- Charging excessive initiation fees
- Practicing "featherbedding" or requiring payment of wages for services not performed

In 1974, Congress extended coverage of Taft-Hartley to private nonprofit hospitals and nursing homes. Provisions require unions representing hospital employees to give 90 days' notice before terminating a labor agreement—30 days more than in other industries. Also, a labor dispute in a healthcare facility is automatically subject to mediation efforts of the Federal Mediation and Conciliation Service (Sloane & Witney, 1996).

Landrum-Griffin Act

The Landrum-Griffin Act, also known as the Labor-Management Reporting and Disclosure Act, was passed in 1959 because the provisions in the Taft-Hartley Act did not cover labor racketeering. These investigations revealed that a few labor organizations and employers were denying employees' rights to representation and due process within their labor organizations.

The act requires that labor organizations hold periodic elections for officers, that members be entitled to due process both within and outside the union, that copies of labor agreements be made available to covered employees, and that financial dealings between union officials and companies be disclosed to the U.S. Department of Labor. The law tightens the Taft-Hartley Act restrictions against secondary boycotts. A secondary boycott occurs when a union asks firms or other unions to cease doing business with someone handling a product affected by a strike. The union can ask only that the product not be used.

Contract Negotiations

After a union wins recognition as the employees' bargaining agent, negotiations begin on a contract with management. Bargaining is a difficult and sensitive proceeding.

In local negotiations, the union side is represented by its local negotiating committee and may include a field representative of the national union with which it is affiliated. The management team may be made up of organizational staff, financial or operations managers, and perhaps an attorney. Negotiations at the national level are led by top-level officials from the national union and by top-level corporate managers and staff.

Major bargaining issues usually fall into five areas:

- *Economic issues* deal with such provisions as base pay, shift differentials, overtime pay, length of service increases, and cost of living allowances and with benefits such as pension plans, insurance, holidays, and vacations.
- *Job security* means an entitlement to work or, in lieu of work, to income protection. Procedures for handling layoffs, promotions, transfers, assignments to specific jobs, unemployment benefits, severance pay, and call-in pay are among the job security provisions generally included in union contracts.
- *Working conditions issues* include work rules, relief periods, work schedules, and health and safety.

- *Management rights issues* detail the rights of management to give direction and discipline employees.
- *Individual rights issues* concern establishment of grievance procedures for employees.

Careful preparation for negotiations is critical and should include planning the strategy and assembling data to support bargaining proposals or positions. The conditions under which negotiations take place vary widely among organizations and are dependent on the goals that each side seeks to achieve and the strength of their relative positions. First-time negotiations are usually more difficult than subsequent negotiating sessions after a union has become established. The negotiation of an agreement often takes on the characteristics of a game in which each side attempts to determine its opponent's position while not revealing its own.

Proposals or positions of each side generally are divided into those that it feels it must achieve, those on which it will compromise, and those submitted primarily for trading purposes. Throughout the negotiations, all proposals or positions submitted by both sides must be disposed of if agreement is to be reached; that is, each proposal or position must be accepted by the other side, either in its entirety or in some compromised form, or it must be withdrawn. To achieve its bargaining proposals or positions, each side presents arguments and evidence necessary to support it and exerts pressure on the other side. The outcome of this power struggle determines which side will make the greater concession to avoid a bargaining deadlock.

The power of the union may be exercised by striking, picketing, or boycotting the employer's product. An employer who is struck by a union may continue to operate or opt for a *lockout,* which entails shutting down operations. If a strike or lockout occurs, both parties are affected. Therefore, both participants usually are anxious to achieve a settlement.

When the two parties are unable to resolve a deadlock, a third party, either a mediator or an arbitrator, may be called in to provide assistance. A **mediator** attempts to establish a channel of communication between the union and management but has no power to force a settlement. In mediation, compromise solutions are only offered by the third party. If the contract calls for binding arbitration, an **arbitrator** is called in to render a decision that is binding on both the union and the employer. After an agreement has been reached through collective bargaining, the written document must be signed by representatives of both parties.

PRODUCTIVITY IMPROVEMENT

Productivity improvement is a phrase widely used in newspapers, on television, in news magazines, and in the trade and professional literature of almost all fields. Productivity concerns the efficient use of human, equipment, and financial resources and often is expressed mathematically as the ratio of output to input (Olson & Meyer, 1987). Productivity measures, principles of work design, work measurement, and the use of quality improvement teams have helped improve the level of productivity in a foodservice organization.

Productivity Measures

Measurement of productivity is important for foodservice managers. Productivity can be evaluated by means of the ratio of one or more inputs compared with a variety of outputs. Formulas for frequently used measures of labor productivity are shown in Figure 12.18. Recordkeeping systems should be established to record data on a systematic basis to determine the productivity measures that the management of the foodservice organization has selected for analysis. These measures can be used to examine trends over time within a particular operation, to compare various operations within an organization, or to compare the results from a specific foodservice operation with available industry data.

Productivity is the ratio of output to input, or the ratio of goals to resources of the foodservice system. Productivity can be increased by reducing input, by increasing output, or by doing both at the same time. High turnover, rising costs, and the relatively high percentage of the revenue dollar devoted to labor indicate that foodservice managers should focus their efforts on increasing productivity.

According to Konz and Johnson (1995), labor inefficiency may be the result of several factors:

- Poor product design
- Work methods
- Management
- Workers
- Material waste
- Improper tools
- Inadequate maintenance
- Poor production scheduling
- Absences without cause
- Carelessness

Work Measurement in Foodservice

Work measurement is a method of establishing an equitable relationship between the amount of work performed and the human input used to do that work. Several productivity measures were outlined previously in this chapter. In any production operation, work measurement is necessary for effective use of human resources.

Data from work measurement studies can aid in evaluating alternative production/service subsystems, determining and controlling cost, staffing, scheduling work, deciding whether to make or buy, planning facilities and layout, identifying needs for changes in employee assignments, and timing or sequencing tasks. David (1978) indicated that work measurement data also are needed for developing useful managerial aids, such as production time standards. Activity analysis, activity or occurrence sampling, elemental standard data, and predetermined motion time are the primary techniques of work measurement for analysis in foodservice operations.

Activity analysis involves continuous observation for a chronological record of the nature of activities performed by individual workers, work performed at one workstation, work units produced, or the amount of time that equipment is used and for what purpose. Data are used to establish standards for short- or long-cycle

$$\text{Meals/labor hour} = \frac{\text{total meals served/day}}{\text{labor hours/day}}$$

$$\text{Minutes/meal} = \frac{\text{labor minutes/day}}{\text{total meals served/day}}$$

Payroll cost/day = Σ of hourly rate of each employee $\times$ hours worked by each employee

$$\text{Payroll cost/meal served} = \frac{\text{total daily payroll cost}}{\text{meals served/day}}$$

Labor cost/day = total payroll cost/day + total of all other direct labor costs (fringe benefits, etc.)/day

$$\text{Labor cost/meal served} = \frac{\text{total labor cost/day}}{\text{meals served/day}}$$

Figure 12.18. Selected productivity measures.

work by persons in motion. A simplified technique has been developed that involves employee recording of activities at periodic intervals, usually between 5 and 15 minutes. This technique has been referred to as an employee time log reporting system (DenHartog, Carlson, & Romisher, 1978). Figure 12.19 shows a time sheet and the codes for reporting foodservice activities on an *employee time log* in a school foodservice operation. Employees are asked to enter data on the form every 10 minutes during each day of a time study, according to the function and project from the code sheet. Data permit an analysis of time devoted to various operations within the school foodservice unit.

Occurrence sampling, or activity sampling, are terms used in the literature to describe a method for measuring working time and nonworking time of people employed in direct and indirect activities, and to measure operating time and downtime of equipment. The term *work sampling* also may be used to refer to this technique. Konz and Johnson (1999) stated that occurrence sampling is a more accurate term than work sampling, however, because what is being sampled is the occurrence of the various events that may invoke direct work, indirect work, and delays. By means of intermittent, randomly spaced, instantaneous observations, estimates can be made of the proportion of time spent in a given activity over a specified time.

Occurrence sampling has been widely used for studying work in foodservice operations. Several of these studies are cited in the reference list at the end of this chapter. Much of the classic work has been done at the University of Wisconsin, where the *Methodology Manual for Work Sampling: Productivity of Dietary Personnel* (University of Wisconsin–Madison, 1967) was developed. In the manual, methods for conducting activity sampling studies in foodservice operations are described. Work functions and classifications, including direct work, indirect work, and delays, are defined in Figure 12.20.

TIME REPORTING SHEET FOR THE LINCOLN PUBLIC SCHOOLS' FOODSERVICE STUDY

LINCOLN PUBLIC SCHOOLS
DIVISION OF BUSINESS AFFAIRS
TIME MANAGEMENT REPORTING FORM

____
DEPARTMENT

• EMPLOYEE NAME WORK WEEK

Day	Unit	Function	Project	Day	Unit	Function	Project	Day	Unit	Function	Project

CODES FOR REPORTING FOODSERVICE ACTIVITIES: EMPLOYEES, FUNCTIONS, AND PROJECTS

Coding for Foodservices Work Analysis

Code/Function		Code/Project	
1	Washing	1	Salad Fruit/Veg
2	Chopping	2	Fruit Can/Fresh
3	Dipping	3	Dessert
4	Stirring	4	Vegetables
5	Mixing	5	Entree
6	Panning	6	Juice
7	Shaping	7	Milk
8	Weighing	8	Breads
9	Buttering	9	Restroom
10	Cutting	10	Dishes
11	Sorting	11	Trays
12	Opening	12	Silverware
13	Peeling	13	Carts
14	Supplying	14	Tables
15	Cleaning	15	Refrig./Freezers
16	Scraping	16	Pots & Pans
17	Shredding	17	Steam Tab/Count
18	Pouring	18	Stove
19	Planning	19	Ovens
20	Ordering	20	Mixer
21	Counting	21	Work Schedules
22	Wrapping	22	Qualify Planner
23	Accounting	23	Deposit Slip
24	Deliming	24	Floors
25	Putting Away	25	Steam Equipment
		26	Dish Machine
		27	Dish Ally
		28	Food Order
		29	Nonfood Order
		30	Clothes

Figure 12.19. Form and codes for employee time log system.
Source: "Employee Logs as a Basis for Time Analysis" by R. DenHartog, H. Carlson, and J. M. Romisher, 1978, *School Food Service Research Review, 2*(2), p. 99. Used by permission.

Direct Work Functions

Any essential activity contributing directly to the production of the end product (end product is total number of meals served per day).

- **Processing**
  Act of changing the appearance of a foodstuff by physical or chemical means.
  - **Preparation or preliminary processing**
    Preliminary act or process of making ready for preparation, distribution, or service.
  - **Preparation or cooking**
    Final act or process of making ready for distribution or service.
- **Service**
  Act of preparing facilities for distribution and of portioning and assembling prepared food for distribution to patients and to cafeteria customers (to coffee shop also if dietary is responsible for operation of coffee shop).
- **Transportation**
  Act of transporting food, supplies, or equipment from a location in one functional area to a designated location in another area within the department or to patients' wards.
  - **Transportation of food**
    Act of moving food from a location in one functional area to a designated location in another area within the department.
  - **Transportation of equipment, supplies, and other**
    Act of moving equipment, supplies, and other items from a location in one functional area to a designated location in another area within the department.
  - **Delivery of trays to patients** (if this function is performed by dietary)
    Act of removing patients' trays from food trucks, dumbwaiter or trayveyor, and carrying to patients' bedside.
  - **Return of trays from patients** (if this function is performed by dietary)
    Act of removing trays from patients' bedside to food trucks; dumbwaiter on the ward.
  - **Transportation empty**
    Act of moving without carrying or guiding anything from a location in one functional area to a designated location in another area within the department.
- **Clerical (routine)**
  Act of receiving, compiling, distributing, and storing of routine records of data and information necessary for operation of the department.
- **Cleaning**
  Act of removing soil or dirt to provide sanitary conditions for the use of equipment, facilities, and supplies.
  - **Pot and pan washing**
    Act of scraping, washing, or rinsing quantity food containers and cooking utensils.

Figure 12.20a. Work function classification and definitions.
Source: Adapted from *Methodology Manual for Work Sampling, Productivity of Dietary Personnel* by the University of Wisconsin–Madison, 1967.

- **Dishwashing**
  Act of preparing for or removal of soil or dirt to provide sanitary conditions for use of tableware (china, silverware, glassware, and trays).
- **Housekeeping**
  Act of removing soil or dirt to provide sanitary conditions for the use of installed and mobile equipment and facilities.

- **Receiving**
  Act of acquiring, inspecting, and storing food and/or supplies from an area outside the department.

Indirect Work Functions

Any catalytic activity which contributes to production of the end product.

- **Instruction or teaching**
  Act of directing or receiving direction by oral or written communication in a training or classroom situation or on the job.
- **Appraisal**
  Act of judging or estimating the value or amount of work in order to make decisions for future planning.
- **Conference**
  Act or oral communication with one or more persons in the form of a scheduled meeting.
- **Clerical (original or nondelegable)**
  Act of compiling and formulating management control records of data and information necessary for the operation of the department.

Delays

All time when an employee is scheduled to be working and is not engaged in either a direct or an indirect work function.

- **Forced delay**
  The time an employee is not working due to an interruption beyond his or her control in the performance of a direct or an indirect work function.
- **Personal and idle delays**
  The time an employee is not working due to personal delays or avoidable delays.
  - **Personal delays**
    The time an employee is not working due to time permitted away from the work area.
  - **Idle time**
    Any avoidable delay (other than forced or personal delay) that occurs for which the employee is responsible.

Figure 12.20b. Continued.

The number of observations required in occurrence sampling depends on the type of study, the type of operation, and the number of personnel. Data from an occurrence sampling study are used to calculate labor minutes per meal equivalent or labor minutes for some other specific activity. For example, Block, Roach, and Konz (1985) used occurrence sampling to study cleaning times for vegetables in a university residence hall foodservice, and Oyarzan and colleagues (2000) used it to study time spent by diet clerks. One advantage of occurrence sampling is that several workers in a specific area can be studied simultaneously by a single observer.

Elemental standard data are time values that have been determined for many elements and motions common to a wide variety of work (David, 1978). From these values, total times for specific tasks can be synthesized. David (1978) stated that job variables significantly affecting normal time for a given type of operation must first be hypothesized; then time data collected on the number and variety of jobs in that operation. Data are used to determine the relationship between normal time and each of the variables believed to affect normal time significantly.

Predetermined motion time includes techniques in which tasks are broken down into basic motions with known normal time values (David, 1978). The purpose is to establish cycle time for a specific operation without actually performing the task. Instead, the predetermined time for the basic motions that make up the cycle are synthesized. One technique, *Methods Time Measurement (MTM),* is widely used in industry but is time-consuming; David (1978) concluded that MTM is usually not applicable to long-cycle work or work with limited repetition, such as that in foodservice operations. An alternative technique has been developed, called **master standard data (MSD),** in which seven basic elements of work are combined into larger, more condensed elements.

Montag, McKinley, and Kleinschmidt (1964) were among the first to apply MSD to foodservice operations. They concluded that the method was applicable for developing coded standard elements with universal application in foodservice operations. Several studies listed in the references for this chapter cite other studies that used MSD for examining production times in foodservice facilities. One of the studies (Ridley, Matthews, & McProud, 1984) used MSD to develop labor times for the assembling and microwave heating of menu items in a hospital galley. They also found that the technique could be used effectively for developing standard labor times because data from their study indicated that total labor time under actual conditions in a hospital galley was similar to MSD-predicted time. However, use of these work measurement techniques is limited in the foodservice industry. The amount of time and expertise needed to interpret these data offer are not available.

CHAPTER SUMMARY

This summary is organized by the learning objectives.

1. Human resources management is an integrative process that includes recruiting, selecting, orienting, training, and maintaining the workforce needed to achieve an organization's goals. Recruiting is the process of locating and encouraging potential applicants to apply for a job opening. Selection is the process of comparing applicants' knowledge skills, and abilities to those required of a position and choosing the applicant most qualified. Orientation is a formal process of

Professional Organization Profile

American Correctional Food Service Association

MISSION

The mission of the American Correctional Food Service Association (ACFSA) is to develop and promote cooperative programs and activities that will improve the level of professionalism, provide a vehicle for broadening the knowledge within the correctional food service segment of the food service industry, and efficiently manage a quality association within a budget that returns maximum benefits for all members.

WHO BELONGS TO THE ORGANIZATION

Members include nearly 1,000 innovative foodservice professionals that work in or with a correctional institution. These professionals include Jail or Prison Administrators, Dietitians, Food Service Directors, Managers, Supervisors, Dietary Officers, Bakers, Cooks, and Line Staff. Our Membership also includes Professional Partners who are the Food and Equipment Suppliers that sell to the Correctional Institutions. Student memberships are not available.

ADVANTAGES OF MEMBERSHIP

Membership benefits include opportunities for networking and sharing information with the other members and professional partners at state, regional, or the annual International Exposition and Training Conference and Vendor Trade Show; scholarships for conference travel; quarterly magazine, *INSIDER;* and educational programs for continuing education and certification credit.

WEBSITE

www.ACFSA.org

MEET THE PRESIDENT

Ellen White, president 2005–2006, is the Correctional Dietary Supervisor at the Pre-Release Center for the Montgomery County Department of Corrections and Rehabilitation. Her 30+ year foodservice career began as a Dietary Aide in a nursing home while in high school where she was promoted to Cook and then Assistant Manager.

familiarizing new employees to the organization, job, and work unit. Training is the ongoing process of updating skills of an employee. Performance appraisal is the comparison of an individual's performance with established standards.
2. Many laws have been passed that impact the management of human resources in an organization. Laws such as the Equal Pay Act, Title VII and the Civil Rights Act, the age and pregnancy discrimination acts, the Equal Employment Opportunity Act, and the Americans with Disabilities Act all impact the selection and hiring process. Other acts such as the Family and Medical Leave Act and the Health Insurance Portability and Accountability Act influence and organization's benefits programs. Labor relations are covered by the Norris-LaGuardia Act, Wagner Act, Taft-Hartley Act, and the Landrum-Griffin Act.
3. There are many terms that are unique to labor relations. Collective bargaining is the negotiation that occurs between management and the union regarding the collective agreement that will govern their relationship. The union steward is a union employee who is elected to represent employees in their relationship with management. Mediation is the process of having an objective third party offer a nonbinding solution to a situation. Arbitration is the use of an objective third party whose solution is binding and must be implemented.
4. Productivity is the outputs achieved using given inputs. Productivity often can be improved by changing work methods, improving scheduling, and proper training of employees.

TEST YOUR KNOWLEDGE

1. What are the steps to make human resources planning a success in the foodservice industry?
2. How can a foodservice manager control labor costs through selection, orientation, training, and performance appraisal?
3. What variables in the organization will affect staffing and scheduling?
4. What is the difference between part-time and temporary workers?
5. How can productivity be improved?
6. What is labor relations? How does it relate to human resources management?

CLASS PROJECTS

1. Form groups of three or four students. Have each student describe the recruitment and selection process for a position he or she was hired for. Make a chart detailing similarities and differences in the process each experienced.
2. Invite a human resources manager from a local foodservice operation to class to discuss current issues in human resources management in his or her operation.
3. Obtain a copy of a union contract for a local foodservice operation. Identify the types of practices covered by the contract.

WEB SOURCES

http://stats.bls.gov	Bureau of Labor Statistics
www.hr-guide.com	Human Resources Management Website
www.eeoc.gov	Equal Employment Opportunity Commission
http://online.onetcenter.org	O*NET, Occupational Information Network
www.dol.gov	U.S. Department of Labor
www.nlrb.gov	National Labor Relations Board

BIBLIOGRAPHY

Aaron, T., & Dry, E. (1992). Sexual harassment in the hospitality industry. *Cornell Hotel and Restaurant Administration Quarterly, 33*(2), 92–95.

ADA employment rules to take effect July 26 for many employers. (1993). *Washington Weekly, 12*(20), 1, 4.

Batty, J. (1993). Preventing sexual harassment in the restaurant. *Restaurants USA, 13*(1), 30–34.

Beasley, M. A. (1991). The story behind quality improvement. *Food Management, 26*(5), 52, 56, 58, 60.

Beasley, M. A. (1993). Better performance appraisals. *Food Management, 28*(9), 38.

Blackburn, R., & Rosen, B. (1993). Total quality and human resources management: Lessons learned from Baldrige Award-winning companies. *Academy of Management Executive, 7*(3), 49–66.

Block, A. A., Roach, F. R., & Konz, S. A. (1985). Occurrence sampling in a residence hall foodservice: Cleaning times for selected vegetables. *Journal of the American Dietetic Association, 85*(2), 206–209.

Boyle, K. (1987). Effective employee discipline requires keeping in close touch. *Restaurants USA, 7*(10), 26–28.

Byers, B. Shanklin, C. W., Hoover, L. C. & Puckett, R. (1994). *Food service manual for health care instituitions.* Chicago, IL: American Hospital Publishing.

Chase, R. B. & Aquilano, N. J. (1992). *Production and operations management: A life-cycle approach* (4th ed). Homewood, IL: Richard D. Irwin.

Cheney, K. (1993). Avoid the Zoë Baird trap. *Restaurants & Institutions, 103*(11), 153–154.

Conrad, P. J., & Maddux, R. B. (1997). *Guide to affirmative action: A primer for supervisors and managers.* Los Altos, CA: Crisp Publications.

Cross, E. W. (1992). AIDS: Legal implications for managers. *Journal of the American Dietetic Association, 92,* 74–77.

Cross, E. W. (1993). Implementing the Americans with Disabilities Act. *Journal of the American Dietetic Association, 93,* 272–275.

David, B. D. (1978). Work measurement in food service operations. *School Food Service Research Review, 2*(1), 5–11.

Davis, H. A. (1998). Sex and the workplace: Part II. *Food Management, 33*(8), 82, 84, 86.

Dear National Restaurant Association member . . . (Issue on the Family and Medical Leave Act). (1993). *Washington Weekly, 13*(28), 1–12.

Deinhart, J. R., & Gregoire, M. B. (1993). Job satisfaction, job involvement, job security, and customer focus of quick-service restaurant employees. *Hospitality Research Journal, 16*(2), 29–43.

DenHartog, R., Carlson, H., & Romisher, J. M. (1978). Employee logs as a basis for time analysis. *School Food Service Research Review, 2*(2), 98–101.

Department of Labor (1999). The Family and Medical Leave Act of 1993. [Online] Available: http://www.dol.gov.

Doherty, R. F. (1989). *Industrial and labor relations terms: A glossary* (5th ed.). ILR Bulletin No. 44. Ithaca, NY: Cornell University Press.
Doyle, F. (1992). Hiring your way to a great staff. *Restaurants USA, 12*(6), 20–22.
Drucker, P. F. (1993). *Managing for the future: The 1990s and beyond.* New York: Truman Talley Books/Dalton.
Employee focus groups criticize immigration paperwork. (1993). *Washington Weekly, 13*(34), 4.
Employers should brush up on teen-labor laws for summer hire. (1993). *Washington Weekly, 13*(25), 4.
Foodservice accounts for about 11% of ADA charges at Justice Department. (1993). *Washington Weekly, 13*(4), 1.
Forrest, L. C. (1996). *Training for the hospitality industry* (2nd ed.). East Lansing, MI: Educational Institute of the American Hotel and Motel Association.
Freedman, D. H. (1992). Is management still a science? *Harvard Business Review, 70*(6), 26–28, 30, 32–38.
Gómez-Mejía, L. R., Balkin, D. B., & Cardy, R. L. (2001). *Managing human resources* (3rd ed.). Upper Saddle River, NJ: Prentice Hall.
Goodgame, D. (1993). Ready to operate. *Time, 142*(12), 54–58.
Griffin, R. W. (2001). *Management* (7th ed.). Boston: Houghton Mifflin.
Hackman, J. R. (1977). Work design. In J. R. Hackman and J. L. Sattle (Eds.), *Improving life at work: Behavioral sciences approaches to organizational change* (pp 86–162). Glenview, IL: Scott, Foresman.
Hamilton, A. J., & Veglahn, P. A. (1992). Sexual harassment: The hostile workplace. *Cornell Hotel and Restaurant Administration Quarterly, 33*(2), 88–92.
Heneman, H. G., Schwab, D. P., Fossum, J. A., & Dyer, L. D. (1989). *Personnel/human resource management* (4th ed.). Homewood, IL: Richard D. Irwin.
Hogan, J. J. (1992). Turnover and what to do about it. *Cornell Hotel and Restaurant Administration Quarterly, 33*(1), 40–45.
Industry at a Glance. (2002). Available online at www.restaurant.org
Ingalsbe, N., & Spears, M. C. (1979). Development of an instrument to evaluate critical incident performance. *Journal of the American Dietetic Association, 74*(2), 134–140.
Iverson, K. M. (2001). *Managing human resources in the hospitality industry.* Upper Saddle River, NJ: Prentice Hall.
Iwamuro, R. (1992). Disabled workers in foodservice. *Restaurants USA, 11*(6), 36–38.
Iwamuro, R. (1993). Foodservice employee profile. *Restaurants USA, 13*(6), 37–41.
Jenks, J. M., & Zevnik, B. L. P. (1989). ABCs of job interviewing. *Harvard Business Review, 67*(4), 38–39, 42.
Jesseph, S. A. (1989). Employee termination. II: Some dos and don'ts. *Personnel, 66*(2), 36–38.
Kansas Department of Administration: Division of Personnel Services. (1999). About the State Affirmative Action Plan for Equal Employment Opportunity. [Online]. Available: http://da.state.ks.us/ps/subject/eeoc.htm.
Kazarian, E. A. (1979). *Work analysis and design for hotels, restaurants and institutions* (2nd ed.). Westport, CT: AVI.
Keiser, J. R., DeMicco, F., & Grimes, R. N., (2000). *Contempory management theory: Controlling and analyzing costs in food service operations* (4th ed.). Upper Saddle Review, NJ: Prentice Hall.
Kohl, J. P., & Greenlaw, P. S. (1992). The ADA, part II: Implications for managers. *Labor Law Journal, 42,* 52–56.
Konz, S., & Johnson, S. (1999). *Work design: Industrial ergonomics* (5th ed.). Scottsdale, AZ: Holcomb Hathaway.
Koral, A. M. (1994). *Conducting the lawful employment interview: How to avoid charges of discrimination when interviewing job candidates* (4th ed.). New York: Executive Enterprises Publications.

Krone, C., Tabacchi, M., & Farber, B. (1989b). Manager burnout. *Cornell Hotel and Restaurant Administration Quarterly, 30*(3), 58–63.

Lorenzini, B. (1992). The accessible restaurant. Part II: Employee accommodation. *Restaurants & Institutions, 102*(12), 150–151, 154, 158, 162, 166, 168, 170.

Lundberg, D. E., & Armatas, J. P. (1982). *The management of people in hotels and restaurants* (5th ed.). New York: Richard Irwin.

Marvin, B. (1992). How to hire the right people. *Restaurants & Institutions, 102*(14), 60–61, 68, 73, 75, 77.

Mayo, C. R., & Olsen, M. D. (1987). Food servings per labor hour: An alternative productivity measure. *School Food Service Research Review, 11*(1), 48–51.

McCool, A. C., & Stevens, G. E. (1989). Older workers: Can they be motivated to seek employment in the hospitality industry? A preliminary report: Factors influencing older persons to retire or to work beyond the usual retirement age. *Hospitality Education and Research Journal, 13*(3), 569–572.

McLauchlin, A. (1992). Take note: ADA's employment rules take effect in July. *Restaurants USA, 12*(5), 9–10.

Mondy, R. W., Noe, R. M., Premeaux, S. R. (2002). *Human resource management* (8th ed.). Upper Saddle River, NJ: Prentice Hall.

Montag, G. M., McKinley, M. M., & Kleinschmidt, A. C. (1964). Predetermined motion times: A tool in food production management. *Journal of the American Dietetic Association, 45,* 206–211.

Muczyk, J. P., & Reimaon, B. C. (1989). MBO as a complement to effective leadership. *Academy of Management Executive, 3*(2), 131–138.

National Restaurant Association. (1987). *A primer on how to recruit, hire, and retain employees.* Washington, DC: Author.

National Restaurant Association. (1992). *Americans with Disabilities Act: Answers for foodservice operators.* Washington, DC: Author.

National Restaurant Association. (1998). *Employee profile.* Conducted by the Research Department at National Restaurant Association. Washington, DC: Author.

Ninemeier, J. D. (1999). *Planning and control for food and beverage operations* (4th ed.). East Lansing, MI: Educational Institute of the American Hotel and Motel Association.

Olson, M. D., & Meyer, M. K. (1987). Current perspectives on productivity in food service and suggestions for the future. *School Food Service Research Review, 11*(2), 87–93.

Oyarzan, V. N., Lafferty, L. J., Gregoire, M. B., Sowa, D. C., Dowling, R. A., & Shott, S. (2000). Evaluation of efficiency and effectiveness measurements of a foodservice system that included a spoken menu. *Journal of the American Dietetic Association, 100,* 460–463.

Palmer, J., Kasavana, M. L., & McPherson, R. (1993). Creating a technological circle of service. *Cornell Hotel and Restaurant Administration Quarterly, 34*(1), 81–87.

Panitz, B. (1999). Year of the restaurant. *Restaurants USA, 19*(2), 26–30.

Pesci, P. H., Spears, M. C., & Vaden, A. G. (1982). A method for developing major responsibilities and performance standards for food service personnel. *NACUFS Journal, 4:* 17–21.

Pickworth, J. R. (1987). Minding the Ps and Qs: Linking quality and productivity. *Cornell Hotel and Restaurant Administration Quarterly, 8*(1), 40–45.

Puckett, R. P. (1982). Dietetics: Making or breaking the new employee. *Contemporary Administrator, 5*(10), 14–16.

Quick, R. C. (1989). Employee assistance programs: Beating alcoholism in the dish room and the board room. *Cornell Hotel and Restaurant Administration Quarterly, 29*(4), 63–69.

Ridley, S. J., Matthews, M. E., & McProud, L. M. (1984). Labor time code for assembling and microwave heating menu items in a hospital galley. *Journal of the American Dietetic Association, 84*(6), 648–654.

Ruf, K. L., & Matthews, M. E. (1973). Production time standards. *Hospitals, 47*(9), 82, 84, 86, 88, 89, 90.

Sherer, M. (1990). If you don't have a literacy program READ THIS NOW! *Restaurants & Institutions, 100*(27), 75, 78, 82, 84, 88, 92.

Sherman, A. W., Snell, S., & Bohlander, G. W. (1997). *Managing human resources* (11th ed.). Cincinnati: South-Western Publishing.

Shuit, D,P. (2005). Yum does a 360. *Workforce Management.* [Online]. Available: http://www.workforce.com/section/10/feature/23/99/39/

Sloane, A. A., & Witney, F. (1996). *Labor relations* (9th ed.). Englewood Cliffs, NJ: Prentice Hall.

Smart, B. D. (1990). *The smart interviewer.* New York: Wiley.

Smith, J. H., Jr. (1992). A study of handicapped employment in the hospitality industry. *Hospitality and Tourism Educator, 4*(3), 16–25.

Sneed, J., & Kresse, K. H. (1989). *Understanding foodservice financial management.* Rockville, MD: Aspen Publishers.

Spertzel, J. K. (1992). The Americans with Disabilities Act: Employment doors open a little wider for the disabled. *School Food Service Journal, 46*(4), 36–37.

This, L., & Lippitt, G. (1966). Learning theories and training. *Training and Development Journal, 20*(4), 2.

University of Wisconsin–Madison. (1967). *Methodology manual for work sampling: Productivity of dietary personnel.* Madison: Author.

Ursin, C. (1999). It's your move: stop sexual harassment before it degrades your restaurant. *Restaurant USA, 19*(1), 34–38.

VanDyke, T., & Strick, S. (1988). New concepts to old topics: Employee recruitment, selection and retention. *Hospitality Education and Research Journal, 12*(2), 347–360.

Weinstein, J. (1992a). Here's how smart restaurant companies turn training and development into . . . personnel success. *Restaurants & Institutions, 102*(29), 92–93, 96, 101, 104, 108, 113.

Weinstein, J. (1992b). Initial ADA complaints scarce. *Restaurants & Institutions, 102*(25), 105–108.

What the new family and medical leave law will require. (1993). *Washington Weekly, 13*(7), 2.

Wheelhouse, D. R. (1989). *Managing human resources in the hospitality industry.* East Lansing, MI: Educational Institute of the American Hotel and Motel Association.

Woods, R. H., & Kavanaugh, R. R. (1992). Here comes the ADA—are you ready? *Cornell Hotel and Restaurant Administration Quarterly, 33*(1), 24–32.

Woods, R. H., & Macaulay, J. F. (1987). Exit interviews. How to turn a file filler into a management tool. *Cornell Hotel and Restaurant Administration Quarterly, 28*(3), 38–46.

Zemke, R. (1989a). Employee orientation: A process, not a program. *Training, 26*(8), 33–35, 37–38.

Zemke, R. (1989b). Workplace illiteracy. Shall we overcome? *Training, 26*(6), 33–39.

CHAPTER 13

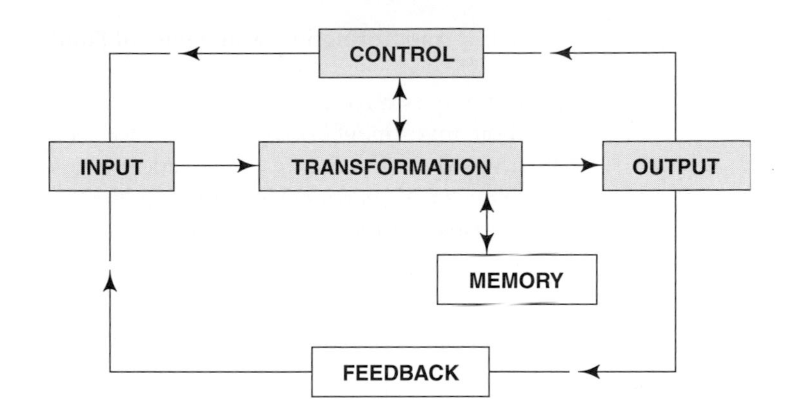

Management of Financial Resources

Enduring Understanding

- Maximizing revenue while minimizing cost is central to financial management.
- The effectiveness of a foodservice manager will be determined in part by his or her success in managing financial resources.

Learning Objectives

After reading and studying this chapter, you should be able to

1. Define selected accounting principles, such as the business entity concept, the going concern concept, cost principle, cost versus accrual bases of accounting, consistency principle, and conservatism.
2. Evaluate information presented in balance sheets and income statements.
3. Differentiate among liquidity, solvency, profitability, and operating ratios.
4. Describe break-even analysis and its component parts, fixed and variable costs.
5. Calculate the payback period and net present value of a proposed equipment purchase.
6. Describe strategies for pricing a menu.

Effective management of financial resources is critical to the success of any foodservice operation. In this chapter, you will learn how managers use accounting and financial management concepts to analyze financial performance. You will be introduced to key reports that help guide a manager's decision making related to use of financial resources.

USERS OF FINANCIAL STATEMENTS

Seven groups of users of financial statements have been identified in profit organizations: owners, boards of directors, managers, creditors, employees, governmental agencies, and financial analysts. Each user group has a different need for financial data. Owners have invested in the business, and their primary concern is the state of their investment. Boards of directors, elected to oversee operations and make business decisions for an organization, use financial statements to determine the effectiveness of managers supervising the daily operation. Managers are concerned with assessing the daily and long-term success of their decisions, and they use financial data to evaluate plans. Creditors, whose concern is the likelihood that payment obligations will be met, are those lending money or goods on credit to the operation. Employees have an interest in financial information to help assess the company's ability to meet wage and benefit demands. Governmental agencies are concerned with financial data as they relate to taxation and regulation. Financial analysts are persons outside the firm who desire information about a firm for their own or a client's purpose.

Not-for-profit organizations, which provide goods or services that fulfill a social need and are not operating for anyone's personal financial gain, have many of the same users of their financial information. In not-for-profit operations, such as hospitals and schools, excess revenues over expenditures are not distributed to those who contributed to support through taxes or voluntary donations but are used to further the purposes of the organization. In tax-supported organizations, the public is viewed as the owner, and its interest is the efficiency of the organization in using public monies. Boards of directors often oversee operations in not-for-profit organizations; they use financial statements to determine the effectiveness of operations. Suppliers are concerned with the ability of an organization to pay bills. Managers use financial statements to provide information for controlling expenditures in relation to available funds. Some not-for-profit organizations are established to provide services to a specific clientele for a fee that closely approximates the cost of providing the service. This clientele can be categorized as an additional user because of its concern about the organization's efficiency to ensure that costs for services are maintained as low as possible.

SYSTEMS APPROACH TO MANAGING FINANCIAL RESOURCES

The management of financial resources involves many components of the foodservice system as shown in Figure 13.1. Money is an input into the foodservice system. Foodservice managers transform that input into financial accountability, an output to the system. Budgets are plans and as such are controls of the system. The storing of financial data occurs as part of the memory function within the system. Preparation of financial statements and the interpretation of these statements are critical responsibilities of the foodservice manager.

KEY ASPECTS OF ACCOUNTING

Auditing is an area of accounting concerned with an independent review of accounting records. An audit involves examination of records that support financial reports and formulation of an opinion regarding the fairness and reliability of these reports.

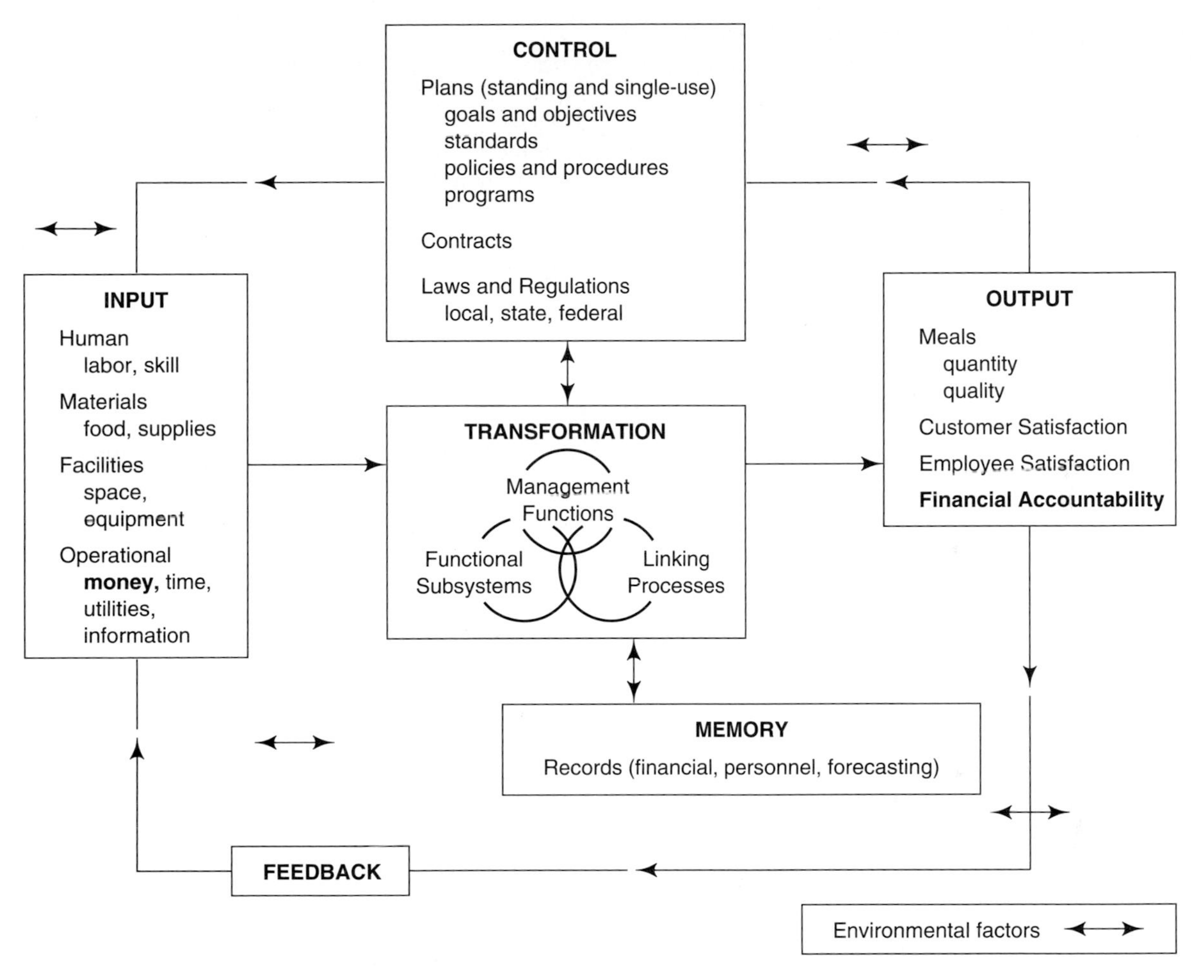

Figure 13.1. Foodservice systems model.
Source: Adapted from *A Model for Evaluating the Foodservice System* by A. G. Vaden, 1980. Manhattan, KS: Kansas State University. © 1980. Used by permission.

Cost accounting involves the determination and control of cost. It focuses on assembling and interpreting cost data for use by management in controlling current operations and planning for the future. Traditionally, the cost of production processes and products has been cost accounting's emphasis, but increasing attention is being given to distribution costs. In a foodservice organization, service costs also are of particular concern.

Financial accounting is concerned with the reporting of transactions for an organization and the periodic preparation of various reports from these records. Income statements and balance sheets are examples of statements prepared by the financial accountant.

Managerial accounting uses historical and estimated financial data to assist management in daily operations and in planning future operations. Identifying the cost of alternative courses of actions is one important function of managerial accounting. Another is budgeting and the reports comparing performance to budget.

SELECTED ACCOUNTING PRINCIPLES

Users of financial statements need a basic understanding of the principles underlying the preparation of these statements for proper interpretation. These generally accepted accounting principles (often referred to as GAAP) help provide consistency to the preparation of financial statements. Organizations that have been involved in the formulation of these principles include the American Institute of Certified Public Accountants, the American Accounting Association, and the Financial Accounting Standards Board.

Business Entity Concept

The business entity concept assumes that a business enterprise is separate from the person or persons who supply its assets, and the financial records of each are distinct. Without this distinction, determining the organization's true performance and current status would be impossible. For example, if individual restaurateurs make personal or family purchases from funds available in the cash register without recording them, they violate good business procedures and also destroy the integrity of the business as separate from the owner's personal affairs. For accounting purposes, only those resources that are set aside for the business and those debts directly related to it are properly included in the accounts.

The Fundamental Equation

In accounting terminology, the resources, debts, and ownership interests of an organization are referred to as its assets, liabilities, and owner's equity, respectively. The relationship among them, "assets equal liabilities plus owner's equity," is known as the fundamental accounting equation. For accounting records to be in balance, each increase in assets must be accompanied by a corresponding decrease in another asset or an increase in liabilities or owner's equity.

Going-Concern Concept

One of accounting's basic assumptions is that an organization will continue to operate for an indefinite time. This is not to imply that a given organization will exist permanently, but simply that it will continue. This concept implies that the value of a company's assets is its ability to generate revenue rather than the value the assets would bring in liquidation.

Money as a Unit of Measure

Money is the basis for business transactions and is the unit of measure commonly referred to as revenues. To lend uniformity to financial data, all business transactions are recorded as dollar amounts. Using this concept, only information that can be stated in monetary terms is included on a company's financial statement. Thus, other information, such as the value of good employees, is not included on financial statements because it cannot be expressed in monetary terms.

Cost Principle

The cost principle involves recording transactions or valuing assets in terms of dollars at the time of the transaction. Cost is the amount measured in dollars expended for goods or services.

Cash versus Accrual Bases of Accounting

Two distinct methods can be used to determine when to record a transaction. The *cash basis* of accounting recognizes a transaction at the time of cash inflow or outflow. The *accrual basis* of accounting, which is used in most organizations, recognizes revenues when earned (regardless of when the actual cash is received) and expenses when incurred (regardless of when cash is dispersed).

Matching Revenues and Expenses

The matching concept involves matching revenues with all applicable expenses during the accounting period in which they occur. For example, a restaurant owner may purchase food in one accounting period and sell it in the following period. If the matching concept is not used, the cost of the food would be recorded in the accounting period prior to when the sale was recorded, thereby overstating cost in the first period and profit in the next. This matching concept is the basis for the accrual basis of accounting.

Depreciation

Depreciation, an aspect of accrual accounting, is a systematic means by which costs associated with the acquisition and installation of a fixed asset are allocated over the estimated useful life of the asset. A number of methods are used for depreciation, one being the straight-line method. Using this method the cost of the asset is depreciated equally over the life of the asset. The formula for calculating the yearly amount of depreciation using the straight-line method is

$$(\text{Cost of the asset} - \text{Salvage value})/\text{Years of useful life}$$

Because some assets frequently decline in value faster during the first few years, accelerated depreciation methods may be used, which give larger amounts of depreciation in the early years and lesser amounts in the later years.

Adequate Disclosure

Financial statements and their accompanying footnotes or other explanatory materials should contain full information on all data believed essential to a reader's understanding of the financial statement. Such disclosures might include accounting methods used, changes in accounting methods, and any unusual or nonrecurring issues pertinent to accurate interpretation of the financial statement.

Consistency Principle

The consistency principle states that once an organization chooses an accounting method, it should be used from one period to another to make financial data comparable. Without consistent methods, financial statements could be interpreted incorrectly. For example, a change in deprecation methods would affect the amount of depreciation expense, and net income from one period to the next would be affected.

Materiality Principle

Absolute accuracy and complete full disclosure may be neither practical nor economically feasible in presenting accounting information. The materiality principle means that events or information must be accounted for if they "make a difference" to the user of financial statements. Generally, an item is considered as material if its inclusion or omission would change or influence the judgment of a reasonable person.

Conservatism

Conservatism refers to the concept of moderation in recording transactions and assigning values. Historically, accountants have tended to be conservative, favoring the method or procedure that yielded the lesser amount of net income or of asset value.

BASIC FINANCIAL STATEMENTS

Primary financial statements used by foodservice managers are the balance sheet and income statement. Each provides unique information necessary for analyzing the effectiveness of operations.

Balance sheet: Statement of assets, liabilities or debts, and capital or owner's equity at a given time or at the end of the accounting period.

Income statement: Financial report that presents the net income or profit of an organization for the accounting period.

The **balance sheet** is a statement of assets, liabilities or debts, and capital or owner's equity at a given time or at the end of the accounting period. The **income statement** is the financial report that presents the net income or profit of an organization for the accounting period. It provides information about the revenues and expenses that resulted in the net income or loss. The balance sheet is considered a static statement because it presents the financial position at a specific date or time. The income statement is considered a flow or dynamic statement because operating results over time are presented.

Standard methods of accounting and presentation of financial statements, termed *uniform systems of accounts,* have been established in a number of industries, including certain segments of the hospitality services industries. These uniform systems of accounts within a particular industry provide for the uniform classification, organization, and presentation of revenues, expenses, assets, liabilities, and equity. They include a standardized format for financial statements, which permits comparability of financial data within an industry. Financial statements can be prepared using computer spreadsheet or customized software programs.

Balance Sheet

The balance sheet, or statement of financial condition, is a list of assets, liabilities, and owner's equity of a business entity at a specific date, usually at the close of the last day of a month, quarter, or year. This statement is designated a balance sheet because it is based on a fundamental equation that shows that assets equal liabilities plus owner's equity. An example of a balance sheet for the Midtown Deli is shown in Figure 13.2.

Midtown Deli
BALANCE SHEET
December 31, 2005

Assets		Liabilities and Owner's Equity	
CURRENT ASSETS		**CURRENT LIABILITIES**	
Cash		Accounts payable	$ 28,585
on hand and in checking	$ 34,156	Accrued expenses	8,590
in savings	5,521	Mortgage payable, current	24,000
Accounts receivable	67,278	Total Current Liabilities	$ 61,175
Inventory	5,330		
Prepaid expenses	5,688		
Total Current Assets	$117,973		
FIXED ASSETS		**LONG-TERM LIABILITIES**	
Land	$375,000	Mortgage payable	$390,047
Building	402,950	Less current portion	(24,000)
Furniture, fixtures, equipment	67,000	Total Liabilities	$366,047
China, glassware, silver	28,856		
Less depreciation	(117,000)		
Total Fixed Assets	$756,806		
OTHER ASSETS	$ 6,083	**OWNER'S EQUITY**	
		Capital stock	$ 47,000
		Additional paid in capital	365,503
		Retained earnings	41,137
		Total Owner's Equity	$453,640
TOTAL ASSETS	$880,862	**TOTAL LIABILITIES & OWNER'S EQUITY**	$880,862

Figure 13.2. Example of a balance sheet.

Assets

Assets: Resources of a company.

The first section of the balance sheet is a list of **assets,** which are generally categorized as current or fixed. The example in Figure 13.2 shows three categories of assets for the Midtown Deli: current, fixed, and other assets. *Current assets* include cash and all assets that will be converted into cash in a short period of time, generally 1 year. The cash accounts are cash on hand and in checking accounts and cash in savings. Other current assets include accounts receivable, inventory, prepaid expenses, and entrance fees receivable. Any marketable securities held by the deli would have been included as a current asset.

Fixed, or *long-term, assets* are those of a permanent nature, most of which are acquired to generate revenues for the business. Fixed assets are not intended for sale and include land, buildings, furniture, fixtures, and equipment, in addition to small equipment such as china, glassware, and silver. Because fixed assets generally lose value over their expected life, their value is reduced by a value called *accumulated depreciation.*

Liabilities

Liabilities: Debts of a company.

Liabilities are categorized as current and long term. *Current liabilities* represent those that must be paid within a period of 1 year, including such items as accounts payable for merchandise, accrued expenses, and annual mortgage payment. *Accrued expenses* are due but not paid at the end of the accounting period, such as salaries, wages, or interest. *Fixed,* or *long-term, liabilities,* in contrast, are obligations that will not be paid within the current year. An example of a long-term liability is a mortgage for building and land; annual mortgage payments due during the current year are current liabilities and reduce the long-term mortgage liability, as shown in Figure 13.2.

Owner's Equity

Owner's equity: Money value of a company in excess of its debts that is held by the owners.

The **owner's equity** or capital section of the balance sheet represents that portion of the business that is the ownership interest, along with earnings retained in the business from operations. In profit-oriented enterprises, the ownership may be one of three kinds: a *proprietorship,* a business owned by a single individual; a *partnership,* a business owned by two or more people; or a *corporation,* a business incorporated under the laws of the state with ownership held by stockholders. In a not-for-profit corporation, the members may be the owners. As shown on the balance sheet from the Midtown Deli (Figure 13.2), the owner's equity section includes the value of the members' capital stock, additional paid-in capital, and the retained earnings.

Income Statement

The income statement (also known as the statement of income or the profit-and-loss statement) is a primary managerial tool reporting the revenues, expenses, and profit or loss as a result of operations for a period of time. An example of an income statement for a commercial operation such as a restaurant is shown in Figure 13.3. On-site operations such as hospitals and schools are more likely to use monthly operating statements as shown in Figure 13.4.

Sales or revenues include the cash receipts or the funds allocated to the operation for the period. For example, in a college residence hall, revenues might include monies paid by residents for meal plans and cash sales to nonresidents who may eat in the foodservice facility.

In a foodservice establishment, the cost of sales section of the income statement reflects the cost of products sold that generated the revenue. Cost of sales often is calculated in the following manner:

Inventory at beginning of period	$ XXX
+ Purchases during the period	+XXX
Total value of available food	$ XXX
− Inventory at end of period	−XXX
Cost of goods sold during period	$ XXX

Gross profit or income is determined by subtracting cost of goods sold from sales or revenue.

North Star Restaurant December 21, 2005		
Income		
Food		$ 884,250
Beverage		379,150
Total income		$1,263,400
Undistributed operating expenses		
Administrative and general	$252,000	
Marketing	69,200	
Property operation and maintenance	108,300	
Energy costs	73,000	502,500
Income before fixed charges		$ 760,900
Fixed charges		
Property taxes	$101,200	
Insurance	23,800	
Interest	84,900	
Depreciation	170,100	380,000
Income before income tax		$ 380,900
Income tax		182,832
Net income		$ 198,068

Figure 13.3. Example of an income statement.

Net profit or loss is determined by subtracting expenses from gross profit. In a not-for-profit organization, net profit often is referred to as excess revenues over expenditures.

TOOLS FOR COMPARISON AND ANALYSIS

The foodservice manager should use a variety of tools to analyze financial data, such as ratio analysis, trend analysis, common-size statements, and break-even analysis. The resulting operational indicators help managers understand financial information and compare performance to earlier periods. The manager should make several kinds of comparisons in reviewing and analyzing financial information. These comparisons can be categorized as internal and external. Managers can complete these comparisons more efficiently when using a computer to perform the calculations and comparisons.

Internal standards of comparisons include a review of current performance in relation to budgeted performance, past performance, or preestablished department standards. Because the budget represents the plan for financial operations for a period of time, comparisons with the budget indicate whether or not operations are proceeding as planned. Comparison of current performance in relation to past performance provides information on changes occurring in the operation over time.

Riverside Regional Medical Center April 2004						
	Patient Foodservice	Employee Cafeteria	Visitor Café	Vending	Catering	TOTAL
Income		$271,442	$45,431	$20,610	$43,991	$381,474
Expense						
Food						
Meat	$30,301	$39,590	$5,088	$1,803	$5,389	$82,171
Dairy	$21,032	$14,570	$1,890	$938	$2,002	$40,432
Grocery	$17,794	$15,569	$2,035	$1,001	$2,156	$38,555
Fresh/Frozen	$22,435	$18,296	$2,326	$721	$1,463	$45,241
Bakery	$14,571	$16,248	$1,894	$936	$1,694	$35,343
Supplements	$11,822	$0	$0	$0	$0	$11,822
Other	$6,940	$8,542	$1,615	$1,804	$1,548	$20,449
Total	$124,895	$112,815	$14,848	$7,203	$14,252	$274,013
Labor						
Salaried						$165,478
Hourly	$175,469	$157,390	20,437	$3,770	$16,314	$373,380
Cleaning supplies	$400	$425	$187	$90	$176	$1,278
Paper	$1,260	$1,629	$405	$250	$598	$4,142
Repairs/Maint	$2,256	$0	$0	$0	$0	$2,256
Office Supplies	$230	$169	$97	$0	$83	$579
China/silver	$382	$258	$0	$0	$0	$640
Books/Mag	$59	$0	$0	$0	$25	$84
Other	$479	$258	$84	$25	$49	$895
Total Expense	$305,430	$272,944	$36,058	$11,338	$31,497	$822,745

Figure 13.4. Example of hospital foodservice department monthly operating statement.

Having preestablished department standards provides objective benchmarks for comparison of actual performance with expected performance.

External standards of comparison include a review of performance in relation to similar operations or comparisons with industry performance. For example, companies such as Solucient and professional organizations such as Health Care Foodservice Management offer healthcare foodservice managers the opportunity to compare these new data with similar facilities from across the country.

Ratio Analysis

Most financial information is presented as a collection of totals or balances of accounts, the meaning or significance of which may not be readily apparent. A **ratio analysis,** or analysis of financial data in terms of relationships, facilitates interpretation and understanding. Computation of various types of financial ratios is an important tool in analysis.

A *ratio* is a mathematical expression of the relationship between two items that may be expressed in several ways:

- **As a common ratio.** The ratio between x and y may be stated as x to y ($x:y$). For example, an operation that has 12 full-time employees and 4 part-time employees would have a full-time to part-time employee ratio of 3 to 1 (3:1).
- **As a percentage.** The ratio may be expressed as a percentage, such as a percentage of sales.
- **As a turnover.** Some relationships are best expressed as a turnover or the number of times x must be "turned over" to yield the value of y, calculated by dividing y by x. The number of times inventory is turned over in a month is one example.
- **On a per unit basis.** The relationship may also be expressed in dollars per unit, such as sales dollars per stool at a restaurant counter.

Ratios frequently are categorized according to primary use. The major categories include the following:

- Liquidity
- Solvency
- Activity
- Profitability
- Operating

Examples of each of these major types are listed in Table 13.1.

Table 13.1 Ratios Categorized by Primary Use

Types	Examples
Liquidity ratios	Current ratio Acid-test ratio
Solvency ratios	Solvency ratio Debt to equity ratio Debt to assets ratio
Activity ratios	Inventory turnover Occupancy percentage
Profitability ratios	Profit margin Return on equity Return on assets
Operating ratios	Analysis of the revenue mix Average customer check Food cost percentage Labor cost percentage Food cost per customer Meals per labor hour Meals per full-time equivalent Labor minutes per meal

Liquidity Ratios

Liquidity ratios indicate the organization's ability to meet current obligations—in other words, its ability to pay bills when due. An organization may be making a profit but have insufficient cash to pay current bills. Several ratios are used to analyze the liquidity of a business, two of the most common being the current ratio and the acid-test ratio. Data from the balance sheet are used in both calculations. The *current ratio* is the relationship between current assets to current liabilities and is computed as follows:

$$\text{Current ratio} = \frac{\text{Current assets}}{\text{Current liabilities}}$$

Creditors may consider a high current ratio as assurance they will receive payment for products and services, but managers may wish to maintain a lower current ratio by avoiding excess buildup of cash or inventories.

The *acid-test ratio,* also called the *quick ratio,* is another comparison of current assets and current liabilities, but it yields a more accurate measure of bill-paying capability. Current liabilities are measured against cash and other assets readily convertible to cash, such as accounts receivable and marketable securities. The acid-test ratio is calculated as follows:

$$\text{Acid-test ratio} = \frac{\text{Cash} + \text{Accounts receivable} + \text{Marketable securities}}{\text{Current liabilities}}$$

Solvency Ratios

Solvency ratios are used to examine an establishment's ability to meet its long-term financial obligations and its financial leverage. The basic solvency ratio is the relationship between total assets and total liabilities, calculated as follows:

$$\text{Solvency ratio} = \frac{\text{Total assets}}{\text{Total liabilities}}$$

Other solvency ratios examine the relationship between liabilities and equity (debt to equity ratio) and between liabilities and assets (debt to asset ratio).

Activity Ratios

Activity ratios are designed to examine how effectively an organization is using its assets. These ratios usually are expressed as either percentages or turnovers. The *inventory turnover ratio,* one of the most widely used activity ratios, shows the number of times the inventory is used up and replenished during a period. Inventory turnover is calculated as follows:

$$\text{Inventory turnover} = \frac{\text{Cost of goods sold}}{\text{Average inventory value}}$$

A high ratio indicates that a limited inventory is being maintained by a foodservice organization, and a low ratio indicates that larger amounts of money are tied up in inventories.

Percentage of occupancy is another activity ratio important in many segments of the industry. Hotels, hospitals, college or university residence halls, and nursing homes compute occupancy percentages regularly. These percentages indicate the relationship between the number of beds or rooms available and the number being used by the clientele of the operation:

$$\text{Percentage of occupancy} = \frac{\text{No. of beds/rooms occupied}}{\text{No. of beds/room available}}$$

Profitability Ratios

Profitability ratios measure the ability of an organization to generate profit in relation to sales or the investment in assets. Profit or net income is an absolute term expressed as a monetary amount of income remaining after all expenses have been deducted from income or revenue; profitability is a relative measure of the profit-making ability of an organization. Most foodservice operations, whether profit or not-for-profit, must generate some level of net income; disposition of profit is a primary difference between these two types of operation. Also, the not-for-profit organization may have a lower expectation with regard to level of profit. Three major profitability ratios are used:

$$\text{Profit margin} = \frac{\text{Net profit}}{\text{Sales}}$$

$$\text{Return on equity} = \frac{\text{Net profit}}{\text{Equity}}$$

$$\text{Return on assets} = \frac{\text{Net profit}}{\text{Total assets}}$$

The *profit margin* is the most commonly used measure of operating profitability; it uses information from the income statement to assess overall financial efficiency. The *return on equity* measures the adequacy of profits in providing a return on owners' investments; the *return on assets* is a measure of management's ability to generate a return on the assets employed in generating revenue. Both calculations combine data from the income statement and balance sheet to assess financial efficiency.

Operating Ratios

Operating ratios are primarily concerned with analysis of the success of the operation in generating revenues and in controlling expenses. *Analysis of the revenue mix* of the operation is one type of sales analysis. For example, a restaurant may analyze its food and beverage sales to determine the relative proportion of revenues generated by these two aspects of the operation; percentage of revenues generated by food and beverage sales would be calculated by dividing the sales in each area by total sales. In a student union, the relative percentage of sales from catering, vending, and cafeteria operations may be of interest and similarly would be calculated by dividing the sales from each area by the total sales.

The *average customer check* is another measurement of generation of sales dollars; it is calculated by dividing total sales by the number of customer checks. Sales per seat in a restaurant, per meal, per waitress, and per menu item are other sales ratios that may be used in a foodservice establishment.

As stated initially in this chapter, control of costs is a primary responsibility of foodservice managers; their organizations' profitability is strongly related to their effectiveness in such control. In foodservice organizations, two commonly used in indicators to assess operational effectiveness are

$$\text{Food cost percentage} = \frac{\text{Cost of food}}{\text{Sales}}$$

and

$$\text{Labor cost percentage} = \frac{\text{Cost of labor}}{\text{Sales}}$$

A percentage breakdown of the food dollar spent on various food categories also can be calculated by dividing the cost of food for each category by the total spent on food. Such analysis may reveal possible cost problems. If an analysis of expenditures indicates that dollars spent on meat have increased for two successive months, for example, the manager may need to examine menu offerings, purchase prices, inventory control procedures, production methods, portion controls, or selling prices.

Another performance indicator is *food cost per customer* (or per patient, or student, or the like), which often is used for analyzing food cost in healthcare or university foodservices. This indicator is determined by dividing total food cost by the number of customers (patients/students) served.

Also, various types of ratios are computed to analyze labor cost:

$$\text{Meals per labor hour} = \frac{\text{Total number of meals served}}{\text{Total labor hours to produce meals}}$$

$$\text{Meals per full-time equivalents} = \frac{\text{Total number of meals served}}{\text{Total full-time equivalents to produce meals}}$$

$$\text{Labor minutes per meal} = \frac{\text{Total labor minutes to produce meals}}{\text{Total number of meals served}}$$

Using Ratio Analysis

Ratio analysis constitutes an effective tool for evaluating financial stability and operating effectiveness, providing managers with the information they need to make decisions and control operations. Ratios selected for analysis of financial results must be appropriate to the operation, however, and the foodservice manager must understand the relationships measured.

Managers should use a variety of ratios in analyzing operations because a single ratio is insufficient for sound decision making. Ratios should be used in combination with trend analysis to compare changes over a period of time. Consistency in accounting methods, however, is required to yield comparable data for comparisons over a period of time.

Trend Analysis

Trend analysis is a comparison of results over several periods of time; changes may be noted in either absolute amounts or percentages. It also is used to forecast future revenues or levels of activity. Trend analysis may use several of the various types of ratios discussed in the preceding section.

Often a graphic analysis of the financial data over a period of time will help managers detect and understand changes. For example, many foodservice managers regularly plot on a graph the percentage of sales spent on food, labor, and other operating expenses. The percentage is depicted on the vertical axis and the period of time (days, months, years) on the horizontal axis. Figure 13.5 shows the graphic record of food cost percentage for a 6-month period.

Common-Size Statements

Comparison among financial statements for various periods or from different departments within an organization may differ because of varying levels of volume. If the financial data are expressed as percentages, however, meaningful comparisons can be made because data have a common base.

Common-size statements: Financial statement in which data are expressed as percentages for comparing results from one accounting period to another.

Financial statements in which data are expressed as percentages are called **common-size statements**. These are especially useful in comparing results of the income statement of an operation from one accounting period to another or for comparing results among units of a multiunit operation, such as a chain restaurant or food centers in a campus dining operation. To prepare a common-size income statement, each figure on the statement is calculated as a percentage of total sales by dividing each figure by the total sales. A common-size balance sheet sets total assets equal to 100% and individual asset categories as percentages of the total. Similarly, total liabilities and owner's equity are taken as 100% and individual categories become percentages of that 100%. Comparisons with industry performance are facilitated by expressing the financial statements in percentages.

Break-Even Analysis

Break-even analysis: Technique for assessing financial data to determine the point at which profit is not being made and losses are not being incurred.

Break-even analysis is another tool for analyzing financial data. The name of this technique is derived from the term break-even point, or the point at which an operation is just breaking even financially, making no profit but incurring no loss. In other words, total revenues equal total expenses.

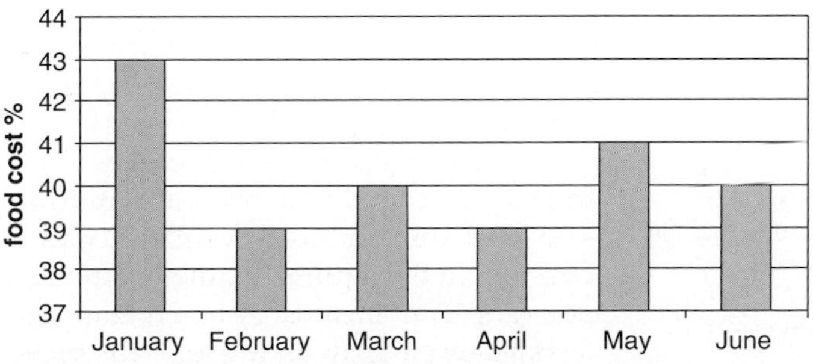

Figure 13.5. Food cost percentage trend analysis.

Fixed costs: Costs that do not vary with changes in the volume of sales.

Variable costs: Costs that vary directly with changes in sales.

Break-even analysis requires classification of costs into fixed and variable components. **Fixed costs (FCs)** are those costs required for an operation to exist, even if it produces nothing. These costs do not vary with changes in the volume of sales but stay fixed, or constant, within a range of sales volume. The size of the existing physical plant and the equipment capacity define this volume. If either is expanded, however, fixed costs will increase. Insurance, rent, and property taxes are examples of fixed costs.

Variable costs (VCs) are those costs that change in direct proportion to the volume of sales. As the volume of sales increases, a proportionately higher amount of these costs is incurred, as with direct materials or food cost.

A complication in defining fixed and variable costs is that some costs cannot be clearly classified as either entirely fixed or entirely variable. These costs are termed *semivariable* because a portion of the total cost will remain fixed regardless of changes in sales volume, and a portion will vary directly with changes in sales volume. Such costs often include labor, maintenance, and utilities. Managers need to divide these semivariable costs into their fixed and variable components before they can use break-even analysis for decision making. Once semivariable costs have been divided into their fixed and variable portions, the break-even point can be calculated by the following formula:

$$\text{Break-even point} = \frac{\text{Fixed costs}}{1 - \dfrac{\text{Variable costs}}{\text{Sales}}}$$

To illustrate, the following data are available for XYZ cafeteria:

$$\text{Fixed costs} = \$28{,}000$$

$$\text{Variable costs} = \$60{,}000$$

$$\text{Sales} = \$100{,}000$$

Substituting the figures in the formula, the break-even point for XYZ cafeteria is calculated as follows:

$$\begin{aligned}\text{Break-even point} &= \frac{\$28{,}000}{1 - \dfrac{\$60{,}000}{\$100{,}000}}\\ &= \frac{\$28{,}000}{1 - .60}\\ &= \frac{\$28{,}000}{.40}\\ &= \$70{,}000\end{aligned}$$

Contribution margin: Proportion of sales that can contribute to fixed costs and profits after variable costs have been covered.

The denominator in this formula, or the ratio of variable cost to sales subtracted from 1, is referred to as the **contribution margin.** It represents the proportion of sales that can contribute to fixed costs and profit after variable costs have been covered. In the illustration, the contribution margin is 0.40, or 40%. The break-even point for XYZ cafeteria is shown graphically in Figure 13.6.

Break-even analysis is a tool for projecting income, expense, and profit under several assumed conditions. It can assist the foodservice manager in understanding

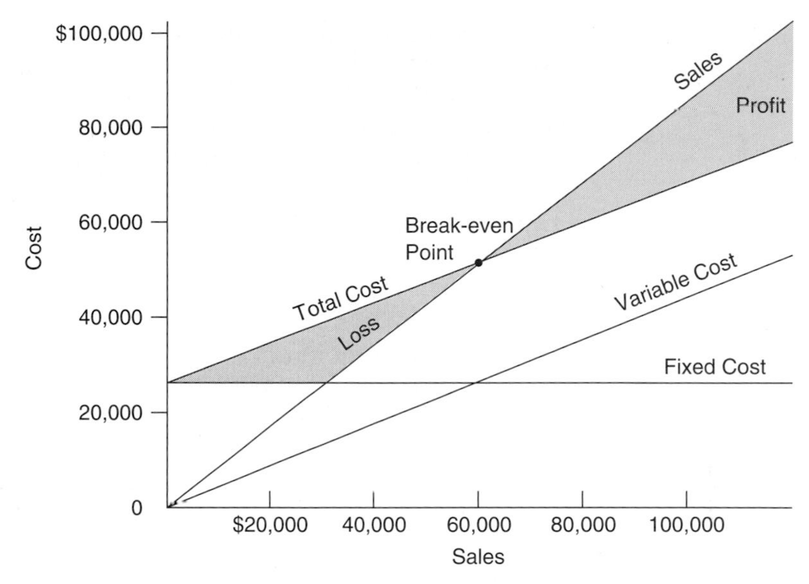

Figure 13.6. Break-even chart for XYZ cafeteria.

the interrelationships among volume, cost, and profits, but it requires that the operational costs of an organization be known and that they can be segmented into fixed and variable classifications. An expansion of this break-even concept is known as cost-volume-profit (CVP) analysis. Using the CVP formula, a manager can determine the volume required for a given level of profit. The CVP formula is as follows:

$$\text{Sales or revenue level} = \frac{\text{Fixed costs} + \text{Profit desired}}{1 - \dfrac{\text{Variable costs}}{\text{Sales}}}$$

BUDGETING

Budget: Plan for operating a business expressed in financial terms.

A **budget** is a plan for operating a business expressed in financial terms or a plan to control expenses and profit in relation to sales. Budgeting is the process of budget planning, preparation, control, reporting, utilization, and related procedures. Budget control involves the use of budgets and performance reports throughout the planned period to coordinate, evaluate, and control operations in accordance with the goals specified in the budget plan.

In a governmental operation, however, the budget represents something different: a schedule of authorizations for a given period or purpose and the proposed means of financing it. Funds are established, representing a sum of money or other resources segregated for carrying on specific activities or attaining certain objectives in accordance with specific regulations, restrictions, and limitations. These funds constitute an independent fiscal entity.

Budgeting provides an organized procedure for planning and for development of standards of performance in numerical terms. Planning, coordination, and control are the three primary objectives of budgeting. A well-constructed operating budget shows the overall structure of the operation and enables staff to visualize its place in it.

Budgets provide a basis for control, but they must be implemented by operational personnel within the organization. Budgets are a tool for periodic examination, restatement, and establishment of guidelines. A budget is a plan that should be well constructed and capable of attainment. To be effective, all staff members responsible for implementing the budget should have input.

Types of Budgets

Foodservice operations often will have several different budgets. Most common are the operating and capital budgets. Cash budgets also are important.

Operating Budget

The first step in preparing the operating budget in most organizations is the development of the *sales* or *revenue portion of the budget*. Various internal and external influences must be considered in constructing the sales budget, which includes an estimate of revenues expected during the budget period. Past experience should be examined, including both past performance and past budgets, and any unusual aspects clarified.

Changes in pricing also must be considered. For example, if board rates are increased for a college residence hall, the proportion allocated per student for the foodservice operation should be reviewed. National and local indicators also should be considered, such as new competitors, other new businesses in the area, industry trends, and changes in the national economy.

The profit objective of the organization should be considered. The projected profit from a given volume of sales should be estimated and the ability to attain the profit objective determined. Seasonal variations in sales also should be projected, and segmentation of sales in various units of the organization forecast. For example, in a large medical center, the sales forecast for the employee cafeteria catering, a coffee shop, and vending may be appropriate.

Once sales have been forecast, the *expenditures* are budgeted. Expenditures in relation to the projected level of revenues should be estimated for food, labor, and other operational expenses. In estimating food costs, anticipated changes in food prices should be considered.

Increments in salaries and wages and in payroll taxes are key elements in estimating the *labor budget*. Also, the foodservice manager must project the impact of menu changes or other operational changes that may affect the labor needed in the operations. Similar factors should be considered in projecting expenditures for other operating expenses.

Capital Budget

The development of the *capital expenditure budget* usually is completed at the same time the operating budget is being planned. Improvements, expansions, and replacements in building, equipment, and land are the major capital expenditures. These major capital investments may be for purposes of expanding or improving facilities.

Careful planning is particularly important in capital budgeting because of the magnitude of funds and the long-term commitments involved. Capital expenditures may be prorated over several budget periods.

Cash Budget

Many operations also will develop a cash budget, which is a detailed estimate of anticipated cash receipts and disbursements throughout the budget period. The cash budget will assist managers in coordinating cash inflow and outflow and in synchronizing cash resources with need. Seasonal effects on cash position thus become apparent and the availability of cash for taking advantage of discounts is identified. Also, the cash budget assists management in planning financial requirements for large payments, such as tax installments or insurance premiums, and indicates availability of funds for short-term investments.

Pro Forma Statement

Pro forma statement: Statement that projects expected income and expenditures.

Managers in profit generating organizations usually compile a forecast income statement, often termed a pro forma statement, as part of their budget planning. This **pro forma statement** is a composite of the sales and expenditure budgets and includes a projection of profit. The budget planner should compute various ratios to determine if the appropriate relationship of expenditures to sales or to revenues has been projected.

Budgeting Concepts

The basic steps of the budget process may vary in different organizations, and the foodservice manager may be involved in only certain aspects of budget planning. For example, in a very large organization, the cash budget and compilation of the total budget are often the responsibility of the controller and other top-management staff.

After the budget planning process has been completed, various reviews and approvals are needed before the budget document is finalized; these vary widely, depending on the size and type of the organization. The approved budget then becomes the standard against which operations are evaluated. Periodic reports are issued comparing operating results with budget estimates. The manager's job becomes one of using these comparative reports to bring about operational changes conforming to budget plans unless, however, data suggest that budget amendments are in order.

A *fixed budget* is one that is prepared at one level of sales or revenues. Incremental budgeting involves using the existing budget as a base and projecting changes for the ensuing year in relation to the current budget.

The *flexible budget* is adjusted to various levels of operation or sales; it is useful for operations with varying sales or revenues throughout the year. A foodservice organization may develop budgets at two or three possible levels of sales to assist in adjusting expenditures to actual sales volume. For example, resort hotels and hospitals have wide variations in occupancy rates. A flexible budget will enable the management of these organizations to adjust expenditures in relation to occupancy levels.

Investment Decisions

Requests for renovation of space or the purchase of supplies/equipment valued at more than $5,000 often are termed capital expenditures. These expenditures usually are identified in the capital budget and reflect the purchase of assets assumed to have a relatively long life for the organization. When submitting a capital budget request, foodservice directors usually will be asked to justify the benefits to the organization of this capital expenditure.

There are several methods that can be used to help evaluate capital budget requests. Most commonly used are payback period and net present value (NPV).

Payback period is an easy-to-compute indicator of the time it will take an organization to recover the money invested in a particular project or piece of equipment. Calculation of the payback period involves dividing the cost of the project or equipment by the expected annual income of cash savings. This result gives an estimate of the number of years that it would take for the income or savings from the project or equipment to equal the investment in that project or piece of equipment.

For example, a hospital foodservice manager may be considering the purchase of a soft serve machine for the employee cafeteria. The machine will cost $8,000. The manager has estimated profits of $2,500 per year from the machine. The payback period for this investment would be 3.2 years.

$$\frac{\text{Initial investment}}{\text{Expected yearly income}} \qquad \frac{\$8{,}000}{\$2{,}500} = 3.2$$

Time value of money: Money has a differing value over time; having $1 today is worth more than receiving $1 in the future.

Although the payback period is easy to compute, it is criticized because it does not consider the **time value of money**. The basic premise of the time value of money is that having $1 today is worth more than receiving that dollar one year or more from now because money in hand today could be invested and should be worth more in the future. The future worth of money is dependent on the rate of return achieved by investing the money.

Net present value (NPV) is considered a more sophisticated and preferred method for evaluating a proposed capital expenditure because it considers the time value of money. The NPV method determines the present value of expected future cash inflows and outflows related to a capital expenditure.

Calculation of the present value of cash flows is a critical component of the NPV method. This calculation requires knowledge of the following:

- Expected initial investment for a capital expenditure (renovation and/or equipment) 12,500
- Expected income or cost savings each year for the expected useful life of the project or piece of equipment
- The cost of capital used by the organization for evaluating capital expenditures
- A table of discounted cash flows factors (see Table 13.2)

The *initial investment* for a capital expenditure is the dollar amount of money that the organization will need to spend to renovate a space and/or purchase a piece of equipment. The *expected income* or *cost savings* is the dollar amount that a foodservice director expects to have earned (income) or saved each year as a result of this capital expenditure. The *cost of capital* is the minimum return a company expects on an investment. In essence, a company executive will ask, "How much would I make if I invested this amount of money in the bank, or in the stock

Table 13.2 Table of Discounted Cash Flow Factors

Period	5%	6%	7%	8%	9%	10%	11%	12%	13%	14%	15%	16%	17%	18%	19%	20%	25%	30%
1	0.9524	0.9434	0.9346	0.9259	0.9174	0.9091	0.9009	0.8929	0.8850	0.8772	0.8696	0.8621	0.8547	0.8475	0.8403	0.8333	0.8000	0.7692
2	0.9070	0.8900	0.8734	0.8573	0.8417	0.8264	0.8116	0.7972	0.7831	0.7695	0.7561	0.7432	0.7305	0.7182	0.7062	0.6944	0.6400	0.5917
3	0.8638	0.8396	0.8163	0.7938	0.7722	0.7513	0.7312	0.7118	0.6931	0.6750	0.6575	0.6407	0.6244	0.6086	0.5934	0.5787	0.5120	0.4552
4	0.8227	0.7921	0.7629	0.7350	0.7084	0.6830	0.6587	0.6355	0.6133	0.5921	0.5718	0.5523	0.5337	0.5158	0.4987	0.4823	0.4096	0.3501
5	0.7835	0.7473	0.7130	0.6806	0.6499	0.6209	0.5935	0.5674	0.5428	0.5194	0.4972	0.4761	0.4561	0.4371	0.4191	0.4019	0.3277	0.2693
6	0.7462	0.7050	0.6663	0.6302	0.5963	0.5645	0.5346	0.5066	0.4803	0.4556	0.4323	0.4104	0.3898	0.3704	0.3521	0.3349	0.2621	0.2072
7	0.7107	0.6651	0.6228	0.5835	0.5470	0.5132	0.4817	0.4524	0.4251	0.3996	0.3759	0.3538	0.3332	0.3139	0.2959	0.2791	0.2097	0.1594
8	0.6768	0.6274	0.5820	0.5403	0.5019	0.4665	0.4339	0.4039	0.3762	0.3506	0.3269	0.3050	0.2848	0.2660	0.2487	0.2326	0.1678	0.1226
9	0.6446	0.5919	0.5439	0.5003	0.4604	0.4241	0.3909	0.3606	0.3329	0.3075	0.2843	0.2630	0.2434	0.2255	0.2090	0.1938	0.1342	0.0943
10	0.6139	0.5584	0.5084	0.4632	0.4224	0.3855	0.3522	0.3220	0.2946	0.2697	0.2472	0.2267	0.2080	0.1911	0.1756	0.1615	0.1074	0.0725
11	0.5847	0.5268	0.4751	0.4289	0.3875	0.3505	0.3173	0.2875	0.2607	0.2366	0.2149	0.1954	0.1778	0.1619	0.1476	0.1346	0.0859	0.0558
12	0.5568	0.4970	0.4440	0.3971	0.3555	0.3186	0.2858	0.2567	0.2307	0.2076	0.1869	0.1685	0.1520	0.1372	0.1240	0.1122	0.0687	0.0429
13	0.5303	0.4688	0.4150	0.3677	0.3262	0.2897	0.2575	0.2292	0.2042	0.1821	0.1625	0.1452	0.1299	0.1163	0.1042	0.0935	0.0550	0.0330
14	0.5051	0.4423	0.3878	0.3405	0.2993	0.2633	0.2320	0.2046	0.1807	0.1597	0.1413	0.1252	0.1110	0.0986	0.0876	0.0779	0.0440	0.0254
15	0.4810	0.4173	0.3625	0.3152	0.2745	0.2394	0.2090	0.1827	0.1599	0.1401	0.1229	0.1079	0.0949	0.0835	0.0736	0.0649	0.0352	0.0195
16	0.4581	0.3937	0.3387	0.2919	0.2519	0.2176	0.1883	0.1631	0.1415	0.1229	0.1069	0.0930	0.0811	0.0708	0.0618	0.0541	0.0281	0.0150
17	0.4363	0.3714	0.3166	0.2703	0.2311	0.1978	0.1696	0.1456	0.1252	0.1078	0.0929	0.0802	0.0693	0.0600	0.0520	0.0451	0.0225	0.0116
18	0.4155	0.3503	0.2959	0.2503	0.2120	0.1799	0.1528	0.1300	0.1108	0.0946	0.0808	0.0691	0.0592	0.0508	0.0437	0.0376	0.0180	0.0089
19	0.3957	0.3305	0.2765	0.2317	0.1945	0.1635	0.1377	0.1161	0.0981	0.0829	0.0703	0.0596	0.0506	0.0431	0.0367	0.0313	0.0144	0.0068
20	0.3769	0.3118	0.2584	0.2146	0.1784	0.1486	0.1240	0.1037	0.0868	0.0728	0.0611	0.0514	0.0433	0.0365	0.0308	0.0261	0.0115	0.0053

market, and how does this compare to what I will make if I invest in this foodservice capital expenditure." Cost of capital is given as a percentage figure based on the percent a company would expect an investment to earn each year. The discounted cash flows factors (Table 13.2) are factors that are used to discount a future cash flow to a current value based on cost of capital percentage figures.

Table 13.3 shows the calculations of NPV for the soft serve machine described earlier. The machine is estimated to cost $8,000. This is shown as a negative value or a cash outflow in the NPV calculation. The cost of capital for the hospital in this example is assumed to be 7%. The foodservice director expects an income (sales minus expenses) of $2,500 each year for the 10-year useful life of the machine. Table 13.3 shows how the discount factors (taken from the column headed 7% in Table 13.2) are applied to the future expected cash flows (income) to achieve a NPV of $9,558. There is no target comparison number for deciding whether this NPV figure is good or bad. Typically, organizations will want a positive NPV value when evaluating whether to commit funds to a capital expenditure.

Inherent in the use of NPV for evaluation of a capital expenditure is the assumption that the cost of capital figure accurately reflects the return a company could expect to make on their money if it were invested elsewhere. Also assumed is that the cost of capital figure remains constant for the life of the project. Although these assumptions may not always be valid, the NPV method does provide a more sophisticated way than the payback period to evaluate capital expenditures and commonly is used, especially in foodservice operations that have a profit goal.

MENU PRICING

Pricing menu items follows planning the menu and can be one of the most difficult decisions management makes. Menu pricing should cover the cost of food and labor and additional operating costs, including rent, energy, and promotional advertising. Other important factors to be considered when menu prices are set are

Table 13.3 Calculation of Net Present Value for Soft Serve Machine

	Cash Flow	Discount Factor	Present Value
Initial Investment	−$8,000		−$8,000
Year 1 income	$2,500	.9346	$2,336
Year 2 income	$2,500	.8734	$2,183
Year 3 income	$2,500	.8163	$2,041
Year 4 income	$2,500	.7629	$1,907
Year 5 income	$2,500	.7130	$1,782
Year 6 income	$2,500	.6663	$1,666
Year 7 income	$2,500	.6228	$1,557
Year 8 income	$2,500	.5820	$1,455
Year 9 income	$2,500	.5439	$1,360
Year 10 income	$2,500	.5084	$1,271
NPV			$9,558

perception of value and competition (Keiser, DeMicco, & Grimes, 2000). Perception of value is what a customer believes the menu is worth. The foodservice manager also needs to be aware of what the competition is doing, for example, McDonald's often sets local standards for hamburger prices.

Setting menu prices is a continual challenge for foodservice managers (Keiser, DeMicco, & Grimes, 2000). A restaurant manager wants to make the highest possible profit and retain repeat customers. Onsite foodservice managers want to serve the best food possible within the budget. Unless they are subsidized, onsite foodservice managers may need to develop a pricing strategy that covers all costs.

Menu pricing today generally is computerized; most food management software programs include a menu-pricing component. Menu items and portion count can be entered into the program, and it will calculate portion sizes, selling prices, item costs, and the raw food and markup percentages. Sales can be calculated for any period of time, as can total or per-customer costs and profit margins (Keiser, DeMicco, & Grimes, 2000).

Pricing Methods

Various methods are used to price menus; the one most often used is based on establishing a percentage of the selling price for food and labor. The three most often used in foodservice operations are discussed here: factor, prime cost, and actual cost methods.

Factor

The **factor pricing method** is also known as the *markup method* (Keiser, DeMicco, & Grimes, 2000). *Markup,* the difference between cost and selling price, varies among types of foodservice operations. First, the desired percentage of food cost must be selected and divided into 100 to give a pricing factor. By multiplying the raw food cost by this factor, a menu sales price will result:

$$\text{Raw food cost} \times \text{Pricing factor} = \text{Menu sales price}$$

If the operator chooses a 40% food cost, the pricing factor would be 100/40, or 2.5. For a raw food cost of \$2.00, the selling price would be \$2.00 × 2.5 = \$5.00.

The factor method often is used by foodservice managers because simple mathematics are involved. The principal disadvantage is that costs other than food are not known until the end of the month, and this method disregards perception of value and the fact that customers will not pay a uniform markup on all menu items.

Prime Cost

Prime cost consists of raw food cost and direct labor cost of those employees involved in preparation of a food item but not service, sanitation, or administrative costs. An accurate determination of prime cost for each menu item would require calculating the raw food cost and direct labor cost for pricing. In addition to cost records on raw food purchased for each menu item, time studies of the amount of direct labor would be required. This would be a gigantic task for an entire menu and difficult to justify because it is so labor intensive. The total process seems thoroughly impractical, especially in commercial operations, because pricing would have to be done almost daily and menu prices changed accordingly.

To make this cost method practical, some assumptions must be made on the percentage of prime cost attributable to raw food, direct labor, and operating margin. Each restaurateur would need to decide what percentage of the selling price would be assigned to the raw food and direct labor costs to give a prime cost total. As an example, in examining financial records, the foodservice manager finds that for every $2.00 spent for raw food, 9.5 cents are spent for direct labor, making the prime cost total $2.095. The manager decides that 40% of the selling price of a food item would be for raw cost and 8% for direct labor, leaving an operating margin of 52%. The relationship of selling price with a margin of 52% is 100 divided by 52, or 1.923, as a multiplier for the prime cost to yield the selling price. In the example, the prime cost of $2.095 multiplied by 1.923 gives a selling price of $4.028. In actuality, this would probably be changed to $4.25. The multiplier, 1.923, would be used for pricing most menu items.

Actual Cost

Actual cost is used in operations that keep accurate cost records. The initial step, as usual, is to establish the food cost from standardized recipes and labor costs, which are the principal variable costs. Other variable costs, fixed costs, and profit can be obtained as a percentage of sales from the profit-and-loss statement. The menu price consists of the actual food cost + actual labor cost + other variable costs + fixed cost + profit. The actual-cost method has the advantage of including all costs and the desired profit in the selling price of the menu item.

Pricing Psychology

Foodservice operators consider **pricing psychology** in determining what and how to charge. Psychological aspects of pricing affect customer perceptions, which then influence the purchase decision (Pavesic, 1988). Many schemes have been devised to entice the customer to buy; these schemes are apparent in all merchandising. Menu items are priced by the same general considerations that are apparent in other items for sale. Some of the schemes in use by foodservice operators are odd-cents pricing, cost by the ounce, two-tier foodservice, à la carte, and table d'hôte.

Odd-cents pricing follows the basic philosophy of creating an illusion of a bargain. The so-called magic numbers supposedly stimulate the consumer to buy. Kreul and Scott (1982) identified three methods practiced in odd-cents pricing:

- Price ends in an odd number (e.g., $4.75).
- Price ends in a number other than zero (e.g., $4.77).
- Price is just below a zero (e.g., $4.99).

Pricing by the ounce is used in some onsite foodservice operations, especially for salad and sandwich bars. This concept also is popular for salad bars in supermarkets. Increasing menu prices causes dissatisfied customers, but when customers can weigh their own portions, the sense of control reduces complaints. The initial step in using this pricing system requires estimation of the total amount of raw food needed to serve an anticipated number of customers to determine the raw food cost per ounce. A markup factor per ounce of raw food to cover labor, other expenses, and profit must be established. The selling price per ounce is computed by multiplying the cost per ounce by a markup factor. The cost-per-ounce pricing scheme has satisfied many customers because they like paying only for what they eat.

Two-tier foodservice is being used in a number of healthcare centers around the country as hospitals compete for patients. Upscale amenities and menus are being prepared for patients who are willing to pay for special food items and elegant service. Special kitchens, private dining rooms, and suites have been added to many hospitals, with chefs preparing fine restaurant menu items to respond to individual desires.

The à la carte menu is actually not a separate type of menu but a method of pricing a menu. Menu items are priced, offered, and selected separately by the customer. The à la carte menu is typical in many commercial foodservices, ranging from upscale gourmet to limited-menu restaurant operations. Cafeterias, both onsite and commercial, generally offer à la carte pricing of menu items.

Table d'hôte menus group several food items together and offer them at a fixed price. The menu is a complete meal with several courses, and the only choices might be soup, salad, or dessert.

CONTROLLING COSTS

Controlling labor and food costs in a foodservice operation is a primary responsibility of the manager of that operation. Pavesic and Magnant (2005) suggested that effective cost control programs:

- Provide information for daily decision making
- Monitor individual and departmental efficiency
- Inform management of income, expenses, and variations from budget
- Prevent fraud and theft
- Are the basis of knowing where the company is going, not where it has been
- Emphasize prevention, not correction
- Maximize profit, not minimize loss

Labor Costs

Controlling and reducing labor costs and simultaneously increasing labor productivity have been growing challenges for many years among managers of all types of foodservice operations. Many different approaches have been used, a number of which have been helpful, such as increasing the use of convenience foods, decreasing the number of items on a menu, improving the efficiency of the layout and equipment, and increasing employee benefits. Also needed for streamlining a labor force are the following seven steps: personnel policies, job analysis, work simplification, work production standards, workload forecasting, scheduling, and control reports. According to Keiser, DeMicco, and Grimes (2000), labor cost is the most important cost in hospitality operations. In the past, when labor was abundant and cheap, and there were few government controls, food was considered the highest cost item. That period has passed. If management is to generate a profit for commercial operations or provide the best services resources permit for onsite foodservice enterprises, control of labor costs is inevitable.

One technique that is being used to help reduce labor costs is programable equipment. Computerized French fry equipment and self-service ordering technology already are being tested by limited-menu chains.

The foodservice industry has some unique labor control problems. The most competent personnel in the workforce will choose industries in which financial compensation is higher than in foodservice. According to the National Restaurant Association (2006), the typical foodservice employee is

- Female
- Under 30 years old
- Single
- Living in a household with two or more wage earners
- A part-time employee with weekly hours averaging 25

The number of African American–owned and female-owned eating-and-drinking businesses increased at double-digit rates over the past decade, with sales also rising dramatically. Almost 6 out of 10 employees in foodservice occupations are women (57%), 12% are African American, and 16% are of Hispanic origin. Of the 1.5 million managers of foodservice and lodging establishments in 2001, 68% were women, 13% were African American, and 12% were of Hispanic origin (Industry at a Glance, 2002).

For many years, the foodservice industry has been the biggest employer of teenagers, who consider their jobs temporary; however, that source of supply is diminishing because of population trends. In addition, foodservice employees' work hours may be long and scheduled at times that teenagers study or have social activities. The prediction is that the elderly and persons with disabilities might take up this employment slack. Scheduling and staffing become top priorities for many foodservice managers. Lowering labor costs by improving productivity and providing satisfactory wages that attract and retain competent employees is a monumental challenge for managers.

Ninemeier (1999) offered a number of points that justify efforts to control labor cost.

- Direct labor costs are increasing due to elevation of minimum wage scales and competition for qualified people.
- Increases in labor costs often are not balanced by increases in productivity.
- Fringe benefit packages are often very creative and unfortunately expensive.
- People with different attitudes, beliefs, problems, goals, and personalities, as are commonly found among hospitality staff, are difficult to manage.
- Managers often recognize the need to help employees find meaning in their jobs and get more than wages from their work.

Foodservice is labor intensive, and generally food industry technology has not been able to replace people with equipment.

Cost-containment pressures and the prospective payment plans in the healthcare segment of the industry, for example, have caused a major emphasis to be placed on improved productivity. Because labor represents a major component of the total operational cost in a foodservice establishment, controlling labor cost is a key aspect of attaining financial objectives in foodservice operations.

Effective use of labor in the foodservice industry is especially difficult because of the following unique characteristics:

- Many 7-day-a-week operations, some of which may require staff coverage 24 hours a day
- Peaks in service demand requiring additional staff at those times
- Seasonal variations in patronage of establishments
- The highly perishable nature of food items both before and after production
- The labor-intensive aspect of most production and service activities
- The large number of unskilled and semiskilled personnel in the industry

CHAPTER SUMMARY

This summary is organized by the learning objectives.

1. A variety of accounting principles exist to help ensure accuracy and consistency in financial reporting. The business entity concept means that the accounting for the business will be kept separate from that of the owner(s). The going concern concept is the assumption that a business will continue. The cost principle means that business transaction will be measured in terms of dollars. The cost basis of accounting recognizes a transaction at the time of the cash inflow or outflow; whereas the accrual basis of accounting recognizes revenues when earned and expenses when incurred. The consistency principles states that once a company chooses an accounting method, that method should be used consistently. The practice of conservatism implies use of moderation in recording transactions and assigning value.
2. The balance sheet gives a picture of a company's financial status at one point in time; the income statement reports operating results over a period of time. Financial data presented in these statements can be analyzed using ratio analysis, trend analysis, and common-size statement comparison. Data also can be compared with data from similar operations.
3. Liquidity ratios, such as the current ratio and acid-test ratio, indicate a company's ability to meet current obligations. Solvency ratios, such as the solvency ratio, debt to equity ratio, and debt to asset ratio, are used to examine a company's ability to meet its long-term obligations. Profitability ratios, such as profit margin, return on equity, and return on assets, measure the ability of a company to generate profits. Activity ratios, such as inventory turnover, are designed to help evaluate how well a company is using its assets. Operating ratios, such as food cost percentage, food cost per customer, and meals per labor hour, help evaluate the success of the operation in generating revenue and controlling expenses.
4. Break-even analysis is used to determine the point at which a company sales and expenses are equal. Costs are divided into fixed costs (those that do not change with variations in sales) and variable costs (those that change directly with sales).
5. The payback period is calculated by dividing the initial cost of the piece of equipment by the expected annual savings or income generated by having that piece of equipment. Net present value (NPV) is the present value of all future cash flows decreased by the amount of the initial cost.

Professional Organization Profile

Consultant Dietitians in Health Care Facilities Dietetic Practice Group of the American Dietetics Association

MISSION

Committed to strengthening members with food, nutrition, and management skills to promote quality of life for the people we serve

WHO BELONGS TO THE ORGANIZATION

Members are also members of the American Dietetic Association and work in a variety of settings including long-term care, assisted living, home health, corrections, hospice, hospitals, and physician's offices. Student membership available through the American Dietetic Association.

ADVANTAGES OF MEMBERSHIP

Members have access to a members-only online network; receive a quarterly newsletter, *The Consultant Dietitian;* earn continuing education credit by attendance at association sponsored seminars and workshops; and have access to resources developed specifically for consultant dietitians.

WEBSITE

www.cdhcf.org

MEET THE PRESIDENT

Carolyn Breeding, RD, LD, FADA; President 2005–2006; is president and owner of Dietary Consultants, Inc., a dietary consulting company; president of NAPA Health Care, a group purchasing organization; and owner of Quality Provider Services, a contract billing company. Carolyn was recipient of the Better Life Award from the Kentucky Association of Health Care Facilities and the American Dietetic Association Award for Excellence in Consultation Business Practice. Carolyn offers the following advice for students: "Many jobs in dietetics today require broad knowledge in a variety of areas. Do not short change your education in foodservice management. These skills can be as critical to your success as clinical skills."

6. Several strategies exist for pricing a menu; all involve computations using the food cost for that menu item. The factor pricing method involves multiplying the food cost by a predetermined factor to determine the price to be charged for the item. In the prime cost method, the food and direct labor costs are totaled and multiplied by a factor. In the actual-cost method, the actual costs (fixed and variable) and the desired profit are added together to give a suggested selling price.

TEST YOUR KNOWLEDGE

1. Write definitions for these terms: auditing, depreciation, materiality principle, asset, liability, capital budget, and time value of money.
2. Why is the current ratio termed a liquidity ratio, whereas the debt to asset ratio is termed a solvency ratio?
3. What are examples of operating ratios? Why would these be used by a foodservice manager?
4. How do changes in fixed costs impact the break-even point?
5. Why might the chief financial officer of a hospital prefer that the foodservice director analyze a proposed equipment purchase using NPV rather than the payback period?

CLASS PROJECTS

1. Obtain financial reports from a local foodservice operation. Calculate ratios on the data and discuss the viability of that operation based on its financial data.
2. Divide into groups of two to three students. Have each group conduct an interview with a local foodservice manager to determine what financial reports they review each month and how they analyze the data they receive. Share results of the interviews with fellow classmates.

WEB SOURCES

www.nfsmi.org	National Food Service Management Institute
www.restaurant.org	National Restaurant Association
www.foodservice.com	Foodservice management information
www.restaurantresults.com	Restaurant management information

BIBLIOGRAPHY

Anthony, R. N., & Young, D. W. (1998). *Management control in nonprofit organizations* (6th ed.). Homewood, IL: Richard D. Irwin.

Baker, M. M. (1988). Meeting the challenge in foodservice management: Enhanced quality at less cost. *Journal of the American Dietetic Association, 88,* 441–442.

Barwise, P., Marsh, P. R., & Wensley, R. (1989). Must finance and strategy clash? *Harvard Business Review, 67*(5), 85–90.

Coltman, M. M. (1997). *Hospitality management accounting* (6th ed.). New York: Van Nostrand Reinhold.

DeMicco, F. J., Dempsey, S. J., Galer, F. F., & Baker, M. (1988). Participative budgeting and participant motivation: A review of the literature. *FIU Hospitality Review, 6*(1), 77–94.

DeYoung, R., & Gregoire, M. B. (1993). Use of capital budgeting techniques by foodservice directors in for-profit and not-for-profit hospitals. *Journal of the American Dietetic Association, 93,* 67–69.

Dittmer, P. R., & Griffin, G. G. (1997). *Principles of food, beverage, and labor cost controls for hotels and restaurants* (5th ed.). New York: Van Nostrand Reinhold.

Educational Institute of the American Hotel and Motel Association. (1987). *A uniform system of accounts and expenses dictionary for small hotels, motels, and motor hotels* (4th ed.). East Lansing, MI: Author.

Everett, M. D. (1989). Managerial accounting systems: A decision-making tool. *Cornell Hotel and Restaurant Administration Quarterly, 30*(1), 46–51.

Fromm, B., Moore, A. N., & Hoover, L. W. (1980). Computer-generated fiscal reports for food cost accounting. *Journal of the American Dietetic Association, 77,* 170–174.

Horngren, C. T., & Foster, G. (1991). *Cost accounting: A managerial emphasis* (7th ed.). Englewood Cliffs, NJ: Prentice Hall.

Industry at a Glance. (2002). Available online at www.restaurant.org

Educational Institute of the American Hotel and Motel Association. (1996). *A uniform system of accounts for the lodging industry* (9th ed.). New York: Author.

Keiser, J. R., DeMicco, F., & Grimes, R. N. (2000). *Contemporary management theory: Controlling and analyzing costs in food service operations* (4th ed.). Upper Saddle Review, NJ: Prentice Hall.

Kreul, L. M., & Scott, A. M. (1982). Magic numbers: Psychological aspects of menu pricing. *Cornell Hotel and Restaurant Administration Quarterly, 23*(2), 70–75.

Kwansa, F., & Evans, M. R. (1988). Financial management in the context of the organizational life cycle. *Hospitality Education and Research Journal, 12*(2), 197–214.

Moncarz, E. S., & O'Brien, W. G. (1990). The powerful and versatile spreadsheet. *The Bottom Line, 5*(4), 16–21.

National Restaurant Association. (1995). *Uniform system of accounts for restaurants* (3rd ed.). Washington, DC: Author in cooperation with Laventhol & Horwath.

National Restaurant Association (2006). Fact Sheet. [Online]. Available: www.restaurant.org.

Ninemeier, J. D. (1999). *Planning and control for food and beverage operations* (4th ed.). East Lansing, MI: Educational Institute of the American Hotel and Motel Association.

Pavesic, D. (1988). Taking the anxiety out of menu pricing. *Restaurant Management, 2*(2), 56–57.

Pavesic, D. V., & Magnant, P. E. (2005). *Fundamental principles of restaurant cost control* (2nd ed). Upper Saddle River, NJ: Prentice Hall.

Rose, J. (Ed.). (1984). *Handbook for healthcare foodservice management.* Rockville, MD: Aspen Publishers.

Schmidgall, R. S. (1997). *Hospitality industry managerial accounting* (4th ed.). East Lansing, MI: Educational Institute of the American Hotel and Motel Association.

Schmidgall, R. S., Geller, A. N., & Ilvento, C. (1993). Financial analysis using the statement of cash flows. *Cornell Hotel and Restaurant Administration Quarterly, 34*(1), 47–53.

Schmidgall, R. S., & Ninemeier, J. D. (1986). Food-service budgeting: How the chains do it. *Cornell Hotel and Restaurant Administration Quarterly, 26*(4), 51–55.

Sneed, J., & Kresse, K. H. (1989). *Understanding foodservice financial management.* Rockville, MD: Aspen Publishers.

Stinson, J. P., & Guley, H. M. (1988). Use of the microcomputer to determine direct costs of menu items. *Journal of the American Dietetic Association, 88,* 586–590.

CHAPTER 14

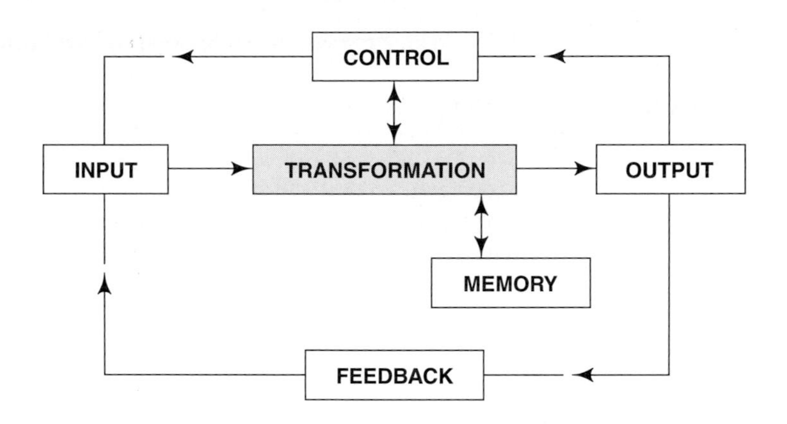

Marketing Foodservice

Enduring Understanding

- Marketing is more than advertising and promoting products or services.
- Marketing involves determining customer wants and needs and designing products/services that will be profitable and meet those wants/needs.

Learning Objectives

After reading and studying this chapter, you should be able to

1. Define core marketing terms such as marketing mix, target market, market segmentation, and promotion.
2. Develop a marketing plan for a foodservice operation.
3. Describe ways a market may be segmented, based on demographic, geographic, or psychographic characteristics.

Marketing has become a philosophy and a way of doing business in many industries, including the foodservice industry. In this chapter, you will see that marketing is much more than advertising and selling. You will be introduced to basic marketing concepts and definitions and will learn strategies and techniques that foodservice managers can use to develop a strong marketing program for their operation.

SYSTEMS APPROACH TO MARKETING MANAGEMENT

Managers in all types of foodservice operations, whether commercial or onsite, recognize marketing as a component of management. Marketing is an integral component of the management functions that transform inputs into outputs (see Figure 14.1). Marketing, long defined in terms of a product, now includes a much stronger emphasis on the consumer. Achieving customer satisfaction as an output of the foodservice system depends in part on the success of a foodservice manager's ability to apply marketing principles in his or her operation.

DEFINITION OF MARKETING

Marketing has been defined in many ways. The American Marketing Association, the professional association for those interested in the marketing discipline, defines **marketing** as "an organizational function and a set of processes for creating, communicating, and delivering value to customers and for managing customer relationships in ways that benefit the organization and its stakeholders" (see www.marketingpower.com).

Marketing Products

Many activities are needed to market products or goods. Producers, sellers, and buyers of products are all involved in marketing. Because of escalating costs in products and labor in the past decade, noncommercial foodservice managers have become cognizant of the value of using marketing principles. Competition for survival has become a priority for healthcare organizations, universities, and other institutions. Commercial foodservice managers learn early to be competitive because of the high failure rate of restaurants.

For an *exchange* to occur between two or more individuals or organizations, each must be willing to give up "something of value" for "something of value" (Pride & Ferrell, 1997). Both the buyer and seller have to communicate with each other to make their "something of value" available, as shown in Figure 14.2. In most situations, the seller has products, and the buyer has financial resources such as money or credit. In an exchange, products are traded for other products or money. The exchange must be satisfying to both the buyer and the seller. The

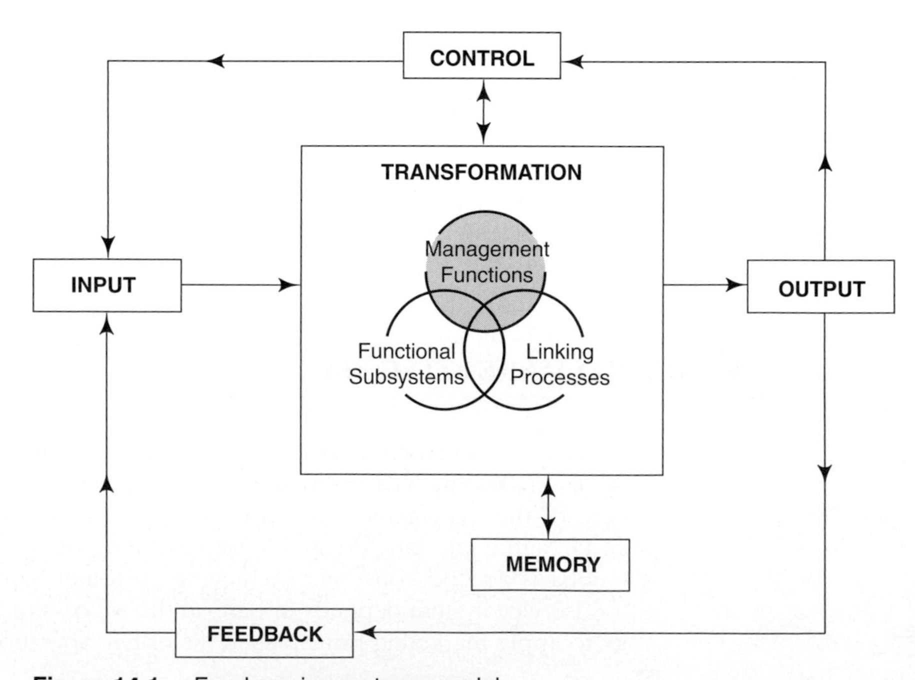

Figure 14.1. Food service systems model.

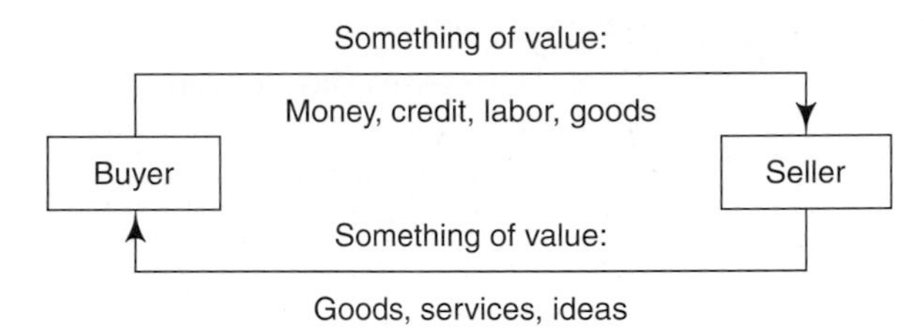

Figure 14.2. Exchange between buyer and seller.
Source: Pride, William, & Ferrell, O. C. *Marketing: Concepts and Strategies,* 10th edition. Copyright © 1997 by Houghton Mifflin Company. Used by permission.

buyer must be pleased with the product received from the seller and the seller with the reimbursement.

MARKETING CONCEPT

The **marketing concept** is a management philosophy about how an organization views customers and the sale of their product. In marketing terms, a "product" is a good, service, or idea. Kotler, Bowen, and Makens (1999) categorize marketing philosophies into five concepts:

- **Manufacturing/production concept**—based on the belief that customers favor products that are available and highly affordable; thus, companies should focus on production and distribution efficiency.
- **Product concept**—based on belief that customers prefer existing products and product forms; thus, companies should develop good versions of these products.
- **Selling concept**—based on the belief that customers will not buy enough of the organization's product unless the organization undertakes a large selling and promotion effort.
- **Marketing concept**—based on the belief that organizations should determine the needs and wants of target markets and deliver desired satisfaction more effectively and efficiently than competitors.
- **Societal marketing concept**—based on the belief that organizations should determine needs and wants of target markets and deliver desired satisfaction more effectively and efficiently than competitors in a way that maintains or improves the consumer's and society's well-being.

Evolution of Marketing

Satisfying customers has not always been the philosophy of business. This concept was preceded by the production and sales eras. During the late 1800s, the Industrial Revolution marked the beginning of the modern concept of marketing. With mass production, better transportation, and more efficient technology, products could be manufactured in greater quantities and sold at lower prices. In the initial stages of the Industrial Revolution, output was limited, and marketing was devoted to the physical distribution of products. Because demand was high and competition low, businesses did not have to conduct consumer research, modify products, or otherwise adapt to consumer needs. The production era of marketing had the goal of increasing production to keep up with demand (Evans & Berman, 1996).

Once a company was able to maximize its production capabilities, it hired a sales force to sell its inventory. The sales era began in the 1920s, as businesses sought to alter customers' desires to accept the products being produced. Business executives decided that advertising and sales were the major means of increasing profits.

By the early 1950s, business executives recognized that efficient production and promotion of products did not guarantee that customers would buy them. The marketing era began with the creation of the marketing department to conduct consumer research and advise management in how to design, price, distribute, and promote products. During the past 20 years, marketing managers have been represented on organizations' decision-making teams because of their ability to conduct consumer research.

Competition is intense, and companies must draw sophisticated customers to their products and retain them (Evans & Berman, 1996). According to Pine (1993), today's competitive edge is found in the concept of mass customization, which is the development, production, marketing, and delivery of affordable goods and services with enough variety that most people will find exactly what they want. For example, Burger King's advertising campaigns have included anti-mass-production slogans such as "Have It Your Way!" and "Sometimes You've Gotta' Break the Rules." The marketing concept, with its emphasis on satisfying the customer, thus forms the basis of the marketing era.

Marketers must have a good understanding of customer satisfaction and loyalty to be successful. Harrell (2002) defined **customer satisfaction** as a customer's positive, neutral, or negative feelings about the value received from a product. **Customer loyalty** refers to the frequency with which a customer consistently purchases a specific brand.

Customer satisfaction: Positive, neutral or negative feelings about the value received from a product.

Customer loyalty: Frequency with which a customer consistently purchases a specific brand.

Implementation of Marketing

Once the management of an organization has adopted a marketing philosophy, the development and implementation of that philosophy is based on the marketing concept. According to Lewis, Chacko, and Chambers (1997), the marketing concept derives from the premise that the customer is king, has a choice, and does not have to buy the product. Thus the best way to earn a profit is to serve the customer better.

The marketing concept affects all types of business activities and should be adopted entirely by top-level management. These executives must incorporate the marketing concept into their personal philosophies of business management so completely that customers become the most important concern in the organization. Support of managers and employees at all levels of the organization is required for implementation of the marketing concept (Pride & Ferrell, 1997).

First, management must establish an information system to determine the customers' real needs and use the information to develop products that satisfy them. This is expensive and requires money and time to make the organization customer oriented. Second, the organization must be restructured to coordinate all activities.

The head of the marketing department should be a member of the top-level management team in the organization (Pride & Ferrell, 1997).

Problems can occur with this new marketing approach. Most operations cannot make products specific to the needs of each customer in our mass-production economy. Regardless of the great amount of time and money spent for research, products still are produced that do not sell. Occasionally, satisfying one segment of the population makes another dissatisfied. Limited-menu restaurants realized early that one menu would not appeal to all members of a family. Applebee's Neighborhood Grill and Bar and Denny's have capitalized on this fact by recognizing that young families will become repeat customers if children are occupied and prices are low (Chaudhry, 1993).

MARKETING MANAGEMENT

Marketing management is a process of planning, organizing, implementing, and controlling marketing activities to facilitate and expedite exchanges effectively and efficiently (Pride & Ferrell, 1997). The managerial functions in the transformation element of the foodservice system have an important role in marketing management. **Effectiveness** refers to the degree to which an exchange helps to achieve an organization's objectives; the quality of the exchanges may range from highly desirable to highly undesirable. **Efficiency** refers to the minimization of resources that an organization must spend to achieve a specific level of desired exchanges. Pride and Ferrell (1997) summarized these definitions by stating that the overall goal of marketing management is to facilitate highly desirable exchanges and to minimize, as much as possible, the costs of doing so.

To achieve the goal of facilitating and expediting desirable exchanges, marketing management is responsible for developing and managing marketing strategies. Strategy encompasses key decisions for reaching an objective. A **marketing strategy** pertains to the selection and analysis of a group of people, identified as a target market, which the organization wants to reach; it includes the creation and maintenance of an appropriate marketing mix that will satisfy those people.

Marketing Mix

Marketing mix: Combination of product, price, place, and promotion to satisfy target market.

Target market: Customer with common characteristics for which an organization creates products/services.

To manage marketing activities, managers must deal with variables relating to the marketing mix and the marketing environment. The **marketing mix** is defined as the specific combination of marketing elements used to achieve an organization's objectives and satisfy the target market (Evans & Berman, 1996). The marketing mix decision variables are product, price, place, promotion, and other factors over which an organization has control. These variables are constructed based on buyer preferences. The **target market** is a group of persons for whom an organization creates a marketing mix that specifically meets the needs of that group (Pride & Ferrell, 1997). The marketing environment variables are political, legal, regulatory, societal, economic, competitive, and technological forces. Many decisions must be made concerning the activities required for each element included in the mix.

Product

A **product** can be a good, a service, or an idea (Hsu & Powers, 2002). Even though the manufacturing of products is not a marketing activity, research on customer needs and product designs is. Product decisions focus on which products to develop, which current products to promote, and which products to discontinue. The term **new product** means it is a genuine innovation because it has not been served commercially (such as the McGriddle® syrup-infused pancake). Products referred to as **new to the chain** such as Chicken McNuggets® are really an imitation of a successful product offered by another chain, such as KFC's chicken nuggets.

Price

Price is the amount of money charged for a product. Price competition has become very common in foodservice operations. In 1989, Taco Bell introduced value pricing. More than half of Taco Bell's menu was priced under $1.00 and company sales tripled. Once the success of value pricing became clear, McDonald's and many others followed. Marketing managers usually are involved in establishing pricing policies for various products because consumers are concerned about the value obtained in the exchange. Price is a critical competent of the marketing mix and often is used as a competitive tool. Price also helps establish a product's image. The goal is to set the price at a point that customers perceive value yet the company achieves the volume and profit it desires.

Promotion

This element is used to facilitate exchanges by informing prospective customers about an organization and its products. **Promotion** is used to increase public awareness about a new product or brand; also, it is used to renew interest in a product that is waning in popularity. Upscale restaurants spend less on advertising than mid-scale and quick-service restaurants (Powers, 1995). The foodservice industry is one of the largest advertisers in the United States. The strategic change in foodservice promotion took place in the 1980s, when advertising increased four times over the previous decade. The level of advertising in the quick-service restaurant has become quite large—McDonald's spends over $1 billion on marketing each year. In upscale foodservice operations, promotion plays a less crucial role.

Place

In marketing, **place** refers to the location, the place where food or services are offered. Increasingly, food is prepared elsewhere. Food manufacturers are preparing, packaging, and distributing menu items for restaurants and contract companies. Each is intruding on the turf of others, making foodservice competitive. Many foodservice operations are downsizing traditional distribution operations to fit into kiosks and mobile carts that are much less expensive. Baskin Robbins ice cream is being sold from kiosks in malls and airports, and Pizza Hut is delivering pizza to several thousand school lunch programs and hospitals. Customers have many options to purchase food when they are away from home.

Environmental Forces

As shown in Figure 14.3, the **marketing environment** surrounds the buyer and the marketing mix (Pride & Ferrell, 1997). Political, legal, regulatory, societal, economic, competitive, and technological forces in the environment affect the marketing manager's ability to facilitate and expedite change. The marketing environment influences customers' preferences and needs for products. These forces also directly influence how a marketing manager should perform certain marketing activities. Finally, a manager's decisions may be affected by environmental forces that influence customers' reactions to the organization's marketing mix.

Political forces influence the country's economic and political stability and decision making, which in turn affects domestic matters, negotiation of trade agreements, and determination of foreign policy. Political trends can have tremendous impact on the hospitality and healthcare industries. Organizations such as the National Restaurant Association, American Hotel and Lodging Association, American Dietetic Association, and School Food Service Association maintain lobbyists in Washington, D.C.

Legal forces are responsible for legislation and interpretation of laws. Marketing is controlled by numerous laws designed to preserve competition and protect the consumer; interpretation of laws by the marketers and courts has a great effect on marketing mix components.

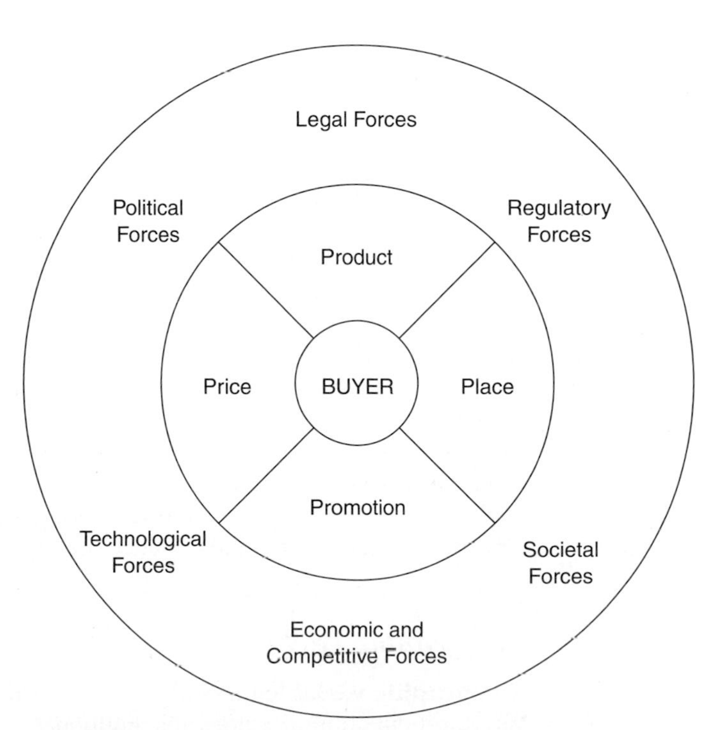

Figure 14.3. Components of the marketing mix and marketing environment.
Source: Pride, William, & Ferrell, O. C. *Marketing: Concepts and Strategies,* 10th edition. Copyright © 1997 by Houghton Mifflin Company. Used by permission.

Local, state, and federal regulatory forces develop and enforce regulations that can affect marketing decisions. Quite often, regulatory agencies, especially at the federal level, encourage industries to develop guidelines to stop questionable practices. Industry leaders usually cooperate to avoid government regulations. Individual industries and trade organizations also put regulatory pressures on themselves and their members.

Societal forces cause marketers to be responsible for decisions. Thousands of consumer groups have been formed to discuss such issues as environmental pollution and the use of pesticides on fruit and vegetable crops. Consumer groups also have been active in discussions on food labeling.

Economic forces have a major influence on competition, which is affected by the number of industries controlling the supply of a product, the ease by which a new operation can enter the industry, and the demand for the product relative to the supply. Demand is determined by buyers' abilities and willingness to purchase.

Technological forces have an impact on everyday living, influencing consumers' desires for products and the stability of the marketing mix. The technologies of communication, transportation, computers, and packaging influence the types of products being produced.

Managers must be able to adjust marketing strategies to major changes in the environment. If they want to develop effective strategies, managers must recognize the dynamic environmental forces that cause marketing problems and opportunities.

Market Segmentation

Market segmentation: Division of total market into groups of customers who have similar needs, wants, values, and buying behavior.

Swinyard and Struman (1986) stated that separating customers into "natural" market groups provides the basis for successful strategy development in marketing a restaurant. **Market segmentation** is the process of dividing a total market into groups of people with similar needs, wants, values, and buying behvaiors (Harrell, 2002). A market, in the context used here, is not a place but rather a group of people; as individuals or organizations, the group needs products and possesses the ability, willingness, and authority to purchase them. A market segment is a mixture of individuals, groups, or organizations that share one or more characteristics, which causes them to have similar product needs.

In a homogeneous market, which consists of individuals with similar product needs, a marketing mix is easier to design than one for a heterogeneous group with dissimilar needs. Choosing the correct variable for segmenting a market is important in developing a successful strategy. Variables have been grouped into four categories for the segmentation process: geographic, demographic, psychographic, and behavioristic. Examples of each are shown in Table 14.1.

Geographic Variables

Geographic variables include climate, terrain, natural resources, population density, and subcultural values that influence customers' product needs. In addition, the size of the region, city, county, or state and whether the area is urban, suburban, or rural have an effect on the market. Population in these areas, customer preferences, and spending patterns also need to be considered in marketing decisions. As an example of geographical variables, Cajun food, which is part of the

Table 14.1 Major Segmentation Variables for Consumer Markets

Variable	Typical Breakdown
Geographic	
Region	Pacific, Mountain, West North Central, West South Central, East North Central, East South Central, South Atlantic, Middle Atlantic, New England
City or metro size	Under 5000; 5000–20,000; 20,000–50,000; 50,000–100,000; 100,000–250,000; 250,000–500,000; 500,000–1,000,000; 1,000,000–4,000,000; 4,000,000 or over
Density	Urban, suburban, rural
Climate	Northern, southern
Demographic	
Age	Under 6, 6–11, 12–19, 20–34, 35–49, 50–64, 65+
Gender	Male, female
Family size	1–2, 3–4, 5+
Family life cycle	Young, single; young, married, no children; young, married, youngest child under 6; young, married, youngest child 6 or over; older, married, with children; older, married, no children under 18; older, single; other
Income	Under $10,000; $10,000–$15,000; $15,000–$20,000; $20,000–$30,000; $30,000–$50,000; $50,000–$100,000; $100,000 and over
Occupation	Professional and technical; managers, officials, and proprietors; clerical, sales; craftspeople, foremen; operatives; farmers; retired; students; housewives; unemployed
Education	Grade school or less; some high school; high school graduate; some college; college, graduate
Religion	Catholic, Protestant, Jewish, Muslim, Hindu, other
Race	White, Black, Asian
Nationality	American, British, French, German, Italian, Japanese
Psychographic	
Social class	Lower lowers, upper lowers, working class, middle class, upper middles, lower uppers, upper uppers
Lifestyle	Straights, swingers, longhairs
Personality	Compulsive, gregarious, authoritarian, ambitious
Behavioral	
Occasions	Regular occasion, special occasion
Benefits	Quality, service, economy, speed
User status	Nonuser, ex-user, potential user, first-time user, regular user
Usage rate	Light user, medium user, heavy user
Loyalty status	None, medium, strong, absolute
Readiness stage	Unaware, aware, informed, interested, desirous, intending to buy
Attitude toward product	Enthusiastic, positive, indifferent, negative, hostile

Source: Marketing for Hospitality and Tourism, 2nd edition, by Kotler, Bowen, and Makens. Reprinted by permission of Pearson Education, Inc., Upper Saddle, NJ.

culture in Louisiana, has been accepted in New York City and in California. The astute foodservice manager knows when and where to introduce such food items on the menu by doing a market analysis before making a decision.

Demographic Variables

Demographic variables consist of population characteristics that might influence product selection, such as age, gender, race, ethnicity, income, education, occupation, family size, family life cycle, religion, social class, and price sensitivity. Only those demographic variables pertinent to the population segment under consideration must be ascertained.

Psychographic Variables

Psychographic variables: Motives and lifestyle characteristics.

Psychographic variables include many factors that can be used for segmenting the market, but the most common are motives and lifestyles. When a market is segmented according to a *motive,* it recognizes the reason a customer makes a purchase. *Lifestyle* segmentation categorizes people according to what is important to them and their mode of living.

A useful classification system for segmenting customers in terms of a broad range of lifestyle factors is the **Values and Life-Styles (VALS)** research program, which is sponsored by SRI International in Menlo Park, California (see www.sric-bi.com). The current version of the program seeks to predict customer behavior by defining segments, based on resources and primary motivation.

Resources include the psychological, physical, demographic, and materials means and capabilities that people have. Resources are viewed on a continuum from low to high. Motivation is viewed on a continuum from low to high innovation and is divided into three components in the VALS model:

- **Ideals**—Consumers make choices based on their knowledge and principles.
- **Achievement**—Consumers make choices based on what they perceive will show their success to their peers.
- **Self-expression**—Consumers make choices based on a desire for social or physical activity, variety, or risk.

Combining the dimensions of resources and primary motivation results in the categorization of consumers into one of eight groups defined by SRI (see www.sric-bi.com) as follows:

- **Innovators**—Highest level of resources and motivation. They are characterized as change leaders who are active consumers and receptive to new ideas. They tend to prefer upscale, niche products and services.
- **Thinkers**—These consumers are characterized as mature, reflective, well-informed individuals who are motivated by ideals. They value order, knowledge, and responsibility and base decisions on their principles. Although their income allows abundant choices, they tend to look for durability, functionality, and value in products they purchase.
- **Achievers**—These individuals have high levels of resources and are motivated by achievement. They are characterized as work-oriented, conservative, and value-conscious consumers who favor established, prestige products that demonstrate success to their peers.
- **Experiencers**—These consumers have high levels of resources and are motivated by self-expression. They seek variety, excitement, new and risky activities. They tend to be young, impulsive, and rebellious individuals who spend much of their income on products such as fashion, entertainment, and socializing.
- **Believers**—These consumers have low amounts of resources and are motivated by ideals. They tend to be conservative, follow established routines, and organize their lives around home, family, and social or religious organizations. They favor American products and established brands.

- **Strivers**—These consumers have limited resources and are motivated by achievement. They often emulate those who own more impressive possessions, but usually lack the financial resources to purchase what they would like.
- **Makers**—These consumers have limited resources and are motivated by self-expression. They are characterized as practical, self-sufficient, and traditional. They experience the world by working on it (i.e., by building a house, fixing a car, canning vegetables, etc.). They purchase products that have practical or functional purpose.
- **Survivors**—These consumers have the lowest resources and motivation. They are very cautious consumers whose chief concerns are for security and safety. They are loyal to favorite brands purchased at a discount.

Behavioristic Variables

Behavioristic variables are the basis of some feature of consumer behavior toward and use of a product. They include variables such as purchase volume, purchase readiness, loyalty, and shopping behavior. To satisfy a specific group of customers, a special product might need to be produced, such as caffeine-free diet cola. How the customer uses a product also may determine segmentation. Frozen menu items are being packaged in single servings to meet the needs of people living alone.

Niche Marketing and Micromarketing

Newer strategies for market segmentation are niche marketing and micromarketing. Niche is a narrowly defined portion of the market with unique interests or characteristics. **Niche marketing** focuses on identifying small but profitable segments of the market and making products specifically for this segment. **Micromarketing** is marketing to the single customer, the smallest niche.

SERVICE MARKETING

In the definition of marketing used in this chapter, a *product* refers to a good, a service, an idea, or any combination of the three. A *good* is a tangible product that a customer can physically touch; a *service* is the application of human or mechanical efforts to people or objects. The most satisfactory definition of service is that given by Kotler (1996): A service is any act or performance that one party can offer to another that is essentially intangible and does not result in the ownership of anything.

Manufacturing firms still exist, but services have replaced goods in the U.S. economy, accounting for 67% of the **Gross Domestic Product (GDP)** and 57% of consumer expenditures (www.bea.gov, 2002). Several reasons for this growth in the services sector are apparent. The United States now is an information rather than an industrial society; services are the primary material of an information society. Services are no longer byproducts in a manufacturing or production process but often are the products. A third reason is the prosperity in the United States that has led to growth in financial, travel, entertainment, and personal-care services. Lifestyle changes, such as more women in the workforce, have created increased demand for child care, domestic, and other time-saver services including meal preparation. The elderly require increased healthcare services, and at the same time the younger population demands increased fitness and health maintenance services. The technology explosion has created more complex goods that require servicing and repair; the business environment is more specialized, justifying more business and industrial services.

Lewis, Chacko, and Chambers (1997) emphatically stated that marketing and management in a service business, such as the hospitality industry, are one and the same. Many marketing experts argue that service marketing is different from goods marketing and requires different strategies and tactics; they contend that a pure good without some elements of service attached to it is impossible.

Characteristics of Services

Zeithaml, Parasuraman, and Berry (1985) stated that the problems of service marketing are not the same as those of goods marketing. A look at the four basic characteristics of service marketing—intangibility, inseparability of production and consumption, perishability, and heterogeneity—explains why. Services generally are sold before they are produced, and goods generally are produced before they are sold (Berry & Parasuraman, 1991). Moreover, services marketing has a more limited influence on customers before purchase than goods marketing.

Intangibility

Intangibility of services: Inability of services to be seen, touched, tasted, smelled, or possessed before buying.

Atmospherics: Physical elements in an operation's design that appeal to customers' emotions.

Generally, **intangibility of services** is defined in terms of what services are not; they cannot be seen, touched, tasted, smelled, or possessed (Pride & Ferrell, 1997). Services are performances and, therefore, intangible; products are tangible. Services, however, have a few tangible attributes. Atmospherics is an example of a tangible attribute and has been used as a marketing tool for many years.

Atmospherics describes the physical elements in an operation's design that appeal to customers' emotions and encourages them to buy (Pride & Ferrell, 1997). Berry and Parasuraman (1991) stated that the principal responsibility for the service marketer is to manage tangibles to convey the proper signals about service.

The atmosphere of the exterior and interior of the operation may be friendly, exciting, quiet, or elegant. *Exterior atmospherics* is important to new customers who often judge a restaurant by its outside appearance. If windows are foggy or the grounds unkempt, customers might decide that service would be unacceptable, too. *Interior atmospherics* includes lighting, wall and floor coverings, furniture, and rest rooms. A pleasing and clean interior probably would indicate to the customer that the service would be impeccable. Sensory elements also contribute to atmosphere. *Color* can attract customers. Many limited-menu restaurants, for example, use bright colors, such as red and yellow, because they have been shown to make customers feel hungrier and eat faster, thus increasing turnover. *Sound* is important; a noisy restaurant probably would not be chosen by customers who are celebrating a wedding anniversary. *Odor* also might be relevant; the scent of freshly baked bread makes customers feel cared for and wanted. Again, these tangible sensory elements might reflect on the customer's perception of the intangible service (Pride & Ferrell, 1997).

Levitt (1981) stated that the most important thing to know about intangible service is that "customers don't know what they're getting until they don't get it." A customer is more precious than the tangible assets shown on a balance sheet, and companies that understand this concept continually tailor products and services to their customers. They look beyond the value of a single transaction to the customer's lifetime value to the company (Treacy & Wiersema, 1993).

Inseparability

Inseparability: Inability to separate production and service.

According to Pride and Ferrell (1997), **inseparability** of production and consumption is related to intangibility. Services are normally produced at the same time they are consumed. In a commercial foodservice operation, the waitstaff, bartender, and maître d'hôtel are producing services at the same time the customer is consuming them. The knowledge and efficiency of the waitstaff in taking the order and serving the meal, the desire of the bartender to mix a drink exactly the way the customer wants it, and the concern of the maître d'hôtel that the customer is satisfied are examples of inseparability of production and consumption. In hospitals, foodservice personnel deliver trays to patients who are either satisfied or dissatisfied immediately with the attitude or concern of the delivery person. Likewise, the warmth of a smile from a cook serving a child a school lunch is strongly associated with the acceptance of the meal.

Perishability

Perishability of services: Services cannot be stored for future sale.

The **perishability of services** means that those services cannot be stored for future sale (Evans & Berman, 1996). Unused capacity cannot be shifted from one time to another. Because service is produced and consumed simultaneously, it is perishable. The service supplier must try to regulate customer usage to develop consistent demand throughout various periods. The service operation must have the capacity and capability to produce when demand occurs; if demand does not occur, however, that capacity and capability are lost and wasted, resulting in losses in the bottom line (Lewis, Chacko, & Chambers, 1997). For example, if overstaffing occurs, the labor cost is too high; if understaffing occurs and demand increases, service becomes too slow.

One alternative is to charge prices high enough to permit overstaffing. Restaurant managers of operations known for a high level of service often use this scheme. Most foodservices cannot afford such a solution to the problem, and the result is irate customers. Reduction of staff is both a marketing and management decision, and the impact on the customer must be the first consideration. If service is being marketed, it becomes an expectation of the customer, and management must accept the risk of overstaffing and the higher cost to the customer. A customer, however, often makes a sacrifice beyond cost, which is time. Waiting for room service, for lunch in a restaurant, for a bottle of wine to be served, or for the check to come causes the customer to become irritated. An alternative is the limited-menu restaurant, which has been very successful because it capitalizes on the time saved. If marketing creates expectations, makes promises, offers value, and reduces risk, management needs to understand that the cost of keeping a customer is far less than that of creating a new one.

Heterogeneity

Heterogeneity of service: Variation and lack of uniformity in the performance by different service employees.

Heterogeneity of service is concerned with the variation and lack of uniformity in the performance of people. This is different from the poor service caused by an insufficient number of staff; rather, it is fluctuations in service caused by unskilled employees, customer perceptions, and the customers themselves. Variations might occur between services within the same organization or in the service provided by one employee from day to day or from customer to customer. Most services are labor intensive, and the performance of each employee is different. Managers have difficulty in predicting how employees with different backgrounds and personalities will react in various circumstances.

Marketers of services who make promises to the customer never know how employees will handle a situation or how the customer will perceive a service. Lewis, Chacko, and Chambers (1997) stated that the consequence is that good service may equal bad service. One customer may be pleased that the waitstaff never permits the coffee cup to become empty, but another customer may be annoyed because the cup never becomes empty. Sometimes, less service is more service. An example is the popular salad bar. Many people like the idea that they can select what they want, although this is less service because they have to get their own food.

Components of Service Products

Customers are concerned with the components of goods, services, and environment when purchasing the hospitality product (Lewis, Chacko, & Chambers, 1997). *Goods* are mostly physical factors over which management has direct, or almost direct, control and are usually tangible. The manager's expertise determines the quality level of goods. Lewis, Chacko, and Chambers (1997) define price as being tangible, although it is a cost of services and goods. To the customer, however, price is tangible in any purchase decision.

Service includes nonphysical, intangible attributes that management should control. Personal elements provided by employees, such as friendliness, speed, attitude, and responsiveness, are very important components of service. In the environment category, items over which management may have some control may be included. These items may or may not be tangible but are something the customer feels. That feeling is what the manager is marketing. Décor, atmosphere, comfort, ambience, and architecture are attributes included in this category.

Service Marketing Mix

Managers desiring to market service must provide benefits that satisfy the needs of the customer. Target markets should be defined and the marketing mix identified before a marketing strategy can be finalized. The four elements required for a marketing mix for goods are applicable to service: product, distribution, promotion, and price.

Services are intangible products and thus difficult for customers to evaluate. If a limited-menu restaurant chain can standardize a service and market it more effectively than other chains can market the service, the chain generally will gain a greater share of the market. An example has been the serving of certain food items, such as baked-to-order pizza within 15 minutes of placing the order.

Distribution in the service context refers to making services available to prospective users. Instead of taking the goods to the customer, customers must come to the service. Pizza, hamburger, chicken, fish, and other specialty limited-menu restaurants distribute products that are the same in many locations.

Promotion: Use of communication to inform and influence consumers.

Lewis, Chacko, and Chambers (1997) define **promotion** as marketing communication that serves specifically as an incentive to stimulate sales on a short-term basis. Promotions are frequently used to stimulate business in off periods when normal business flow has decreased.

Establishing a price for service can be difficult because of its intangibility. The more standardized service becomes, the easier pricing is for the manager. Pricing service in a limited-menu restaurant with well-defined procedures for employees to follow is much easier than pricing in an upscale restaurant in which employees are encouraged to satisfy individual customer's desires.

STRATEGIC MARKETING

Marketing strategy pertains to the selection and analysis of a target market and the creation and maintenance of an appropriate marketing mix. Any strategic planning process begins with the organization's mission statement and objectives and ends with a marketing plan.

Strategic Planning Process

Evans and Berman (1996) identified seven interrelated steps in the strategic planning process for marketing:

- Define the organizational mission.
- Establish strategic business units.
- Set marketing objectives.
- Perform a situation analysis.
- Develop a marketing strategy.
- Implement tactical plans.
- Monitor results.

The organizational **mission statement** is a summation of the organization's purpose, competition, target market, product, and service and of the recipients of the service, including consumers, employees, owners, and the community. For example, the mission of a hospital could be to provide both inpatient and outpatient medical service to the people in the community within budgetary limitations. Each department would have a mission compatible with that of the organization. The mission of the foodservice department thus would be to provide food within the departmental budget to the patients, employees, and visitors.

After defining a mission, an organization should establish **strategic business units (SBUs).** Each SBU is a separate component of the organization and has a specific market focus and a manager with responsibility for placing all functions into a strategy. SBUs are the basis for a strategic marketing plan. In business terminology, the hospital is the corporate level, and the foodservice department is the strategic business unit. Every SBU has a clearly defined market segment with a strategy consistent with that of the corporation, its own mission, and its own competitors.

A **marketing objective** is a statement of what is to be accomplished through marketing activities. Each SBU in an organization needs to set its own objectives in clear, simple terms for marketing performance. Objectives generally are described in both quantitative terms (dollar sales, percentage profit growth, market share) and qualitative terms (image, uniqueness, customer service). Many restaurants and other food-service organizations combine quantitative and qualitative goals, such as dollar sales and uniqueness based on a new theme.

Situation analysis: Identification of marketing opportunities and challenges.

Situation analysis is the identification of marketing opportunities and potential problems confronting an organization. The manager needs to know the current condition of the organization and the direction in which it is going.

A technique often used in situation analysis is a **SWOT analysis.** A SWOT analysis focuses on identifying the *S*trengths, *W*eaknesses, *O*pportunities and *T*hreats to an organization. When using a SWOT analysis for marketing, the focus typically is on factors related to market share, customer loyalty, customer satisfaction, and previous marketing success. Strengths are defined as the unique resources that a company can provide. Weaknesses or constraints are those aspects of a company that limit the company's ability to achieve its goals. Opportunities are areas where competitive advantage exists or where new markets could be developed. Threats are those elements that might prevent accomplishment of objectives.

A **marketing strategy** encompasses selecting and analyzing a target market and creating and maintaining an appropriate marketing mix that will satisfy that market (Pride & Ferrell, 1997). A strategy should be as specific as possible. For example, to increase dessert sales, a poor strategy might be something imprecise like, "The addition of low-fat frozen yogurt to the menu will be advertised." A better strategy would provide more guidance: "Dessert sales will be increased by 10% within 3 months by adding low-fat frozen yogurt to the menu and increasing advertising to health-conscious consumers."

Tactic: Specific action.

The marketing strategy is implemented through a series of **tactics,** which are specific actions. According to Lewis, Chacko, and Chambers (1997), strategy is the way to gain and keep customers; tactics are the step-by-step procedures on how to do it. The objective could be "to be perceived as the restaurant of choice," and the strategy, "to give customers better value." Some of the tactics could include having a table ready for customers who have made reservations, calling customers by name, having the print on the menu large enough to read, and offering a selection for customers with special dietary needs. Tactics flow from strategy, which means the appropriate strategy must be developed first.

Monitoring results involves the comparison of performance standards against actual performance over a definite time. Budgets, timetables, sales, and cost analyses may be used to analyze results. If actual performance does not meet the standards, corrective action should be taken in problem areas. Many organizations have contingency plans if performance standards are not met.

Marketing Research

The foundation of a successful marketing plan is research (Yesawich, 1987). Only through research can proper judgments be made about the best combination of product, distribution, promotion, and price. Market research can help an establishment succeed, rather than merely survive, by attracting new customers, keeping up with trends, and tailoring menus to meet customer needs (Stern, 1990). Intuition and past experience, rather than scientific decision making, often govern marketing decisions, however.

The American Marketing Association defines **marketing research** as the function that links the consumer, customer, and public to the marketer through information that is used to identify and define marketing opportunities and problems; generate, refine, and evaluate marketing actions; monitor marketing performance; and improve understanding of marketing as a process (see www.marketingpower.com). To be effective, marketing research must be systematic and not haphazard or disjointed. Marketing research involves a series of steps including data collection, recording, and analysis (Evans & Berman, 1996). Data may be available from different sources: the organization itself, an impartial marketing research company, or a research specialist working for the organization.

Objectivity, accuracy, and thoroughness are important when conducting research.

- **Objectivity**—conducted in an unbiased, open-minded manner; conclusion based on data and analysis
- **Accuracy**—use of research tools that are carefully constructed and implemented
- **Thoroughness**—ensuring that the sample represents the population; a questionnaire, if used, is pretested; and the analysis of data is statistically correct.

The marketing research process, as shown in Figure 14.4, consists of five steps for logically solving a problem: problem definition, data collection, data analysis, recommendations, and preparation of the report. Foodservice managers conducting marketing research should think about each of these steps and tailor them to fit the problem.

The Marketing Plan

Marketing plan: Written document or blueprint governing an organization's marketing activities.

Pride and Ferrell (1997) define **marketing plan** as a written document or blueprint governing an organization's marketing activities, including the implementation and control of those activities. **Marketing planning** is a systematic process involving the assessment of marketing opportunities and resources, the determination of marketing objectives, the development of a marketing strategy, and planning for implementation and control. A marketing plan needs to be integrated and evaluated.

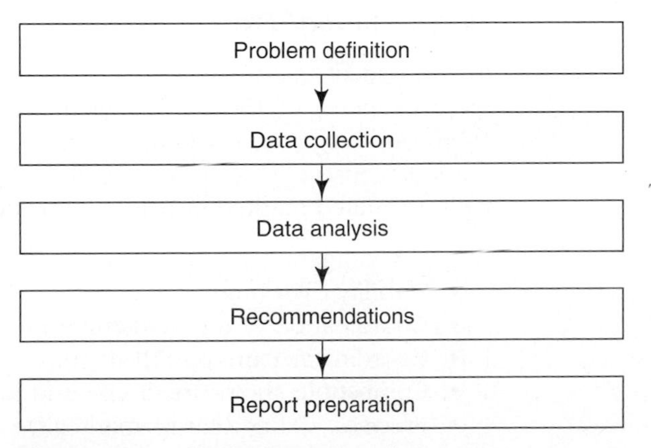

Figure 14.4. Marketing research process.

Development of Plan

Strategic planning should be done before a marketing plan is developed. The marketing should define the target market and marketing mix variables. The plan also should identify resources needed and objectives to be achieved.

A marketing plan should be real and workable and should be easy to execute (Lewis, Chacko, & Chambers, 1997). It also should be flexible but have a certain amount of stability. Specific responsibilities with times and dates for accomplishment should be designated in the plan. Finally, a marketing plan needs to be constantly reviewed and evaluated to keep it current.

Types of Plans

Marketing plans can be categorized according to duration, scope, and method of development (Evans & Berman, 1996). Marketing plans typically are developed for one year and are considered short range; medium-range plans from two to five years are sometimes used; and those over five years, long-range marketing plans, are seldom developed. Short- and medium-range plans are more detailed and more geared to the operation than long-range plans.

The scope of marketing plans varies tremendously. Separate marketing plans may be developed for individual menu items and special services.

Finally, the method of development of plans may be bottom-up, top-down, or a combination of the two. In the *bottom-up approach,* information from employees is used to establish objectives, budgets, forecasts, timetables, and marketing mixes. Bottom-up plans are realistic and good for morale. Coordination of each bottom-up plan into one integrated plan may be difficult to achieve because of conflicts, for example, in estimates of the impact of marketing a new menu item. In the *top-down approach,* top management directs and controls planning activities. Top-level managers understand the competition and environment and provide direction for marketing efforts. If input from lower-level managers is not sought, however, morale may diminish. A combination of these two approaches could be the best solution; top management could set the overall objectives and policy, and lower-level managers could establish the plans for implementing the policy (Evans & Berman, 1996).

Integration of Plans

Integration of marketing is necessary if the product, distribution, promotion, and price elements of the marketing mix are to be synchronized (Evans & Berman, 1996). An integrated marketing plan is one in which all the various components are unified, consistent, and coordinated. According to Evans and Berman (1996), a well-integrated marketing plan will include the following:

- Clear organizational mission
- Stability over time
- Coordination of the marketing mix
- Coordination among SBUs
- Compatible short-, medium-, and long-range plans
- Precisely defined target market(s)
- Long-term competitive advantages

A clear organizational mission defines an organization's type of business and place in the market. The mission is involved each time products or services are added or deleted or new target markets are sought or abandoned. A marketing plan must show stability over time to be implemented and evaluated correctly.

The product, distribution, promotion, and price components of the marketing mix need to be coordinated within each department (Evans & Berman, 1996). Coordination among individual departments or units is increased when the strategies and resources allocated to each are described in short-, medium-, and long-range plans. Compatible plans for each department or unit become the broad marketing plan for the organization. The target market needs to be identified in a marketing plan to guide marketing efforts and future direction. If two or more distinct target markets are present, each should be clearly defined. For example, a foodservice organization catering to children and parents probably will have separate strategies for each of these segments.

Customer expectations and quality of the product and service need to be emphasized in developing long-term plans for successful competition in the market. Internal customers, like the maître d', chef, waitperson, and cashier, all need to work together while keeping the external customer in focus.

Control and Evaluation

Control. Control is as necessary in marketing as in managing all facets of the foodservice organization. The manager should establish performance standards for marketing activities based on goals of the organization. Internal standards generally are expressed as profits, sales, or costs. Most organizations use external individuals or organizations, such as consultants or marketing research firms, for marketing assistance.

When foodservice managers attempt to control marketing activities, they frequently have problems because information is not always available. Even though controls should be flexible enough to allow for environmental changes, the frequency and intensity of changes may curtail effective control. Because marketing overlaps other activities, the precise costs of marketing are difficult to define.

Evaluation. Sales analysis can be used for evaluating the actual performance of marketing strategies (Pride & Ferrell, 1997). A *sales analysis* is the detailed study of sales data, either volume or market share. Dollar volume sales are frequently used because the dollar is the common denominator of profits, sales, and costs. Price increases and decreases, however, affect total sales figures and need to be considered in the analysis. For example, if a restaurant increases prices by 10% this year and its sales volume is 10% greater than last year's, it has not had any increase in unit sales. A restaurant marketing manager should factor out the effects of price changes.

Market share: Percentage of industry sales for a product.

Market share is stated as the percentage of industry sales for a product. The rationale for using market share is to estimate if sales changes occurred because of the organization's marketing strategy or from uncontrollable environmental factors. The assumption is that industry sales decrease when restaurant sales decrease and market share remains constant. If a restaurant suffers a decrease in both sales and market share, however, the marketing strategy is not effective. Market share analysis should be interpreted with caution because it is based on uncontrollable factors, such as differing objectives among companies.

Professional Organization Profile

School Nutrition Services Dietetic Practice Group of the American Dietetic Association

MISSION

To be recognized leaders in "keeping nutrition in our schools"

WHO BELONGS TO THE ORGANIZATION

Members are also members of the American Dietetic Association engaged in the management of school food service and nutrition education programs at the local, state, and national levels or employed by companies providing products and services to these programs. Student membership available through the American Dietetic Association.

ADVANTAGES OF MEMBERSHIP

Members receive a quarterly newsletter, *The Nutrition Link;* earn continuing education credit by attendance at association sponsored seminars and workshops; have access to scholarships designated for members; and receive a membership directory for networking.

WEBSITE

www.eatright.org

MEET THE PRESIDENT

Linda Godfrey, RD; President 2005–2006; is an adjunct instructor at Samford University in Homewood, AL. Her career has included foodservice director positions in healthcare and schools, most recently as foodservice director for the Shelby County Schools in Alabama.

CHAPTER SUMMARY

This summary is organized by the learning objectives.

1. Marketing is a complex process that involves planning, promoting, and distributing goods, services, and ideas that satisfy customers and meet organizational goals. There are several terms that are used in marketing. Marketing mix is the combination of product, price, place, and promotion. Target market is a group of consumers with common characteristics. Market segmentation is the process of dividing a total market into groups based on consumer characteristics. Promotion is the use of various techniques to increase public awareness or renew interest in a product or service.
2. The marketing plan should define the target market, describe the marketing mix, and identify resources needed to execute the plan. The plan should be workable, easy to execute, and flexible, and it should clearly identify specific responsibilities and time/date deadlines for accomplishment.
3. There are many ways to segment a market. Segmentation by demographic characteristics would involve grouping consumers based on attributes such as age, gender, race, ethnicity, income, education, occupation, religion, social class, and so on. Segmentation by geographic characteristics would involve grouping consumers based on indicators such as city, state, region of the country or world, population of area, and so on. Psychographic segmentation would involve grouping of consumers based on factors such as lifestyle, motives, values, and so on.

TEST YOUR KNOWLEDGE

1. What is the marketing environment? How does it impact the development and execution of a marketing plan?
2. Define the term marketing concept and describe how it applies to a foodservice operation.
3. What are variables that influence the marketing mix and market segmentation?
4. Explain the four characteristics of services and describe how they impact marketing in foodservice operations.
5. Describe how strategic planning might be used to enhance foodservice marketing.

CLASS PROJECTS

1. Ask a manager from a local foodservice operation that has an active marketing program to speak to the class about his or her program; have the manager include description of target market, types of data collected, and how a marketing plan is developed and evaluated.
2. Divide into groups of two to three students. Develop a marketing plan for a local foodservice operation.

WEB SOURCES

www.bea.doc.gov	Bureau of Economic Analysis
www.marketingpower.com	American Marketing Association
www.hsmai.org	Hospitality Sales and Marketing Association International
www.demographics.com	American Demographics
www.sric-bi.com	SRI Consulting Business Intelligence
www.quirks.com	Quirk's Marketing Research Review
www.str-online.com	Smith Travel Research
www.census.gov	U.S. Census Bureau

BIBLIOGRAPHY

Albrecht, K. (1995). *At America's service: How your company can join the customer service revolution*. Homewood, IL: Dow Jones-Irwin.

Albrecht, K., & Zemke, R. (1995). *Service America! Doing business in the new economy*. Homewood, IL: Dow Jones-Irwin.

Allen, R. L. (1992). The world of target marketing. *Nation's Restaurant News, 26*(11), 25, 62.

Berry, L. L., & Parasuraman, A. (1991). *Marketing services: Competing through quality*. New York: Free Press.

Chaudhry, R. (1993). Food for tot. *Restaurants & Institutions, 103*(7), 131, 134.

Congram, C. A., & Friedman, M. L. (Eds.). (1991). *The AMA handbook of marketing for the service industries*. New York: AMACOM.

Davidow, W. H., & Uttal, B. (1989). Service companies: Focus or falter. *Harvard Business Review, 67*(4), 77–85.

Dodd, J. (1992). President's page: The fifth P. *Journal of the American Dietetic Association, 92,* 616–617.

Evans, J. R., & Berman, B. (1996). *Marketing* (7th ed.). Upper Saddle River, NJ: Prentice Hall.

Forgac, J. (1999). Marketing. In J. Martin & M. Conklin (eds.). *Managing child nutrition programs*. Gaithersburg, MO: Aspen, pp 611–626.

Gronroos, C. (1990). *Service management and marketing: Managing the moments of truth in service competition*. Lexington, MA: Lexington Books.

Hart, C. W., Casserly, G., & Lawless, M. J. (1984). The product life cycle: How useful? *Cornell Hotel and Restaurant Administration Quarterly, 25*(3), 54–63.

Harrell, G. B. (2002). *Marketing* (2nd ed.). Upper Saddle River, NJ: Prentice Hall.

Heskett, J. L. (1986). *Managing in the service economy*. Boston: Harvard Business School Press.

Houston, F. S. (1986). The marketing concept: What it is and what it is not. *Journal of Marketing, 50*(2), 81–87.

Hsu, C. H. C., & Powers, J. (2002). *Marketing hospitality* (3rd ed.). New York: John Wiley.

Jacobs, P. (1993). Staying focused: Strategic plans need constant care. *Nation's Restaurant News, 27*(20), 22.

Kanter, R. M. (1992). Think like the customer: The global business logic. *Harvard Business Review, 70*(4), 9–10.

Kashani, K. (1989). Beware the pitfalls of global marketing. *Harvard Business Review, 67*(5), 91–98.

Kotler, P. (1977). From sales obsession to marketing effectiveness. *Harvard Business Review, 55*(6), 67–75.

Kotler, P. (1996). *Marketing management: Analysis, planning and control* (9th ed.). Englewood Cliffs, NJ: Prentice Hall.

Kotler, P., Bowen, J., & Makens, J. (1999). *Marketing for hospitality and tourism* (2nd ed.). Upper Saddle River, NJ: Prentice Hall.

Levitt, T. (1981). Marketing intangible products and product intangibles. *Harvard Business Review, 59*(3), 94–102.
Lewis, R. C. (1984). Theoretical and practical considerations in research design. *Cornell Hotel and Restaurant Administration Quarterly, 24*(4), 25–35.
Lewis, R. C. (1989). Hospitality marketing: The internal approach. *Cornell Hotel and Restaurant Administration Quarterly, 30*(3), 41–45.
Lewis, R. C., Chacko, H. E., & Chambers, R. E. (1997). *Marketing leadership in hospitality: Foundations and practices* (2nd ed.). New York: Van Nostrand Reinhold.
Lovelock, C. H. (1980). Why marketing management needs to be different for services. In J. H. Donnelly & W. R. George (Eds.), *Marketing of services* (pp. 708–719). Chicago: American Marketing Association.
Lovelock, C. H. (1983). Classifying services to gain strategic marketing insights. *Journal of Marketing, 47*(3), 9–20.
Lovelock, C. H. (1996). *Services marketing* (3rd ed.). Englewood Cliffs, NJ: Prentice Hall.
McKenna, R. (1988). Marketing in an age of diversity. *Harvard Business Review, 66*(5), 88–95.
McKenna, R. (1991). Marketing is everything. *Harvard Business Review, 69*(1), 65–79.
National income and product accounts. (1992). *Survey of Current Business, 72*(9), 5–47.
Pine, B. J., II. (1993). *Mass customization: The new frontier in business competition*. Boston: Harvard Business School Press.
Powers, T. (1995). *Introduction to management in the hospitality industry* (5th ed.). New York: John Wiley.
Pride, W. M. & Ferrell, O. C. (1997). *Marketing: Concepts and strategies* (10th ed.). Boston: Houghton Mifflin.
Solomon, J. (1993). Homemade marketing strategies. *Restaurants USA, 13*(1), 17–19.
Star, S. H. (1989). Marketing and its discontents. *Harvard Business Review, 67*(6), 148–154.
Stern, G. M. (1990). The case for marketing research. *Restaurants USA, 10*(7), 26–29.
Sullivan, J. (1991). Market your restaurant as you work the floor. *Nation's Restaurant News, 25*(16), 22.
Swinyard, W. R., & Struman, K. D. (1986). Market segmentation: Finding the heart of your restaurant's market. *Cornell Hotel and Restaurant Administration Quarterly, 27*(1), 89–96.
Treacy, M., & Wiersema, F. (1993). Customer intimacy and other value disciplines. *Harvard Business Review, 71*(1), 84–93.
Uhl, K. P., & Upah, G. D. (1986). The marketing of services. In J. N. Sheth & D. E. Garrett (Eds.), *Marketing management: A comprehensive reader* (pp. 999–1026). Cincinnati: South-Western Publishing.
Vance, D. E. (1992). Capture your market—then work to keep it. *Nation's Restaurant News, 26*(6), 52.
West, J. J., & Olsen, M. D. (1989). Competitive tactics in foodservice: Are high performers different? *Cornell Hotel and Restaurant Administration Quarterly, 30*(1), 68–71.
Yesawich, P. C. (1987). Hospitality marketing for the '90s: Effective marketing research. *Cornell Hotel and Restaurant Administration Quarterly, 28*(1), 49–57.
Yesawich, P. C. (1988). Planning: The second step in market development. *Cornell Hotel and Restaurant Administration Quarterly, 28*(4), 71–81.
Yesawich, P. C. (1989). The final steps in market development: Execution and measurement of programs. *Cornell Hotel and Restaurant Administration Quarterly, 29*(4), 83–91.
Zeithaml, V. A., Parasuraman, A., & Berry, L. L. (1985). Problems and strategies in services marketing. *Journal of Marketing, 49*(2), 33–46.
Zeithaml, V. A., Parasuraman, A., & Berry, L. L. (1990). *Delivering quality service: Balancing customer perceptions and expectations*. New York: Free Press.

CHAPTER

15

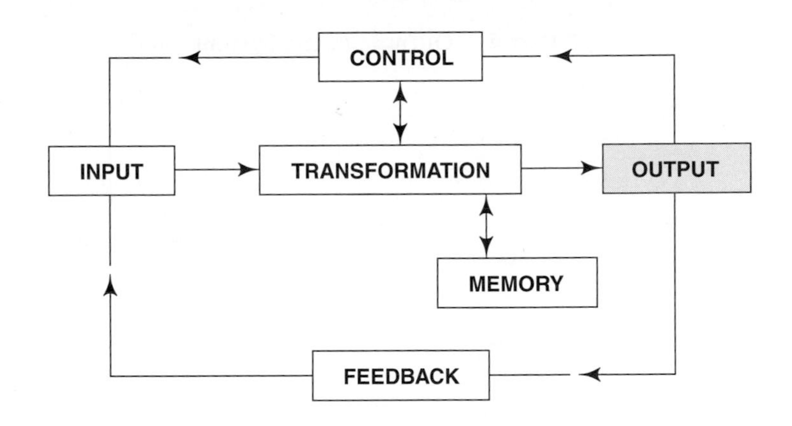

Meals, Satisfaction, and Accountability

Enduring Understanding

- Successful foodservice organizations will serve quality food in the correct quantity, will have satisfied employees and customers, and will be financially accountable.

Learning Objectives

After reading and studying this chapter, you should be able to

1. Describe outcomes of the foodservice systems model.
2. Measure system outcomes.

A key component of the foodservice systems model is the outputs of the system. In this chapter, you will learn about the outputs of the foodservice systems model. You will be introduced to techniques that can be used to measure the quality of the system outputs. We will describe ways companies are recognized for the quality of their outputs.

OUTPUTS IN THE FOODSERVICE SYSTEMS MODEL

As shown in Figure 15.1, the primary outputs of the foodservice systems model are the quantity and quality of meals produced, customer satisfaction, employee satisfaction, and financial accountability. All are important outcomes, and the foodservice manager will be responsible for ensuring their achievement.

QUANTITY OF MEALS

Most foodservice operations will have an objective of producing meals in sufficient quantity to meet the needs of their customers. Quantity of meals refers to the number of meals that are prepared for service.

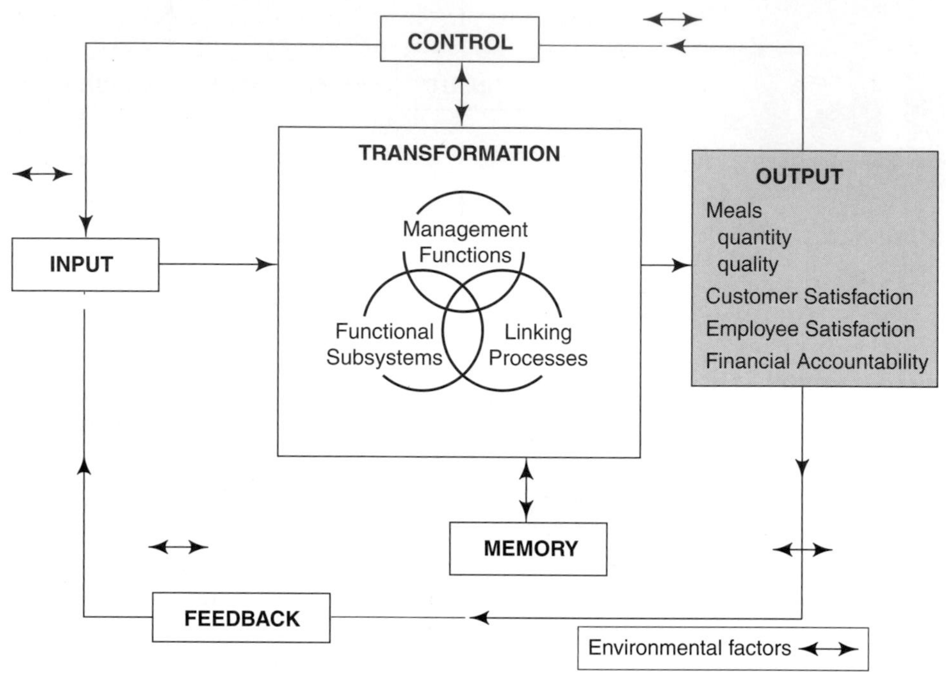

Figure 15.1. Outcomes of the foodservice system.
Source: Adapted from *A Model for Evaluating the Foodservice System* by A. G. Vaden, 1980. Manhattan, KS: Kansas State University. © 1980. Used by permission.

The quantity of meals is an important output because it impacts both the financial performance of an organization and the satisfaction of its customers. Overproduction (production of too great a quantity) of menu items results in a loss of income because often the excess food items cannot be used and are discarded. The organization incurred a cost for the product and for the production done to these products. When these products are discarded rather than being sold, no income is earned to offset this cost. Conversely, underproduction (production of not enough quantity) can result in a decrease in customer satisfaction. Customers who expect a foodservice operation to have a specified menu item usually are disappointed when they learn that an item on the menu is not available that day. Such dissatisfaction can lead to a loss of customers.

QUALITY OF MEALS

The quality of meals focuses on the ingredients and the preparation techniques used to prepare those products. According to Sanders and Hill (2001), quality of meals should be determined according to

- Taste
- Quality of ingredients
- Portion size
- Methods of preparation
- Service

Quality ingredients and product standards are important components for helping ensure that food products meet the organization's quality goals. The quality of ingredients is identified in the food specifications written by the foodservice manager. Appendix A contains example specifications for food products and Appendix B provides reference sources for product specifications. The quality of the ingredient specified often depends on the menu item for which the ingredient will be used, the price to be charged for the food product, and the atmosphere in which it will be served.

Product standards define what is expected in a food product. They provide the basis for monitoring performance of the organization and taking corrective action deemed necessary.

The word *quality* often is used in combination with standards and is central to the traditional statement of the objective of a foodservice: the production of the highest possible quality food. The primary quality attributes of food are microbiological, nutritional, and sensory, which require controls throughout the procurement/production/service cycle to maintain them. The increased time lapse between the preparation and distribution of food to the customer in some foodservices is critical and requires monitoring for nutritional value.

Although the customer's interpretation of quality is certainly important in the foodservice operation, quality can also be defined from a scientific standpoint. Thorner and Manning (1983) stated that the analyst or technologist refers to quality as an index or measurement obtained by grading or classifying a product's chemical and physical characteristics in accordance with explicit, predetermined specifications. The essential elements that must be evaluated in establishing the quality of a product are flavor, texture, appearance, consistency, palatability, nutritional value, safety, ease of handling, convenience, storage stability, and packaging.

Thorner and Manning (1983) identified two dominant factors in the evaluation of quality:

- The actual chemical or physical measurement of the product
- The acceptance of the product by consumers

In addition, management will be concerned with the relationship of quality to cost, profit, and selling price, in particular to cost of the product, profits generated, and consumer acceptance.

Many factors are responsible for poor-quality food, among them poor sanitation, improper handling, malfunctioning equipment, incorrect pre-preparation or preparation, and carelessness. Thorner and Manning (1983) compiled the following list of the prime factors responsible for significant quality changes:

- Spoilage due to microbiological, biochemical, physical, or chemical factors
- Adverse or incompatible water conditions
- Poor sanitation and ineffective ware washing
- Improper and incorrect precooking, cooking, and postcooking methods
- Incorrect temperatures
- Incorrect timing
- Wrong formulations, stemming from incorrect weight of the food or its components
- Poor equipment maintenance
- Presence of vermin and pesticides
- Poor packaging

MENU

Cheeseburger on Bun

Hot Dutch Potato Salad

Sliced Tomatoes Dill Pickle Spears

Fudge Brownie

Fresh Fruit Cup

STANDARDS FOR MENU ITEMS

Cheeseburger. The hamburger patty should be evenly browned, juicy, and glossy with a moist and tender interior. The interior cooked temperature should exceed 160°F. The appearance should be pleasing and the flavor typical of beef. Processed American cheese covering the grilled patty will be melted without becoming stringy. The color will be yellow orange, typical of the type of cheese, and the flavor will be distinctive but will not dominate the meat.

Bun. The top crust should be golden brown and tender. The interior texture should be fine, even grained, free from large air bubbles, and have thin cell walls. The white crumb should be moist and silky. The bun will be fresh.

Hot Dutch Potato Salad. The cooked potato slices will be soft but firm. Diced celery, chopped onion, pimento, parsley, and crisp ⅛-inch-wide bacon pieces will be identifiable both in flavor and color. The tart vinegar sauce will be adequately sweet and thick enough to coat all ingredients. The salad will be served hot.

Fudge Brownie. The exterior should be dark brown and smooth with a darker interior. Texture should be slightly chewy and free of crumbling. The brownie should be slightly moist and tender, not doughy or pasty. Nut pieces should be evenly dispersed and identifiable. Each brownie should be approximately 3 inches square and ¾-inch uniform height.

Fresh Fruit Cup. Fruit should be selected for color, texture, flavor, and shape—e.g., golden peaches, blue plums, red watermelon, green grapes, white bananas. The fruit should be cut into bite-size pieces without predominance of any one type. A simple syrup flavored with lemon, lime, or other fruit juices should slightly coat the fruit.

Figure 15.2. Example of quality standards for a lunch menu in a university residence hall foodservice.

Any of these factors, either singly or in combination, will contribute to poor quality and cause changes affecting the flavor, texture, appearance, and consistency of food.

In Figure 15.2, a menu for a university residence hall foodservice is given, together with the quality standards for evaluating each of the menu items. The assumption is made that the items will be prepared using a standardized recipe defining the ingredient amounts, precise production methods, product yield, and serving directions. Too frequently, foodservice managers have failed to define standards of quality for menu items, and therefore the basis for control and evaluation is not available. In Appendix A, product standards for several menu items for a university foodservice are included. These standards are good examples of the efforts of a management team to define quality for foodservice operations.

Product evaluation is an important component of assessing the quality of meals served. Evaluating a product entails comparing to both the specification requirements for the ingredients and the determined standard for the finished product.

Food quality is evaluated by sensory, chemical, and physical methods. Sensory methods are used to determine if foods differ in such qualities as taste, odor, juiciness, tenderness, or texture and to define the extent and direction of the differences (Palmer, 1972). Chemical and physical methods for testing food often are used with sensory analysis to identify reasons for differences in color, texture, and flavor. These methods are usually more reproducible and less time consuming than sensory analysis, but they are limited to areas in which they have been shown to measure the quality that is apparent to the senses.

Sensory analysis: Science that measures the texture, flavor, and appearance of food products through human senses.

In this text, **sensory analysis** is defined as a science that measures the texture, flavor, and appearance of food products through human senses. Commonalities in all the definitions are measurement, food characteristics, and human senses.

In very large foodservice organizations, competent personnel in quality control laboratories routinely conduct sensory, physical, and chemical tests during the initial development of new menu items. Large multiunit operations often have a sophisticated quality control program centralized at corporate headquarters.

Physical and chemical testing of foods may be limited in smaller operations in which a full-scale quality control program is not justified. These tests are in the domain of food science and are beyond the scope of this text. Because sensory evaluation of new menu items by panels and consumer testing can be a regular part of any foodservice operation, practical applications will be discussed.

Skelton (1984) contended that to ensure continuing success, foodservice operations should use sensory evaluation for new menu items and for maintaining the quality of existing items. The cost of an error in a menu innovation can be extremely high, involving such costs as reprinting menus, training staff, and loss of business from dissatisfied patrons.

The necessary sensory evaluation of new menu items requires a sensory or consumer panel before serving them to customers. *Sensory panels* are relatively small, ranging from 6 to 12 persons trained to judge quality characteristics and differences among food items. Panel members must be experienced in the use of score cards and in the vocabulary of food description; they also must be able to distinguish among various levels of the basic tastes (sweet, salt, sour, and bitter) and to repeat their assessments with reasonable precision.

In contrast to the trained panel, the *consumer panel* usually includes 50 to 100 persons who are reasonably representative of the target market. The objective in using consumer panels is to evaluate acceptance of, or preference for, a menu item.

Analytical and affective sensory tests are used for product evaluation and defined as follows:

- **Analytical sensory test**—differences and similarities of quality and quantity of sensory characteristics that are evaluated by a panel of specially trained persons, commonly identified as a trained panel

- **Affective sensory test**—preference, acceptance, and opinions of a product that are evaluated by consumers who have no special sensory training, commonly identified as an untrained panel

Skelton (1984) outlined three general purposes for sensory tests in foodservice operations: discriminating among food items, describing characteristics, and determining acceptance and preference.

- **Discrimination sensory test**—This test determines detectable differences among food items. For example, judges may be presented three samples and asked to choose the one that is different. If the panel has difficulty identifying the odd sample, the conclusion can be made that the product was not altered by the ingredient or process change.
- **Descriptive sensory test**—Quality control and recipe development both depend on descriptive tests to provide information about certain sensory characteristics. Adjectives, numerical scales, and rankings are used to evaluate such attributes as taste, aroma, texture, tenderness, and consistency. Figure 15.3 gives a list of terms that are useful in describing food products.
- **Acceptance and preference sensory test**—This test, used with a consumer panel, is intended to answer the questions of whether or not people will like the menu item. Using this test, preference for certain characteristics may be rated, or an overall preference score may be attained. The conclusion can be made that one recipe is not better than the other if consumers indicate no difference in overall preference.

Properly designed, executed, and analyzed sensory tests can be used to assist the foodservice manager in developing products that are more likely to succeed in the marketplace and in setting standards for ingredients and menu items (Setser, 1992). Table 15.1 outlines typical problem areas and the type of panel and category of sensory test appropriate to each.

Appearance (optical properties)

- Color

blueness	dark	greenness	redness
bright	dull	light	white
brownness	grayness	pale	yellowness

- Other optical properties[a]

clear	irridescent	scum	translucent
cloudy	lustrous	sediment	transparent
frothy	muddy	shiny	uniform pigment distribution
glossy	opaque	sparkling	

Appearance (physical form)[b]

broken	large	rounded	smooth
crumbly	loose particles	shriveled	stringy
curdled	medium	shrunken	uniform shape
flat	rough	small	uniform size
irregular			

Figure 15.3a. Terms used in judging food products.

Aroma and Flavor by Mouth[c]

burned	eggy	medicinal	scorched
buttery	fishy	nutty	spicy
caramelized	floral	putrid	starchy
dairy-like	fruity	rancid	

Quality Judgments

acceptable	fresh	low	optimum
appealing	full-bodied	mellow	pleasing
delicate	good	normal	poor
desirable	high	objectionable	rich
excellent	ideal	obnoxious	stale

Taste and Chemical Feeling Factors

astringent	burning	sharp	sweet
biting	coolness	sour	tart
bitter	salty		

Textural Attributes

- Consistency—resistance to deformation through continuous changes of form

liquid	stiff	thick	viscous
runny	syrupy	thin	watery
slimy			

- Geometrical properties—size, shape, and orientation of particles perceived by tactile nerves

abrasive	crystalline	granular	rough
aerated	even	irregular	sharp
amorphous	fine	lumpy	smooth
beady	flaky	porous	stringy
broken	flat	powdery	uneven
cellular	fluffy	puffy	uniform
coarse	foamy	pulpy	
creamy	grainy	regular	

- Mechanical deformation—reaction of food to stress

adhesive	elastic	mealy	stiff
bouncy	firm	pasty	sticky
brittle	fracturable	plastic	tacky
chewy	friable	rubbery	tender
cohesiveness	gooey	short	tooth-packing
crisp	gummy	soft	tough
crumbly	hard	solid	
crunchy	limp	springy	

- Mouth feel—physical feel of moistness and oiliness of food as it is broken down

dehydrating	greasy	moist	saliva-inducing
dry	juicy	mouth-coating	soggy

[a]attributes related to transmission or reflectance of light but not related to pigmentation
[b]visual perceptions related to dimensions and adherence between particles
[c]sensations produced by volatile substances through nasal and oral cavities that provide characteristic flavors perceived in foods and beverages

Figure 15.3b.

Table 15.1 Problems Solved by Sensory Evaluation

Problem	Type of Panel	Category of Test
(1) Recipe development: Maximizing quality	Trained	Discrimination and description
(2) Shelf life: Storage time and temperature	Trained	Discrimination and description
(3) Acceptance: Likelihood of purchase	Consumer	Acceptance or preference
(4) Convenience food: Best substitution	Trained or Consumer	Description or acceptance
(5) Quality control: Product consistency	Trained	Discrimination and description

Source: "Sensory Evaluation of Food" by M. Skelton, 1984, *Cornell Hotel and Restaurant Administration Quarterly, 24*(4), 51. Used by permission.

Sensory Analysis Instruments

Before introducing a new menu item, the recipe must be standardized. The final step in recipe standardization is evaluation of the product for acceptability by a taste panel that usually consists of cooks, supervisors, and managers. Standards for evaluating a recipe for blueberry muffins and stir-fried chicken and vegetables are shown in Figure 15.4.

Sensory analysis can be a useful tool for recipe standardization. In the early stages of this process, cooks and managers together can help in determining which sensory attributes are most affected by changes in formulation and ingredients and in indicating general trends in acceptability for those attributes. The two basic tests in this type of sensory evaluation are discrimination and acceptance. For *discriminatory tests,* such as difference tests and ranking, panelists do not need extensive training and large panels are not required. The *paired comparison test* can be used to differentiate between a pair of coded samples on the basis of some specified characteristic, such as sweetness, crumbliness, moistness when chewing,

Blueberry Muffins

A blueberry muffin should be scored on overall acceptability, including appearance, texture, tenderness, and flavor. The crust should be crisp, shiny, pebbly, and golden brown with a well-rounded top free from knobs. It should be large in volume compared to weight. The interior crumb should be moist, light, and tender with a coarse, even grain and no tunneling. The whole blueberries should be evenly distributed through the muffin. The blueberries should be moist but not discolor the muffin. The muffin flavor should be delicate, not bready or too sweet, and the blueberries should have a natural taste.

Stir-Fried Chicken and Vegetables

The chicken and vegetables should be scored on overall acceptability, including appearance, texture, tenderness, and flavor. The appearance should be pleasing and the chicken and vegetables identifiable. The bite-size pieces of chicken should be tender and juicy. The fresh vegetables should retain some of the original crispness and natural color. Fresh ginger and garlic should enhance, but not overpower, the flavors of the other ingredients. The sauce should be clear and ingredients glazed.

Figure 15.4. Example of quality standards for standardizing recipes.

Name of Panelist ______________________ Date __________

Product ______________________________

Evaluate the sweetness of the two samples of blueberry muffins. Taste the muffin on the left first. Indicate which is sweeter.

Code number 581 —　　*Code number 716* —

Comments:

Figure 15.5. Example of paired comparison test form.

lightness, or degree of browning. A sample evaluation form for a paired comparison test is found in Figure 15.5.

The *ranking test* extends the paired comparison test to three or more coded samples, and panelists are asked to rank them by intensity of the characteristics that differentiate the products. For example, if moistness of crumb obviously is affected by changing the formulation of the recipe, this characteristic is one that should be ranked. An example of this type of questionnaire for a ranking test is provided in Figure 15.6.

Panelists should receive the samples for all tests in a random order to avoid any order biases in the testing. Ten to twelve panelists should be used. Results of the discrimination testing can be compared to affective testing of acceptance or preference to determine how panelists like the changes. For *affective tests,* the number and type of panelists are an important issue. Such tests are aimed at determining the response of the consumer to the product. If too few representative or unbiased panelists are used, the results can be questionable. In the laboratory, 20 to 40 panelists generally can establish relative desirability; for hedonic evaluations of products, 50 to 100 judgments usually are necessary. Larger panels are needed only when unusually high precision is required.

Name of Panelist ______________________ Date __________

Rank the muffins for crumb moistness as you chew them. The least moist muffin is ranked first and the most moist sample is ranked fourth. Place the code numbers on the appropriate lines. Test the samples of the coded muffins in the following order: *212, 336, 471, 649.*

____ **1**　____ **2**　____ **3**　____ **4**

Figure 15.6. Example of ranking test form.

For the relative desirability information, the use of the paired comparison or ranking test is also appropriate. In this type of test, the panelists' response to the following statement is requested: "Two samples of stir-fried chicken and vegetables are presented. Taste each sample in the order specified and indicate which saltiness level you prefer." Further information can be obtained by using a *hedonic scale* for individual attributes. An example of such a scale for three characteristics of interest in the blueberry muffins is given in Figure 15.7. A comparison of the attribute intensity information from the discrimination testing and the relative preference data might suggest changes for the next test run, which likely would involve adjustments in ingredient quantities, cooking times, procedures, or serving methods. Generally, only one change should be made at a time in systematically evaluating the effect of the various changes.

Overall acceptability using the hedonic scale of two or more different muffins could also be obtained after recipe modifications are considered complete. Such evaluations of the product likely would include assessments from customers. To perfect some recipes, many tests might be needed. If test products are acceptable, the recipe then is put into the format being used in the operation and placed in the permanent file.

Name of Panel Member: John Green

Date: 02/05/99

PRODUCT: Blueberry Muffin

Please rate the muffin by checking one point on the following scale to indicate your evaluation of each attribute.

SWEETNESS	SALTINESS	CRUMBLINESS
☐ Like extremely	☐ Like extremely	☐ Like extremely
☐ Like very much	☐ Like very much	☑ Like very much
☐ Like moderately	☐ Like moderately	☐ Like moderately
☐ Like slightly	☐ Like slightly	☐ Like slightly
☐ Neither like, nor dislike	☑ Neither like, nor dislike	☐ Neither like, nor dislike
☑ Dislike slightly	☐ Dislike slightly	☐ Dislike slightly
☐ Dislike moderately	☐ Dislike moderately	☐ Dislike moderately
☐ Dislike very much	☐ Dislike very much	☐ Dislike very much
☐ Dislike extremely	☐ Dislike extremely	☐ Dislike extremely

COMMENTS: Explain your decisions

Muffin not sweet enough for tart blueberries

Figure 15.7. Example of a hedonic scale for rating the acceptance of blueberry muffin attributes.

CUSTOMER SATISFACTION

Customer satisfaction is critical output of the foodservice system. It involves the perceptions of customers related to the food they were served, the service that was provided to them, and the atmosphere in which it was provided. The foodservice industry is very customer-satisfaction sensitive; customers decide while they are still consuming the food product whether they are satisfied with the food and their dining experience.

Techniques to Assess Customer Satisfaction

There are a variety of techniques a foodservice manager can use to help determine the level of satisfaction of their customers. Gathering input from guests as they are finishing their dining experience provides the most accurate information because the dining experience is still very real to the guest. Techniques commonly used to gather customer satisfaction information include (Sanders & Hill, 2001)

- Walk-through audits
- Talking with guests
- Customer comment cards
- Exit interviews
- Mystery shopper reports
- Customer surveys
- Focus groups

A *walk-through audit* involves the manager walking through the operation, observing and evaluating the food, the service, and the appearance of the operations. Typically a standard check sheet will be developed based on the standards for the operation, which the manager can complete during the walk-through.

Talking to guests and carefully listening to their responses is one of the best ways for a manager to assess customer satisfaction. Often managers will elicit the greatest amount of input from guests by asking open-ended questions such as, "How was your meal?" "What could we do to improve your dining experience?" Open-ended questions such as these allow the guest to comment on a variety of attributes related to the food, the service, or the dining atmosphere.

Many foodservice operations use *customer comment cards* as a means for gathering input from their customers. These cards may be given to guests by the person serving the meal or might be available on the dining table or meal tray. Usually the comment cards are limited to a few key questions and space for guests to write additional comments. One of the greatest concerns with using comment cards is the representativeness of the information collected. Often, only guests who were extremely satisfied or extremely dissatisfied complete the cards, which may give a distorted view of actual overall customer satisfaction.

Exit interviews provide another way to gather customer satisfaction data. Exit interviews often are difficult to conduct because the customer is ready to leave and may not appreciate being asked to stay to provide information related to their dining experience. The form used for exit interviews varies. Sometimes closed-ended questions are used in which a customer is asked to provide a specific response such as yes or no or excellent, good, poor. In other cases, more open-ended questions are used to allow the customer to provide information about the questions being asked.

Mystery shoppers are individuals hired to dine in a foodservice operation as customers and then to prepare a report evaluating their experience. Often the evaluation completed by a mystery shopper is a fairly comprehensive evaluation form. The identity of the mystery shopper is unknown to the foodservice staff who serves the customers; sometimes such a shopper is hired by the corporate level and the person's identity is unknown to the managers as well. The University of Massachusetts hires students to serve as mystery shoppers in its dining centers. These mystery shoppers complete weekly evaluations on many aspects of food and service quality.

Customer surveys may be developed by the manager of a foodservice operation or may be purchased from a consulting firm. The Press Ganey, Inc. survey is an example of a purchased survey used with hospital patients to evaluate patient satisfaction. It includes a section assessing patient satisfaction with foodservice. Customer surveys may be completed by mail or through a telephone interview. Typically the surveys include a list of questions to gather information on the guest's satisfaction with specific aspects of the foodservice operation. Two national level surveys are conducted to provide data on restaurant customer satisfaction: the CREST® customer satisfaction report and the American Customer Satisfaction Index (ASCI). Both provide an overall measure of customer satisfaction with the industry.

Focus groups involve discussions with a group of 5 to 15 individuals to gather more specific information. Focus group members might be repeat customers or randomly selected potential customers. Conducting focus group meetings can be very time consuming but provide an excellent way to gather detailed information from those external to the organization.

The content of any of these techniques will vary depending on what is of most importance to the foodservice manager. Table 15.2 contains possible topic areas that might be included to help assess customer satisfaction.

Customer Satisfaction and Organizational Profitability

Heskett and colleagues (1994), in a much quoted research paper, demonstrated empirically the linkages among employee satisfaction, quality service, customer satisfaction, and operational profitability. The authors proposed a Service-Profit Chain to model the relationships they found among these variables. The Service-Profit Chain suggests that when there is internal service quality through workplace design, job design, and employee recognition, the organization will have more satisfied employees. When employees are more satisfied, they will be more productive and are more likely to stay with the organization. These satisfied employees give better customer service, which results in more satisfied and loyal customers. Having satisfied and loyal customers will increase an organization's revenue and profitability. Sherman (1999) also stressed the importance of customer satisfaction for organizational success. The author recommended a list of 18 "commandments" for healthcare professionals to follow to deliver total customer satisfaction (Table 15.3).

Table 15.2 Topics for Customer Satisfaction Evaluations

Category	Possible Areas to Address
Facilities	Inside and outside appearance Lighting Noise level Cleanliness Atmosphere/ambience Location
Food	Appearance Taste/flavor Temperature Price/value Menu selections Size of portions
Service	Attentiveness of staff Knowledge of staff Timeliness of service Accuracy of order Bill preparation/payment Responsiveness of staff Friendliness of staff Appearance of staff Appropriateness of interaction with customers

Table 15.3 Commandments of Total Customer Satisfaction in Health Care

1. Constantly measure, monitor and share customer feedback
2. Accept no customer defection as inevitable or negligible
3. Evangelize customer satisfaction at every level in the organization
4. Every customer contact is an opportunity to measure, manage and master customer satisfaction
5. Use Big D (dissatisfaction) to drive improvements in customer satisfaction
6. Commit to customer satisfaction BHAGs (Big, Hairy, Audacious Goals)
7. Be critical in you interpretation of customer satisfaction ratings
8. Listen to the special needs of every customer
9. Associate Satisfaction = Customer Satisfaction = Success
10. Manage associate turnover to manage customer satisfaction
11. Accept no excuses for poor customer satisfaction
12. Reward, recognize, and reinforce excellent associate performance for customer satisfaction
13. Balance the seven areas of key business results (customer satisfaction, productivity, economics, quality, organization climate, people growth and innovations)
14. Establish uniform standards of performance for customer satisfaction and accept nothing less
15. Benchmark against only the most excellent companies and don't limit your options to the health care industry
16. Free associates to problem solve and create customer satisfaction
17. Link up with physicians
18. Value and measure customer share first, market share second

Source: Sherman (1999); American Dietetic Association (2005).

EMPLOYEE SATISFACTION

Employee satisfaction: Beliefs and feelings an employee has about his or her job.

Employee satisfaction is the beliefs and feelings an employee has about his or her job. As shown in Figure 15.8, from George and Jones (2002), satisfaction with one's job is impacted by four factors: personality, values, work situation, and social influences. *Personality* influences how a person thinks and feels about his or her job. *Values* include both intrinsic (related to the nature of the job itself) and extrinsic (related to the consequences of work) components. Values impact an employee's convictions about what outcomes to expect from work and what behavior is appropriate at work. The *work situation* often is believed to have the greatest influence on an employee's level of job satisfaction. The work situation includes tasks employees perform, the individuals with whom they interact, the physical environment in which they work, and the way in which the organization treats employees as evidenced by things such as pay, work hours, job security, and so on. *Social influence* is the influence that others and the organizational culture have on a person's attitudes and behaviors.

There are a variety of measurement instruments that have been developed to help managers assess employee satisfaction. The most commonly used are the Minnesota Satisfaction Questionnaire (Weiss et al., 1967) and the Job Descriptive Index (JDI) (Balzer et al., 1997). The Minnesota Satisfaction Questionnaire measures satisfaction across 20 dimensions of work: ability utilization, achievement, activity, advancement, authority, company policies, compensation, co-workers, creativity, independence, moral values, recognition, responsibility, security, social service, social status, supervision-human relations, supervision-technical, variety, and working

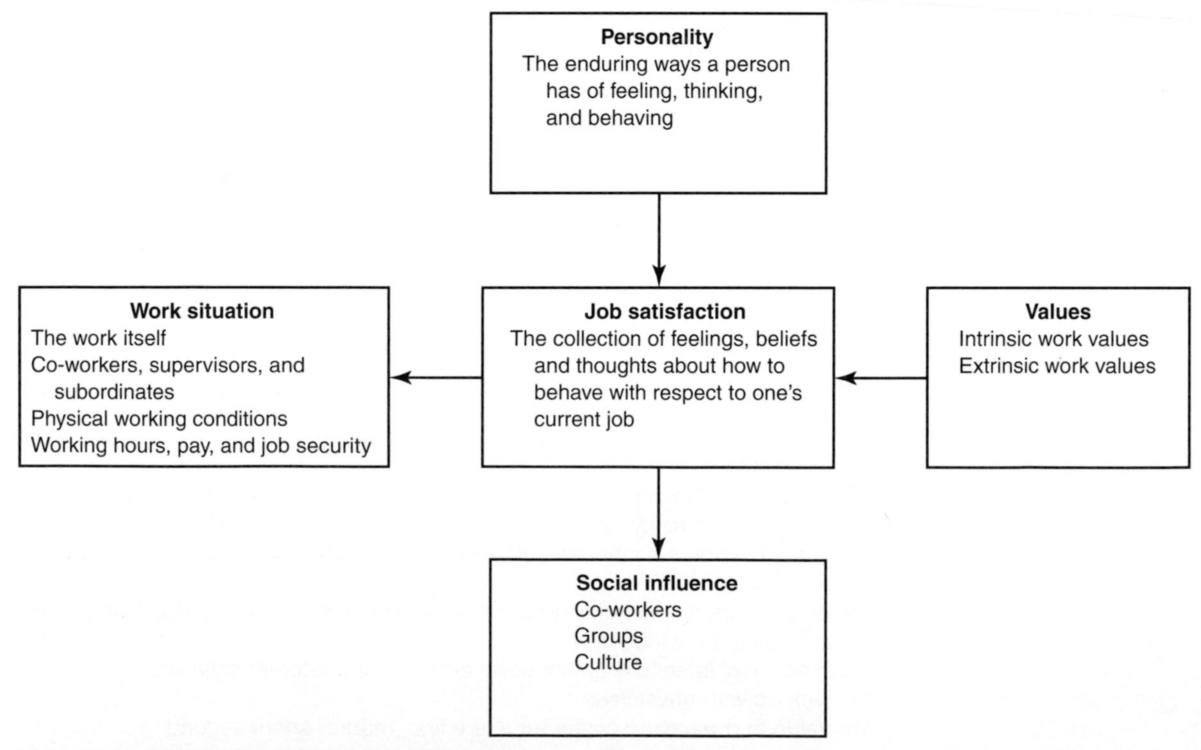

Figure 15.8. Determinants of job satisfaction.
Source: Reprinted with permission from George, J. M., & Jones, G. R. (2002). *Organizational Behavior.* Upper Saddle River, NJ: Prentice Hall.

conditions. The JDI measures satisfaction on five dimensions of a job: work on present job, present pay, opportunities for promotion, supervision, and co-workers.

Employee satisfaction is important to managers because it impacts an employee's behavior on the job. Satisfied employees have a more positive attitude on the job which is readily seen by customers and often results in increased customer satisfaction.

FINANCIAL ACCOUNTABILITY

A final outcome of the foodservice system model is **financial accountability.** Foodservice managers will be expected to control costs and generate revenue regardless of whether the operation is a for-profit or not-for-profit organization. Chapter 13 provided details on the types of financial records that an organization might have to help manage their finances.

Financial accountability includes not only keeping the proper financial records but also communicating appropriately the financial status of the operation to the appropriate stakeholders. Uniform systems of accounts (national restaurant association, 1995) typically are used to provide the framework for financial reporting.

Menu engineering, introduced by Kasavana and Smith in 1982, is a management information tool that focuses on both the popularity and the contribution to profit of menu items. The menu engineering model segments menu items into one of four segments as shown in Figure 15.9 based on the item's popularity relative to

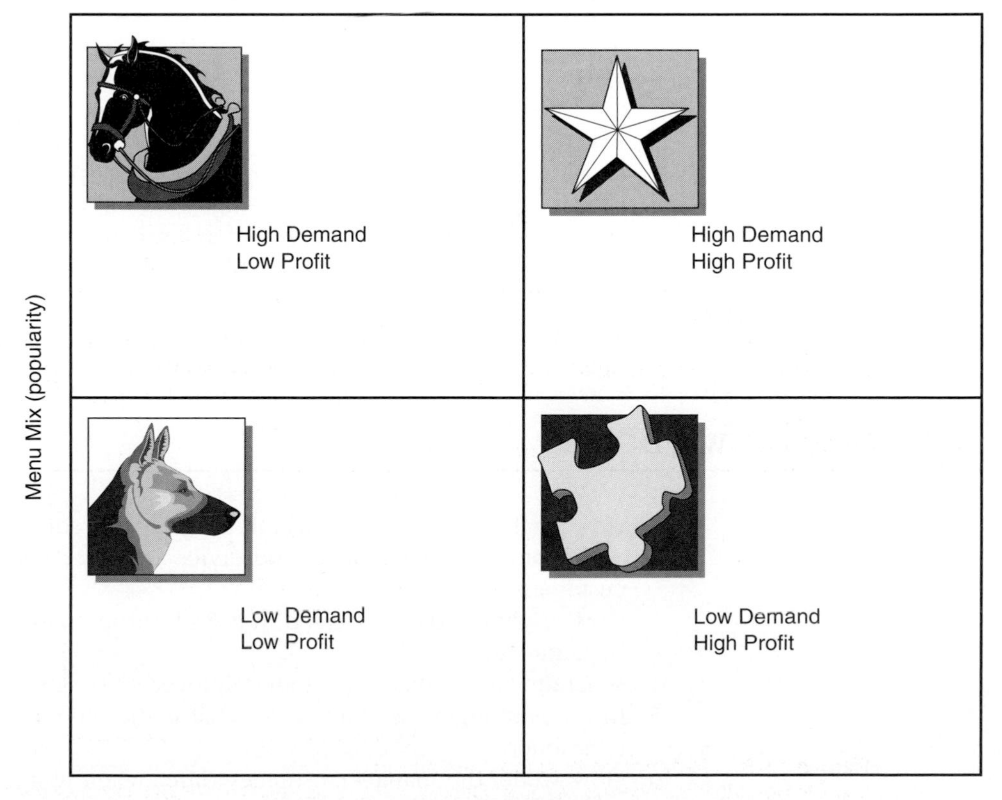

Figure 15.9. Menu engineering matrix. *Source:* Adapted from Kasavana, M. L., & Smith, D. I. (1982). *Menu Engineering: A Practical Guide to Menu Analysis.* Lansing, MI: Hospitality Publications.

other menu items (menu mix) and its contribution to profit (contribution margin). Menu mix is calculated by dividing the number sold of a particular menu item by the total number of all menu items sold. Contribution margin is calculated by subtracting a menu item's food cost from its selling price.

Items are categorized into high and low menu mix using the formula based on having popularity greater or less than 70% of total sales. Items are categorized into high and low contribution margin based on whether they have a contribution margin above or below the average contribution margin for the entire menu. This categorization results in four categories of menu items:

- **Stars**—These menu items are highly profitable and popular; managers should maintain standards for these items and promote them.
- **Plow Horses**—These menu items are very popular but not very profitable; managers might consider trying a price increase for these items to make them more profitable.
- **Puzzles**—These items are very profitable but not very popular; managers might consider whether to continue to offer these items.
- **Dogs**—These items are not profitable nor are they popular; mangers should consider eliminating these items from their menu.

CHAPTER SUMMARY

This summary is organized by the learning objectives.

1. Outcomes of the foodservice systems model include the quantity and quality of meals, customer satisfaction, employee satisfaction, and financial accountability.
2. System outcomes can be measured in a variety of ways. The quantity of meals can be measured by keeping accurate production records. Meal quality can be assessed through evaluations of food product attributes, portion sizes, the quality of the ingredients, and comparison against product standards. Customer satisfaction might be evaluated using walk-through audits, talking with guests, customer comment cards, exit interviews, mystery shopper reports, customer surveys, and focus groups. Employee satisfaction might be measured using standardized forms such as the Minnesota Satisfaction Questionnaire or the Job Diagnostic Instrument. Financial accountability will be measured based on financial reports and organizational goals.

TEST YOUR KNOWLEDGE

1. How might you determine whether quality ingredients are being used?
2. What information might a foodservice manager collect to determine whether customers are satisfied?
3. How might the data collected in a walk-through audit differ from that collected by talking to guests?
4. What are factors that impact an employee's job satisfaction?
5. In a menu engineering analysis, what is the difference between menu mix and contribution?
6. How would the manager's strategy differ when working with a menu item classified as a plow horse compared to one classified as a puzzler?

Professional Organization Profile

Management in Food and Nutrition Systems Dietetic Practice Group of the American Dietetic Association

MISSION

To advance leadership for dietitians practicing in food and nutrition systems management

WHO BELONGS TO THE ORGANIZATION

Members are also members of the American Dietetic Association. Most manage foodservice operations in a variety of health care, college and university, and other on-site foodservice operations; others work as foodservice management educators. Student membership is available through the American Dietetic Association.

ADVANTAGES OF MEMBERSHIP

Members have access to a members-only online network; receive a quarterly newsletter, *MARKET LINK;* earn continuing education credit by attendance at association sponsored seminars and workshops; and have access to resources and scholarships specifically for members.

WEBSITE

www.rdmanager.org

MEET THE PRESIDENT

Pam Ferguson, RD; President 2005–2006; is a foodservice consultant. She has worked in food and foodservice equipment sales and as a hospital foodservice director.

CLASS PROJECTS

1. Develop a customer satisfaction survey and a walk-through audit form for a local foodservice operation. Collect data using these forms and write a report for the manager of the foodservice operation at which the data were collected.
2. Request financial and sales data from a local foodservice operation for their menu items. Use the menu engineering technique to analyze the menu. Present results of the analysis and present your recommendations to the foodservice manager.

WEB SOURCES

www.busreslab.com	The Business Research Lab—restaurant customer satisfaction surveying
www.pressganey.com	Press Ganey, Inc.—hospital patient satisfaction
www.theacsi.org	American Customer Satisfaction Index
www.npd.com	NPD Foodservice Information—CREST® report
www.bgsu.edu/departments/psych/JDI	Job Descriptive Index
www.psych.umn.edu/psylabs/vpr/msqinf.htm	Minnesota Satisfaction Questionnaire

BIBLIOGRAPHY

American Dietetic Association (2005). Putting the WOW! in Dietetics: An Introdution to Customer Satisfaction. [Online]. Available: www.eatright.org

Balzer, W. K., Kihm, J. A., Smith, P. C., Irwin, J. L., Bachiochi, P. D., Robie, C., Sinar, E. F., & Parra, L. F. (1997). *Users' manual for the Job Descriptive Index (JDI, 1997 Revision) and the Job In General scales.* Bowling Green, OH: Bowling Green State University.

Bartlett, A. L., Probber, J. K., & Scerbo, F. B. (1999). Student employees in university foodservice: Job design, job characteristics, and job satisfaction. *NACUFS Journal, 21,* 14–29.

Duke K. M., & Sneed J. (1989). A research model for relating job characteristics to job satisfaction of university foodservice employees. *Journal of the American Dietetic Association, 89,* 1087–1091.

George, J. M., & Jones, G. R. (2002). *Organizational behavior* (3rd ed.). Upper Saddle River, NJ: Prentice Hall.

Heskett, J. L., Lones, T. O., Loveman, G. W., Sasser, W. E., & Schlesinger, L. A. (1994). Putting the service-profit chain to work. *Harvard Business Review,* March–April: 164–174.

Kasavana, Michael, & Smith, D. (1982). *Menu engineering—A practical guide.* Lansing, MI: Hospitality Publishers.

National Restaurant Association. (1995). *Uniform system of accounts for restaurants* (3rd ed.). Washington, DC: Author.

Palmer, N. H. (1972). Sensory methods in food-quality assessment. In Paul, P. C., & Palmer, H. H. (Eds.), *Food theory and application* (pp. 727–738). New York: John Wiley.

Sanders, E. E., & Hill T. H. (2001). *Foodservice profitability: A control approach.* Upper Saddle River, NJ: Prentice Hall.

Setser, C. S. (1992). Sensory evaluation. In Karmel, B. S., & Stauffer, C. E. (Eds.). *Advances in baking technology* (pp. 254–291). New York: VCH Publishers.

Sherman, S. C. (1999) *Total customer satisfaction: A comprehensive approach for health care providers.* San Francisco, CA: Jossey-Bass.

Skelton, M. (1984). Sensory evaluation of food. *Cornell Hotel and Restaurant Administration Quarterly, 24*(4), 51–57.

Sneed, J. (1988) Job characteristics and job satisfaction of school foodservice employees. *School Food Service Research Review, 12,* 65–68.

Sneed J., & Herman C. M. (1990). Influence of job characteristics and organizational commitment on job satisfaction of hospital foodservice employees. *Journal of the American Dietetic Association, 90,* 1072–1076.

Thorner, M. E., & Manning, P. B. (1983). *Quantity control in foodservice.* Westport, CT: AVI.

Weiss, D. J., Dawis, R. V., England, G. W., & Lofquist, L. H. (1967). *Manual for the Minnesota Satisfaction Questionnaire.* Minneapolis: University of Minnesota Industrial Relations Center.

Sample Specifications for Food Products

EP—edible portion; **AP**—as purchased; **IQF**—individually quick frozen

FRESH FRUITS AND VEGETABLES

Grapefruit

Origin: Texas preferred when in season. If not available, substitute with Florida or California. (Check with buyer.)

Use: breakfast fruit (halves will be served), EP 47%

Preservation form: fresh, delivery in refrigerated truck

Quality: U.S. No. 1, golden pink or white flesh

Tolerance: not more than one-third of the surface may be discolored

Pack: 35 pounds/carton, 48 count/carton

Price: by the box

Lettuce, Iceberg

Origin: California; if quality not good, substitute Arizona, Florida, or Mexico

Use: liner for salads. (Heads need to be fairly well trimmed unless specified as closely trimmed.)

Preservation form: delivery in refrigerated truck at 41°F or lower

Quality: U.S. #1, fresh and green and free of decay or damage

Tolerance: 40% hard heads and 60% firm heads

Pack: 40–50 pounds/carton, 24 count per carton

Price: by carton

PROCESSED FRUITS & VEGETABLES

Beans, Green

Origin: Oregon blue lake or Northwest

Form: canned

Use: green bean and onion casserole

Quality: U.S. Grade A

Size: 3–4 sieve, cut approximately 2″ in length

Pack: 6/#10 cans per case

Price: per case

Cherries

Form: red sour, pitted

Preservation form: canned, water packed

Use: cherry pie

Quality: U.S. Grade B

Pack: 6/#10 cans per case

Price: per case

Potatoes, French Fried, Frozen

Form: crinkle cut, 3⁄8″ × 3⁄8″, long

Use: vegetable choice on menu

Quality: U.S. Grade A

Brand: Ore-Ida or Simplot

Preservation form: frozen, 0°F or lower inside carton on delivery in refrigerated truck

Pack: 6/5-pound cartons per case

Price: per carton or case

Green Peas and Pearl Onions, Frozen

Use: vegetable choice on cafeteria line

Quality: U.S. Grade A for peas; U.S. Grade No. 2 or better for pearl onions

Size: 3–4 sieve for peas. Less than 1⁄2″ in diameter for pearl onions, minimum of 100 onions per 2 1⁄2-pound carton

Preservation form: frozen, 0°F or lower inside carton on delivery in refrigerated truck

Pack: 12/2 1⁄2-pound cartons per case

Price: by carton or case

Strawberries, Frozen

Form: IQF, whole

Size: range from 3⁄4″ to 1 diameter

Quality: U.S. Grade A

Preservation form: 0°F or lower inside case on delivery in refrigerated truck

Pack: polyethylene bags inside fiberboard cartons, 6/6½ pounds per case

Price: by carton or case

Peaches

Form: yellow cling, halves

Use: dessert and salad

Preservation form: canned

Syrup density: light 14°–18° Brix

Quality: U.S. Grade B

Count: 30–40 maximum

Pack: 6/#10 cans per case

Price: per case

Cranberry Juicc Cocktail

Form: crystal clear with cranberry red color

Use: beverage

Quality: U.S. Grade A

Enriched: Vitamin C

Ingredients: water, cranberry juice, sugar and inverted sugar, dextrose, vitamin C enriched

Brand: Ocean Spray or Sysco Classic

Pack: 12/46-ounce bottles per case

Price: per case

Jelly

Use: baked products

Variety: 2 apple, 2 blackberry, 2 grape per case

Pack: 6/#10 cans per case

Price: per case

Jam, Strawberry

Use: individual service

Brand: Smuckers

Pack: 200/½-ounce packets per case

Price: per case

DAIRY PRODUCTS

Milk

Form: fluid milk, 2% reduced fat

Use: hospital individual patient service

Preservation form: fresh, 41°F or lower on delivery in refrigerated truck

Quality: U.S. Grade A, pasteurized

Milkfat: not less than 2%

Milk solids: not less than 8.25% not fat

Vitamin A: restored to not less than 2,000 IU per quart

Vitamin D: restored to not less than 400 IU per quart

Pack: half-pint cartons, 50/case

Price: by the half-pint carton

Yogurt

Form: plain nonfat yogurt

Use: salad dressings, Mexican menu items

Preservation form: 41°F or lower on delivery in refrigerated truck

Brand: Dannon plain nonfat yogurt

Quality: meets National Yogurt Association criteria for live and active culture yogurt

Milkfat: not more than 0.5%

Milk solids: not less than 8.25% without fat

Pack: 32-ounce carton, 6 cartons/case

Price: by case

Cheese, American

Form: lowfat American pasteurized processed

Use: hot entrées, salad bar

Preservation form: 38°F–40°F on delivery in refrigerated truck

Quality: must rate a 3 in USDA melt test and meet FDA composition requirements for pasteurized processed American cheese

Pack: 6/5-pound loaves per case

Price: by loaf or case

Butter

Form: sweet cream, unsalted

Use: customer service

Preservation form: 41°F or lower on delivery in refrigerated truck

Quality: USDA Grade AA (score 93)

Milkfat: not less than 80%

Pack: 72 pats per pound individually wrapped in foil, 5 pounds/case

Price: by the case

DEHYDRATED PRODUCTS

Raisins

Form: Muscat, seedless

Use: bread and sweet rolls

Quality: U.S. Grade B (Choice)

Pack: 30-pound polyethylene bag inside fiberboard carton

Price: by carton

Orange Juice

Form: frozen, concentrated, unsweetened, calcium enriched

Use: breakfast and snack beverage

Preservation form: frozen, minimum 0°F inside case on delivery in refrigerated truck

Quality: U.S. Grade A

Enrichment: 300 mg tricalcium phosphate and calcium phosphate per 8 fluid ounces

Dilution: ratio 3:1

Brand: Minute Maid, Coca Cola Co.

Pack: 12/32-ounce cans per case

Price: by case

POULTRY AND EGGS

Chicken Breast Fillets

Form: boneless, skinless, Cajun marinated breast halves with no rib meat

Use: grilled Cajun flavored chicken breast

Class: broiler/fryer

Preservation form: frozen, 0°F on delivery in refrigerated truck

Quality: U.S. Grade A

Weight: 3½ to 4½ ounces

Pack: 10-pound carton

Price: by the carton

Eggs

Form: shell

Use: breakfast items—fried, poached, soft cooked, omelets

Preservation form: 41°F or lower when delivered in refrigerated truck
Quality: U.S. Grade AA
Color of shell: white
Delivery time: within 14 days of laying
Size: large, 24 ounce per dozen
Pack: 30 dozen moisture-proof crate
Price: by crate

Dried Egg Whites

Form: dried egg whites
Use: baking, especially meringues and cakes
Preservation form: spray dried, glucose, removed
Pack: 6/#10 cans per case
Price: by case

GRAIN FOODS

Flour

Origin: Northern United States and Canada
Form: hard wheat bread flour
Use: specialty breads and rolls
Composition: minimum of 11% gluten, enriched, bleached
Pack: 50-pound sack
Price: by sack

Cereal

Form: dry
Use: breakfast food and snacks out of a dispenser
Brand: Kellogg's Frosted Flakes
Pack: 4/2½-pound polyethylene bags per case
Price: by case

Pasta

Form: precooked lasagna noodles
Use: baked lasagna casserole
Composition: 100% durum wheat semolina, enriched
Pack: 10-pound carton, handpacked
Price: by the carton

SEAFOOD

Rainbow Trout

Origin: trout farms on the freshwater Snake River, Idaho

Form: dressed (eviscerated and head, tail, fins, and skin removed)

Use: grilled whole fish to be boned at table

Preservation form: fresh, 41°F or lower on delivery in refrigerated truck

Quality: USDA Grade A or equivalent

Weight: 8–10 ounces drawn

Pack: 10-pound carton

Price: by the carton

Orange Roughy

Origin: New Zealand waters

Form: boneless and skinless fillets

Use: broiled

Preservation form: frozen, 0°F or lower on delivery in refrigerated truck

Quality: USDC grade A or equivalent

Certification: PUFI seal

Weight: 6–8-ounce fillet without glaze. Tolerance ± 0.5 ounce. Glaze shall not be over 6% of total weight.

Pack: 20-pound carton

Price: by the carton

Breaded Fish (Pollock) Sticks

Origin: Northern Atlantic Ocean

Form: boneless and skinless fish sticks, lightly breaded IQF

Use: oven-baked

Preservation form: frozen, −10°F on delivery in refrigerated truck

Quality: Van de Kamp's breaded fish sticks or equal

Weight: 0.67-ounce/stick, 65% solid fish, 35% breading

Pack: 10-pound carton

Price: by the carton

Mussels

Origin: farm raised

Form: live, in-shell, debearded

Use: steamed

Preservation form: fresh, 35°F–40°F on delivery in refrigerated truck

Quality: shells should be clean and reasonably free of dirt and debris and should close tightly when tapped

Weight: 45–55 pounds/bushel

Pack: bushel

Price: by the bushel

Appendix B Resources for Writing Specifications

GOVERNMENT PUBLICATIONS

Institutional Meat Purchase Specifications (IMPS). Available from: Livestock and Meat Standardization Branch, USDA, AMS, LS, SB, 1400 Independence Avenue, SW, Stop 0254, Washington, DC 20250-0254, (202) 720–4486. Available online at *www.ams.usda.gov/LSG/stand/imps.htm*

Gunn, M. (1995). *First Choice: A purchasing systems manual for school food service.* Hattiesburg, MS: National Food Service Management Institute.

BOOKS

Almanac of the canning, freezing, preserving industries. (Annual). Westminister, MD: Edward E. Judge & Sons.

Dietary Managers Association. (1991). *Professional procurement practices: A guide for dietary managers.* Lombard, IL: Author. (Developed by J. D. Ninemeier.)

Kotschevar, L. H., & Donnelly, R. (1999). *Quantity food purchasing* (5th ed.). Upper Saddle River, NJ: Merrill/Prentice Hall.

North American Meat Processors Association. (2002). *The meat and poultry buyers guide.* Available on CD-ROM at *www.namp.com*

Produce Marketing Association. (1994). *Foodservice guide to fresh produce, 1994.* Available from the Produce Marketing Association. P.O. Box 6036, Newark, Delaware, 19714-6036. $7.00.

Produce Marketing Association. (2002). *Fresh produce manual.* Available at *www.pma.com*

Produce Marketing Association. (2002). *Profitability in fresh produce.* Available at *www.pma.com*

Reed, L. S. (1993). *SPECS: The Comprehensive Foodservice Purchasing and Specification Manual.* New York: Van Nostrand Reinhold.

Spears, M. C. (1999). *Foodservice procurement: Purchasing for profit.* Upper Saddle River, NJ: Merrill/Prentice Hall.

Stefanelli, J. M. (1997). *Purchasing: Selection and procurement for the hospitality industry* (4th ed.). New York: John Wiley.

Warfel, M. C., & Cremer, M. L., & Hug, R. J. (1996). *Purchasing for food service managers* (3rd ed.). Berkeley, CA: McCutchen Publishing.

Standards for Food Products

BEVERAGES

Cocoa Cocoa or hot chocolate should be a rich brown color. Flavor should be delicately sweet with the rich, aromatic flavor of cocoa predominant. Cocoa powder should be well mixed in the milk with no visible flecks of undissolved cocoa. Cocoa should have a cooked starch flavor, be free of skin, and be slightly thicker than milk (Kotschevar & Donnelley, 1999).

Coffee Brewed coffee, no more than 30 minutes old, should be evaluated for flavor, including taste and aroma, which should be balanced and sufficiently strong to give a pleasing flavor. Taste should be properly balanced between bitterness (astringency), acidity, and sweetness. Aroma should be fragrant, mellow, heavy, and rich with coffee bouquet. Brewed coffee should be bright and clear and the color rich and deep brown. It should be full bodied and sensed in the mouth as denser than water. Coffee should be served at 160° to 200°F (Spears, 1999). Iced coffee should have the same characteristics, but the original brew should be stronger to permit dilution by ice.

Tea Tea is judged by flavor, strength, clarity, and color. Tea generally has a slightly bitter flavor caused by tannin, some sweetness, and acidity. Aroma should be fruity and fragrant. The term *brisk* often is used to describe the zestful, stimulating quality of tea. The type of tea determines the color of the brew. For example, green tea produces a delicate, greenish yellow pale liquor and rather fruity flavor; oolong is darker, less bitter, still fruity, but has a softer flavor than green tea; black tea is a copper-colored brew with a soft, mild flavor, but slightly acidic (Kotschevar & Donnelly, 1999). Iced tea should have the same characteristics, but the original brew should be stronger to permit dilution by ice.

BREADS, QUICK

Biscuits, Baking Powder A baked biscuit should be round in shape with vertical sides and a fairly smooth level top. Size should be two to three times that of an unbaked biscuit. Crust should be light brown in color with some evidence of flour. Interior can be peeled in thin layers and should be creamy white with a fine even-grained cell structure and a moist, fluffy crumb. Flavor should be mild with a slight taste of baking powder.

Biscuits, Cinnamon Raisin Addition of cinnamon, sugar, and plumped raisins to the baking powder biscuit dough yields a light brown biscuit with a spicy, sweet taste. Flecks of brown cinnamon are evident in the biscuits. Raisins should be whole, moist, and evenly distributed. Cell structure is slightly more compact than in a plain biscuit. A thin powdered sugar glaze on the baked biscuit gives additional sweetness and a smooth top.

Bread, Corn Crust is crisp, shiny, pebbly, and slightly brown on top with darker sides and bottom. Interior should have a coarse, moist, light, and tender crumb and be free of tunneling. When cut, slight crumbling may occur. Color should be characteristic of type of cornmeal used, that is, white, yellow, or blue. Flavor should have a well-balanced corn taste. Sweetness depends on regional preferences.

Bread, Loaf The muffin batter bread should be loaf-shaped with an even, medium brown crust, and a crack down the middle. Interior grain, from coarse to very fine, depends on the formulation. The bread should be free from tunnels and have a moist crumb. Flavor should be slightly sweet, due to the sweetener, and mild with a characteristic flavor of any additions (e.g., bananas, cranberries, nuts, dates, spices). Color depends on added ingredients, which should be evenly dispersed and identifiable.

Cake Doughnuts Exterior should have a rich, golden brown, crisp crust and an inner core that resembles a baked product more than a fried food. Interior should be pale yellow. The doughnut should be evenly rounded with a slightly wrinkled or star-shaped center hole. Crumb should be fine, tender, and moist with uniform-sized cells. Volume should be large in proportion to the weight of the doughnut. Flavor should be delicate and well-balanced. Glaze or coating, if used, should be light to enhance but not dominate the flavor.

Coffee Cake Golden brown crust appears sugary or shiny and may be slightly crisp. Top may be uneven depending on topping. Cell size is slightly larger than in a butter cake, yet smaller than in a muffin. Flavor, aroma, and color should be characteristic of ingredients, especially additional ones, such as extracts, fruit, nuts, and spices. Any addition should be evenly dispersed and identifiable. Coffee cake should be served warm because freshness is lost rather rapidly.

BREADS, YEAST

Bread, French Exterior surface should be relatively thin, hard and crusty, and an even, golden brown color with three to five diagonal slashes across the top. The loaf shape should be a cylinder, a minimum of 12 inches long and 3 inches in diameter. Interior is snowy white with a moist, soft, elastic crumb. Cell walls are uniformly thin and of varying sizes. Air pockets are desirable. Flavor is bland with slight yeast emphasis.

Bread, Plain The bread should have a symmetrical loaf shape with a golden brown top crust and uniformly brown sides, ends, and bottom. Crust should be tender and easily cut. Interior texture should be fine, even grained, free from large air bubbles, and have oval-shaped cells with thin walls. The creamy white crumb should be moist, silky, and slightly resistant to bite with no trace of being doughy. Flavor should be pleasant, nutlike, and have a hint of yeast. Density and color of bread will vary depending on type of flour used (such as cracked wheat, rye, white). Color also is dependent on added ingredients (for example, brown sugar, molasses, raisins, spices), which should be evenly dispersed.

Bread Sticks Characteristics are identical to those of plain bread. Outer surface should be golden brown and crisp. Sticks should be approximately 5 inches in length and 1 inch in diameter.

Croissants Exterior surface is golden brown, flaky, and tender. Interior is very moist, light, and flaky and easily peels off in thin sheets. Croissants have a buttery flavor and color. Shapes vary from crescent to round to straight, depending on use.

Kolache Characteristics are identical to those of sweet rolls with added flavor of grated lemon peel. Kolaches are round with an indentation in the middle filled with prune or apricot, cottage or cream cheese, or poppy seed filling. Crust is golden brown, and filling is moist but solid. Top is covered with sifted powdered sugar.

Rolls, Cinnamon The rolls should have characteristics of sweet rolls and good volume in proportion to weight. Shape is distinctive owing to basic preparation from a sheet of dough sprinkled with cinnamon and sugar, rolled, and cut in slices of uniform thickness. Rolls should have a golden brown crust and are generally topped with a powdered sugar glaze. Flavor is sweet, wheaty, mild, and enhanced with cinnamon.

Rolls, Dinner Characteristics are identical to those of loaf bread. Size and shape may vary. Top crust should be golden brown, smooth, and round, and the bottom light brown.

Rolls, Pan Characteristics are similar to those of dinner rolls. Pan rolls should be brown on top and bottom and also on the sides of the rolls placed on the edge of the baking pan. Other sides resemble the interior of the rolls.

Rolls, Sticky Pecan Characteristics are similar to cinnamon rolls except for the baked-on brown sugar syrup topping with bite-sized pieces of pecans. The soft syrup topping should be shiny, clear, and smooth. Pecans should be evenly dispersed over the rolls and in the interior.

Rolls, Sweet Characteristics are similar to loaf bread except for the tinge of yellow egg color in the interior, moister crumb, and sweeter and more full-bodied flavor. Addition of other ingredients like fruit, nuts, and spices enhances the flavor. Glazes, icings, or powdered sugar add to flavor and appearance of the rolls.

CEREAL, PASTA, RICE

Cereal, Cooked Color of the cereal depends on the color of the grain: tan for oatmeal, white for farina, and yellow for cornmeal. Cooked cereal should be tender with a cooked starch flavor. Consistency should be thick, and texture, smooth. Cooked cereal should be served hot.

Pasta, Cooked Pasta has an ivory color or a color characteristic of added ingredients: green from spinach, pink from tomatoes, black from squid, golden yellow from corn or saffron, medium yellow from egg yolks, or light tan from whole wheat flour. Cooked pasta should be firm to the bite, al dente, and have a bland, cooked starch flavor that blends with various sauces.

Pasta Sauce, Creamy The white sauce base should be smooth and mild with a cooked starch flavor. If cream is one of the ingredients, the rich flavor is evident. Addition of chicken broth, garlic, and/or herbs should blend with the white sauce. Ingredients, including cheese, poultry, seafood, and vegetables, should be identifiable and add characteristic flavor and color to the sauce.

Pasta Sauce, Light Sauce is light in color and almost clear. Basil, garlic, olive oil, and parsley are the predominant flavors and enhance the mild-flavored pasta. Addition of herbs and freshly grated Parmesan cheese adds piquancy to the finished product. Minced clams add a characteristic flavor and chewy texture to the sauce.

Pasta Sauce, Red The sauce should be an orange-red tomato color. Consistency should be sufficiently thick to preclude running when served over hot pasta. The tomato flavor should be predominant but enhanced with chopped garlic, green pepper, onions, and herbs. Browned ground meat or meatballs can be cooked in the sauce, giving a meaty flavor.

Rice, Boiled, Steamed, Baked Rice should be pure white, and each grain should be intact and separate. Brown rice has a light brown color. Rice grains should be tender but firm and have a bland, cooked starch flavor.

DESSERTS

Cake

Angel Food Rough, light brown crust is dry to the touch. Interior is snowy white. Crumb should be soft, moist, light, feathery, and slightly elastic. Cell size is somewhat uneven and cell walls are uniformly thin. Flavor should be mildly sweet and well balanced and have a cooked egg white flavor.

Butter Crust should be an even golden brown with a dull matte finish. Top should be slightly rounded, puffy, and free of cracks. Light yellow interior has a soft, moist, tender, velvety crumb and very small cells with uniformly thin walls. Flavor is sweet, mild, and buttery. Flavor and color depend on added ingredients (for example, brown sugar, chocolate, spices).

Carrot Crust is soft, and browning is barely noticeable. Interior is a cinnamon brown color and is very moist and compact. Grated raw carrots are difficult to detect but add body and moisture to the product. Chopped nuts and cinnamon flavor are identifiable. The cake usually is topped with a mild-flavored cream cheese icing, which blends with the positive cinnamon flavor.

Chiffon Rough, light brown crust is dry to the touch. Texture should be delicate, springy, and light. Small, fine, uniform cells with thin walls are desirable. Crumb should be moist and soft. Color is a pale yellow and flavor is mild, but both may be changed by additional ingredients and coloring and flavoring agents, such as chocolate, maraschino cherry juice, spices.

Chocolate Characteristics except for color and crust are identical to those of a butter cake. Color may vary from reddish to deep brown. Browning of the crust is barely noticeable. Flavor intensity depends on type of chocolate, which may vary from mild sweet to strong bittersweet.

Icing Cream icing should be smooth and easy to spread. It should be fluffy, yet maintain its shape, and adhere to the cake. Color and flavor depend on ingredients. Fat content for cake icings ranges from 10% to 25% to permit ease in cutting.

Pound The loaf cake should have a light brown crust with a narrow split down the middle of the slightly rounded top. Interior is yellow with a smooth crumb. Texture is firm with fine, compact cells. Well-blended flavor is buttery and sweet.

Cookies

Bar Bars should be approximately 3-by-2½ inches with a uniform height of ½ to ¾ inches. Top and bottom crusts should be uniformly medium brown. Cut edges should be straight. Texture may vary from chewy to cakelike. Any additions to the bars (e.g., chocolate chips, coconut, dried fruits, nuts, spices) should be evenly dispersed and identifiable. Flavor should be rich, sweet, and characteristic of additions. A topping of glaze, icing, or powdered sugar may be added.

Brownie Characteristics are similar to those of bar cookies. Exterior is dark brown and smooth. Interior is darker than the exterior. Brownie is tender and slightly moist with a moderately chewy texture free of crumbling. Nuts should be evenly dispersed and identifiable. A topping of icing may be added.

Drop Shape is symmetrically round, approximately 3 inches in diameter. Height may vary from ¼ to ½ inch depending on whether the cookie is chewy or cakelike. Top and bottom should be evenly browned with no evidence of sugar crystals unless added for specific varieties (e.g., sugar cookies). Flavor should be rich, sweet, and characteristic of any additions (e.g., butterscotch or chocolate chips, chocolate chunks, coconut, dried fruits, nuts, oatmeal, spices), which should be evenly dispersed and identifiable. Cookies may be topped with a glaze, icing, or powdered sugar.

Pie

Cream The filling should be opaque, creamy, smooth, and free of lumps. Filling should retain its shape when cooled and sag only slightly at the cut edge. Quantity of filling should be adequate to fill the shell. Flavor and color are characteristic of the additional ingredients and coloring and flavoring agents, such as banana, butterscotch, chocolate, coconut, and vanilla. Cream pies must be refrigerated immediately after preparation.

Custard Top surface film should be yellow with flecks of brown nutmeg. Pale yellow filling should be creamy, smooth, shiny, and tender. Gel should be uniformly coagulated and free of porosity. Egg and nutmeg flavors predominate. The pie should have a crisp crust with an adequate amount of filling without overflow. Custard pies must be served chilled.

Fruit Thickened juice should be clear, bright, smooth, and have a slight cooked starch flavor. Each piece of bite-sized fruit should be distinct and identifiable. Filling should sag only slightly at the cut edge. Flavor is moderately sweet and characteristic of the specific fruit—apple, apricot, blueberry, cherry, gooseberry, peach, pineapple, and rhubarb.

Lemon The filling should be a translucent gel with a color characteristic of lemon rind. Filling should be viscous and retain its shape when cut. Flavor is a pleasing blend of tart and sweet. Quantity of filling should be adequate to fill the crisp shell.

Meringue Top is a pale golden color, and interior is snowy white. Meringue should be tender, fluffy, moist, and have a sweet flavor. It should adhere to and completely cover the pie filling, meeting the crust at the edge. Beads of liquid or weeping should not be present.

Pastry Crust should be thin. Top surface should be rough and blistery with a soft luster. Center of the top crust should be light, golden brown with color deepening slightly toward the edges. Texture of both top and bottom crusts should be crisp, flaky, and fork tender, but not crumbly. Crust should cut easily. Flavor depends on the amount of salt and type of fat. Top crust may be brushed with egg wash and sprinkled with granulated sugar before baking.

Pumpkin Color should be orangish brown with noticeable flecks of spices. Filling should be viscous and retain its shape when cut. Texture should be slightly coarse. Flavor is a mild blend of pumpkin with pleasing notes of cinnamon, cloves, ginger, and nutmeg.

Shell, Graham Cracker Shell should retain the shape of the pie pan. Shell should be crisp with no evidence of sugar crystals. Flavor is buttery, sweet, and characteristic of graham crackers.

Other

Cheesecake The top is covered with a creamy white sour cream mixture, and sides and bottom are the brown color of a graham cracker shell. Interior has the smooth consistency of cream cheese and is uniformly coagulated. Cheesecake is pale yellow and has a very mild cheese, egg, and vanilla flavor. Variations in color and flavor depend on additional ingredients (e.g., chocolate, lemon, mint, pumpkin, spices). The baked cheesecake often is topped with a cherry, lemon, raspberry, or strawberry glaze.

Cream Pudding Pudding should be glossy, creamy smooth, and free of lumps and surface skin. It should mound slightly when dipped with a spoon. Color and flavor depend on added ingredients and coloring and flavoring agents (e.g., butterscotch, chocolate, vanilla).

Fruit Crisp A crunchy topping should be evenly distributed over a slightly thickened and sweetened fruit mixture (e.g., apples, blueberries, or sour cherries). Topping should be delicate brown and have a coarse, crumbly consistency with a flavor characteristic of oatmeal, brown sugar, and butter. Fruit should be identifiable by color and flavor.

EGGS

Fried Appearance should be bright, glossy, and compact with rounded edges. A sunny-side-up egg should have a bright yellow or orange-yellow, well-rounded yolk. If the egg is turned over, the yolk is covered with a thin film of coagulated white. The egg should be shiny and soft underneath. Coagulated areas should be firm, yet tender.

Omelet Bottom is a delicate light brown, and the top is pale yellow. The omelet should be moist and tender and have a delicate flavor. Omelets generally are filled with various ingredients (such as diced tomatoes, sautéed mushrooms, shredded cheese, and small cubes of ham). The folded or rolled shape should be even and uniform.

Poached The egg should have some shiny white adhering closely to the bright yellow yolk. It should be compact with smooth edges, rather than spread out. Coagulated areas should be slightly firm, yet tender.

Scrambled Eggs should be bright and clear with a soft sheen and have no evidence of browning. The uniform pale yellow eggs should be in small pieces that are tender and moist. Flavor should be delicate and mild.

Soft or Hard Cooked Appearance should be bright and fresh with the colors natural and clear. Yolk of a soft-cooked egg should be warm throughout but liquid with one-half to three-fourths of the white coagulated, depending on degree of doneness. A hard-cooked egg has a solid, mealy yolk that is bright yellow or orange-yellow and is uniformly coagulated. White of a hard-cooked egg is firm and glossy.

ENTRÉES, ACCOMPANIMENTS

Entrées

Chicken Fried Steak Exterior should be golden brown with a crisply fried, well-seasoned, batter coating. The tenderized beef cutlet interior should be moist. Flavor should be characteristic of fresh beef. It is typically served with a milk gravy made from meat drippings.

Chicken Tetrazzini Melted orange-yellow or grated Parmesan cheese tops spaghetti and chicken chunks in a pale yellow chicken sauce with identifiable ingredients: chopped green pepper, onion, parsley, pimiento, and sliced mushrooms. Spaghetti is tender but al dente with a cooked starch flavor. Sauce blends with the mild-flavored spaghetti. A portion flows slightly when spooned on a plate.

Chili con Carne Color of coarsely ground meat, tomato, and bean mixture should be a reddish brown. Well-blended flavor should be characteristic of ingredients and seasonings, including chili powder, cumin, garlic, and onion. Consistency when served should be thick.

Fajita Color of the meat or poultry strips in the filling is light brown, contrasted with sautéed red and green pepper strips and sliced onions. Tender strips of meat or poultry are flavored with various marinades. Filling typically is rolled into a steamed flour tortilla and served with condiments, such as cheese, guacamole, salsa, and sour cream.

Grilled Ground Beef Patty (Hamburger) Exterior should be an even dark brown color and glossy. Interior should reflect desired degree of doneness ranging from slightly pink to light brown. Patty should be moist and tender with characteristic beef flavor. Fat content of cooked patty should reflect raw meat specification for lean/fat ratio. Patty should be served immediately.

Lasagna Layers of ingredients should be identifiable: ribbon-edged lasagna noodles, pale yellow Parmesan and mozzarella cheese, white ricotta or cottage cheese, and tomato red sauce with flecks of browned meat and dark green herbs. Noodles are tender but al dente; the cheese has a soft, smooth consistency; and the sauce is thick. Lasagna should have a spicy tomato beef flavor blended with the mild mozzarella and white cheese and the piquancy of Parmesan. A lasagna serving has sharply cut edges and retains its shape.

Macaroni and Cheese Macaroni should be al dente but tender and have a cooked starch flavor. Sauce should have a medium thick consistency with an American or cheddar cheese flavor. Each piece of the macaroni should be completely coated with the light yellow creamy sauce. A portion should spread slightly when spooned on a plate.

Meat Loaf Ground meat mixture should have a dark brown, thin crust with an even brown interior. Meat loaf should be moist, yet firm enough to retain its shape when sliced. Flavor should be characteristic of ingredients (beef base, catsup, dried onion soup, garlic, onions).

Pizza Color of baked pizza should be characteristic of ingredients. Yeast leavened, slightly blistered crust may be thin and crisp or thick and chewy, and should be well browned on the bottom and edges. Typically, sauce is thick and tomato based with well-blended seasonings of basil, fennel, garlic, and oregano, for example. Toppings may include any combination of cheese (usually mozzarella), fruit, meat, seafood, and vegetables. Portions can be individual round, wedge shaped, or square, depending on shape of pan.

Stew Fork-tender, bite-sized meat or poultry cubes should be of uniform size and lightly browned. Bite-sized vegetables should be firm, yet tender, while maintaining characteristic flavor and shape. Meat or poultry and vegetables are covered with smooth, mild-flavored natural gravy of medium consistency. A variation is pot pie, in which stew is covered with a rich pastry or drop biscuits and baked until golden brown.

Tuna and Noodles Lightly browned melted orange-yellow cheese tops ivory-colored noodles and tuna chunks in a chicken-based white sauce. Noodles should be tender but al dente and have a cooked starch flavor. White sauce should be smooth and mild enhanced by the characteristic flavor of tuna. Chopped celery and onion add crunchiness and also flavor. A portion should have an overall semisolid consistency and flow slightly on a plate.

Accompaniments

Bread Dressing Baked dressing should be light in weight for volume and tender. Moistness should complement the meat or poultry, dry with rich sauces and moist with dry meats. It should hold together when served. Chopped celery, onion, and seasoning should blend with the flavor of the meat or poultry. Day-old white bread is the major ingredient, but cornbread also is popular. Color of the dressing is characteristic of type of bread.

Fried Apples Apple rings, approximately ½ inch thick, should be tender, glossy, and lightly browned. Tart apples should have caramelized or brown sugar and butter flavor. Apple flavor should predominate with only enough sweetness to complement the entrée.

Fried Rice Color and flavor are characteristic of soy sauce, which blends well with the bland steamed rice. Chopped onions, frozen peas, shredded carrots, small pieces of tender but firm scrambled eggs are identifiable and give additional color and flavor to the rice.

Fritters Crisp crust is golden brown with a yellow interior that resembles a baked product. Fritters usually are spherical in shape. Flavor is slightly sweet and mild and complements characteristic flavors of additional ingredients (such as apples, bananas, corn, peaches, or pineapple).

Rice Pilaf Rice is pale yellow and tender with a cooked starch flavor. Grains are light, fluffy, and remain separate. Finely chopped onions and margarine or butter enhance the bland flavor of the rice in addition to the chicken broth and seasonings. Flavor also may be enhanced by additional ingredients (such as green peppers, mushrooms, pimiento, and water chestnuts).

FISH, MEAT, POULTRY

Fish, Cooked Color of cooked fish should be reminiscent of the raw state (i.e., white for orange roughy, pinkish orange for salmon, pale tan for swordfish). Flavor and odor should be mild and characteristic of fresh fish. Flesh should be moist and should flake easily. Seasonings and sauces enhance the mild flavor of fish.

Meat, Cooked Meat should have a pleasing appearance when cooked. Meat should be moist and drippings rich and of good color. Flavor and color should be natural to type of meat, style of cooking, and degree of doneness. Beef and lamb roasts should be well browned; veal should have a reddish brown surface, and pork should have a uniformly rich brown surface. Outer surfaces of lamb and pork roasts should be crisp. Sliced roasted meat should be firm, juicy, tender, and retain its shape. Broiled meats should have a rich brown color with well-developed flavor and aroma. Braised meat should be tender and juicy with a rich, glossy surface. Pieces of braised meat should be uniform, even, and symmetrical.

Poultry, Cooked Poultry flesh should be moist, and drippings pale yellow to light brown. Flavor and color should be natural to type of poultry and style of cooking. Skin on roasted poultry should be crisp. Sliced roasted poultry should be firm, juicy, tender, and retain its shape. Flavor of simmered poultry is delicate and the color ivory. Flesh should be tender and moist and retain its shape. Color of the stock should be pale yellow.

Seafood, Cooked Color should be typical of seafood type (pink exterior and white interior for shrimp, lobster, and crab; gray oysters; white scallops). Shellfish should be free of sand, tender, and moist. Flavor should be mild, characteristic of type, and blend well with seasonings and sauces.

SALADS

Fruit Fresh fruits should be ripe but firm and selected for variety in flavor, texture, and color. Flavors should be zestful and well blended. Color should be clear, fresh, and natural. Fruit that oxidizes quickly should be dipped into lemon or other juices to prevent darkening. Fruit should be identifiable and neatly cut into bite-sized pieces. The leafy greens used as underliners should be chilled and crisp. Dressing for fruit salads is usually slightly sweet and flavored with fruit juices or honey. Canned fruit should be well drained before using in salads.

Molded Commercial dessert gelatins generally are used as the base. Gelatin should have bright sparkle, clear color, good setting ability, and a true, pleasing flavor. It should be firm enough to hold its shape, especially when cut, yet tender, quivery, and transparent. Ingredients should complement the slightly sweet, natural, fruit flavor of the gelatin. Ingredients should be cut into bite-sized pieces and evenly distributed through the gelatin, not exceeding 35% of total recipe weight. Underliner greens should be cold and crisp.

Tossed Greens should be distinct, bite-sized, neatly cut or torn, and of good proportion. More than one type of greens should be used to give variety in texture, color, and flavor (e.g., curly endive, escarole, iceberg or leaf lettuce, radicchio, romaine, and spinach). Greens should be chilled, crisp, and dry. Other ingredients also may be added for interest (e.g., avocado, broccoli florets, carrot shreds, mushroom slices, tomato wedges, radish slices). Dressing and seasoning, added just before serving, should complement ingredients. Garnishes will give additional flavor and crispness to the salad (e.g., alfalfa sprouts, crisp bacon, seasoned croutons, sunflower seeds, toasted sesame seeds, and toasted wheat germ).

Vegetable Raw and cooked vegetables should be selected for flavor, texture, and color. Flavors should be zestful, pleasing, and complement each other. Color should be clear, fresh, and natural. Raw vegetables should be crisp and cooked tender and firm. Vegetables should be neatly cut into bite-size pieces. Marinating in a piquant dressing adds flavor.

SANDWICHES

Cold

Cheese, Meat, Poultry Bread should be fresh, firm, and moist with a close, smooth crumb, and good flavor. Bread should cut with a clean edge. Meat, poultry, or cheese should be thinly sliced or shaved. Lettuce, sliced pickles, and thin slices of tomato add crispness and flavor.

Soft Filling Bread should be fresh, firm, and moist with a close, smooth crumb, and good flavor. Chicken, egg, ham, tuna, and other fillings should be identifiable, easy to spread, but firm enough to retain shape. Chopped celery, green pepper, and pickles add color and crispness to the filling. Salad dressing should coat all ingredients and serve as a binding agent. Various seasonings such as lemon juice for tuna and dry mustard for eggs complement the mild flavor of basic ingredients. Bread should cut with a clean edge. Sandwich should stay together when picked up, without losing filling.

Hot

Bacon, Lettuce, and Tomato Bread should be fresh, firm, and moist with a close, smooth crumb, and good flavor. Golden brown toast should be hot, and bacon crisp and hot. Lettuce should be cold and crisp, and the thinly sliced tomato cold. Toast and filling should cut with a clean edge. Mayonnaise usually is served separately.

Barbecued Meat Bread or roll should be fresh, firm, and moist with a close, smooth crumb, and good flavor. Meat should be tender, thinly sliced, and covered with a reddish brown, spicy barbecue sauce. Flavor of meat should be characteristic of type, and flavor of sauce should complement it.

Cheeseburger Round hamburger bun should be freshly baked and have a golden brown exterior and a soft white interior. Grilled hamburger should have a fresh beef flavor and be juicy and tender with a brown and glossy exterior. It should be covered with melted cheese.

Grilled Cheese Bread should be fresh, firm, and moist with a close, smooth crumb, and good flavor. Grilled bread should be crisp and have a buttery flavor. Cheese should be kept within limits of the bread and completely melted.

Roast Beef Sandwich may be either closed or open-face and consists of one or two slices of bread, roast beef, and gravy. Beef is juicy, thinly sliced, hot, and the flavor characteristic. Gravy, often served separately, is either au jus or thickened meat juice.

SOUPS

Hot

Bean The slightly thick soup is light tan with small cubes of pink ham. Beans and ham should be very tender but identifiable. Ham flavor is predominant and blends well with the mild navy or Great Northern beans. Seasonings of black pepper, chopped celery, onion enhance the flavor.

Beef Rice Transparent brown beef stock has body and minute globules of fat on the surface. Each grain of rice should be separate. Chopped celery and onions add characteristic flavor to the broth. Chicken rice soup, pale yellow in color, is a variation.

Bisque Bisque is a variation of a cream soup and contains chopped shellfish, such as crab, lobster, or shrimp. The slightly thick soup is a pale orange color. Flavor is characteristic of type of seafood, cream, and various seasonings, quite often dry sherry wine.

Bouillon Bouillon is a clear, light soup, with a rich meat, poultry, or vegetable flavor. Color depends on the basic ingredient. Surface may have a few minute globules of fat.

Cheese Thin white-sauce-based soup is the orange-yellow color characteristic of cheddar cheese. Chopped onions and finely diced carrots and celery are tender but crisp and enhance the flavor of the chicken stock and cheese. Consistency of soup is similar to moderately heavy cream and lightly coats a spoon.

Clam Chowder, Manhattan Chowder should be a tomato red color. Chopped clams and small pieces of sautéed onion and golden brown salt pork cracklings or bacon should be identifiable and give a characteristic flavor. Chopped potatoes and celery and diced carrots should be tender but firm. Flavor should be spicy but not overpowering, using bay leaves, black pepper, thyme, and Worcestershire sauce.

Clam Chowder, New England The chowder should be a light cream color with the consistency of heavy cream. Chopped clams and small pieces of lightly sautéed chopped onion and golden brown salt pork or bacon should be identifiable and give a characteristic flavor. One-half-inch cubes of potato should be tender but firm.

Consommé Consommé is a clear, concentrated broth, pale yellow or light brown with a subtle chicken, fish, game, meat, or vegetable flavor. Surface may have a few minute globules of fat.

Cream Color and flavor depend on type of mashed, strained, or finely chopped meat, poultry, or vegetables added to a white sauce base. Consistency should be similar to moderately heavy cream and thick enough to delicately coat a spoon.

French Onion Dark brown beef broth and thin slices of sautéed onions are topped with a slice of toasted French bread, completely covered with melted Swiss and Parmesan cheeses. Soup has a slightly thick consistency and is a blend of strong and spicy flavors, such as black pepper, sautéed onion, and Worcestershire sauce.

Oriental Chicken Mildly flavored and transparent stock should be pale yellow and the color of vegetables distinct and bright. Tender chicken should be in one-half-inch cubes and green onions in one-eighth-inch slices. Other vegetables (e.g., celery, mushrooms, and spinach) should be crisp and in bite-sized pieces. Thin noodles or chicken wontons should be tender.

Potato The creamy white soup is thickened primarily with pureed potatoes. Chopped celery and onions blend well with the mild flavor of the potatoes and milk. Butter and cream may be added to give a rich flavor. Seasonings, especially white pepper, enhance the flavor.

Split Pea Soup is a green pea color with small cubes of pink ham. Split peas should be very soft or pureed and the ham tender. Characteristic flavor of ham is predominant and blends with the mild-flavored peas. Chopped carrots, onions, and potatoes are identifiable and add flavor to the soup. Variations include black beans, lentils, and yellow split peas.

Vegetable Beef Soup should have a reddish brown color with chopped or one-half-inch cubes of beef and numerous identifiable vegetables. Soup traditionally has chopped celery, onions, and tomatoes and cubed carrots and potatoes, which are tender and firm with a characteristic flavor.

Cold

Beet Borscht Chilled soup is a beet-red color topped with a dollop of sour cream. Ingredients are chopped, and the flavor is a combination of beets, celery, onions in a consommé stock. Sour flavor is derived from lemon juice or red wine vinegar and sour cream.

Fruit Color and predominant flavor of the chilled soup depend on type of fruit or berry puree (i.e., blueberry, peach, raspberry, or strawberry). Flavor is slightly sweet with a piquancy provided by lemon juice and a dollop of sour cream. Consistency is smooth and slightly thick.

Gazpacho Chilled soup is a tomato-red color with chopped pieces of celery, cucumbers, green peppers, parsley, sautéed mushrooms, and onion. Characteristic flavor of tomato is predominant and blends well with the other ingredients. Sour flavor is derived from red wine vinegar and spicy flavor from Worcestershire and tabasco sauces.

Vichyssoise Chilled soup is creamy white and topped with chopped chives or parsley. Flavor is a blend of chicken broth, white potatoes, and onions. Texture is creamy smooth.

VEGETABLES

General

Deep-Fat Fried Batter coating or breading should be golden brown and crisp and the vegetable soft but not saturated with oil. Vegetables most often deep-fat fried are eggplant, mushrooms, okra, onions, and zucchini squash.

Cauliflower, carrots, parsnips, and sweet potatoes are steamed before frying. Fat or oil should be bland. Dipping sauces enhance the flavor of the vegetables.

Grilled Identifiable pieces of vegetables, such as green onions, eggplant, mushrooms, peppers (green, orange, red, yellow), and zucchini squash, should be soft and slightly browned showing darker grill marks. Oil, most often olive, for basting should be mild flavored. Herbs may be added to the oil to enhance flavor.

Steamed Color should be natural and clear and the form distinct and uniform. Flavor should be natural, sweet, pleasant, and free of any trace of rawness. Seasonings should mildly contrast but blend with the flavor of the vegetable. Texture varies with the type. High-moisture vegetables, like green beans and cabbage, should be slightly underdone. Legumes and potatoes should be soft.

Stir-Fried Bite-sized pieces of vegetables are identifiable and retain natural color and some original crispness. Flavors should be characteristic of vegetable type (e.g., broccoli, carrots, cauliflower, celery, mushrooms, onions, peppers, snow peas, tomatoes, and zucchini squash). Seasoning enhances the flavor of the vegetables (fresh ginger, garlic, sesame oil, and soy sauce). Vegetables can be coated with a light, clear, cornstarch glaze.

Legumes

Beans, Baked Cooked navy beans should be soft and tender and the color of the added ingredients (catsup, molasses, mustard, onion, and/or salt pork). Flavor of the added ingredients should permeate the beans. Baked beans are moist and retain shape.

Beans, Refried The product should be pinkish brown, the color of pinto beans. Seasonings should enhance, but not overpower, the natural flavor of the beans (chili powder, garlic, jalapeño peppers, onion, and tabasco sauce). Flavor of cooking oil should be unobtrusive. Refried beans should have a mashed, slightly lumpy consistency and be slightly moist.

Cooked The natural shape of legumes should be preserved in cooking. Cooked legumes should be soft and tender, but retain identity, and the color typical of the fresh product (such as black, pinto, red, white beans; black-eyed and split green or yellow peas; and brown or red lentils). Flavor varies with the type of legume and should be distinctive. Various seasonings and ingredients complement the flavor.

Potatoes

Au Gratin Surface should have a lightly browned soft yellow sheen characteristic of melted cheese. Cooked cubed or sliced potatoes should be uniform in size and tender but firm. Cheese sauce should be smooth and velvety with a light yellow color and thick enough to coat the potatoes. The product should retain its shape when served.

Baked Skin should be the typical brown potato color and should be crisp, with doneness indicated by touch. Interior should be white with a soft and mealy texture. The potato should be split open immediately after removing from the oven to let steam escape and prevent sogginess. Flavor is mild and blends well with toppings, such as bacon, butter, cheese, chives, meat sauces, vegetables, or sour cream.

Boiled or Steamed White, slightly opaque potatoes should have a mild flavor that blends well with seasonings, sauces, and other ingredients. Potato should be tender, but firm, and retain its shape when cubed, shredded, or sliced.

Deep-Fat Fried Potatoes cut into varying lengths and shapes may be crinkle-cut or smooth. Exterior should be golden brown and crisp and interior white and soft. Cooking oil should be bland with no distinctive taste.

Mashed Color should be creamy white and texture soft and smooth throughout. The blend of salt, butter, and milk should enhance the natural potato flavor without being predominant.

Potato Salad, German Cooked potato slices should be soft but firm. Chopped onion, crisp bacon pieces, diced celery, pimiento, and parsley are identifiable in color and flavor. The tart vinegar sauce should be adequately sweet and thick enough to coat all ingredients. Salad should be served hot.

BIBLIOGRAPHY

Kotschevar, L. H., & Donnelly, R. (1999). *Quantity food purchasing* (5th ed.). Upper Saddle River, NJ: Merrill/Prentice Hall.

Spears, M. C. (1999). *Foodservice procurement: Purchasing for profit.* Upper Saddle River, NJ: Merrill/Prentice Hall.

Glossary

ABC inventory method Tool for classifying products as A, B, or C according to value.

Acceptance and preference sensory test Consumer panel answers questions about whether or not people will like the menu item.

Acceptance authority Authority based on employee's acceptance of that authority.

Accident Unexpected event resulting in injury, loss, or damage.

Accountability State of being responsible to one's self, to some organization, or even to the public.

Achievement-power-affiliation McClelland's theory emphasizing needs that are learned and socially acquired as the individual interacts with the environment.

Activity analysis Continuous observation for a chronological record of the nature of activities performed by individual workers.

Activity sampling *See* Occurrence sampling.

Actual cost Pricing method using actual costs plus profit to determine the selling price of the menu item.

Actual purchase price Inventory valuation method involving pricing the inventory at the exact price of each product.

Administrative intensity Degree to which managerial positions are concentrated in staff positions.

Aerobic bacteria Bacteria that need oxygen to grow.

Affective commitment Commitment to an organization because one is happy to be working for the organization, believes in the organization, and wants to do what is best for the organization.

Affective sensory test Preference, acceptance, and opinions of a product are evaluated by consumers who have no special sensory training.

Affirmative action Employers are required to analyze their workforce and develop a plan of action to correct areas of past discrimination.

AFL-CIO Merger of the American Federation of Labor (AFL), composed primarily of craft unions, and the Congress of Industrial Organizations (CIO), made up of industrial unions.

Agency Business relationship between the agent and principal.

Agent Individual who has been authorized to act on behalf of another party, known as the principal.

À la carte Method of pricing menu items individually.

All-or-nothing bid Suppliers are required to quote the best price on a complete list of items to be awarded the order.

American Dietetic Association (ADA) Professional organization serving the public through promotion of optimum nutrition, health, and well-being.

American Public Health Association (APHA) Professional organization representing all disciplines and specialties in public health.

American School Food Service Association (ASFSA) National association for federally sponsored child nutrition programs.

Americans with Disabilities Act (ADA) Comprehensive legislation that creates new rights and extends existing rights for Americans with disabilities.

Anaerobic bacteria Bacteria that reproduce without oxygen.
Analytical sensory test Differences and similarities of quality and quantity of sensory characteristics are evaluated by a panel of specially trained persons.
Approved brand specifications Specification that indicates quality by specifying a brand name or label.
Argyris's maturity theory Argyris's theory that a number of changes take place in the personality of individuals as they develop over the years into mature adults.
Artificial intelligence Computer program that attempts to duplicate the thought processes of experienced decision makers.
As purchased (AP) Amount of food before processing.
Assembly/serve Menu items are purchased preprepared and require minimal cooking before service.
Assets Resources of a company.
Atmospherics Physical elements in an operation's design that appeal to customers' emotions.
Authority Delegation from top to lower levels of management and the right of managers to direct others and take action because of their position in the organization.
Authority of competence Authority based on a manager's competence or expertise.
Balance Managerial adaptation to changing economical, political, social, and technological conditions.
Balance sheet Statement of assets, liabilities or debts, and capital or owner's equity at a given time or at the end of the accounting period.
Barbecue smoker Compact-size oven with racks to smoke, using wood chips, up to 100 pounds of meat at one time.
Bargaining unit Group of jobs represented in the union contract negotiations.
Batch cooking Cooking smaller quantities of menu items as needed for service.
Behavioristic variables Market segmentation by customers' attitudes and needs for a product.
Benchmarking Comparison against best performance in field.
Bid A price quotation from seller.
Bid buying Buyer decides which supplier will be chosen for the order based on bids submitted by the seller.
Biochemical spoilage Type of spoilage caused by natural food enzymes, which are complex catalysts that initiate reactions in foods.
Boundaries Limits of a system that set the domain of organizational activity.
Brand Product trade name, which may be used to indicate quality level in food specifications.
Break-even analysis Technique for assessing financial data to determine the point at which profit is not being made and losses are not being incurred.
Broiling Similar to grilling except the heat source is above the rack with the food item.
Broker Independent sales and marketing representative who contracts with manufacturers, processors, or prime source producers to both sell and conduct local marketing programs with wholesalers, suppliers, or foodservice operations.
Brunch Hybrid of breakfast and lunch.
Budget Plan for operating a business expressed in financial terms.
Burnout Emotional exhaustion.
Business ethics Self-generating principles of moral standards to which a substantial majority of business executives gives voluntary assent.
Business representative Hired by the local union to manage the union and also to settle a member's grievance if the steward was not successful.
Buyer A person who purchases goods or services.
Carry-over cooking The increase in internal meat temperature after removal from the heat source.

Causal forecasting model Based on the assumption that an identifiable relationship exists between the item being forecast and other factors, such as selling price, number of customers, and market availability.
Center-of-the-plate Most prominent menu item, usually the entrée, on which the menu is based.
Centralized service/centralized meal assembly Plates or trays are assembled in area close to production.
Centralized purchasing Purchasing activity is done by one person or department.
Chain of command Clear and distinct lines of authority in the organization, "who reports to whom."
Change Movement from one state to another.
Change agent Person who initiates change.
Charbroiler Either gas or electric equipment with a bed of ceramic briquettes above the heat source and below the grid.
Charrette Collaborative planning session.
Chemical spoilage Type of food spoilage of certain ingredients in a food or beverage resulting from interaction with oxygen or light.
Clamshell Hinged or removable top with its own heat source has been added to a gas or electric grill that permits cooking both sides of the food at one time.
Clean Free of physical soil and with an outwardly pleasing appearance.
Code of ethics Set of rules for standards of professional practice or behavior established by a group.
Collective bargaining Negotiation between management and the union on terms of the collective agreement between them.
Combi-oven Flow of both convected air and steam through oven cavity producing superheated, moist, internal atmosphere.
Combustion Form of solid waste recycling in which the energy value of combustible waste materials is recovered.
Composting Controlled application of the natural process of organic degradation.
Commercial foodservice Foodservice operations in which sale of food is the primary activity and a profit is desired.
Commissary foodservice Centralized procurement and production facilities with distribution of prepared menu items to several remote areas for final preparation and service.
Common-size statement Financial statement in which data are expressed as percentages for comparing results of the income statement from one accounting period to another.
Communication Oral, written, or computer-generated information whereby decisions and other information are transmitted.
Compensation Financial remuneration given by the organization to its employees in exchange for their work.
Competitor Another organization selling a similar product/service to the same market segment.
Competitive advantage Characteristic(s) of a company that distinguish it from others.
Conceptual framework Loosely organized set of ideas, some simple and some complex, that provides the fundamental structure of an organization.
Conceptual skill Ability to view the organization as a whole, recognizing how various parts depend on one another and how changes in one part affect other parts.
Conditions of certainty Adequate information is available to assure results of decision.
Conditions of risk Results of decision are uncertain, and probability techniques are necessary for estimating the likelihood of events occurring in the future.
Conditions of uncertainty Occurrence of future events cannot be predicted.
Conduction Transfer of heat through direct contact from one object or substance to another.
Constraint Something that limits an organization from reaching its goals.

Continuing care retirement communities (CCRCs) Living centers for retired persons who want quality residential services with healthcare available.

Continuous commitment Commitment to an organization only because the cost of leaving is too great.

Continuous quality improvement (CQI) Focused management philosophy for providing leadership, structure, training, and an environment to improve all organizational processes continuously.

Contribution margin (CM) Proportion of sales that can contribute to fixed costs and profits after variable costs have been covered.

Controlling Management function of ensuring that plans are being followed.

Convection Distribution of heat by the movement of liquid or vapor; may be either natural or forced.

Convection oven Fan on the back wall of an oven creates currents of hot air within the cooking chamber.

Convection/microwave oven Multifunction equipment that can use convection air alone, microwave energy alone, or a combination of the two.

Convenience store Retail business with primary emphasis placed on providing the public a convenient location to purchase quickly from a wide array of consumable products (predominantly food) and gasoline.

Conventional foodservice Foods are purchased in various stages of preparation for an individual operation, and production, distribution, and service are completed on the same premises.

Conveyor oven A gas or electric oven, often called a pizza oven in quick-service operations, with a belt running through a hot chamber on which products are baked.

Cook-chill Method in which menu items are partially cooked, rapidly chilled, held in chilled storage, and reheated just prior to service.

Cook-freeze Method in which menu items are partially cooked, rapidly frozen, held in freezer storage, and reheated just prior to service.

Cooking loss Decrease in yield of many foods in production primarily because of moisture loss.

Coordination Process of linking activities of various departments in the organization.

Corporate culture, or organizational culture Shared philosophies, values, assumptions, beliefs, expectations, attitudes, and norms that knit an organization together.

Cost center Department that has expenses but does not generate a profit to cover those expenses.

Cost-effectiveness Technique that provides a comparison of alternative courses of action in terms of their cost and effectiveness in attaining a specific objective.

Critical control points (CCPs) Locations in the food product flow where mishandling of food is likely to occur.

Critical incident Technique for identifying when an employee has or has not done something that results in unusual success or failure.

Cross-training Technique being used to familiarize employees with other jobs in the organization.

Customer Anyone who is affected by a product or service.

Customer satisfaction Positive, neutral, or negative feelings about the value received from a product.

Customer loyalty Frequency with which a customer consistently purchases a specific brand.

Cycle menu Series of menus offering different items daily on a weekly, biweekly, or some other basis, after which the menus are repeated.

Debit card Form of customer payment that looks like a credit card but allows owner to purchase using money already in an account.

Decentralized service/Decentralized meal assembly Food is transported in bulk to a location separate from production and plates or trays of food are assembled in that location.

Decisional roles According to Mintzberg, a manager can commit the organization to new courses of action and determine strategy; includes entrepreneur, disturbance handler, resource allocator, and negotiator roles.

Decision making Selection by management of a course of action from a variety of alternatives.

Decision tree Technique for assessing consequences of a sequence of decisions with reference to a particular problem.

Deck oven Oven that comes in basic sections, each with two shelves and its own source of heat, and that may be stacked on each other.

Deep-fat fryer Tank of oil or fat heated by gas or electricity into which foods are immersed; fryer has thermostatic control and fast recovery of fat temperature.

Delegation Process of assigning job activities and authority to a specific employee within the organization.

Delphi group Panel of experts who individually complete questionnaires on a chosen topic until a consensus is reached.

Demographic variables Population characteristics that might influence product selection, such as age, sex, race, income, education, occupation, and religion.

Demotion Change in job to one at a lower level in the organization.

Departmentalization Process of grouping jobs according to some logical arrangement.

Descriptive sensory test Information about certain sensory characteristics, such as taste, aroma, texture, tenderness, and consistency, that is needed for quality control and recipe development.

Design Defining the size, shape, style, and decoration of a space.

Development Training to prepare employees for increased responsibility.

Dietary Guidelines for Americans Recommendations for good health developed by the U.S. Department of Agriculture (USDA) and the U.S. Department of Health and Human Services (HHS).

Direct energy Energy expended within the foodservice operation to produce and serve meals.

Direct lighting Lighting aimed at a certain place.

Direct reading measurement tables Recipe adjustment method that uses printed tables for adjusting weight and volume ingredients for recipes that are divisible by 25.

Directing Management function of directing human resources for the accomplishment of objectives.

Discipline Action against an employee who fails to conform to the policies or rules of an organization.

Discrimination sensory test Detectable differences among food items are determined.

Distribution and service The third subsystem in the foodservice system.

Distributor Wholesaler responsible for transferring products from the processor manufacturer to the supplier.

Downsizing *See* Rightsizing.

Dry heat method Heat is conducted by dry air, hot metal, radiation, or a minimum amount of hot fat.

Dry storage Holding of food not requiring refrigeration or freezing but requiring protection from the elements, insects, rodents, and theft.

Dynamic equilibrium Continuous response and adaptation of a system to its internal and external environment.

Economic order quantity (EOQ) Inventory concept derived from a sensible balance of ordering cost and holding cost.

Edible portion (EP) Amount of food available for eating after preparation and/or cooking.

Effectiveness "Doing the right things."

Efficiency "Doing things right."

Emotional intelligence Extent to which a person is in tune with his/her own feelings and the feelings of others.

Employee assistance program (EAP) Program that provides diagnoses, counseling, and referral for advice or treatment for problems related to alcohol or drug abuse, emotional difficulties, and marital or family difficulties.

Employee satisfaction Beliefs and feelings an employee has about his or her job.

Empowerment Level or degree to which managers allow employees to act within their job descriptions to make decisions.

Environment All internal and external conditions, circumstances, and influences affecting the system.

Environmental factors Things outside the system that can impact the operation of the system.

Environmental scanning Search for and acquisition of information about events and trends external to the organization.

Equal employment opportunity Policy of equal employment (nondiscrimination) for all.

Equifinality Same or similar output can be achieved by using different inputs or by varying the transformation process.

Ergonomics Study of how workers interact with their work environment.

Ethics Principles of conduct governing an individual or a business.

Expectancy Theory based on the belief that people act in such a manner as to increase pleasure and decrease displeasure.

Expert systems Computer programs that build knowledge bases for making decisions.

Exponential smoothing forecasting model Time series model in which an exponentially decreasing set of weights is used, giving recent values more weight than older ones.

External audit Inspection program performed by governmental or nongovernmental agencies.

External control Consists of local, state, and federal regulations and contracts with outside companies.

External customers Recipients of food and services that do not belong to the organization that produces them.

Factor pricing method Also known as the mark-up method, the difference between cost and selling price.

Factor recipe adjustment method Ingredients are changed from measurements to weights and multiplied by a conversion factor determined by dividing the desired yield by the current recipe yield.

Fair Labor Standards Act Federal law concerned with minimum wage rates and overtime payments, child labor, and equal rights.

Fast-food restaurant Establishment primarily selling limited lines of refreshments and prepared food.

Feedback Those processes by which a system continually receives information from its internal and external environment.

FIFO (first in, first out) Inventory valuation method based on the assumption that pricing closely follows the physical flow of products through the operation.

Fixed costs Costs that do not vary with changes in the volume of sales.

5 a day program Program sponsored by the Produce for Better Health Foundation and the National Cancer Institute to increase consumption of fruits and vegetables to at least five servings a day.

Focus group Qualitative marketing technique bringing a group of people together for a one-time meeting to discuss a specific aspect of an operation.

Flow Movement of product or people through an operation.

Food and Drug Administration (FDA) Agency within the U.S. Department of Health and Human Services that enforces the Federal Food, Drug and Cosmetic Act, which ensures that foods other than meat, poultry, and fish are pure and wholesome.

Foodborne infection Caused by activity of large numbers of bacterial cells carried by the food into the gastrointestinal tract.

Foodborne intoxication Caused by toxins formed in food prior to consumption.

Foodborne pathogen Virus, microorganism, or other substance that causes disease; an infecting agent.

Food Guide Pyramid Illustration of recommended food intake.

F.O.B. Free on board.

F.O.B. destination Buyer pays freight charges, but seller owns product during transit.

F.O.B. origin Buyer selects freight company, pays freight charges, and owns product during transit.

Food product flow Alternative paths within foodservice operations that food and menu items may follow, initiating with receiving and ending with service to the customer.

Food production Preparation of menu items in the needed quantity and the desired quality at a cost appropriate to the particular operation.

Food Safety and Inspection Service (FSIS) Responsible for ensuring that meat and poultry products destined for interstate commerce and human consumption do not pose any health hazards.

Foot-candle Measurement of illumination equal to 1 lumen of light on 1 square foot of space.

Forecasting Art and science of estimating events in the future, which provides a database for decision making and planning.

Forecasting model Mathematical formula to predict future needs, which is used as an aid in determining quantity.

Formal authority Authority that exists because of position in the organization.

Franchisee Person who is granted a franchise.

Franchising Right granted to an individual or group to market a company's concepts.

Franchisor Person who grants a franchise.

Frying Cooking in fat or oil; includes sautéing, pan frying, and deep frying.

Full or broadline wholesalers Generally, wholesalers who carry large amounts of stock, permitting the buyer to purchase everything from frozen and canned products to kitchen equipment and furniture.

Full-service restaurants Establishments that sell food and service to customers who prefer to eat at a table on premises; includes casual, theme, family dining, and fine-dining restaurants.

Full-time equivalent (FTE) Number of total hours worked in a week divided by 40 (the number of hours worked by one full-time employee) to determine the number of full-time-equivalent employees.

Functional foods Often used interchangeably with nutraceuticals, this broad term refers to foods containing significant levels of biologically active compounds that impart health benefits beyond basic nutrition.

Functional manager Responsible for only one area of organizational activity.

Game theory Competition in decision making is accomplished by bringing into a simulated decision situation the actions of an opponent.

Gauge The weight of the material per square foot; a measurement characteristic of metal.

General manager Responsible for all activities of a unit.

Genetically engineered food Food whose genetic structure has been altered by adding or eliminated genes to enhance qualities of the product.

Geographic variables Climate, terrain, natural resources, population density, and subcultural values that influence customer product needs.

Goal Aim or purpose of the foodservice system.

Grade Level of quality of agricultural products as defined by the USDA.

Grade standards Requirements, which are voluntary, that must be met by a product, such as fruits, vegetables, eggs, dairy products, poultry, and meat, to obtain a particular USDA grade.

Grapevine Informal communication channel in which information is passed from one employee to another.

Griddle Extra-thick steel plate with a ground and polished top surface with raised edges, either gas or electric, usually set on a countertop.

Grilling Cooking on an open grid over a heat source, which may be an electric or gas-heated element, ceramic briquettes, or exotic woods and flavored chips.

Gross Domestic Product (GDP) Total national output of goods and services valued at market prices.

Gross National Product (GNP) Measurement of all currently produced final goods and services; the term is being phased out in favor of GDP.

Group decision making More than one person is needed to solve complex problems for which specialized knowledge is required from several fields.

Group purchasing Bringing together managers from different operations for joint purchasing.

Groupthink Situation where reaching agreement becomes more important to group members than arriving at a sound decision.

Halo effect When a single trait dominates the assessment of another individual.

Handling loss Decrease in the yield of a recipe because of preparation process.

Hazard Unacceptable contamination of food.

Hazard analysis Identification of specific foods that are at risk of contamination.

Hazard Analysis Critical Control Point (HACCP) Systematic analysis of all process steps in the foodservice subsystems, starting with food products from suppliers and ending with consumption of menu items by customers.

Heat recovery time Amount of time for oil in fryer to return to optimum cooking temperature after food is added.

Heat transfer rate The amount of heat from one substance to another in a given amount of time in a given space measured in British thermal units (Btu's).

Heterogeneity of service Variation and lack of uniformity in the performance by different service employees.

Hierarchy Characteristic of a system that is composed of subsystems of a lower order and a suprasystem of a higher order.

Historical records Past purchasing and production records provide the base for forecasting quantities.

Home meal replacement Meal similar to one that is prepared in the home that can be purchased from a foodservice operation and taken or delivered to the home.

Horizontal division of labor Division of work among employees at the same level in the organization.

Human resources Skills, knowledge, and energies of people required for the system to function.

Human resources management Management of all aspects of human resources from recruitment through termination.

Human resources planning Process of making provision for the movement of people into, within, and out of an organization.

Human skill, or interpersonal skill Concerns working with people and understanding their behavior.

Immigration Reform and Control Act (IRCA) Control of unauthorized immigration by making it unlawful for a person or organization to hire persons not legally eligible for employment in the United States.

Impinger Conveyorized gas-fired or electric oven that toasts the bottom of pizza crust and makes it flaky.

Income statement Financial report that presents the net income or profit of an organization for the accounting period.

Independent purchasing Unit or department of an organization that has been authorized to purchase.

Indirect energy Energy expended to facilitate functions that use energy directly.

Indirect lighting Lighting shining over a space rather than at a certain place.

Induction Use of electrical magnetic fields to excite the molecules of metal cooking surfaces.

Induction-heat grill Cooks magnetically and has no open flame or thermostat for control.

Informational roles According to Mintzberg, communication may be the most important aspect of a manager's job and includes monitor, disseminator, and spokesperson roles.

Infrared waves Type of radiation used in food production that has a longer wavelength than visible light.

Ingredient room Ingredient assembly area designed for measuring ingredients to be transmitted to the various work centers.

Input Any human, physical, or operational resource required to accomplish objectives of the system.

Inseparability Inability to separate production and service.

Institutional foodservice Foodservice operations in which sale of food is secondary to the goal of the organization; typically not-for-profit, but the foodservice must break even financially while satisfying customers. *See also* noncommercial and onsite foodservice.

Intangibility of services Inability of services to be seen, touched, tasted, smelled, or possessed before buying.

Integration Parts of the system sharing objectives of the entire organization.

Interacting groups Decision-making technique in which members discuss, argue, and agree upon the best alternative.

Interdependency Each part of the system affects performance of other parts of the system.

Interface Area where two systems or subsystems come in contact with each other.

Internal audit Self-inspection program.

Internal control Consists of plans including the goals and objectives of the organization, standards, and policies and procedures.

Internal customers Employees and suppliers of products who also belong to the organization that produces them.

Interpersonal roles According to Mintzberg, the focus is on relationships and includes figurehead, leader, and liaison roles.

Interpersonal skill *See* Human skill.

Inventory Record of material assets owned by an organization.

Inventory control Technique of maintaining assets at desired quantity levels.

Invoice Document prepared by the supplier that contains product, name, quantity, and price.

Irradiation Exposure of foods to gamma rays or radient energy to reduce harmful bacteria.

Job analysis Process of determining duties and tasks of a job.

Job characteristics Job dimensions that could improve the efficiency of organizations and job satisfaction of employees.

Job description List of duties, working conditions, and tools and equipment needed to perform the job.

Job design Concerned with structuring jobs to improve organization efficiency and employee job satisfaction.

Job duties Statements that usually are arranged in order of importance that should indicate the weight or value of each job.

Job enlargement Increase in the total number of tasks employees perform.
Job enrichment Increase in the variety and number of tasks and control the employee has over the job.
Job identification Such information as location of job, to whom the employee reports, and often the number of employees in the department and the Directory of Occupations code number.
Job satisfaction An individual's feelings and beliefs about their job.
Job specification List of abilities, skills, and other credentials needed to do the job.
Job title Indication of level in the organization and degree of authority the job possesses.
Joint Commission on Accreditation of Healthcare Organizations (JCAHO) Regulatory agency to determine the degree to which the organization complies with established control standards.
Just-in-time (JIT) Philosophy and strategy that involves purchasing products.
Just-in-time purchasing Purchasing of products for immediate production and consumption by the customer without having to record it in inventory records.
Labor relations Interaction between management and a labor union.
Latest purchase price Inventory valuation method in which the last price paid for a product is used.
Law of agency Buyer's authority to act for the organization, the obligation each owes the other, and the extent to which each may be held liable for the other's actions.
Law of contract Agreement between two or more parties requiring the buyer to be certain that each agreement is legally sound.
Law of warranty Guarantee by the supplier that an item will perform in a specified way.
Layout Detailed arrangement of equipment, floor space, and counter space.
Leader-member relations Nature of the relationship between the leader and work group.
Leadership Process of influencing activities of an individual or group toward achieving organisational goals.
Lead time Interval between the time that a requisition is initiated and receipt of the product.
Lean production Superior way for humans to make things by providing better products in wider variety at lower cost.
Liabilities Debts of a company.
LIFO (last in, first out) Inventory valuation method based on the assumption that current purchases are largely made for the purpose of meeting current demands of production.
Limited-menu restaurant Classification based on the Standard Industrial Classification (SIC), Eating and Drinking Places, and used by the NRA for a fast-food restaurant.
Line and staff Organizational relationship of line personnel, those with a linear responsibility, to staff personnel, those who serve in an advisory capacity to line managers.
Linear programming Technique used in determining an optimal combination of resources to obtain a desired objective.
Line-item bidding Each supplier quotes a price on each product on the buyer's list, and the one offering the lowest price receives the order for that product.
Line position A position in the direct chain of command.
Linking process Coordination of the characteristics of a system in the transformation of resources into goals.
Low-temperature cooking and holding ovens Cooking temperatures are from 100°F to 325°F, giving a 25% yield increase for meat; holding temperatures are from 60°F to 200°F.
Low-temperature storage Holding perishable food in refrigerated or frozen storage for preservation of quality and nutritive value immediately after delivery and until use.
Lumen Amount of light generated when 1 foot-candle of light shines from a source.
Maître d'hôtel Master of the house, head steward, majordomo, or head waiter.

Make-or-buy decisions Procedure for deciding whether to purchase from oneself (make) or purchase from suppliers (buy).

Managed care Relatively new term coined originally to refer to the prepaid healthcare sector where care is provided under a fixed budget.

Management Process integrating resources for accomplishment of objectives.

Management by objectives Technique sometimes used for performance appraisal in which managers and their superiors agree on the objectives to be achieved, usually for a 1-year period.

Management development Programs designed to improve the technical, human, and conceptual skills of managers.

Management functions Integral component of the transformation element, including planning, organizing, staffing, leading, and controlling, performed by managers to coordinate subsystems in accomplishing goals.

Manufacturers' representatives Distributors who represent a manufacturing company and informs suppliers of products by this manufacturer.

Market Medium through which a change in ownership moves commodities from producer to customer.

Market segmentation Process of dividing a total market into groups of customers who have similar needs, wants, values, and buying behavior.

Market share Percentage of industry sales for a product.

Marketing Activities that facilitate satisfying customers in a dynamic environment through the creation, distribution, promotion, and pricing of goods, services, and ideas.

Marketing channel Exchange of ownership of a product from the producer through the processor or manufacturer and the distributor to the customer.

Marketing cost analysis Classifies costs to determine which are associated with specific marketing activities.

Marketing environment Consists of all forces outside the organization that influence marketing activities and exchanges.

Marketing mix Combination of product, price, place, and promotion to satisfy target market.

Marketing objective Statement of what is to be accomplished through marketing activities.

Marketing plan Written document governing an organization's marketing activities.

Marketing research Systematic gathering, recording, and analyzing of data about problems relating to the marketing of goods and services.

Marketing strategy Encompasses selecting and analyzing a target market and creating and maintaining an appropriate marketing mix that will satisfy that market.

Markup Difference between cost and selling price.

Master schedule Overall plan for employee scheduling, including days off.

Master standard data (MSD) Seven basic elements of work are combined into larger, more condensed elements.

Meal equivalent Conversion of number of snacks, nourishments, and paid meals into a common number of meals.

Memory All stored information that provides historical records of a system's operations.

Mentor Teacher, tutor, or a coach, depending upon the relationship between the mentor and the protégé.

Menu List of items available for selection by a customer and the most important internal control of the foodservice system.

Menu engineering Management information tool that focuses on comparing the popularity of the item with its profit.

Menu evaluation Continuing process that should be conducted during production and service of each meal and after major menu planning sessions.

Menu mix (MM) Concerned with the popularity of the menu item, as measured by customer demand.
Menu pricing Cost of a menu item to a customer, including food and labor, operating, and profit.
Menu psychology Designing and layout a menu in such a way as to influence the sale of foods served on that menu.
Mesquite grill Smoke from mesquite wood, the source of heat, gives a smoky and somewhat subtle tangy flavor to grilled meats.
Metacognition Ability to learn and adapt to change.
Method Details for one step of a procedure.
Microbiological spoilage Type of food spoilage caused by microorganisms, such as bacteria, molds, yeast, viruses, rickettsiae, protozoa, and parasites.
Microbreweries Place where beer is brewed, which may be sold on or off premises.
Microwave oven An oven in which food is cooked by the absorption of microwave energy by water molecules in the food.
Microwaves Very short wavelengths that penetrate partway into food and agitate water molecules, causing friction to create heat.
Mini-max (minimum-maximum) method Tool for controlling inventory by establishing lower and upper levels for each product in storage.
Mission statement Summation of the purpose of the organization, competition, target market, product and service, and recipients of the service including customers, employees, owners, and the community; differentiates it from others.
Model Conceptual simplification of a real situation in which extraneous information is excluded and analysis is simplified.
Moist heat method Heat is conducted to the food product by water or steam.
Motive Something that prompts a person to action.
Motivation An inner force that activates a person toward achievement of a goal.
Moving average forecasting model Time series model that uses a repetitive process for developing a trend line by averaging the number of servings for a specified number of times for the first point on the line and then dropping the oldest and adding the newest number of servings for subsequent points.
Multidepartment management More than one department is managed by one manager.
MyPyramid Illustration of nutrition and physical activity recommendations.
Mystery shoppers Persons unknown to customers and employees who eat at a restaurant and evaluate their own experiences and those of other customers.
National Association of College and University Food Services (NACUFS) Professional organization that defines standards and criteria to be used for self-monitoring or for a voluntary peer review program.
National Council of Chain Restaurants (NCCR) Professional organization that has an active role in shaping food safety policy for the future.
National Marine Fisheries Service (NMFS) A voluntary inspection program for seafood safety.
National Sanitation Foundation International (NSF International) Nonprofit, noncommercial organization that develops minimum sanitation standards for foodservice equipment.
Need hierarchy Maslow's theory that states people are motivated by their desire to satisfy specific needs in the following ascending order: physiological, safety, social, esteem, and self-actualization.
Negotiation Process of working out an agreement, mutually satisfactory to both parties.
Network Graphic representation of a project, depicting the flow and sequence of defined activities and events.
New organization Organizations in which employers are challenged to improve the quality of work life and to develop a corporate culture.

Nominal group Structured technique for generating creative and innovative alternatives or ideas.

Noncommercial foodservice Foodservice operations in which sale of food is secondary to the goal of the organization; see also institutional foodservice and onsite foodservice.

Nonprogrammed decision Relatively unstructured decision that takes a higher degree of judgment.

Nutraceuticals Combinations of natural ingredients that provide a proactive, natural approach to promote and maintain health.

Nutrition labeling Regulations issued by the FDA and the USDA's Food Safety and Inspection Service (FSIS) state that any nutritional health claims made about food must be documented on labels and menus.

Observation Method requiring trained observers to estimate visually the amount of plate waste.

Occurrence sampling Method for measuring work and nonwork time of employees in direct and indirect activities and for measuring operating and downtime of equipment.

Offer-versus-serve A meal plan in school foodservice programs in which students must choose three of five menu components for the lunch to be reimbursed by the federal government.

Off-the-job training Occurs outside the workplace.

Omnibus Budget Reconciliation Act (OBRA) of 1987 Federal nursing home regulation designed to improve the quality of life and care for residents in nursing homes.

One-stop shopping Customers can purchase almost anything from food products, service supplies and equipment, and office supplies to furniture from one distributor.

Onsite foodservice Foodservice operations in which sale of food is secondary to the goal of the organization; typically not-for-profit.

On-the-job training Training an employee by having them work with another employee.

Open systems Organizations that are in continual interaction with the environment.

Operating in the black When revenue minus expenses is a positive value.

Operational resources Money, time, utilities, and information.

Organization Group of people working together in a structured and coordinated way to achieve goals.

Organization chart Graphic portrayal of the organization structure that depicts basic relationships of positions and functions while specifying the formal authority and communication network.

Organizational change Any substantive modification to some part of the organization.

Organizational citizenship behaviors (OCB) Positive, voluntary behaviors that enhance organizational efficiency.

Organizing Management function of grouping activities, delegating authority, and providing for coordination of relationships, horizontally and vertically.

Orientation Formal process of familiarizing new employees to the organization, job, and work unit.

Output Result of transforming input into achievement of a system's goal.

Overproduction Production of more food than is needed for service.

Owner's equity Money value of a company in excess of its debts that is held by the owners.

Paradigm Mental frame of reference that dominates the way people think and act.

Paradigm shift Changing the way one views a situation.

Pareto analysis 80–20 rule—approximately 80% of total sales volume comes from approximately 20% of customers.

Participative management Involving employees in the decision-making process.

Percentage method Recipe adjustment method in which measurements for ingredients are converted to weights and then the percentage of the total weight for each ingredient is computed.

Performance Degree of accomplishment of tasks that make up an individual's job.

Performance appraisal Comparison of an individual's performance with established standards for the job.

Performance specifications Specification that indicates quality by functioning characteristics of the product.

Performance standards Desired results at a definite level of quality for a specified job.

Perishability of services Services cannot be stored for future sale.

Permeability of boundaries Characteristic of an open system that allows the system to be penetrated or affected by the changing external environment.

Perpetual inventory Purchases and issues continuously are recorded for each product in storage, making the balance in stock available at all times.

Personal bias When personal preferences alter objective decision making.

Personal ethics Person's religion or philosophy of life that is derived from definite moral standards.

Personnel management Performance of basic functions such as selection, training, and compensation of employees.

pH value Degree of a food's acidity or alkalinity.

Physical inventory Periodic actual counting and recording of products in stock in all storage areas.

Physical resources Materials and facilities.

Physical spoilage Type of food spoilage caused by temperature changes, moisture, and dryness.

Pilferage Inventory shrinkage.

Planning Management function of determining in advance what should happen.

Planning for production Establishment of a program of action for transformation of resources into products and services.

Plate waste Amount of food left on a plate, which often is used as a measure of food acceptability.

Policy General guide to organizational behavior developed by top-level management.

Portion control Service of same size portion to each customer.

Potenially hazardous food (time/temperature control for safety food) Food items that require temperature control because they are capable of capable of supporting growth of pathogenic microorganisms or toxin formation.

Power Resource that enables a leader to induce compliance or influence others.

Power Soak Pot- and pan-washing equipment that capitalizes on the natural scouring abilities of high-turbulence, heated water.

Preferred futuring Process that solidifies many of the changes that already have taken place.

Prepotent need Need that is dominant over all others.

Pressure steamers Equipment that works by trapping and removing air from its cavity, causing steam pressure to build.

Pressureless convection steamer Equipment in which heat is transferred from steam to food by forced convection caused by a fan inside that encircles the food, thus cooking it without pressure.

Preventive maintenance Keeping equipment and facilities in a good state of repair.

Pricing psychology Psychological aspects of pricing affect customer perceptions, which then influence the purchase decision.

Prime cost Pricing method which includes raw food cost and direct labor cost.

Principal Person who needs an agent to work on his or her behalf.

Prix fixe Fixed price for a complete meal; compare to à la carte.

Procedure Chronological sequence of activities.

Procurement Managerial responsibility for acquiring material for production; the first subsystem in the transformation element of the foodservice system.

Producers Farmers or ranchers who produce raw food to sell to processors who sell to distributors or directly to the foodservice operation.

Production Second subsystem in the transformation element of the foodservice system, defined as the process by which products and services are created.

Production schedule Written plan for production for a specific data and/or meal.

Production scheduling Time sequencing of events required by the production subsystem to produce a meal.

Productivity Ratio of output to input.

Productivity level Meals per labor hour in foodservice operations.

Profit center Department that is generates revenues greater than its expenses creating profit for the organization.

Pro forma statement Statement that projects expected income and expenditures.

Programmed decision Decision reached by following established policies and procedures.

Promotion Change in job to one at a higher level in the organization.

Promotion Use of communication to inform and influence consumers.

Psi Pounds per square inch, used to measure steam pressure.

Psychographic variables Motives and lifestyles characteristics.

Pulper Water-filled tank in which solid waste is broken down into a slurry by a shredding device and then water is pressed out of it.

Purchase order Document, based on information in the requisition, completed by the buyer, who gives it to the supplier.

Purchasing Activity concerned with the acquisition of products.

Quality Features and characteristics of a product or service that bears on its ability to satisfy stated or implied needs and a product or service that is free or defects.

Quality assurance (QA) Procedure that defines and ensures maintenance of standards within prescribed tolerances for a product or service.

Quality attributes Characteristics of a food, such as microbiological, nutritional, and sensory, that require control throughout the procurement/production/service cycle to maintain them.

Quality circles Problem-solving process by small groups of employees who choose a project and focus on quality improvement.

Quality control Continuous process of checking to determine if standards are being followed, and, if not, taking corrective action.

Quality improvement process (QIP) Structured problem-solving approach that focuses on operating processes and involves staff in analyzing current situations and developing recommendations for problem resolution.

Quality of work life How work is organized by the manager and how jobs are designed.

Queuing theory Cost of waiting in lines is balanced against the cost of expanding facilities.

Radiation Generation of heat energy by wave action within an object.

Range oven Part of a stove, generally called a range, located beneath the cooking surface.

Rating scales Number of dimensions on which an employee is rated on a three- to five-point scale for measuring quality of performance.

Ratio analysis Mathematical expressions of the relationship between two items are categorized according to primary use, such as liquidity, solvency, activity, profitability, and operating, to facilitate understanding.

Ready prepared foodservice Menu items are produced and held chilled or frozen until heated for serving.

Receiving Activity for ensuring that products delivered by suppliers are those that were ordered.

Recipe Formula by which weighed and measured ingredients are combined in a specific procedure to meet predetermined standards.

Recipe format Definite pattern or style for recording a recipe.

Recipe standardization Process of tailoring a recipe to suit a particular purpose in a specific foodservice operation.

Recommended Dietary Allowance Recommendations for dietary intake of nutrients for healthy growth.

Recruitment Process of locating and encouraging potential applicants to apply for a job opening.

Recycling Act of removing materials from the solid waste stream for reprocessing into valuable new materials and useful products.

Regression analysis forecasting models Causal models that are based on the assumption that the linear relationship between variables will continue for a reasonable time in the future.

Reinforcement Skinner's theory that people behave in a certain way because they have learned that certain behaviors are associated with positive and others with negative outcomes.

Reorder point Lowest stock level that safely can be maintained to avoid emergency purchasing.

Requisition First document in the purchasing process that may have originated in any one of a number of units in a foodservice operation.

Resources Human and physical inputs into the foodservice system that are transformed to produce outputs.

Responsibility Obligation to perform an assigned activity or see that someone else performs it.

Restaurant-type menu Static menu in which the same items are offered every day.

Retailers Final persons in the marketing channel who sell products to the ultimate buyers, the customers.

Rightsizing Eliminating unneeded managers and empowering employees can increase productivity in a time of economic slowdown.

Risk Possibility of loss or injury.

Risk management Discipline dealing with possibility that some future event will cause harm to an organization.

Risky shift Tendency of individuals to take more risk in groups than as individuals.

Rotisserie An appliance fitted with a spit on which food is rotated over a source of heat.

Rules Specification of action, stating what must or must not be done.

Safety stock Backup supply to ensure against sudden increases in product usage rate.

Sales analysis Detailed study of sales data for the purpose of evaluating the appropriateness of a marketing strategy.

Sanitarian Health official or inspector who is trained in sanitation principles and methods.

Sanitary Free of disease-causing organisms and other contaminants.

Sanitation and maintenance Fourth subsystem in the foodservice system.

Scheduling Assigning employees to specific work hours and days.

Scientific management Theory developed by Frederick W. Taylor in the early 1900s emphasizing the systematic approach for improving worker efficiency by using performance standards and time studies.

Scrapping Disposing of fragments of discarded or leftover food in the dishwashing process.

Selection Process of comparing applicant skills, knowledge, and abilities with the requirements of a position and choosing the most qualified.

Self-reported consumption Technique for measuring plate waste in which individuals are asked to estimate their plate waste using a scale similar to one used by trained observers.

Sensory analysis A science that measures the texture, flavor, and appearance of food products through human senses.

Separation Voluntary or involuntary termination of a job.

Service Wide variety of intangible factors influencing the satisfaction of the buyer.

SERVSAFE® Training program developed by the Educational Foundation of National Restaurant Association that concentrates on food safety.

Sexual harassment Form of gender discrimination in violation of Title VII of the Civil Rights Act of 1964.

Shift schedule Staffing pattern for each block of time, usually 8 hours, regardless of idle time in the operation.

Simulation Use of a model for imitating a real-life occurrence and studying its properties, behavior, and operating characteristics.

Single-use menu Menu that is planned for service on a particular day and not used in the exact form a second time.

Situation analysis Identification of marketing opportunities and challenges.

Skill As defined by Katz, an ability that can be developed and is manifested in performance.

Skills inventory Information on each employee's education, skills, experience, and career aspirations.

Smoker oven Electric, compact oven with racks to smoke up to 100 pounds of meat at a time.

Social caterers Establishments primarily engaged in serving prepared food and beverages at another facility rather than at a fixed business location, such as a restaurant.

Social intelligence Ability to determine the requirements for leadership in a particular situation and select the appropriate response.

Social Security Act Insurance program to protect covered employees against loss of earnings resulting from retirement, unemployment, disability, or from death or disability of the person supporting them.

Sociocultural factors Customs, mores, values, and demographic characteristics of the society in which the organization functions.

Solid waste Products and materials discarded after use in homes, commercial establishments, and industrial facilities (U.S. Environmental Protection Agency).

Source reduction Design and manufacture of products and packaging with minimum toxic content, minimum volume of material, and a longer useful life (U.S. Environmental Protection Agency).

Sous vide A process of sealing raw, fresh food items in plastic pouches to allow chilled storage and then cooking in boiling water prior to service.

Span of management (span of control) Number of employees that can be effectively supervised by one manager.

Special breed distributors Purchasing and product movement specialists that purchase food directly from processors and hire a distributor to deliver the products to their restaurant chain clients.

Specialty wholesalers Distributors that specialize in a particular product category, such as meat, produce, dairy, paper, or detergent.

Specification Statement understood by buyers and suppliers of the required quality of products, including allowable limits of tolerance.

Split shift scheduling Assigning employees to work with extra staff at peak times of business.

Spoilage Denotes unfitness for human consumption due to chemical or biological causes.

Spoken menu Menu that is presented by the technician orally to the patient.

Staff position Intended to provide expertise, advice, and support for line positions.

Staffing Management function that determines the appropriate number of employees needed by the organization for the work that must be accomplished.

Staffing tables Pictorial representation of all jobs with the number of employees in those jobs and future employment requirements.

Staggered schedule Employee work hours that begin at varying times to meet the needs of the foodservice operation.

Stakeholders Individuals or groups who are significantly affected by or can significantly influence a company's decisions.

Standard Definition of what is expected to happen.

Standard Operating Procedures (SOPs) Written, step-by-step instructions for routine tasks.

Standardized recipe Recipe that consistently delivers the same quantity and quality of product when followed precisely.

Statement of changes in financial position Information about the source and use of funds.

Static menu Same menu items are offered every day, that is, a restaurant-type menu.

Steam-jacketed kettle One kettle in a larger kettle with an open space, or jacket, between the two into which steam is introduced.

Storage Holding of products under proper conditions to ensure quality at time of use.

Strategic business units (SBUs) Division, product-line, or product department that sells specific products or services to a target market and competes against other businesses in that market.

Strategic planning A continuous and systematic process in which people make decisions about intended future outcomes, how outcomes are to be accomplished, and how success is measured and evaluated.

Strategies Decisions and actions to assist a company meet its objectives.

Subjective forecasting model Based on the idea that little relationship exists between the past and long-term future and, therefore, forecasters must rely on opinions and quantitative information that might be related to the item being forecast.

Subsystem Complete system in itself but not independent; an interdependent part of the whole system.

Superdistributor Person who offers an extremely wide variety of products, thus giving the customer the opportunity to engage in one-stop shopping.

Supplier Fourth component of the marketing channel, often identified as a seller or vendor; the person who offers products for sale.

Sweet spot Size or capacity of equipment from which the buyer gets the best value and lowest cost per menu item produced.

Synergy Working together to create greater outcome than individually.

System Collection of interrelated parts or subsystems unified by design to obtain one or more objectives.

Systems analysis Decision-making process aiding the manager in making the best choice among several alternatives.

Systems approach Keeping the organization's objectives in mind throughout the performance of all activities.

Systems management Application of systems theory to managing organizations.

Systems philosophy Way of thinking about an event in terms of parts or subsystems with emphasis on interrelationships.

Table d'hôte (French for "host's meal") Several food items grouped together and sold for a fixed price (prix fixe).

Tactic Short-term action.

Target market Customer with common characteristics for which an organization creates products/services.

Team Group of workers that serves as a unit, often with little or no supervision.

Team-based leadership Two or more persons who interact to accomplish a common purpose or goal regularly enough to be considered a group.

Technical skill Understanding of, and proficiency in, a specific kind of activity, particularly one involving methods or techniques.

Technical specifications Applicable to products for which quality may be measured objectively and impartially by testing instruments.

Theft Premeditated burglary.

Theory X McGregor's theory that motivation is primarily through fear and the supervisor is required to maintain close watch of employees if organizational goals are to be met.

Theory Y McGregor's theory that emphasizes managerial leadership by permitting employees to experience personal satisfaction and to be self-directed.

Thermal death time Time required at a specific temperature to kill a specified number of vegetative cells or spores.

Tilting skillet Floor-mounted rectangular pan with a gas or electric-heated flat bottom, pouring lip, and hinged cover.

Time and temperature Critical elements in quantity food production that must be controlled to produce a high-quality product.

Time series forecasting model Based on the assumption that actual occurrences follow an identifiable pattern over time.

Time value of money Money has a differing value over time; having $1 today is worth more than receiving $1 in the future.

Total customer value Combination of the tangible and intangible experienced by customers that become their perception of doing business with an organization.

Total quality management (TQM) Management philosophy directed at improving customer satisfaction while promoting positive change and an effective cultural environment for continuous improvement of all organizational aspects.

Total quality service (TQS) Based on the assumption that all quality standards and measures should be customer referenced and help employees guide the organization to deliver outstanding value to customers.

Traditional organization Organization in which lines of authority, which create order, are established.

Training Ongoing process of updating of employees.

Transactional leadership Leadership that focuses on clarifying roles and responsibilities and use of rewards and punishment to achieve goals.

Transfer Change in job to one at approximately the same level elsewhere in the organization.

Transformation Any action or activity used in changing inputs into outputs.

Transformational leadership Leadership that inspires followers to become motivated to work toward organizational rather than personal gain.

Trend analysis Comparison of results over several periods of time to forecast future revenues or levels of activity.

Two-factor Herzberg's theory of work motivation focusing on outcomes of performance that satisfy needs.

Underwriters Laboratory, Inc. (UL) An organization responsible for the compliance of equipment with electrical safety standards.

Underproduction Production of less food than is needed for service.

Union salting Union organizers who apply for positions in nonunionized companies and when hired, work to form a union of the employees

Union steward (shop steward) Union employee who is elected to represent other union members in their relations with an immediate supervisor or other managers.

United States Department of Agriculture (USDA) Federal department with an important role in the food regulatory process by grading, inspecting, and certifying all agricultural products.

United States Department of Health and Human Services (USHHS) Federal department that administers the Food, Drug and Cosmetic Act and the Public Health Service Act.

Universal Product Code (UPC) System for uniquely identifying thousands of different suppliers and millions of different products that are warehoused, sold, delivered, and billed throughout retail and commercial channels of distribution.

Usage rate Determined by past experience and forecasts of a product.

Valence Value an employee places on rewards offered by the organization.

Value Perceived relationship between quality and price.

Value-added Increase in value caused by processing, manufacturing, marketing, or distributing, exclusive of the cost of material, packaging, and overhead.

Value analysis Methodical investigation of all components of an existing product or service with the goal of discovering and eliminating unnecessary costs without interfering with the effectiveness of the product or service.

Values and Life-Styles (VALS) program Classification system for segmenting customers by demographic and lifestyle factors.

Variable costs Costs that vary directly with changes in sales.

Vegan-approved menu items Those with no animal meat or byproducts for strict vegetarians.

Ventilation The circulation of fresh air in a space.

Vertical division of labor Based on lines of authority.

Vision Statement of where a company wants to be in the future.

Warehouse club purchasing No-frills approach to purchasing because the wholesale warehouse is a self-service, cash and carry operation.

Ware washing Process of washing and sanitizing dishes, glassware, flatware, and pots and pans either manually or mechanically.

Weighted average Inventory valuation method in which a weighted unit cost is used and is based on both the unit purchase price and the number of units in each purchase.

Wholesaler Distributor that purchases from various manufacturers or processors, provides storage, sells, and delivers products to suppliers.

Work design Industrial engineering term for productivity improvement to assist the worker to work more efficiently without expending more effort.

Work measurement Method of establishing an equal relationship between the amount of work performed and the human input to do the work.

Yield Amount of product resulting at the end of the procurement/production/process.

Index

Page numbers followed by a *f* indicate figure.
Page numbers followed by a *t* indicate table.

J

K

L

M

N

O